Synthesis and Applications of Nanoparticles

Atul Thakur • Preeti Thakur • S. M. Paul Khurana
Editors

Synthesis and Applications of Nanoparticles

Springer

Editors
Atul Thakur
Centre of Nanotechnology
Amity University Haryana
Gurugram, Haryana, India

Preeti Thakur
Department of Physics, Amity School of Applied Sciences
Amity University Haryana
Gurugram, Haryana, India

S. M. Paul Khurana
Science Instrumentation Centre
Amity University Haryana
Gurugram, Haryana, India

ISBN 978-981-16-6818-0 ISBN 978-981-16-6819-7 (eBook)
https://doi.org/10.1007/978-981-16-6819-7

This Springer imprint is published by the registered company Springer Nature Singapore Pte Ltd.
The registered company address is: 152 Beach Road, #21-01/04 Gateway East, Singapore 189721, Singapore

Preface

In this fast-pacing world of science and technology, nanotechnology is making its presence felt in everyday life. This multidisciplinary science is filling the gap and bridging all the major branches of science and providing a common platform for scientists and technologists. As a result, new innovative nanoproducts are being developed making our lives all the more easier.

Synthesis and Applications of Nanomaterials provides a comprehensive introduction to nanomaterials, their properties, processing techniques, and applications. This book covers the basics of nanotechnology and provides a solid understanding of the subject. Starting from the basics of nanomaterials to their types, structure and synthesis techniques, the book gradually gives an insight to the properties of nanomaterials. It includes chapters on the various applications of nanoscience and nanotechnology. It is written in a simple form, making it useful for students of physical and material sciences.

The editors wish to thank all the authors for their efforts in writing their chapters and, also Springer Nature Publishing who have rendered every possible help for the successful completion of this book.

Gurugram, India

Atul Thakur
Preeti Thakur
S. M. Paul Khurana

Contents

1 **Introduction to Nanotechnology** 1
Preeti Thakur and Atul Thakur

2 **Nanomaterials, their Types and Properties** 19
Preeti Thakur and Atul Thakur

3 **Synthesis of Nanoparticles by Physical Route** 45
Lucky Krishnia, Preeti Thakur, and Atul Thakur

4 **Synthesis of Nanomaterials by Chemical Route** 61
Shilpa Taneja, Pinki Punia, Preeti Thakur, and Atul Thakur

5 **Synthesis of Nanomaterials by Biological Route** 77
Atul Thakur, Deepika Chahar, and Preeti Thakur

6 **Structural and Morphological Characterization of Nanomaterials** 121
Munirah A. Almessiere, Yassine Slimani, Alex V. Thurkanov, and Abdulhadi Baykal

7 **Optical Characterization of Nanomaterials-I** 139
Vladimir Pavelyev

8 **Optical Characterization of Nanomaterials-II** 151
Vladimir Pavelyev and Nishant Tripathi

9 **Magnetic Characterization of Nanomaterials** 177
Yassine Slimani, Sadik Guner, Munirah A. Almessiere, Essia Hannachi, Ayyar Manikandan, and Abdulhadi Baykal

10 **Electrical Characterization of Nanomaterials** 239
Artem Kozlovskiy, Inesh Kenzhina, Maxim V. Zdorovets, Kayrat K. Kadyrzhanov, and Alex V. Trukhanov

11 **Electromagnetic Characterizations of Nanomaterials** 255
Atul Thakur, Preeti Thakur, Dinesh Kumar, and P. B. Sharma

12 Electromagnetic Characterization of Nanomaterials: Preliminary Study of 60 GHz Millimetre Wave Li-NGD Circuit in Microstrip Technology 267
Fayu Wan, Xiaoyu Huang, Samuel Ngoho, Kishore Ajay Kumar Ayyala, Preeti Thakur, M. S. Prasad, Atul Thakur, Sébastien Lallèchère, Wenceslas Rahajandraibe, Nour Mohammad Murad, and Blaise Ravelo

13 Nanomaterials for Antenna Applications 297
Anwer S. Abd El-Hameed, Asmaa I. Afifi, Moustafa A. Darwish, and Trukhanov Alex

14 Advancement in Crops and Agriculture by Nanomaterials 319
Sandeep Kaur, Kajal Sharma, Rajat Singh, and Naveen Kumar

15 Micronutrient Nanoparticles: Synthesis, Properties and Application in Agriculture 337
Kartik Pal, S. M. Paul Khurana, Nitai Debnath, and Sumistha Das

16 Phytosynthesis of Silver Nanoparticles and Their Role as Antimicrobials 357
Alka Yadav and Mahendra Rai

17 Microbicidal Nanoparticles 371
Nishant Srivastava and Shailendra K. Saxena

18 Advancement in Water Purification by Nanomaterials 387
Andrei Ivanets, Vladimir Prozorovich, and Tatyana Kouznetsova

19 Nanotechnology in Healthcare 405
Deepak Kala, Shagun Gupta, and Ankur Kaushal

20 Graphene Quantum Dots and Their Hybrid Hydrogels: A Multifaceted Platform for Theranostic Applications 417
Sujata Sangam, Piyush Garg, Trinanjana Sanyal, Siddhartha Pahari, S. M. Paul Khurana, and Monalisa Mukherjee

21 Functionalized Nanoparticles in Drug Delivery: Strategies to Enhance Direct Nose-to-Brain Drug Delivery via Integrated Nerve Pathways 455
Fakhara Sabir, Qurrat Ul Ain, Abbas Rahdar, Zhugen Yang, Mahmood Barani, Mauhammad Bilal, and Nikhil Bhalla

22 Environmental Nanomedicine 487
Nishant Srivastava, Gourav Mishra, and Shailendra K. Saxena

23 Development of Nanostructured Material by Severe Plastic Deformation 503
Sanjeev Sharma and P. B. Sharma

24 Nanotechnology for Functional/High-Performance/Smart Textiles 525
Sapna Gautam, Anupama Mishra, and Pooja Koundal

25 Toxicity of Nanomaterials: An Overview 535
Preeti Thakur, Yeshvandra Verma, and Atul Thakur

Correction to: Functionalized Nanoparticles in Drug Delivery: Strategies to Enhance Direct Nose-to-Brain Drug Delivery via Integrated Nerve Pathways C1
Fakhara Sabir, Qurrat Ul Ain, Abbas Rahdar, Zhugen Yang, Mahmood Barani, Mauhammad Bilal, and Nikhil Bhalla

Editors and Contributors

About the Editors

Atul Thakur is professor and director of the Amity Institute of Nanotechnology & Amity School of Applied Sciences, Amity University, Haryana. His research interests are spinel ferrites for antenna miniaturization and sensor applications, magnetic nanoferrites for water purification and agriculture applications, synthesis and characterization of nanoferrites for high-frequency applications, and radar-absorbing materials. He has published more than 150 research articles in journals of high repute and filed 18 patents till date. He has worked on several funded research projects from MNRE, Jal Shakti, MeitY, DST, DAE, DRDO and is recipient of Royal Academy of Engineering UK award.

Preeti Thakur is a professor and head of the Department of Physics, Amity School of Applied Sciences, Amity University, Haryana. She is Gold Medalist in electronics. Her areas of interest are spinel ferrites synthesis and characterization of RADAR-absorbing materials, sensor, high-frequency materials, and application of magnetic nanoferrite materials for wastewater treatment and agriculture purposes. Till now, she has published more than 100 research articles in journals of high repute and filed 15 patents till date. She has worked on several funded research projects from DST, DAE, DRDO and is recipient of Royal Academy of Engineering UK award.

S. M. Paul Khurana PhD is a retired professor of biotechnology and head of the University Science Instrument Centre, Amity University, Haryana, Gurgaon. Earlier, he served as the vice chancellor of Rani Durgavati University, Jabalpur, from 2004 to 2009, project coordinator AICRP (Potato) from 1994 to 2004, and director of the Central Potato Research Institute, Shimla, from 2002 to 2004. He is the founder director of Amity Institute of Biotechnology, Gurgaon Haryana; Dean, Science, Engineering, and Technology; consultant at CIP/FAO, 1992, 1996, and 1997. He has more than 54 years of experience in pathology, virology, and nanotechnology. He has published more than 230 research papers, 135 reviews/chapters, and authored and/or edited 25 books.

Contributors

Anwer S. Abd El-Hameed Center for Northeast Asian Studies, Tohoku University, Sendai, Miyagi, Japan
Electronics Research Institute, Giza, Egypt

Asmaa I. Afifi Electronics Research Institute, Giza, Egypt

Trukhanov Alex National University of Science and Technology, MISiS, Moscow, Russia
SSPA "Scientific and Practical Materials Research Centre of NAS of Belarus", Minsk, Belarus
South Ural State University, Chelyabinsk, Russia

Munirah A. Almessiere Department of Biophysics, Institute for Research and Medical Consultations (IRMC), Imam Abdulrahman Bin Faisal University, Dammam, Saudi Arabia
Department of Physics, College of Science, Imam Abdulrahman Bin Faisal University, Dammam, Saudi Arabia

Kishore Ajay Kumar Ayyala Centre of Nanotechnology, Amity University Haryana, Gurugram, India

Mahmood Barani Medical Mycology and Bacteriology Research Center, Kerman University of Medical Sciences, Kerman, Iran

Abdulhadi Baykal Department of Nanomedicine Research, Institute for Research and Medical Consultations (IRMC), Imam Abdulrahman Bin Faisal University, Dammam, Saudi Arabia

Nikhil Bhalla Nanotechnology and Integrated Bioengineering Centre (NIBEC), School of Engineering, Ulster University, Newtownabbey, Northern Ireland, UK
Healthcare Technology Hub, Ulster University, Newtownabbey, Northern Ireland, UK

Mauhammad Bilal School of Life Science and Food Engineering, Huaiyin Institute of Technology, Huaian, China

Deepika Chahar Department of Physics, Amity School of Applied Sciences, Amity University Haryana, Gurugram, India

Moustafa A. Darwish Physics Department, Faculty of Science, Tanta University, Tanta, Egypt
National University of Science and Technology MISiS, Moscow, Russia

Sumistha Das Amity Institute of Biotechnology, Amity University Haryana, Gurugram, India

Nitai Debnath Amity Institute of Biotechnology, Amity University Haryana, Gurugram, India

Piyush Garg Amity Institute of Biotechnology, Amity University, Noida, Uttar Pradesh, India
Amity Institute of Click Chemistry Research and Studies, Amity University, Noida, Uttar Pradesh, India

Sapna Gautam Department of Textiles & Apparel Designing, College of Home Science, CSK HP Agricultural University, Palampur, India

Sadik Guner Institute of Inorganic Chemistry, RWTH Aachen University, Aachen, Germany

Shagun Gupta Department of Biotechnology, Maharishi Markandeshwar University, Mullana, Ambala, Haryana, India

Essia Hannachi Department of Nuclear Medicine Research, Institute for Research and Medical Consultations (IRMC), Imam Abdulrahman Bin Faisal University, Dammam, Saudi Arabia

Xiaoyu Huang School of Electronics and Information Engineering, NUIST, Nanjing, China

Andrei Ivanets Institute of General and Inorganic Chemistry, National Academy of Sciences of Belarus, Minsk, Belarus

Kayrat K. Kadyrzhanov L.N. Gumilyov Eurasian National University, Nur-Sultan, Kazakhstan

Deepak Kala Department of Biotech Engineering and Food Technology, Chandigarh University, Mohali, Punjab, India

Sandeep Kaur Department of Life Sciences, RIMT University, Mandi Gobindgarh, Punjab, India

Ankur Kaushal Department of Biotechnology, Maharishi Markandeshwar University, Mullana, Ambala, Haryana, India

Inesh Kenzhina L.N. Gumilyov Eurasian National University, Nur-Sultan, Kazakhstan
The Institute of Nuclear Physics, Almaty, Kazakhstan

S. M. Paul Khurana Science Instrumentation Centre, Amity University Haryana, Gurugram, Haryana, India

Pooja Koundal Department of Fashion and Textile Design, NBSCFF, Subharti University, Meerut, Uttar Pradesh, India

Tatyana Kouznetsova Institute of General and Inorganic Chemistry, National Academy of Sciences of Belarus, Minsk, Belarus

Artem Kozlovskiy L.N. Gumilyov Eurasian National University, Nur-Sultan, Kazakhstan
The Institute of Nuclear Physics, Almaty, Kazakhstan
Research School of Chemistry and Applied Biomedical Sciences National Research Tomsk Polytechnic University, Tomsk, Russia

Lucky Krishnia Centre of Nanotechnology, Amity University Haryana, Gurugram, India

Dinesh Kumar Centre of Nanotechnology, Amity University Haryana, Gurugram, India

Naveen Kumar Amity Institute of Biotechnology, Amity University Rajasthan, Jaipur, India

Sébastien Lalléchère Université Clermont Auvergne (UCA), CNRS, SIGMA Clermont, Institut Pascal, Aubière, France

Ayyar Manikandan Department of Chemistry & Center for Catalysis and Renewable Energy, Bharath Institute of Higher Education and Research (BIHER), Bharath University, Chennai, Tamil Nadu, India

Anupama Mishra Central Agricultural University, Tura, Meghalaya, India

Gourav Mishra Department of Biotechnology, Meerut Institute of Engineering and Technology, Meerut, India

Monalisa Mukherjee Amity Institute of Biotechnology, Amity University, Noida, Uttar Pradesh, India
Amity Institute of Click Chemistry Research and Studies, Amity University, Noida, Uttar Pradesh, India

Nour Mohammad Murad PIMENT Lab, Network and Telecom Lab, Institut Universitaire de Technologie, University of La Reunion, Saint Pierre, France

Samuel Ngoho Association Française de Science des Systèmes (AFSCET), Paris, France

Siddhartha Pahari Amity Institute of Click Chemistry Research and Studies, Amity University, Noida, Uttar Pradesh, India

Kartik Pal Amity Institute of Biotechnology, Amity University Haryana, Gurugram, India

Vladimir Pavelyev Samara National Research University, Samara, Russia

M. S. Prasad Amity Institute of Space Science and Technology, Amity University Noida, Noida, India

Vladimir Prozorovich Institute of General and Inorganic Chemistry, National Academy of Sciences of Belarus, Minsk, Belarus

Pinki Punia Department of Physics, Guru Jambheshwar University, Hisar, India

Qurrat Ul Ain Institute Teknologi Bandung, Bandung, Indonesia

Wenceslas Rahajandraibe Aix-Marseille University, CNRS, University of Toulon, IM2NP UMR7334, Marseille, France

Abbas Rahdar Department of Physics, Faculty of Science, University of Zabol, Zabol, Iran

Mahendra Rai Department of Biotechnology, Sant Gadge Baba Amravati University, Amravati, India
Department of Microbiology, Nicolaus Copernicus University, Torun, Poland

Blaise Ravelo School of Electronics and Information Engineering, NUIST, Nanjing, China

Fakhara Sabir University of Szeged, Faculty of Pharmacy, Institute of Pharmaceutical Technology and Regulatory Affairs, Szeged, Hungary

Sujata Sangam Amity Institute of Biotechnology, Amity University, Noida, Uttar Pradesh, India
Amity Institute of Click Chemistry Research and Studies, Amity University, Noida, Uttar Pradesh, India

Trinanjana Sanyal Amity Institute of Biotechnology, Amity University, Noida, Uttar Pradesh, India

Shailendra K. Saxena Centre for Advanced Research (CFAR), Faculty of Medicine, King George's Medical University (KGMU), Lucknow, India

Kajal Sharma Chitkara School of Health Sciences, Chitkara University, Rajpura, Punjab, India

P. B. Sharma Centre of Nanotechnology, Amity University Haryana, Gurugram, India
Department of Mechanical Engineering, Amity University, Manesar, Haryana, India

Sanjeev Sharma Department of Mechanical Engineering, Amity University, Manesar, Haryana, India

Rajat Singh Department of Food Technology, Eternal University, Sirmour, Himachal Pradesh, India

Yassine Slimani Department of Biophysics, Institute for Research and Medical Consultations (IRMC), Imam Abdulrahman Bin Faisal University, Dammam, Saudi Arabia

Nishant Srivastava Department of Biotechnology, Meerut Institute of Engineering and Technology, Meerut, India

Shilpa Taneja Department of Physics, Amity School of Applied Sciences, Amity University Haryana, Gurugram, India

Atul Thakur Centre of Nanotechnology, Amity University Haryana, Gurugram, India

Preeti Thakur Department of Physics, Amity School of Applied Sciences, Amity University Haryana, Gurugram, India

Alex V. Thurkanov Department of Physics, College of Science, Imam Abdulrahman Bin Faisal University, Dammam, Saudi Arabia
South Ural State University, Chelyabinsk, Russia
SSPA "Scientific and Practical Materials Research Centre of NAS of Belarus", Minsk, Belarus
L.N. Gumilyov Eurasian National University, Nur-Sultan, Kazakhstan

Nishant Tripathi Samara National Research University, Samara, Russia

Alex V. Trukhanov National University of Science and Technology MISiS, Moscow, Russia
SSPA "Scientific and Practical Materials Research Centre of NAS of Belarus", Minsk, Belarus
South Ural State University, Chelyabinsk, Russia

Yeshvandra Verma Department of Toxicology, C.C.S. University Campus, Meerut, India

Fayu Wan School of Electronics and Information Engineering, NUIST, Nanjing, China

Alka Yadav Department of Biotechnology, Sant Gadge Baba Amravati University, Amravati, India

Zhugen Yang Cranfield Water Science Institute, Cranfield University, Cranfield, Bedfordshire, UK

Maxim V. Zdorovets L.N. Gumilyov Eurasian National University, Nur-Sultan, Kazakhstan
The Institute of Nuclear Physics, Almaty, Kazakhstan
Ural Federal University, Yekaterinburg, Russia

Introduction to Nanotechnology 1

Preeti Thakur and Atul Thakur

Abstract

Nanoscience and Nanotechnology are very vast and very old sciences that people know from the very beginning. Life started with the big bang, where all atomic and molecular phenomena happened, which falls under nanoscience. It is right to say nanoscience as the science of God. Knowingly or unknowingly, we come across many events in our daily life that are influenced by nano science and technology. Our "Rishis" used "Bhasma" for medication, which is a classic example of nanotechnology. 0D, 1D, 2D, and 3D nanoparticles are described and explained in detail. In this chapter, structural, optical, chemical, electronic, mechanical, thermal, and magnetic properties of nanoparticles in general are also discussed.

Key words

Nanoscience · Nanotechnology · Atomic · Size

1.1 Introduction

The term nanotechnology is made up of two words; in which, Greek word "nano" means billionth and the second word is technology. Nanotechnology considers the objects that are of the size below 100 nm. As an outcome, nanotechnology or nanoscale technology (Ghazi et al. 2018; Wang et al. 2017; Ghaffari et al. 2012; Gleiter 2009; Bhushan 2016; Nouailhat 2010) is generally considered to be at a size below 100 nm (a nanometer is one billionth of a meter, 10^{-9} m). Nanotechnology, in

P. Thakur (✉)
Department of Physics, Amity School of Applied Sciences, Amity University Haryana, Gurugram, India

A. Thakur
Centre of Nanotechnology, Amity University Haryana, Gurugram, India

A. Thakur et al. (eds.), *Synthesis and Applications of Nanoparticles*,
https://doi.org/10.1007/978-981-16-6819-7_1

short, is called as "nanotech," and it is defined as the study of controlling matter on an atomic and molecular scale. In general, nanotechnology is the name of the science which deals with structures of the size 100 nm or smaller in at least one dimension, and also it involves developing materials or devices within that size (Baer et al. 2003; Whitesides 2005). The range of nanotechnology is very wide, from the smallest particles which cannot be seen by the naked eye to the food we eat and clothes we wear. The definition of nanotechnology can be stated as:

- The development in the field of research and technology at various levels such as atomic, molecular, or macromolecular levels, having a length scale of approximately 1–100 nm.
- Creation and use of devices and structures with novel properties due to small size.
- Controlling or manipulating matter at atomic scale.

Nanotechnology is becoming very popular with time. Firstly, Michael Faraday in 1857 mentioned during a lecture on the optical properties of gold noticeable that "a mere variation in size of the particles gave rise to a variety of resultant colors." The invention of atomic force microscopy (AFM) in 1986 and first electron transistor in 1987 and then creation of carbon nanotube in 1991 depict the growth in the field of nanotechnology. Here some of the historical contents of nanotechnology development are listed in the flowchart below.

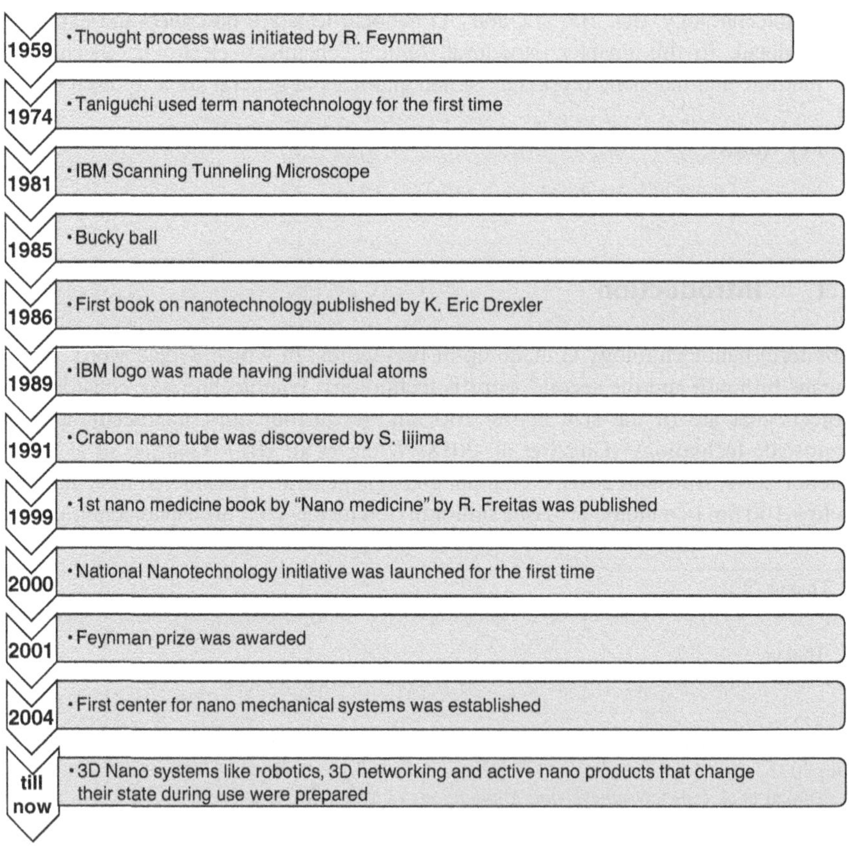

Nanoparticles in the range 1–100 nanometer are the fundamental components of nanotechnology and made up of metal, carbon, organic matter, or metal oxides. The physical, chemical, and biological properties of nanoparticles are unique at nanoscale in comparison with their respective bulk counterparts. This is due to larger surface area to the volume ratio, high reactivity or chemical stability, increased mechanical strength, etc. (Sui et al. 1996). Due to these properties, nanoparticles are used in many applications. The nanoparticles are different in terms of dimensions, shapes, and sizes. The classification of nanoparticles on the basis of dimensions can be done as zero dimensional, one dimensional, two dimensional, and three dimensional. In case of a zero-dimensional nanomaterial, the length, breadth, and height are fixed at a single point like nanodots. In case of one-dimensional nanomaterial, it can possess only one parameter, for example, carbon nanotubes, and for two dimensional, it has length and breadth in nanoscale, for example, graphene. For three-dimensional nanomaterial, it has all the parameters such as length, breadth, and height of nanorange, for example, gold nanoparticles. The shape, size, and structure of nanoparticles are different as it may be cylindrical, spherical, conical, tubular, hollow core, spiral, flat, etc. The surface may have variation or it may be uniform. Nanoparticles are classified into crystalline or amorphous based on whether the single or multicrystal solids are loose or agglomerated (Rashad et al. 2009).

There are numerous synthesis techniques that are developed to improve the properties of nanomaterials and to reduce the cost of production. The modification in some methods is done to improve their optical, mechanical, physical, and chemical properties (Chandamma et al. 2017; Gao et al. 2013; Yadav et al. 2016). The characterizations are improved due to vast development in the instrumentation. The use of nanoparticles is in every field like in cooking vessel, electronics to renewable energy, and aerospace industry. So, it can be said that nanotechnology is the key for a clean and sustainable future. The nanostructured materials and nanosystems can be invented by discovering new materials and processes at the nanoscale and the advancement of novel theoretical and experimental techniques. There are much current and expected advancement in nanoscale science and nanotechnology in terms of its applications in agriculture, medicine, energy, electronics, etc. The developments in the domain of nanotechnology are increasing day by day, and it plays a very important role in creating new products, substituting present production equipments, and reformulating novel materials to improve the performance and reduce the consumption of energy and is also used for remediation of environment (Melo et al. 2015; Mallesh and Srinivas 2019). Although it is beneficial for the environment that consumption of matter and energy is decreased, a more sustainable route can be offered by nanotechnology to remediate the problems. Nanotechnology can be used to develop solutions to environmental problems, measures to address both the ensuing problems from interactions of material and energy with the environment and the risks associated with nanotechnology (Harris and Šepelák 2018).

Nanotechnology can be stated as the complex interdisciplinary science that includes (Ghazi et al. 2018; Mehta 2017; Thakur and Hsu 2011; Thakur et al. 2014; Sawant et al. 2016; Manikandan et al. 2018) the study of nanophysics, nanochemistry, nanomaterial science, nanoelectronics, nanobionics, and nanometrology. This is illustrated as under:

- Nanophysics including spintronics, quantum physics, and photonics is destined for assembling and fabricating nanostructures artificially and doing research about the external size effects.
- Nanochemistry including sol-gel, nanocolloid, and quantum chemistry is destined for the nanoparticle synthesis and doing research about their intrinsic size effects.
- Nanomaterial science including nanoceramic compounds, nanotribology, nanopowder technology, nanosintering, etc. is about developing and producing nanostructured materials and nanocomposites having unique properties.
- Nanobionics is the branch of nanotechnology that is about developing nanobiochips, nanobiorobots, etc.
- Nanoelectronics is the development of nanomotors, nanodevices, ultra-large integrated circuits (ULCI), micro-optoelectronic-mechanical systems (MEMS, MOEMS), nanorobots, etc.
- Nanometrology is building and developing special nanotools, information, instrumentations, and computational systems.

1.2 Classification of Nanoparticles

The classification of nanoparticles can be done in accordance with organic, inorganic, and carbon based as shown in Fig. 1.1. Some organic nanoparticles or polymers are micelles, dendrimers, ferritin, liposomes, etc. The features of these nanoparticles are listed as:

- Biodegradable.
- Nontoxic.
- Some particles among them like micelles and liposomes have a hollow core (Fig. 1.2), known as nanocapsules.
- Sensitive to thermal and electromagnetic radiation such as heat and light.

These features make these materials capable for application in drug delivery. The drug carrying capacity, delivery systems, its stability, and entrapped or adsorbed drug system decide the field of applications and their efficiency apart from their normal characteristics like the surface morphology, size, composition, etc. The use of these organic nanoparticles is in the field of biomedicines. In case of drug delivery systems, these are efficiently injected on particular parts of the body called as targeted drug delivery.

The particles that are not made up of carbon are called inorganic nanoparticles. Inorganic nanoparticles are metal- and metal oxide-based nanoparticles. Metal-based nanoparticles are the nanoparticles that are prepared from metals to nanometric sizes by using either destructive or constructive methods. It is possible to synthesize almost all the metals into their nanoparticles (Sivakumar et al. 2011). The metals

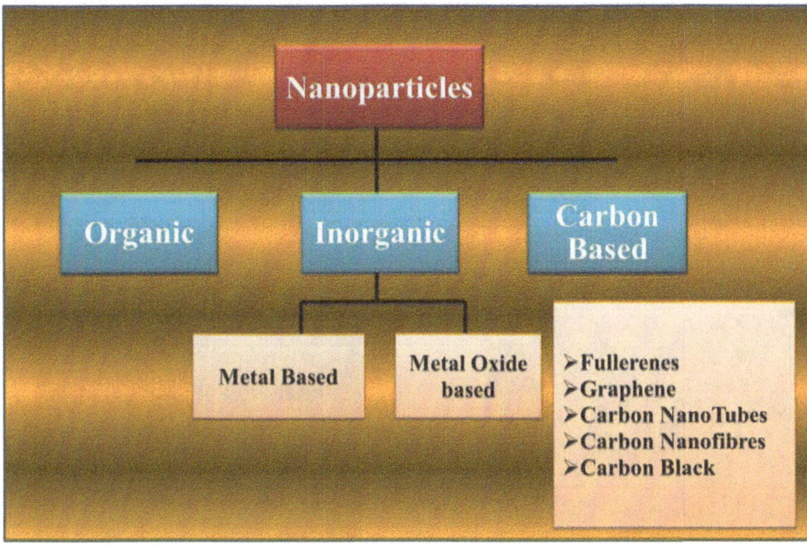

Fig. 1.1 Classification of nanoparticles according to organic, inorganic, and carbon based

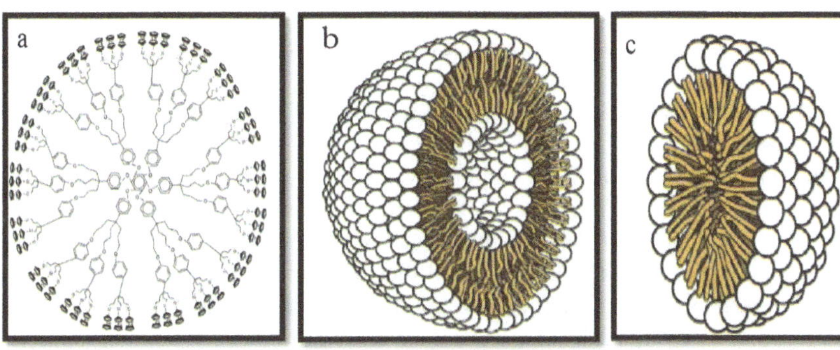

Fig. 1.2 Organic nanoparticles: (**a**) Dendrimers; (**b**) liposomes; (**c**) micelles

that are commonly used for the synthesis of nanoparticles are cobalt (Co), aluminum (Al), copper (Cu), cadmium (Cd), iron (Fe), gold (Au), silver (Ag), lead (Pb), and zinc (Zn). The nanoparticles have properties like sizes as low as 10–100 nm, high surface area to volume ratio, surface charge, pore size, surface charge density, color, shapes like spherical and cylindrical, crystalline and amorphous structures, and sensitivity and reactivity to environmental factors like heat, air, sunlight, moisture, etc. The metal oxide-based nanoparticles have modified properties in comparison to their respective metal-based nanoparticles, for example, in the presence of oxygen, iron nanoparticles oxidize to iron oxide (Fe_2O_3) at room temperature due to which its reactivity gets increased compared to iron nanoparticles. Due to increased reactivity

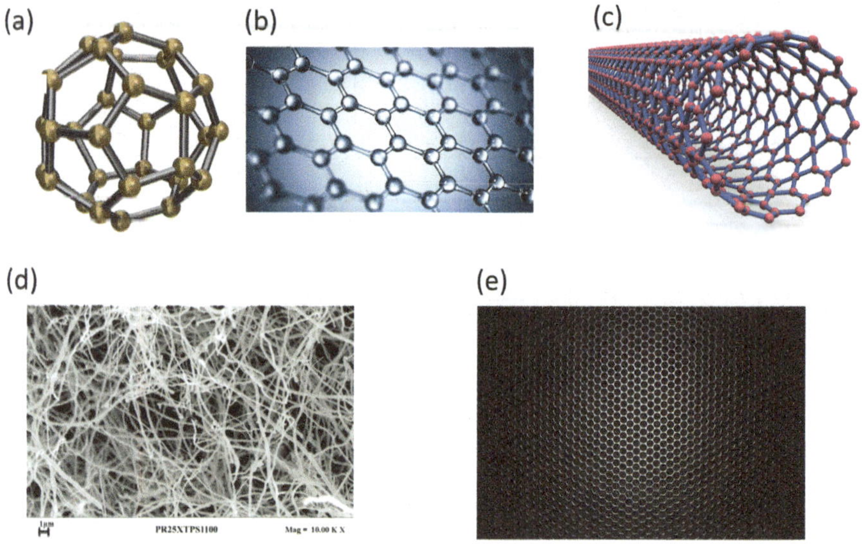

Fig. 1.3 Carbon-based nanoparticles: (**a**) fullerenes; (**b**) graphene; (**c**) carbon nanotubes; (**d**) carbon nanofibers; and (**e**) carbon black

and efficiency, metal oxide nanoparticles are synthesized (Baer et al. 2003). Some commonly synthesized are aluminum oxide (Al_2O_3), iron oxide (Fe_2O_3), silicon dioxide (SiO_2), cerium oxide (CeO_2), titanium oxide (TiO_2), magnetite (Fe_3O_4), and zinc oxide (ZnO).

The nanoparticles which are completely made of carbon are known as carbon based nanoparticles (Whitesides 2005). These materials can be classified into graphene, fullerenes, carbon nanofibers, carbon nanotubes (CNT), carbon black, and activated carbon in nanosize and are illustrated in Fig. 1.3. Fullerene (C_{60}) is spherical in shape and is a carbon molecule that is made up of carbon atoms that are held together by sp^2 hybridization. About 28–1500 carbon atoms build the spherical structure having diameters up to 8.2 nm for single-layered and 4–36 nm for multi-layered fullerenes. Graphene is called as an allotrope of carbon. Graphene shape is hexagonal having honeycomb lattice that is made up of carbon atoms in a 2D planar surface. Graphene sheet has a thickness of 1 nm. A Carbon nanotube (CNT) is a graphene nanofoil having a honeycomb lattice made up of carbon atoms wound into hollow cylinders to build nanotubes with diameters of measurement 0.7 nm for single-layered and 100 nm for multilayered carbon nanotubes and length in the range of a few micrometers to several millimeters. The ends can be open or closed by a half fullerene molecule. The particles have high interaction such that the bound inaggregates and around 500 nm agglomerates are formed.

In simple words, nanotechnology may be defined as a branch of science which deals with materials or structures in nanoscale range varying from subnanometer to several nanometers. This field is quite similar to quantum mechanics and is a new

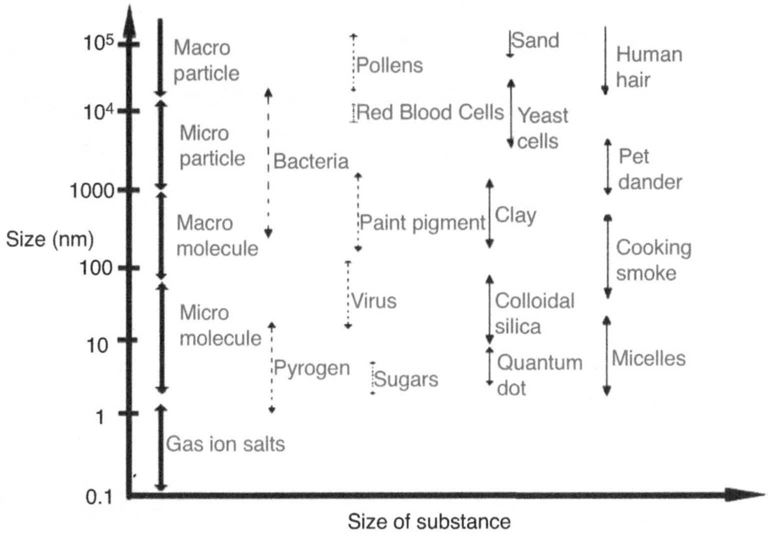

Fig. 1.4 Zero-dimensional representation of nanostructures with their typical ranges of dimensions

scientific domain. The zero-dimensional representation of nanomaterials having typical ranges of dimensions is shown in Fig. 1.4. In the nanometer scale, the synthesized materials or structures have some new physical properties from which some properties are known. For example, by varying the dimension of material, band gap of the semiconductor can be tuned. Still, there are certain properties which are not in the knowledge till now. These new physical properties are capable of satisfying human beings and also proved to bring new advancements in the field of science and technology.

1.2.1 Classification of Nanomaterials on the Basis of Size

On the basis of size, the classification of nanoparticles can be done into zero, one, two, and three dimensional as shown in Fig. 1.5. The materials which have structures in the range 1–100 nm are called nanostructured materials. The size and nature of the nanostructures define the properties of the nanostructured materials. Large changes in the material properties in comparison with a non-nanostructured material can be observed, if the characteristic length scale of the microstructure is comparable to the associated lengths to fundamental physical phenomena. A great variety of nanostructures can be produced using crystallites of nanometer size of elements like sodium chloride and gold, depending on the chemical composition of the mixture, the crystallographic orientation, and the possibility to have nonequilibrium structures that have certain property advantages. Nanostructured materials can be made by using nanoparticles as building blocks. The nature of nanostructured material can vary, and these may be nanocrystallites, fullerenes, nanofibers, nanotubes, etc. Nanomaterials are the simplest building blocks of nanostructured

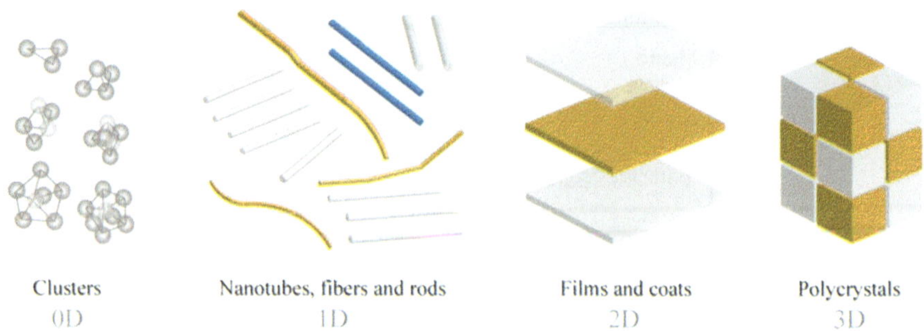

Fig. 1.5 Nanostructured materials

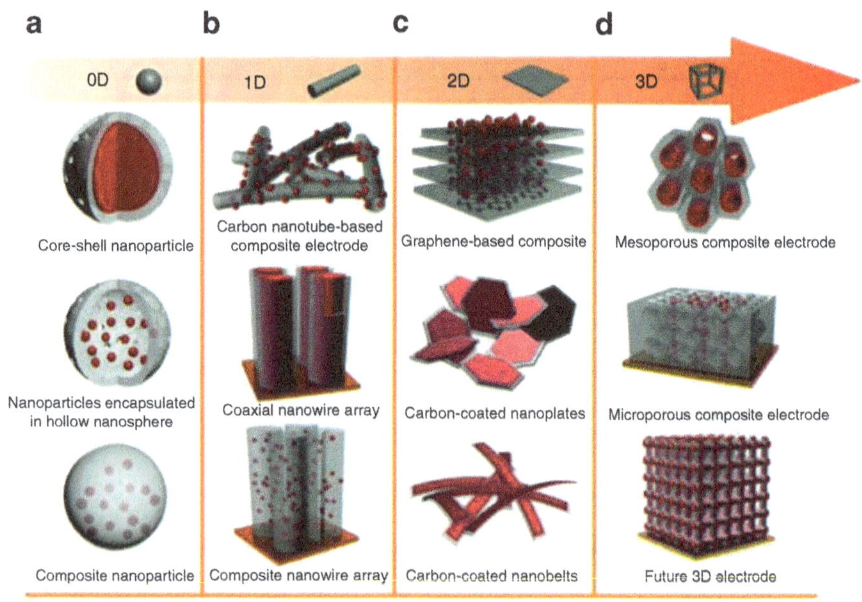

Fig. 1.6 Different types of (**a**) 0D (**b**) 1D (**c**) 2D, and (**d**) 3D nanostructured materials

nanomaterials, and it is possible to use more complicated elementary structures to make nanocomposites. Nanoparticles are the very simple objects which can be used to make nanomaterials. However, it is not an easy task to self-assemble nanoparticles according to a given template. There are only few examples such as sulfides or selenides combining with success soft templates, oriented attachment resulting in 1D structures, and self-alignment of nanoparticles by dipolar interactions. Different techniques like electron lithography can be used to design hard nanotemplates. Nanoparticles having controlled size and shape can be synthesized using soft nanotemplates, for example, mesophases and micellar systems. Different types of 0D, 1D, 2D, and 3D nanostructured materials are shown in Fig. 1.6.

Various types of nanostructures can be differentiated on the basis of dimensionality. The word "nano" is originated from a Greek word "nanos," which means dwarf. This word "nano" is meant for a number 10^{-9}, i.e., one billionth of a unit.

A significant progress has been made in the field of zero-dimensional nanostructured materials in the past 10 years. The zero-dimensional nanostructured materials can be synthesized using a variety of physical and chemical methods. Recently, zero-dimensional nanostructures like quantum dots, core-shell quantum dots, hollow spheres, heterogeneous particle arrays, onions, and nanolenses have been prepared by several research groups. Also, these materials like quantum dots have been extensively studied in single-electron transistors (Nayak et al. 2011), light-emitting diodes (Harzali et al. 2018), lasers (Zhang et al. 2009), and solar cells (Saeedi Afshar et al. 2018).

Due to the importance in research and having a variety of potential applications, in one-dimensional nanostructured materials, the interest of researchers is increasing in these materials. A large number of novel phenomena can be explored at the nanoscale using one-dimensional nanostructured materials. Also, these materials are useful in investigating the size and dependence on dimensions of functional properties. They are also used to play the role of interconnects and as a key unit in fabrication of optoelectronic, electronic, and EEDs having nanoscale dimensions. After useful work, significant attention has been attained by one-dimensional nanostructured materials like nanotubes. There is a great impact of one-dimensional nanostructured materials in nanodevices, nanoelectronics, alternative energy resources, nanosystems, nanocomposite materials, and national security. One-dimensional nanostructured materials include hierarchical nanostructures, nanotubes, nanowires, nanoribbons, nanorods, and nanobelts (Amer 2017; Sertkol et al. 2010).

The two-dimensional nanostructured materials have two dimensions outside of the nanoscale range. Also, these nanomaterials have many low-dimensional characteristics that are different from their bulk counterparts. Some unique shape-dependent properties are exhibited by these two-dimensional nanostructured material geometries. Also, these materials are the key components for synthesizing nanodevices (Chithra et al. 2017; Rafiq et al. 2015). The mechanism for the growth of nanostructures, investigation, and developing applications in the field of nanoreactors, sensors, photocatalysts, and nanocontainers can be easily understandable by fabricating two-dimensional nanostructured materials (Narang and Pubby 2021). Two-dimensional nanostructured materials include nanoprisms, junctions (continuous islands), branched structures, nanodisks, nanoplates, nanosheets, and nanowalls.

Due to many superior properties and large specific surface area over the bulk materials, researchers are taking great interest in three-dimensional nanostructured materials, and these materials are being synthesized from the past 10 years (Ling et al. 2010; Costa et al. 2008; Pei and Wang 2018). As it is a well-known fact that the behavior of nanostructured materials is strongly affected by size, shape, morphology, and dimensionality which are the key factors for the applications and ultimate performance of the nanomaterials. Hence, three-dimensional nanostructured materials having controlled structure and morphology are being synthesized by the researchers. The range of applications of these materials is very wide, for example, in the area of electrode material for batteries, catalysis, and magnetic material. Due to supply enough absorption sites for all involved molecules in a small space and higher surface area, the three-dimensional nanostructured materials are attracting intensive interest by researchers. Also, better transport of the molecules is possible due to porosity of these materials in three dimensions. Some examples of three-dimensional nanomaterials are nanocones, nanoballs, nanopillers, nanocoils, and nanoflowers.

1.3 Properties of Nanomaterials

1.3.1 Structural Properties

The changes in the spacing between interatoms can lead to an increase in the surface area and surface energy with a decrease in particle size. This is because of the compressive strain caused by the internal pressure by the small radius of curvature in the nanoparticle. It is evident that interatomic spacing increases with a decrease in particle size for semiconductors and metal oxides. One more effect is the stability of metastable structures in small clusters and nanoparticles, and due to this there is loss in all traces of the usual bulk atomic arrangement. Metallic nanoparticles, for example, gold, adopt polyhedral shapes like multiply twinned icosahedra, cube octahedra, and multiply twinned decahedra. These nanoparticles may be considered as multiply twinned crystalline particles (MTPs) in which understanding of shapes in terms of surface energies of various crystallographic planes, the growth rates along various crystallographic directions, and the energy required for the formation of defects such as twin boundaries can be made possible. But it is evident that these particles are crystalloids or quasiperiodic crystals. The growth of nanocluster, up to a size where they will switch into a more regular crystalline packing, is possible by these icosahedral and decahedral quasicrystals. Crystalline solids are different from amorphous solids because they possess long-range periodic order and the patterns and symmetries correspond to 230 space groups. Such long-range periodic order is not possessed by quasiperiodic crystals, and more differently five-fold symmetry is exhibited by them, which is forbidden in the 230 space group. In the hexagonal close packed and cubic close packed structures, that is exhibited by many metals in which each atom is coordinated by 12 neighboring atoms. These all coordinating atoms are in contact, but these are not evenly distributed around the central atom. Each atom situated at the apex of icosahedra is in contact only with the central atom in the alternative arrangement. The body of the material gains shape and point group symmetry of regular icosahedra by relaxing the rigid atmospheric model, allowing the central atom to decrease in diameter by 10%, and bringing the coordinating atoms in contact. This symmetry indicates the presence of 20 threefold, 12 fivefold, and 30 twofold axes of symmetry. This geometry depicts a quasiperiodic crystal nucleus which may grow in the form of pentagonal dodecahedra or icosahedra.

These are dual solids having identical symmetry in which the apices of one are replaced by the faces of the other. There is difficulty in understanding the characteristics that are related to size instability of quasiperiodic crystals. The process of multiple twinning is a frequently observed process, and such crystals are differentiated from quasiperiodic crystals by their electron diffraction patterns. Here, the five triangular faces of the fivefold symmetric icosahedra can be mimicked by five twin-related tetrahedra (with a close-packed crystalline structure) through relatively small atomic movements.

1.3.2 Optical Properties

The optical properties are greatly affected by reducing the dimension of materials. There are two groups when the size dependence is classified. One is because of the increase in energy level spacing as the system becomes more confined, and the other is because of surface plasmon resonance. The band gap increases with decrease in size due to quantum size effect in the semiconductor nanoparticles in which the interband transition shifts to higher frequencies. In a semiconductor, there is a rapid increase in energy separation (the energy difference between the completely filled valence band and the empty conduction band) with a decreasing size, and this energy separation is of the order of a few electron volts. A blue shift in the band gap is produced by quantum confinement and also in the appearance of discrete subbands attributed to quantization along the direction of confinement. The optical properties of the nano semiconductors can be modified by varying the size and keeping the same chemical composition. The variation in the nanoparticle size can lead to luminescent emission from the semiconductor nanostructures. The nature of electronic density of states and carrier confinement of semiconductor nanostructures make it more efficient for devices that are operating at lower threshold currents than lasers. The size-dependent emission spectra of quantum dots, quantum wells, and quantum wires make the lasing media attractive. The quantum dot lasers show less dependence on temperature than conventional semiconductor lasers. The same quantum size effect is also known in metal nanoparticles. However, in order to observe the localization of the energy levels, there is requirement of very small size so that the level spacing exceeds the thermal energy (~26 MeV). An in-phase oscillation is caused by surface plasmon resonance which is the coherent collective excitation of all the free electrons within the conduction band. A surface plasmon resonance is generated when the size of a metal nanocrystal is smaller than the wavelength of incident radiation. Biomedicine, photocatalysis, optical detectors, imaging, lasers, sensors and solar cells are some of the prominent applications based on the optical properties of the nanomaterials.

1.3.3 Chemical Properties

Chemical reactivity of the materials has a link with the size effects. Nanoscale structures, for example, nanolayers and nanoparticles, have potentially different crystallographic structures and very high surface area to volume ratios that causes a radical alteration in chemical reactivity. Nanoparticles generally show new chemistry which are different from their particular large counterparts; for example, in the form of micron-sized particles, there are many new medicines which are insoluble in water, but in a nanostructure form, they get dissolved easily. Hence, it is important to chemically identify the nanomaterials and characterize them. A few of the chemical properties that are essential for characterizing nanomaterials are composition, structure, chemical bonding, reactivity, stability, melting and boiling points.

1.3.4 Electronic Properties

The changes in electronic properties during decrease in the system length scale are mainly related to the increasing influence of the electrons' wavelike property, i.e., quantum mechanical effects and lack of scattering centers. The discrete nature of the energy states becomes apparent when the size of the system becomes comparable with the de Broglie wavelength of the electrons. But to observe a fully discrete energy spectrum, the system should be confined in all three dimensions. Below a critical length scale, conducting materials behave as insulators due to overlapping of the energy bands. Due to their intrinsic wavelike nature, quantum mechanical tunneling of electrons is possible between two closely adjacent nanostructures. Resonant tunneling occurs when a voltage is applied between two nanostructures due to which discrete energy levels are aligned in the density of state causing an increase in the tunneling current. The impurities, scattering with phonons, and scattering at rough surfaces determine the electronic transport in macroscopic systems. There is diffusive transport and path of every electron relates a random walk. In inelastic scattering, when system has dimensions smaller than the electron mean free path, electrons travel through the system without phase randomization of wave functions. This gives rise to additional localization phenomena related to phase interference. If due to small system, all scattering centers are to be eliminated completely, and if boundary reflections are purely specular due to smooth sample boundaries, then the electron transport is purely ballistic, and the sample acts as a waveguide for the electron wave function. Conduction in highly confined structures like quantum dots is very sensitive to the presence of other charge carriers and thus to the charge state of the dot. The conduction processes involving single electrons are caused by these Coulomb blockade effects due to which very substantial amount of energy is required by them to operate a transistor, switch, or memory element. Different types of components for information processing applications, electronic, and optoelectronic can be produced by utilizing all these phenomenas.

1.3.5 Mechanical Properties

The mechanical properties of the nanomaterials (hardness, fracture toughness, scratch resistance, elastic modulus, fatigue strength, etc.) are different from the bulk materials because of the nanometer size. This modification may result in an enhancement of mechanical properties of nanomaterials that often results from structural perfection of the materials. The small size either renders them free of internal structural imperfections such as dislocations, impurity precipitates, and micro twins. It is not possible to cause mechanical failure due to few defects or impurities. The highly energetic imperfections within the nano dimension will migrate to the surface to relax themselves under annealing, thereby causing purification of the material and leaving perfect material structures inside the nanomaterial. Moreover, the external surface of nanomaterials are free of defects in comparison to the bulk materials, causing enhancement in the mechanical properties of nanomaterials.

1.3.6 Thermal Properties

There is low progress in study of the thermal properties of nanomaterials due to the difficulties encountered in measuring experimentally and controlling the thermal transport in nanoscale dimensions. The introduction of atomic force microscopy (AFM) to measure the thermal transport of nanostructures within nanometer scale with high spatial resolution has provided a promising way to probe the thermal properties of nanostructures. The availability of the definition of temperature is in question when the dimensions go down into nanoscale. Phonons carry the thermal energy in nonmetallic material system which has a wide variation in mean free path and frequency. Generally, at room temperature, the phonons that carry heat have large mean free path and wave vectors in nanoscale range. Due to this, the nanostructure dimensions are comparable to the wavelength and mean free path of phonons. However, average energy of a material system defines the temperature. In case of macroscopic systems, a local temperature in each region within the materials is defined by the dimension, and there is variation in this local temperature from region to region, so that thermal transport properties based on certain temperature distributions can be investigated. But in case of nanomaterial systems, a local temperature sometimes can't be defined by just the dimensions because dimensions are too small to define. Also, the concept of temperature defined in equilibrium conditions is difficult or problematic to use for theoretical analysis of thermal transport in nanoscale. In nanomaterial systems, various factors like the large interfaces, the special shape, and the small size do modification in the thermal properties of the nanomaterials, rendering them a quite different behavior in comparison to the macroscopic materials. The size of the nanomaterials become comparable to the mean free path and wavelength of the phonons. When the dimension goes down to nanoscale, there is a significant change in phonon transport within the material due to the phonon confinement and quantization of phonon transport, which results in modified thermal properties. The thermal properties are also affected by the special structure of nanomaterials. For example, carbon nanotubes due to their tubular structures have extremely high thermal conductivity in axial directions, leaving high anisotropy during the heat transport in the materials. The thermal properties of nanomaterials are also determined by interfaces. The thermal properties of nanomaterials have another promising application in the use of nanofluid to enhance the thermal transport. The nanofluids consist of nanomaterials of size in the range 1–100 nm which are suspended in a liquid generally referred to as the solid-liquid composite materials. The increase in thermal conductivity in comparison to liquids not containing nanomaterials is an important feature of nanofluids.

1.3.7 Magnetic Properties

Magnetic nanoparticles have a wide range of applications, such as ferrofluids, refrigeration, bioprocessing, and color imaging, as well as high storage density magnetic memory media. The large surface area to volume ratio results in a substantial proportion of atoms (those at the surface which have a different local environment) having a different magnetic coupling with neighbouring atoms, leading to differing magnetic properties. When the particle size decreases below a certain value, ferromagnetic particles become unstable as domains are spontaneously switched in polarization directions by gaining surface energy; due to which, ferromagnetic becomes paramagnetic. However, this ferromagnetic which is nanometer-sized ferromagnetic turned to paramagnetic has a different behavior than the conventional paramagnetic and is known as superparamagnetic. While multiple magnetic domains are formed by bulk ferromagnetic materials, only one domain is formed by small magnetic nanoparticles exhibiting a phenomenon known as superparamagnetism. The overall magnetic coercivity in this case is then lowered, and there is random distribution of magnetizations of the various particles due to thermal fluctuations and only get aligned in the presence of an applied magnetic field. The nanoscale multilayers show giant magnetoresistance (GMR) which consist of a strong ferromagnet (e.g., Fe, Co) and a weaker magnetic or nonmagnetic buffer (e.g., Cr, Cu).

1.4 Conclusions

Enormous progress is made by nanotechnology in the past decades. In summary, the requirements of nanotechnology are the fabrication of matter on the scale of atoms and molecules, prediction, and measurement. Hopefully, there is revolutionary impact of the atomic scale nanotechnology in the way of doing, designing, and producing things in the future. Nanotechnology can be defined as an atomic or molecular approach using which physically, chemically, and biologically stable structures can be built of one atom, or one molecule, at a time. Nanomaterials can be classified as organic, inorganic and carbon based. According to their size 0D, 1D, and 2D nanostructures are defined. Due to large surface to volume ratio and size in nanometer range, these materials exhibit unique physical, chemical, optical, thermal and magnetic properties. The results of developments and investigations in nanotechnological fields are entering into all areas of our lives, like aerospace, agriculture, materials science, energy, medicine, defense, and environmental science. There are some active research areas which include nanodevices, nanolithography, nanopowders, nanorobotics, nanostructured catalysts, nanocomputers, nanoporous materials and molecular nanotechnology, nanolayers, molecular manufacturing, medicines and nanobiology (e.g., prediction, prevention, and treatment of diseases), and some organic nanostructures. Researchers have come to know from many years that current technologies are depending on processes which take place at the nanoscale. Some instances of these technologies are adsorption, lithography, catalysis, plastics, drug design, composites, and ion exchange.

References

Amer MA (2017) Structural , elastic and magnetic studies of the as-synthesized co 1 À x Sr x Fe 2 O 4 nanoparticles. J Alloys Compd 690:293–303. https://doi.org/10.1016/j.jallcom.2016.08.135

Baer DR, Burrows PE, El-Azab AA (2003) Enhancing coating functionality using nanoscience and nanotechnology. Prog Org Coat 47:342–356. https://doi.org/10.1016/S0300-9440(03)00127-9

Bhushan B (2016) Introduction to nanotechnology: history, status, and importance of nanoscience and nanotechnology education. https://doi.org/10.1007/978-3-319-31833-2_1

Chandamma N, Manohara BM, Ujjinappa BS, Shankarmurthy GJ, Santhosh Kumar MV (2017) Structural and electrical properties of zinc doped nickel ferrites nanoparticles prepared via facile combustion technique. J Alloys Compd 702:479–488. https://doi.org/10.1016/j.jallcom.2016.12.392

Chithra M, Anumol CN, Sahu B, Sahoo SC (2017) Structural and magnetic properties of ZnXCo1−XFe2O4 nanoparticles: nonsaturation of magnetization. J Magn Magn Mater 424:174–184. https://doi.org/10.1016/j.jmmm.2016.10.064

Costa ACFM, Diniz APA, de Melo AGB, Kiminami RHGA, Cornejo DR, Costa AA, Gama L (2008) Ni-Zn-Sm nanopowder ferrites: morphological aspects and magnetic properties. J Magn Magn Mater 320:742–749. https://doi.org/10.1016/j.jmmm.2007.08.011

Gao P, Hua X, Degirmenci V, Rooney D, Khraisheh M, Pollard R, Bowman RM, Rebrov EV (2013) Structural and magnetic properties of Ni1-xZnxFe 2O4 (x=0, 0.5 and 1) nanopowders prepared by sol-gel method. J Magn Magn Mater 348:44–50. https://doi.org/10.1016/j.jmmm.2013.07.060

Ghaffari M, Huang H, Tan OK, Shannon M (2012) CrystEngComm Band gap measurement of SrFeO 3 2 d by ultraviolet photoelectron spectroscopy and photovoltage method. https://doi.org/10.1039/c2ce25751c

Ghazi N, Mahmoudi Chenari H, Ghodsi FE (2018) Rietveld refinement, morphology analysis, optical and magnetic properties of magnesium-zinc ferrite nanofibers. J Magn Magn Mater 468:132–140. https://doi.org/10.1016/j.jmmm.2018.07.084

Gleiter H (2009) Nanoscience and nanotechnology: the key to new studies in areas of science outside of nanoscience and nanotechnology. MRS Bull 34:456–464. https://doi.org/10.1557/mrs2009.122

Harris VG, Šepelák V (2018) Mechanochemically processed zinc ferrite nanoparticles: evolution of structure and impact of induced cation inversion. J Magn Magn Mater 465:603–610. https://doi.org/10.1016/j.jmmm.2018.05.100

Harzali H, Marzouki A, Saida F, Megriche A, Mgaidi A (2018) Structural, magnetic and optical properties of nanosized Ni0.4Cu0.2Zn0.4R0.05Fe1.95O4 (R = Eu3+, Sm3+, Gd3+ and Pr3+) ferrites synthesized by co-precipitation method with ultrasound irradiation. J Magn Magn Mater 460:89–94. https://doi.org/10.1016/j.jmmm.2018.03.062

Ling W, Zhang H, He Y, Wu Y, Yang K, Li Y, Li S (2010) Low temperature fired Ni-cu-Zn ferrite with bi 4 Ti 3 O 12. J Magn Magn Mater 322:819–823. https://doi.org/10.1016/j.jmmm.2009.11.010

Mallesh S, Srinivas V (2019) A comprehensive study on thermal stability and magnetic properties of MnZn-ferrite nanoparticles. J Magn Magn Mater 475:290–303. https://doi.org/10.1016/j.jmmm.2018.11.052

Manikandan V, Denardin JC, Vigniselvan S, Mane RS (2018) Structural, dielectric and enhanced soft magnetic properties of lithium (Li) substituted nickel ferrite (NiFe2O4)nanoparticles. J Magn Magn Mater 465:634–639. https://doi.org/10.1016/j.jmmm.2018.06.059

Mehta RV (2017) Synthesis of magnetic nanoparticles and their dispersions with special reference to applications in biomedicine and biotechnology. Mater Sci Eng C 79:901–916. https://doi.org/10.1016/j.msec.2017.05.135

Melo RS, Silva FC, Moura KRM, De Menezes AS, Sinfrônio FSM (2015) Magnetic ferrites synthesised using the microwave-hydrothermal method. J Magn Magn Mater 381:109–115. https://doi.org/10.1016/j.jmmm.2014.12.040

Narang SB, Pubby K (2021) Nickel spinel ferrites: a review. J Magn Magn Mater 519:167163. https://doi.org/10.1016/j.jmmm.2020.167163

Nayak RR, Pradhan N, Behera D, Pradhan KM, Mishra S, Sukla LB, Mishra BK (2011) Green synthesis of silver nanoparticle by Penicillium purpurogenum NPMF: the process and optimization. J Nanopart Res 13:3129–3137. https://doi.org/10.1007/s11051-010-0208-8

Nouailhat A (2010) An introduction to nanoscience and nanotechnology. Nanotechnology:1–206. https://doi.org/10.1002/9780470610954

Pei J, Wang Z (2018) Effect of bi-co co-doping on the microstructure and magnetic properties of NiMgCuZn ferrites. J Magn Magn Mater 465:598–602. https://doi.org/10.1016/j.jmmm.2018.06.045

Rafiq MA, Khan MA, Asghar M, Ilyas SZ, Shakir I, Shahid M, Warsi MF (2015) Influence of Co2+ on structural and electromagnetic properties of Mg–Zn nanocrystals synthesized via co-precipitation route. Ceram Int 41:10501–10505. https://doi.org/10.1016/j.ceramint.2015.04.141

Rashad MM, Hessien MM, El-midany A, Ibrahim IA (2009) Journal of magnetism and magnetic materials effect of synthesis conditions on the preparation of YIG powders via co- precipitation method. J Magn Magn Mater 321:3752–3757. https://doi.org/10.1016/j.jmmm.2009.07.033

Saeedi Afshar SR, Hasheminiasari M, Masoudpanah SM (2018) Structural, magnetic and microwave absorption properties of SrFe12O19/Ni0.6Zn0.4Fe2O4 composites prepared by one-pot solution combustion method. J Magn Magn Mater 466:1–6. https://doi.org/10.1016/j.jmmm.2018.06.061

Sawant VJ, Bamane SR, Shejwal RV, Patil SB (2016) Comparison of drug delivery potentials of surface functionalized cobalt and zinc ferrite nanohybrids for curcumin in to MCF-7 breast cancer cells. J Magn Magn Mater 417:222–229. https://doi.org/10.1016/j.jmmm.2016.05.061

Sertkol M, Köseoğlu Y, Baykal A, Kavas H, Toprak MS (2010) Synthesis and magnetic characterization of Zn0.7Ni0.3Fe2O4 nanoparticles via microwave-assisted combustion route. J Magn Magn Mater 322:866–871. https://doi.org/10.1016/j.jmmm.2009.11.018

Sivakumar P, Ramesh R, Ramanand A, Ponnusamy S, Muthamizhchelvan C (2011) Preparation and properties of nickel ferrite (NiFe2O 4) nanoparticles via sol-gel auto-combustion method. Mater Res Bull 46:2204–2207. https://doi.org/10.1016/j.materresbull.2011.09.010

Sui X, Scherge M, Kryder MH, Snyder JE, Harris VG, Koon NC (1996) Barium ferrite thin-film recording media. J Magn Magn Mater 155:132–139. https://doi.org/10.1016/0304-8853(95)00722-9

Thakur A, Hsu J (2011) Novel magnetodielectric nanomaterials with matching permeability and permittivity for the very-high-frequency applications. Scr Mater 64:205–208. https://doi.org/10.1016/j.scriptamat.2010.09.045

Thakur A, Thakur P, Hsu JH (2014) Structural, magnetic and electromagnetic characterization of In3+ substituted Mn-Zn nanoferrites. Z Phys Chem 228(6–7):663–672. https://doi.org/10.1515/zpch-2014-0477

Wang HG, Liu W, Jia N, Zhang M, Guo M (2017) Facile synthesis of metal-doped Ni-Zn ferrite from treated Zn-containing electric arc furnace dust. Ceram Int 43:1980–1987. https://doi.org/10.1016/j.ceramint.2016.10.164

Whitesides GM (2005) Nanoscience, nanotechnology, and chemistry. Small 1:172–179. https://doi.org/10.1002/smll.200400130

Yadav RS, Havlica J, Masilko J, Kalina L, Wasserbauer J, Hajdúchová M, Enev V, Kuřitka I, Kožáková Z (2016) Impact of Nd^{3+} in $CoFe_2O_4$ spinel ferrite nanoparticles on cation distribution, structural and magnetic properties. J Magn Magn Mater 399:109–117. https://doi.org/10.1016/j.jmmm.2015.09.055

Zhang E, Yin D, Xu L, Yang L, Yang K (2009) Microstructure, mechanical and corrosion properties and biocompatibility of Mg-Zn-Mn alloys for biomedical application. Mater Sci Eng C 29:987–993. https://doi.org/10.1016/j.msec.2008.08.024

Nanomaterials, their Types and Properties 2

Preeti Thakur and Atul Thakur

Abstract

In this modern era, nanotechnology may be called as an important field that deals with the synthesis, design, and manipulation of particle structures that have dimensions ranging from 1 to 100 nm. The classification of nanoparticles can be done on the basis of their different properties, shapes, or sizes. Due to their nanoscale size and high surface area, they possess unique physical and chemical properties. Even their properties like reactivity, toughness, and some other factors are also dependent on their unique shape, size, and structure. This is the reason of diverse applications of nanoparticles in different areas like optoelectronics, electronics, pharmaceuticals, biomedical sciences, health care, drug delivery, environmental health, cosmetics, chemical industries, food industry, optics, nonlinear optical devices, space technology, and energy sciences. Ferrites are magnetic nanoparticles of oxides of iron and are of much interest because of their biological compatibility, chemical stability, relative ease of preparation, and various other applications associated with them. So, various types of nanoparticles and their properties are covered in this chapter. Also, the structure, types, applications, and properties of ferrite nanoparticles will also be discussed.

Keywords

Nanoparticles · Properties · Applications · Ferrites

P. Thakur (✉)
Department of Physics, Amity School of Applied Sciences, Amity University Haryana, Gurugram, India

A. Thakur
Centre of Nanotechnology, Amity University Haryana, Gurugram, India

A. Thakur et al. (eds.), *Synthesis and Applications of Nanoparticles*,
https://doi.org/10.1007/978-981-16-6819-7_2

2.1 Introduction to Nanomaterials

Nanotechnology (Mathew and Juang 2007) is defined as the engineering, science, and technology conducted at nanoscale (1–100 nanometers) in which the word "nanometer" refers to one billionth of meter or 10^{-9} m in size, and expressed as 1 nm.

2.1.1 Starting of Nanotechnology

A renowned physicist Richard Feynman's talk entitled "There's Plenty of Room at the Bottom" at the American Physical Society meeting at the California Institute of Technology (CalTech) on December 29, 1959, includes the ideas and concepts behind nanoscience and nanotechnology. Feynman explained a mechanism, how individual atoms and molecules can be manipulated and controlled by scientists. After a decade, Professor Norio Taniguchi coined the term nanotechnology in his explorations of ultra-precision machining. In 1981, scanning tunneling microscope was developed, and it was made possible to "see" individual atoms, with that modern nanotechnology began. The meaning of prefix "nano" is one billionth which originated from the Greek word "dwarf" and is about 1 ten thousandths (10^{-4}) the diameter of a human hair. One nanometer (1 nm) is defined as 1/1000,000,000 of a meter. To get a sense of the nanoscale, a human hair measures 50,000 nanometers across, and a bacterial cell measures a few hundred nanometers across. The unaided human eye can see smallest things which measure 10,000 nanometers across. One nanometer can be made with ten hydrogen atoms arranged in a line, and really it is very small. Many atoms and molecules are of diameter of a few nanometers or less. As an example, the smallest atom of hydrogen has a diameter of 0.078 nm. Thousands of atoms are consisted by some biological molecules that have a diameter in 10's of nm or more. For instance, the diameter of RBC is about 7000 nm and of water molecule is about 0.3 nm. Nanotechnology may be defined as the "Engineering of functional systems at the molecular scale." In simple words, "Nanotechnology is the creation, use or manipulation of matter on the atomic scale." Nanotechnology is an emerging, interdisciplinary field that combines principles of physics and chemistry with the engineering principles of electrical engineering, mechanical design, computer science, structural analysis, and system engineering. Nanotechnology is the technology of preference to make things small, light, and cheap. Nanomaterials are materials with morphological features smaller than a micron in at least one dimension. The term nanotechnology includes nano particles, nano powders, nano clusters, and nano crystals. Nanotechnology refers to a field of applied science and technology whose theme is construction and fabrication of matter or devices or materials on the atomic and molecular scale in the range of 1–100 nm.

2.1.2 Classification of Nanomaterials

Nanomaterial size is extremely small, that is, materials in the nanoscale exist in one dimension, two dimensions, and three dimensions. Nanomaterials can also occur in different shapes and phases. Generally, nanomaterials are categorized into four types on the basis of their origin, dimensions, and structure.

2.1.2.1 Zero-Dimensional Nanomaterials

In zero-dimensional nanomaterials, the electron movement is confined in all the three dimensions, for example, quantum dots, gold spheres, and silver clusters. These types of nanomaterials generally have spherical morphology (average diameter in the range of 1–50 nm), cubic structures, and polygon-shaped materials.

2.1.2.2 One-Dimensional Nanomaterials

In one-dimensional nanomaterials, free electron movement is confined only in one dimension. Nanotubes, nanorods, nanowires, and nanofibers are a few examples. The diameter of such materials falls in the nanoscale range, whereas the length is out of nanoscale range in several micrometers.

2.1.2.3 Two-Dimensional Nanomaterials

In two-dimensional nanomaterials, free electron movement is confined in two dimensions, for example, thin films, nanosheets, multilayer film, nanowalls, etc. Such nanomaterials may have larger surface area and may be of several micrometers, and the thickness is in the nanoscale range.

2.1.2.4 Three-Dimensional Nanomaterials

In three-dimensional nanomaterials, the free electrons can move in all the three dimensions, for example, bulk materials, particles, and hollow spheres consisting of materials of nanoscale range as building blocks.

2.1.3 Synthesis of Nanoparticles

The two bottom-up and top-down approaches are used in nanotechnology. In bottom-up approach, material and devices are built from molecular atoms which assemble themselves chemically, and in top-down approach, nano objects are constructed from larger entities without atomic level control. Nanotechnology allows the generation of a great variety of matters, devices, and products with properties unobtainable by usual invention techniques. Top-down approach creates smaller objects using the larger objects, begins with a pattern generated on a larger scale and then reduced to nanoscale, and is relatively an expensive and time-consuming technique, whereas bottom-up approach arranges smaller components into more complex, and fabrication is much less expensive (Fig. 2.1).

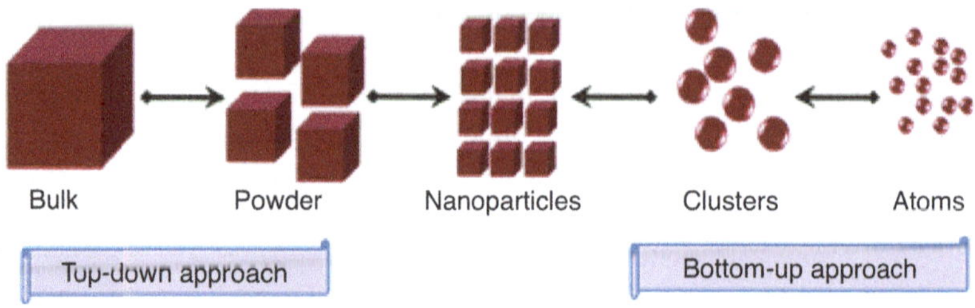

Fig. 2.1 Synthesis approaches for nanoparticles

2.2 Properties of Nanomaterials

Nanomaterials are different from other materials on the basis of various factors like quantum effects, reduced imperfections, surface area, surface energy, surface atoms, etc. which are responsible for the enhancement or alteration of the characteristic properties of the materials like optical, mechanical, electrical, and magnetic properties. The major advantage of nanomaterials is that many atoms are available on the material's surface with the decrease in particle size. Different properties of nanomaterials are discussed below:

2.2.1 Optical Properties

The study of nanomaterials is very important due to their novel optical properties in comparison with their bulk counterparts. The optical properties of the nanomaterials are dependent on several parameters like shape, dimensions, doping, surface properties, etc. Also, the surface to volume ratio increases with a decrease in the size of the materials.

2.2.2 Magnetic Properties

The magnetic materials are the materials that exhibit permanent magnetization in the absence of an external magnetic field. The terms used to measure the strength of the magnets are coercivity and saturation magnetization. There is an increase in the saturation and decrease in coercivity, with the decrease in size of nanoparticles. For example, in the bulk form, Au, Pt, and Pd are nonmagnetic, but in nano size, they behave as magnetic materials.

2.2.3 Size-Dependent Properties

Under a certain condition, how the material acts is described by properties of a material which are often measured by looking at large ($\sim 10^{23}$) aggregation of atoms or molecules. The properties that often change are electrical (e.g., conductivity), chemical (e.g., reactivity, reaction rates), optical (e.g., color, transparency), and physical (e.g., hardness, boiling point) among others.

2.2.4 Chemical Properties

The nanoscience and nanotechnology have a great impact on the chemical properties of nanomaterials. In case of nanomaterials, due to large surface area and small size, chemical activity is very high that make these materials useful for a wide range of practical applications. Hence, these materials can be used as catalysts for water decomposition and thereby produce energy, which is clean, efficient, and environmental-friendly preventing pollution.

2.3 Ferrites

The history of ferrites, that is, magnetic oxides, began with the discovery of stones that would attract iron, centuries before the birth of Christ. In the district of Magnesia in Asia Minor, the most plentiful deposits of these stones were found, and thus the mineral's name became magnetite (Fe_3O_4). Much later, the first application of magnetite was to locate magnetic North as "Lodestones" used by early navigators. In 1600, William Gilbert published the foremost scientific study on magnetism and called it *De Magnete*. In 1819 Hans Christian Oersted observed that a magnetic compass needle affects an electric current in a wire. Later Faraday, Hertz, Maxwell, and many others developed the new science of electromagnetism. Naturally occurring magnetite is a weak "hard" ferrite. Permanent magnetism is possessed by "hard" ferrites. After that, hard ferrites that were man-made were developed with superior properties, but producing an analogous "soft" magnetic material in the laboratory proved elusive. During the 1930s research on "soft" ferrites continued in Japan and the Netherlands. However, it was not until 1945 that a "soft" ferrite for commercial applications was produced by J. L. Snoek from the Philips Research Laboratories of the Netherlands. Originally manufactured soft ferrites for a multitude of uses have proliferated into countless sizes and shapes for inductor and antenna applications. Ferrites are used in power applications, multilayer inductor chip (MLIC) applications, and electromagnetic interference (EMI) suppression. In 1956, garnet ferrite class of materials was discovered by Neel. This type of ferrite material has three sub-lattices and is also referred to as rare-earth iron garnets. These materials have a magnetization lower than spinel ferrite. The application range of ferrites in electronic circuitry has been continuously growing. This continuing improvement in material characteristics, a wide range of possible geometries, and their relative cost-effectiveness make ferrite components the choice for both innovative and conventional applications.

The term "ferrites" has a different meaning for the different scientific community. For metallurgists, the meaning of ferrite is pure iron. For geologists, the term ferrites stand for a set of iron oxide-based minerals. For an electrical engineer also, ferrite means a group of iron oxide-based materials having some specific magnetic as well as dielectric properties. The meaning of magnetic properties is that iron-based materials are attracted by a piece of ferrite, and magnets of opposite polarity are attracted and magnets of like polarity are repelled. Both geologists and engineers consider magnetite or lodestone a ferrite that is a naturally occurring iron oxide. The strange properties of lodestone were recognized over 2000 years ago by the ancient Greeks, and Chinese used it to invent the magnetic compass almost 1000 years ago. The meaning of dielectric properties is that ferrites do not readily conduct electricity even though electromagnetic waves can pass through them. So, they are advantageous over nickel, iron, and other transition metals that show useful magnetic properties in several applications as these metals are found to also conduct electricity. Ferrites are termed as magnetic materials due to magnetic moments of the molecules of the material. The magnetic field is produced when they all line up. There is a distinctive arrangement of parallel and perpendicular magnetic moments in "ferrimagnetic" ferrites as compared to "ferromagnetic" metals. This effect can be achieved through several different crystal structures. Both the ferromagnetic and ferrimagnetic materials are found to lose their magnetism when heated at a very high temperature or are subjected to mechanical stresses. This is due to the fact that their magnetism depends on an orderly crystal structure.

Ferrites are found to exhibit diverse uses. Ferrites are used whensoever a fixed magnet, instead of an electromagnet, is required. The use of ferrites is in electric generators and electric motors, transformer, and inductor cores. Ferrite coating on top of the plastic base for recording of the signal was used in videotapes and cassettes. And many computers up to the 1970s employed magnetic core memory with ferrite cores (ferrite core memories were even used in the Space Shuttle until 1990 due to their dependability). Ferrites have recently been used as a medium for transmitting microwaves, which is perhaps their most important application. This is because some ferrites display a nonreciprocal effect at very high frequencies (starting at 500 MHz and very strongly in the microwave range of 1–30 GHz). That implies electromagnetic waves flowing through them go in different directions and react differently. One-way transmission allows for the creation of structures that can manage microwave "traffic" and other microscopic traffic and other microwave control devices. Without ferrites, our contemporary telecommunications system would not be conceivable. In 1956, discovery of garnet ferrite class of materials is done by Neel. This ferrite substance is also known as rare-earth iron garnets because it has three sub-lattices. Despite having a lower magnetism than spinel ferrite, these materials are useful.

2.3.1 Soft Ferrites

Soft ferrites with a low coercivity are ferrites that contain nickel, zinc, and/or manganese compounds and are utilized in transformer or electromagnetic cores. Low coercivity means that the direction of a material's magnetization can be easily reversed without wasting a lot of energy (hysteresis losses). These materials have high resistivity which is another source of energy loss that prevents eddy currents in the core. Because of their low losses at high frequencies, they are commonly employed in RF transformers and inductors cores, as well as in applications such as switched mode power supply. Manganese-zinc ferrite is the most prevalent soft ferrite (MnZn, with the formula $Mn_xZn_{1-x}Fe_2O_4$). MnZn has higher permeability and saturation induction than NiZn. Nickel-zinc ferrite (NiZn, with the formula $Ni_aZn_{(1-a)}Fe_2O_4$). NiZn ferrites have a higher resistivity than MnZn ferrites, making them better suited to frequencies beyond 1 MHz.

2.3.2 Hard Ferrites

In contrast, hard ferrites are used in making permanent ferrite magnets having high coercivity and high remanence magnetization. These are made up of iron and barium or strontium oxides. The materials are very resistant to becoming demagnetized due to their high value of coercivity, which is an essential characteristic for a permanent magnet. Also the magnetic permeability is high, and they also conduct magnetic flux well. Hence, stronger magnetic fields than iron can be stored in these so-called ceramic magnets. They are widely used in household products and are very cheap. The value of maximum magnetic field B is about 0.35 Tesla and of magnetic field strength H is about 30–160 kilo ampere turns per meter (400–2000 Oersted). The value of density of ferrite magnets is about 5 g/cm^3. The examples of most common hard ferrites are barium ferrite, $BaFe_{12}O_{19}$ ($BaO{\cdot}6Fe_2O_3$), which is a common material for permanent magnet applications, and strontium ferrite, $SrFe_{12}O_{19}$ ($SrO{\cdot}6Fe_2O_3$), a common material for permanent magnet applications. Barium

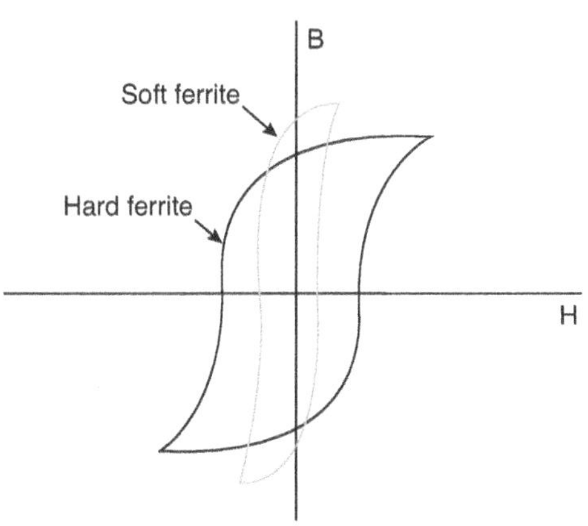

Fig. 2.2 Hysteresis loop of soft and hard ferrites

ferrites are corrosion resistant and stable to moisture. They are used in, for example, subwoofer magnets and as a medium for magnetic recording, on magnetic stripe cards; cobalt ferrite, $CoFe_2O_4$ ($CoO{\cdot}Fe_2O_3$), is used in some media for magnetic recording (Fig. 2.2).

2.4 Ferrite Structure

Ferrites contain iron oxide and metal oxides as their main constituents. Ferrites are classified into the following types depending upon the crystal structure.

2.4.1 Spinel Ferrite

Spinel ferrites are also called cubic ferrite as shown in Fig. 2.3. It is the most widely used family of ferrite. These materials are ideal for use at microwave frequencies due to high values of electrical resistivity and low eddy current losses. Bragg and Nishikawa determined the spinel structure of ferrite firstly as possessed by mineral spinel $MgAl_2O_4$ in 1915. The formula of a spinel ferrite can be written in general as MFe_2O_4 where M is a divalent metal ion such as Zn^{2+}, Co^{2+}, Mg^{2+}, Fe^{2+}, Cu^{2+}, Ni^{2+}, Cd^{2+}, or a combination of these ions. The structure of unit cell of spinel ferrite is FCC having eight formula units per unit cell. The formula can be written as $M_8Fe_{16}O_{32}$. The anions form an FCC lattice and are the greatest. There are two types of interstitial positions within these lattices, and metallic cations occupy these positions. One unit cell has 96 interstitial sites in which 64 are tetrahedral (A) and 32 are octahedral (B) sites as depicted in Fig. 2.4. Ni-Cu-Zn is the class of soft ferrite and chemically symbolized as MFe_2O_4. $NiFe_2O_4$ and $CuFe_2O_4$ are inverse spinel and $ZnFe_2O_4$ ferrite has a normal spinel structure. Ni^{2+} and Cu^{2+} ions show their strong preference to the octahedral B site because of favorable fit of charge

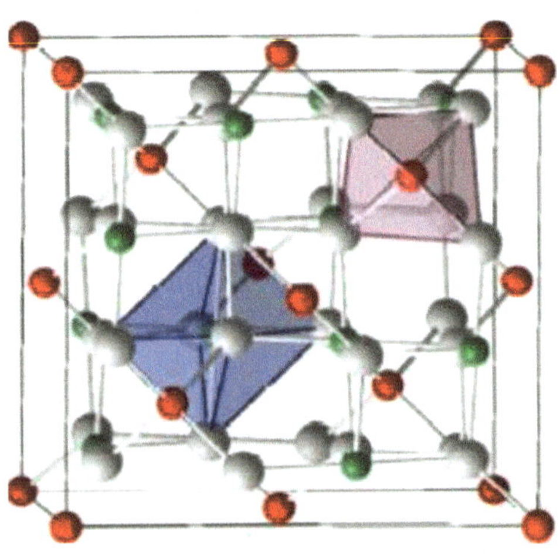

Fig. 2.3 Spinel unit cell structure

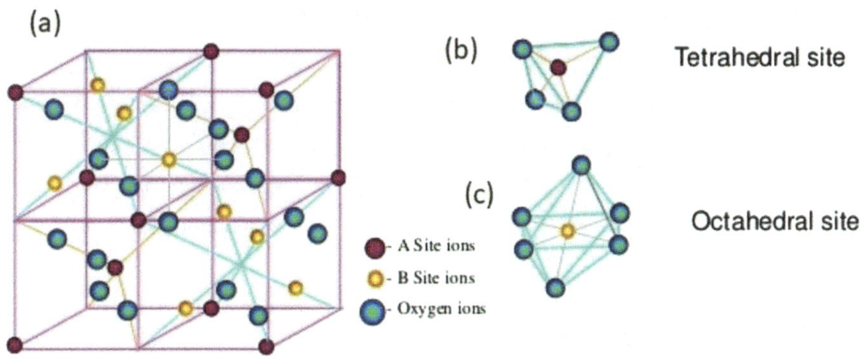

Fig. 2.4 (**a**) Spinel structure, (**b**) octahedral interstice (B site, 32 per unit cell, 16 occupied) and (**c**) tetrahedral interstice (A site, 64 per unit cell, 8 occupied)

distribution. Zn^{2+}ions show a strong preference for tetrahedral A site because of its electronic configuration. These ferrites are used in the surface-mount devices (SMD) and multilayer chip inductors (MLCI) due to their excellent soft magnetic properties at high frequencies and their high electrical resistivity. The spinel ferrite has been classified into three categories based on the distribution of cations on tetrahedral (A) and octahedral (B) sites: normal spinel ferrite, inverse spinel ferrite, and random spinel ferrite.

2.4.1.1 Normal Spinel

The spinel is normal if there is only one kind of cations on octahedral [B] site. In these ferrites the tetrahedral (A) sites are occupied by divalent cations while the octahedral [B] sites by trivalent cations. To indicate the ionic distribution of the octahedral [B] sites, square brackets are used. The formula for representing the normal spinel is $(M^{2+})_A[Me^{3+}]_BO_4$ (M represents divalent ions and Me for trivalent ions). Bulk $ZnFe_2O_4$A is a typical example of normal spinel ferrite. The cation disorder is defined in terms of a “normal” spinel structure, for example, for ideal $MgAl_2O_4$, all the Mg resides on tetrahedral sites coordinated with oxygen, and all the Al resides on octahedral sites coordinated with oxygen. The inversion parameter is defined as the ratio of the atomic fraction of Al on tetrahedral sites to the atomic fraction of Al on octahedral sites. The inversion parameter is 0.0 for a perfect normal spinel. Normal spinel structures are usually cubic closed-packed oxides having one octahedral and two tetrahedral sites per oxide. The tetrahedral points are smaller than the octahedral points. The octahedral holes are occupied by B^{3+} ions because of a charge factor but can only occupy half of the octahedral holes. 1/eighth of the tetrahedral holes are occupied by A^{2+} ions. If the ions are similar in size, this maximizes the lattice energy. A common example of a normal spinel is $MgAl_2O_4$.

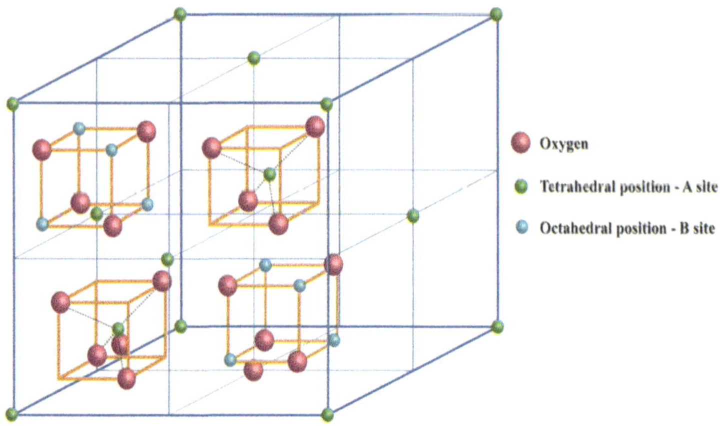

Fig. 2.5 Inverse spinel

2.4.1.2 Inverse Spinel

In inverse spinel structure, tetrahedral (A) sites and half octahedral [B] sites are occupied by half of the trivalent ions; the remaining cations are randomly distributed among the octahedral [B] sites as shown in Fig. 2.5. This formula for representing these ferrites is given by $(Me^{3+})_A$ $[M^{2+}Me^{3+]}{}_BO^4$. Fe_3O_4 is a typical example of inverse spinel ferrite which has divalent cations of Fe occupied at the octahedral [B] sites. In an ideal "inverse" spinel structure (e.g., $MgFe_2O_4$), octahedral sites have all of the Mg ions, and Fe is distributed equally over all of the tetrahedral sites and the remaining octahedral sites. The inversion parameter would be 1.0 in this case. Due to the crystal field stabilization energies (CFSE), of the transition metals present, inverse spinel structures are slightly different. Some ions dependent on the d-electron count may have a distinct preference on the octahedral site. If the A^{2+} ions have a strong preference for the octahedral site, they will force their way into it, and displacement of half of the B^{3+} ions from the octahedral sites to the tetrahedral sites takes place. If the B^{3+} ions have no preference if they have a low or zero octahedral site stabilization energy (OSSE), then they will adopt the tetrahedral site. A common example of an inverse spinel is Fe_3O_4, if the Fe^{2+} $(A^{2+)}$ ions are d^6 high-spin and the Fe^{+3} (B^{+3}) ions are d^5 high-spin.

2.4.1.3 Random Spinel

The spinel which has ionic distribution intermediate between normal and inverse is called mixed spinel structure. The cations are equally distributed over the two sites in ratios $[M^{2+}{}_x\ Fe^{3+}{}_{1-x}]A[M^{2+}{}_{1-x}\ Fe^{3+}{}_{1\ +\ x}]B\ O^{2-}{}_4$ for a "random" spinel structure, proportional to their stoichiometry and the site ratios. The inversion parameter is (2/3), or 0.667 for a random spinel structure. The spinel ferrite is called random spinel if the divalent metal ions and trivalent Fe^{3+}ions are distributed randomly over the tetrahedral and octahedral B sites.

2.4.2 Garnet Structure

The trivalent cations (including rare earth and Fe3+) occupy tetrahedral (d), octahedral (a), or dodecahedral—a 12-sided deformed polyhedron—(c) sites in garnets, which have an orthorhombic crystal structure (oxygen polyhedra encircling the cations). The tetrahedral and octahedral sites interact in an antiparallel manner, and the net magnetic moment is oriented in the opposite direction as the rare-earth ions on the c sites. It's the most intricate crystal structure, and drawing it in 2 dimensions with all 160 ions visible in the unit cell is difficult. The garnet structure is composed of a combination of octahedral (trivalent cation surrounded by 6 oxygen ions), tetrahedral (trivalent cations surrounded by 4 oxygen ions), and 12-sidedpolyhedral—dodecahedral—(trivalent cations surrounded by 8 oxygen atoms) sites. $3Me_2O_3 \cdot 5Fe_2O_3$ is the chemical formula for garnets in which Me represents the trivalent rare-earth ions like nonmagnetic yttrium or a magnetic rare earth such as from lanthanum through ytterbium.

2.4.3 Orthoferrites

Rare-earth orthoferrites are classified as ferrites, although they are called antiferromagnets. The magnetic oxides having perovskite structure are an exception in the group of oxides. The figure shown below depicts the perovskite structure. The corners of a cube are occupied by large divalent or trivalent ions (A), and small trivalent or tetravalent metal ions (B) occupy the center of the cube. The oxygen ions are situated centrally on the faces of the cube. The chemical formula of orthoferrites is ABO_3, where A represents yttrium or a rare earth. The inside of a sub-lattice is usually collinear ferromagnetic structure, but the different sub-lattices are coupled antiferromagnetically. Because of the different number of magnetic ions in different sub-lattices, there is a net resulting magnetic moment, giving rise to ferrimagnetism. The nature of the superexchange interaction depends not only on the type of the magnetic ion but rather strongly on the bond length and bonding angle (Fig. 2.6).

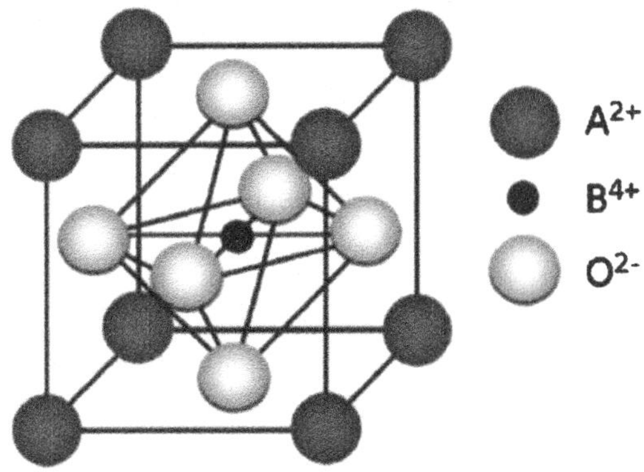

Fig. 2.6 An orthoferrite with perovskite structure

2.4.4 Hexagonal Ferrites

The crystal structure of regular barium hexaferrite is of the mineral magnetoplumbite. It belongs to the group of M-type hexagonal ferrite. The crystallographic unit cell is related to the space group P63/mmc and has two molecules with chemical composition $BaFe_{12}O_{19}$. The dimensions of the unit cell are approximately $a = b = 6$ Å and $c = 23$ Å. The unit cell has the basic structure made up of ten layers of oxygen ions that are formed by alternate close packing of cubic or hexagonal stacked layers. In every fifth layer, one O^{-2} ion is replaced by barium. The crystal structure of M-type barium can be divided into various blocks. The symbolic representation of M-type barium ferrite structure is described as RSR*S*, where R is a one-layer block having composition $Ba^{2+}Fe_3^{3+}O_3^{2-}$ and S is a four-O^{4-} layer block which has composition $Fe_9^{3+}O_{16}^{2-}$, where the asterisk means that the corresponding block has been turned 180° around the hexagonal c-axis. The interstices of oxygen atom have all the metal ions arranged in it. There are in total five Fe sites: two ions have tetrahedral surroundings ($4f_1$), and one Fe ion is located in a trigonal bipyramid (2b) with fivefold coordination (Ishino and Narumiya 1987); three octahedral positions (2a, 4f1, and 12k) are occupied by one, two, and six Fe ions.

2.5 Properties of Ferrites

From the application point of view, the structural, electrical, and magnetic properties of ferrite are very much important. By using proper method of preparation, sintering temperature, choosing appropriate type and amount of dopant, etc., the electrical and magnetic properties of ferrites can be modified.

2.5.1 Magnetic Properties

The magnetic characteristics of ferrite are determined by the distribution of metal cations over the available sites. Magnetic characteristics of ferrite are classified as intrinsic or structure-insensitive and extrinsic or structure-sensitive. Two prominent structure-insensitive properties are saturation magnetization and Curie temperature. On the other hand, there are a number of structure-sensitive properties, and the attribute can be characterized as static or dynamic depending on whether or not it has frequency dependence. Structure-dependent static qualities include induction, permeability, coercive force, hysteresis loop and associated energy loss, and remanence. Structure-sensitive dynamic features include spin resonance, eddy current loss, and domain barriers (Mathew and Juang 2007).

The properties which are unaffected by the microstructure are called as **intrinsic properties,** for example, grain size and crystal orientation of grains. The saturation magnetization (M_S) of ferromagnetic material is defined as the maximum attainable intensity of magnetization per unit volume. It is dependent on the strength of the dipole moment on the atom that makes the material and how densely they are packed together. The nature of the atom and the overall electronic structure within the compound affect the atomic dipole moment. The crystal structure, i.e., the spacing of the moments and the presence of any nonmagnetic elements within the structure, determine the packing density of the atomic moments. M_S will also depend on how well these moments are aligned at finite temperatures in case of ferromagnetic materials, as thermal vibration of the atoms causes misalignment of the moments and M_S is reduced. All of the moments are not aligned parallel for ferromagnetic materials even at zero Kelvin, and hence relative alignment of the moments and the temperature affect the M_S. Curie temperature is the characteristic temperature exhibited by ferromagnetic materials. The magnetic moments are partially aligned within magnetic domains in ferromagnetic materials at temperatures below the Curie point. Thermal fluctuations increasingly destroy this alignment as the temperature is increased from below the Curie point, until the net magnetization reached zero at and above the Curie point. The material is purely paramagnetic above the Curie point (Beatrice et al. 2008). An applied magnetic field has a paramagnetic effect on the magnetization, at temperatures below the Curie point, but the combination of paramagnetism with ferromagnetism leads to the magnetization followed by a hysteresis curve with the applied field strength. The destruction of magnetization at the Curie temperature is a second-order phase transition and a critical point where the magnetic susceptibility is theoretically infinite.

The extrinsic properties or structure-sensitive properties are listed under:

(a) Static properties.
(b) Dynamic properties.

Static properties include magnetic induction. If the performance of a magnetic component or device is determined by the magnetization of the ferromagnetic material involved, magnetization is used to evaluate the performance instead of induction.

This is because of the definition of induction, $B = \mu_0 (H + M)$.

Because the μ_0, H term has inherently made it a structure-sensitive static feature, induction is more beneficial than magnetization in practice. Magnetic permeability (μ) is defined as the ratio of magnetic induction (B) to magnetic field (H) or applied field strength. The ratio of flux density to applied field is known as absolute permeability. The closed B-H curve is called hysteresis loop because B lags behind H in the B-H curve. As illustrated in the diagram below, this loop appears in the four quadrants (Fig. 2.7).

In the B-H plot, an energy is given by $W = \int H.\ dB$ is represented by the area enclosed by hysteresis loop. This magnetic energy in the form of heat dissipates immediately into the lattice upon generation and lost permanently. Thus, hysteresis loss is designated by having units of Joules per m^3 per cycle. Hysteresis loss is very undesirable in soft magnetic materials, not only because it wastes energy, but

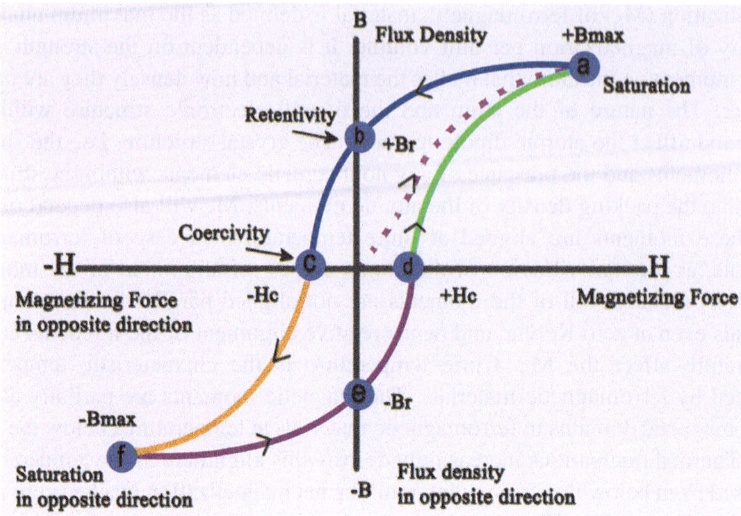

Fig. 2.7 Typical hysteresis loop

problems are also created due to heat involved in the application of the material. The ability of a ferromagnetic material to withstand an external magnetic field without being demagnetized is called as coercive force and coercivity. Coercivity is the criterion to determine whether a ferromagnetic material is soft or hard. A hard ferrite and a soft ferrite are differentiated based on their respective coercive force. H_c is large (~100 Oe) for a hard ferrite and small (<20 Oe) for a soft ferrite. The major goal of soft magnetic material preparation is to keep coercivity to a minimum. Coercivity is intimately related to magnetic properties that are structure-sensitive, as evidenced by experimental data. Remanence or retentivity is the amount of induction that remains after the field H is made zero. Remanence or retentivity is the name for this type of induction.

Eddy currents and the energy losses they cause, are examples of dynamic characteristics. An electromagnetic force (e.m.f.) is induced in a ferromagnetic body when it is magnetized by an alternating field. If the material is a good conductor, such as metals and alloys, the induced e.m.f. produces a significant quantity of currents in various places; these currents are known as eddy currents, and their presence causes energy loss (W). At high frequency, the resonance of Bloch walls is a prominent phenomenon found in ferrites. It plays an important role in the application of soft magnetic materials. It is found that the eddy current power loss is proportional to the square of the velocity of the domain walls. Thus, we must try to keep the wall velocity at the lowest possible value to minimize the loss. Some applications of ferrites at microwave frequency also involve the resonance of domain walls.

2.5.2 Electrical Properties

Due to their high electrical resistivity, ferrites play a useful role in many technological applications. Electrical properties of ferrites are dependent on the chemical composition as well as on the various heat treatments during the course of preparation. The method of preparation also affects these properties (Coey 2009). The Verwey hopping mechanism explained the variation in dc electrical resistivity (Jonker 1959) according to which, the electrical conduction in ferrites is due to the hopping of electrons between the ions of different valence states present at the octahedral sites but of the same element. The hopping of electrons between Fe^{3+} and Fe^{2+} ions present at octahedral sites causes the conduction in ferrites (Elwell et al. 1966). Generally, with the increase of temperature, the electrical resistivity of ferrites decreases. This shows semiconductor-like behavior of ferrites (Verwey et al. 1950). Ferrites are applicable for microwave applications due to their very high resistivity (Snoek 1947). For ferrites, resistivity values vary from 10^{-2} ohm-cm to 10^{11} ohm-cm depending on chemical composition of the material at room temperature (Mathew and Juang 2007).

2.5.3 Dielectric Properties

Many factors decide the dielectric properties such as the method of preparation, grain structure or size, and chemical composition. The valence state changes when a ferrite is sintered under slightly reducing condition, and the individual cation found in the sample leads to high conductivity. When cooling of such a material is done in an oxygen atmosphere, films of high resistivity are formed over the constituent grain. Such ferrites behave as inhomogeneous dielectric material in which the individual grains are separated by air gaps or low-conducting layers. This piqued people's curiosity in ferrite's dielectric behavior in the low-frequency region (10^2 Hz–10^5 Hz).

2.6 Application of Ferrites

Due to lower cost, high resistivity, superior magnetization properties, and easier manufacture, ferrites are considered as better magnetic materials as compared to pure metals. The application area of ferrites includes radar, bubble devices, audio-video and digital recording, microwave devices, satellite communication, and memory cores (Mathew and Juang 2007; Beatrice et al. 2008; Coey 2009). Application area of ferrites is very vast ranging from microwave to radio frequencies. They are used in flyback transformer in TV picture tube, mechanical filter, antenna cores in radio receivers, moderators, broadband transformer, phase shift, ultrasonic generators, and isolators. Ferrites are used in computers, telephone exchange, and control equipments nowadays. Ferrites are generally of two types—soft ferrite and hard ferrite. Soft ferrites are a class of magnetic materials used in telecommunication computer, transformer cores primarily in television, medical, and also in electronics.

Hard ferrites are being used in making permanent magnets, in loudspeakers and micromotors. The use of low loss polycrystalline ferrites should be in a high frequency range. For the good performance in application area and classified by the initial permeability, for the low and high-frequency applications, the most important technological properties are coercive force (H_C), initial permeability, saturation magnetization (M_S), and losses. It is generally impossible to get the best combination of these properties for any specific application. Most of the parameters can be controlled either by varying the compositions or by varying the synthesis technique or adding additives. Thin films of defect spinel ferrites have application as write-once read many media working with blue wavelengths. Also, due to the metastable nature of nonstoichiometric ferrites, they can be transformed by a laser spot into corundum phases at moderate temperatures. In contrast to the starting ferrite film, the transformed regions have different optical indices from which makes the readout process possible (Jonker 1959). Temperature is controlled by using magnetic sensors which are prepared by using ferrite materials having definite and sharp Curie temperature. Proximity switches can also be fabricated using ferrites. Also, ferrites are used to control pollution in which precipitation of ferrite precursors is used to scavenge pollutant materials such as lead and cadmium from waste streams. In addition, the produced ferrites are easily magnetically separated along with the pollutant. Due to their high resistance to corrosion, ferrites which have suitable conductivities can be used as electrode in applications, for example, chromium plating. Ferrites have applications in radio and television circuits such as flyback transformers, SMPS transformer for power applications, and deflection Yokes. Soft ferrites are used for computer memory hard disc, floppy disc audio-video cassette, high-frequency transformer core, and recorder head. Hard ferrites are used in permanent magnets in motor, generator, telephone, and loudspeaker.

2.6.1 Some Technically Important Ferrites

Due to high permeability at high frequency, mechanical hardness, remarkably high electrical resistivity, reasonable cost, and chemical stability, Ni-Zn ferrites having spinel crystal structure are extensively used in a number of electronic devices (Yamashita and Kurusawa 1958). Ni ferrite is a model inverse ferrite, and Zn ferrite is a model normal ferrite because of strong preferences of Zn and Ni for the tetrahedral and octahedral sites, respectively (Ishino and Narumiya 1987). The cation distribution of Ni-Zn ferrite is represented by $(Zn_xFe_{1-x})[Ni_{1-x}Fe_{1+x}]O_4$. This system has been extensively studied for various properties including structural issues. The studies on magnetic and dielectric properties in Ni-Zn ferrites synthesized by conventional technique have been reported by many researchers (Smit and Wijn 1959; Murthy and Ramaiah 2000; Ajmal and Maqsood 2008; Yadoji et al. 2003; Akther Hossaina et al. 2007). The most common method for preparing ferrites is conventional method from so decades. The dielectric properties of nickel-zinc ferrites $Ni_xZn_{1-x}Fe_2O_4$, where x changed from 0.2 to 1.0 were investigated by Mohan et al. (Abdeen 1998) using standard ceramic technique for synthesis. The investigation of frequency, temperature, and composition dependence of NiZn ferrites was done. The dielectric constant and loss tangent were found to decrease with an increase in zinc content up to $x = 0.4$. Beyond $x = 0.4$, these parameters increased progressively. This was due to the reason that the variation of dielectric constant depends linearly on the variation of available ferrous ions on octahedral sites. With increasing frequency dielectric constant declined. Maximum dielectric dispersion was seen at $x = 0.8$, which can be explained by the availability of ferrous ions on octahedral sites. The electron exchange between Fe2+ and Fe3+ in an n-type semiconducting ferrite and hole exchange between Ni3+ and Ni2+ in a p-type semiconducting ferrite explained the difference in dispersion for mixed nickel-zinc ferrites. The dielectric constant and DC resistivity are inversely proportional to one another. Additionally, the dielectric loss reflected in resistivity that the lower loss exhibited higher resistivity and vice versa. With an increase in temperature up to a particular temperature, there is gradual increase in dielectric constant, which is designated as the dielectric transition temperature Ts. However, there was a continuous decrease in the values of the dielectric constant for all the samples beyond this temperature. The conductivity of $Ni_{1-x}Zn_xFe_2O_4$ as a function of composition and temperature was studied by Ranga and Ravinder (Ranga et al. 1999). They observed an increase in conductivity with zinc content and temperature. The charge carrier concentration was calculated, and the higher charge carrier in higher temperature up to magnetic transition temperature was observed. There is a decrease in the concentration of charge carrier beyond this temperature. El-Sayed (El-Sayed et al. 2003) did further research on the electrical conductivity of $Ni_{1-x}Zn_xFe_2O_4$ ($x = 0.1$, 0.3, 0.5, 0.7, and 0.9) prepared by using ceramic processing technique. For $x = 0.3$, they found the lowest conductivity and higher activation energy in nickel-zinc ferrite. The substitution of Zn in the $Ni_{1-x}Zn_xFe_2O_4$ system was investigated by Ajmal and Maqsood (Ajmal et al. 2007), and they checked the effect on the physical properties. The variation of zinc content on DC resistivity, dielectric constant, and loss factor was observed. The dielectric constant was found to increase with the increase in Zn concentration. The resistivity values were calculated in the range of 1.629×10^6–3.0×10^3 Ω-cm. With the variation in frequency from 80 Hz to 1 MHz, loss factor remained in the range of 9.057–0.456. It is observed by various workers that the magnetization having the highest values for $x = 0.3$ /0.4 in $Ni_{1-x}Zn_xFe_2O_4$ (Murthy and Ramaiah 2000; Ajmal and Maqsood 2008; Ajmal and Maqsood 2007; Jadhav et al. 2008; Kakatkar et al. 1996). So it was decided to conduct further research with different ion substitution by keeping $Ni_{0.7}Zn_{0.3}Fe_2O_4$ as a parent compound. There are some inherent drawbacks in conventional methods, for example, poor compositional control, coarser particle, chemical inhomogeneity, and introduction of some impurities during milling. Thus, some voids and low-density areas are formed in the green compact due to the coarser and nonuniform particles (Li et al. 2000). However, a variety of wet methods such as ball milling, sol-gel, hydrothermal, coprecipitation, and microemulsions have been used to synthesize ferrite materials. There are unique advantages of each method. But, still it is a matter of interest to try to improve the physical properties of ferrites by new designing in synthesize. Recently sol-gel methods have been used to prepare ultrafine, reproducible, and homogenous powder (Kumar et al. 2011; Patil et al. 2002; Popovici et al. 2003; VijayaBhasker Reddy et al. 2010; Shirsath et al. 2010). In recent years, it has been discovered that the sol-gel approach has considerable advantages over traditional dry processing. The ultimate goal is to create ferrites with improved physical and magnetic properties that can be used in a range of industrial applications. The synthesis of nickel-zinc ferrites was done by Elsa E. Sileo (Mane et al. 2011) using citrate precursor method. The combustion method was used for the preparation of $Ni_{1-x}Zn_xFe_2O_4$ (0.2, 0.4, 0.5, 0.6, and 0.7). Metal nitrates and citric acid were used to obtain precursors by sol process and then heated at 200 °C. The heating of residues was done at 1000 °C for 2 h. Insertion of small amounts of different R (III) cations (R = yttrium, ruthenium, and rare-earth cations) into $Ni_{0.5}Zn_{0.5}$ Fe_2O_4 has also been studied. The structure and magnetic properties of Ni-Zn ferrites can be modified from rare-earth addition. The auto-combustion method was used to prepare $Ni_{0.5}$ $Zn_{0.5}$ $Fe_{2-y}R_YO_4$ samples. Different proportions of nickel and zinc oxalates and iron nitrates were weighed by authors and then diluted in water ([Fe (III)] + [Ni (II)] + [Zn (II)] = 1 M) followed by addition of 3 M citric acid solution (50 ml) to each metal solution. The heating was done at 40 °C with continuous stirring for 30 min. A highly viscous gel formation took place after evaporation which was heated at approx. 200 °C. Then calcination of the final residue was done at 1000 °C for 2 h. For the preparation of R-inserted sample, $Ni_{0.5}$ $Zn_{0.5}Fe_2O_4$ was used as reference compound. For XRD analysis, D5000 diffractometer was used, keeping 2-theta collection range 16. 125° with scanning time of 15 s and step size 0.02° with CuK_α radiation. The gel was heated in the temperature range 25–450 °C for the characterization of auto-combustion process at a heating rate of 10 °C min^{-1}. It was observed from XRD analysis that the only product in auto-combustion method was nanometer-sized spinels. There is an increment in crystallite size (10.1–12.0–13.6–24.7–33.4 nm) for $Ni_{0.5}$ $Zn_{0.5}$ Fe_2O_4 sample calcined for 1 h at temperatures (300, 500, 700, 1000 °C). There is a regular increase in lattice constant for Zn-substituted Ni ferrites with Zn(II) content. The decomposition range for the citrate precursor was found in the range of 197–369 °C for $NiFe_2O_4$ and 178–328 °C for $Ni_{0.5}Zn_{0.5}Fe_2O_4$. Pure ferrites were obtained for Ru ($y = 0.01$ and 0.02), Eu ($y = 0.02$), and Y ($y = 0.01$ and 0.02) preparations. There are many applications of Ni-Zn ferrites, and they are commercially used as high-frequency ferrites for radio-frequency coils and transformer cores.

Coprecipitation method was used by S.J. Azhagushanmugam (Azhagushanmugam et al. 2013) to synthesize nanoferrite powder followed by heat treatment at different sintering temperatures (130, 600, and 900°C). The structural, SEM, and FTIR investigations were done on the sintered Ni-Zn ferrite powders. From the line broadening in XRD pattern, the average crystallite was calculated. XRD patterns confirmed the single-phase cubic spinel structure of Ni-Zn ferrite. Particle size increased with an increase in sintering temperature from 45.59 nm at 130 °C to 47.21 nm at 600 °C and 50.47 nm at 900 °C.With increase in sintering temperature, lattice parameter also increased (at 130 it is found to be 8.359, at 600 °C is 8.360 and at 900 °C is 8.364). With increase of temperature, all the XRD peaks become sharper and narrower indicating an increase in particle size and enhancement of crystallinity. IR spectra were measured in the range of 400–4000 cm^{-1}. The characteristic bands at intensities (3400 cm^{-1}, 1500 cm^{-1} and 1380 cm^{-1} approx.) were observed. This is due to O-H stretching vibration. There is drastic increase in intensities of the bands with increase in sintering temperature due to the loss of residual water in the sample. From SEM it was observed that the microstructure consisted of clusters of about 1 μm in size.

Nickel ferrite is a very attracting class because of its interesting and important properties. This class of soft ferrite is also useful in many technical applications, like in catalysis, sensor, and so on. In the referred paper (Nejati and Zabihi 2012), $NiFe_2O_4$ nanoparticles were synthesized by the hydrothermal method, and investigation of the inhibition of surfactant (glycerol or sodium dodecyl sulfate) on the particle growth was done. The samples were prepared in the presence of glycerol and sodium dodecyl sulfate, for investigation of the inhibition effect of surfactant $NiFe_2O_4$ particle growth. The samples were characterized using X-ray diffraction (XRD), Fourier-transform infrared spectroscopy (FTIR), transmission electron microscopy (TEM), vibrating sample magnetometer, and inductively coupled plasma atomic emission spectroscopy (ICP-AES) techniques. The author obtained that the products were pure $NiFe_2O_4$ from XRD and ICP-AES and with increasing the temperature (from 30 nm at 100 °C to 53 nm at 150 °C), the growth in nanoparticles was also noticed, while under the same condition, surfactant prevents the particle growth. The Scherrer equation was used to calculate the average particle size, and TEM micrographs revealed the size of nanoparticles in the range of 50–60 nm in the presence of surfactant decreased up to 10–15 nm. The nanoparticle at the room temperature exhibited a superparamagnetic behavior.

The assisted hydrothermal methods were used for nanosized nickel ferrite particle synthesis with and without surfactant. The results show that the crystallinity of nanoparticles is increased with increasing of temperature. In comparison with surfactant-free prepared samples, the crystallinity of $NiFe_2O_4$ nanoparticles decreased in the presence of surfactant. At room temperature, all of the nickel ferrite nanoparticles were superparamagnetic. In his research Binu P Jacob (Jacob et al 2011) synthesized nickel ferrites by sol-gel and coprecipitation techniques. Attractive properties were shown by nanosized nickel ferrites for the application of soft magnets, core materials in power transformers, and low loss material at high frequencies. High permeability in the radio-frequency region, high Curie temperature, high electrical resistivity, and low eddy current loss are important properties of nickel ferries due to which these ferrites are suitable for many applications. XRD patterns of $NiFe_2O_4$ confirmed the formation of single-phase cubic $NiFe_2O_4$ nanoparticles with hematite phase. It is observed that nanoparticles with less crystallite size were obtained as compared with the sol-gel synthesis and the crystallite size was found to increase with the temperature in both processes. EDX analysis was done to check the stoichiometry of the powdered sample. In the samples prepared by coprecipitation, Ni deficiency and excess of oxygen were observed, while in the case of sol-gel-derived sample, expected stoichiometry was shown. The particle size and morphology of the sample were estimated by using TEM samples, and slightly agglomerated particles with almost spherical shape were observed in case of sol-gel annealed at 400 °C, and a particle size of 18–25 nm was obtained. An irregular shape with a wide particle size distribution of 8–20 nm was observed in coprecipitation-derived samples. FTIR spectral analysis confirmed the spinel structure in ferrite samples. The vibrational frequencies of IR bands ν_1 and ν_2 by coprecipitation ($\nu_1 = 401$, $\nu_2 = 597.5$) and by sol-gel ($\nu_1 = 407$, $\nu_2 = 584$) were observed. The spectra showed prominent bands near 3400 and 1600 cm^{-1}corresponding to stretching modes of H-O-H bending vibration absorbed water. VSM was used to do magnetic characterization of samples at room temperature with max applied field of 15 kOe. With the increase in annealing temperature, the saturation magnetization was found to increase. Less hysteresis loss with almost the same saturation magnetization was exhibited by coprecipitation at the same annealing temperature. It is suitable for synthesis of ferrites having very small size and superparamagnetic nature. For the preparation of homogenous, pure, and compositionally stoichiometric ferrites, sol-gel method is better. The effect of Fe^{3+} replacement by RE (Yb, Er, Sm, Tb, Gd, Dy, and Ce) ions on the properties of $(Ni_{0.7}Zn_{0.3})Fe_{2-x}RE_xO_4$ ferrite was investigated by Rezlescu et al. (Rezlescu et al. 1993). The electrical resistivity of a ferrite increased by substitution of a small quantity of Fe_2O_3 with RE_2O_3. The effects of rare-earth ions on the properties of $(Ni_{0.5}Zn_{0.5})$ $Fe_{1.98}RE_{0.02}O_4$ (RE = Y, Eu, or Gd) nominal composition were investigated by Sun et al. (Sun at el. 2004). The partial substitution of Fe^{3+} with a small amount of RE ions increased the electrical resistivity and relative loss factor, whereas it slightly decreased the Curie temperature.

Ni-Zn ferrite and Mg-Zn ferrite nanoparticles were synthesized by Kumar et al. (Kumar et al. 2013) and (Dora et al. 2014) by adapting green route via modified sol-gel method in which aloe vera plant extract was used as a precursor. X-ray diffraction (XRD), transmission electron microscopy (TEM), Fourier-transform infrared spectroscopy (FTIR), and vibrating-sample magnetometer (VSM) were used to determine the structural characteristics, size of the particles, and magnetization measurements. The aloe vera plant extract as a precursor is not only inexpensive and user-friendly, but it also serves as high production having a perfect structure and particle size. Ni-Zn ferrites were fabricated through sol-gel technique by Jacob et al. (Jacob et al. 2011) with the doping of terbium. XRD showed the formation of a single-phase FCC spinel structure. Investigators demonstrated that the properties, size, structure, and AC conductivity of prepared sample are influenced by the doping of Tb^{3+}.

Among other ferrites, zinc ferrite shows a superparamagnetic behavior. It is being used in various commercial applications, for example, as absorbents, sensors, photocatalysts, catalysts, and Li-ion batteries, because of its unique electrical and magnetic properties. As nanoferrites exhibit magnificent physical and chemical properties in comparison with bulk counterparts, they were widely examined. Zinc ferrite set a marvelous example of direct relation between structure, composition, and properties of nanoparticle (Jacob et al. 2012). The zinc ferrite has a spinel structure AB_2O_4 with A site occupied by Zn^{2+} ions and B site occupied by Fe^{3+} ions when prepared as a bulk material. Spinel ferrites can be classified as normal spinel like zinc ferrite or inverse spinel having half of the trivalent ions in the A site and the other half together with the divalent ions in the B sites. Various synthesizing methods have been reported to prepare high electromagnetic zinc ferrites, like in situ precipitation method, facile polyol process, high-energy ball milling, hydrothermal technique, reduction roasting, coprecipitation, advanced combustion route, solvothermal method, ultrasonic cavitation, microwave-assisted synthesis, microemulsion process, thermal plasma, sol-gel auto-combustion method, combustion front quenching method, chemical precipitation method, auto-combustion technique, self-propagating low-temperature combustion method, conventional method, etc. But, the above methods have some problems like nonuniform particle size and contain impurities that impose further advancement in the achievement of the products. Spinel ferrites are now widely used as sensing elements. Ferrites could be used as humidity/gas sensor materials due to very sensitive nature with these parameters. An important parameter for the sensing property is porosity. The size of pores affects and also serves as adsorption sites for humidity/gases.

2.6.2 $ZnFe_2O_4$ as Corrosion Inhibitor

Nowadays nanoparticles are playing a vital role in inhibiting corrosion. In actual, there are certain chemical substances which when applied on corrosive surface act as a corrosion inhibitor. These suitable inhibitors because of their ease in production and low cost possess properties such as low toxicity and high inhibition efficiency. High corrosion inhibition efficiency was shown by iron oxide nanoparticles due to their superior stabilizing proficiency. For green synthesis of iron oxide nanoparticles, olive oil was used as a natural agent. Therefore, olive oils are used as natural stabilizing agents wherein iron oxide nanoparticles are coated on mild steel and after storage of 6 months at room temperature so as to obtain good corrosion protection. The results indicated that the olive oil stabilized nanoparticles showed high inhibition proficiency having greater anticorrosion behavior (Naidu et al. 2011).

2.6.3 Cytotoxicity Effect of $ZnFe_2O_4$ on Cancer Cells

The effect of cytotoxicity of magnetic cobalt-zinc nanoferrite particles on the human prostate cancer cell lines was studied (Palanisamy et al. 2014). The cytotoxicity of nanoparticles is recently being studied for various biomedical applications. Dimercaptosuccinic acid (DMSA)-coated cobalt-zinc ferrite nanoparticles were studied on human prostate cancer cell lines, HPCs, (PC3 and DU145). It was also exhibited that in certain concentration, these nanoparticles are not only nontoxic but also lead to proliferation of cancer cells.

2.6.4 Zinc Ferrite for Dye Degradation

Nowadays due to the presence of dye in many industrial applications, most of the nations are confronting a problem of wastewater treatment. Due to complex aromatic structure, these dyes are very hard to degrade. In several processes in water, these dyes undergo hydrolysis and oxidation due to which they generate carcinogenic by-products. Hence, for the safety purposes, degradation of these dyes is essential. For this, nanoparticles have generated greater interest among researchers. Zinc nanoferrite particles are found to be efficient in eliminating organic and inorganic pollutants present in water (Gahrouei et al. 2013). The photocatalytic dye degradation in the presence of hydrogen peroxide and mineralization ability of zinc ferrite nanoparticles were studied wherein Reactive Red 120 and Reactive Red 198 were used as model dyes. The effects of $ZnFe_2O_4$ dose, salt, and initial dye concentration on the dye degradation were evaluated. It was hence validated that zinc nanoferrite ($ZnFe_2O_4$) could be proficiently used for dye degradation from colored wastewater.

2.7 Characterization Techniques

2.7.1 X-Ray Diffraction (XRD)

A schematic representation of powder X-ray diffractometer is shown in Fig. 2.8. X-rays are electromagnetic radiation having wavelength ranging from 0.01 to 10 nm. In diffraction experiments it is used in the typical wavelength range 0.5–1.5 A. The spacing in the grating should be of the same order as the wavelength for electromagnetic radiation to be diffracted. Therefore, to study the crystal structures, X-rays can be used.

The primary X-rays are made to fall on the sample substance to be investigated in this technique. X-ray gets diffracted to a certain angle due to its wave nature, like light waves. The information regarding the crystal nature of the substance is given by the angle of diffraction differing from the incident beam. By using a grating plate, the wavelength of X-rays can also be varied for any specific application. The interference due to X-rays which are scattered by the electrons in atoms located at various positions in the unit cell is represented by diffraction. Bragg's law given by $n\lambda = 2d\cdot\sin\theta$ is used to describe the diffraction from the crystal where n is the diffraction order, λ is the wavelength of the X-rays, d is the spacing between the

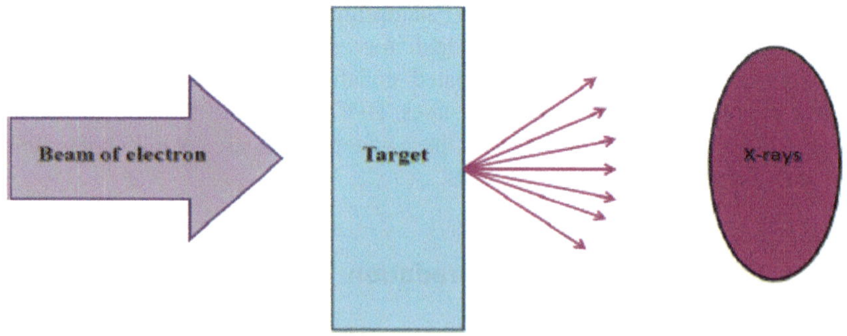

Fig. 2.8 Powdered X-ray diffractometer

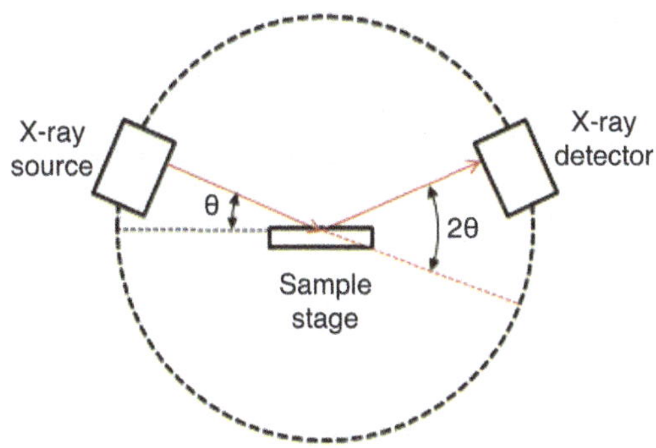

Fig. 2.9 Powder X-ray diffractometer

planes in an atomic lattice and θ is the angle between the incident ray and scattering planes. A powder X-ray diffractometer consists of an X-ray source (generally an X-ray tube), a detector, a sample stage, and also the provision for changing angle θ. At some angle θ, the X-ray is focused on the sample, while the detector opposite to the source reads the intensity of the X-ray which receives at 2θ away from the source path. The incident angle is increased over time, while the detector angle always remains 2θ above the source path (Fig. 2.9).

X-ray diffraction (XRD) method is employed to know the structural properties of materials and get information like lattice parameters, crystal structure/phase, orientation of single crystals, crystallite size, defects, preferred orientation of polycrystals, strains, and so on (Wu et al. 1987). This technique is suitable for bulk, nanomaterials, and thin films. The variation in lattice parameter w.r.t. bulk gives an idea of the type of strain that is present in the film in case of nanostructures (Fig. 2.10).

The intense peaks corresponding to planes 220, 311, 400, 420, 511, and 440 are shown by XRD spectra of sample of $ZnFe_2O_4$. These peaks confirmed cubic spinel structure without impurities. A single-phase spinel structure is noticed.

2.7.2 Scanning Electron Microscope (SEM)

An electron beam is focused on the sample surface kept in a vacuum by electromagnetic lenses in this technique (since electron shows properties of both particle and wave, hence an electron beam can be focused or condensed like an ordinary light). The beam is then scanned over the surface of the sample. Then through an amplifier, the scattered electron from the sample is fed to the detector and then to a cathode ray tube. The images are formed there, giving the information of the sample (Wu et al. 1987). It comprises a heated filament as a source of electron beam, aperture, condenser lenses, evacuated chamber for placing the sample, amplifier, electron detector, CRT with image forming electronics, etc. A largely magnified image by using electrons instead of light to form an image is produced by SEM. The figure shows a schematic diagram of the FE-SEM. A beam of electrons is produced at the top of the microscope by an electron gun. Then through the microscope, the electron

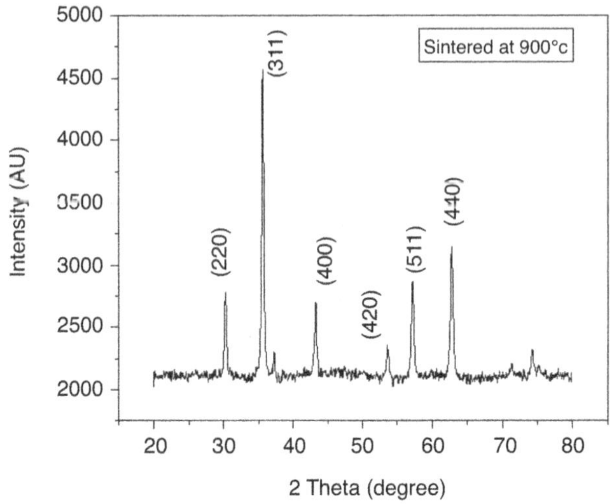

Fig. 2.10 XRD patterns of $ZnFe_2O_4$

beam follows a vertical path. The beam travels through electromagnetic fields and lenses, which focus the beam down toward the sample. Electrons and X-rays are ejected from the sample when the beam hits the sample. Detectors collect these X-rays, secondary electrons, and backscattered electrons and convert them into a signal that is sent to a screen similar to a television screen. This produces the final image.

Scanning electron microscopy has been used to study the surface properties of polymers, metals, composites, ceramics, and biological materials for both topography and compositional analysis. Electron probe microanalysis (EPMA) is an extension of this technique, wherein when the sample surface is exposed to a beam of high-energy electrons, the emission of X-rays takes place. This method is classified into two depending on the type of detectors: energy-dispersive spectrometry (EDS) and wavelength-dispersive spectrometry (WDS). The technique is widely used in the inclusions in polymeric materials, analysis of metallic and ceramic inclusions, and diffusion profiles in electronic component (Fig. 2.11).

The SEM is used to investigate the morphological properties of zinc ferrites. The SEM micrograph shows a compact arrangement of homogenous particles having almost spherical shapes. Most of the particles are aggregated, with some pores and voids.

2.8 Conclusions

Nanoparticles have a large surface area making them suitable material for diverse applications. XRD and SEM are used to investigate the structure and morphology of the nanoparticles. At nanometric size, their structural, electrical, magnetic, and

Fig. 2.11 SEM micrograph of $ZnFe_2O_4$

dielectric properties are dominant, which further increase the importance of these materials for multiple applications. Synthetic techniques can be useful to control the specific morphology, size, and magnetic properties of nanoferrites.

References

Abdeen AM (1998) J Magn Magn Mater 185:199

Ajmal M, Maqsood A (2007) Mater Sci Eng B 139:164

Ajmal M, Maqsood A (2008) Mater Lett 62:2077

Akther Hossaina AKM, Mahmuda ST, Sekib M, Kawaib T, Tabata H (2007) J Magn Magn Mater 312:210

Azhagushanmugam SJ, Suriyanarayanan N, Jayaprakash R (2013) Synthesis and characterization of Nano crystalline $Ni_{(0.6)}Zn_{(0.4)}Fe_2O_4$ spinel ferrite magnetic material. Physics Procedia 49:44–48

Beatrice C, Filorillo F, Landgraf FJ, Lazaro-Colan V, Janasiand S, Leicht J (2008) J Magn Magn Mater 320:865

Coey JMD (2009) Magnetism and magnetic materials. Cambridge University press

Dora BB, Kumar S, Sahu MC (2014) Size controlled synthesis and magnetic behaviour of mg-Zn nano ferrites by using Aloe vera extract solution. Int J Pharm Sci Rev Res 29(2014):307–311

El-Sayed AM (2003) Mater Chem Phys 82:583

Elwell D, Griffiths BA, Parker R (1966) Br J Appl Phys 17:249

Gahrouei D, Ghasemian Z, Abdolahi M, Manouchehri S, Javanmard SH et al (2013) *In vitro* evaluation of cobalt-zinc ferrite nanoparticles coated with DMSA on human prostate cancer cells. J Mol Biomark

Ishino K, Narumiya Y (1987) Ceram Bull 66:1469

Jacob BP, Thankachan S, Xavier S, Mohammed EM (2012) Dielectric behavior and AC conductivity of Tb^{3+} doped $Ni_{0.4}Zn_{0.6}Fe_2O_4$ nanoparticles. J Alloys Comp 541:29–35

Jacob BP, Ashok Kumar RP, Pant SS, Mohammed EM (2011) Influence of preparation method on structural and magnetic properties of nickel ferrite nanoparticles. Bull Mater Sci 24:1345–1350

Jadhav SS, Shirsath SE, Toksha BG, Shukla SJ, Jadhav KM (2008) Chin J Chem Phys 21:381

Jonker SJGH (1959) Phys Chem Solids 9:105

Kakatkar SV, Kakatkar SS, Patil RS, Sankpal AM, Suryawanshi SS, Bhosale DN, Sawant SR (1996) Phys Stat Sol (b) 198:853

Kumar G, Chand J, Verma S, Singh M (2011) J Phys D Appl Phys
Kumar S, Sharma A, Singh M, Sharma SK (2013) Simple synthesis and magnetic properties of nickel-zinc ferrites nanoparticles by using Aloe vera extract solution. Arch Appl Sci Res 5(2013):145–151
Li W, Fa-Shen L, Qing-Guo Z (2000) Chin Phys 9:685
Mane DR, Birajdar DD, Patil S, Shirsath SE, Kadam RH (2011) J Sol-Gel Sci Tech
Mathew DS, Juang R-S (2007) An overview of the structure and magnetism of spinel ferrite nanoparticles and their synthesis in microemulsions. Chem Eng J 129:51–65
Mohan GR, Ravinder D, Reddy AV, Boyanov BS (1999) Mater Lett 40:39
Murthy SR, Ramaiah B (2000) J Mater Sci Lett 19:703
Naidu V, Ahamed Kandu Sahib SKA, Suganthi M, Prakash C (2011) Study of electrical and magnetic properties in Nano sized Ce-Gd doped magnesium ferrite. Int J CompuAppl 27:40–45
Nejati K, Zabihi R (2012) Preparation and magnetic properties of nano size nickel ferrite particles using hydrothermal method. Chem Cent J 6:23
Palanisamy KL, Devabharathi V, Sundaram NM (2014. -2015) Corrosion inhibition studies of mild steel with carrier oil stabilized of iron oxide nanoparticles incorporated into paint. Int J ChemTech Res 7:1661–1664
Patil KC, Aruna ST, Mimani T (2002) Curr Opin Solid State Mater Sci 6:507
Popovici M, Savii C, Niznanskya D, Subrta J, Bohaceka J, Becherescub D, Caizerc C, Enache C, Ionescu C (2003) J Opto Adv Mater 5:251
Ranga Mohan G, Ravinder D, Ramana Reddy AV, Boyanov BS (1999) Mater Lett 40:39
Rezlescu N, Rezlescu E (1993) J Solid State Commun 88(2):139
Shirsath SE, Toksha BG, Kadam RH, Patange SM, Mane DR, Jangam GS, Ghasemi A (2010) J Phys Chem Solids 71:1669
Sileo EE, Rotelo R, Jacobo SE (2002) Nickel zinc ferrites prepared by the citrate precursor method. Physica B 320:257–260
Smit J and Wijn HPJ (1959) Ferrites-physical properties of ferrimagnetic oxides in relation to their technical applications (N. V. Philip's Gloeilampenfabrieken, Eindhoven, Holland, Chap. VIII, pp. 136–176
Snoek SL (1947) New development in ferromagnetic material. Elsevier publishing Co., New York, Amsterdam
Sun GL, Li JB, Sun JJ, Yang XZ (2004) J Magn MagnMater 281:173
Verwey EJW, Haaijman PW, Romeyn FR, Van Oostehout GW (1950) J Phil Res Rep 5:173
VijayaBhasker Reddy P, Ramesh B, Gopal Reddy C (2010) Physica B: Cond Matter 405:1852
Wu XD, Dijkamp D, Ogale S, Inam A, Chase EW, Miceli PF, Chang CC, Tarascon JM, Venkateshan T (1987) Appl Phys Lett 51:861
Yadoji P, Peelamedu R, Agrawal D, Roy R (2003) Mater Sci Engg B 98:269
Yamashita J, Kurusawa (1958) J Phys Chem Solids 5:34

Synthesis of Nanoparticles by Physical Route

3

Lucky Krishnia, Preeti Thakur, and Atul Thakur

Abstract

Development of nanoscience has evoked new technologies both in sample preparation and device fabrication. Synthesis and development of nanoparticles that are synonyms to quantum-confined atom is an important milestone in this pursuit. In recent years, a significant development with advanced improvements has been been made in the synthesis methods of nanomaterials. The nanoparticles can be prepared by using two well-known approaches, i.e., top-down approach and bottom-up approach. This chapter gives an overview of the various physical routes of synthesis of nanoparticles. Various methods of preparing nanomaterials including mechanical milling, sputtering, laser pyrolysis, laser ablation, electron beam evaporation, and nanolithography are discussed in this chapter. As a summary, this chapter describes the main physical routes for nanoparticle synthesis.

Keywords

Physical synthesis route · Laser ablation · Mechanical milling · Sputtering · Lithography

L. Krishnia · A. Thakur
Centre of Nanotechnology, Amity University Haryana, Gurugram, India

P. Thakur (✉)
Department of Physics, Amity School of Applied Sciences, Amity University Haryana, Gurugram, India

A. Thakur et al. (eds.), *Synthesis and Applications of Nanoparticles*,
https://doi.org/10.1007/978-981-16-6819-7_3

3.1 Introduction

There are numerous techniques available for synthesis of nanomaterials which can be in the form of powders, tubes, wires, rods, spheres, or thin films. Nanomaterials can be synthesized by either physical, chemical, biological, or hybrid routes. These different routes yield a different variety of nanomaterials which may be dependent on the route, material used, catalyst, enzyme, starting chemicals, and many more. In this chapter, we have discussed some of the physical routes which are available to synthesize nanoparticles for commercial scale production and have been optimized parametrically for synthesis of nanomaterials.

3.2 Approaches for Synthesis of Nanomaterials

Basically, the synthesis of nanomaterials can be done using two main approaches: top-down and bottom-up (Fig. 3.1).

Top-down approaches: The bulk materials are processed to produce nanomaterials of desired parameters in top-down approach. The physical methods which include top-down approach are physical vapor deposition, mechanical milling, electrospinning, sputtering, laser ablation, electro-explosion, lithography (e-beam), arc discharge, and thermal evaporation.

Bottom-up approaches: The building up of a material from the bottom, i.e., atom by atom, molecule by molecule, or cluster by cluster. The physical methods

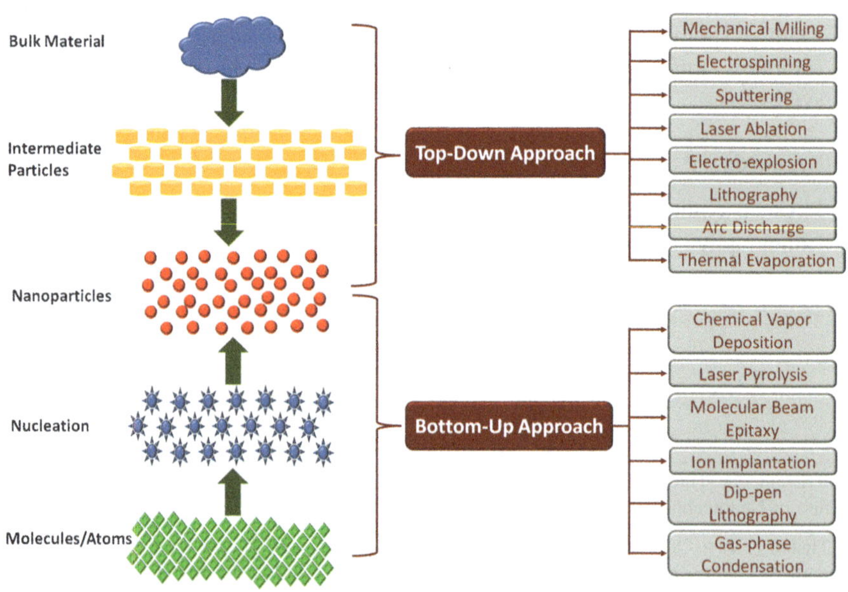

Fig. 3.1 The two approaches, i.e., top-down and bottom-up approaches, for synthesis of nanoparticles

which include bottom-up approach are chemical vapor deposition, laser pyrolysis, molecular beam epitaxy, ion implantation, dip-pen lithography, and gas phase condensation.

3.2.1 Physical Vapor Deposition

Physical vapor deposition (PVD) is a commonly used technique for the fabrication of thin films and surface coatings. Physical vapor deposition is characterized by a process in which the material to be deposited is converted into vapor by high-temperature vacuum or gaseous plasma, and then transported to a region of low pressure from its source to the substrate where the vapor undergoes condensation on the substrate to form a thin film. PVD technique results in the formation of coatings with improved properties as compared to the substrate material. All types of inorganic materials and some types of organic materials can be used in PVD technique.

3.2.2 Mechanical Milling

Mechanical milling is one of the simplest and the most cost-effective method of producing nanoparticles of some metals and alloys from their bulk materials. The material development by high-energy ball milling of powders was first established by John Benjamin (1970) and his co-workers at the International Nickel Company in the late 1960s wherein complex oxide dispersion-strengthened (ODS) alloys for high-temperature structural applications were produced (Benjamin 1970). This mechanical alloying method produced fine, uniform dispersions of oxides of Al_2O_3, Y_2O_3, and ThO_2 in nickel-based superalloys. Mechanical milling technique involves placing a suitable powder charge (typically, a blend of elemental) in a high-energy mill, along with a suitable milling medium. The mechanical milling helps to reduce the particle size and also to achieve new phases. This method produces nanomaterials of different phases which makes it suitable for production of nanocomposites. There are different types of mills being used in this technique such as vibratory, planetary, rod, tumbler, etc. (Kulkarni 2015). The size of the container used for milling depends upon the quantity of sample to be prepared. The kinetics of mechanical milling depends on the energy transferred to the powder from the balls during milling which is governed by many parameters such as the type of milling (dry or wet), the powder supplied to drive the milling chamber, milling speed, temperature, and duration of milling. Either tungsten carbide or hardened steel balls are put in the container with the desired bulk material. The initial/bulk material being used can be of arbitrary size and shape. After putting the balls and the bulk material in the container, it is closed with the use of tight lids. The mass ratio of balls to material which is most suitable to get nanomaterials is 2:1, i.e., if the container is more than half filled, then the efficiency of the milling is reduced. Use of heavy milling balls increases the impact energy of collision, but at the same time the defect density also increases. The temperature during milling depends on

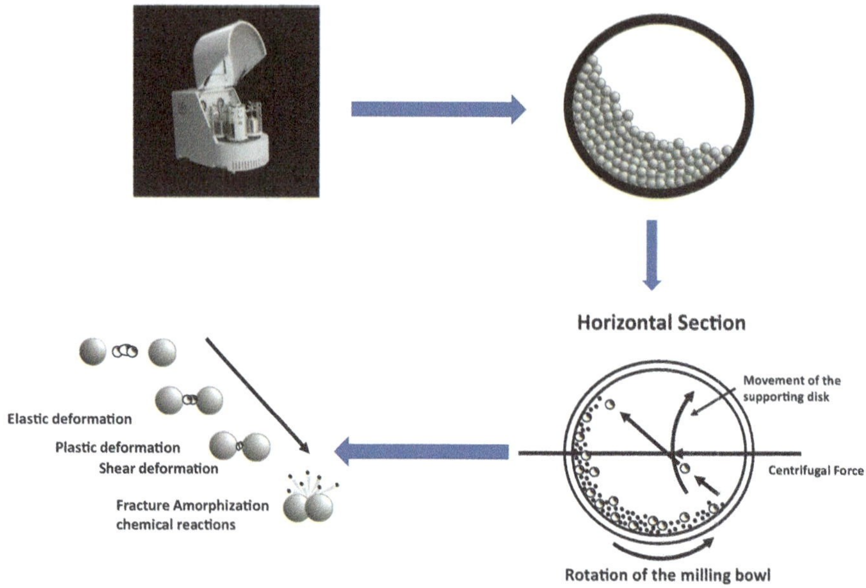

Fig. 3.2 The principle of the ball milling method. Reprinted with permission from ref. Zhuang et al. (2016). Copyright: ©2016, John Wiley & Sons, Ltd.

powder used, milling media, and kinetic energy of the ball. The temperature of the powder during milling influences the diffusivity and defect concentration in the powder inducing the phase transformations. Higher temperature is likely to result in intermetallic phases, while lower temperature results in the formation of amorphous phases. Sometimes, there are chances of addition of some kind of impurities from balls. The gases being used to make the inert atmosphere may also serve as impurities if they are not of high purity. The basic principle of ball milling is displayed in Fig. 3.2 (Zhuang et al. 2016). This technique is mainly used to produce oxide-like ferrites and carbide-strengthened aluminum alloys, wear-resistant spray coatings, aluminum-/nickel-/magnesium-/copper-based nano-alloys, and many other nanocomposite materials (Yadav et al. 2012). Carbon nanomaterials prepared through ball milling are considered to be a novel class of nanomaterials as they have varied applications in the fields of energy storage, energy conversion, and environmental remediation (Lyu et al. 2017; Kammakakam and Falath 2021). Different types of nanoparticles can be synthesized using ball milling like ZnO nanoparticles of size 5–110 nm which were prepared depending on different milling speed and milling time (Salah et al. 2011; Giri et al. 2007; Damonte et al. 2004). CuO nanoparticles of size 11–20 nm were synthesized from metallic powder of size 60 micrometer (Khayati et al. 2013; Yang and Chen 2017). TiO_2 nanoparticles in the range of 10–37 nm were prepared by ball milling (Khayati et al. 2013; Yang and Chen 2017; Carneiro et al. 2014; Salari et al. 2008; Yadav et al. 2015), and Ag_2O powder (5–40 μm) was ball-milled to nanoparticles of size 14 nm in 95 h (Khayati and Janghorban 2013).

3.2.3 Electrospinning

Electrospinning is another one of the simplest top-down methods for synthesis of nanomaterials which are usually fibers of a variety of materials which are mainly polymers (Ostermann et al. 2011). Development of coaxial spinning method was one of the breakthroughs in the electrospinning technique. Coaxial electrospinning is an effective technique for synthesis of core-shell nanofibers at a large scale. The length of the ultrathin nanofibers synthesized using this technique may go up to several centimeters. The spinneret in this technique has two coaxial capillaries, i.e., one of the capillaries with viscous liquid is used to generate shell, and the other capillary with nonviscous liquid is used to generate shell. A schematic diagram of coaxial electrospinning is shown in Fig. 3.3 (Du et al. 2012). This technique is highly helpful in development of core-shell as well as hollow organic, inorganic, polymeric, and hybrid nanomaterials (Kumar et al. 2014). Nanocomposite fibers by electrospinning can be divided into blending, post-modification, and posttreatment methods. In the blending method, metal nanoparticles are mixed with the polymer solution to form a uniform precursor solution. This mixed solution is directly electrospun and various materials interfaces are formed. The process is beneficial in terms of ease of preparation and high yield and therefore is widely used in preparation of fluorescent and electrochemical sensing interfaces. In post-modification, metal nanoparticles are adsorbed or modified onto the nanofibers to obtain the metal nanoparticle-based interface. Posttreatment method is conducted on the obtained metal nanoparticle-based interfaces, such as calcining to obtain the interface of the new structure.

3.2.4 Sputtering

Sputtering is a technique which is being widely used for producing thin films of nanomaterials (Kulkarni 2015; Ayyub et al. 2001). The main advantage of using this

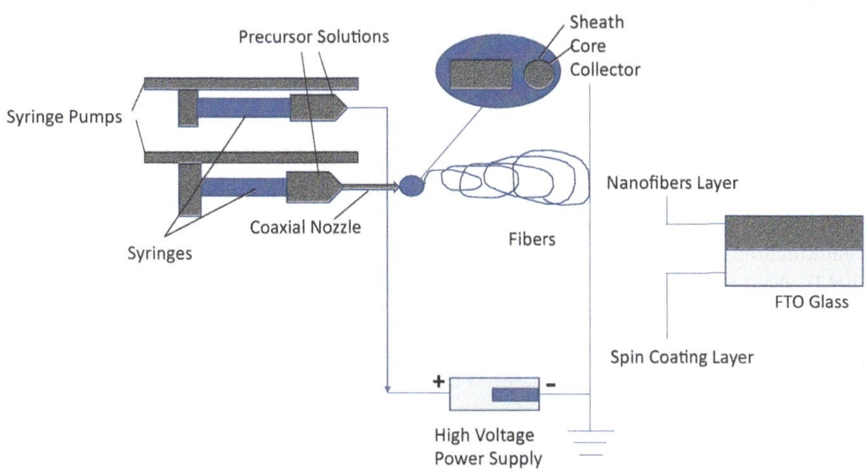

Fig. 3.3 A schematic diagram of coaxial electrospinning. Reprinted with permission from ref. Du et al. (2012). Copyright: ©2012, Elsevier Ltd. All rights reserved

technique is development of stoichiometric thin films which makes the technique more cost-effective. Thin films/nanoparticles are obtained by bombarding the solid surfaces by high-energy particles, i.e., gas and plasma. It is the phenomena of deposition of nanoparticles when highly energetic gaseous ions are bombarded on the target surface resulting in physical ejection of small atomic clusters (Son et al. 2017; Wender et al. 2013; Shah and Gavrin 2006; Bharti et al. 2020). It is generally carried out in an evacuated chamber, in which sputtering gas is introduced. The cathode (target) is supplied with a high voltage, and free electrons collide with the gas to produce gas ions. The positively charged ions are strongly accelerated in the electric field toward the target resulting in the ejection of atoms from the surface (Munoz-Garcıa et al. 2009; Nam et al. 2020). Some important factors which are responsible for determining the size and shape of the nanoparticles are substrate temperature, energy of particles, and annealing duration (Bharti et al. 2020). There can be many ways through which the high-energy particles can be bombarded on the substrate and depending on change of source material according to which it can be classified into magnetron, radio-frequency diode, and DC diode sputtering (Wender et al. 2013). Development of multilayer thin films/magnetic films for spintronic applications is also possible using this technique (Kulkarni 2015). A schematic representation of DC magnetron sputtering is shown in Fig. 3.4 (Son et al. 2017). Different nanoparticles of silver, gold, iron, ferrites, copper, copper oxide, and zinc oxide have been synthesized using sputtering (Asanithi et al. 2012; Hu et al. 2013; Xing et al. 2016; Peng et al. 2003; Gunnarsson et al. 2015; Jaiswal et al. 2015; Das et al. 2016; Rashid et al. 2015).

3.2.5 Laser Ablation

Laser ablation technique involves synthesis of nanoparticles by usage of a powerful laser beam which hits the target material and ablates the surface. The target/source/ precursor material is vaporized due to the high energy of the laser irradiation which results in nanoparticle formation. The laser beam condenses a plasma which produces nanoparticles by irradiating the different metal substrates (Amendola and Meneghetti 2009). This technique is very useful for reduction of metal to nanoparticles. The stable nanoparticles are synthesized using laser ablation techniques and do not need any stabilizing agent or chemicals. Different types of nanoparticles of silver, gold, copper, copper oxide, tin oxide, etc. can be manufactured by using laser ablation (Maciulevičius et al. 2013; Wender et al. 2011; Al-Azawi and Bidin 2015; Khumaeni et al. 2017; Valverde-Alva et al. 2015; Tajdidzadeh et al. 2014; Pyatenko et al. 2004; Hajiesmaeilbaigi et al. 2005; Gondal et al. 2009; Al-Dahash et al. 2018; Mintcheva et al. 2018; Boutinguiza et al. 2013; Singh et al. 2016; Abdulateef et al. 2016; Gondal et al. 2013).

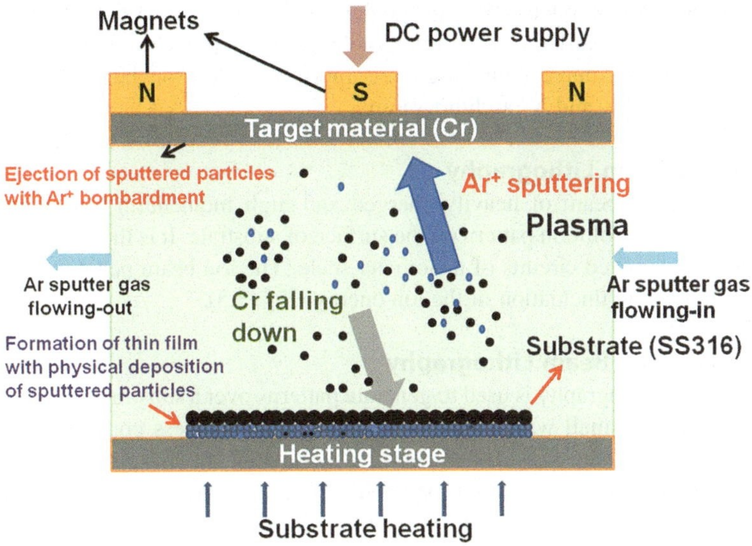

Fig. 3.4 A schematic diagram of the DC magnetron sputtering process. Reprinted with permission from ref. Son et al. (2017). Copyright: ©2017, Elsevier Ltd.

3.2.6 Electrical Explosion

Electrical explosion of metal wires is another upcoming technology for the synthesis of nanoparticles with increased activity. In this method, a high-density current pulse is passed through a metal wire resulting in heating up of the wire quickly. The wire explodes with the formation of explosion products, which form nanoparticles while passing a gas atmosphere. This technique results in the possibility of obtaining nanopowders of metals, alloys, and oxides and nitrides of metals (Lerner et al. 2016; Llyin et al. 2012). Also, this method shows stability in properties of nanopowders obtained with high activity in chemical processes. Ease of changes in process parameters, small-sized nanoparticles, and high energy efficiency of the process characterized by low energy losses for heating the environment are some of the other advantages of this method.

3.2.7 Lithography

This term lithography can be defined in two distinct ways. First is the general definition which is by Google "lithography," seven of the first ten hits will show the art world and a process invented by Aloys Senefelder in 1788 (Britannica Concise Encyclopedia 2005). And second is "the process of printing from a plane surface (as a smooth stone or metal plate) on which the image to be printed is ink-receptive and the blank area ink-repellent" given by the Merriam-Webster

Online Dictionary (Mollenstedt and Speidel 1960). The second definition gives a much clearer and boarder view of lithography. Lithography can be of many types depending on the source being used for imprinting, i.e., ion beam lithography, e-beam lithography, and X-ray lithography.

3.2.7.1 Ion Beam Lithography

In this process a beam of heavily charged and high momentum ions is used to fabricate a high-resolution pattern on the surface of substrate. It is therefore useful in developing integrated circuits of nanometer scale. The ion beam penetration power can be adjusted by fluctuation of the ion energy (Fig. 3.5).

3.2.7.2 Electron Beam Lithography

Electron beam lithography is used to generate patterns over a substrate using focused electron beam of small wavelength. This lithographic process comprises of three steps: exposure of the sensitive material, development of the resist, and pattern transfer. An electron-sensitive film or resist is placed on the sample; thereafter it is exposed by spin coating spun at 1000–6000 rpm to form a coating. The energy deposited during the exposure creates a latent image that materializes in the course of chemical development. When exposed to the electron beam, the resistance solubility changes to the point where it becomes selectively more soluble in the exposed part or in unexposed parts depending upon the type of resist. The resist can be removed by

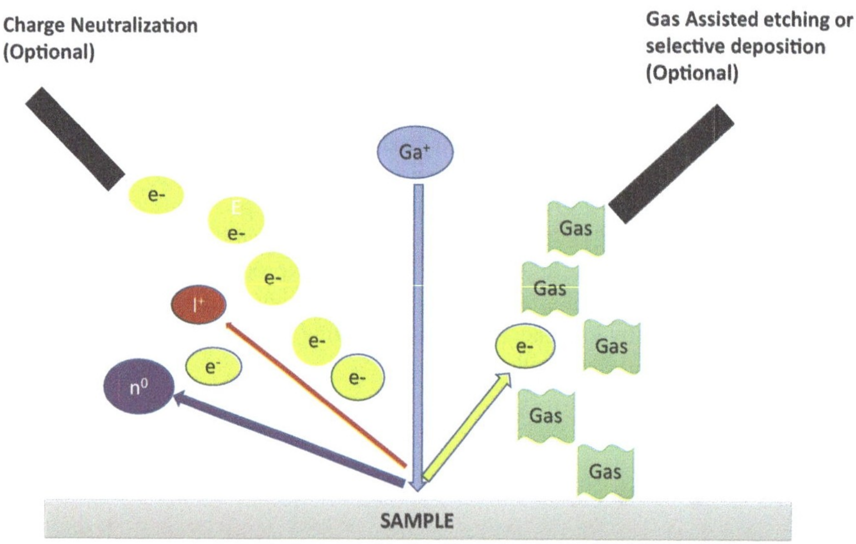

Fig. 3.5 Focused ion beam lithography. Reprinted with permission from ref. Sutter et al. (2008). Copyright: ©2017, Elsevier Ltd. License Number 5134251485537

immersing it in a solvent which is often termed as developing. The intensity and energy of the electron beam, thickness of the resist, and irradiation time are some of the important parameters to be taken into account. Etching (chemical removal of the layer from the surface of substrate) and liftoff (creating a photoresist profile that ensures separation between the thin film coating in desired and undesired areas of the pattern) are the main pattern transfer methods that can be applied. Polymethyl methacrylate (PMMA) is the commonly used resist material due to its advantages of high resolution and easy processing (Fig. 3.6).

3.2.7.3 X-Ray Lithography

X-ray lithography is a masked lithographic process used to transfer patterns from a mask to a resist on the surface of substrate by X-rays. X-ray lithography is a masked lithographic technique that uses X-rays to transfer patterns from a mask to a resist on a substrate's surface. The procedure begins with the application of a mask and resist to the wafer surface, followed by exposure. The photoresist is penetrated by X-rays that emanate from a synchrotron and are selectively passed or inhibited by the patterned mask absorber, depositing energy to expose the resist. In this procedure, gold is commonly utilized as an absorber. To generate nano-patterns or nanostructures on the substrate, the same etching and developing technique as photolithography is used (Betancourt et al. 2021; Sebastian et al. 2020) (Fig. 3.7).

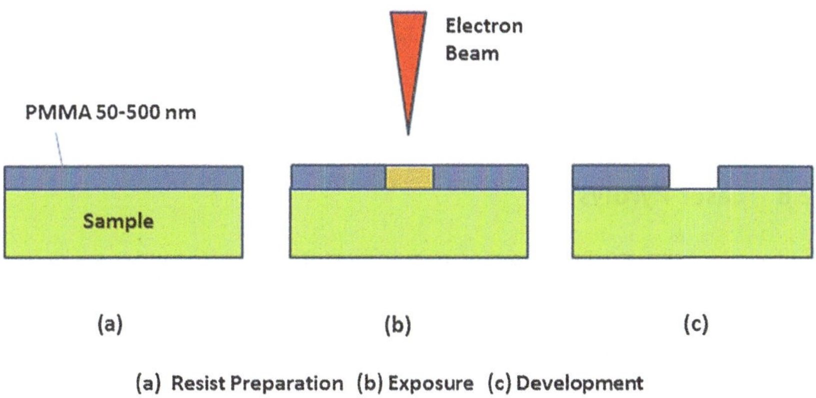

Fig. 3.6 Electron beam lithography. Reprinted with permission from ref. Llyin et al. (2012). Copyright: ©2017, Elsevier Ltd. License Number 5134251485537

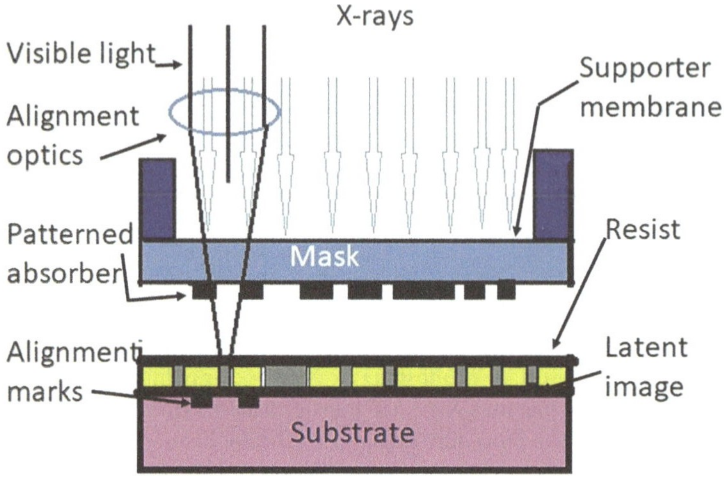

Fig. 3.7 Process of X-ray lithography. Reprinted with permission from ref. Sutter et al. (2008). Copyright: ©2017, Elsevier Ltd. License Number 5134251485537

Bottom-up approach: The basic molecular and atomic units are being used for synthesis of nanoparticles in bottom-up approach. The physical methods which include bottom-up approach are chemical vapor deposition, laser pyrolysis, molecular-beam epitaxy (MBE), ion implantation, dip pen lithography, and gas-phase condensation.

3.2.8 Laser Pyrolysis

Laser pyrolysis is one of the important, easy, and effective vapor phase nanopowder synthesis methods. In this technique, a laser beam is used selectively to heat a gas stream of nanoparticle precursors. The reaction between the laser beam and gaseous precursors will enhance the temperature, and hence the precursor gets decomposed, thereby inducing nucleation of nanoparticles. The powders are deposited where they will be obtained. One of the most important advantages of laser pyrolysis is its flexibility to synthesize nanoparticles of diverse materials using appropriate precursors of 15–20 nanometers at 100 g/h. Some other advantages of this technique are flow of reaction, high level of purity, and excellent chemical and physical properties. A study reported the laser pyrolysis synthesis of zinc sulfide and zinc oxide nanoparticles with an average diameter below 10 nm using a low-power CO_2 laser and controlling the operating parameters (Malekzadeh et al. 2020).

3.2.9 Chemical Vapor Deposition

Chemical vapor deposition (CVD) is a vacuum-based deposition process for producing high-quality, high-performance solid materials. CVD is a method for depositing nonvolatile solid thin films on substrates that involves chemical reactions between an organometallic or halide compound and other gases. CVD differs from PVD in that it uses a multidirectional deposition method to deposit material onto the substrate, whereas PVD uses a line-of-site impingement method. CVD is frequently used in microfabrication techniques to deposit materials in a variety of morphologies, such as monocrystalline, polycrystalline, amorphous, and epitaxial.

In contrast to PVD, in CVD, a mixture of gases interacts chemically with the bulk surface of the material, causing chemical breakdown of some of the specific gas elements and the formation of a solid coating on the base material's surface (Jones and Hitchman 2008; Shah and Tali 2016).

CVD is an important material preparation technology that is used to make precious metal thin films and coatings. A variety of CVD processes exist, including atmospheric pressure chemical vapor deposition (APCVD), low-pressure chemical vapor deposition (LPCVD), plasma-enhanced chemical vapor deposition (PECVD) or plasma-assisted chemical vapor deposition (PACVD), and laser-enhanced chemical vapor deposition (LECVD). Hybrid techniques, which combine physical and CVD properties, have also arisen (Fig. 3.8).

3.2.10 Molecular-Beam Epitaxy (MBE)

In thermal evaporation, the molecular-beam epitaxy (MBE) technique is the most reliable deposition process. A typical MBE system is depicted in Fig. 3.9. The system is a controlled MBE process in which a computerized process control unit controls the evaporation rate of the source materials in situ. Esaki has successfully deposited a man-made superlattice structure formed of thin alternating layers of GaAs and GaAlAs, as illustrated in Fig. 3.9 (Adachi and Wasa 2012).

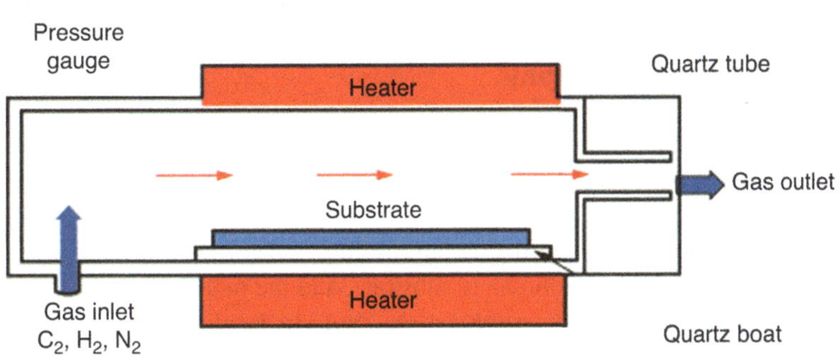

Fig. 3.8 CVD setup. Reprinted with permission from ref. Jones and Hitchman (2008). Copyright: ©2017

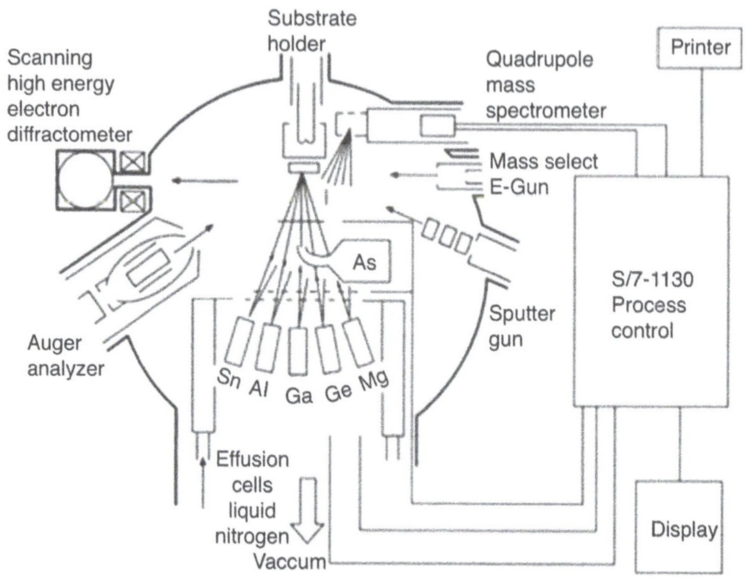

Fig. 3.9 A typical MBE system. Reprinted with permission from ref. Adachi and Wasa (2012). Copyright: ©2017

Fig. 3.10 Schematic representation of DPN. Reproduced with permission from ref. Piner et al. (1999). Copyright 1999 American Association for the Advancement of Science

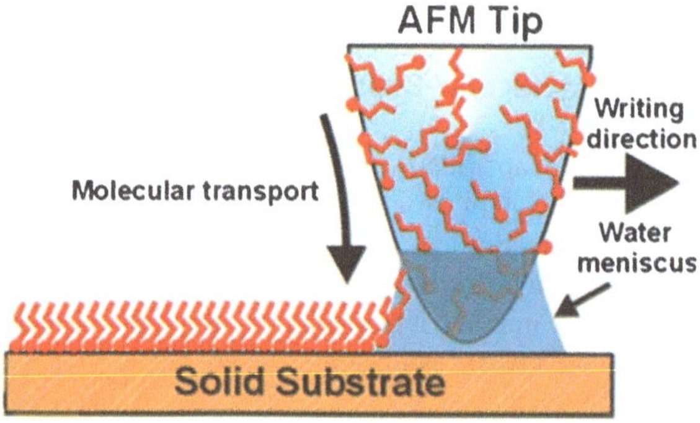

3.2.11 Dip Pen Lithography

Figure 3.10 (Piner et al. 1999) depicts the invention of DPN, which was published in a Science article in 1999. The Mirkin group employed AFM tips to transport molecules (particularly, alkanethiols, ODT, and MHA) directly onto a gold thin film in a manner similar to a dip pen with 30 nm line width resolution. The "nib" in this case was an AFM tip, the "paper" was a solid-state substrate, and the molecules with a chemical affinity for the substrate were employed as "ink" to be directly written into the "paper" by the "nib." The feature size was dependent on the tip scanning speed and the ink transport rate. The meniscus and humidity were employed to aid molecule transport and build chemisorbed nanostructures on a surface. DPN uses a positive printing mode to easily deposit different types of molecules at particular spots without the use of a stamp, resist, or specialized tool.

3.3 Conclusion

A large number of nanoparticles and nanocomposites can be fabricated by using the methods and techniques described in this chapter. Different physical methods have been developed to obtain nanoparticles of various shapes and sizes, including mechanical milling, laser ablation, e-beam evaporation, electric deposition, and lithographic techniques. But to know the optimum conditions to improve the reproducibility and the quantity is an important parameter in nanoparticle synthesis.

References

Abdulateef SA, et al. (2016) *Preparation of CuO nanoparticles by laser ablation in liquid. AIP Conference Proceedings*. Vol. 1733. No. 1. AIP Publishing

Adachi H, Wasa K (2012) 1—thin films and nanomaterials. In: Wasa K, Kanno I, Kotera H (eds) Handbook of sputtering technology, 2nd edn. William Andrew Publishing, pp 3–39

Al-Azawi MA, Bidin N (2015) Gold nanoparticles synthesized by laser ablation in deionized water: influence of liquid layer thickness and defragmentation on the characteristics of gold nanoparticles. Chin J Phys 53(4):201–209

Al-Dahash G, Khilkala WM, Abd Alwahid SN (2018) Preparation and characterization of ZnO nanoparticles by laser ablation in NaOH aqueous solution. Iran J Chem Chem Eng (IJCCE) 37(1):11–16

Amendola V, Meneghetti M (2009) Laser ablation synthesis in solution and size manipulation of noble metal nanoparticles. Phys Chem Chem Phys 11(20):3805–3821

Bharti et al (2020) Potential of magnetic nanoferrites in removal of heavy metals from contaminated water: mini review. J Supercond Nov Magn 33:3651–3665

Asanithi P, Chaiyakun S, Limsuwan P (2012) Growth of silver nanoparticles by DC magnetron sputtering. J Nanomater 2012:79

Ayyub P, Chandra R, Taneja P, Sharma AK, Pinto R (2001) Appl Phys A Mater Sci Process 73:67–73

Benjamin JS (1970) Metal Trans 1:2943

Betancourt AP, Goswami DY, Bhethanabotla VR, Kuhn JN (2021) Catal Commun 149:106213

Boutinguiza M et al (2013) Synthesis of titanium oxide nanoparticles by ytterbium fiber laser ablation. Physics Procedia 41:787–793

"lithography." Britannica Concise Encyclopedia. 2005. Encyclopædia Britannica Premium Service 14 Feb. 2005 http://www.britannica.com/ebc/article?tocId=9370361

Carneiro JO et al (2014) Synthesis of iron-doped TiO2 nanoparticles by ball-milling process: the influence of process parameters on the structural, optical, magnetic, and photocatalytic properties. J Mater Sci 49(21):7476–7488

Damonte LC et al (2004) Nanoparticles of ZnO obtained by mechanical milling. Powder Technol 148(1):15–19

Das A et al (2016) Copper oxide nano-particles film on glass by using sputter and chemical bath deposition. Adv Mater Lett 7(8):600–603

Du P, Song L, Xiong J, Li N, Xi Z, Wang L, Jin D, Guo S, Yuan Y (2012) Electrochim Acta 78: 392–397

Giri PK et al (2007) Correlation between microstructure and optical properties of ZnO nanoparticles synthesized by ball milling. J Appl Phys 102(9):093515

Gondal MA et al (2009) Synthesis of ZnO2 nanoparticles by laser ablation in liquid and their annealing transformation into ZnO nanoparticles. Appl Surf Sci 256(1):298–304

Gondal MA et al (2013) Synthesis of cu/Cu2O nanoparticles by laser ablation in deionized water and their annealing transformation into CuO nanoparticles. J Nanosci Nanotechnol 13(8): 5759–5766

Gunnarsson R, Helmersson U, Pilch I (2015) Synthesis of titanium-oxide nanoparticles with size and stoichiometry control. J Nanopart Res 17(9):353

Hajiesmaeilbaigi F et al (2005) Preparation of silver nanoparticles by laser ablation and fragmentation in pure water. Laser Phys Lett 3(5):252

Hu XL, Takai O, Saito N (2013) Synthesis of gold nanoparticles by solution plasma sputtering in various solvents. J Phys Conf Ser 417:1. IOP Publishing

Jaiswal J, Chauhan S, Chandra R (2015) Influence of sputtering parameters on structural, optical and thermal properties of copper nanoparticles synthesized by dc magnetron sputtering. Int J Sci Technol Manage 4(01):678–688

Jones AC, Hitchman ML (2008) In: Jones AC, Hitchman ML (eds) Chemical vapour deposition. Royal Society of Chemistry, Cambridge, pp 1–36

Kammakakam I, Falath WS (2021) Nanomaterials: a review of synthesis, properties, recent progress, and challenges. Mater Adv 2021(2):1821–1871

Khayati GR, Janghorban K (2013) Preparation of nanostructure silver powders by mechanical decomposing and mechanochemical reduction of silver oxide. Trans Nonferrous Met Soc China 23(5):1520–1524

Khayati GR et al (2013) Synthesis of cuprous oxide nanoparticles by mechanochemical oxidation of copper in high planetary energy ball mill. Adv Powder Technol 24(1):301–305

Khumaeni A, Budi WS, Sutanto H (2017) Synthesis and characterization of high-purity gold nanoparticles by laser ablation method using low-energy Nd: YAG Laser 1064 nm. J Phys Conf Ser 909:1

Kulkarni SK (2015) Nanotechnology: principles and practices. Springer International Publishing. 978-3-319-09171-6

Kumar PS, Sundaramurthy J, Sundarrajan S, Babu VJ, Singh G, Allakhverdiev SI, Ramakrishna S (2014) Energy Environ Sci 7:3192–3222

Lerner MI, Pervikov AV, Glazkova EA, Svarovskaya NV, Lozhkomoev AS, Psakhie SG (2016) Powder Technol 288:371–378

Llyin AP, Nazarenko OB, Tikhonov DV (2012) J Nanosci Tech 12(10):8137–8142

Lyu H, Gao B, He F, Ding C, Tang J, Crittenden JC (2017) ACS Sustain Chem Eng 5:9568–9585

Maciulevičius M et al (2013) On-line characterization of gold nanoparticles generated by laser ablation in liquids. Phys Procedia 41:531–538

Malekzadeh M, Rohani P, Liu Y, Raszewski A, Ghanei F, Swihart MT (2020) Laser pyrolysis synthesis of zinc-containing nanomaterials using low-cost ultrasonic spray delivery of precursors. Powder Technol 376:104–112

Mintcheva N et al (2018) Laser-ablated ZnO nanoparticles and their photocatalytic activity toward organic pollutants. Materials (Basel) 11(7):1127

Mollenstedt G, Speidel R (1960) Physik Blatter 16:100

Munoz-Garcıa J, Vazquez L, Cuerno R, Sanchez-Garcıa JA, Castro M, Gago R (2009) Toward functional nanomaterials. Springer US, New York, NY, pp 323–398

Nam JH, Jang MJ, Jang HY, Park W, Wang X, Choi SM, Cho B (2020) J Energy Chem 47:107–111

Ostermann R, Cravillon J, Weidmann C, Wiebcke M, Smarsly BM (2011) Chem Commun 47:442–444

Peng Y, Park C, Laughlin DE (2003) Fe 3 O 4 thin films sputter deposited from iron oxide targets. J Appl Phys 93(10):7957–7959

Piner RD, Zhu J, Xu F, Hong S, Mirkin CA (1999) Dip-pen″ nanolithography. Science 283:661–663

Prasad Yadav T, Yadav RM, Pratap Singh D (2012) Nanosci Nanotechnol 2:22–48

Pyatenko A et al (2004) Synthesis of silver nanoparticles by laser ablation in pure water. Appl Phys A Mater Sci Process 79(4–6):803–806

Rashid J et al (2015) ZnO-nanoparticles thin films synthesized by RF sputtering for photocatalytic degradation of 2-chlorophenol in synthetic wastewater. J Ind Eng Chem 23:134–139

Salah N et al (2011) High-energy ball milling technique for ZnO nanoparticles as antibacterial material. Int J Nanomedicine 6:863

Salari M et al (2008) Effect of milling time on mechanochemical synthesis of TiO2 nanoparticles. Int J Mod Phys B 22(18n19):2955–2961

Sebastian EM, Jain SK, Purohit R, Dhakad SK, Rana RS (2020) Nanolithography and its current advancements. Mater Today Proc 26(2):2351–2356

Shah KA, Tali BA (2016) Mater Sci Semicond Process 41:67–82

Shah P, Gavrin A (2006) Synthesis of nanoparticles using high-pressure sputtering for magnetic domain imaging. J Magn Magn Mater 301(1):118–123

Singh A et al (2016) Pulsed laser ablation-induced green synthesis of TiO2 nanoparticles and application of novel small angle X-ray scattering technique for nanoparticle size and size distribution analysis. Nanoscale Res Lett 11(1):447

Son HH, Seo GH, Jeong U, Shin DY, Kim SJ (2017) Int J Heat Mass Transf 113:115–128

Sutter E, Ozturk B, Sutter P (2008) Nanotechnology 19(43):435607

Tajdidzadeh M et al (2014) Synthesis of silver nanoparticles dispersed in various aqueous media using laser ablation. Sci World J 2014:324921

Valverde-Alva MA et al (2015) Synthesis of silver nanoparticles by laser ablation in ethanol: a pulsed photoacoustic study. Appl Surf Sci 355:341–349

Wender H, Migowski P, Feil AF, Teixeira SR, Dupont J (2013) Coord Chem Rev 257:2468–2483

Wender H et al (2011) Synthesis of gold nanoparticles by laser ablation of an Au foil inside and outside ionic liquids. Nanoscale 3(3):1240–1245

Xing L et al (2016) Synthesis and morphology of iron–iron oxide Core–Shell nanoparticles produced by high pressure gas condensation. Nanotechnology 27(21):215703

Yadav BC, Singh S, Yadav TP (2015) Titania prepared by ball milling: its characterization and application as liquefied petroleum gas sensor. Synth React Inorg Met-Org Nano-Met Chem 45(4):487–494

Yang B, Chen D (2017) Synthesis of CuO nanoparticles for catalytic application via ultrasound-assisted ball milling. Process Appl Ceram 11(1):39–44

Zhuang S, Lee ES, Lei L, Nunna BB, Kuang L, Zhang W (2016) Int J Energy Res 40:2136–2149

Synthesis of Nanomaterials by Chemical Route

4

Shilpa Taneja, Pinki Punia, Preeti Thakur, and Atul Thakur

Abstract

Nanoferrites are the most frequently used ceramic nanomaterials due to their excellent physical and chemical properties. The properties of nanomaterials are more significant than their bulk counterparts. Selection of specific synthesis route plays a vital role in the preparation of nanoparticles. The reaction conditions and other physical parameters can affect the properties of particles to be developed. Several wet chemical methods have been introduced in the past few decades. Sol-gel, polyol, electrochemical, citrate precursor, sonochemical, solvothermal, co-precipitation, hydrothermal, etc. are some trendy methods to synthesize nanoparticles. Various synthesis routes are explained briefly in this chapter. Various factors such as reaction conditions, energy utilization, reagent compositions, and costs of production are discussed as well.

Keywords

Nanomaterials · Synthesis methods · Sol-gel method · Chemical properties

S. Taneja · P. Thakur (✉)
Department of Physics, Amity School of Applied Sciences, Amity University Haryana, Gurugram, India

P. Punia
Department of Physics, Guru Jambheshwar University, Hisar, India

A. Thakur
Centre of Nanotechnology, Amity University Haryana, Gurugram, India

A. Thakur et al. (eds.), *Synthesis and Applications of Nanoparticles*,
https://doi.org/10.1007/978-981-16-6819-7_4

4.1 Introduction

Science and engineering have made rapid advances in the synthesis of nanomaterials to attain unique properties that differ from those of bulk materials. The particle has fascinating features below 100 nm, mostly due to two physical factors. The quantization of electronic states appears, resulting in very sensitive size-dependent effects such as optical and magnetic properties. The high surface-to-volume ratio modifies the thermal, mechanical, and chemical properties of materials. Because of their unique physical and chemical properties, nanoparticles are ideal for a variety of specialized applications. The top-down and bottom-up approaches to synthesize metal nanoparticles have been identified. Milling, lithography, and repeated quenching are examples of top-down methods. The bottom-up method, which involves building a material, atom by atom, molecule by molecule, and cluster by cluster (Hazra and Ghosh 2014; Shaikh et al. 2020), is the method most commonly used by researchers in the synthesis of nanoparticles. Using chemical reductants in solvents, numerous chemical processes have been recognized to synthesize colloidal metal nanoparticles from various precursors (aqueous and nonaqueous). Electrochemical method (Wang et al. 2008), sonochemical method (Gogate and Pandit 2004), radiolytic method (Uttayarat et al. 2015), and photochemical method (Parra et al. 2002) are some of the chemical processes which are used for industrial applications.

4.2 Synthesis Methods

4.2.1 Chemical Methods

4.2.1.1 Polyol Method

In the polyol method, nonaqueous liquid (polyol) is used as a reducing agent and solvent for the benefit of reducing surface oxidation and agglomeration. This method regulates the size, texture, and shape of nanomaterials. It can also be used in the mass production of nanomaterials (Zhao et al. 2010). If the synthesis is carried out at an elevated temperature with precise particle growth, this process can be used as a sol-gel method for synthesis of oxide (Fi et al. 2018). Due to its strong reducing ability, high dielectric constant and boiling point, ethylene glycol is used as solvent in this method. This solvent is also used as a cross-linking substance to form metal glycolate, which leads to oligomerization (Woei and Ying 2012). When as-synthesized glycolate precursors are calcined in air, they can be converted to metal oxide derivatives (Quievryn et al. 2014). Metallic alloys and core-shell size nanoparticles have also been synthesized using the polyol synthesis process (Kyun et al. 2007; Hoon and Kim 2012; Chang and Chen 2005; Dong et al. 2015). Yang et al. used this method to form icosahedral and cubic gold particles in the range of 100–300 nm (Kim et al. 2004). To control the molar ratio between silver nitrate and PVP, Xia et al. studied the management of morphologies like nanocubes and nanowires (Xia et al. 2009).

4.2.1.2 Microemulsions

An emulsion is a fluid that is dispersed in another fluid. As long as the polymer solution is liquid, it can produce emulsions. Emulsions are classified as macro, mini, or microemulsions based on the size of the droplet (Drmota et al. 2012). Using the microemulsion synthesis, the nanoparticles are formed. Immiscible liquids are separated into two phases when mixed (Foroughi et al. 2016; Mathew and Juang 2007). To create water-oil, energy is required to mix the two phases. In order to associate both phases, energy is required to create a water-oil connection; afterward it replaces the water-water and oil-oil contacts. Surfactants can reduce the interfacial tension between two liquids. Hydrophilic and lipophilic groups are found in surfactants (Puliová et al. 2013). If there are enough surfactant molecules, the interface between oil and water can be aligned and established by lowering the interfacial tension. The most common method for preparing nanomaterials in both phases is shown in Fig. 4.1. After mixing two microemulsions, due to the collision between micelles, Brownian motion occurs. Good collisions result in the reactants coalescing, fusing, and mixing well. The reaction between solubilizates produces metal nuclei. From the nucleation stage, Bönnemann et al. studied zerovalent metal atoms (Bönnemann and Richards 2001).

The nucleation point is a collision between a reverse micelle moving a nucleus and moving product monomer due to intermicellar conversation throughout the growth stage. The size and shape of nanodroplets, as well as the type of surfactant, determine the size and morphology of nanomaterials. Surfactants are commonly used to stabilize particles and prevent them from growing (Malik et al. 2012).

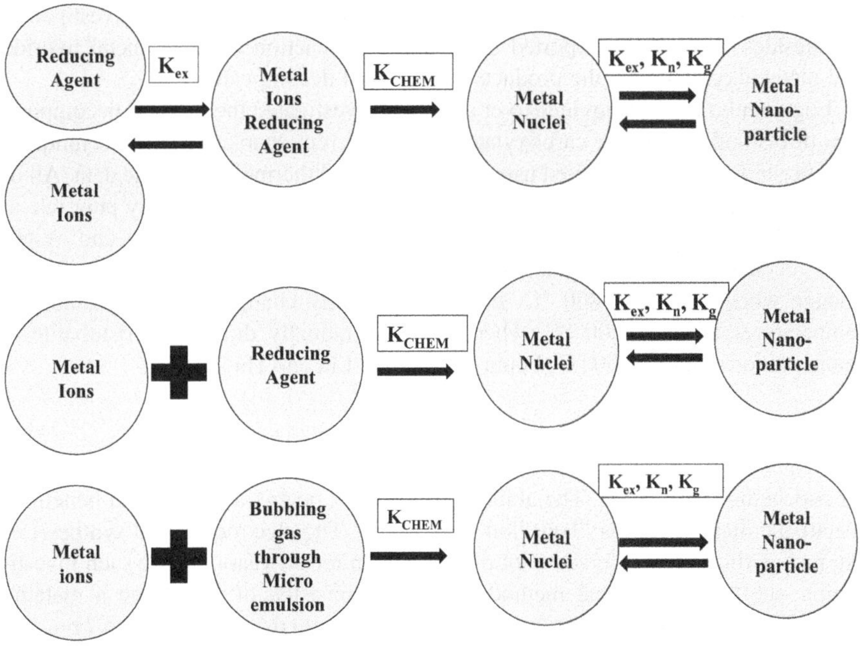

Fig. 4.1 Schematic diagram of nanoparticle preparation (Zhi et al. 2006)

Wongwailikhit et al. investigated the creation of Fe_2O_3 by adding water in oil microemulsion. Spherical nanoparticles were obtained with a diameter of approx. 50 nm. The size of the particles was determined by the amount of water in the microemulsion system. In an oil microemulsion, increasing the water fraction resulted in a larger particle size (Wongwailikhit 2011).

Sarkar et al. used water-in-oil microemulsion method to synthesize pure zinc oxide nanomaterials in a variety of shapes (Sarkar et al. 2011). Maitra was the first to use the microemulsion technique to create chitosan nanoparticles. Chitosan nanoparticles cross-linked with glutaraldehyde in the aqueous core of reverse micellar droplets (Mitra et al. 2001).

4.2.1.3 Thermal Decomposition

Thermal decomposition is a chemical method in which heat is needed to break chemical bonds in the compound being decomposed in this method, and the reaction is endothermic. A positive feedback loop is created when decomposition is sufficiently exothermic (Parra et al. 2002). Arshad et al. used TG-DTA-DTG techniques to study the thermal decomposition of metal complexes in an air atmosphere. To investigate the thermal behavior and mode of decomposition, thermo-analytical techniques were used in a static air atmosphere. When heated to 740°, the complexes and ligands disintegrated in two steps. The residue corresponded to metal oxide at temperatures above 740°; the thermal stability improves in the order Co(II) < Cu (II) < Zn(II) < Cd(II) (Arshad and Qureshi 2008).

Patil et al. investigated metal acetates and dicarboxylates to study TG-DTA-DTG to regulate the metal acetate bonding (Patil et al. 1968). George et al. studied the formation of thermal decomposition of n-butyl copper. This investigation (Whitesides et al. 1970) reported an example of reaction between metal hydride and metal alkyl to create the products of a thermal decomposition.

Logvinenko et al. (Logvinenko et al. 2007) investigated the thermal decomposition of bismuth and silver carboxylates using different characterization techniques. Kinetic studies were conducted using non-isothermal thermogravimetric data. All of the decomposition methods were multistep. Ewell et al. studied nearly pure talc at different temperatures, both unheated and after heating. The heat effects and weight losses occurred when heating talc was calculated. The crystal structure of talc did not change when heated to 800 °C. Enstatite gradually changed to clinoenstatite at temperatures around 1200 °C, whereas silica gradually directs to cristobalite at temperatures around 1300 °C (Punia et al. 2021; Liu and Hu 2014).

4.2.1.4 Electrochemical Synthesis

The combination of chemical compounds in an electrochemical cell is known as electrochemical synthesis. The ability to accept the potential is the main benefit of electrochemical synthesis (Tourillon et al. 2000). The electrochemical synthesis of Ag nanoparticles has received a lot of attention in recent years. In one such investigation, the electrochemical method employed consisted of dissolving a metallic anode in a solvent. The formation of Ag nanoparticles in the size range 2 to 7 nm was observed using this method. By varying the current density, the particle size was determined. The impact of various parameters on the size of nanoparticles was investigated. Two different silver clusters were detected in the UV-Vis spectra (Balakumaran et al. 2016).

Dobre et al. described the synthesis of colloidal silver solutions by the "sacrificial anode" method, which was carried out with a stirrer and a current pulse generator. The researchers created Ag particles of spherical geometry with a diameter of 10–55 nm. The absorption band at 420 nm was visible in the UV-Vis spectra, indicating the presence of Ag nanoparticles (Hajos et al 2011). The synthesis of Ag nanoparticles in aqueous PVA solution has also been studied which is a low-cost synthetic polymer with numerous mechanical properties. The average diameter of Ag nanoparticles was found to be 15 $\pm$ 9 nm (Starowicz et al. 2006).

The synthesis of silicon (Si) nanoparticles under UV excitation is useful for optics and other applications (Choi et al. 2014). More research was completed on the electrochemical method for silver nanoparticles in aqueous solutions. The size distribution of the produced silver nanoparticles ranged from 2 to 20 nm. On the surface of the cathode, Ag crystals are obtained with size less than 40 nm (El-sherbiny et al. 2012).

The study focused on synthesizing highly pure silver nanoparticles using an electrochemical method which was chosen because it is simple to control at room temperature and does not require the use of hazardous chemicals. The anode oxidation and cathode reduction were brought up by the experimental setup. The geometry of silver nanoparticles was found to be with a size of less than 50 nanometers (Khodashenas and Ghorbani 2015). Islam et al. investigated the electrochemical deposition method for the synthesis of platinum nanoparticles. Variations in electrolysis parameters were used to control particle size, and the composition of electrolytic solutions was used to improve platinum particle homogeneity. The particle sizes of platinum nanoparticles were larger than 10 nm (Islam and Islam 2013).

4.2.1.5 Precursor Method

Decomposition of precursor compounds is used to make complex oxides (Assar and Abosheiasha 2012). The hydrate of Li [Cr (C2O4)2 can be used to make LiCrO2. Metal oxides are typically synthesized using alkoxides and carboxylates as precursors. A variety of oxide metals have been prepared using hydrazine precursors. The thermal decomposition of ammonium oxalate precursors yielded ceramic composites. Semiconducting compounds such as GaAs and InP have been synthesized using organometallic precursors.

4.2.1.6 Combustion Synthesis

Combustion synthesis is a well-known procedure for producing a wide range of solids (Ehi-eromosele et al. 2015). This method has been used to make borates, carbides, oxides, etc. The mixture of reactants must be highly dispersed for combustion to occur. The combustion method can be used to produce the product by combining a fuel and an oxidizer. The powder mixture of reactants (0.1–100 m particle size) is commonly used in combustion synthesis. Then, select a gas medium that promotes the ignition of an exothermic reaction (Cheruku et al. 2012). Depending on the reaction, the combustion temperature ranges from 1500 to 3000 K because the desired products are obtained quickly after combustion. The

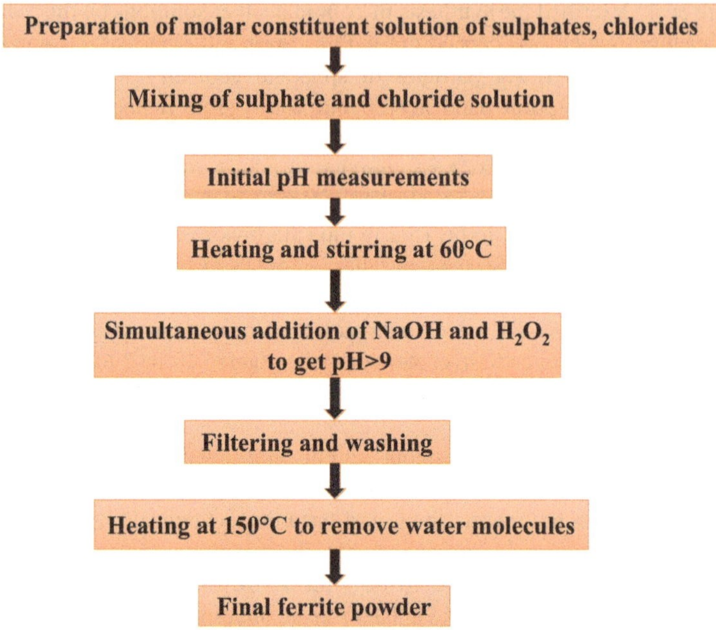

Fig. 4.2 Flowchart of wet chemical coprecipitation method

synthesis of silicates and carbides does not require the use of a gas medium. This method can be used to make superconducting cuprates, ferrites, and various oxides.

4.2.1.7 Wet Chemical Coprecipitation Method

Chemical coprecipitation can be used to make wet ferrites at different properties at low temperatures (55 °C). The oxidation method for making ferrite powders comprises ferrous ion and ferrite powders (Thakur et al. 2011). Many researchers have studied wet chemically prepared ferrites in depth. (Kumar and Kumar 2015; Sheikh and Jain 2016; Chahar et al. 2022; Rana et al. 2018). Ferrites are made by air oxidizing an aqueous suspension containing stoichiometric proportions of constituent cations. The starting solutions are made by stoichiometrically mixing 50 ml of aqueous solution with the appropriate sulfate. As a precipitant, a two-molar (2 M) solution of NaOH is prepared. Continuous stirring of all the precipitates has converted into brownish oxides of soft ferrites, and finally all the samples have been filtered, washed, and dried. The flowchart for the wet chemical method (Ataie et al. 1995) is shown in Fig. 4.2.

4.2.1.8 Sol-Gel Method

Sol-gel method is a wet chemical method used for preparing inorganic oxides by combining both chemical and physical processes (Bel-hadj-tahar and Mohamed 2014). The rise in viscosity is the conventional characteristic of sol-gel formation. The important characteristics of the sol-gel method are good size, homogeneity, morphology, high purity, lower cost, and temperature. The six steps in sol-gel method are as follows:

1. **Hydrolysis** – In this process, a mixture of metal alkoxide and water in a solvent is stirred at the elevated temperature.
2. **Polymerization –** In this step, condensation of adjacent molecules was eliminated, and metal oxide linkages are designed to form liquid (sol) state.
3. **Gelation:** Then, the polymer networks join up to arrange a 3-D network throughout the liquid. The system becomes rigid due to removing the solvent from the sol.
4. **Drying:** H_2O and alcohol are removed at modest temperatures by reducing a hydroxylated metal oxide with some residual content.
5. **Dehydration:** In this step organic residues and chemically bound water exit out.
6. **Densification:** For compression temperature in the range of 1200–1400 K is used to form the dense oxide product.

The sol-gel procedure has been used to prepare metal oxide powders with a less particle size distribution and different particle shapes. Metal-ceramic composites and organic-inorganic composites have been ready by the sol-gel process.

Sol-gel route comprises three different techniques, namely, (a) autoignition, (b) auto-combustion, and (c) Pechini synthesis.

4.2.1.8.1 Sol-Gel Autoignition Method

In this method, the starting compounds were reserved in nitrates to form homogeneous powders. The flowchart is given in Fig. 4.3. The figure shows the detailed

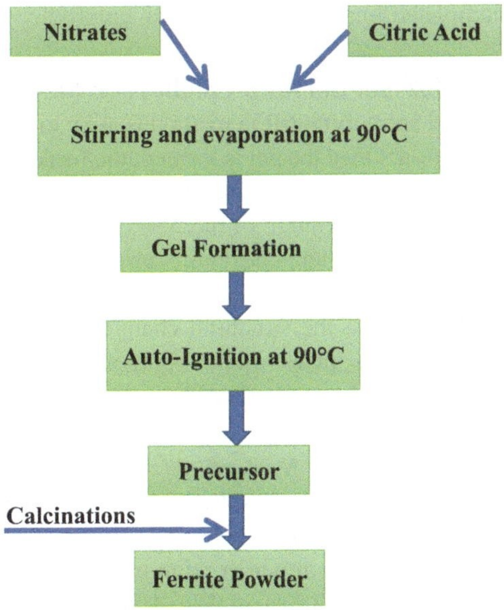

Fig. 4.3 Flowchart of sol-gel autoignition method

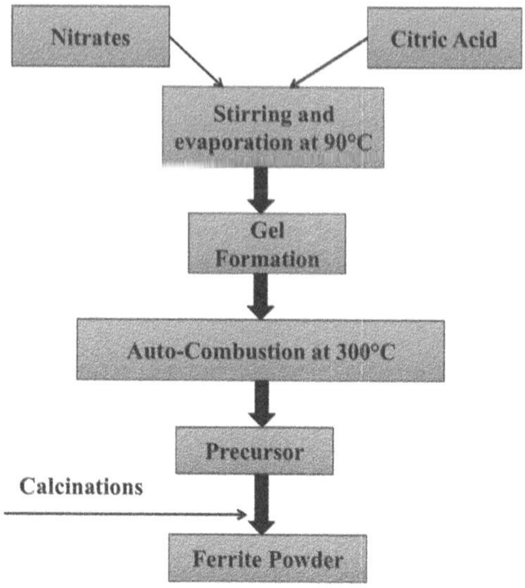

Fig. 4.4 Flowchart of sol-gel auto-combustion method

process of obtaining the required ferrite powders by sol-gel autoignition method (Raghavender et al. 2011).

The nitrates were used as starting materials and citric acid as chelating material. The metal nitrates were dissolved together in a deionized water to get a pure solution. To adjust the pH at 7, an aqueous solution of citric acid was mixed with metal nitrate solution. After full mixing of the chemical solutions, the mixed solution was placed onto a hot plate with nonstop stirring at 90 °C. After evaporation, the solution became viscous and finally is shaped to viscous brown gel. After several minutes the gel mechanically ignited and burnt with shining flints. The autoignition was completed finally with the brown colored.

4.2.1.8.2 Sol-Gel Auto-Combustion Method

Sol-gel auto-combustion method is alike to the method as described above till the gel formation. Once the gel is formed, the beaker with gel is moved onto the mantle, and the temperature is increased to 300 °C (Fig. 4.4).

As the temperature of the beaker reaches high, the entire gel is transformed into glowing flints, and the entire process would not stop till the citric acid is not consumed. The obtained precursor powders will also show some interesting properties, but the structural changes, which are taking place at low temperature, i.e., the initial phase of the compound formation, cannot be investigated. This is because the obtained powders by this method are presintered at 300 °C.

4.2.1.8.3 Pechini Method

This is also one of the sol-gel techniques employed by Pechini (Parvin et al. 2019; Massoudi et al. 2020; Jebeli Moeen et al. 2010). The metal nitrate mixture was heated to 90 °C, at which point ethylene glycol was added at a mass ratio of 4060 with respect to citric acid (Fig. 4.5). The temperature was maintained constant up to gel formation, which polymerized at 300 °C.

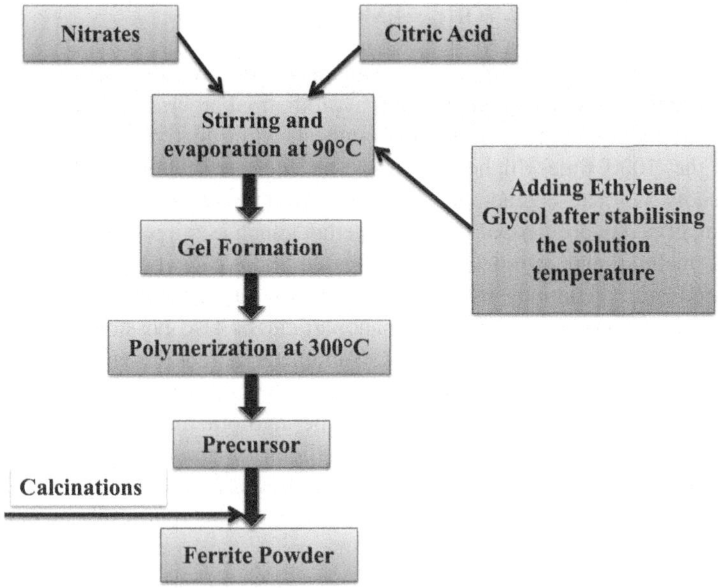

Fig. 4.5 Flowchart of Pechini method

4.2.1.9 Hydrothermal Synthesis

Among all the synthesis routes, hydrothermal synthesis is one of the most frequently used solution reaction-based procedure for synthesizing nanomaterials. This process requires very high temperature for the formation of nanoparticles. The process could be employed at a very low or very high pressure for the controlled morphology, and the process is nondestructive at high vapor pressure (Gan et al. 2020). The size of nanomaterials depend on hydrolysis rate and solubility of the metal compounds. First of all, the precursor solutions are poured into a Teflon container and then loaded in an autoclave reactor. Then, the solution is heated at specific temperatures up to 300 °C for several hours. Quenching with cold water is used to end up the reaction, and the prepared samples are washed with acetone or water many times to remove impurities (Dunne et al. 2015). The residue material is then dried at low temperature (60–80 °C) before calcination. Several metal nanoparticles like metal oxides, metal sulfides, metal phosphates, metal ferrites, etc. can be fabricated using this route. For nanoparticle formation the hydrated metal compounds undergo hydrolysis process, and then precipitates of metal nanoparticles are formed using dehydration (Hayashi and Hakuta 2010). Hydrolysis and dehydration processes during metal oxide nanoparticle formation take place according to the following reactions:

$$MA_x + xH_2O \rightarrow M(OH)_x + xHA$$

$$M(OH)_x \rightarrow MO_{x/2} + x/2H_2O$$

Hydrolysis is also the combination of an electrostatic reaction between metal and hydroxyl ions. Though the conventional hydrothermal process is much effective for nanomaterial fabrication, hydrothermal synthesis with supercritical water is superior due to the 1000 times higher reaction and high crystallinity. This process is eco-friendly as it can degrade alkaline concentrations and other toxic by-products. Other integrated hydrothermal synthesis methods such as microwave-assisted hydrothermal method and template-free self-assembling catalytic synthesis are also introduced with better results for the formation of nanomaterials (Machmudah et al. 2014). Controlled synthesis of nanoparticles is also possible through liquid phase or multiphase chemical reactions. Nanomaterials generated via hydrothermal process have low stability at high temperatures. Beyond these great advantages, the process holds some drawbacks also. As the hydrothermal process needs high temperature and high pressure, it is quite difficult to attain the required values of temperature and pressure. Second and the most serious issue is that the process takes much more time than some other conventional synthesis routes like precursor and sol-gel method.

4.2.1.10 Solvothermal Method

Solvothermal synthesis is basically a process having chemical reactions within any solvent at specific physical parameters like pressure and temperature. Usually, temperature above the boiling point of the solvent and pressure more than 1 bar is significant for the process to fabricate metal nanoparticles, but the parameters may vary according to the required properties of nanomaterials to be formed. Solvothermal synthesis is a different process from hydrothermal synthesis as it consists of chemical reactions within nonaqueous solutions at comparatively high temperatures (Peh and Zhao 2020). Numerous organic and inorganic solvents or alcohol, etc. can be used as medium for solvothermal process. In this process temperature and pressure are the key parameters like in hydrothermal synthesis for crystallization of the precursor solution. Further addition of solvents gives rise to the mobility of dissolved ions, and a fine mixture of all the reagents can be formed. The technique is useful for complex materials also as it gives a single-step reaction route for complex elements. Different reagents such as chlorides, acetates, nitrates, etc. undergo solvothermal process; form precipitates in the form of hydroxide or any other compound depending upon the specific nanomaterial to be prepared. Sodium hydroxide, potassium hydroxide, or any other base could be mixed with the reagent solution for precipitation. The properties of the solvent like dielectric constant, thermal conductivity, density, etc. change with temperature varying toward higher values (Shaikh et al. 2020). Thus, desirable characteristics of the resultant product can be attained by controlling the reaction parameters. Autoclave reactors are used for the heat treatment of the prepared solutions. These reactors are usually made up of strong materials or alloys like steel so that the reactor can uphold the pressure created inside the autoclave. To get a fine chemically inert vessel and prevent it from corrosion, a layer of Teflon is used inside the walls of reactor (Nunes et al. 2019).

Batch reactors are also very strong candidates for uninterrupted nanomaterial formation via solvothermal route. In one such reaction, Graphene oxide (GO) was prepared by mixing concentrated H_2SO_4 and $NaNO_3$ to the flake graphite powder while $KMnO_4$ was added to control the temperature of the solution. Graphene oxide nanoparticles were formed after the solution was kept in an ice bath for a few hours prior to the heat treatment, and residue was washed away with HCl solution and water many times (Yang et al. 2007). On the other hand, silver nanoparticles were successfully synthesized by one-step seedless solvothermal reduction route by using silver nitrate as a reagent and dimethylformamide (DMF) as solvent with polyvinylpyrrolidone (PVP) in a Teflon autoclave reactor at different temperatures (Liu et al. 2014). These solvents are mild reductant used to stop agglomeration of particles and growth mechanism of crystalline faces. Other than the conventional solvothermal technique, some integrated synthesis routes like microwave-assisted solvothermal synthesis have also shown excellent results in the formation of nanoparticles (Chella et al. 2015; Zhang et al. 2014).

4.2.1.11 Sonochemical Method

Nanoparticles could be synthesized by countless procedures, but controlled chemical reaction is the key factor to prepare nanomaterials with desired properties. The chemical reaction during the synthesis process strongly depends on many physical parameters like temperature, feed inlet, pressure, time, etc. Among all these factors, the feed inlet or energy supply is one of the most important parameter to control the reaction (Hankare et al. 2011; Patil and Bhanage 2016). Though each and every energy type has specific properties along with their reaction conditions, ultrasonic radiations give unique and most effective reaction conditions than other conventional energy source techniques (Xu et al. 2012). The effect of ultrasonic waves in liquids was first observed by Robert Williams Wood. The interaction between acoustic wave and matter was nil at atomic level that shows the zero interaction between chemical species and ultrasound at molecular level (Ali Dheyab et al. 2021). Further the acoustic cavitation process involves formation of bubbles and their disintegration that gives rise to intense heat emission. This intense heat is generated with high pressure abruptly and leads to the high-energy chemical reactions. This technique of sonochemistry is effectively applicable to the formation of nanostructured materials. Numerous nanomaterials including metal nanoparticles, metal oxide nanoparticles, nanostructured carbides, etc. have been synthesized using sonochemical synthesis route till now. Kumar et al. prepared silver nanoparticles by sonochemical route using starch and silver nitrate in specific amounts. Further for complete dissolution the mixture was stirred and agitated under sonication, and ultrasonic processor (DAIGGER GE 505, 500 W, 20 kHz) was used for radiation feed under effective operating conditions. The reduction of silver ions was observed at different time intervals for 22 days using a UV-visible spectrophotometer (Kumar et al. 2014). Different kind of solvents are used to dissolve the reagents before sonication in sonochemical synthesis, but some traditional solvents include dichloromethane, hexane, hexadecane, isopropyl ether, diethyl ether, pentane, etc. D. Mahajan and R. R. Adzic used hexadecane and hexane in equal amounts as solvent and polyvinylpyrrolidone (PVP) as stabilizer in formation of molybdenum and palladium nanoparticles. In addition sodium borohydride ($NaBH_4$) was mixed with the solution as a reducing agent, and monitoring of sonolysis was completed through CO gas extraction for 3 h (Okoli et al. 2018). Sonochemical synthesis method is proven to be very effective in formation of nanoparticles of Ag, copper, and cobalt oxide using distilled water as solvent. The prepared mixture was sonicated for 10 min using high-intensity ultrasonic horn (Kis-csitari et al. 2008). The sonochemical process is a simple, easy, and rapid path to form nanostructured materials, but the main problematic step is to control the reactions among the precursors due to the ultrasonic radiations. The process basically involves two-step reactions, first is the reaction between reagents, and second is the actual formation of nanoparticles. The properties of final product can be specified if these two reactions are controlled.

4.3 Conclusion

The synthesis of uniformly sized nanoparticles is increasing day by day due to their unique and modified physical properties. These properties of nanomaterials are much more efficient than the properties of the bulk materials. The electrical, structural, optical, and magnetic characteristics of the nanoparticles are dependent on the synthesis procedure. Several synthesis methods have been evolved according to the required properties of materials and ease of operation. Among all the synthesis methods, sol-gel, polyol, electrochemical, citrate precursor, sonochemical, coprecipitation, hydrothermal, etc. are some most frequently used methods to synthesize nanoparticles. A number of synthesis routes are discussed in this chapter including the abovementioned methods. Selection of a particular synthesis process for nanoparticles depend on the required properties of final product and availability of resources. Energy consumption, reaction conditions, reagent compositions, and cost-effectiveness are some factors which should be taken care of during the selection of synthesis route. Some integrated techniques have shown better results than the conventional methods and proven to be of great advantage.

References

Ali Dheyab M, Aziz AA, Jameel S (2021) Recent advances in inorganic nanomaterials synthesis using sonochemistry: a comprehensive review on iron oxide, gold and iron oxide coated gold nanoparticles. Molecules 26(2453):1–19. https://doi.org/10.3390/molecules26092453

Arshad M, Qureshi AH (2008) Thermal decomposition of metal complexes of type MLX 2 (M = Co (II), Cu (II), Zn (II), and Cd (II); L = DIE; X = NO 1 3) by TG-DTA-DTG techniques in Air Atmosphere. Turk J Chem 32:593–604

Assar ST, Abosheiasha HF (2012) Structure and magnetic properties of Co-Ni-Li ferrites synthesized by citrate precursor method. J Magn Magn Mater 324:3846–3852. https://doi.org/10.1016/j.jmmm.2012.06.033

Ataie A, Piramoon MR, Harris IR, Ponton CB (1995) Effect of hydrothermal synthesis environment on the particle morphology, chemistry and magnetic properties of barium hexaferrite. J Mater Sci 30:5600–5606. https://doi.org/10.1007/BF00356692

Balakumaran MD, Ramachandran R, Balashanmugam P, Mukeshkumar DJ, Kalaichelvan PT (2016) Mycosynthesis of silver and gold nanoparticles: optimization, characterization and antimicrobial activity against human pathogens. Microbiol Res 182:8–20. https://doi.org/10.1016/j.micres.2015.09.009

Bel-hadj-tahar R, Mohamed AB (2014) Sol-gel processed indium-doped zinc oxide thin films and their electrical and optical properties. New J Glass Ceram 4:55–65. https://doi.org/10.4236/njgc.2014.44008

Bönnemann H, Richards RM (2001) Nanoscopic metal particles—synthetic methods and potential applications. Eur J Inorg Chem 2001(10):2455–2480. https://doi.org/10.1002/1099-0682(200109)2001:10<2455::aid-ejic2455>3.0.co;2-z

Chahar D, Thakur P, Kumar R, Thakur A (2022) Influence of Mg doping on the structural, electrical and dielectric properties of Co-Zn nanoferrites. J Magn Magn Mater 544:168726. https://doi.org/10.1016/j.jmmm.2021.168726

Chang YC, Chen DH (2005) Adsorption kinetics and thermodynamics of acid dyes on a carboxymethylated chitosan-conjugated magnetic nano-adsorbent. Macromol Biosci 5:254–261. https://doi.org/10.1002/mabi.200400153

Chella S, Kollu P, Komarala EVPR, Doshi S, Saranya M, Felix S, Ramachandran R, Saravanan P, Koneru VL, Venugopal V, Jeong SK, Grace AN (2015) Solvothermal synthesis of MnFe 2 O 4 -graphene composite-investigation of its adsorption and antimicrobial properties. Appl Surf Sci 327:27–36. https://doi.org/10.1016/j.apsusc.2014.11.096

Cheruku R, Vijayan L, Govindaraj G (2012) Electrical relaxation studies of solution combustion synthesized nanocrystalline Li 2NiZrO 4 material. Mater Sci Eng B Solid-State Mater Adv Technol 177:771–779. https://doi.org/10.1016/j.mseb.2012.04.005

Choi J, Kim K, Han H, Hwang MP, Lee KH (2014) Electrochemical synthesis of red fluorescent silicon nanoparticles. Bull Korean Chem Soc 35(1):35–38. https://doi.org/10.5012/bkcs.2014.35.1.35

Dong H, Chena YC, Feldmann C (2015) Polyol synthesis of nanoparticles: status and options regarding metals, oxides, chalcogenides, and non-metal elements. Green Chem 00(5):1–23. https://doi.org/10.1039/C5GC00943J

Drmota A, Drofenik M, Koselj J, Nidari A (2012) Microemulsion method for synthesis of magnetic oxide nanoparticles. In: Microemulsions – An Introd. to Prop. Appl. https://doi.org/10.5772/36154

Dunne PW, Munn AS, Starkey CL, Huddle TA, Lester EH (2015) Continuous-flow hydrothermal synthesis for the production of inorganic nanomaterials. Philos Trans R Soc A Math Phys Eng Sci 373. https://doi.org/10.1098/rsta.2015.0015

Ehi-eromosele C, Ita B, Iweala E (2015) Magneto-structural properties of Ni–Zn nanoferrites synthesized by the low-temperature auto-combustion method. Bull Mater Sci 38(5):1–8. https://doi.org/10.1007/s12034-015-1038-1

El-sherbiny I, Reicha FM, Sarhan A, Abdel-hamid MI, El-sherbiny IM (2012) Preparation of silver nanoparticles in the presence of chitosan by electrochemical method preparation of silver nanoparticles in the presence of chitosan by electrochemical method. Carbohydr Polym 89:236–244. https://doi.org/10.1016/j.carbpol.2012.03.002

Fi F, Brayner R, Chau F, Giraud M, Mammeri F, Peron J, Piquemal J, Sicard L, Viau G (2018) The polyol process: a unique method for easy access to metal nanoparticles with tailored sizes, shapes and compositions. Chem Soc Rev 47:5187–5233. https://doi.org/10.1039/c7cs00777a

Foroughi F, Hassanzadeh-Tabrizi SA, Bigham A (2016) In situ microemulsion synthesis of hydroxyapatite-MgFe2O4 nanocomposite as a magnetic drug delivery system. Mater Sci Eng C 68:774–779. https://doi.org/10.1016/j.msec.2016.07.028

Gan YX, Jayatissa AH, Yu Z, Chen X, Li M (2020) Hydrothermal synthesis of nanomaterials. J Nanomater 2020. https://doi.org/10.1155/2020/8917013

Gogate PR, Pandit AB (2004) A review of imperative technologies for wastewater treatment II: hybrid methods. Adv Environ Res 8:553–597. https://doi.org/10.1016/S1093-0191(03)00031-5

Hajos M, Stypuła B, Starowicz M, Kasprzyk D (2011) Electrochemical synthesis of silver nanoparticles in alcoholic electrolytes. Arch Metall Mater 56(1). https://doi.org/10.2478/v10172-011-0016-x

Hankare PP, Patil RP, Jadhav AV, Pandav RS, Garadkar KM, Sasikala R, Tripathi AK (2011) Synthesis and characterization of nanocrystalline Ti-substituted Zn ferrite. J Alloys Compd 509: 2160–2163. https://doi.org/10.1016/j.jallcom.2010.10.173

Hayashi H, Hakuta Y (2010) Hydrothermal synthesis of metal oxide nanoparticles in supercritical water. Materials (Basel) 3:3794–3817. https://doi.org/10.3390/ma3073794

Hazra S, Ghosh NN (2014) Preparation of nanoferrites and their applications. J Nanosci Nanotechnol 14:1983–2000. https://doi.org/10.1166/jnn.2014.8745

Hoon J, Kim Y (2012) Ultrasonics Sonochemistry a novel polyol method to synthesize colloidal silver nanoparticles by ultrasonic irradiation. Ultrason – Sonochemistry 19:209–215. https://doi.org/10.1016/j.ultsonch.2011.06.004

Islam MA, Islam MS (2013) Electro-deposition method for platinum nano-particles synthesis. Eng Int 1(2):9–18. https://doi.org/10.18034/ei.v1i2.120

Khodashenas B, Ghorbani HR (2015) Synthesis of silver nanoparticles with different shapes. Arab J Chem 1(1):1–16. https://doi.org/10.1016/j.arabjc.2014.12.014

Kim F, Connor S, Song H (2004) Platonic gold nanocrystals. Angew Chem Int Ed 43:3673–3677. https://doi.org/10.1002/anie.200454216

Kis-csitari J, Konya Z, KIiricsi I (2008) Sonochemical synthesis of inorganic nanoparticles. NATO Sec Sci Ser B Phys Biophys:369–372. https://doi.org/10.1007/978-1-4020-8903-9

Kumar B, Smita K, Cumbal L, Debut A, Pathak RN (2014) Sonochemical synthesis of silver nanoparticles using starch: a comparison. Bioinorg Chem Appl 2014:1–8. https://doi.org/10.1155/2014/784268

Kumar NS, Kumar KV (2015) Synthesis and structural properties of bismuth doped cobalt nanoferrites prepared by sol-gel combustion method. World J Nano Sci Eng 5:140–151. https://doi.org/10.4236/wjnse.2015.54016

Kyun B, Jeong S, Kim D, Moon J, Lim S, Sub J (2007) Synthesis and size control of monodisperse copper nanoparticles by polyol method. 311:417–424. https://doi.org/10.1016/j.jcis.2007.03.039

Liu X, Liu X, Hu Y (2014) Investigation of the thermal decomposition of talc. Clays Clay Miner 62: 137–144. https://doi.org/10.1346/CCMN.2014.0620206

Liu YF, Yuan GH, Jiang ZH, Yao ZP (2014) Solvothermal synthesis of Mn3O4 nanoparticle/graphene sheet composites and their supercapacitive properties. J Nanomater 2014:190529. https://doi.org/10.1155/2014/190529

Logvinenko V, Polunina O, Mikhailov Y, Mikhailov K, Bokhonov B (2007) Study of thermal decomposition of silver acetate. J Therm Anal Calorim 90:813–816

Machmudah S, Widiyastuti W, Prastuti OP, Nurtono T, Winardi S, Wahyudiono H, Kanda MG (2014) Synthesis of ZrO2 nanoparticles by hydrothermal treatment. AIP Conf Proc 1586:166–172. https://doi.org/10.1063/1.4866753

Malik MA, Younus M, Hashim MA (2012) Microemulsion method : a novel route to synthesize organic and inorganic nanomaterials. Arab J Chem 5:397–417. https://doi.org/10.1016/j.arabjc.2010.09.027

Massoudi J, Smari M, Nouri K, Dhahri E, Khirouni K, Bertaina S, Bessais L, Hlil EK (2020) Magnetic and spectroscopic properties of Ni-Zn-Al ferrite spinel: from the nanoscale to microscale. RSC Adv 10:34556–34580. https://doi.org/10.1039/d0ra05522k

Mathew DS, Juang RS (2007) An overview of the structure and magnetism of spinel ferrite nanoparticles and their synthesis in microemulsions. Chem Eng J 129:51–65. https://doi.org/10.1016/j.cej.2006.11.001

Mitra S et al (2001) Tumor targeted delivery of encapsulated dextran-doxorubicin conjugate using chitosan nanoparticles as carrier. J Control Release 74(1-3):317–323. https://doi.org/10.1016/s0168-3659(01)00342-x

Moeen J, Vaezi MR, Yousefi AA (2010) Chemical synthesis of nano-crystalline nickel-zinc ferrite as a magnetic pigment. Prog Color Color Coat 3:9–17

Nunes D, Pimentel A, Santos L, Barquinha P, Pereira L, Fortunato E, Martins R (2019) Synthesis, design, and morphology of metal oxide nanostructures, Metal oxides series. Elsevier, Amsterdam, pp 21–57. https://doi.org/10.1016/B978-0-12-811512-1.00002-3

Okoli CU, Kuttiyiel KA, Cole J, McCutchen J, Tawfik H, Adzic RR, Mahajan D (2018) Solvent effect in sonochemical synthesis of metal-alloy nanoparticles for use as electrocatalysts. Ultrason Sonochem 41:427–434. https://doi.org/10.1016/j.ultsonch.2017.09.04

Parra S, Olivero J, Pulgarin C (2002) Relationships between physicochemical properties and photoreactivity of four biorecalcitrant phenylurea herbicides in aqueous TiO2 suspension. Appl Catal B Environ 36:75–85. https://doi.org/10.1016/S0926-3373(01)00283-1

Parvin F, Rikta SY, Tareq SM (2019) Application of nanomaterials from wastewater. Elsevier Inc., pp 137–158. https://doi.org/10.1016/B978-0-12-813902-8.00008-3

Patil AB, Bhanage BM (2016) Sonochemistry: a greener protocol for nanoparticles synthesis. In: Handbook of nanoparticles. Springer, Cham, pp 143–166. https://doi.org/10.1007/978-3-319-15338-4

Patil KC, Chandrashekhar GV, George MV (1968) Infrared spectra and thermal decompositions of metal acetates and dicarboxylates. Can J Chem 46(2):257–265. https://doi.org/10.1139/v68-040

Peh SB, Zhao D (2020) Synthesis and development of metal–organic frameworks. In: Nanoporous materials for molecule separation and conversion. Elsevier Inc. https://doi.org/10.1016/B978-0-12-818487-5.00001-7

Puliová P, Kováč J, Voigt A, Raschman P (2013) Structure and magnetic properties of Co and Ni nano-ferrites prepared by a two step direct microemulsions synthesis. J Magn Magn Mater 341: 93–99. https://doi.org/10.1016/j.jmmm.2013.04.003

Punia P, Bharti MK, Chalia S, Dhar R, Ravelo B, Thakur P, Thakur A (2021) Recent advances in synthesis, characterization, and applications of nanoparticles for contaminated water treatment-a review. Ceram Int 47:1526–1550. https://doi.org/10.1016/j.ceramint.2020.09.050

Quievryn C, Bernard S, Miele P (2014) Polyol-based synthesis of praseodymium oxide nanoparticles. Nanomater Nanotechnol 4(7):1–8. https://doi.org/10.5772/58458

Raghavender AT, Bili N, Skoko Ž (2011) XRD and IR analysis of nanocrystalline Ni–Zn ferrite synthesized by the sol–gel method. Mater Lett 65:677–680. https://doi.org/10.1016/j.matlet.2010.11.071

Rana K, Thakur P, Tomar M, Gupta V, Thakur A (2018) Investigation of cobalt substituted M-type barium ferrite synthesized via co-precipitation method for radar absorbing material in Ku-band (12–18 GHz). Ceram Int 44:6370–6375. https://doi.org/10.1016/j.ceramint.2018.01.028

Sarkar D, Tikku S, Thapar V, Srinivasa RS, Khilar KC (2011) Colloids and surfaces A : physico-chemical and engineering aspects formation of zinc oxide nanoparticles of different shapes in water-in-oil microemulsion. Colloids Surf A Physicochem Eng Asp 381:123–129. https://doi.org/10.1016/j.colsurfa.2011.03.041

Shaikh SF, Ubaidullah M, Mane RS (2020) Types, synthesis methods and applications of ferrites. Elsevier Inc., pp 51–82. https://doi.org/10.1016/B978-0-12-819237-5.00004-3

Sheikh A, Jain P (2016) A thorough study of zinc ferrite nanoparticles with reference to green synthesis. Int J Nanomed Nanosurg 2(3). https://doi.org/10.16966/2470-3206.115

Starowicz M, Stypuła B, Banas J (2006) Electrochemical synthesis of silver nanoparticles. Electrochem Commun 8:227–230. https://doi.org/10.1016/j.elecom.2005.11.018

Thakur A, Thakur P, Hsu J (2011) Enhancement in dielectric and magnetic properties of substituted Ni-Zn nano-ferrites by coprecipitation method. IEEE Trans Magn 47:4336–4339

Tourillon G, Pontonnier L, Levy JP, Langlais V (2000) Electrochemically synthesized co and Fe nanowires and nanotubes. Electrochem Solid-State Lett 3:20–23. https://doi.org/10.1149/1.1390946

Uttayarat P, Eamsiri J, Tangthong T, Suwanmala P (2015) Radiolytic synthesis of colloidal silver nanoparticles for antibacterial wound dressings. Adv Mater Sci Eng A 2015:1–6. https://doi.org/10.1155/2015/376082

Wang XW, Yuan ZH, Sun SQ, Duan YQ, Bie LJ (2008) Electrochemically synthesis and magnetic properties of Ni nanotube arrays with small diameter. Mater Chem Phys 112:329–332. https://doi.org/10.1016/j.matchemphys.2008.05.077

Whitesides GM, Stedronsky ER, Casey CP, Filippo JS (1970) Mechanism of thermal decomposition of ylphosphine cop n-butyl (tri-n-but per (I)). Am Chem Soc 1426:1–3

Woei B, Ying Y (2012) Synthesis of ZnO nanoparticles by modified polyol method. Mater Lett 73:78–82. https://doi.org/10.1016/j.matlet.2012.01.004

Wongwailikhit K (2011) The preparation of iron (III) oxide nanoparticles using W/O microemulsion. Mater Lett 65:2820–2822. https://doi.org/10.1016/j.matlet.2011.05.063

Xia Y, Xiong Y, Lim B, Skrabalak SE (2009) Shape-controlled synthesis of metal nanocrystals: simple chemistry meets complex physics? Angew Chem Int Ed 48:60–103. https://doi.org/10.1002/anie.200802248

Xu H, Zeiger BW, Suslick KS (2012) Sonochemical synthesis of nanomaterials. Chem Soc Rev 42:2555–2567. https://doi.org/10.1039/c2cs35282f

Yang Y, Matsubara S, Xiong L, Hayakawa T, Nogami M (2007) Solvothermal synthesis of multiple shapes of silver nanoparticles and their SERS properties. J Phys Chem C 111:9095–9104. https://doi.org/10.1021/jp068859b

Zhang Y, Feng J, Xu J, Chen G (2014) Solvothermal synthesis and visible-light driven photocatalytic properties of YFeO3 nanoparticles, integrated ferroelectrics. An Int J 151 (1):108–115. https://doi.org/10.1080/10584587.2014.899876

Zhao T, Sun R, Yu S, Zhang Z, Zhou L, Huang H, Du R (2010) Size-controlled preparation of silver nanoparticles by a modified polyol method. Colloids Surf A Physicochem Eng As 366:197–202. https://doi.org/10.1016/j.colsurfa.2010.06.005

Zhi J, Wang Y, Lu Y, Ma J, Luo G (2006) In situ preparation of magnetic chitosan/Fe3O4 composite nanoparticles in tiny pools of water-in-oil microemulsion. React Funct Polym 66:1552–1558. https://doi.org/10.1016/j.reactfunctpolym.2006.05.006

Synthesis of Nanomaterials by Biological Route

5

Atul Thakur, Deepika Chahar, and Preeti Thakur

Abstract

Nanotechnology has become one of the most important and emerging technologies in all areas of science. Various metallic and nonmetallic nanoparticles synthesized using nanotechnology have received global attention due to their potential applications in the biomedical and physiochemical areas. From the past years, the synthesis of nanoparticles using plants, bacteria, microorganisms, and various other biosources have been extensively studied and have been accepted as a green and efficient way for producing nanostructures. Various naturally occurring biodegradable materials like vitamins, sugars, and tea- or polyphenol-rich agricultural residues, which act as reducing and capping agents, are used for synthesizing nanoparticles with no toxicity. The green synthesis of nanostructures means plant- and other biosource (like fungi, flower, fruit, bacteria, starch, etc.)-mediated synthesis of nanoparticles. Various physiochemical and greener methods are available for the eco-friendly synthesis of nanoparticles that also require use of synthetic compounds for the assembly of nanostructures. The green approach of nanoparticle synthesis has many advantages like low cost, the lack of dependence on the use of any toxic materials, and the environmental friendliness for the sustainable assembly of stable nanostructures. The nanoparticles synthesized using green technology have

A. Thakur (✉)
Centre of Nanotechnology, Amity University Haryana, Gurugram, India
e-mail: athakur1@ggn.amity.edu

D. Chahar · P. Thakur
Department of Physics, Amity School of Applied Sciences, Amity University Haryana, Gurugram, India

A. Thakur et al. (eds.), *Synthesis and Applications of Nanoparticles*,
https://doi.org/10.1007/978-981-16-6819-7_5

numerous applications in fields like health care, food and feed, cosmetics, biomedical science, energy science, drug-gene delivery, environmental health, and so on. This technique is economic and sustainable, and hence an ideal way of production of natural nanoparticles. This chapter highlights the synthesis of biosource-mediated nanostructures and their applications in various fields. The aim is to provide insight into the use of plants, fruits, flowers, fungi, etc. as a bio-renewable, sustainable, diversified resource and platform for the production of useful nanoparticles having applications in various fields, including medicine, industry, defense, water purification, agriculture, and pharmaceuticals.

Keywords

Green synthesis · Flowers · Plants · Agriculture · Environment

5.1 Introduction

The green synthesis of nanoparticles by using plant extracts has drawn a lot of attention in recent years because it is environment friendly, highly efficient, and economical and is a simple synthesis method that can produce nanoparticles at industrial scale (Khan et al. 2017a; Khan et al. 2018a; Ambika and Sundrarajan 2015; Momeni and Nabipour 2015; Singh et al. 2016; Jagtap and Bapat 2013; Ghoreishi et al. 2011; Song and Kim 2008; Rao et al. 2016; Moteriya and Chanda 2016; Giljohann et al. 2010; Pereira et al. 2013). The nanoparticles that are synthesized based on secondary metabolites of plants like phenolics, alkaloids, saponins, terpenes, lipids, and carbohydrates are used for the treatment of toxic organic contaminations existing in the environment, and used for environmental remediation (Bhainsa and Souza 2006; Bagher et al. 2018; Gholami et al. 2018). Among the various eco-friendly methods available for the synthesis of nanomaterials, the use of medicinal/herbal plant extracts seems to be a better choice for chemical factories (Abdel-aziz et al. 2013; Dipankar and Murugan 2012; Khatami et al. 2016; Patil et al. 2012; Ramar et al. 2014). Natural compounds extracted from medicinal/herbal plants are used for capping of nanoparticles and exhibited a suitable platform for the development of newly discovered treatment selection switch with improved features (Patil and Kim 2016; Ahn et al. 2016; Khatami 2018; Dhanuskodi and Prabukumar 2018; Karthik et al. 2018a). There are many advantages of using natural plant products, for example, these are cheap, safe, and suitable primary materials for the synthesis of nanoparticles (Singh et al. 2016; Gnanasangeetha and Saralathambavani 2013; Singh et al. 2015; Zhang et al. 2011; Taylor et al. 2015; Golinska et al. 2014; Seshadri et al. 2011; Noruzi 2014; Elango and Roopan 2015). A number of secondary metabolites can be exploited in medicinal plant extracts regarding the green synthesis of nanoparticles (Sayed et al. 2018; Khatami 2018). This does not require the expensive and harmful chemicals; moreover in comparison with other microbial synthesis techniques, the rate of synthesizing nanoparticles by plant extracts is much higher and faster (Oh et al.

2017; Devi et al. 2017; Hamedi et al. 2016; Miri and Sarani 2018). Industrialization and urbanization lead to generation of excess of harmful and unwanted substances causing damage to the environment. A variety of such pollutants in the environment (including water and soil) are exposed by the microorganisms. Therefore, the microbial growth is inhibited due to persistence metal ions as they are nonbiodegradable in nature and often cause toxicity. However, microorganisms can survive even at high metal ion concentration and grow due to their ability to fight against the metal stress. The mechanisms include efflux systems; alteration of solubility and toxicity via reduction or oxidation; biosorption; bioaccumulation; extracellular complexation or precipitation of metals, and lack of specific metal transport systems (Bruins et al. 2000; Beveridge et al. 1997). The area of applications of these metal-microbe interactions is very vast in the field of biotechnology such as in bioremediation, biomineralization, bioleaching, and microbial corrosion (Prasad et al. 2018; Rajeshkumar 2016; Dhand et al. 2016). The microorganisms can be used in synthesis of metallic nanoparticles such as cadmium sulfide, gold, and silver. The use of nanoparticles in medical applications is limited due to chemical methods of synthesis that usually happen in the presence of toxic-reducing materials that may get attached with the surface of the prepared nanoparticles (Khatami et al. 2018a; Khan et al. 2012). Also, the physical and chemical methods of synthesis are expensive and require much energy (Dhanuskodi and Prabukumar 2018; Jana et al. 2001; Gontero et al. 2017; Dhanuskodi and Prabukumar 2017a; Jamdagni et al. 2018; Safaei et al. 2019; Torkzadeh-mahani et al. 2019; Khan et al. 2017b). This problem can be solved by synthesizing NPs using natural resources by which environment can be protected from toxic substances (Thg-ls et al. 2018; Singh et al. 2017; Ur et al. 2019; Javaid et al. 2018; Bharathi et al. 2018; Karthiga 2017; Goutam et al. 2017; Wongpreecha et al. 2018; Stone 2002; Phull et al. 2016). Various natural biosources used to synthesize nanoparticles are shown in Fig. 5.1. Nanoparticles play a very important role in development of sustainable technology for the future. Plant extracts are used for the synthesis of nanoparticles that connects nanotechnology and plant biotechnology. Nanoparticles are formed by plant extracts by bioreduction of metal ions. Various metabolites present in plant extracts like sugars, terpenoids, polyphenols, alkaloids, phenolic acids, and proteins play a significant role in metal ion reduction into nanoparticles. If nanoparticles are synthesized by using chemical methods like citrate precursor, coprecipitation, autocombustion, etc. in which various chemicals are used, then it may be a serious issue for the environment because of their general toxicity and harmful fumes liberated during burning. Hence, to avoid this problem, green synthesis using biological route is a best option as the molecules of extracts derived from plant sources and many other sources is preferred over other chemical methods (Nithyaja et al. 2012; Kasyanenko et al. 2016; Lok et al. 2007; Dai et al. 2005; Verma et al. 2010; Bindhu and Umadevi 2015; Sonker et al. 2017; Marulasiddeshwara et al. 2017; Bonilla et al. 2017; Carmen et al. 2018; Nin 2008). Green synthesis of nanoparticles using various plant extracts are presented in Table 5.1. Nanoparticles synthesized by green approach have potential to fight against all types of cancer, neurodegenerative disorders, and other diseases. The bio–/green-synthesized nanomaterials have been efficiently controlling the various

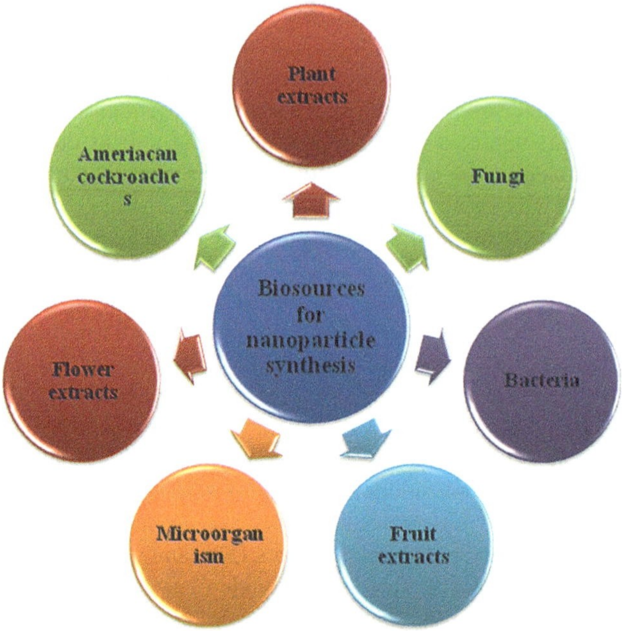

Fig. 5.1 Various types of biosources used to synthesize nanoparticles

endemic diseases with less adverse effect. The trend of using natural products has increased lately, and the active biosource extracts are frequently used for discovering new drugs. Thus, green synthesis of nanoparticles using biological molecules derived from plants, bacteria, microorganisms, flowers, etc. in the form of extracts is found to be superior, economic, and environmental friendly over chemical methods.

From the past many years, a lot of studies have proven that the biosources play a significant role of a potential precursor for the synthesis of nanomaterials in environmental friendly ways. Various biosources are used for the synthesis of several greener nanoparticles such as cobalt, copper, silver, gold, palladium, platinum, zinc oxide, iron-based magnetite/hematite, etc. Recently, various biological systems that include plants and algae (Govindaraju and Khaleel 2008), diatoms (Scarano and Morelli 2002; Lengke et al. 2007), bacteria (Kowshik et al. 2002), yeasts (Rautaray et al. 2003), fungi (Anshup et al. 2005), and human cells have proved their ability to transform inorganic metal ions into metal nanoparticles by the reductive capacities of the proteins and metabolites present in these organisms (Jeevanandam et al. 2018). Synthesis/preparation of metallic/nonmetallic nanoparticles using biological entities has been of great interest due to their unusual optical (Kudelski et al. 2003) and chemical properties (Kumar et al. 2003). There are various varieties of plants like geranium (*Pelargonium graveolens*) (Shankar et al. 2003), leaf extracts of lemongrass (*Cymbopogon flexuosus*) (Shankar et al. 2005), *Cinnamomum camphora* (Huang et al. 2007), neem (*Azadirachta indica*) (Shankar et al. 2004), aloe vera

Table 5.1 Green synthesis of nanoparticles using various plant extracts

Nanoparticle	Source of green synthesis	Name of source	Characterization	Size of the nanoparticles	Applications of nanoparticles	Reference
ZnO	Plant extract	*Coriandrum sativum*	XRD, SEM, FTIR, EDAX		Bimolecular detection, diagnostics, microelectronics, water remediation	Gnanasangeetha and Saralathambavani (2013)
Gold nanoparticles	Plant leaf extract	*Panax ginseng*			Therapeutic agent for cure of inflammation	Ahn et al. (2016)
ZnS	Plant extract	*Stevia rebaudiana*	XRD, EDAX, SEM, TEM, FTIR	8.35 nm	Capping and stabilizing agent	Alijani et al. (2018)
$NiFe_2O_4$ nanorods	Plant extract	Rosemary extract	XRD, HRTEM, FESEM, XPS, FTIR, VSM	10–28 nm	Anticancer agent	Alijani et al. (2020)
Ni-cu-Zn ferrite	Plant extract	Aloe vera	SEM, TEM, XRD, FTIR, VSM	60 nm		Duong et al. (2008)
$CoFe_2O_4$ and $AgCoFe_2O_4$	Plant extract of flower and leaf	*Hibiscus rosa-sinensis*	XRD, SEM, FTIR, VSM	18.8 nm & 15.8 nm	Antimicrobial agent	Gholami et al. (2018)
CdO	Plant extract	*Andrographis paniculata*	XRD, FESEM, HRTEM, FTIR	22 nm	Antibacterial agent	Dhanuskodi and Prabukumar (2017b)
Fe_3O_4	Plant extract	Rosemary	UV-visible spectroscopy, XRD, TEM, FTIR	4 nm		Khatami et al. (2017)
Cu/Cu_2O	Plant extract	*Stachys lavandulifolia*	XRD, TEM, UV-visible spectroscopy, FTIR	80 nm	Antibacterial agent for *Pseudomonas aeruginosa*	Khatami et al. (2017)
Silver nanoparticles	Plant	Dried grass	UV-visible spectroscopy, XRD, TEM	15 nm	Anticancer, antifungal, and antibacterial agent	Khatami et al. (2018b)

(continued)

Table 5.1 (continued)

Nanoparticle	Source of green synthesis	Name of source	Characterization	Size of the nanoparticles	Applications of nanoparticles	Reference
$CuFe_2O_4$, $NiFe_2O_4$, $ZnFe_2O_4$	Plant extract	Aloe vera	XRD, FTIR, SEM, TEM, VSM	15–70 nm		Laokul et al. (2011)
Gold and silver nanoparticles	Plant extract	*Panax ginseng* fresh leaves	XRD, FESEM, EDX, SAED		Drug delivery	Singh et al. (2016)
Ceria nanoparticles	Plant	Pullulan	XRD, TGA/DTG, FESEM, FTIR	12 nm	Biological and medical applications	Bagher et al. (2018)
$CuFe_2O_4$	Plant extract	*Hibiscus rosa-sinensis*	XRD, FTIR, EDX, SAED, HRTEM, HRSEM, VSM	17.16 nm		Durka and Antony (2015)
$CoFe_2O_4$	Plant extract	Aloe vera	XRD, EDX, SAED, TEM, UV-visible spectroscopy, VSM	14–26 nm		Manikandan et al. (2014)
CeO_2	Plant extract	*Prosopis farcta*	PXRD, EDX, FESEM, TEM, UV-visible spectroscopy, Raman spectroscopy, FTIR	21–28 nm	Drug delivery	Miri and Sarani (2018)
Gold and silver nanoparticles	Plant extract	*Hibiscus rosa-sinensis*	XRD, FTIR, TEM, SAED, UV-visible spectroscopy			Philip (2010)
MFe_2O_4 (M = Ni, co, Mn, mg, Zn)	Plant extract	Aloe vera	XRD, SEM, TEM, SAED, VSM	8.2 nm 8.5 nm 15.9 nm 45.3 nm 17.9 nm		Phumying et al. (2013)
Silver nanoparticles	Plant extract	*Pseudomonas*	XRD, FESEM, DLS, FTIR	10–40 nm	Antibacterial and antimicrobial agent	Thg-ls et al. (2018)
Gold nanoparticles	Plant extract	*Hibiscus rosa-sinensis*	FTIR, TEM, UV-visible spectroscopy, FTIR	16–30 nm	Medical applications	Yasmin et al. (2014)

Hydroxyapatite (HAp) nanoparticles	Plant extract	Aloe vera	XRD, FTIR, TEM, Raman spectroscopy	43–171 nm	–	Klinkaewnarong et al. (2010)
Gold nanotriangles and silver nanoparticles	Plant extract	Aloe vera	UV-visible, TEM, EDAX, AFM, FTIR	15.2 ± 4.2 nm	Cancer hyperthermia, optical coating	Chandran et al. (2006)
Silver nanoparticles	Plant extract	*Zingiber officinale*	UV-visible, TEM, DLS, AFM	6–20 nm	Drug delivery and gene delivery	Kumar et al. (2012)

(Chandran et al. 2006), and tamarind (*Tamarindus indica*) (Ankamwar et al. 2005b) that are used for efficient and rapid extracellular synthesis of gold and silver nanoparticles. Also, fruit extracts of *Emblica officinalis* (Ankamwar et al. 2005a) have been effectively used for synthesizing gold nanoparticles. Biomasses of wheat (*Triticum aestivum*) and oat (*Avena sativa*), alfalfa (*Medicago sativa*) (Gardea-torresdey et al. 2003), native and chemically modified hop biomass (Lo et al. 2005), and remnant water collected from soaked Bengal gram bean (*Cicer arietinum*) (Ghule et al. 2006) have also been used for various nanoparticle synthesis. Also, gold nanoparticles have been synthesized from alfalfa (*Medicago sativa*), *Chilopsis linearis*, and Sesbania seedlings, and silver and Ag-Au-Cu alloy nanoparticles have been synthesized from alfalfa (*Medicago sativa*) (Gardea-Torresdey et al. 2002) sprouts and *Brassica juncea* germinating seeds. There are various advantages of synthesis of nanoparticles using greener approach, for example, green techniques eliminate the use of expensive chemicals, consume less energy, and generate environmentally benign products and by-products. Green nanobiotechnology is a promising alternate route for synthesis of biocompatible stable nanoparticles (Narayanan and Sakthivel 2011).

5.2 Nanoparticle Synthesis

Nanoparticles can be synthesized by two different ways called "bottom-up" approach and the "top-down" approach. The two methods can be differentiated in the sense that bottom-up approach is meant for creating nanoparticles by grouping atoms and molecules. This grouping happens in a clear and managed way which results in an increase in the functionality of the structure of such materials. The top-down approach is clear reduction or breaking down of systems in their current state by making existing technologies more efficient. This causes reduction in the size into nanoscale aspects.

The bottom-up approach is more advantageous than the top-down approach because it has a better chance of producing nanostructures having less defects, more homogenous chemical composition, and better short- and long-range ordering. In bottom-up synthesis technique, the nanoparticles are synthesized onto the substrate by stacking atoms onto each other, which gives rise to crystal planes. Then, the crystal planes further stack onto each other, resulting in the synthesis of the nanostructures. A bottom-up approach can thus be viewed as a synthesis approach where the building blocks are added onto the substrate to form the nanostructures. In top-down synthesis approach, the nanostructures are synthesized by removing out crystal planes which are present on the substrate. A top-down approach can thus be viewed as an approach where the building blocks are removed from the substrate to form the nanostructure. Green synthesis of nanoparticles using other biosources are presented in Table 5.2.

There are various techniques of top-down approach for the nanoparticle synthesis. These techniques include chemical etching, laser ablation, mechanical milling/ball milling, sputtering, electroexplosion, etc. The bottom-up approach includes

Table 5.2 Green synthesis of nanoparticles using other biosources such as flower, fruit, fungi, etc.

Nanoparticle	Source of green synthesis	Name of source	Characterization	Size of the nanoparticles	Applications of nanoparticles	Reference
Silver nanoparticles	Fungus	*Aspergillus fumigatus*	XRD, TEM, UV-visible	5–25 nm		Bhainsa and Souza (2006)
Gold nanoparticles	Biosynthesis	Glucose and starch solution	UV-visible spectroscopy, TEM, AFM, electrochemistry		Biological and medical application	Engelbrekt et al. (2009)
Silver nanoparticles	Microorganism	*Neurospora intermedia*	UV-visible spectroscopy, DLS, SDS-PAGE, SEM			Hamedi et al. (2016)
Silver nanoparticles	Fruit extract	Longan fruit peel	UV-visible spectroscopy, XRD, EDX, HRTEM, FTIR	21 nm	Anticancer agent against breast cancer	Khan et al. (2018a)
Silver nanoparticles	American cockroach wings	*Periplaneta americana*	UV-visible spectroscopy, XRD, TEM	50 nm	Anti-insect agent on *Aphis gossypii*	Khatami et al. (2019)
$ZnFe_2O_4$	Wood apple juice	*Limonia acidissima*	XRD, FTIR, SEM, EDAX, TEM, UV-DRS	20 nm	Degradation of MB dye and antibacterial agent	Naik et al. (2019)
Gold and silver nanoparticles	Fruit extract	*Chaenomeles sinensis*	XRD, FETEM, UV-visible spectroscopy, DPPH	20–40 nm & 5–20 nm	Antimicrobial and anticancer agent	Oh et al. (2017)
CeO_2	Flower extract	*Hibiscus sabdariffa*	XRD, EDs, HRTEM, Raman spectroscopy, FTIR, XPS	3.9 nm		Thovhogi et al. (2015)
ZnO nanoparticles	Biosource	Gelatin	XRD, FESEM, TGA/DTA, AFM	–	Cosmetic and medicinal applications, optical and electrical devices	Darroudi et al. (2013)
Gold nanowires	Biosource	Sugar beet pulp	UV-visible, TEM, FTIR, EDS	–	Capping agent	Castro et al. (2011)
Gold and silver nanoparticles	Fruit extract	*Tanacetum vulgare*	UV-visible, XRD, EDX, FTIR	11 nm 16 nm	Biotechnological and biomedical applications	Prabha et al. (2010)

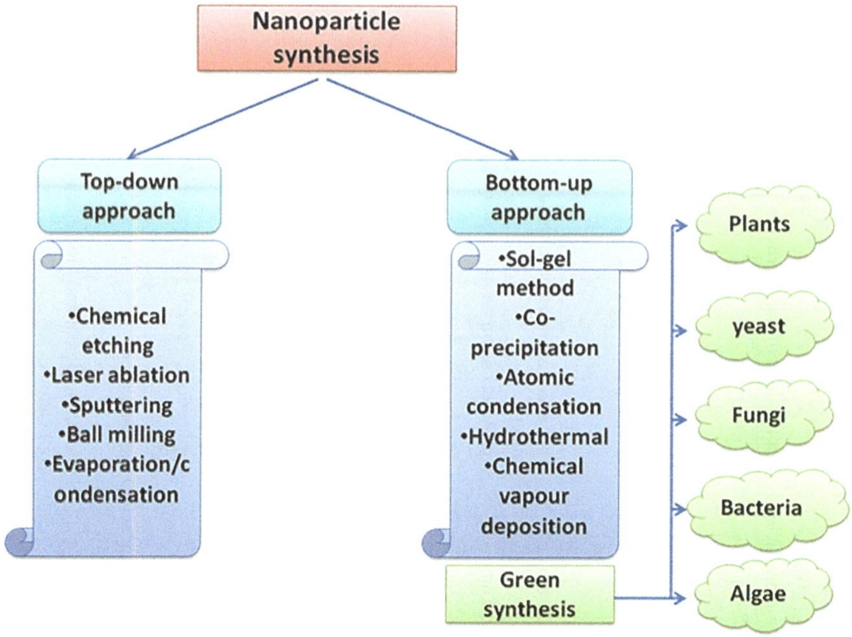

Fig. 5.2 Methods of nanoparticle synthesis

chemical vapor deposition, sol-gel processes, laser pyrolysis, spray pyrolysis, atomic/molecular condensation, and aerosol processes. The green synthesis of nanoparticles is also a bottom-up approach. Figure 5.2 shows various methods of nanoparticle synthesis. The green synthesis of nanoparticles is a cost-effective and environmental friendly method to synthesize nanoparticles in place of various chemical and physical methods. Green synthesis connects nanotechnology with nature. The synthesis happens at ambient temperature, neutral pH, and low cost and in environmentally friendly way. Synthesis of nanoparticles using plants is the best green synthesis technique as plants are nature's "chemical factories." They are cost-efficient and require low maintenance.

5.3 Green Synthesis Using Plant Extract

From the many past years, many chemical and physical methods have been used by researchers to synthesize nanomaterials that include electrochemical technique, chemical reduction, and photochemical reduction (Chen et al. 2001). It is verified from the previous studies that experimental conditions, kinetic interaction between metals and reducer agents, and stabilizer agent behavior influence the size, morphology, stability, and chemical and physical properties of the synthesized nanoparticles (Knoll and Keilmann 1999; Sengupta et al. 2005). Therefore, it is a great challenge to find a method to control nanoparticles' properties (Wiley and Sun 2007).

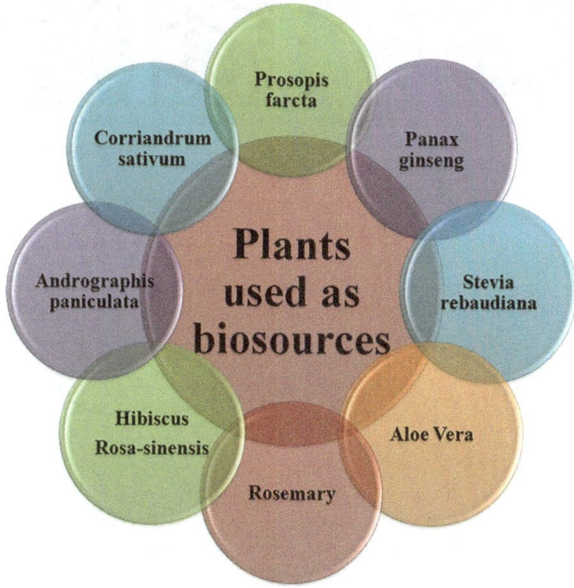

Fig. 5.3 Various plants used to synthesize nanoparticles

Nanoparticle synthesis using plants is one of the environmental friendly, affordable, nontoxic, and clean method (Dwivedi and Gopal 2010; Khatami et al. 2015; Castro et al. 2011; Kumar et al. 2012; Darroudi et al. 2013; Ponnuchamy and Jacob 2016; Prabha et al. 2010). Various plants can be used as natural sources for nanoparticle synthesis as shown in Fig. 5.3. Green chemistry gains its inspiration from nature through plants, yeast, fungi, and bacteria. In the nanoscience research, the integration of green chemistry principles is a key issue (Philip 2010). Figure 5.4 shows the green synthesis of nanoparticles by using different plants. The nanostructured magnetic ferrites can be prepared by using plant extracts from leaves, flowers, roots, or seeds in place of several chemical pathways using benign reagents and hence reduce the risk of hazardous substances (Manikandan et al. 2014; Phumying et al. 2013; Laokul et al. 2011). Figure 5.5 shows various plants' digital photographs whose extracts can be used for green synthesis of nanoparticles. A variety of metabolites that are released from the plants and also contained in plants such as carbohydrates, polysaccharides, phenols, amino acids, and vitamins. These metabolites can act as capping agents, reducing agents, and stabilizing and/or chelating agents for "capturing" the metal ions, and also they can also act as fuel. Size, shape, and morphology of the nanoparticles get influenced by the use of plant extracts during synthesis. Nanoparticles with high dispersity, high stability, and narrow size distribution can be produced by these metabolites (Laokul et al. 2011). Nowadays, metal oxides and mixed oxide nanoparticles can be obtained in bulk scale from many plant extracts like *Aloe vera* leaves, ginger roots, and *Hibiscus rosa-sinensis* flowers/leaves (Phumying et al. 2013; Laokul et al. 2011). Miri et al. (Miri et al. 2018) synthesized gold nanoparticles by using *Prosopis farcta* extract and studied its application for the decay of colon cancer cells.

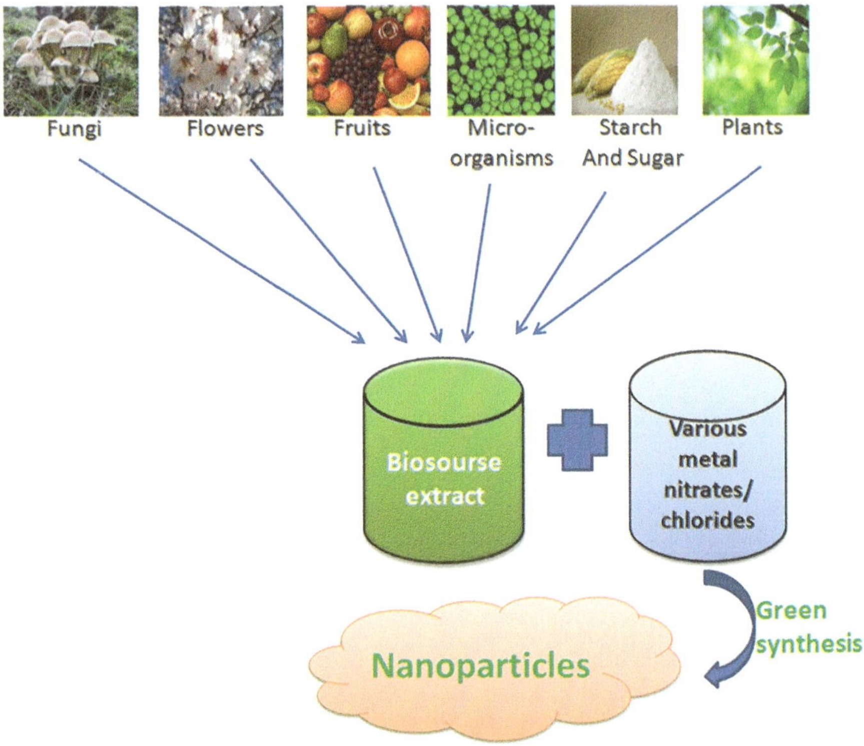

Fig. 5.4 Green synthesis of nanoparticles

Fig. 5.5 Digital images of various plants used for green synthesis of nanoparticles (Reproduced by permission from Ref. Mittal et al. (2013), License No. 4830671099554, Copyright 2013, Elsevier)

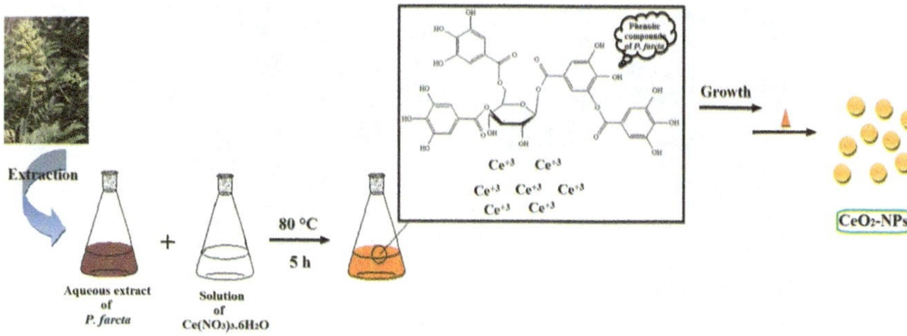

Fig. 5.6 Synthesis of CeO_2 NPs by using an aqueous extract of *P. farcta* (Reproduced by permission from Ref. Miri and Sarani (2018), License No. 4830660768290, Copyright 2018, Elsevier)

The various compounds present in the *Prosopis* are quercetin (flavonoids), tryptamine, apigenin 5-hydroxytryptamine (alkaloids), l-arabinose, and lectin (Jannet 2005; Gulalp and Karcioglu 2008). This plant named *Prosopis farcta* contains various medicinal properties to cure many diseases like gastric ulcers, fetus abortion, dysentery, arthritis, larynx inflammation, heart pains, and asthma, and it grows in the area of the Middle East (Al-Qura 2008). It contains phenolic compounds including tannins, vicenin-2, caffeic acid derivative, and apigenin c-glycoside (Jannet 2005) that can be helpful in the reduction process of Au^+ to Au^0. For the preparation of extract of *Prosopis farcta*, 5 mg of leaf powder was added to 50 mL of water, and prepared mixture was shaken at 150 rpm for 4 h, and resultant mixture was filtered to get a brown solution. This solution was then kept at 4 °C. To synthesize the gold nanoparticles, 5 mL of leaf extract was increased to 50 mL by addition of gold chloride solution, and that final mixture was then shaken at 150 rpm at 25 °C for 30 min. The prepared gold nanoparticles were found to have a very good potential for the decaying of cancer cells. Also, CeO_2 nanoparticles were synthesized using the extracts of *P. farcta* (Miri and Sarani 2018). The detailed synthesis process is shown in Fig. 5.6. The synthesized CeO_2 nanoparticles have found useful for application in decay of cancer cells. Figure 5.7 shows the TEM images and morphology of HT-29 cancer cells before and after treatment of nanoparticles.

Andrographis paniculata is a herbaceous plant having a height of 30–110 cm and is grown in moist shady areas. It can be found in South Asia, China, and Europe. This plant is effectively used to cure body illness like body heat, common cold, and upper respiratory tract infections including sinusitis and fever and is useful to dispel toxins from the body. Karthik et al. (Dhanuskodi and Prabukumar 2017b) synthesized CdO nanoparticles using *Andrographis paniculata* extract. For preparing the extract of *Andrographis paniculata,* 10 g leaves were washed using double-distilled water at 303 K and then immersed in water to extract the green-colored dye. Thereafter it was heated for 30 min at 393 K and filtered to remove residual solids. Then, 0.5 M of cadmium acetate was added in 100 mL of *A. paniculata* solution. The precipitate from the solution was washed with ethanol and double-distilled water

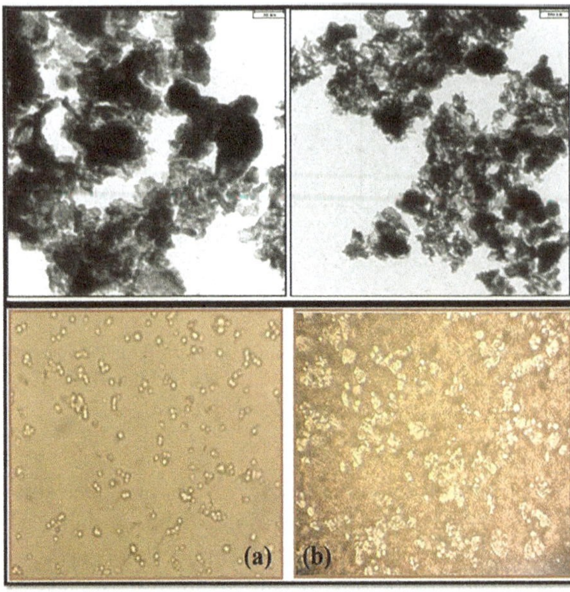

Fig. 5.7 (**A**) TEM images of synthesized CeO_2 NPs at 400 °C and (**B**) Morphology of HT-29 cells (a) before treatment and (b) after treatment with 200 μg/ml synthesized CeO_2 NPs (Reproduced by permission from Ref. Miri and Sarani (2018), License No. 4830660768290, Copyright 2018, Elsevier)

after stirring for 1 h and then dried at 303 K. After that, the resultant was sintered at 673 K for 4 h, and finally brown CdO powder was obtained. The antibacterial activity of green-mediated CdO nanoparticles was carried out against *Escherichia coli*, *Staphylococcus aureus*, *Aeromonas hydrophila*, *Vibrio cholerae*, and *Rhodococcus rhodochrous*. From the antibacterial studies, CdO nanoparticles exhibit an antibacterial activity that can be helpful in the food packaging industries.

5.3.1 Synthesis of Nanoparticles Using *Hibiscus Rosa-Sinensis*

Hibiscus rosa-sinensis is found throughout India, and known as shoe flower plant or Chinese hibiscus. It is an evergreen woody, showy shrub and glabrous plant found in India. Hibiscus is a medicinal herb that plays a role in many medical applications as it can be used as a native cure for hypertension, liver disorder, and pyrexia (Chen et al. 2003). It also prohibits adipogenesis and is used against the problem of dandruff, thereby promoting hair growth. Hibiscus plant extract can be used in the cure of diseases like diarrhea, fatigue and skin problems, gonorrhea, and menorrhagia and as a medicine against diabetes (Venkatesh et al. 2008; Sachdewa and Khemani 2003). Figure 5.8 depicts the digital photograph of hibiscus leaf and TEM images of the synthesized gold nanoparticles using hibiscus leaf extract. The ingredients in the hibiscus leaf extract are proteins, vitamin C, organic acids,

Fig. 5.8 (**A**) Digital image of the *Hibiscus* leaf (**B**) TEM images of gold nanoparticles (a) and (b) under different magnification (c) single multibranched gold nanoparticle. (Reproduced by permission from Ref. Philip (2010), License No. 4830661263292, Copyright 2010, Elsevier)

flavonoids, and anthocyanins. It is evident from the literature that *Hibiscus rosa-sinensis* flower/leaf extracts were often used to obtain Au and Ag nanoparticles (Bhainsa and Souza 2006; Muzammil 2013), ZnO, and CeO_2 nanoparticles (Yasmin et al. 2014; Thovhogi et al. 2015). The first study on the synthesis of spinel copper ferrite ($CuFe_2O_4$) using hibiscus flower extract was done by the Manikandan et al. (Durka and Antony 2015). Many diseases can be cured by *Hibiscus rosa-sinensis*. Its antibacterial activity is known for more than 50 years (Atwan and Saiwan 2010; Missoum 2018; Mak et al. 2013). The organic and phenolic acids such as citric, malic, succinic, lactic, gallic, hibiscus, and homogentisic acids form the chemical composition of this plant. Also, flavonoids such as luteolin, quercetin, and gossypetin and their glycosides are also present. The bright color of the flowers is due to anthocyanins. The chemical composition varies with the species, origin, age, and color. The antioxidant and antimicrobial activities are due to total phenolic compounds and flavonoids (Alaga et al. 2014). Gingasu et al. (Gingasu et al. 2016) synthesized $CoFe_2O_4$ and $AgCoFe_2O_4$ nanoparticles using *Hibiscus rosa-sinensis* flower and leaf extract and studied its potential for the antimicrobial actions. For the synthesis of $CoFe_2O_4$ nanoparticles, firstly 5 g of dried flowers were taken in 100 mL distilled water under continuous stirring. The mixture was then boiled for 15 min. The obtained bright red extract was cooled to room temperature and filtered to get the hibiscus flower extract. To obtain the hibiscus leaf extract, 5 g of fresh leaves were taken and placed in 100 mL of distilled water after cutting under continuous stirring followed by boiling the mixture for 45 min till the formation of yellow-colored solution with pH = 6. The obtained solution was then cooled at room temperature and filtered. The cobalt ferrites were prepared using the self-combustion process. For this, the metal nitrates were added to the extract of *Hibiscus rosa-sinensis* flower under constant slow stirring. The obtained gel was put on a heater at 250–300 °C. Initially, the melting of gel took place followed by spontaneous decomposition by self-ignition, leaving behind voluminous foam. This was then annealed at 800 °C for 1 h to improve the degree of crystallization of cobalt ferrite. After that for the wet fertilization reaction, the metal nitrates in a proper ratio were added under stirring to the aqueous extract of *Hibiscus rosa-sinensis* leaf. The pH of the solution was raised to 10 by adding NH_4OH followed by separation of a dark brown precipitate. The suspension became magnetic after 4 h when maintained at 80 °C. A thermal treatment at 800 °C for 1 h led to the formation of a well-crystallized cobalt ferrite. For the synthesis of Ag-Co nanoferrite particles, the abovementioned self-combustion process was performed. The prepared ferrite nanoparticles showed good potential for the antimicrobial and antifungal activities. Philip et al. (Philip 2010) synthesized gold and silver nanoparticles using *Hibiscus rosa-sinensis* plant extract. For collecting the hibiscus leaf extract, firstly the leaf of the plant was washed many times using deionized water and cut followed by stirring at 300 K with 200 mL deionized water for 1 min. After that it was filtered to get the required extract. This filtrate acts as a reducing agent and stabilizer. In it 5 mL of hibiscus extract is added stirring vigorously for 1 min. Slow reduction happens and got completed in 1.5 h shown by the change in color to light violet of the solution. For the synthesis of silver nanoparticles, 20 mL of the hibiscus extract is added to a vigorously stirred 25 mL aqueous solution of $AgNO_3$ continuously stirring for 1 min. The pH of the solution was adjusted to 6.8 using NaOH. Reduction takes place rapidly, and it was indicated by golden yellow color of the solution. These colloids are best able for 4 months.

5.3.2 Synthesis of Nanoparticles Using *Stevia Rebaudiana* Bertoni

Stevia rebaudiana Bertoni is a herbaceous plant and sweet steviol glycosides can be produced (Karaköse et al. 2015). Until 2013, 34 sweet steviol glycosides had been discovered in leaf extracts (stevioside; steviolbioside (trace); rebaudiosides A, B (trace), C, D (trace), and E (trace); and dulcoside A) including 8 isomers and glycosylated forms of oxidized steviol derivatives (Vashist et al. 2017). These glycosides contain different glucose units (Melis et al. 2009). The gold, ZnS, Ag, and CdS nanoparticles can be synthesized by using glucose and biodegradable materials (Dhanuskodi and Prabukumar 2017b; Engelbrekt et al. 2009; Karthik et al. 2018b; Salari et al. 2018). This can help in reducing the cytotoxicity of the nanoparticles for their bioanalysis applications in the quantitation tags, signal transducers, encoded substrates, functional tags, and drug-gene delivery (Dhanuskodi and Prabukumar 2018). Because of its natural sweetness and therapeutic effects of diterpene steviol glycosides, this plant is considered highly economical and scientific. This compound was found to be 250–300 times sweeter than sucrose (Karthik et al. 2018a). Its applications are found in the fields of biomedicine (Das 2008), food industry, and biotechnology (Karak et al. 2011). Alijani et al. (Alijani et al. 2018) synthesized zinc sulfide nanoparticles using extract of *Stevia rebaudiana*. For the synthesis of ZnS nanoparticles, the aqueous crude extract of Stevia and Na_2S were used as sources of glucose and sulfur, respectively. For the preparation of *Stevia rebaudiana* leaf extract, 15 g of dried leaf powder was added to 105 mL of deionized water at room temperature. After that, the flasks were shaken with a rotation rate of 90 rpm for 48 h at a temperature of 37 °C. Then, centrifugation was done at 4500 rpm to separate the resultant mixture for a period of 10 min. Thereafter, 100 mL of 1 M $Zn(NO_3)_2$ solution was added dropwise to 100 mL of 1 M Na_2S stirring continuously. Then, the white-colored solution was stirred for 16 h followed by adding 100 mL of this resultant crude extract dropwise to the solution and achieved light green-colored solution. This was then incubated at 70 °C for 6 h. Then, centrifugation was done to separate the resultant mixture, and finally the obtained light green-colored product was dried at 50 °C for 4 h. The particle size of the synthesized nanoparticles ranged from 1 to 40 nm. FTIR study confirmed the role of the prepared nanomaterial as capping and stabilizing agents.

5.3.3 Synthesis of Nanoparticles Using *Aloe vera* Extracts

Recently, biosynthesis has emerged as an alternative synthesis technique to prepare nanocrystalline inorganic materials. The metal and semiconductor nanoparticles are synthesized using fungi, actinomycetes, and plant extracts. Aloe vera is a native plant in Thailand and several other countries and contains 99.5% water content in the leaves. The rest is solid material containing over 75 different ingredients including vitamins, minerals, enzymes, sugars, anthraquinones or phenolic compounds, lignins, saponins, sterols, amino acids, and salicylic acids. Cosmetic industries use aloe vera gel as a hydrating ingredient in liquids, creams, sun lotions, lip balms, healing ointments, etc. Also, the gel is used in pharmacology for wound healing and anti-inflammatory and burn treatment. Properties of *Aloe vera* and *Panax ginseng* are shown in Fig. 5.9. Phumying et al. (Phumying et al. 2013) synthesized magnetic

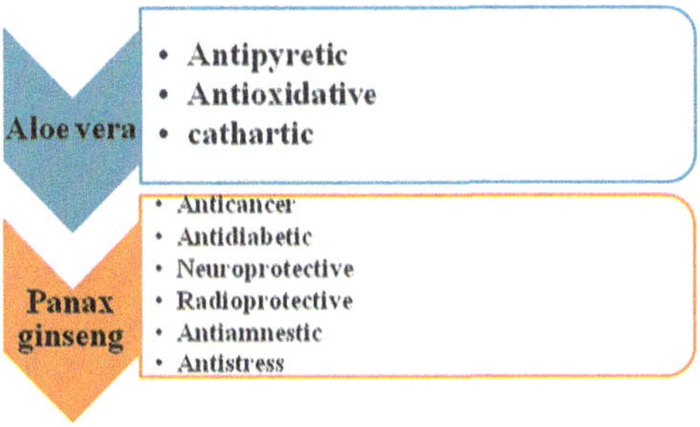

Fig. 5.9 Various properties of *Aloe vera* and *Panax ginseng*

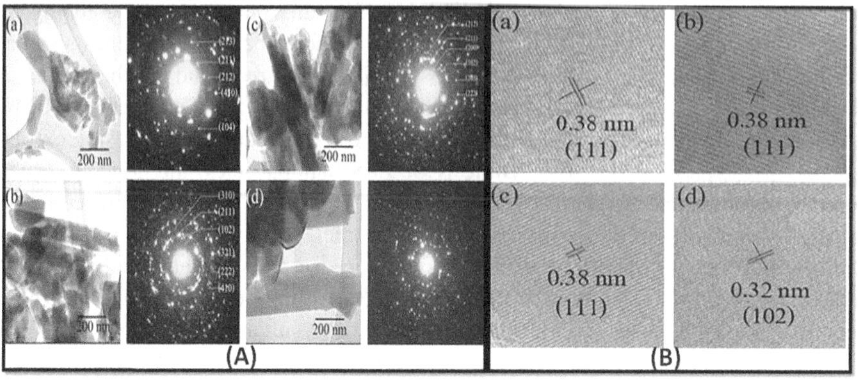

Fig. 5.10 (**A**) TEM images with corresponding selected area electron diffraction (SAED) patterns of HAp powders calcined in air for 2 h at (a) 500 °C, (b) 600 °C, (c) 700 °C, and (d) 800 °C. (**B**) HRTEM images of HAp powders calcined in air for 2 h at (a) 500 °C, (b) 600 °C, (c) 700 °C, and (d) 800 °C. (Reproduced by permission from Ref. Klinkaewnarong et al. (2010), License No. 4830681177416, Copyright 2010, Elsevier)

ferrites by using aloe vera extract. Aloe vera-extracted solution was prepared from 35 g *Aloe vera* leaves after cutting and boiling in 100 ml of deionized water. The resulting extract was used as an aloe vera extract. In the preparation of MFe_2O_4 (M = Ni, Co, Mn, Mg, Zn) samples, all the starting materials were first dissolved in 50 ml aloe vera extract solution and stirred for 1 h at room temperature. The solution was sealed in a Teflon-lined autoclaved of 100 mL capacity and heated at 200 °C for 2 h. It was then gradually cooled to room temperature, filtered and washed with deionized water and ethanol several times. The resultant was then dried in air at 85 °C. The hydrothermal synthesis using aloe vera extract is a relatively new method to produce precursors of nanocrystalline spinel ferrite powders at a low temperature and in short time. Synthesis of nanomaterials using aloe vera extract is a simple, efficient, and green method (Chandran et al. 2006; Phumying et al. 2013; Laokul et al. 2011; Klinkaewnarong et al. 2010). Figure 5.10 shows TEM and HRTEM images of the HAp nanoparticles synthesized using aloe vera gel (Klinkaewnarong et al. 2010). There are many advantages of using aloe vera extract solution for the synthesis of nanomaterials, for example, aloe vera plant extract is an environment friendly, nonpolluting solvent system, an eco-friendly reducing agent, and a non-hazardous agent for the stabilization of the nanostructures (Visinescu et al. 2011; Varma 2012). Aloe vera plant contains water ranging from 97.5% to 99.5% of fresh matter. Also, the components of this plant are water, fat-soluble vitamins, minerals, enzymes, polysaccharides, phenolic compounds, and organic acids. About 60% of the left out solid is made up of polysaccharides (Boudreau et al. 2017). The metal oxides can be prepared with aloe vera extract by using it as a bio-reducing agent due to the long-chain polysaccharides present in the Aloe vera plant extract that affords the homogeneous distribution of metal oxides. In one such study, Manikandan et al. (Manikandan et al. 2014) synthesized ferrite nanoparticles using aloe vera extract.

For the preparation of extract solution, a 5 g portion of thoroughly washed *Aloe vera* leaves was finely cut, and the obtained gel was added in 10 ml of deionized water, clear solution was then obtained after stirring for 30 min. For preparing $CoFe_2O_4$ samples, firstly ferric nitrate (10 mmol) and cobalt nitrate (5 mmol) were dissolved in the prepared extract solution under vigorous stirring at room temperature for 1 h to obtain a clear transparent solution. Aloe vera plant extract solution acts both as a reducing and gelling agent for the synthesis of mixed metal oxides. Then, the precursor mixture of metal nitrates in Aloe vera-extracted solution was put in a domestic microwave oven and exposed to the microwave energy in a 2.45 GHz multimode cavity at 850 W for 10 min. Initially, boiling of the precursor mixture took place followed by evaporation and then decomposition with the evolution of gases. After reaching the point of spontaneous combustion, it vaporized and instantly became a solid. The obtained solid was then washed with ethanol and dried at 70 °C for 1 h. Therefore, aloe vera plant-extracted microwave combustion synthesis can be used as an eco-friendly method to produce precursors for nanocrystalline spinel ferrite powders with low temperature, energy consuming, and short time. Also Laokul et al. (Laokul et al. 2011) synthesized copper ferrite, nickel ferrite, and zinc ferrite nanoparticles using aloe vera extract, wherein, the modified sol-gel method was used to synthesize MFe_2O_4 using various appropriate metal nitrates such as $Ni(NO_3)_2.6H_2O$, $Cu(NO_3)_3.3H_2O$, $Zn(NO_3)_2.6H_2O$, $Fe(NO_3)_3.9H_2O$, and aloe vera extract solution as starting materials. For this, firstly 250 mL aloe vera plant extract solution was dissolved with 50 g of nitrate with stirring on a hot plate for 60 min at room temperature. After that, the temperature was increased to 90 °C so that a dried solid precursor was obtained. Finally, the calcination was done at different temperatures in the range of 600–900 °C for 2 h in a furnace with heating rate of 5 °C/min. This is a simple method with the use of cheap precursors of Aloe vera plant extract that gives very high-yield nanomaterials having a well crystalline structure and acceptable magnetic properties.

5.3.4 Nanoparticle Synthesis Using *Panax Ginseng*

The meaning of ginseng is the essence of man, and it is a very important pharmacological herbal medicinal plant. *P. ginseng* was found in China over 5000 years ago, and since that time, its roots are used as a very important medicinal herb in traditional Chinese medicine (Radad et al. 2006). This plant which resembles the "human body" is a very slow growing perennial herb. The roots of the plant grow from the third year, and an increase in its diameter can be seen in the fourth year. By the sixth year, the length of the plant reaches to 7–10 cm and breadth 3 cm (Immer 1996). The components of the *P. ginseng* plant include ginsenosides, sugar residues, flavonoids, proteins, etc. having ginsenoside as the major pharmacological ingredient (Kim et al. 2014). It has been proposed that the pharmacological efficacies of ginseng plants and their products are due to the ginsenosides. However, ginsenosides are found mainly in roots of the plant, but leaves also contain a large amount of it and can be used easily as compared to roots (Wang et al. 2006). Gold and silver nanoparticles can be synthesized using this plant. Gold nanoparticles of anti-inflammatory nature were synthesized using *Panax ginseng* leaves (Ahn et al. 2016). The experiment was performed with decreased availability of inflammatory mediators in macrophages. The synthesized gold nanoparticles were found to have a potential application for therapeutic application for inflammatory diseases by blocking of NF-jB via p38 MAPK.

Singh et al. (Singh et al. 2016) synthesized gold and silver nanoparticles using *Panax ginseng* fresh leaves. Although *P. ginseng* roots, leaves, and other products have the capability of preparing the extract, leaves are found to have very rapid, facile, stable, economical source and result in nanoparticles of high pharmacological importance and effect. Hence, gold and silver nanoparticles synthesized using *P. ginseng* leaves are used in in vitro trial for biocompatibility, anticancer, and anti-inflammatory efficacies. The characterization done of the synthesized silver and gold nanoparticles was in terms of biological active groups, surface charge, and temperature stability, and further applied for antioxidant efficacy, cell viability on normal cells lines, anticancer efficacy on lung cancer and skin cancer cell lines (melanoma cancer), and anti-inflammation effect on RAW 264.7. Hence, the green synthesis using *P. ginseng* is an advantageous and cost-effective method for the development of herbal medicinal plant-mediated, low-cost, and safe nano-drug carriers in targeted drug delivery systems, cancer diagnostic, photothermal therapy, biosensing, and medical imaging. The use of such plants that have medical application with therapeutic importance can create a new platform for effective and green nanoparticle synthesis that have many applications on medical platform.

5.4 Green Synthesis of Nanoparticles Using Flower Extracts

5.4.1 Synthesis of Nanoparticles Using *Hibiscus sabdariffa*

Hibiscus sabdariffa (*H. sabdariffa*) is a shrub that belongs to the family Malvaceae with red flowers in the form of calyces. From the study of the composition of this plant, it is evident that various phenolic compounds are present in this plant that include various organic and phenolic acids like citric acid, hydroxycitric acid, and hibiscus acid (Mahadevan and Kamboj 2009). Also this plant contains flavonoids such as quercetin, luteolin, or gossypetin and their glycosides. The red color in the flowers is due to anthocyanins that are present in high amount. The anthocyanins present in flowers contain cyanidin-3-glucoside, delphinidin-3-glucoside, cyanidin-3-sambubioside, and delphinidin-3-sambubioside (Alaga et al. 2014; Qourzal et al. 2005; Pacôme et al. 2014; Guardiola and Mach 2014; Patel 2013; Borrás-linares et al. 2015). Thovhogi et al. (Thovhogi et al. 2015) synthesized CeO_2 nanoparticles using *Hibiscus sabdariffa* flower extract. Firstly, the washing of the dried red flowers was done to remove dust. To synthesize the required nanoparticles, 10.0 g of clean *H. sabdariffa* flowers was weighed in a beaker, and 400 ml distilled water was added to it and left for 2 h at room temperature. Filtering of the obtained solution was done twice to remove residual solids followed by adding 2.0 g of $Ce(NO_3)_3.6H_2O$ in 100 mL of the prepared solution. The solution was mixed homogeneously and heated for ~2 h. The precipitate formed is presumably CeO_x and $Ce(OH)_x$ mixture. Then, after the centrifugation at ~10,000 rpm, it was then dried in oven at ~100 °C. Then, the resultant product was annealed at ~500 °C for 2 h using a high-temperature tubular furnace. The nanoparticles formed are spherical in shape having a diameter of 3.9 nm.

5.4.2 Synthesis of Nanoparticles Using *Stachys Lavandulifolia*

Stachys lavandulifolia flower extract was used for the synthesis of copper and copper oxide nanomaterials (Khatami et al. 2017). In that research, copper chloride was used as a precursor of copper ions, and the extract as the other reagent. The pH of the extract of *S. lavandulifolia* flowers was adjusted to alkali for the synthesis of nanomaterials. Various characterizations were performed like transmission electron microscopy (TEM), X-ray diffraction (XRD), UV-visible spectroscopy, and Fourier-transform infrared spectroscopy (FTIR) for the characterization of the nanoparticles. Finally, agar well diffusion method was used to study the antibacterial activity of the nanoparticles against *Pseudomonas aeruginosa*. For the synthesis, firstly, 5.0 g of *S. lavandulifolia* flowers was added in deionized water, and then the sepals were transferred into an Erlenmeyer flask having 100 mL hot deionized water and boiled for 20 min. Then, the resultant was filtered using a filter paper, and this extract was used to synthesize copper and copper oxide nanomaterials. Then, 50 mL of copper chloride solution was added dropwise to 25 ml herbal extract with rapid stirring at 50 °C. The pH of the solution was adjusted to 10 by adding NaOH solution. Then, centrifugation was done to separate the precipitate, and washing was done with water and ethanol. The resultant was then dried at ambient temperature. The samples were kept at room temperature. The nanomaterials using the *S. lavandulifolia* can be synthesized using no additional surfactants, polymers, or chemical reagents at room temperature and pressure. This is an attempt to further develop the green synthesis approach of nanoparticles (Table 5.3).

5.4.3 Synthesis of Nanoparticles Using Rosemary Extract

Rosmarinus officinalis, a renewable biological resource, has high antioxidants (Genena et al. 2008) and is used as a safe food flavoring because of its favorable taste and aroma (Nieto et al. 2018). Many researchers have used rosemary extract for the synthesis of nanomaterials. Khatami et al. (2017) synthesize Fe_3O_4 nanoparticles using rosemary extract. *R. officinalis* leaves were obtained from the local market of Kerman city. Firstly, 10 g of leaves was taken and washed with double-distilled deionized water and dried at air temperature (28 °C). Then, the leaves were powdered with mortar and added to the Erlenmeyer containing 1000 mL of sterile double-distilled deionized water. This final mixture was heated at 70–80 °C for

Table 5.3 Various biosources used for green synthesis

Plant	Native place	Ingredients	Properties	Synthesized nanoparticles	References
Panax ginseng	China	Ginsenosides, sugar residues, flavonoids, proteins	Anti-inflammatory	Gold and silver nanoparticles	Attele et al. (1999); Meyer (2012)
Aloe vera	Thailand and several other countries	99.5% of water, vitamins, minerals, enzymes, sugars, anthraquinones or phenolic compounds, lignin, saponins, sterols, amino acids, and salicylic acid	Used for wound healing, anti-inflammatory, burn treatment, and cosmetic industry	Nanoferrites	Manikandan et al. (2014); Boudreau et al. (2017)
Hibiscus rosa-sinensis	India and China	Proteins, vitamin C, organic acids, flavonoids, and anthocyanins	Used to cure hypertension, liver disorder, pyrexia, dandruff, diarrhea, fatigue, skin problems, and gonorrhea	Various ferrites, gold nanoparticles, silver nanoparticles	Muzammil (2013); Durka and Antony (2015); Missoum (2018); Mak et al. (2013)
Andrographis paniculata	South Asia, China, and Europe		Cure gastric ulcers, fetus abortion, dysentery, arthritis, larynx inflammation, heart pains, and asthma	CdO nanoparticles	Dhanuskodi and Prabukumar (2017b)
Prosopis farcta	Grows in the Middle East	Quercetin (flavonoids), tryptamine, apigenin 5-hydroxytryptamine (alkaloids), l-arabinose, and lectin	Cure gastric ulcers, fetus abortion, dysentery, arthritis, larynx inflammation, heart pains, and asthma	Gold nanoparticles	Miri et al. (2018); Jannet (2005); Gulalp and Karcioglu (2008)

30 min (Khanna-chopra and Semwal 2011). Then, centrifugation was done for 5 min, and the supernatant was collected for further processing. These all procedures were performed inside the laminar airflow cabinet. Therefore, a sterile condition was maintained during the experiments. After that, the ferric (III) chloride hexahydrate stock solution (0.1 M) was prepared by addition of 1000 mL of sterile double-distilled deionized water to 27.03 g of $FeCl_3.6H_2O$. For the preparation of iron oxide (Fe_3O_4) nanoparticles (NPs), 10 mL of filtered *R. officinalis* extract was mixed to a 1 mM $FeCl_3$ solution with constant stirring at room temperature. After some time, light yellow color was changed to black denoting the synthesis of nanoparticles (Groiss et al. 2016). In this report, nanoparticles were synthesized using a green method and were mainly spherical and homogeneous having an average size of about 4 nm. The green method of synthesis used was an eco-friendly, simple, and inexpensive method. Alijani et al. (Alijani et al. 2020) synthesized nickel ferrite nanoparticles using rosemary extract. $NiFe_2O_4$ nanoparticles were synthesized by green and sustainable process using natural plant extract. Firstly, distilled water containing sodium hypochlorite was used to wash rosemary young leaves, and then the surface moisture was removed by placing at room temperature. For the extract of rosemary leaves, 200 g of rosemary healthy leaves was added to 1000 mL of deionized water and heated at 80 °C for 1 hour. After that the resultant was filtered with filter paper. For synthesis of nickel ferrite nanoparticles, 0.5 g of $FeCl_3.6H_2O$ and 0.1 g of $NiCl_2.6H_2O$ were dissolved in 30 mL of rosemary extract at 70 °C under vigorous stirring. The pH of the solution was maintained at 7.4. Finally, nanoparticles were washed with ethanol and deionized water three times each and then dried in the oven at 60 °C for 12 h. The $NiFe_2O_4$ nanorod particles were synthesized using greener and cost-effective methodologies using rosemary extract. Metallic nanomaterials have their application in biomedicine and environment. This technique helps the industries' application faster to the end products. Various characterizations were performed using HRTEM, XRD, FeSEM, XPS, VSM, and FTIR. Also, $NiFe_2O_4$ nanoparticles had a cytotoxicity effect on MCF-7 cell survival which suggests that $NiFe_2O_4$ nanoparticles can be used as a new class of anticancer agent in designing novel cancer therapy research. These prepared nickel ferrite nanorod particles can be used to increase the level of public health. Also, the nickel ferrite nanoparticles using rosemary extract can be used as medical sensor or antibacterial application.

5.5 Green Synthesis of Nanoparticles Using Fruit Extracts

Among all the methods of synthesis of nanoparticles, biosynthesis comes out to be a good, eco-friendly method to synthesize nanomaterials as in this process, nontoxic reducing and capping agents are used. These capping agents get adsorbed on the surface of nanoparticles. The fruit extracts can be used for the green synthesis of nanomaterials (Isaac et al. 2013). *Solanum torvum* fruit extract is also used for the synthesis of gold and silver nanoparticles (Ramamurthy et al. 2013). The prepared nanoparticles showed good antibacterial and antioxidant properties as shown by

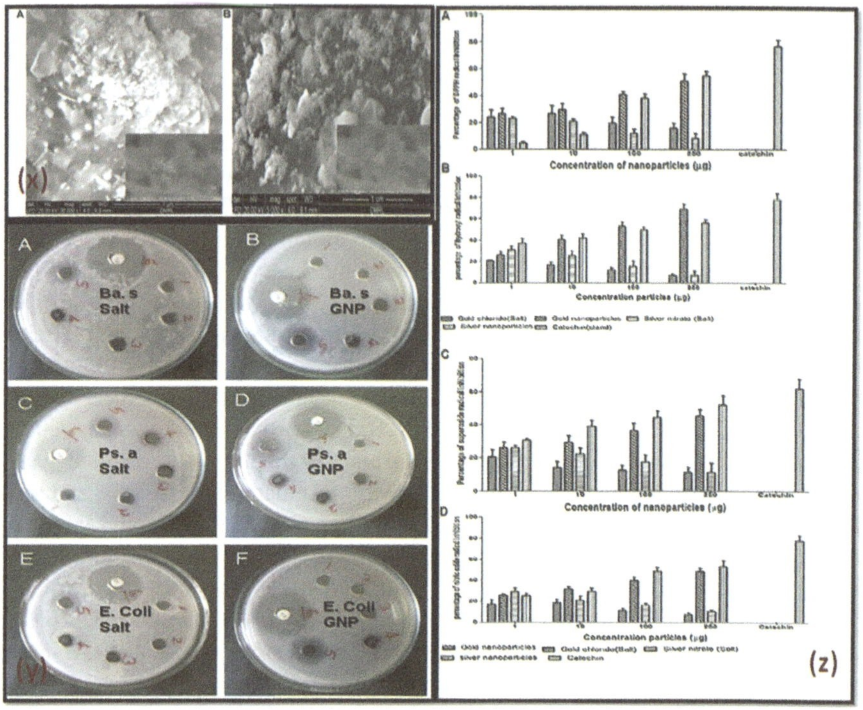

Fig. 5.11 (**X**) SEM photographs of *S. torvum*-synthesized gold nanoparticles and SEM photographs of *S. torvum*-synthesized silver nanoparticles. (**Y**) Antioxidant properties of gold and silver nanoparticles and their respective salts. (A) DPPH antioxidant assay of gold and silver nanoparticles; (B) hydroxyl radical scavenging assay of gold and silver nanoparticles; (C) superoxide radical scavenging assay of gold and silver nanoparticles; (D) nitric oxide radical scavenging activity of gold and silver nanoparticles. Catechin was used as positive control. (**Z**) Antimicrobial activity of gold nanoparticles against (A and B) *Bacillus subtilis*; (C and D) *Pseudomonas aeruginosa* and (E and F) *E. coli* (Reproduced by permission from Ref. Ramamurthy et al. (2013), License No. 4830671457452, Copyright 2013, Elsevier)

Fig. 5.11. Also, tansy is a fruit that is considered a cure for intestinal worms, rheumatism, digestive problems, fevers, etc. This fruit extract was used for the synthesis of silver and gold nanoparticles (Prabha et al. 2010). Figure 5.12 shows the digital image of the tansy fruit and TEM images of the synthesized nanoparticles. The *Limonia acidissima* is a fleshy fruit, and its components include large amounts of carbohydrate (18.1 g), protein (7.1 g), fat (3.7 g), iron (6 mg), and vitamin C (3 mg). These components act as a reducing agent and metal nitrates as oxidizers that are useful for microwave process. Naik et al. (Naik et al. 2019) used *Limonia acidissima* juice for the synthesis of zinc ferrite nanoparticles and tested their application as a photocatalyst and antibacterial agent. The zinc nitrates and ferric nitrates with a ratio of 1:2 in 5 mL of *Limonia acidissima* juice were used as starting materials for the synthesis of zinc ferrite nanoparticles. The *L. acidissima* juice acts

Fig. 5.12 (**a**) Digital image of the tansy fruit extract. (**b**) TEM images of silver nanoparticles. (Reproduced by permission from Ref. Prabha et al. (2010), License No. 4830680693395, Copyright 2010, Elsevier)

as a reducing agent. The whole prepared mixture was taken in 45 mL deionized water. Then, the prepared mixture was stirred for 60 min to get the homogeneous solution. The microwave irradiation was done in a domestic microwave oven (2.54 GHz at 900 W) for 15 min. Then, calcination was done at 600 °C for 4 h. Further, the prepared nanoparticles are used for structural, optical, morphological, magnetic, photocatalytic (Evans blue and methylene blue), and antibacterial (foodborne pathogens) studies. Figure 5.13 shows the SEM, TEM, and EDAX spectra of the synthesized nanoparticles. The synthesized zinc ferrites showed significant photocatalytic activity for Evans blue and methylene blue dyes. The prepared ferrites are ferromagnetic in nature as seen in VSM study. Also, the prepared $ZnFe_2O_4$ nanoparticles show an effective antibacterial activity against foodborne pathogens. The antibacterial activity is shown in Fig. 5.14. The microwave-assisted green synthesis of $ZnFe_2O_4$ nanoparticles are the suitable materials for wastewater treatment and biomedical applications.

5.5.1 Synthesis of Nanoparticles Using Longan Fruit

Longan (*Euphoria longana Lam.*) fruit is a crop in Thailand and used through whole of Asia. The components of this fruit are "Gallic acid, corilagin and ellagic acid" (Rangkadilok et al. 2005). All these compounds are polyphenols, and are good agents to synthesize AgNPs as well as to stabilize them. Khan et al. (Khan et al. 2018a) synthesized silver nanoparticles using the fruit peel extract of *Longan* fruit. *Longan* fruits were firstly hand peeled, dried in shade and powdered using grinder. Then, 20 g of ground peel was soaked in 160 mL of distilled water at 30 °C for 24 h, filtered and centrifuged at 5000 rpm for 10 min at 4 °C to remove the remaining peel debris. The supernatant was then carefully removed after centrifugation and used for

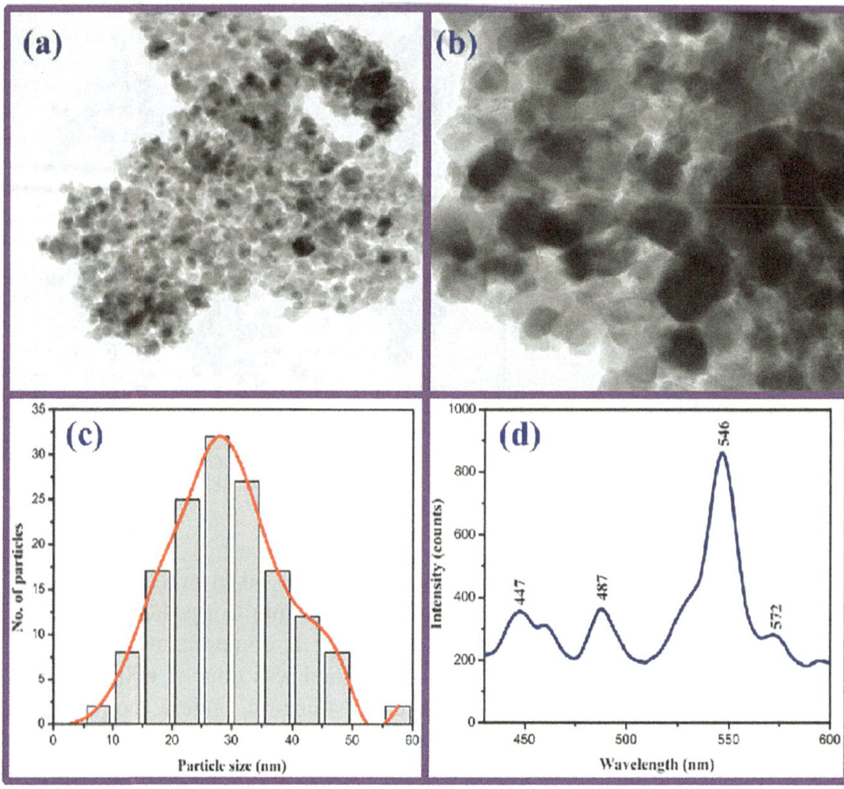

Fig. 5.13 (a) SEM, (b, c) TEM images, (c) typical histogram of the particles' diameter distribution and (d) EDAX spectra of microwave-assisted $ZnFe_2O_4$ nanoparticles. (Reproduced by permission from Ref. Naik et al. (2019), License No. 4830660251110, Copyright 2019, Elsevier)

the synthesis and stabilization of AgNPs. For the synthesis of AgNPs, 10 mL of 3 mM aqueous solution of AgNO3 was added to 10 mL aqueous extract of *Longan* fruit peel in a 100 mL beaker and stirred at 300 rpm (30 °C). The color change from brownish to dark black in 55 min showed the formation of nanoparticles. These prepared silver nanoparticles were found to be very good anticancer agents in the treatment of breast cancer.

5.5.2 Synthesis of Nanoparticles Using *C. Sinensis*

C. sinensis (*Thouin*) *Koehne*, commonly known as Chinese quince or "Guang Pi Mu Gua," is found in Korea, China, and Japan. This fruit is commonly used in traditional Chinese medicine and Korean traditional medicine to cure inflammation, vitalize digestion, and reduce cholesterol and sugar levels (Zhang et al. 2009). Also, this fruit is used to cure and alleviate rheumatoid arthritis, cough, common cold, and diarrhea

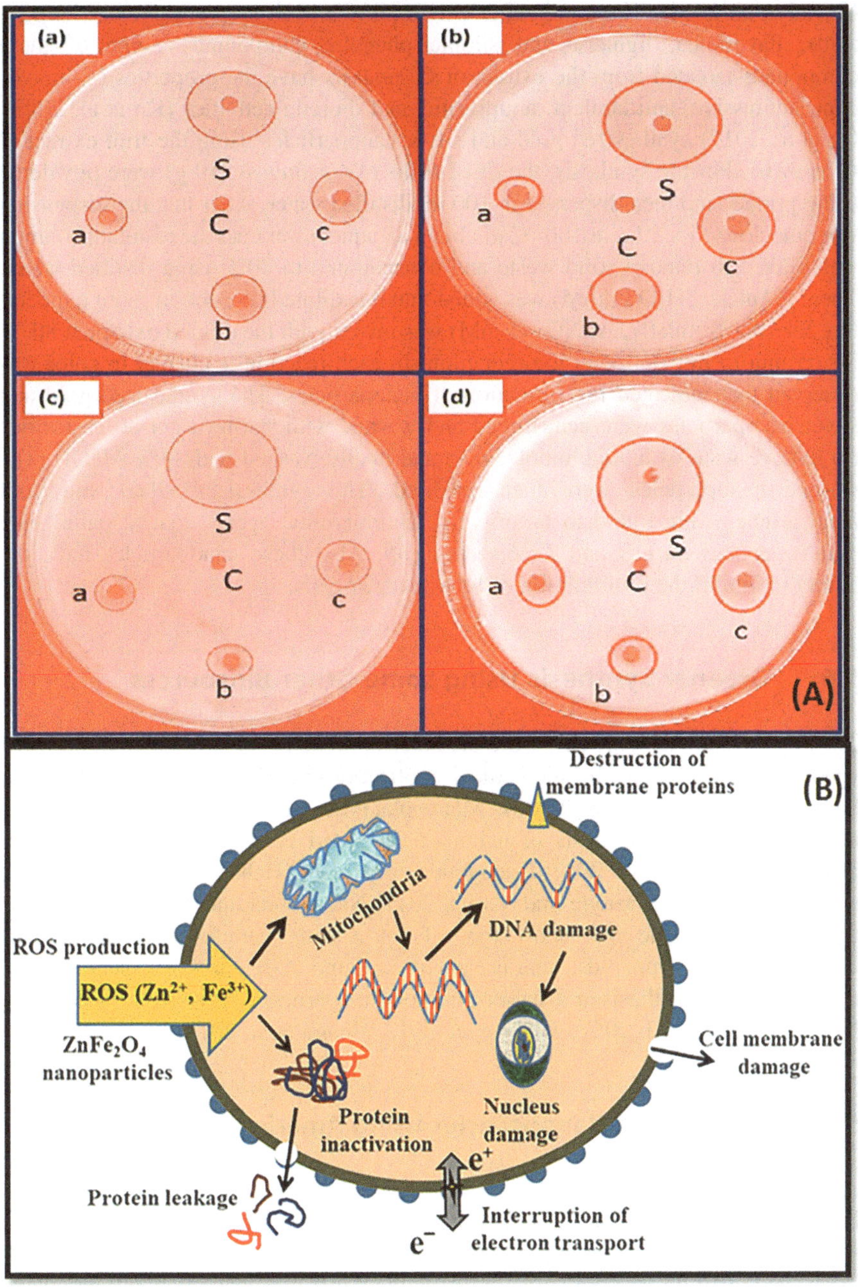

Fig. 5.14 (**A**) Antibacterial activity of microwave-assisted green synthesis of $ZnFe_2O_4$ nanoparticles against pathogenic bacterial strains such as (a) *Staphylococcus aureus*, (b) *Escherichia coli*, (c) *Pseudomonas desmolyticum*, and (d) *Klebsiella aerogenes* (S = standard antibiotic; C = control; a, b, and c are the different concentrations of nanoparticles: 50, 100, and 150 μg/μL). (**B**) Antibacterial activity mechanism of microwave-assisted green synthesis of $ZnFe_2O_4$ nanoparticles. (Reproduced by permission from Ref. Naik et al. (2019), License No. 4830660251110, Copyright 2019, Elsevier)

(Sawai et al. 2008). The compounds present in this fruit are pentacyclic triterpene acids, flavonoids, lignans, and simple phenolic compounds. Some of these compounds isolated from the extract of *C. sinensis* have the properties of antioxidant, antitussive, antiflatulent, antipruritic, and diuretic activities (Ku et al. 2003). Oh et al. 2017) synthesized gold and silver nanoparticles using the fruit extract of *C. sinensis*. For the synthesis, the dried fruits of *C. sinensis* (10 g) were powdered using grinder and then dispersed in 100 ml distilled water. After that the suspension was autoclaved for 1 hr. at 100 °C to obtain an aqueous extract the resultant solution was filtered to remove solid waste and concentrated to 70% using distilled water. Then, $HAuCl_4$. $3H_2O$ (1 mM) was added into this diluted solution in room temperature for 10 s. Similarly, $AgNO_3$ (1 mM) was mixed with the diluted extract at 80 °C for 65 min to initiate metal ion reduction. In each reaction, a change in color was observed that indicated the formation of nanoparticles. The synthesized gold and silver nanoparticles were centrifuged and washed with sterile water at least thrice (to remove water-soluble biomolecules) and finally washed with 80% MeOH. The washed nanoparticles were then air-dried. The synthesized silver and gold nanoparticles were found to be effective antimicrobial agents against pathogenic *Staphylococcus aureus* and *Escherichia coli*. Also these nanoparticles have the ability to inhibit the proliferation of breast cancer cells.

5.6 Greener Synthesis Using some Other Biosources

The synthesis of nanoparticles using the green approach is of very much importance when they are used in the biomedical applications such as anticancer (Miri and Sarani 2018; Khan et al. 2018b). Many plants and even enzymes are used by researchers for the synthesis of nanoparticles (Miri et al. 2018; Darroudi et al. 2014; Muhammad et al. 2017; Nadagouda et al. 2014; Hebbalalu et al. 2013; Kou and Varma 2012; Virkutyte and Varma 2011; Nadagouda and Varma 2006). The reports are available that show the role of a wide variety of phytochemical compounds like terpenoids, tannins, glucose, amino acids, proteins, phenols, and alkaloids in the synthesis and stabilization of nanoparticles (Engelbrekt et al. 2009; Mirzaei and Darroudi 2017; Mittal et al. 2013; Huang et al. 2011).

5.6.1 Synthesis of Nanoparticles Using American Cockroaches

American cockroaches are found to be one of the most widespread insects, and besides that they have a role in medicines. In some parts of Asia, for example, in the southwestern part of China, the diseases such as hepatitis, trauma, stomach ulcers, burns, and heart disease are cured by using the American cockroach extract (Wang et al. 2011; Patterson and Slater 2002). Khatami et al. (2019) synthesized silver nanoparticles by using *Periplaneta americana* wings' extract. Chitin-rich, *Periplaneta americana* (American cockroach) wings' extract has been studied as a novel biomaterial to synthesize silver NPs, and cockroach ball was used to synthesize water-soluble silver NPs in a size range less than 50 nm and was examined for the insecticidal applications. For this, the American cockroach wings were sterilized in 70% ethanol for 10 s, and after that washing is done using deionized water and left in air. Then, 10 cc of deionized water having 1 drop of acetic acid is added to the wings after breaking them into pieces. Then, it was heated for 1 h at 60 °C and kept in the dark. After that the solution was filtered, and centrifuged at 2500 rpm for 20 min. Then, the extract was strained using Whatman filter paper No. 40. For preparing 0.1 M solution of silver nitrate, 50 ml distilled water was added to 0.899 g of silver nitrate powder. Thus, 1 mM of the solution made was then added to 10 mL of the extract of the wings. The results show a significant fatal effect of silver NPs at a concentration of 100 μg/mL on *Aphis gossypii*. However, concentrations of 1–40 μg/mL did not significantly effect on the mortality of immature aphids. It is clear from all results that silver nanoparticles have the lethal effect of green-synthesized silver NPs on *A. gossypii*, in vitro.

5.6.2 Synthesis of Nanoparticles Using Fungi

Since many years, various biological sources have been used for the nanoparticle synthesis. These biosources include bacteria, fungi, plants, etc. and are effective alternate sources for nanoparticle synthesis (Ramamurthy et al. 2013; Wei et al. 2012; Rajasekharreddy and Usha 2010; Gou et al. 2015; John et al. 2020; Rai and Yadav 2013; Kora and Arunachalam 2013). Among these, fungi are the more efficient candidates for synthesis of nanoparticles because of high protein secretion capacity, higher productivity, and easy handling in large-scale production. In addition, extracellular biosynthesis using fungi can also make downstream processing much easier than when employing bacteria (Honary et al. 2013; Musarrat et al. 2010). Various types of fungi were used by researchers like *Fusarium oxysporum* (Ahmad et al. 2003; Masumeh et al. 2012), *Cladosporium cladosporioides* (Balaji et al. 2009), *Penicillium* sp. (Kumar et al. 2010; Ranjan and Nilotpala 2011), and *Aspergillus flavus* (Vigneshwaran et al. 2007) to synthesize nanoparticles. Hamedi et al. (Hamedi et al. 2016) synthesized silver nanoparticles using *N. intermedia*. For the biosynthesis of the nanoparticles, *N. intermedia* fungus was grown aerobically in culture media containing PDB medium supplemented with 0.5% yeast extract. HCL was used to adjust the final pH of the medium to 5.8. After 72 h of fermentation, the culture broth was separated using centrifugation for 20 min at 6000 rpm. For the production of silver nanoparticles, the resultant supernatant and 2 mM silver nitrate solution were mixed together with volume ratios of 1:1. Finally, the mixtures were put in incubators at 28 °C and agitated at 200 rpm to reach the nanoparticle formation. All reactions were performed in the presence of light. After that filtration was done to separate the fungal mycelia, and washing of the resultant mycelia was done three times with sterilized distilled water. Then, 10 g of harvested mycelia was submerged in a 100 ml sterilized distilled water and incubated on an orbital shaker operating at 200 rpm and 4 °C for 72 h. This suspension was filtered, and finally, the obtained filtrate (50 ml) was mixed with 50 ml of a 2 mM aqueous silver nitrate. The reaction mixtures were incubated at 200 rpm, 28 °C until the maximum absorbance at λ_{max} was attained. These experiments were conducted in triplicate under light conditions. After the cultivation period, the mycelial mass was separated from the culture broth by sterile filter paper, and washing of the mycelia was done three times with sterile distilled water. Subsequently, 10 g of harvested mycelia was submerged in 100 ml of a 1 mM aqueous silver nitrate. Finally, the reaction mixtures were incubated until reaching a maximum absorbance at λ_{max} in a shaker that operated at 200 rpm and at 28 °C.

Bhainsa et al. (Bhainsa and Souza 2006) synthesized silver nanoparticles using *Aspergillus fumigatus. A. fumigatus* (NCIM 902) was obtained from the National Chemical Laboratory, Pune, India, and maintained on potato dextrose agar slants. The fungus was grown aerobically in a liquid medium containing (g/l) KH_2PO_4, 7.0; K_2HPO_4, 2.0; $MgSO_4 \cdot 7H_2O$, 0.1; $(NH_4)_2SO_4$, 1.0; yeast extract, 0.6; and glucose, 10.0 to prepare biomass for biosynthesis studies. The flasks were inoculated, incubated on orbital shaker at 25 °C, and agitated at 150 rpm. The biomass was sieved using a plastic sieve and then harvested after 72 h of growth. After that, extensive washing with distilled water was done to remove any medium component from the biomass. Typically 20 g of biomass (fresh weight) was brought in contact with 200 ml of Milli-Q deionized water for 72 h at 25 °C in an Erlenmeyer flask and agitated. After the incubation, the cell filtrate was obtained by passing it through the Whatman filter paper. For the synthesis of silver nanoparticles, $AgNO_3$, 1 mM final concentration was mixed with 50 ml of cell filtrate in a 250 ml Erlenmeyer flask and agitated at 25 °C in dark conditions. A sample of 1 ml was taken at different time intervals, and the UV-visible spectrophotometer was used to measure the absorbance at a resolution of 1 nm. After 72 h of incubation, the cell filtrates containing nanoparticles were characterized using transmission electron microscopy (TEM). Balaji et al. (Balaji et al. 2009) synthesized silver nanoparticles from *Cladosporium cladosporioides* fungus. The TEM images of the synthesized nanoparticles are shown in Fig. 5.15.

5.6.3 Synthesis of Nanoparticles from Glucose and Starch Solution

To solve the problem of cellular toxicity, so that the nanoparticles can be used in biomedical applications, the synthesis of nanoparticles using starch and glucose is a good option. Engelbrekt et al. (Engelbrekt et al. 2009) synthesized the gold nanoparticles using glucose and starch solution. However, gold nanoparticles may be synthesized using other biosources such as lemongrass plant, tea, seaweed, human cells, fungi, microorganisms, protein, but in that cases sometimes it is very difficult to control chemical composition and purity. Hence, glucose and starch were used as the reducing and stabilizing agents, respectively. For the synthesis of Au nanoparticles, 0.5–2.0 mM of AuC14, 10–30 mM glucose, and 0.6–1% (w/w) starch were mixed in 20–100 mL of 10–30 mM buffer. It was observed that the solution immediately changed to a strong red color at room temperature in MES buffer. In all other buffers, pronounced colors were visible only after the solutions had been

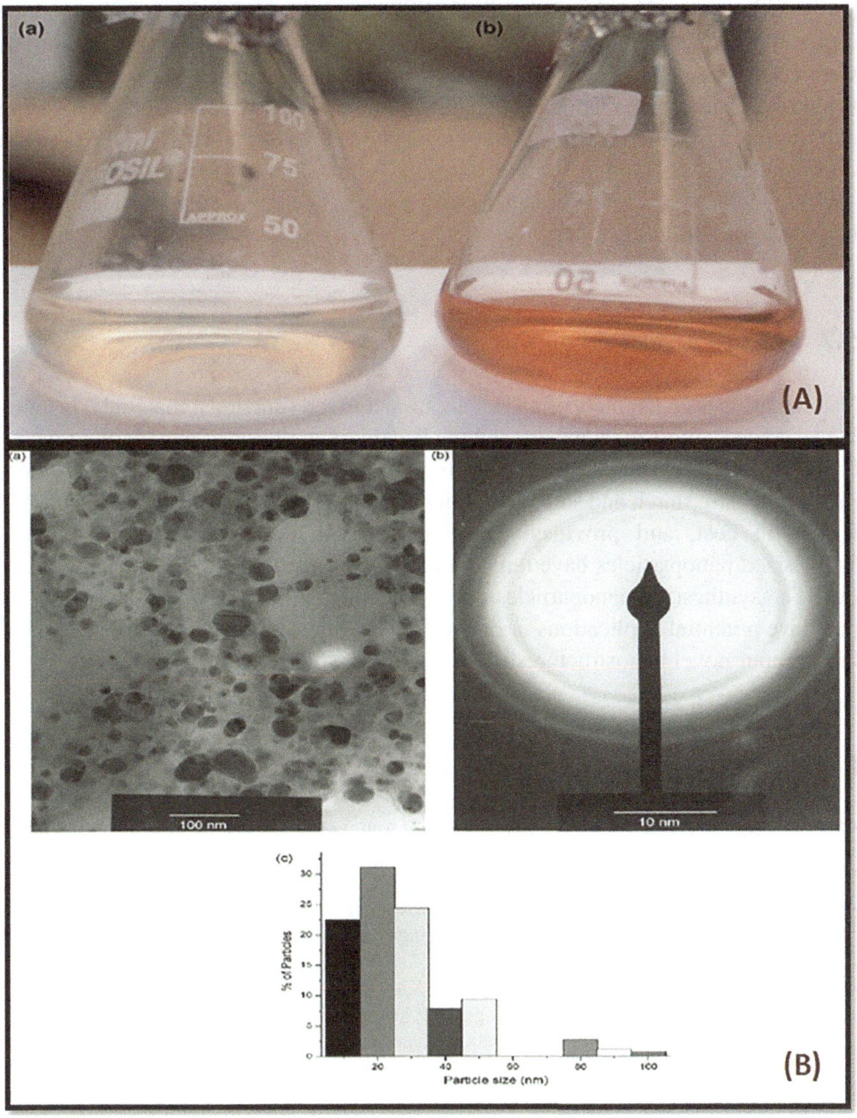

Fig. 5.15 (**A**) Picture of conical flasks containing the extracellular filtrate of the *Cladosporium cladosporioides* in aqueous solution of $AgNO_3$ at the beginning of the reaction (flask 1) and after 1 day of reaction (flask 2), (**B**) (a) TEM images of AgNP. (b) ED pattern images of AgNP. (c) Particle size distribution histogram of AgNP determined from TEM. (Reproduced by permission from Ref. Balaji et al. (2009), License No. 4830680282082, Copyright 2019, Elsevier)

heated at 80–94 °C for 1–2 h. Starch is a key component, although of substantially variable abundance in a variety of vegetable nutrients. The equivalent of 3 g of dry matter, i.e., 13.3 g fresh potato, 20.6 g fresh carrot, or 22.9 g fresh onion, all peeled and were heated at 95 °C in 200 mL of Millipore water for 3 h. These all contained different amounts of starch. After cooling to room temperature, the liquid solutions were centrifuged to remove insoluble parts, and the supernatant is used for further synthesis. That research work has potential both for larger-scale production of stable and variable-size AuNPs and for biological applications of the AuNPs. Due to starch coating, the layers around the AuNPs may further invoke new properties of the AuNPs, and new strategies for new nanostructures. The prepared nanomaterials may be used in bioelectrochemistry.

5.7 Conclusion

Nature has its own manners of producing miniaturized functional materials. Increasing awareness of green chemistry and the benefit of synthesis of nanoparticles using biosources such as plant extracts, fruits extracts, flower extracts, fungi, bacteria, microorganism, starch and sugar, etc. can be ascribed to the fact that it is ecofriendly, is low in cost, and provides maximum protection to human health. Green-synthesized nanoparticles have remarkable significance in the field of nanotechnology. The synthesized nanoparticles using green technology produce nanostructures that have potential applications in medical field. This study will help researchers to know about novel nanostructures using green technology.

References

Abdel-aziz MS, Shaheen MS, El-nekeety AA (2013) Antioxidant and antibacterial activity of silver nanoparticles biosynthesized using Chenopodium murale leaf extract. J Saudi Chem Soc. https://doi.org/10.1016/j.jscs.2013.09.011

Ahmad A, Mukherjee P, Senapati S, Mandal D, Khan MI, Sastry M (2003) Extracellular biosynthesis of silver nanoparticles using the fungus fusarium oxysporum. Colloids Surf B Biointerfaces 28:313–318

Ahn S, Singh P, Castro-aceituno V, Yesmin S, Kim Y, Mathiyalagan R, Yang D (2016) Gold nanoparticles synthesized using Panax ginseng leaves suppress inflammatory—mediators production via blockade of NF-κB activation in macrophages. Artif Cells Nanomed Biotehnol 45:270–276. https://doi.org/10.1080/21691401.2016.1228661

Alaga TO, Edema MO, Atayese AO, Bankole MO (2014) Phytochemical and in vitro anti-bacterial properties of Hibiscus sabdariffa L (Roselle) juice. 8:339–344. https://doi.org/10.5897/JMPR12.1139

Alijani HQ, Pourseyedi S, Torkzadeh-mahani M, Khatami M (2018) Green synthesis of zinc sulfide (ZnS) nanoparticles using *Stevia rebaudiana* Bertoni and evaluation of its cytotoxic properties. J Mol Struct. https://doi.org/10.1016/j.molstruc.2018.07.103

Alijani HQ, Pourseyedi S, Torkzadeh-mahani M, Khatami M (2020) Bimetallic nickel-ferrite nanorod particles: greener synthesis using rosemary and its biomedical efficiency. Artif Cells Nanomed Biotechnol 48:242–251. https://doi.org/10.1080/21691401.2019.1699830

Al-Qura S (2008) Taxonomical and pharmacological survey of therapeutic plants in Jordan. J Nat Prod 1:10–26

Ambika S, Sundrarajan M (2015) Green biosynthesis of ZnO nanoparticles using Vitex negundo L. extract: spectroscopic investigation of interaction between ZnO nanoparticles and human serum albumin. J Photochem Photobiol B Biol. https://doi.org/10.1016/j.jphotobiol.2015.05.004

Ankamwar B, Chaudhary M, Sastry M (2005a) Gold nanotriangles biologically synthesized using tamarind leaf extract and potential application in vapor sensing. Synth React Inorg Met Org Nano-Met Chem 35:19–26. https://doi.org/10.1081/SIM-200047527

Ankamwar B, Chaudhary M, Sastry M (2005b). Gold Nanotriangles Biologically Synthesized using Tamarind Leaf Extract and Potential Application in Vapor Sensing. Synthesis and Reactivity in Inorganic, Metal-Organic, and Nano-Metal Chemistry Synthesis and Reactivity in Inorganic, Metal-Organic, and Nano-Metal Chemistry 35: 19-26. https://doi.org/10.1081/SIM-200047527

Anshup A, Venkataraman JS, Subramaniam C, Kumar RR, Priya S, Kumar TRS, Omkumar RV, John A, Pradeep T (2005) Growth of gold nanoparticles in human cells. Langmuir 21(25): 11562–11567. https://doi.org/10.1021/la0519249

Attele AS, Wu JA, Yuan C (1999) Ginseng Pharmacology: multiple constituents and multiple actions. Biochem Pharmacol 58:1685–1693

Atwan ZW, Saiwan F (2010) The antibacterial activity of cold aqueous and pigment of hibiscus rosa siensis extracts against gram positive and negative bacteria. Bas J Vet Res 10:109–118

Bagher M, Reza H, Pasdar A, Ghayour-mobarhan M, Riahi-zanjani B, Darroudi M (2018) Role of pullulan in preparation of ceria nanoparticles and investigation of their biological activities. J Mol Struct 1157:127–131. https://doi.org/10.1016/j.molstruc.2017.12.053

Balaji DS, Basavaraja S, Deshpande R, Mahesh DB (2009) Extracellular biosynthesis of functionalized silver nanoparticles by strains of Cladosporium cladosporioides fungus. Colloids Surf B Biointerfaces 68:88–92. https://doi.org/10.1016/j.colsurfb.2008.09.022

Beveridge TJ, Hughes MN, Lee H, Leung KT, Poole RK, Savvaidis I, Silver S, Trevors JT (1997) Metal-microbe interactions: contemporary approaches. Adv Microb Physiol 38:177–243. https://doi.org/10.1016/s0065-2911(08)60158-7

Bhainsa KC, Souza SFD (2006) Extracellular biosynthesis of silver nanoparticles using the fungus aspergillus fumigatus. Colloids Surf B: Biointerfaces 47:160–164. https://doi.org/10.1016/j.colsurfb.2005.11.026

Bharathi D, Josebin MD, Vasantharaj S (2018) Biosynthesis of silver nanoparticles using stem bark extracts of Diospyros montana and their antioxidant and antibacterial activities. J Nanostruct Chem. https://doi.org/10.1007/s40097-018-0256-7

Bindhu MR, Umadevi M (2015) Spectrochimica Acta part A: molecular and biomolecular spectroscopy antibacterial and catalytic activities of green synthesized silver nanoparticles. Spectrochim Acta A Mol Biomol Spectrosc 135:373–378. https://doi.org/10.1016/j.saa.2014.07.045

Bonilla JJA, Guerrero DJP, Sáez RGT, Ishida K, Fonseca BB, Rozental S, López CCO (2017) Green synthesis of silver nanoparticles using maltose and cysteine and their effect on cell wall envelope shapes and microbial growth of Candida spp. J Nanosci Nanotechnol 17:1729–1739. https://doi.org/10.1166/jnn.2017.12822

Borrás-linares I, Fernández-arroyo S, Arráez-roman D, Palmeros-suárez PA (2015) Characterization of phenolic compounds, anthocyanidin, antioxidant and antimicrobial activity of 25 varieties of Mexican Roselle (Hibiscus sabdariffa). Ind Crops Prod 69:385–394. https://doi.org/10.1016/j.indcrop.2015.02.053

Boudreau MD, Beland FA, Boudreau MD, Beland FA (2017) An evaluation of the biological and toxicological properties of aloe Barbadensis (miller), aloe Vera an evaluation of the biological and toxicological properties of aloe Barbadensis (miller), Aloe Vera. 0501. https://doi.org/10.1080/10590500600614303

Bruins MR, Kapil S, Oehme FW (2000) Microbial resistance to metals in the environment. Ecotoxicol Environ Saf 207:198–207. https://doi.org/10.1006/eesa.1999.1860

Carmen S, Ruiz-torres CA, Niño-mart N, Roberto S, Mart GA, Ruiz F (2018) Cytotoxic and bactericidal effect of silver nanoparticles obtained by green synthesis method using annona muricata aqueous extract and functionalized with 5-fluorouracil. Bioinorg Chem Appl 15(2018): 6506381. https://doi.org/10.1155/2018/6506381

Castro L, Blázquez ML, Mu JA, González F, García-balboa C, Ballester A (2011) Biosynthesis of gold nanowires using sugar beet pulp. Process Biochem 46:1076–1082. https://doi.org/10.1016/j.procbio.2011.01.025

Chandran SP, Chaudhary M, Pasricha R, Ahmad A, Sastry M (2006) Synthesis of gold nanotriangles and silver nanoparticles using Aloe wera plant extract. Biotechnol Prog 22:577–583

Chen W, Cai W, Zhang L, Wang G, Zhang L (2001) Sonochemical processes and formation of gold nanoparticles within pores of Mesoporous silica. J Colloid Interface Sci 295:291–295. https://doi.org/10.1006/jcis.2001.7525

Chen CC, Hsu J-D, Wang S-F, Chiang H-C, Yang M-Y, Kao E-S, Ho Y-C, Wang C-J (2003) Hibiscus sabdariffa extract inhibits the development of atherosclerosis in cholesterol-fed rabbits. J Agric Food Chem 51(18):5472–5477. https://doi.org/10.1021/jf030065w

Dai S, Zhang X, Li T, Du Z, Dang H (2005) Preparation of silver nanopatterns on DNA templates. Appl Surf Sci 249:346–353. https://doi.org/10.1016/j.apsusc.2004.12.026

Darroudi M, Sabouri Z, Kazemi R, Khorsand A, Kargar H (2013) Green chemistry approach for the synthesis of ZnO nanopowders and their cytotoxic effects. Ceram Int:1–5. https://doi.org/10.1016/j.ceramint.2013.09.032

Darroudi M, Sarani M, Kazemi R, Khorsand A (2014) Nanoceria : gum mediated synthesis and in vitro viability assay. Ceram Int 40:2863–2868. https://doi.org/10.1016/j.ceramint.2013.10.026

Das K (2008) Wound healing potential of aqueous crude extract of Stevia rebaudiana in mice. Revista Brasileira de Farmacognosia – Braz J Pharmacogn 23:351–357. https://doi.org/10.1590/S0102-695X2013005000011

Devi VB, Arulmozhichelvan P, Murugakoothan P (2017) Synthesis and characterization of Znq2 and Znq2 : CTAB particles. Bull Mater Sci. https://doi.org/10.1007/s12034-017-1460-7

Dhand V, Soumya L, Bharadwaj S, Chakra S, Bhatt D, Sreedhar B (2016) Green synthesis of silver nanoparticles using Coffea arabica seed extract and its antibacterial activity. Mater Sci Eng C 58:36–43. https://doi.org/10.1016/j.msec.2015.08.018

Dhanuskodi KKS, Prabukumar CGS (2017a) Dielectric and antibacterial studies of microwave assisted calcium hydroxide nanoparticles. J Mater Sci Mater Electron. https://doi.org/10.1007/s10854-017-7563-5

Dhanuskodi KKS, Prabukumar CGS (2017b) Andrographis paniculata extract mediated green synthesis of CdO nanoparticles and its electrochemical and antibacterial studies. J Mater Sci Mater Electron. https://doi.org/10.1007/s10854-017-6503-8

Dhanuskodi KKS, Prabukumar CGS (2018) Multifunctional properties of microwave assisted CdO–NiO–ZnO mixed metal oxide nanocomposite : enhanced photocatalytic and antibacterial activities. J Mater Sci Mater Electron. https://doi.org/10.1007/s10854-017-8513-y

Dipankar C, Murugan S (2012) Colloids and surfaces B : biointerfaces the green synthesis, characterization and evaluation of the biological activities of silver nanoparticles synthesized from Iresine herbstii leaf aqueous extracts. Colloids Surf B: Biointerfaces 98:112–119. https://doi.org/10.1016/j.colsurfb.2012.04.006

Duong B, Seraphin S, Laokul P, Masingboon C, Maensiri S (2008) Ni-Cu-Zn ferrite prepared by Aloe vera plant extract or egg white. 14:2007–2008. https://doi.org/10.1017/S1431927608083839

Durka AMM, Antony SA (2015) Hibiscus rosa—sinensis leaf extracted green methods, magneto-optical and catalytic properties of spinel CuFe 2 O 4 nano- and microstructures. https://doi.org/10.1007/s10904-015-0203-8

Dwivedi AD, Gopal K (2010) Colloids and surfaces A : physicochemical and engineering aspects biosynthesis of silver and gold nanoparticles using Chenopodium album leaf extract. Colloids Surf A Physicochem Eng Asp 369:27–33. https://doi.org/10.1016/j.colsurfa.2010.07.020

Elango G, Roopan SM (2015) Spectrochimica Acta part A: molecular and biomolecular spectroscopy green synthesis , spectroscopic investigation and photocatalytic activity of lead nanoparticles OH OH. Spectrochim Acta A Mol Biomol Spectrosc 139:367–373. https://doi.org/10.1016/j.saa.2014.12.066

Engelbrekt C, Sørensen KH, Zhang J, Welinder AC, Jensen PS, Ulstrup J (2009) Green synthesis of gold nanoparticles with starch—glucose and application in bioelectrochemistry. J Mater Chem 42:7839–7847. https://doi.org/10.1039/b911111e

Gardea-Torresdey J L, Parsons J. G., Gomez E, Peralta-Videa J, Troiani H. E. , Santiago P. ,Jose Yacaman M (2002) Formation and growth of Au nanoparticles inside live Alfalfa plants, Nano Letters 2, 4, 397–401

Gardea-torresdey JL, Gomez E, Peralta-videa JR, Parsons JG, Troiani H, Jose-yacaman M (2003) Alfalfa sprouts: a natural source for the synthesis of silver nanoparticles. Langmuir 19:1357–1361

Genena AK, Hense H, Sm A, Junior NIA, Machado S (2008) Rosemary (Rosmarinus officinalis)—a study of the composition , antioxidant and antimicrobial activities of extracts obtained with supercritical carbon dioxide. Food Sci Technol 28:463–469

Gholami L, Oskuee RK, Tafaghodi M, Farkhani AR, Darroudi M (2018) Author ' s accepted manuscript. Ceram Int. https://doi.org/10.1016/j.ceramint.2018.02.137

Ghoreishi SM, Behpour M, Khayatkashani M (2011) Green synthesis of silver and gold nanoparticles using Rosa damascena and its primary application in electrochemistry. Physica E 44:97–104. https://doi.org/10.1016/j.physe.2011.07.008

Ghule K, Ghule AV, Liu J, Ling Y (2006) Microscale size triangular gold prisms synthesized using Bengal gram beans (Cicer arietinum L.) extract and HAuCl 4 3H 2 O: a green biogenic approach:3746–3751. https://doi.org/10.1166/jnn.2006.608

Giljohann DA, Seferos DS, Daniel WL, Massich MD, Patel PC, Mirkin CA (2010) Gold nanoparticles for biology and medicine Angewandte:3280–3294. https://doi.org/10.1002/anie.200904359

Gingasu D, Mindru I, Patron L, Calderon-moreno JM, Mocioiu C, Preda S, Stanica N, Nita S, Dobre N, Gradisteanu G, Chifiriuc MC (2016) Green synthesis methods of CoFe 2 O 4 and Ag-CoFe 2 O 4 nanoparticles using hibiscus extracts and their antimicrobial potential. J Nanomater 2016:2106756

Gnanasangeetha D, Saralathambavani D (2013) One pot synthesis of zinc oxide nanoparticles via chemical and green method. Sci Rep 11:1–8

Golinska P, Wypij M, Ingle AP (2014) Biogenic synthesis of metal nanoparticles from actinomycetes : biomedical applications and cytotoxicity. https://doi.org/10.1007/s00253-014-5953-7

Gontero D, Veglia AV, Bracamonte AG, Boudreau D (2017) Biosensing applications based on metal-enhanced fluorescence. RSC Adv 7:10252–10258. https://doi.org/10.1039/C6RA27649K

Gou Y, Zhang F, Zhu X, Li X (2015) Biosynthesis and characterisation of silver nanoparticles using Sphingomonas paucimobilis sp. BDS1. IET Nanobiotechnol 9(2):53–57. https://doi.org/10.1049/iet-nbt.2014.0005

Goutam SP, Saxena G, Singh V, Yadav AK, Bharagava RN, Thapa KB (2017) Advanced Materials Research Laboratory, Department of Applied Physics (DAP), School for Laboratory for bioremediation and metagenomics research (LBMR), Department of Solar Energy Laboratory for experimental studies, Department of Environmental Sciences. Chem Eng J. https://doi.org/10.1016/j.cej.2017.12.029

Govindaraju K, Khaleel ÆS (2008) Silver, gold and bimetallic nanoparticles production using single-cell protein (Spirulina platensis) Geitler:5115–5122. https://doi.org/10.1007/s10853-008-2745-4

Groiss S, Selvaraj R, Thivaharan V, Ramesh V (2016) Structural characterization, antibacterial and catalytic effect of iron oxide nanoparticles synthesized using the leaf extract of Cynometra ramiflora. J Mol Struct. https://doi.org/10.1016/j.molstruc.2016.09.031

Guardiola S, Mach N (2014) Endocrinología y Nutrición Potencial terapéutico del Hibiscus sabdariffa : una revisión de las evidencias científicas. Endocrinol Nutr 61:274–295. https://doi.org/10.1016/j.endonu.2013.10.012

Gulalp B, Karcioglu O (2008) The first report of Prosopis farcta ingestion in children: is it serious? Trigeminal Neuralgia as a presentation of infective endocarditis 62(5):829–830. https://doi.org/10.1111/j.1742-1241.2006.00941.x

Hamedi S, Abbas S, Shokrollahzadeh S (2016) Mechanism study of silver nanoparticle production using Neurospora intermedia. IET Nanobiotechnol 11(2):157–163. https://doi.org/10.1049/iet-nbt.2016.0038

Hebbalalu D, Lalley J, Nadagouda MN, Varma RS (2013) Greener techniques for the synthesis of silver nanoparticles using plant extracts, enzymes, bacteria, biodegradable polymers and microwaves. Perspective. https://doi.org/10.1021/sc4000362

Honary S, Barabadi H, Gharaei-fathabad E (2013) Green synthesis of Silver nanoparticles induced by the fungus Penicillium citrinum. Trop J Pharm Res 12:7–11

Huang J, Li Q, Sun D, Lu Y (2007) Biosynthesis of silver and gold nanoparticles by novel sundried Cinnamomum camphora leaf. Nanotechnology 18:105104. https://doi.org/10.1088/0957-4484/18/10/105104

Huang X, Wu H, Pu S, Zhang W, Shi B (2011) Green chemistry one-step room-temperature synthesis of Au @ Pd core—shell nanoparticles with tunable structure using plant tannin as reductant and stabilizer †:950–957. https://doi.org/10.1039/c0gc00724b

Immer ELAZ (1996) Phylogeny and biogeography of Panax L. (the Ginseng Genus, Araliaceae): inferences from ITS sequences of nuclear ribosomal DNA. Mol Psylogenet Evol 6:167–177

Isaac RSR, Sakthivel G, Murthy C (2013) Green synthesis of gold and silver nanoparticles using averrhoa bilimbi fruit extract. J Nanotechnol 2013:906592

Jagtap UB, Bapat VA (2013) Green synthesis of silver nanoparticles using Artocarpus heterophyllus Lam. Seed extract and its antibacterial activity. Ind Crops Prod 46:132–137. https://doi.org/10.1016/j.indcrop.2013.01.019

Jamdagni P, Rana JS, Khatri P, Nehra K (2018) Comparative account of antifungal activity of green and chemically synthesized zinc oxide nanoparticles in combination with agricultural fungicides. Int J Nano Dimens 9(2):198–208

Jana NR, Gearheart L, Murphy CJ (2001) Wet chemical synthesis of high aspect ratio cylindrical gold nanorods. J Phys Chem B:4065–4067

Jannet H (2005) Ben: flavonoids diversification in organs of two prosopis farcta (banks & sol.) eig. (leguminosea, mimosoideae) populations occurring in the northeast and the southeast of Tunisia. J Appl Sci Res 1:130–136

Javaid A, Oloketuyi SF, Khan MM, Khan F (2018) Diversity of bacterial synthesis of silver nanoparticles. BioNanoScience 8:43–59

Jeevanandam J, Barhoum A, Chan YS, Dufresne A, Danquah MK (2018) Review on nanoparticles and nanostructured materials : history , sources , toxicity and regulations. Beilstein J Nanotechnol 9:1050–1074. https://doi.org/10.3762/bjnano.9.98

John MS, Nagoth JA, Ramasamy KP, Mancini A, Giuli G, Natalello A, Ballarini P, Miceli C, Pucciarelli S (2020) Synthesis of bioactive silver nanoparticles by a pseudomonas strain associated with the Antarctic psychrophilic protozoon Euplotes focardii. Mar Drugs 18:38

Karak H, Jaiswal R, Kuhnert N (2011) Characterization and quantification of hydroxycinnamate derivatives in Stevia rebaudiana leaves by LC-MS n †. J Agric Food Chem 59(18):10143–10150

Karaköse H, Müller A, Kuhnert N (2015) Profiling and quantification of Phenolics in stevia Rebaudiana leaves. J Agric Food Chem 63(41):9188–9198. https://doi.org/10.1021/acs.jafc.5b01944

Karthiga P (2017) Preparation of silver nanoparticles by Garcinia mangostana stem extract and investigation of the antimicrobial properties. Biotechnol Res Innovation:1–7. https://doi.org/10.1016/j.biori.2017.11.001

Karthik K, Dhanuskodi S, Gobinath C, Prabukumar S (2018b) Multifunctional properties of CdO nanostructures synthesised through microwave assisted hydrothermal method multifunctional properties of CdO nanostructures synthesised through microwave assisted hydrothermal method. Mater Res Innov 8917:1–8. https://doi.org/10.1080/14328917.2018.1475443

Karthik K, Dhanuskodi S, Gobinath C, Prabukumar S, Sivaramakrishnan S (2018a) Journal of physics and chemistry of solids nanostructured CdO-NiO composite for multifunctional applications. J Phys Chem Solids 112:106–118. https://doi.org/10.1016/j.jpcs.2017.09.016

Kasyanenko N, Varshavskii M, Ikonnikov E, Tolstyko E, Belykh R, Sokolov P, Bakulev V, Rolich V, Lopatko K (2016) DNA modified with metal nanoparticles: preparation and characterization of ordered metal-DNA nanostructures in a solution and on a substrate. Hindawi

Khan AU, Khan M, Malik N, Hwan M, Mohammad C, Khan M (2018b) Recent progress of algae and blue—green algae-assisted synthesis of gold nanoparticles for various applications. Bioprocess Biosyst Eng. https://doi.org/10.1007/s00449-018-2012-2

Khan AU, Yuan Q, Ul Z, Khan H, Ahmad A, Khan FU, Tahir K, Shakeel M, Ullah S (2018a) PT US Department of Environmental Sciences, COMSATS Institute of Information. J Photochem Photobiol B Biol. https://doi.org/10.1016/j.jphotobiol.2018.05.007

Khan FU, Chen Y, Khan NU, Ahmad A, Tahir K, Khan ZU, Khan AU, Khan SU, Raza M, Wan P (2017a) SC. Microbial pathogenesis. doi:https://doi.org/10.1016/j.micpath.2017.04.024

Khan I, Saeed K, Khan I (2017b) Nanoparticles: properties, applications and toxicities. Arab J Chem. https://doi.org/10.1016/j.arabjc.2017.05.011

Khan MM, Kalathil S, Lee J, Cho MH (2012) Synthesis of cysteine capped silver nanoparticles by electrochemically active biofilm and their antibacterial activities. Bull Korean Chem Soc 33:2592–2596

Khanna-chopra R, Semwal VK (2011) Superoxide dismutase and ascorbate peroxidase are constitutively more thermotolerant than other antioxidant enzymes in Chenopodium album. Physiol Mol Biol Plants 17:339–346. https://doi.org/10.1007/s12298-011-0083-5

Khatami M (2018) Core @ shell nanoparticles: greener synthesis using natural plant products. Appl Sci 8(3):411. https://doi.org/10.3390/app8030411

Khatami M, Alijani H, Sharifi I, Sharifi F, Pourseyedi S (2017) Leishmanicidal activity of biogenic Fe_3O_4 nanoparticles. Sci Pharm 85(4):36. https://doi.org/10.3390/scipharm85040036

Khatami M, Alijani HQ, Fakheri B, Mobasseri MM, Heydarpour M, Farahani ZK, Khan AU (2018a) Super-paramagnetic iron oxide nanoparticles (SPIONs): greener. J Clean Prod. https://doi.org/10.1016/j.jclepro.2018.10.182

Khatami M, Heli H, Jahani PM, Azizi H, Nobre AL (2017) Copper/copper oxide nanoparticles synthesis using Stachys lavandulifolia and its antibacterial activity. IET Nanobiotechnol 11:709–713. https://doi.org/10.1049/iet-nbt.2016.0189

Khatami M, Iravani S, Varma RS, Mosazade F, Darroudi M (2019) Cockroach wings—promoted safe and greener synthesis of silver nanoparticles and their insecticidal activity. Bioprocess Biosyst Eng. https://doi.org/10.1007/s00449-019-02193-8

Khatami M, Nejad MS, Salari S, Ghasemi P, Almani N (2016) Plant-mediated green synthesis of silver nanoparticles using Trifolium resupinatum seed exudate and their antifungal efficacy on Neofusicoccum parvum and Rhizoctonia solani. IET Nanobiotechnol 10:237–243. https://doi.org/10.1049/iet-nbt.2015.0078

Khatami M, Pourseyedi S, Khatami M, Hamidi H, Zaeifi M, Soltani L (2015) Synthesis of silver nanoparticles using seed exudates of Sinapis arvensis as a novel bioresource, and evaluation of their antifungal activity. Bioresour Bioprocess 2(19). https://doi.org/10.1186/s40643-015-0043-y

Khatami M, Sharifi I, Nobre MAL, Zafarnia N, Aflatoonian MR (2018b) Waste-grass-mediated green synthesis of silver nanoparticles and evaluation of their anticancer, antifungal and antibacterial activity. Green Chem Lett Rev 8253. https://doi.org/10.1080/17518253.2018.1444797

Kim Y, Jeon J, Jang M, Oh JY, Kwon W, Jung S, Yang D (2014) Ginsenoside pro files and related gene expression during foliation in Panax ginseng Meyer. J Ginseng Res 38:66–72. https://doi.org/10.1016/j.jgr.2013.11.001

Klinkaewnarong J, Swatsitang E, Masingboon C (2010) Synthesis and characterization of nanocrystalline HAp powders prepared by using aloe vera plant extracted solution. Curr Appl Phys 10:521–525. https://doi.org/10.1016/j.cap.2009.07.014

Knoll B, Keilmann F (1999) Near-field probing of vibrational absorption for chemical microscopy. Nature 399(6732):134–137

Kora AJ, Arunachalam J (2013) Biosynthesis of silver nanoparticles by the seed extract of Strychnos potatorum: a natural phytocoagulant. IET Nanobiotechnol 7:83–89. https://doi.org/10.1049/iet-nbt.2013.0001

Kou J, Varma RS (2012) RSC Advances beet juice utilization: expeditious green synthesis of noble metal. Advances RSC, pp 10283–10290. https://doi.org/10.1039/c2ra21908e

Kowshik M, Deshmukh N, Vogel W, Urban J, Kulkarni SK, Paknikar KM (2002) Microbial synthesis of semiconductor CdS nanoparticles, their characterization, and their use in the fabrication of an ideal diode. Biotechnol Bioeng 78(5):583–588. https://doi.org/10.1002/bit.1023

Ku HO, Eda YU, Shiguro KI (2003) Antipruritic effects of the fruits of Chaenomeles sinensis. Biol Pharm Bull 26:1031–1034

Kudelski A, Michota A, Bukowska J (2003) SERS studies on the structure of thioglycolic acid monolayers on silver and gold. Surf Sci 535:227–232. https://doi.org/10.1016/S0039-6028(03)00094-3

Kumar A, Mandal S, Selvakannan PR, Pasricha R, Mandale AB, Sastry M (2003) Investigation into the interaction between surface-bound alkylamines and gold nanoparticles. Langmuir 19:6277–6282

Kumar G, Loganathan K, Rao B (2010) Extracellular biosynthesis of silver nanoparticles using the filamentous fungus Penicillium sp. Arch Appl Sci Res 2(6):161–167

Kumar KP, Paul W, Sharma CP (2012) Green synthesis of silver nanoparticles with Zingiber officinale extract and study of its blood compatibility. IET Nanobiotechnol 13:144–152. https://doi.org/10.1007/s12668-012-0044-7

Laokul P, Amornkitbamrung V, Seraphin S, Maensiri S (2011) Characterization and magnetic properties of nanocrystalline CuFe 2 O 4 , NiFe 2 O 4 , ZnFe 2 O 4 powders prepared by the Aloe vera extract solution. Curr Appl Phys 11:101–108. https://doi.org/10.1016/j.cap.2010.06.027

Lengke MF, Fleet ME, Southam G (2007) Biosynthesis of silver nanoparticles by filamentous cyanobacteria from a silver (I) nitrate complex. Langmuir 23:2694–2699

Lo ML, Parsons JG, Videa JRP, Gardea-torresdey JL (2005) An XAS study of the binding and reduction of Au (III) by hop biomass. Microchem J 81:50–56. https://doi.org/10.1016/j.microc.2005.01.011

Lok C-N, Ho C-M, Chen R, He Q-Y, Yu W-Y, Sun H, Tam PK-H, Chiu J-F, Che C-M (2007) Silver nanoparticles: partial oxidation and antibacterial activities. J Biol Inorg Chem 12(4):527–534. https://doi.org/10.1007/s00775-007-0208-z

Mahadevan N, Kamboj P (2009) Hibiscus sabdariffa Linn.—an overview. Nat Prod Radiance 8:77–83

Mak YW, Chuah LO, Ahmad R, Bhat R (2013) Antioxidant and antibacterial activities of hibiscus (Hibiscus rosa-sinensis L .) and Cassia (Senna bicapsularis L .) flower extracts. J King Saud Univ Sci 25:275–282. https://doi.org/10.1016/j.jksus.2012.12.003

Manikandan A, Sridhar R, Antony SA, Ramakrishna S (2014) A simple aloe vera plant-extracted microwave and conventional combustion synthesis: morphological, optical, magnetic and catalytic properties of CoFe 2 O 4 nanostructures. J Mol Struct 1076:188–200. https://doi.org/10.1016/j.molstruc.2014.07.054

Marulasiddeshwara MB, Dakshayani SS, Kumar MNS, Chethana R, Kumar PR, Devaraja S (2017) PT. Materials Science & Engineering C. https://doi.org/10.1016/j.msec.2017.07.054

Masumeh S, Hamedi S, Abbas S (2012) Green synthesis of silver nanoparticles by a novel method : comparative study of their properties. Carbohydr Polym 89:467–472. https://doi.org/10.1016/j.carbpol.2012.03.030

Meyer CA (2012) Advances in study of ginsenoside biosynthesis pathway in Panax ginseng. Acta Physiol Plant 34:397–403. https://doi.org/10.1007/s11738-011-0844-3
Melis MS, Rocha ST, Augusto A (2009) Steviol effect, a glycoside of Stevia rebaudiana, on glucose clearances in rats. Braz J Biol 69(2):371–374
Miri A, Darroudi M, Entezari R, Sarani M (2018) Biosynthesis of gold nanoparticles using Prosopis farcta extract and its in vitro toxicity on colon cancer cells. Res Chem Intermed. https://doi.org/10.1007/s11164-018-3299-y
Miri A, Sarani M (2018) Author's accepted manuscript. Ceram Int. https://doi.org/10.1016/j.ceramint.2018.04.063
Mirzaei H, Darroudi M (2017) Zinc oxide nanoparticles_biological synthesis and biomedical applications. Ceram Int 43:907–914. https://doi.org/10.1016/j.ceramint.2016.10.051
Missoum A (2018) An update review on Hibiscus rosa sinensis phytochemistry and medicinal uses. J Ayurvedic Herb Med 4:135–146
Mittal AK, Chisti Y, Banerjee UC (2013) Synthesis of metallic nanoparticles using plant extracts. Biotechnol Adv 31:346–356. https://doi.org/10.1016/j.biotechadv.2013.01.003
Momeni S, Nabipour I (2015) A simple green synthesis of palladium nanoparticles with sargassum alga and their electrocatalytic activities towards hydrogen peroxide. Appl Biochem Biotechnol 176:1937–1949. https://doi.org/10.1007/s12010-015-1690-3
Moteriya P, Chanda S (2016) Synthesis and characterization of silver nanoparticles using Caesalpinia pulcherrima flower extract and assessment of their in vitro antimicrobial, antioxidant, cytotoxic, and genotoxic activities. Artif Cells Nanomed Biotechnol. https://doi.org/10.1080/21691401.2016.1261871
Muhammad N, Khan AU, Tahir K (2017) Biomedical applications of green synthesized Nobel metal nanoparticles. J Photochem Photobiol B Biol. https://doi.org/10.1016/j.jphotobiol.2017.05.034
Musarrat J, Dwivedi S, Raj B, Al-khedhairy AA, Azam A (2010) Bioresource technology production of antimicrobial silver nanoparticles in water extracts of the fungus Amylomyces rouxii strain KSU-09. Bioresour Technol 101:8772–8776. https://doi.org/10.1016/j.biortech.2010.06.065
Muzammil MS (2013) Silver Nano Scaffold formation by flowers of Hibiscus Rosa. Int J Herb Med 1:169–174
Nadagouda MN, Iyanna N, Lalley J, Dionysiou DD, Varma RS (2014) Synthesis of silver and gold nanoparticles using antioxidants from blackberry, blueberry, pomegranate and turmeric extracts synthesis of silver and gold nanoparticles using antioxidants from blackberry, blueberry, pomegranate and turmeric extracts. ACS Sustain Chem Eng 2:1717–1723. https://doi.org/10.1021/sc500237k
Nadagouda MN, Varma RS (2006) Green and controlled synthesis of gold and platinum nanomaterials using vitamin B 2: density-assisted self-assembly of nanospheres, wires and rods. Green Chem 6:516–518. https://doi.org/10.1039/b601271j
Naik MM, Naik HSB, Nagaraju G, Vinuth M, Naika HR, Vinu K (2019) Green synthesis of zinc ferrite nanoparticles in Limonia acidissima juice : Characterization and their application as photocatalytic and antibacterial activities. Microchem J 146:1227–1235. https://doi.org/10.1016/j.microc.2019.02.059
Nieto G, Ros G, Castillo J (2018) Antioxidant and antimicrobial properties of (Rosemarinus officinalis, L): a review. Medicines 5(3):1–13. https://doi.org/10.3390/medicines5030098
Narayanan KB, Sakthivel N (2011) Green synthesis of biogenic metal nanoparticles by terrestrial and aquatic phototrophic and heterotrophic eukaryotes and biocompatible agents. Adv Colloid Interf Sci 169:59–79. https://doi.org/10.1016/j.cis.2011.08.004
Nin ÆN (2008) Synthesis and antibacterial activity of silver nanoparticles with different sizes. J Nanopart Res 10:1343–1348. https://doi.org/10.1007/s11051-008-9428-6
Nithyaja B, Misha H, Nampoori VPN (2012) Synthesis of silver nanoparticles in DNA template and its influence on nonlinear optical properties. Nanosci Nanotechnol 2:99–103. https://doi.org/10.5923/j.nn.20120204.02

Noruzi M (2014) Biosynthesis of gold nanoparticles using plant extracts. Bioprocess Biocyst Engg 38:1–14. https://doi.org/10.1007/s00449-014-1251-0

Oh KH, Soshnikova V, Markus J, Kim YJ, Chul S, Singh P, Castro-aceituno V, Ahn S, Hyun D, Shim YJ, Kim YJ, Yang DC, Hyun K, Soshnikova V, Markus J, Kim YJ, Chul S (2017) Biosynthesized gold and silver nanoparticles by aqueous fruit extract of Chaenomeles sinensis and screening of their biomedical activities. Artif Cells Nanomed Biotechnol:1–8. https://doi.org/10.1080/21691401.2017.1332636

Pacôme OA, Bernard D, Sékou D, Allico D, David J, Monqomaké K, Hilaire K (2014) Phytochemicals and antioxidant activity of Roselle (Hibiscus sabdariffa L.) petal extracts. Res J Pharm, Biol Chem Sci 5:1453–1465

Patel S (2013) Hibiscus sabdariffa: an ideal yet under-exploited candidate for nutraceutical applications. Biomed Prev Nutr. https://doi.org/10.1016/j.bionut.2013.10.004

Patil MP, Kim G (2016) Eco-friendly approach for nanoparticles synthesis and mechanism behind antibacterial activity of silver and anticancer activity of gold nanoparticles. Appl Microbiol Biotechnol. https://doi.org/10.1007/s00253-016-8012-8

Patil RS, Kokate MR, Kolekar SS (2012) Spectrochimica Acta Part A : molecular and biomolecular spectroscopy bioinspired synthesis of highly stabilized silver nanoparticles using Ocimum tenuiflorum leaf extract and their antibacterial activity. Spectrochim Acta A Mol Biomol Spectrosc 91:234–238. https://doi.org/10.1016/j.saa.2012.02.009

Patterson ML, Slater JE (2002) Characterization and comparison of commercially available German and American cockroach allergen extracts. Clin Exp Allergy 1:1–7

Pereira L, Mehboob F, Stams AJM, Mota MM, Rijnaarts HHM, Alves MM (2013) Metallic nanoparticles: microbial synthesis and unique properties for biotechnological applications, bioavailability and biotransformation. 8551:1–15. https://doi.org/10.3109/07388551.2013.819484

Philip D (2010) Green synthesis of gold and silver nanoparticles using Hibiscus rosa sinensis. Physica E 42:1417–1424. https://doi.org/10.1016/j.physe.2009.11.081

Phull A, Abbas Q, Ali A, Raza H, Zia M (2016) Antioxidant, cytotoxic and antimicrobial activities of green synthesized silver nanoparticles from crude extract of Bergenia ciliata. Future J Pharm Sci. https://doi.org/10.1016/j.fjps.2016.03.001

Phumying S, Labuayai S, Swatsitang E (2013) Nanocrystalline spinel ferrite (MFe 2 O 4 , M = Ni , co , Mn , mg , Zn) powders prepared by a simple aloe vera plant-extracted solution hydrothermal route. Mater Res Bull 48:2060–2065. https://doi.org/10.1016/j.materresbull.2013.02.042

Ponnuchamy K, Jacob JA (2016) Metal nanoparticles from marine seaweeds—a review. Nanotechnol Rev:589–600. https://doi.org/10.1515/ntrev-2016-0010

Prabha S, Lahtinen M, Sillanpää M (2010) Tansy fruit mediated greener synthesis of silver and gold nanoparticles. Process Biochem 45:1065–1071. https://doi.org/10.1016/j.procbio.2010.03.024

Prasad BBVSV, Ramesh KV, Srinivas A (2018) Structural and soft magnetic properties of nickel-substituted Co-Zn nanoferrites. J Supercond Nov Magn 31(10):3223–3237. 15p

Qourzal S, Tamimi M, Assabbane A, Ait-ichou Y (2005) Photocatalytic degradation and adsorption of 2-naphthol on suspended TiO 2 surface in a dynamic reactor. J Colloid Interface Sci 286:621–626. https://doi.org/10.1016/j.jcis.2005.01.046

Radad K, Gille G, Liu L, Rausch W (2006) Critical review use of ginseng in medicine with emphasis on neurodegenerative disorders. J Pharamcol Sci 100:175–186

Rangkadilok N, Worasuttayangkurn L, Bennett RN, Satayavivad J (2005) Identification and quantification of polyphenolic compounds in Longan (Euphoria longana Lam.) fruit. J Agric Food Chem 53(5):1387–1392

Rai M, Yadav A (2013) Plants as potential synthesiser of precious metal nanoparticles : progress and prospects:117–124. https://doi.org/10.1049/iet-nbt.2012.0031

Rajasekharreddy P, Usha P (2010) Qualitative assessment of silver and gold nanoparticle synthesis in various plants: a photobiological approach. J Nanopart Res 12:1711–1721. https://doi.org/10.1007/s11051-010-9894-5

Rajeshkumar S (2016) Synthesis of silver nanoparticles using fresh bark of Pongamia pinnata and characterization of its antibacterial activity against gram positive and gram negative pathogens. Resource-Efficient Technologies 2:30–35. https://doi.org/10.1016/j.reffit.2016.06.003

Ramamurthy CH, Padma M, Daisy I, Mareeswaran R, Suyavaran A, Kumar MS, Premkumar K, Thirunavukkarasu C (2013) Colloids and surfaces B : biointerfaces the extra cellular synthesis of gold and silver nanoparticles and their free radical scavenging and antibacterial properties. Colloids Surf B: Biointerfaces 102:808–815. https://doi.org/10.1016/j.colsurfb.2012.09.025

Ramar M, Manikandan B, Raman T, Arunagirinathan K, Prabhu NM, Babu MJ, Perumal M, Palanisamy S, Munusamy A (2014) Biosynthesis of silver nanoparticles using ethanolic petals extract of Rosa indica and characterization of its antibacterial, anticancer and anti-inflammatory activities. Spectrochim Acta A Mol Biomol Spectrosc. https://doi.org/10.1016/j.saa.2014.10.043

Ranjan R, Nilotpala N (2011) Green synthesis of silver nanoparticle by Penicillium purpurogenum NPMF: the process and optimization. J Nanopart Res 13:3129–3137. https://doi.org/10.1007/s11051-010-0208-8

Rao NH, Lakshmidevi N, Pammi SVN, Kollu P, Ganapaty S, Lakshmi P (2016) Green synthesis of silver nanoparticles using methanolic root extracts of Diospyros paniculata and their antimicrobial activities. Mater Sci Eng C 62:553–557. https://doi.org/10.1016/j.msec.2016.01.072

Rautaray D, Ahmad A, Sastry M (2003) Biosynthesis of CaCO3 crystals of complex morphology using a fungus and an actinomycete. J Am Chem Soc 125:14656–14657

Sachdewa A, Khemani LD (2003) Effect of Hibiscus rosa sinensis Linn. Ethanol flower extract on blood glucose and lipid profile in streptozotocin induced diabetes in rats. 89:61–66. https://doi.org/10.1016/S0378-8741(03)00230-7

Safaei M, Mehdi M, Ebrahimpoor N (2019) Trends in analytical chemistry a review on metal-organic frameworks : synthesis and applications. Trends Anal Chem 118:401–425. https://doi.org/10.1016/j.trac.2019.06.007

Salari S, Seddighi NS, Nejad PG (2018) Science direct evaluation of biofilm formation ability in different Candida strains and anti-biofilm effects of Fe 3 O 4 -NPs compared with fluconazole : an in vitro study. Journal de Mycologie Medicale:2–7. https://doi.org/10.1016/j.mycmed.2018.02.007

Sawai R, Kuroda K, Shibata T, Gomyou R, Osawa K, Shimizu K (2008) Anti-influenza virus activity of Chaenomeles sinensis. J Ethnopharmacol 118:108–112. https://doi.org/10.1016/j.jep.2008.03.013

Sayed R, Saad H, Hagagy N (2018) Silver nanoparticles : characterization and antibacterial properties. Rendiconti Lincei Scienze Fisiche e Naturali. https://doi.org/10.1007/s12210-017-0663-6

Scarano G, Morelli E (2002) Characterization of cadmium- and lead- phytochelatin complexes formed in a marine microalga in response to metal exposure. Biometals 15:145–151

Sengupta S, Eavarone D, Capila I, Zhao G, Watson N, Kiziltepe T, Sasisekharan R (2005) Temporal targeting of tumour cells and neovasculature with a nanoscale delivery system. Nature 436:3–7. https://doi.org/10.1038/nature03794

Seshadri S, Saranya K, Kowshik M (2011) Green synthesis of lead sulfide nanoparticles by the lead resistant marine yeast, Rhodosporidium diobovatum. Biotechnol Prog 27(5):1464–1469. https://doi.org/10.1002/btpr.651

Shankar SS, Ahmad A, Sastry M (2003) Bioreduction of chloroaurate ions by geranium leaves and its endophytic fungus yields gold nanoparticles of different shapes. J Mater Chem 13:1822–1826. https://doi.org/10.1039/b303808b

Shankar SS, Rai A, Ahmad A, Sastry M (2004) Rapid synthesis of Au, Ag, and bimetallic Au core—Ag shell nanoparticles using neem (Azadirachta indica) leaf broth. J Colloid Interface Sci 275:496–502. https://doi.org/10.1016/j.jcis.2004.03.003

Shankar SS, Rai A, Ahmad A, Sastry M, Group N, Chemistry M, Di V, Sciences B, Di V (2005) Controlling the optical properties of Lemongrass extract synthesized gold nanotriangles and potential application in infrared-absorbing optical coatings. Chem Mater 17(3):566–572. https://doi.org/10.1021/cm048292g

Singh H, Du J, Singh P, Yi TH (2017) Ecofriendly synthesis of silver and gold nanoparticles by Euphrasia officinalis leaf extract and its biomedical applications. Artif Cells Nanomed Biotechnol:1–8. https://doi.org/10.1080/21691401.2017.1362417

Singh P, Singh H, Ahn S, Castro-aceituno V, Simu SY, Kim YJ, Yang DC (2016) Pharmacological importance, characterization and applications of gold and silver nanoparticles synthesized by Panax ginseng fresh leaves. Artif Cells Nanomed Biotechnol. https://doi.org/10.1080/21691401.2016.1243547

Singh P, Kim YJ, Wang C, Mathiyalagan R, Yang DC (2015) Microbial synthesis of flower-shaped gold nanoparticles. Artif Cells Nanomed Biotechnol:1–6. https://doi.org/10.3109/21691401.2015.1041640

Song JY, Kim BS (2008) Biological synthesis of bimetallic Au/Ag nanoparticles using persimmon (Diopyros kaki) leaf extract. Biotechnology 25:808–811

Sonker AS, Pathak J, Vinod K, Sinha RP (2017) Characterization and in vitro antitumor , antibacterial and antifungal activities of green synthesized silver nanoparticles using cell extract of Nostoc sp . strain. 01, 26–37. https://doi.org/10.24870/cjb.2017-000103

Stone MO (2002) Biomimetic synthesis and patterning of silver nanoparticles. Nat Mater 1:169–172. https://doi.org/10.1038/nmat758

Taylor P, Alghuthaymi MA, Almoammar H, Rai M, Said-galiev E, Kamel A (2015) Myconanoparticles : synthesis and their role in phytopathogens management. Biotechnol Biotechnol Equip:37–41. https://doi.org/10.1080/13102818.2015.1008194

Thg-ls P, Singh H, Du J, Singh P, Yi TH (2018) Author ' s accepted manuscript extracellular synthesis of silver nanoparticles by reference. J Pharm Anal. https://doi.org/10.1016/j.jpha.2018.04.004

Thovhogi N, Diallo A, Gurib-fakim A, Maaza M (2015) Nanoparticles green synthesis by Hibiscus Sabdariffa fl ower extract: Main physical properties. J Alloys Compd 647:392–396. https://doi.org/10.1016/j.jallcom.2015.06.076

Torkzadeh-mahani R, Mehdi M, Jahani S (2019) Ultrasonics—Sonochemistry the effect of ultrasonic irradiation on the morphology of NiO/Co3O4 nanocomposite and its application to the simultaneous electrochemical determination of droxidopa and carbidopa. Ultrason Sonochem 56:183–192. https://doi.org/10.1016/j.ultsonch.2019.04.002

Ur A, Ullah A, Yuan Q, Wei Y, Ahmad A (2019) Tuber extract of Arisaema flavum eco-benignly and effectively synthesize silver nanoparticles: photocatalytic and antibacterial response against multidrug resistant engineered E. coli QH4. J Photochem Photobiol B Biol 193:31–38. https://doi.org/10.1016/j.jphotobiol.2019.01.018

Varma RS (2012) Greener approach to nanomaterials and their sustainable applications. Curr Opin Chem Eng 1:123–128. https://doi.org/10.1016/j.coche.2011.12.002

Vashist R, Swarnkar MK, Kumar A, Singh S (2017) Molecular dissection of transcriptional reprogramming of steviol glycosides synthesis in leaf tissue during developmental phase transitions in Stevia rebaudiana. Sci Rep 7:(1)1–13. https://doi.org/10.1038/s41598-017-12025-y

Venkatesh S, Thilagavathi J, Shyam D (2008) Anti-diabetic activity of flowers of hibiscus rosasinensis. Fitoterapia 79:79–81. https://doi.org/10.1016/j.fitote.2007.06.015

Verma VC, Kharwar RN, Gange AC (2010) Biosynthesis of antimicrobial silver nanoparticles by the endophytic fungus Aspergillus clavatus research article. Nanomedicine (Lond) 5:33–40

Vigneshwaran N, Ashtaputre NM, Varadarajan PV, Nachane RP, Paralikar KM, Balasubramanya RH (2007) Biological synthesis of silver nanoparticles using the fungus aspergillus flavus. Mater Lett 61:1413–1418. https://doi.org/10.1016/j.matlet.2006.07.042

Virkutyte J, Varma RS (2011) Green synthesis of metal nanoparticles: biodegradable polymers and enzymes in stabilization and surface functionalization. Chem Sci 5:837–846. https://doi.org/10.1039/c0sc00338g

Visinescu D, Jurca B, Ianculescu A, Carp O (2011) Starch—a suitable fuel in new low-temperature combustion-based synthesis of zinc aluminate oxides. Polyhedron 30:2824–2831. https://doi.org/10.1016/j.poly.2011.08.006

Wang X, He ZC, Pharm MS, Song L, Spencer S, Yang LX, Peng F, Pharm MS, Liu G, Pharm MS, Hu M, Li HB, Wu X, Zeng S, Hilgenfeld R, Stöckigt J, Zhao Y, Qian JF, Pharm MS (2011) Chemotherapeutic effects of bioassay-guided extracts of the American cockroach, Periplaneta americana. https://doi.org/10.1177/1534735411413467

Wang C-Z, Wu JA, McEntee E, Yuan C-S (2006) Saponins composition in American Ginseng Leaf and Berry assayed by highperformance liquid chromatography. J Agric Food Chem 54(6): 2261–2266. https://doi.org/10.1021/jf052993w

Wei X, Luo M, Li W, Yang L, Liang X, Xu L (2012) Bioresource technology synthesis of silver nanoparticles by solar irradiation of cell-free bacillus amyloliquefaciens extracts and AgNO 3. Bioresour Technol 103:273–278. https://doi.org/10.1016/j.biortech.2011.09.118

Wiley B, Sun Y (2007) Synthesis of silver nanostructures with controlled shapes and properties. Acc Chem Res 40:1067–1076

Wongpreecha J, Polpanich D, Suteewong T, Kaewsaneha C, Tangboriboonrat P (2018) One pot, large scale green synthesis of silver nanoparticles-chitosan with enhanced antibacterial activity and low cytotoxicity. Carbohydr Polym 199:641–648. https://doi.org/10.1016/j.carbpol.2018.07.039

Yasmin A, Ramesh K, Rajeshkumar S (2014) Optimization and stabilization of gold nanoparticles by using herbal plant extract with microwave heating. Nano Convergence 1:1–7. https://doi.org/10.1186/s40580-014-0012-8

Zhang L, Xu H, Li S (2009) Effects of micronization on properties of Chaenomeles sinensis (Thouin) Koehne fruit powder. Innov Food Sci Emerg Technol 10:633–637. https://doi.org/10.1016/j.ifset.2009.05.010

Zhang X, Yan S, Tyagi RD, Surampalli RY (2011) Chemosphere synthesis of nanoparticles by microorganisms and their application in enhancing microbiological reaction rates. Chemosphere 82:489–494. https://doi.org/10.1016/j.chemosphere.2010.10.023

Structural and Morphological Characterization of Nanomaterials

6

Munirah A. Almessiere, Yassine Slimani, Alex V. Thurkanov, and Abdulhadi Baykal

Abstract

Nanomaterials have drawn latest research attention in different aspects very rapidly such as medical devices, electronic devices, and other industries due to a remarkable enhancement of characteristics. In this chapter we are debating the classification of nanomaterials as "0-D" nanomaterials, "1-D" nanomaterials "2-D", and "3-D" nanomaterials. Also, we investigated the morphological properties of nanomaterials as a key to discover their properties and discuss how structure and morphology will play a role in exploring their behavior.

M. A. Almessiere (✉)
Department of Biophysics, Institute for Research and Medical Consultations (IRMC), Imam Abdulrahman Bin Faisal University, Dammam, Saudi Arabia

Department of Physics, College of Science, Imam Abdulrahman Bin Faisal University, Dammam, Saudi Arabia
e-mail: malmessiere@iau.edu.sa

Y. Slimani
Department of Biophysics, Institute for Research and Medical Consultations (IRMC), Imam Abdulrahman Bin Faisal University, Dammam, Saudi Arabia
e-mail: yaslimani@iau.edu.sa

A. V. Thurkanov
Department of Physics, College of Science, Imam Abdulrahman Bin Faisal University, Dammam, Saudi Arabia

South Ural State University, Chelyabinsk, Russia

SSPA—Scientific and Practical Materials Research Centre of NAS of Belarus, Minsk, Belarus

L.N. Gumilyov Eurasian National University, Nur-Sultan, Kazakhstan

A. Baykal
Department of Nanomedicine Research, Institute for Research and Medical Consultations (IRMC), Imam Abdulrahman Bin Faisal University, Dammam, Saudi Arabia

A. Thakur et al. (eds.), *Synthesis and Applications of Nanoparticles*,
https://doi.org/10.1007/978-981-16-6819-7_6

Keywords

Nanomaterial · 1-D nanomaterials · 2-D nanomaterials · 3-D nanomaterials · TEM · SAED · SEM · STM · AFM

6.1 Classification of Nanomaterials (Structure and Morphology)

The nanostructured materials are categorized based on electron movement. In "0-D" nanomaterials, the presence of electrons is generally fixed, electrons can move along the x-axis in "1-D" nanomaterials (which is commonly less than 100 nm), and electron movements along the *x*-, *y*-, *z*-axis, or *x*- to *y*-axis or respectively in "2-D" and "3-D" nanomaterials (Figs. 6.1 and 6.2) (Khan Editor n.d.; Saleh 2020).

(A) 0-D nanomaterials (all dimensions at nanoscale): nanorod, hollow sphere, spherical NMs, metal, polygon, and cube
(B) 1-D nanomaterials (one dimension at macroscale and two dimensions at nanoscale): nanotube, nanowires, ceramic, nanorod filament or fiber, and metallic, polymeric, and nanofibers
(C) 2-D nanomaterials (one dimension at nano- and two others at macroscale): nanocoating, crystalline, nanoplates, thin films, and single-layered and multilayered
(D) 3-D nanomaterials (all dimensions at macroscale, beyond 100 nm): carbon nanobuds, polycrystals, fullerenes, fibers, honeycombs, foams, nanotubes, pillars, and layer skeletons
Fig. 6.3 presents the classification of nanomaterials based on morphology. The NPS having a high aspect ratio can have nanohelices, nanowires, nanotubes, nanozigzags, nanobelts, or nanopillar shapes. The low-aspect ratio NPs can have cubes and spherical, pillar-like, helical, and pyramidal shapes, among others (Shukla Editor n.d.).
(E) recent rapid advances in,
(F) 2D nanomaterials have raised important and exciting questions about their
(G) interactions with biological moieties. 2D nanoparticles such as carbon-based,

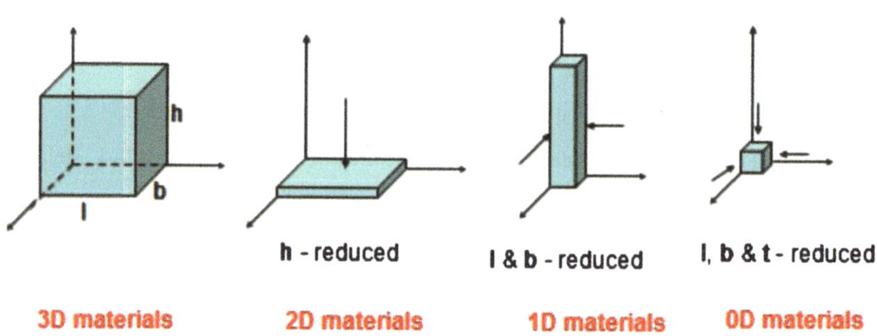

Fig. 6.1 Classification of nanoscale dimensions. (Source: Tallinn University of Technology) (Susi et al. 2017)

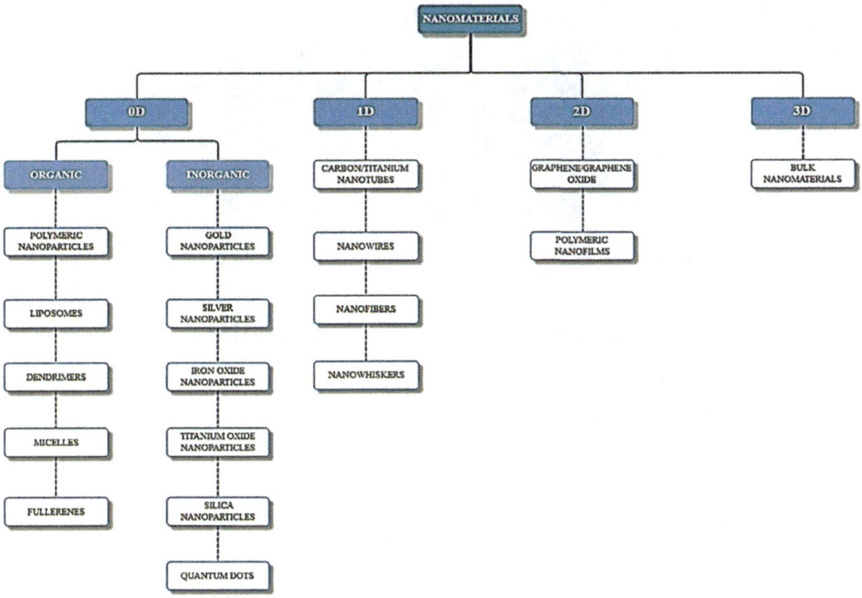

Fig. 6.2 Nanomaterial classification based on their dimensionality (Teleanu et al. 2019)

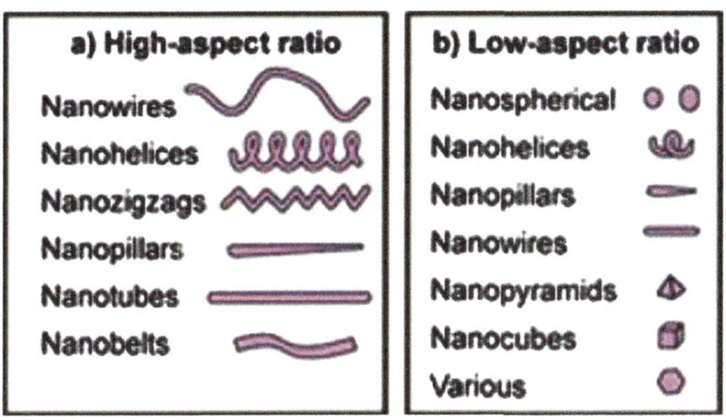

Fig. 6.3 The morphology-based classification of nanomaterials (Shukla Editor n.d.)

(H) 2D materials, silicate clays, transition metal dichalcogenides (TMDs), and
(I) transition metal oxides (TMOs) provide enhanced physical, chemical, and,
(J) biological functionality owing to their uniform shapes, high surface-to-volume,
(K) ratios, and surface charge.

6.1.1 0-D Nanomaterials

Zero-dimensional (0-D) nanomaterials, including inorganic quantum dots (QDs), fullerenes, graphene quantum dots (GQDs), noble metal nanoparticles, magnetic

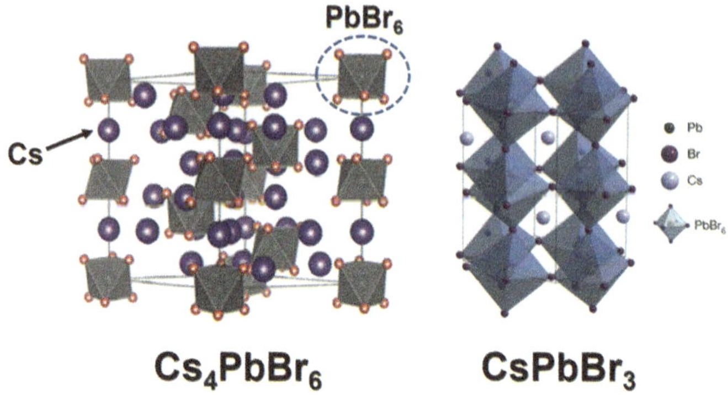

Fig. 6.4 The crystal structure of zero-dimensional perovskite nanocrystals Cs_4PbBr_6 NCs (left) and $CsPbBr_3$ (right) (Becket et al. 2018; Sun et al. 2021)

nanoparticles (MNPs), carbon quantum dots (CQDs), polymer dots (Pdots), and perovskite (Fig. 6.4), are important nanomaterials as biosensor:

Graphene quantum dots (GQDs): 0-D graphene nanomaterials <10 nanometers.
Carbon quantum dots (CQDs): CQDs are quasi-spherical fluorescent particles with sizes.
Fullerenes: Fullerene, having a truncated icosahedron, consisting of five to six sp^2 hybrid carbon rings.
Quantum dots (QDs): QDs generally consist of II–VI or III–V group elements, CdSe, CdTe, and InP.
Magnetic nanoparticles: 0-D magnetic nanoparticles (MNPs) (1–100 nm) with high saturation magnetization, such as alloys (FeCo, permalloy, alnico), pure metals (Fe, Co, Ni), and oxides (Fe_3O_4, $CoFe_2O_4$).

6.1.2 1-D Nanomaterials

One-dimensional nanomaterials (length is larger than width) are in nanoscale in two dimensions (nanowire, nanorod, nanotube, nanofilaments, and nanofiber). They can be single or polycrystalline amorphous and/or ceramic, polymeric, or metallic. However, 1-D nanomaterials can be in doped semiconductor or chemically pure. 1-D nanomaterials can be alone or embedded within another medium (Figs. 6.5 & 6.6).

1-D Nanofillers: The one of dimension of 1-D nanofillers is less than 100 nm (Verdejo et al. 2011). They consist of form of sheets of one to a few nm thick to hundreds and thousands of nm long. The well-known examples of 1-D nanofillers are nanographene platelets, montmorillonite clays, nanodiscs (Schmidt et al. 2012), nanoprisms, nanoplates, nanosheets (Nieto et al. 2012), and nanowalls (Tiwari et al. 2012), ZnO nanodiscs (Umar and Hahn 2006), ZnO nanosheets (Bai et al. 2008),

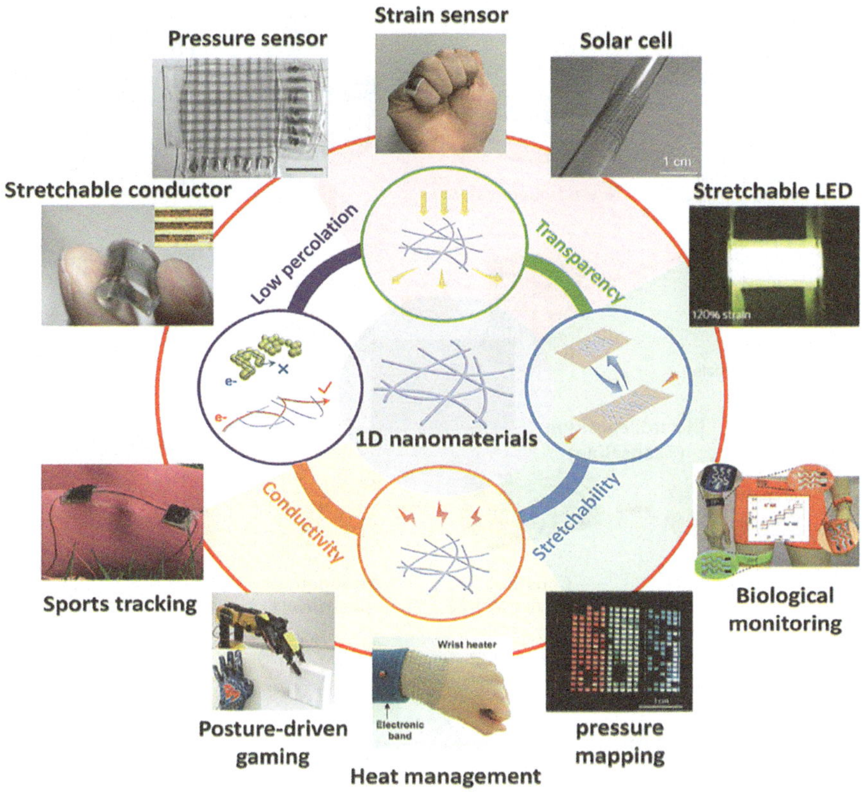

Fig. 6.5 Various applications of 1-D nanomaterials (Gong and Cheng 2017)

ZnO nanoplatelets (Mani and Rayappan 2014), amphiphilic graphene platelets (Shen et al. 2009), carbon nanowalls (Liu et al. 2015), graphite nanoplatelets (GNP) (Li and Zhong 2011), and Fe_3O_4 nanodiscs (Zou et al. 2008). Due to their unique shape-dependent characteristics, 1-D nanofillers have a potential application in nanodevices.

Inorganic Nanowires: Nanowires are playing an increasingly important role in nanoscale electronics and sensors. Nanowires as representative of 1-D nanoscale structures have a large surface-to-volume ratio and unique transport properties which make them as for novel electronics and sensor applications to achieve high performance.

Porous Structure of Nanofibers: Nanofibers, known as superfine fibers, have a diameter smaller than 500 nm and an aspect ratio (length-to-diameter ratio) of at least 100:1. The high surface area of porous nanofibers enhances their performance as conductors, catalysts, sensors, solar cells, and supercapacitors and in tissue engineering, filtration, energy storage, adsorption, etc. (Sabetzadeh and Gharehaghaji 2017).

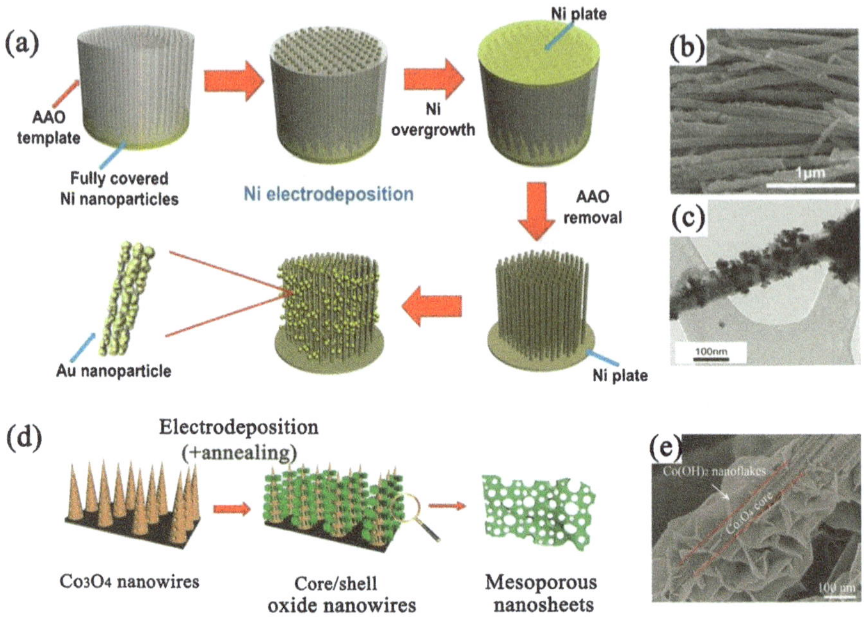

Fig. 6.6 Various applications of 1-D nanomaterials (Jin et al. 2018)

6.1.3 2-D Nanomaterials

2-D can be crystalline or amorphous and composed of polymeric, ceramic, or metallic. They are made of different chemical compositions, deposited on a substrate, or can be integrated into matrix material. Their length is larger than the width. The electrons in 2-D nanomaterials are subject to the electron delocalization and confinement (Liu and Bashir 2015). They are the thinnest nanomaterials due to their dimensions and thickness on nanoscale/macroscale.

2-D nanomaterials have enhanced chemical, physical, and biological functionality due to surface charge, high surface-to-volume ratios, and their uniform shapes (Chimene et al. 2015; Sarmazdeh et al. 2019). On their one of the dimensions, there are only a few atomic layers thick; therefore, they cooperate with biological moieties in a unique way. 2-D nanomaterials can be used in stem cell engineering, medical diagnostics, drug and gene delivery, biosensing, regenerative medicine, basic cell biology, cancer therapy, and bioelectronics. They are the thinnest nanomaterials due to their dimensions and thickness on nanoscale/macroscale.

These nanomaterials have the fast electron transfer kinetics, high electrical conductivity, easy functionalization, and large surface area. Therefore, these properties make them as important materials for biosensor applications. For example, graphene has a high electrical conductivity (~1.0 * 108 S m^{-1}) and theoretically large surface-to-volume ratio (2600 m^2 g^{-1}). For electrochemical applications, actually it has a large effective reaction area for loading a high number of biomolecules and high electron transfer ability. Therefore, the biosensors with 2-D nanomaterials improved the detection performance sensitivity, including stability and linear range. Graphene-based biosensors can detect the substrates in redox proteins (e.g., H_2O_2, glucose) with high sensitivity (Su et al. 2019).

Nowadays, 2-D nanomaterials (graphene (Figs. 6.7–6.10), hBN (Fig. 6.6), and metal dichalcogenides (MX2)) are important due to their important application usage in electronics, energy storage, optoelectronics, solar cells, catalysts, sensors, etc. (Sarmazdeh et al. 2019).

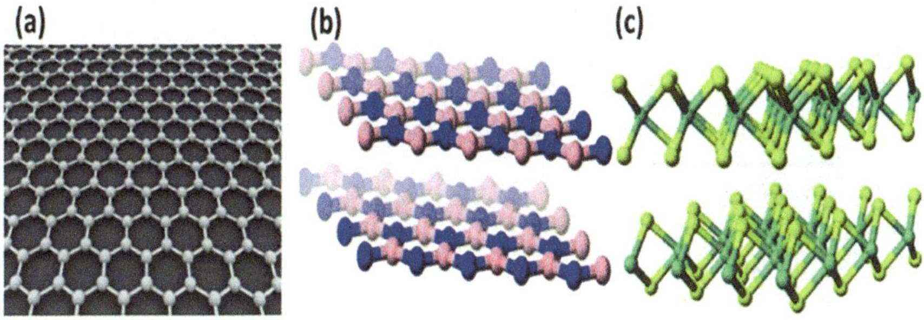

Fig. 6.7 The structure of (**a**) graphene (single layer), (**b**) BN nanosheets, (**c**) WSe_2 (Se in yellow and W in blue) (Latini et al. 2017)

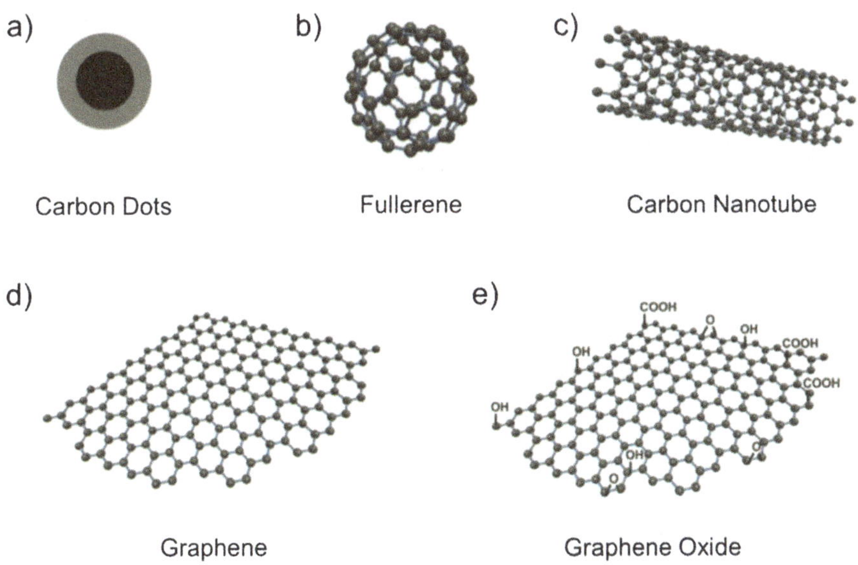

Fig. 6.8 0-D carbon nanomaterials. (**a**) Carbon Dots, (**b**) Fullerene, (**c**) cardon Nanotube, (**d**) Graphene and (**e**) Graphene oxide (Han et al. 2016)

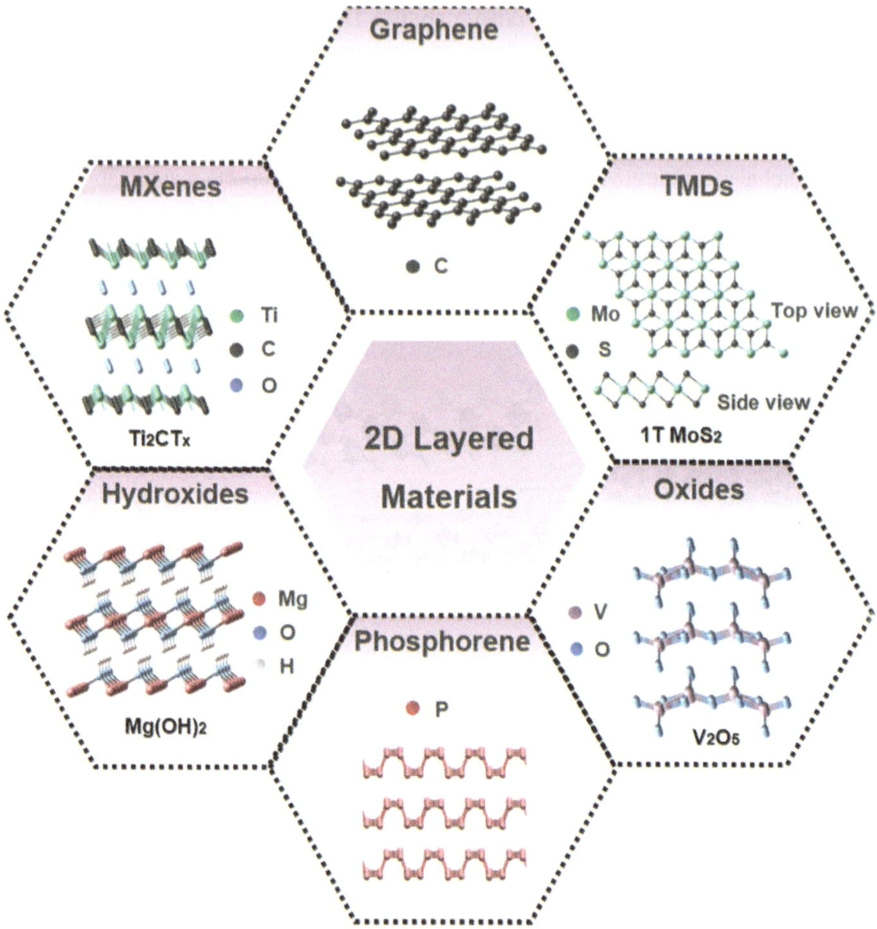

Fig. 6.9 Several typical types of 2-D layered nanomaterials (Li and Wang 2020)

6.1.4 3-D Nanomaterials

3-D nanomaterials are considered as bulk nanomaterials and all dimensions beyond 100 nm. They contain dispersions of nanoparticles, nanotubes, nanowires, and bundles. All electrons in 3-D nanomaterials can move freely within all dimensions (Fig. 6.10).

6.2 Morphological Characterization

Although nanomaterials are important due to their great potential applications in catalysis, harvesting, optics, electronics, magnetic data, and energy storage, the intrinsic sole performance and properties of them depend on the shape, size,

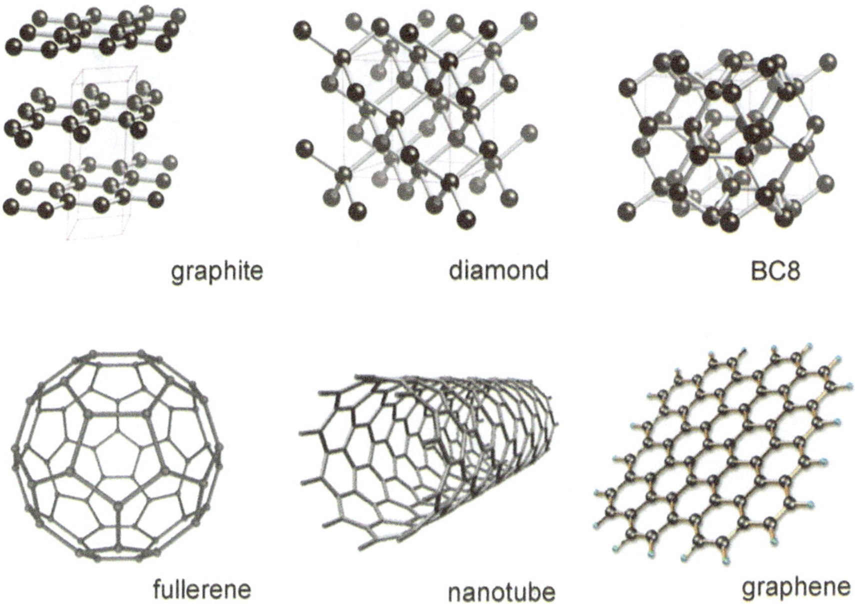

Fig. 6.10 Structure of various allotropes of carbon (Janani et al. 2013)

interparticle interaction, surface structure, etc. They are mainly for the characterization of nanomaterials, which are structural and morphological analyses. It is well-known that morphological characterizations have been carried out using electron microscopy, optical microscopy, etc. Within these techniques, the produced signal at the end of electron sample's atom interaction can provide information about the composition, sample's surface topography, electrical conductivity, etc. TEM can be used to investigate the internal structure of the sample, and SEM deals with the surface (near-surface) (Lieber 1998).

Morphology of nanomaterials is based on their crystal structure, material composition, and synthesis approach based on which nanoparticles with a diversity of shapes (needles, spheres, tubes, rods, octahedrons, cubes, etc.) and sizes can be synthesized. The morphology of a nanomaterial (dimension and shape of geometric) has a considerable impact on its chemical and physical properties. Hence, the determination of the morphology of nanomaterials is crucial for their specific applications. The SEM, TEM, AFM, and STM are powerful microscopic characterization techniques to study the nanomorphology. Among all, TEM is the most powerful technique, and it can identify the intrinsic morphological details of 0-D, l-D, 2-D, and 3-D nanomaterials and can provide more accurate information about atomic arrangements within the crystal lattice (Lieber 1998; Hua et al. 2015). Figure 6.11 provides the comparison of various morphological instruments with eyes.

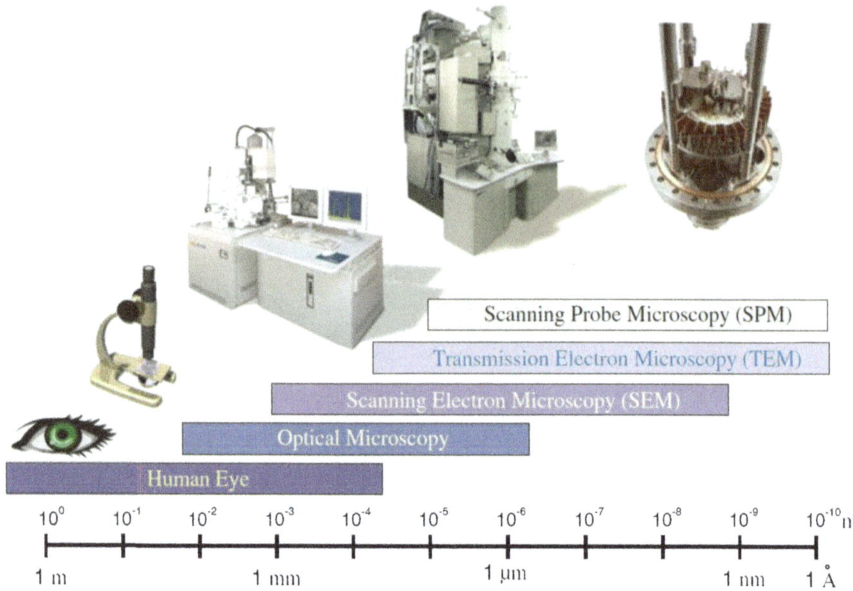

Fig. 6.11 Different microscopy techniques with different imaging ranges (Voigtländer 2015)

6.2.1 High-Resolution TEM

TEM and high-resolution TEM (HR-TEM) can provide information about the grain boundaries and microstructure of interfaces and crystalline defects of nanoparticles. HR-TEM can also resolve the single atoms of the crystalline materials and individual atomic columns. Due to the resistance of polymeric materials to the electron irradiation effects, it is hard to obtain their TEM images. Although TEM has a capacity to detect transmitted electrons and SEM can only detect backscattered electrons, TEM can present 2-D images, and SEM can provide 3-D. TEM can offer the internal composition details (a lattice structure and crystallite size) and greater resolution (Hua et al. 2015; Wang 2000).

6.2.1.1 SAED (Selected Area Electron Diffraction)

TEM-SAED pattern can be obtained by adjusting the magnetic lenses so that the back focal plane of the lens. SAED pattern usually gives an image consisting of rings, a pattern, or dots for the crystalline nanomaterials. Usually dot patterns are obtained for single crystals, and a series of rings was obtained for polycrystalline/amorphous materials. This SAED analysis also provides information about crystal orientation to the beam path and the space group symmetries (Fig. 6.12).

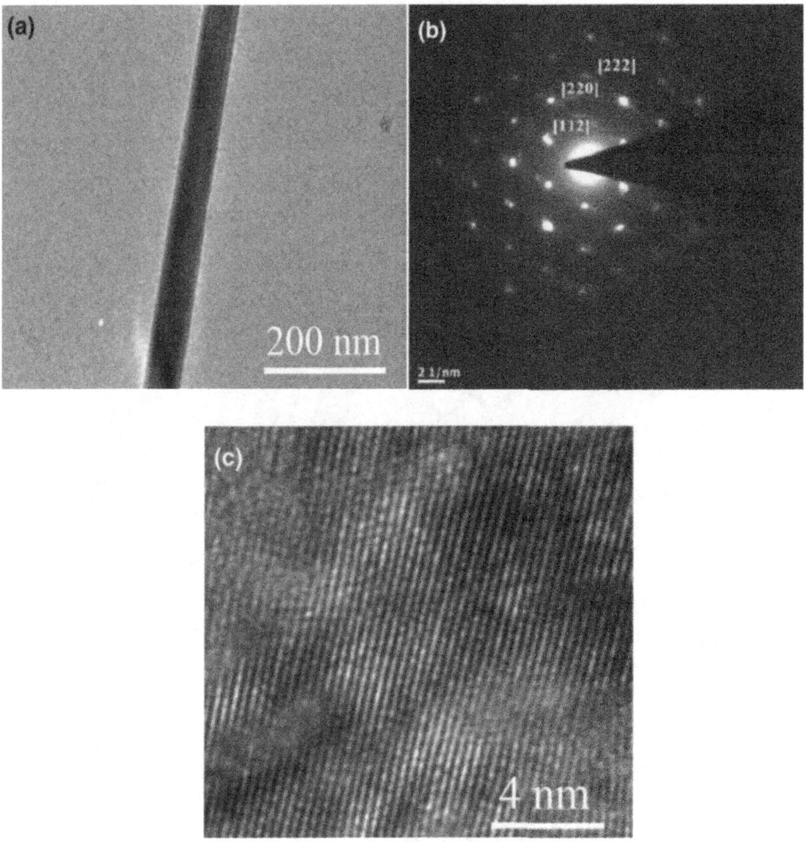

Fig. 6.12 (**a**) TEM image, (**b**) SAED pattern, and (**c**) HR-TEM image of the nanowire (Li et al. 2016)

6.2.2 SEM

To obtain SEM micrographs, the surface of the sample is scanned with a focused beam of energetic electrons. The electrons which are interacting with the atoms in the sample are being used to obtain information about the texture, crystalline arrangement, and chemical composition of sample. SEM provides faster results, more surface information, and easier sample preparation than that of TEM. SEM can also provide information about the microchips for computers, biological specimens, and semiconductors and more. SAED application is also useful for getting qualitative or semiquantitative information about the chemical compositions (Di Giorgio et al. 2017; Bogner et al. 2005).

Morphology of different nanostructures may vary significantly according to crystal structure, material composition, and manufacturing method. Depending upon the synthesis approach, tubes, spheres, octahedrons, wires, needles, rods,

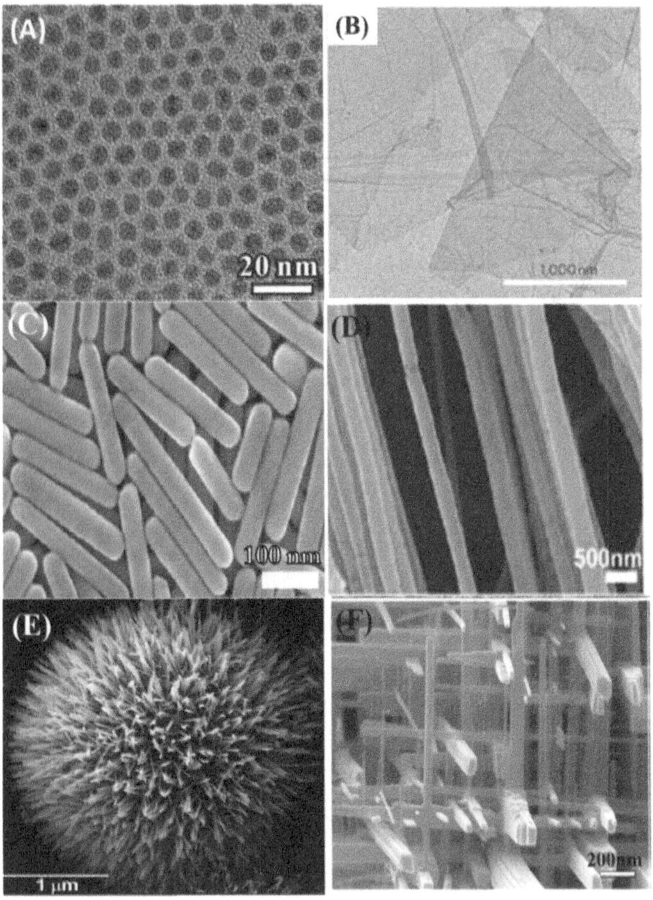

Fig. 6.13 Nanomaterials with different morphologies: (**a**) nonporous Pd NPs (0-D), (**b**) graphene nanosheets (2-D), (**c**) Ag nanorods (1-D) (Umar and Hahn 2006), (**d**) polyethylene oxide nanofibers (1-D), (**e**) urchin-like ZnO nanowires (3-D), (**f**) WO_3 nanowire network (3-D) (Jeevanandam et al. 2018)

cubes, and with varying size nanomaterials can be synthesized. In other words nanomaterials exist in a number of different shapes depending upon the synthesis procedure. Morphological diversity is significant for nanomaterials, although these materials consist of a large number of surface atoms which determine their chemical and physical properties. Based on morphology, nanomaterials can be classified as low- and high-aspect ratio particles (Fig. 6.13). For the NPs having a high aspect ratio, nanohelices, nanowires, nanotubes, nanozigzags, nanopillars, and nanobelts can be given as examples. And the spherical, helical, pillar-like, and pyramidal shapes and cubes are the NPs having a low aspect ratio. In terms of uniformity, NPs can be classified as agglomerate and dispersed. Due to NPs' electromagnetic properties, surface charge magnetism, hydrophobicity, and hydrophilicity are varying. The morphology is important in nanoworld because it affects the NPs' properties. Scanning electron microscope, atomic force microscope, transmission electron microscope, etc. are being used to identify the morphology of nanomaterials (Mayeen et al. 2018a).

6.2.3 SPM (Scanning Probe Microscopy)

SPM forms images of surfaces using a physical probe that scans the specimen, and it provides information about the morphology and local properties of the solid body surface. SPM images of surfaces using a physical probe that touches the sample's surface scan the surface and collect data; hence, two-dimensional grid of data points will be obtained (Mayeen et al. 2018a) (Fig. 6.14).

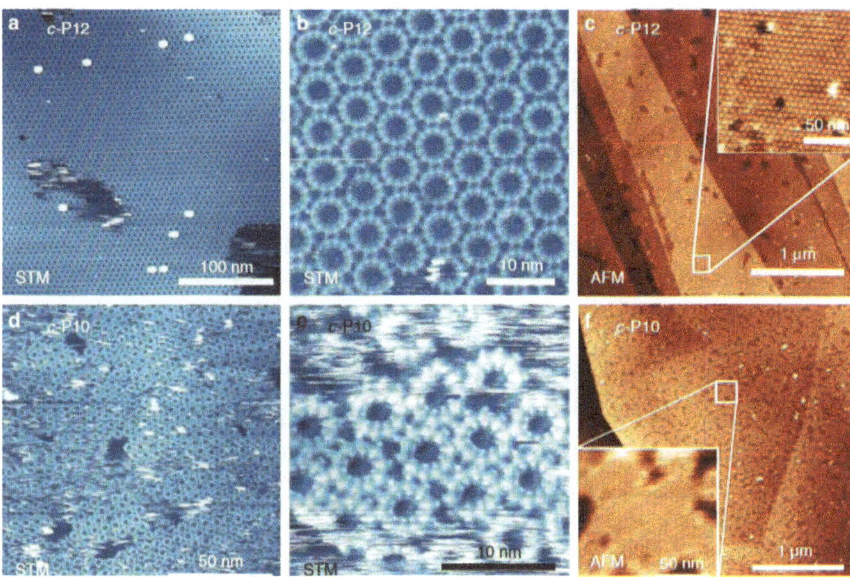

Fig. 6.14 SPM images of porphyrin nanorings deposited on a highly oriented pyrolytic graphite (HOPG) surface. (**a**) STM image of continuous c-P12 domain (the bright spots correspond to stacks of multiple c-P12 rings), (**b**) high-resolution STM image of c-P12 molecules in a single domain, (**c**) AFM image showing large-scale structure of c-P12 domains on HOPG (dark regions correspond to voids in the nanoring monolayer). (Inset) High-resolution image showing hexagonal packing of c-P12 molecules in the region indicated by the white box in the main image, (**d**) STM image of c-P10 domains on HOPG, (**e**) high-resolution image of c-P10 molecules, (**f**) AFM image showing large-scale structure of c-P10 domains on HOPG (Summerfield et al. 2019)

6.2.4 STM (Scanning Tunneling Microscopy)

The STM technique is based on quantum tunneling current to create electron density images for semiconductive or conductive surfaces and biomolecules attached on conductive substrates at the atomic scale (Barth et al. 2005). STM is the most powerful technique for revealing the topographic structures of the surfaces. It can provide defects on or slightly below surfaces, surface roughness, conformation of molecules, and aggregates on the surface (Young et al. 1972). Figure 6.15 represents the STM image Si(111) surface.

6.2.5 AFM (Atomic Force Microscopy)

Atomic force microscopes belong to the scanning probe microscope (SPM) series. It can image almost any type of surface, including biological samples, glass, polymers, composites, and ceramics. It can be used to measure and localize many forces, including mechanical properties, magnetic forces, and adhesion strength, and AFM can be operated in two basic modes, tapping and contact. In the tapping mode, the AFM cantilever is vibrated above the sample surface so that the tip is in contact with the surface only intermittently the contact mode, while the AFM tip is continuously in contact with the surface (Fig. 6.16). The advantages of AFM include:

1. Less requirements for working environment and sample preparation than electron microscope, and it can detect the shape, size, and mechanical properties of

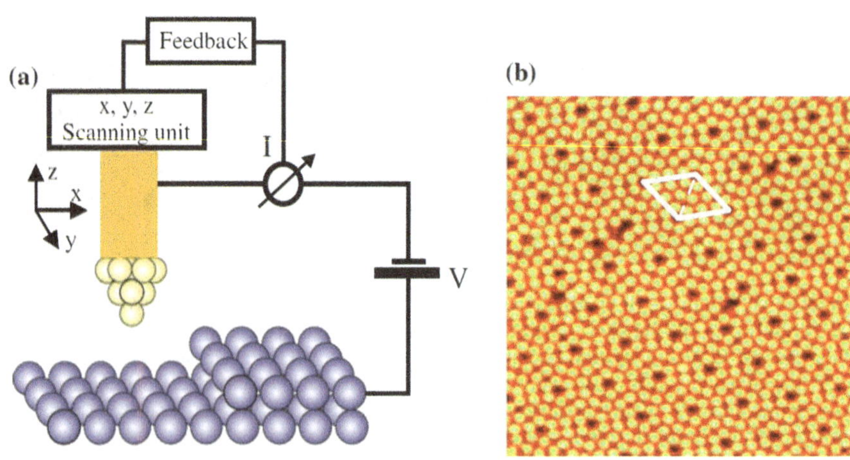

Fig. 6.15 (**a**) Schematic representation of STM, (**b**) STM image of Si(111) surface. Individual atoms are observed as yellow dots. The rhombic unit cell is indicated by white lines. Besides the periodic arrangement of the atoms, also defects such as single missing atoms can be observed (Voigtländer 2015)

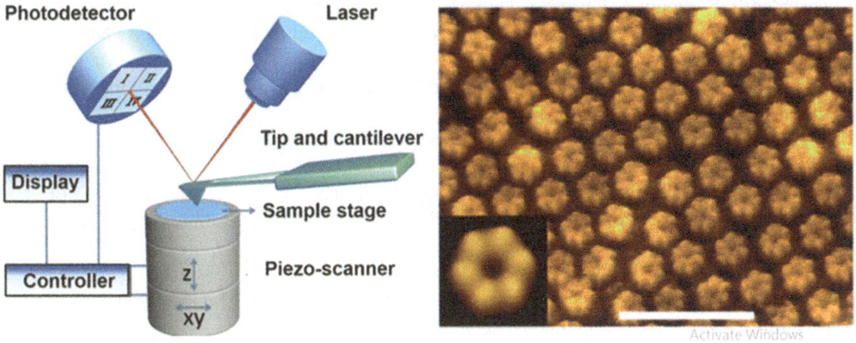

Fig. 6.16 Schematic representation of AFM and high-resolution AFM image of proteins (Shan 2015; Thomas et al. 2017; Mayeen et al. 2018b)

conductors, semiconductors, insulators, and biological samples in the atmosphere, high vacuum, liquid, and other environments.

2. Extremely high resolution. The horizontal resolution is less than 0.1 nm, and the vertical resolution is less than 0.01 nm.
3. Physical and mechanical properties of materials can be measured at the nanometer scale, such as electrical conductivity, stagnation, friction, and lubrication (Fig. 6.13).

References

Bai W, Zhu X, Zhu Z, Chu J (2008) J Appl Surf Sci 254:6483–6488

Barth JV, Costantini G, Kern K (2005) Engineering atomic and molecular nanostructures at surfaces. Nature 437(7059):671–679

Becker MA, Vaxenburg R, Nedelcu G, Sercel PC, Shabaev A, Mehl MJ, Michopoulos JG, Lambrakos SG, Bernstein N, Lyons JL, Stöferle T, Mahrt RF, Kovalenko MV, Norris DJ, Rainò G, Efros AL (2018) Bright triplet excitons in caesium lead halide perovskites. Nature 553 (7687):189–193. https://doi.org/10.1038/nature25147

Bogner A, Thollet G, Basset D, Jouneau PH, Gauthier C (2005) Wet STEM: a new development in environmental SEM for imaging nano-objects included in a liquid phase. Ultramicroscopy 104(3):290–301

Chimene D, Alge DL, Gaharwar AK (2015) Adv Mater 27:7261–7284

Di Giorgio C, Putilov A, Lechner E, Trainer D, Volkova OS, Vasiliev AN, Chareev D, Karapetrov G, Zasadzinski J, Iavarone M (2017) Low temperature scanning tunneling microscopy and spectroscopy investigation of FeSe single crystals. Bull Am Phys Soc 62

Gong S, Cheng W (2017) One-dimensional nanomaterials for soft electronics. Adv Electron Mater 3:1600314

Han X, Li S, Peng Z, Al-Yuobi AO, Bashammakh ASO, El-Shahawi MS, Leblanc RM (2016) Interactions between carbon nanomaterials and biomolecules. J Oleo Sci 65:1–8

Hua X, Liu Z, Bruce PG, Grey CP (2015) The morphology of TiO_2 (B) nanoparticles. J Am Chem Soc 137(42):13612–13623

Janani M, Srikrishnarka P, Nair SV, Nair AS (2013) An in-depth review on the role of carbon nanostructures in dye sensitized solar cells. J Mater Chem A:1–3

Jeevanandam J, Barhoum A, Chan YS, Dufresne A, Danquah MK (2018) Review on nanoparticles and nanostructured materials: history, sources, toxicity and regulations. Beilstein J Nanotechnol 9:1050–1074

Jin T, Han Q, Wang Y, Jiao L (2018) 1D nanomaterials: design, synthesis, and applications in sodium–ion batteries. Small 14(2):1703086

Khan Editor FA (n.d.), Applications of Nanomaterials in Human Health, ISBN 978–981–15-4801-7 ISBN 978–981–15-4802-4 (e-Book) https://doi.org/10.1007/978-981-15-4802-4, This Springer imprint is published by the registered company Springer Nature Singapore Pte Ltd. The registered company address is: 152 Beach Road, #21–01/04 Gateway East, Singapore 189721, Singapore

Li B, Zhong WH (2011) J Mater Sci 46:5595–5614

Li X, Wang J (2020) One-dimensional and two-dimensional synergized nanostructures for high-performing energy storage and conversion. InfoMat 2:3–32

Latini S, Winther KT, Olsen T, Thygesen KS (2017) Interlayer excitons and band alignment in MoS/hBN/WSe van der Waals heterostructures. Nano Lett 17(2):938–945

Li L, Liang L, Wu H, Zhu X (2016) One-dimensional perovskite manganite oxide nanostructures: recent developments in synthesis, characterization, transport properties, and applications. Nanoscale Res Lett 11:121

Lieber CM (1998) One-dimensional nanostructures: chemistry, physics & applications. Solid State Commun 107(11):607–616

Liu JL, Bashir S (2015) Advanced nanomaterials and their applications in renewable energy. Elsevier, Amsterdam, Netherlands

Liu X, Wen N, Wang X, Zheng Y (2015) Nanomater Nanotechnol 5:51–57

Mani GK, Rayappan JBB (2014) RSC Adv 4:64075–64084

Mayeen A, Shaji LK, Nair AK, Kalarikkal N (2018a) Chapter 12—morphological characterization of nanomaterials. In: Characterization of Nanomaterials advances and key technologies micro and Nano Technologies. Woodhead Publishing, pp 335–364

Mayeen A, Shaji LK, Nair AK, Kalarikkal N (2018b) Chapter 12—morphological characterization of nanomaterials. In: Characterization of nanomaterials advances and key technologies, micro and nano technologies. Woodhead publishing, pp 335–364

Nieto A, Lahiri D, Agarwal A (2012) Carbon NY 50:4068–4077

Sabetzadeh N, Gharehaghaji AA (2017) How porous nanofibers have enhanced the engineering of advanced materials: a review. J Text Polym 5(2):57–68

Saleh TA (2020) Nanomaterials: classification, properties, and environmental toxicities. Environ Technol Innovation 20:101067

Sarmazdeh ZR, Dizaji SMZ, Kang AK (2019) Two-dimensional nanomaterials, nanostructures, Sadia Ameen, M. Shaheer Akhtar and Hyung-Shik Shin, IntechOpen, https://doi.org/10.5772/intechopen.85263. https://www.intechopen.com/books/nanostructures/two-dimensional-nanomaterials

Schmidt FP, Ditlbacher H, Hohenester U, Hohenau A, Hofer F, Krenn JR (2012) Nano Lett 12: 5780–5783

Shan E (2015) The structure and function of cell membranes examined by atomic force microscopy and single-molecule force spectroscopy. Chem Soc Rev 44(11):3617–3638

Shen J, Hu Y, Li C, Qin C, Ye M (2009) Small 5:82–85

Shukla Editor AK (n.d.) EMR/ESR/EPR spectroscopy for characterization of nanomaterials, ISSN 1869–8433 ISSN 1869–8441 (electronic) Advanced structured materials ISBN 978–81–322-3653-5 ISBN 978–81–322-3655-9 (eBook) DOI https://doi.org/10.1007/978-81-322-3655-9. This Springer imprint is published by Springer Nature The registered company is Springer (India) Pvt. Ltd. The registered company address is 7th Floor, Vijaya Building, 17 Barakhamba Road, New Delhi 110 001, India

Su S, Sun Q, Gu X, Xu Y, Shen J, Zhu D, Chao J, Fan C, Wang L (2019) Two-dimensional nanomaterials for biosensing applications, rends. Anal Chem 119:1156102

Sun S, Lu M, Gao X, Shi Z, Bai X, Yu WW, Zhang Y (2021) 0D perovskites: unique properties, synthesis, and their applications. Adv Sci *8*(24):2102689. https://doi.org/10.1002/advs.202102689

Susi T, Meyer JC, Kotakoski J (2017) Manipulating low-dimensional materials down to the level of single atoms with electron irradiation. Ultramicroscopy 180:163–172

Summerfield A, Baldoni M, Kondratuk DV, Anderson HL, Whitelam S, Garrahan JP, Besley E, Beton PH (2019) Ordering, flexibility and frustration in arrays of porphyrin nanorings. Nat Commun 10:2932

Teleanu DM, Chircov C, Grumezescu AM, Teleanu RI (2019) Neurotoxicity of nanomaterials: an up-to-date overview. Nano 9:96

Thomas S, Thomas R, Zachariah A, Kumar R (2017) Microscopy methods in nanomaterials characterization, vol 1, 1st edn. Elsevier

Tiwari JN, Tiwari RN, Kim KS (2012) Prog Mater Sci 57:724–803

Umar A, Hahn YB (2006) Nanotechnology 17:2174–2180

Verdejo LJRR, Mar Bernal M, Francisco MAL-M, Tapiador J (2011) Cell Polym 30:45–62

Voigtländer B (2015) Scanning probe microscopy atomic force microscopy and scanning tunneling microscopy. Springer-Verlag, Berlin Heidelberg

Wang ZL (ed) (2000) Characterization of nanophase materials. Wiley-VCH, New York, pp 13–14

Young R, Ward J, Scire F (1972) The topografiner: an instrument for measuring surface microtopography. Rev Sci Instrum 43:999

Zou H, Wu S, Shen J (2008) Chem Rev 108:3893–3957

Optical Characterization of Nanomaterials-I 7

Vladimir Pavelyev

Abstract

The advent of optical nanotechnologies (technologies for structuring optical materials with a resolution above 100 nm) opens the possibility of creating photonics devices, the action of which is based on the effects arising from the interaction of a light wave with a substantially sub-wave structure (visible wavelength range 400–780 nm, infrared wavelength range 0.78–30 мкм). Such devices with wide functional properties and miniature dimensions form a new element base of optical systems for collecting, transmitting, and processing information. At the same time, the methods of nanostructuring optical materials and creating optical nanocomposites allow you to create new metamaterials—that is, composites, the interaction of the light wave with which is described by generalized (effective) physical characteristics. Both nanostructures with linear optical properties and nonlinear ones are successfully used to create optical metamaterials and photonic nanodevices. The first section of this chapter is devoted to the study of functional photonic nanostructures with linear properties, and the second section is devoted to the study of functional photonic nanostructures with nonlinear properties.

Keywords

Photonic nanostructures · Optical metamaterials · Photonic crystals and quasicrystals · Diamond films

V. Pavelyev (✉)
Samara National Research University, Samara, Russia

A. Thakur et al. (eds.), *Synthesis and Applications of Nanoparticles*,
https://doi.org/10.1007/978-981-16-6819-7_7

7.1 Fabrication and Optical Characterization of 2D and 3D Photonic Structures

The monograph (Soifer 2014) discusses the use of nanostructuring techniques to fabricate subwavelength diffractive optical elements (DOE) of the optical range. Lithography techniques (Chaplygin 2005) well known in microelectronics are used to manufacture diffractive optical elements of the optical range with a stepped diffraction microrelief (Soifer 2014). However, optical nanostructures with a more complex topology are difficult to implement with planar lithography technology. Two- and three-dimensional photonic crystal structures are an important example of such structures. Photonic crystals (PhCs) with full photon band gaps, that is, artificial dielectric heterostructures that allow controlling the generation and flow of light, are frequency ranges in which the propagation of electromagnetic waves is banned in all directions and polarizations (Soifer 2014; Yablonovitch 1987; John 1987). Their development has undergone substantial modifications in recent decades, and their unique qualities have found applications in a wide range of fields, including effective radiation sources (Caglayan et al. 2005), telecommunication devices (Knight 2003; Rinnie et al. 2008), and sensors (Chow et al. 2004).The synthesis of three-dimensional (3D) PhCs presents a real technological problem due to the fine features of the 3D configuration, as well as the limited selection of suitable materials (Soifer 2014). Although a number of methods have been proposed to date to solve this problem (Soifer 2014; Ho et al. 1994; Bogomolov et al. 1996; Wijnhoven and Vos 1998; Blanco et al. 2000), they all have significant disadvantages that prevent the successful execution of the concept of photonic crystals in practice. Repurposing classical (2D) lithography is one method. 2D periodic patterns, for example, are first created by selective etching via masks using electron beam lithography, and then new layers are successively manufactured over the old (Ho et al. 1994).

The disadvantage is that the method is laborious and a large number of layers (periods) are difficult to produce. In general, 3D manufacture is limited to only a few layers (Ho et al. 1994). Physicochemical methods of self-assembly using colloidal particles (Bogomolov et al. 1996; Wijnhoven and Vos 1998; Blanco et al. 2000) or separation of the copolymer phase (Kang et al. 2007) are relatively simple and low cost. Nevertheless, the method is not fast, and random defects are unavoidable for large area samples.

Besides, the lattice's potential symmetries are usually confined to hexagonal close-packed and centered cubic configurations. Drilling the material in three directions produces a 3D periodic structure. This method can be used to create the "Yablonovite" structure, which was developed by E. Yablonovitch and is characterized by a diamond-like lattice symmetry (Yablonovich et al. 1991). The Yablonovite structure was created in the near-infrared range by drilling the material with a concentrated ion beam (Chelnokov et al. 2000). Another technology (Deubel et al. 2004; Serbin et al. 2004) uses two-photon stereolithography to generate random 3D structures in the photoresist volume with a precision of 200 nm by sequential (point-by-point) recording with focused femtosecond laser pulses. The similar method was used to produce a flat binary diffractive optics element in Osipov et al. (2010).

A polymerization reaction occurs when light is absorbed by the photoresist at a specific wavelength. Two-photon absorption permits the 3D process to be localized in the beam's waist region rather of dispersing throughout the axis, resulting in 3D structures. Interference lithography (also known as holographic lithography (Campbell et al. 2000)), which exposes the photoresist to the 3D of the interference sample, is a promising industrial technology today (Campbell et al. 2000). As a consequence, an almost flawless periodic structure can be realized. Interference lithography enables to obtain photonic crystal patterns with thickness up to ~50 lattice periods with attenuation less than 5% by film thickness (Campbell et al. 2000). Other advantages of this technology include high fabrication speed (the full lattice volume is exposed at the same time), low cost (no precision positioning device is needed), and the possibility to fabricate rather big specimens. Low resolution, which is restricted by the wavelength used for photoresist exposure, and a lack of options in terms of lattice type are two of the drawbacks. Because of the low refractive index of polymeric PhC lattices, full band gaps are impossible to achieve. A number of authors have recommended that the polymeric lattice be coated with a metal layer to avoid this problem (Lin et al. 2006; Tal et al. 2007; Mizeikis et al. 2007; Kaneko et al. 2008; Walsh et al. 2009). Metallic PhCs and metallodielectric PhCs are two materials with photonic band gaps that have attracted a lot of attention (Sigalas et al. 1995; Fleming et al. 2002; Sedghi et al. 2010; Asadi et al. 2011; Hatef and Singh 2011). Because the dielectric function discontinuities at the metallodielectric interface are much larger in metallodielectric PhCs than in dielectric PhCs, considerable photonic band gap reduction can be accomplished in smaller structures with less lattice periods. A copper layer was placed over a polymeric photonic crystal array created by two-photon polymerization in ref. Tal et al. (2007). The polymeric array produced by two-photon polymerization was electrodeposited with a 300-nm-thick Ni layer in ref. Mizeikis et al. (2007). The authors of ref. Mizeikis et al. (2007) examined the reflection coefficients of such photonic crystals and reported an empirically discovered band gap at a wavelength roughly equivalent to the PhC array period.

In Dyachenko et al. (2011a), a triple exposure of the photoresist film using a two-wave interference route is employed to produce the 3D polymer lattice PhC in the photoresist via interference lithography, as was done previously in Miklyaev et al. (2009), Dyachenko et al. (2011b). The sample is rotated 120 degrees around the axis after each exposure session. The exposure times for all three images must be the same. In Dyachenko et al. (2011a), the angle between interfering waves is identical in each of the three exposure sessions, implying that the response lattice's basic vectors are the same length. In addition, the angles between the primary vectors are the same. As a result, an orthorhombic lattice was used (Dyachenko et al. 2011a). In an SU-8 photoresist, the polymer matrix was recorded using a He-Cd laser at a wavelength of 442 nm (Dyachenko et al. 2011a). According to the manufacturer's instructions, the photoresist was put onto the substrate via centrifugation coating. The aforesaid technique resulted in the production of a 40–50-μm-thick SU-8-50 layer, according to the photoresist specification. To remove the solvent, two phases of drying were performed: 5 min at 60 °C, followed by 20 min at 95 °C. After exposure, the photoresist was baked for 5 minutes at 95 °C, then developed for 5–8 min using PGMEA (propylene glycol methyl ether acetate), and then washed with isopropyl alcohol. The exposure time was consistent, ranging between 10 and 20 min. Due to the polymer's low refractive index, PhC polymer matrices cannot have a complete band gap. By placing a high refractive index material on the polymer surface, the problem of the polymer's insufficient refractive index is solved. Gold was chosen as the deposition medium by the authors (Dyachenko et al. 2011a) because of its low absorption in the visible and infrared range. Using magnetron deposition (SPI-Module Sputter Coater, USA), a 50-nm-thick gold layer was deposited on PhC (Dyachenko et al. 2011a). The thickness of the layer was measured using a quartz thickness sensor. Scanning electron microscopic images of the samples produced are shown in Fig. 7.1.

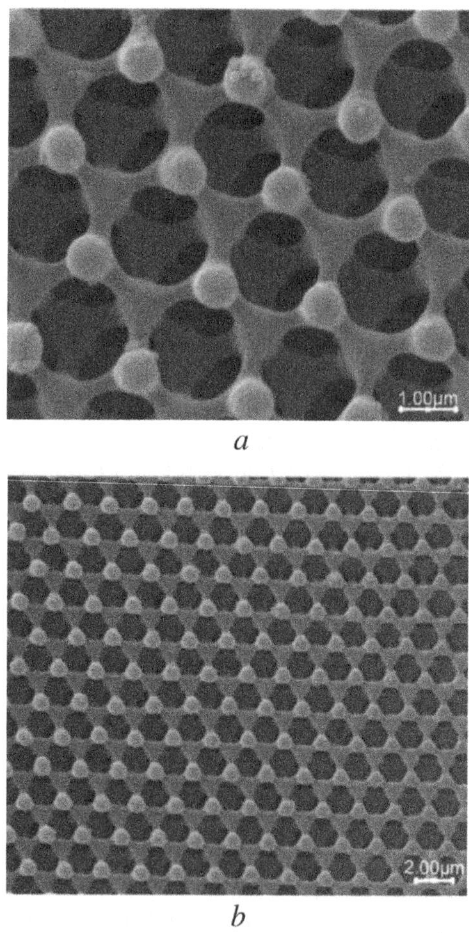

Fig. 7.1 The scanning electron microscopy images ((**a**) and (**b**)) of the PhC with a 50 nm gold film sputtered on surface. Images have been used with the permission of reference Dyachenko et al. (2011a)

The optical properties of PhC were studied on an FT-IR Hyperion 1000 microscope using the FT-IR Tensor 27 spectrometer (Bruker Optics, Germany) and used to measure the reflection coefficients of the structures (Dyachenko et al. 2011a). The PhC reflection coefficients were obtained by normalizing the measured spectra of the structure to the golden mirror spectrum (Dyachenko et al. 2011a). The obtained reflectance spectra are shown in Fig. 7.2. Curve 1 shows the reflection spectrum of PhC obtained using the target angle of 15×. This objective provides integrated reflection spectra for a beam converging in the range of 52°–84° to normal. Curve 2 depicts the PhC reflection spectrum obtained by the Schwarzschild 15 × infrared (IR) lens (Dyachenko et al. 2011a). The objective allows you to obtain integrated reflection spectra for a beam converging in the range from 15° to 30° to normal. Curve 3 shows the reflectance spectrum of the gold film on the photopolymer SU-8. Curve 4 describes the reflectivity spectrum of PhC without the gold layer. Curves 1 and 2 in Fig. 7.2 show that the PhC reflection spectrum has a peak at a wavelength of 2.6–2.8 μm, which corresponds to the period of the PhC array. The said peak of the reflection spectrum indicates the appearance of a photon gap centered at a wavelength of 2.6–2.8 μm. In addition, it is possible to see the difference in peak height in curves 1 and 2, which is due to different light incident angles and nonuniform deposition of the layer in depth (Dyachenko et al. 2011a).

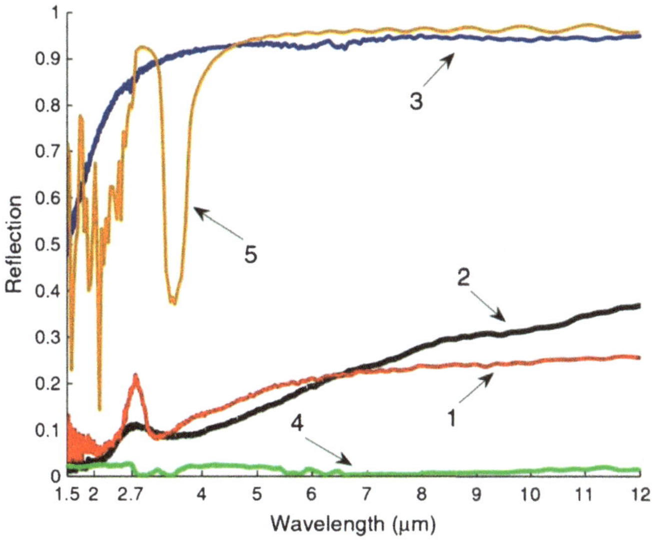

Fig. 7.2 The PhC reflectance spectra with a period of 2.6 μm. 1, with a 15 grazing angle objective; 2, with a 15× IR Schwarzschild objective; 3, the reflectance spectrum of the gold film on SU-8 photopolymer; 4, the reflectance spectrum of the PhC without a gold layer; and 5, the reflectance spectrum of the PhC calculated using the FDTD method. Image has been used with the permission of reference Dyachenko et al. (2011a)

The position of the band gap was observed to be approximately the PhC period. Numerical simulation of PhC optical reflectivity was performed by finite-difference time-domain (FDTD) method (Taflove and Hagness 2000) using freely accessible software package (Oskooi et al. 2010). Reflectivity was determined by simulating the propagation of the plane wave incident on the PhC along the z-axis direction and comparing the intensities of the reflected and incident waves. The FDTD calculation area covers one lattice period in the *x*-*y* plane with periodic boundary conditions. Perfectly consistent layer boundary conditions were applied along the z-axis direction. In the FDTD simulation, the grid dimensions were selected to be 30 nm. The PhC parameters of the model were assumed to be as close to the real one as possible. Authors (Dyachenko et al. 2011a) used realistic parameters (Gansel et al. 2009) for the free electron Drude model explaining the optical properties of gold: the frequency of plasma radiation is assumed to include four lattice periods along the z-axis direction. Curve 5 in Fig. 7.2 shows the PhC reflection spectrum obtained by the FDTD method. As can be seen from Fig. 7.2, the calculated spectra reproduce the essential features of the experimental ones. It is worth to point out a good alignment between spectral positions of reflectance peaks for computed and experimental data. The alteration in peak intensity can be described by deviations between the idealized model, actual samples, and measurement conditions. Nevertheless, the measured reflectivity spectrum is slightly lower than the calculated one. Substantial loss of reflectivity can also happen due to Rayleigh scattering by random inhomogeneities. The loss of reflectivity may be due to "shaded" areas appearing after deposition of the metal layer. These "shaded" regions lead not only to energy dissipation but also to some differences in the reflection spectrum, especially at a wavelength greater than 4 μm (see Fig. 7.2). In addition, the Schwarzschild IR objective used in the IR microscope detects reflection in the conical range of incidence angles from 15° to 30° around the normal to the sample surface. Therefore, the measured reflectance spectrum is the result of angular and spectral averaging, which usually results in a loss of spectral features. Thus, the authors in Dyachenko et al. (2011a) made 3D metal PhC using a mixture of interference lithography and magnetron deposition of a gold nanolayer. The study of the structural and optical properties of the samples carried out during this study makes it possible to draw satisfactory conclusions about the applicability of this technique. Thus, the method provides a flexible new path to new metal dielectric PhC and integrated devices having an optical function in the IR region of the electromagnetic spectrum. Another example of using non-lithographic methods (or direct writing methods) to implement PhC is given in Tukmakov et al. (2013), Tukmakov et al. (2012). Figure 7.3 shows the result of fabrication a photon crystal resonator in a 300-nm-thick diamond film membrane using FIB technology (Chelnokov et al. 2000). It should be noted that the use of focused ion beam technology made it possible to produce a device that seems problematic to produce using lithographic technologies (Chaplygin 2005) (Fig. 7.3). The results of spectral analysis of the obtained diamond photocrystalline resonator are presented in Fig. 7.4.

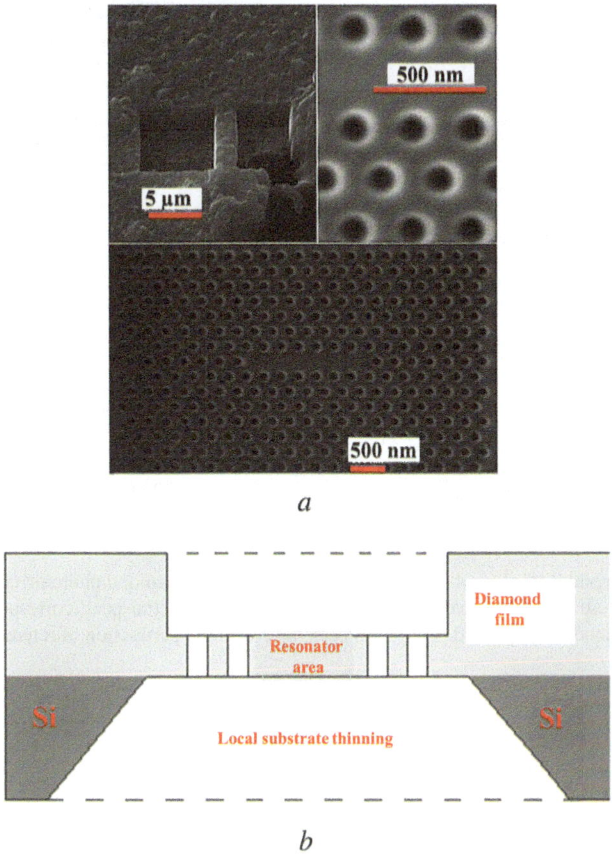

Fig. 7.3 The results of manufacturing a photonic crystal resonator in a diamond film using focused ion beam technology (Tukmakov et al. 2013; Tukmakov et al. 2012): (**a**) images of parts of the manufactured device taken with an electronic microscope, (**b**) device design

Another example of the use of non-lithographic technologies for the manufacture of optical nanostructures is given in Miklyaev et al. (2011). The results of design, fabrication, and investigation of beam splitters with continuous profile made using micromilling technology developed by LIMO company (Miklyaev et al. 2010) are given in the work (Miklyaev et al. 2011). The use of this technology avoided the use of expensive reactive-ion etching (RIE) technology and realized a diffraction microrelief with a continuous profile (Fig. 7.5).With this, the deviation of the realized microrelief from the calculated one was not more than 50 nm. The realization of a microrelief with a continuous profile made it possible to obtain beam splitters with high energy efficiency (Miklyaev et al. 2011). Fig. 7.6 shows the result of investigation of the realized microrelief using a profilometer. Fig. 7.6 shows the result of the measurement of the intensity generated by the 1:3 beam splitter.

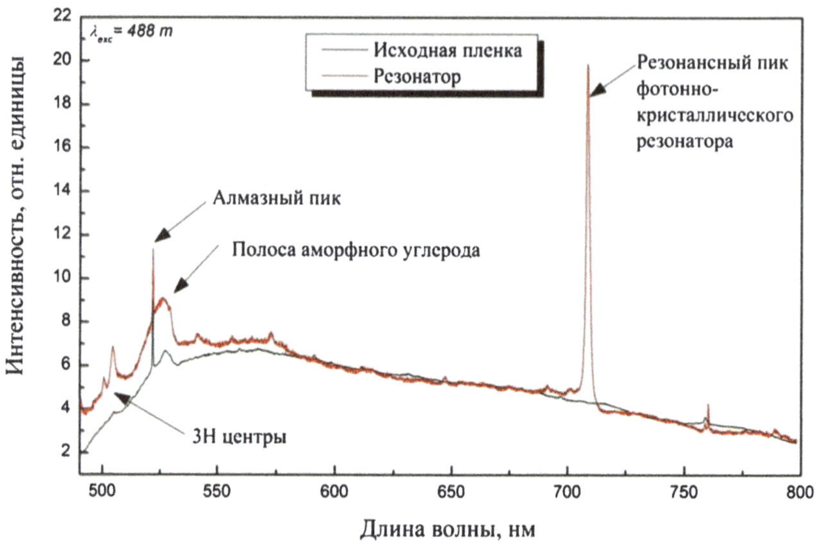

Fig. 7.4 The results of spectral investigation of manufactured diamond photonic crystal resonator (Tukmakov et al. 2013; Tukmakov et al. 2012). The high spectral peak corresponding to PhC resonator frequency is presented. Image has been used with the permission of reference Tukmakov et al. (2012)

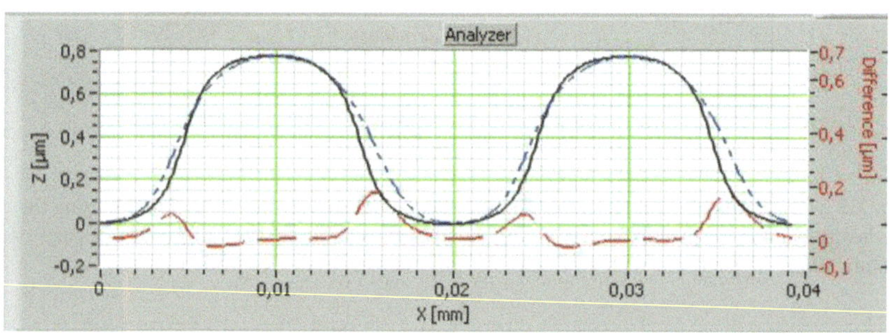

Fig. 7.5 Result of investigation of realized microrelief using profilometer. Image has been used with the permission of reference Miklyaev et al. (2010)

7.2 Conclusion

In conclusion, it is worth noting that the advent of technologies for nanostructuring optical materials and the synthesis of optical nanomaterials, even in the absence of nonlinear effects arising from the interaction of light wave and optical media, allows us to create optical devices with unique capabilities for information collection, transmission, and processing systems.

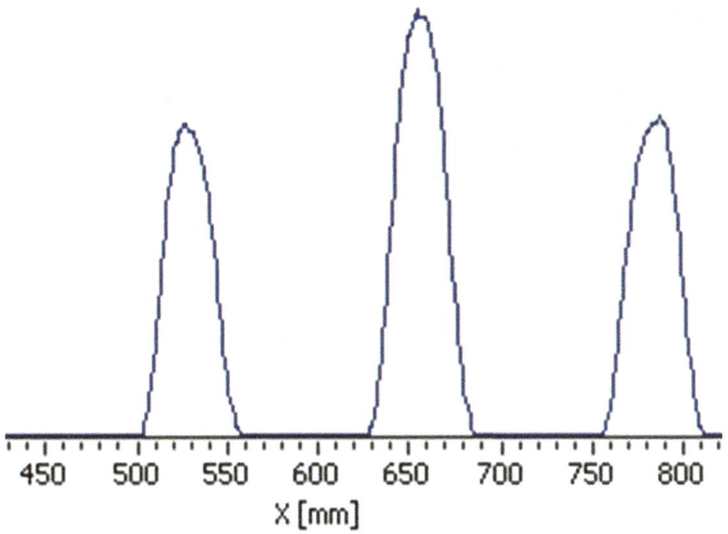

Fig. 7.6 Result of measurement of intensity generated by 1:3 beam splitter. Image has been used with the permission of reference Miklyaev et al. (2010)

In particular, the emergence of possibilities for calculating, optimizing, and manufacturing photon crystal structures of the optical range allows creating efficient photon devices—optical waveguides, resonators, sensors, and elements of integrated photon circuits.

For the technological implementation of such structures, both planar lithography technologies used in microelectronics and direct writing and interference writing technologies are used.

Conventional planar lithography techniques are useful in the manufacture of hybrid optoelectronic or photonic integrated circuits either diffractive optical elements with subwavelength feature size.

Direct writing technologies (focused beam technology, laser ablation, etc.) and interference writing can be used to create photonic nanostructures with a complex three-dimensional topology, for example, in the manufacture of unique or small-scale elements of nanophotonics. In particular, these methods can be used to fabricate elements of classical optics with nanoscale feature sizes (microlenses, microlens arrays, diffraction gratings for UV or visible light).

Another application of non-lithographic methods for creating nanostructures is the synthesis of optical nanocomposites and metamaterials with given optical properties.

Methods such as scanning electron microscopy, atomic force microscopy, interferometry, spectrometry, ellipsometry, etc. are used to study the topology of optical nanostructures and properties of optical materials.

In the next section of the chapter, optical nanostructures with nonlinear features are considered.

References

Asadi R, Malek-Mohammad M, Khorasani S (2011) Opt Commun 284:2230

Blanco A, Chomski E, Grabtchak S, Ibisate M, John S, Leonard SW, Lopez C, Meseguer F, Miguez H, Mondla JP, Ozin GA, Toader O, Van Driel HM (2000) Nature 405:437

Bogomolov VN, Gaponenko SV, Kapitonov AM, Prokofiev AV, Ponyavina AN, Silvanovich NI, Samoilovich SM (1996) Appl Phys A Mater Sci Process 63:613

Caglayan H, Bulu I, Ozbay E (2005) Opt Express 13:7645

Campbell M, Sharp DN, Harrison MT, Denning RG, Turberfield AJ (2000) Nature 404:53

Chaplygin YA (2005) Nanotechnology in electronics. Technosphera, Moscow

Chelnokov A, Wang K, Rowson S, Garoche P, Lourtioz JM (2000) Appl Phys Lett 77:2943

Chow E, Grot A, Mirkarimi LW, Sigalas M, Girolami G (2004) Opt Lett 29:1093

Deubel M, Von Freymann G, Wegener M, Pereira S, Bush K, Soukoulis CM (2004) Nat Mater 3: 444

Dyachenko PN, Karpeev SV, Fesik EV, Miklyaev YV, Pavelyev VS, Malchikov GD (2011b) Opt Commun 284:885

Dyachenko PN, Karpeev SV, Pavelyev VS (2011a) Fabrication and characterization of three-dimensional metallodielectric photonic crystals for infrared spectral region. Opt Commun 284:5381–5383

Fleming JG, Lin S-Y, El-Kady I, Biswas R, Ho KM (2002) Nature 417:52

Gansel JK, Thiel M, Rill MS, Decker M, Bade K, Saile V, von Freymann G, Linden S, Wegener M (2009) Science 325:1513

Hatef A, Singh M (2011) Opt Commun 284:2363

Ho KM, Chen CT, Soukoulis CM, Biswas R, Sigalas M (1994) Solid State Commun 89:413

John S (1987) Phys Rev Lett 58:2486

Kaneko K, Yamamoto K, Kawata S, Xia H, Song J-F, Sun H-B (2008) Opt Lett 33:1999

Kang Y, Walish JJ, Gorishnyy T, Thomas EL (2007) Nat Mater 6:957

Knight JC (2003) Nature 424:6950

Lin SY, Ye D-X, Lu T-M, Bur J, Kim YS, Ho KM (2006) J Appl Phys 99:083104

Miklyaev Y, Imgrunt W, Pavelyev V.S, Kachalov D, Bizjak T, Aschke L, Lissotschenko V (2010) SPIE Advanced Lithography. 764024

Miklyaev YV, Karpeev SV, Dyachenko PN, Pavelyev VS (2009) J Mod Opt 56:1133

Miklyaev YV, Imgrunt W, Pavelyev VS, Soifer VA, Kachalov DG, Eropolov VA, Aschke L, Bolshakov MV, Lissotschenko VN (2011) Synthesis and investigation of diffractive beam splitters with continuous profile. Comput Opt 35(1):42–46

Mizeikis V, Juodkazis S, Tarozaite R, Juodkazyte J, Juodkazis K, Misawa H (2007) Opt Express 15:8454

Osipov V, Pavelyev V, Kachalov D, Zukauskas A, Chichkov B (2010) Realization of binary radial diffractive optical elements by two-photon polymerization technique. Opt Express 18(25): 25808–25814

Oskooi AF, Roundy D, Ibanescu M, Bermel P, Joannopoulos JD, Johnson SG (2010) Comp Phys Commun 181:687

Rinnie SA, Santamaria FG, Braun PV (2008) Nat Photonics 8:52

Sedghi AA, Kalafi M, SoltaniVala A, Rezaei B (2010) Opt Commun 283:2356

Serbin J, Ovsianikov A, Chichkov B (2004) Opt Express 12:5221

Sigalas MM, Chan CT, Ho KM, Soukoulis CM (1995) Phys Rev B 52:11744

Soifer VA (2014) Diffractive nanophotonics. CRC Press, Taylor & Francis Group, CISP, Boca Raton

Taflove A, Hagness SC (2000) Computational electrodynamics: the finite-difference time-domain method. Artech, Norwood MA

Tal A, Chen Y-S, Williams HE, Rumpf RC, Kuebler SM (2007) Opt Express 15:18283

Tukmakov KN, Volodkin BO, Pavelyev VS, Komlenok MS, Khomich AA (2012) Photonic crystal resonator on a diamond film. Bull Samara State Aerosp Univ 7(38):112–116. (in Russian)
Tukmakov KN, Volodkin BO, Pavelyev VS, Komlenok MS, Khomich AA, Ralchenko VG (2013) Fabrication of diamond film-based photonics with focused ion beam. Techn Digest Int Conf Las Appl Technol (LAT), Moscow LAT-04:41–42
Walsh TA, Bur JA, Kim Y-S, Lu T-M, Yu Lin S (2009) J Opt Soc Am B 26:1450
Wijnhoven JEGJ, Vos WL (1998) Science 281:802
Yablonovich E, Gmitter TJ, Leung KM (1991) Phys Rev Lett 67:2295
Yablonovitch Y (1987) Phys Rev Lett 58:2059

Optical Characterization of Nanomaterials-II

8

Vladimir Pavelyev and Nishant Tripathi

Abstract

In the present era of optoelectronics devices, the low-cost, ultra-thin and transparent materials with tunable electronics and optoelectronics properties are in huge demand. The emerging family of nanomaterials, i.e. transition metal dichalcogenides (TMDs), has the potential and properties to fulfil the demand of modern optoelectronics devices. The molybdenum disulphide (MoS_2), an inorganic compound, is a new emerging material of TMD family which shows exceptional electronics as well as optoelectronics properties. In monolayer form, MoS_2 acted as a direct bandgap semiconductor with bandgap energy 1.8 eV. Worldwide researchers developed the MoS_2 with different types of nanostructures such as nanosheets, nanoflowers as well as quantum dots, and all types of nanostructures proven the different types of electronic and optoelectronics properties. This chapter deals with the selective synthesis of MoS_2 nanostructures along with the analysis of their basic construction and respective electronics and optoelectronics properties. We discussed in detail various synthesis technologies for growing different types of MoS_2 nanostructure with their advantages, disadvantages as well as outcomes. The detailed analysis on pristine MoS_2 nanostructures as well as hybrid structure based linear and polarized-light photodetectors has been made. We have discussed different photodetectors (linear and polarized-light) which have been realized using MoS_2 nanostructures. The results show that the MoS_2 nanostructures act as an active material for ultra-broadband detection from the UV-Vis-NIR range of the solar spectrum. Different device architectures and composites were introduced for enhancement in the photodetection properties.

V. Pavelyev (✉) · N. Tripathi
Samara National Research University, Samara, Russia

A. Thakur et al. (eds.), *Synthesis and Applications of Nanoparticles*,
https://doi.org/10.1007/978-981-16-6819-7_8

Keywords

Metal dichalcogenides · Photodetectors · Graphene · Hydrothermal method · Quantum dots

8.1 Introduction

The invention of carbon nanotubes (CNTs), a remarkable achievement of Sumio Iijima, excites worldwide researchers to do work on synthesis of nanomaterials and nanostructures (Tripathi et al. 2018; Tripathi and Islam 2017). Later, the invention of graphene brought a revolutionary impact on the development of optoelectronic nanomaterials and devices (Bera and Banerjee 2019a). The zero bandgap of the graphene disappointed the materials scientists. Sooner, the research interests of worldwide researchers diverted from graphene to the other nanostructures and nanomaterials such as nanocrystalline silicon, TiO_2, reduced graphene oxide and transition metal dichalcogenides (TMDs), etc. (Sharma and Gupta 2016; Sharma et al. 2017; Singh et al. 2019; Kumar et al. 2020; Bera and Banerjee 2019b; Gan et al. 2020). Among all nanostructures and nanomaterials, TMDs show better electronics and optoelectronics properties. The TMD materials were analysed since 1970, but the invention of graphene gave fresh attention towards TMD-based 2D materials (Shi et al. 2019; Wang et al. 2019). TMDs are scientifically interesting and industrially important compounds. Chalcogenides like Se, S and Te (VI A group) are able to make quite firm binary compounds with transition elements. These form layered crystalline structures particularly with transition metals of groups IV–VII B. The layers of TMDs attached with each other by weak van der Waals forces (Bhakhar et al. 2019). Due to weak attachment between the layers, it is very easy to separate layers from their bulk counterparts (Hwang et al. 2019a). Surprisingly, when the monolayer or few layers of TMDs are exfoliated from its bulk counterpart, then the attractive forces between layers either are reduced or eliminated which manipulate the band structure of TMDs; the mentioned property of TMDs is very suitable for manipulation of its electronics as well as optoelectronics properties (Shi et al. 2019). The bandgap energy of TMDs is fallen in the range of 0.2–2 eV, which makes it suitable materials for electronic and optoelectronics devices. The bandgap energy of TMDs highly depends on the elemental combination along with the number of layers, and the existence or lack of a doping atom (Gustavsson et al. 2013). These materials display well-defined structures in contrast with doped graphene, which loses its crystalline structure when covalently doped to tune its bandgap, efficiently creating weakly defined amorphous graphenoid materials (Pumera 2013; Naz et al. 2019). The bandgap tuning is crucial for the making of electronics and optoelectronics devices based on these materials.

In the case of TMD family, two-dimensional (2D) nanostructure of molybdenum disulphide (MoS_2) shows the direct bandgap as well as high light absorption capacity in its single-layer form. The MoS_2 nanostructures are semiconducting materials that show indirect bandgap even in two-layer form, while the monolayer of MoS_2 shows

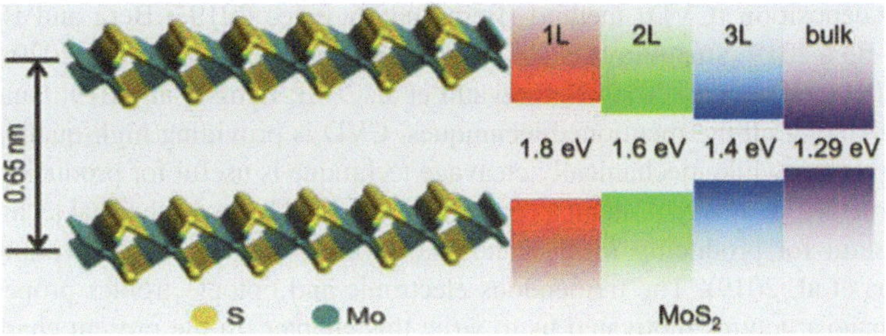

Fig. 8.1 The figure shows the basic structure of MoS_2 as well as its bandgap energy with respect to the number of layers. Images have been used with the permission of references Huang et al. (2020); Zhang et al. (2019c)

the direct bandgap (Bera and Banerjee 2019c). In the nanostructures of MoS_2, the different type of atoms with a strong spin-orbital coupling is situated on the nearest lattice locations; such type of atom arrangements makes MoS_2 different from graphene and also provides it special properties (Shi et al. 2019). In the basic structure of MoS_2, the single molybdenum atom is sandwiched between two sulphide ions (see Fig. 8.1) (Bera and Banerjee 2019a; Meng et al. 2019). The bandgap energy of bulk MoS_2 is estimated around 1.23 eV, while in monolayer form the bandgap energy of 2D-MoS_2 has been found around 1.8 eV (Rahmati et al. 2019). The individual layer thickness of 2D-MoS_2 is in the range of 0.6 Å, and it consists of the covalent bond-attached atoms (see Fig. 8.1) (Bera 2019; Smagulova et al. 2019). Typically, MoS_2 crystals have two types of structures, one is trigonal prismatic and another is rhombohedral (3R) (Deng et al. 2020). Generally, the individual layers of 2D-MoS_2 show the good chemical, mechanical as well as thermal stability (Deng et al. 2020; Subitha et al. 2019; Sangeetha and Madhan 2020; Wei et al. 2019). The above-discussed properties of 2D-MoS_2, and its high in-plane carrier mobility (200–500 $cm^2\ V^{-1}\ s^{-1}$) as well as high conductivity in selective structures, motivate the scholar for development of MoS_2_based highly efficient electronics and optoelectronics devices such as photodetectors, LEDs, spintronic, future electronic switches, transparent electrodes, wearable devices, FETs, saturable absorber and biosensors, electrocatalytic hydrogen evolution (HER), electromagnetic shielding, gas sensors, water purification, lubricant, humidity sensor, energy storage devices, etc. (Bera and Banerjee 2019a; Bera and Banerjee 2019b; Gan et al. 2020; Wang et al. 2019; Naz et al. 2019; Meng et al. 2019; Bera 2019; Subitha et al. 2019; Sangeetha and Madhan 2020; Wei et al. 2019; Tyagi et al. 2019; Zhang et al. 2019a; Hwang et al. 2019b; Ma et al. 2019; Cao et al. 2019; Jia et al. 2020; Huang et al. 2020; Zhang et al. 2019b; Zhou et al. 2019; Chaudhary et al. 2019; Guoa et al. 2020). To synthesize the TMDs, the techniques used for the fabrication of graphene can be implemented. The TMDs can be synthesized by top-down as well as bottom-up techniques (Nethravathi et al. 2013). Specifically for the synthesis of 2D-MoS_2, worldwide researchers utilize hydrothermal process, chemically exfoliation technique, wet chemical reaction, mechanically cleavage process as well as chemical vapour deposition (CVD) method (Bera and Banerjee 2019a; Bera and Banerjee 2019c; Bera 2019; Smagulova et al. 2019; Subitha et al. 2019; Jia et al. 2020; Huang et al. 2020; Zhou et al. 2019; Nethravathi et al. 2013; Leng et al. 2019; Duan et al. 2020). Among all the mentioned techniques, CVD is providing high-quality MoS_2 nanostructure, while mechanically cleavage technique is useful for producing MoS_2 nanostructure on a large scale in a controlled manner. The hydrothermal technique is very useful for producing MoS_2 nanoflakes with a high surface to volume ratio (Subitha et al. 2019). The tremendous electronic and optoelectronics properties of MoS_2 nanostructures motivated us to write this chapter. In the present chapter, we are going to cover the various synthesis processes for the fabrication of different types of MoS_2 nanostructures. The detailed analysis for various properties of different type of MoS_2 nanostructures is also the part of this chapter. For application point of view, a detailed discussion is going to present on MoS_2-based photodetectors.

8.2 Synthesis Techniques for MoS_2 Nanostructures

MoS_2 nanostructures are synthesized in many variants such as nanodiscs, nanoflakes and nanofilms as well as in nanoflower forms. All type of forms of MoS_2 nanostructures shows different properties. So, it is very important to understand all types of MoS_2 nanostructure synthesis methods.

8.2.1 CVD Technique

CVD technique is considered as one of the oldest techniques for depositing high-quality as well as selective forms of 1D and 2D nanostructures in a large scale. In the case of MoS_2, it is utilized for growing monolayer, few layer structures as well as heterostructure of MoS_2 (Li and Zhu 2015). In the growing process of MoS_2 nanostructures, mainly, molybdenum compounds and molybdenum film have been utilized as CVD precursors.

Fazio et al. successfully synthesized monolayer of 2D-MoS_2 on sapphire substrate by utilizing the CVD method. For the same purpose, the annealed and cleaned sapphire substrate was placed inside the CVD chamber on a MoO_3-filled crucible, and another sulphur-filled crucible was placed upstream from the first crucible. The CVD reactor was heated from room temperature to 300 °C in the presence of 200sccm flow of Ar, and after 10 min halted at 300 °C; the temperature of CVD reactor further increased to 700 °C in the presence of 10sccm Ar. The CVD reactor temperature has been maintained at 700 °C for 10 min; then, the furnace was allowed to cool down inflow of 200sccm of Ar gas. The monolayer form of 2D-MoS_2 has been verified by Raman spectroscopy (De Fazio et al. 2016).

Smagulova et al. deposited the monolayer to few layer triangular-shaped MoS_2 nanostructures. To grow the mentioned structures, the powder form mixture of MoO_3 and sulphur was used in the ratio of 1:13 as a precursor. The resultant mixture-filled crucible was placed in CVD chamber, and a cleaned SiO_2 substrate was placed on the crucible. The CVD chamber was heated from room temperature to 300 °C with maintaining the chamber pressure 10^{-5} bar. At 300 °C, the Ar gas supply was turned on with a 100sccm flow rate. After 1 h, the temperature of CVD was increased to 700 °C; the MoS_2 nanostructures had been grown in 30 min with 10sccm Ar flow. The CVD system allowed cooling down naturally after completion of MoS_2 nanostructure synthesis process. The MoS_2 nanostructure-deposited SiO_2 sample was unloaded from the CVD system for its morphological analysis. In the next set of experiment, instead on crucible, the SiO_2 sample was placed far from crucible; the remaining CVD process remains unchanged. It has been observed from microscopic analysis that the largest nanostructures were grown when the SiO_2 substrate placed directly on the crucible and lateral size of grown nanostructure reduced significantly with increased the distance between SiO_2 substrate with the crucible. It has been observed from microscopic analysis that the largest nanostructures were grown when the SiO_2 substrate placed directly on the crucible and lateral size of grown nanostructure reduced significantly with increased the distance between SiO_2 substrate with the crucible. The lateral dimensions of as-grown sample have been found around 250 μm, when the substrate was placed on the crucible. When the SiO_2 substrate is placed 10 cm far from a crucible, the lateral dimensions of MoS_2 decreased to 5–7 μm (Smagulova et al. 2019). It is a well-known fact that 1D nanotubes show different and better properties than their 2D and 3D forms. For analysing the electronic properties of 1D nanostructures of MoS_2, Jia et al. developed the nanotubes of MoS_2. The fabrication process is shown in Fig. 8.2(a). As shown in the figure, first, the SiO_2 nanowires have been prepared by electro-spinning technique. As-prepared SiO_2 nanowires were dispersed on the silicon substrate after annealing. These SiO_2 nanowires worked as templates for MoS_2 nanostructures. The resultant silicon substrate was loaded in CVD chamber for growing the MoS_2 nanostructures.

The standard CVD process has been used for growing MoS_2 nanostructure. The MoS_2 nanostructure formation took place on SiO_2 nanostructures. The resultant sample was dissolved in hydrofluoric solution for removing SiO_2 nanowires. After SiO_2 removal process, the MoS_2 shell remains in the form of nanotubes. It has been observed by HR-TEM and Raman spectroscopy that as-prepared MoS_2 nanotubes are highly crystalline with clear lattice fringes (see Fig. 8.2b–d). As-produced MoS_2 nanotubes have an open end and made with 20 walls. The diameter of MoS_2 nanotubes has been observed around 540 nm, while the length of as-prepared nanotubes is estimated around a few micrometres (Jia et al. 2020).

Sun et al. deposited the thin film of MoS_2 by CVD system with a different approach for photodetection application. For the same purpose, first, a film of a mixture of MoO_3 and ammonia was prepared on SiO_2/Si substrate. As-prepared sample was placed in a boat and on top of the boat another substrate was placed. In another boat, far from the previous boat, sulphur powder was placed. For growing MoS_2 thin-film standard CVD process was run. The MoS_2 thin film was grown by taking 1000sccm argon flow rate and 820 °C CVD temperature. The growth duration was set for 5 min (Suna et al. 2019).

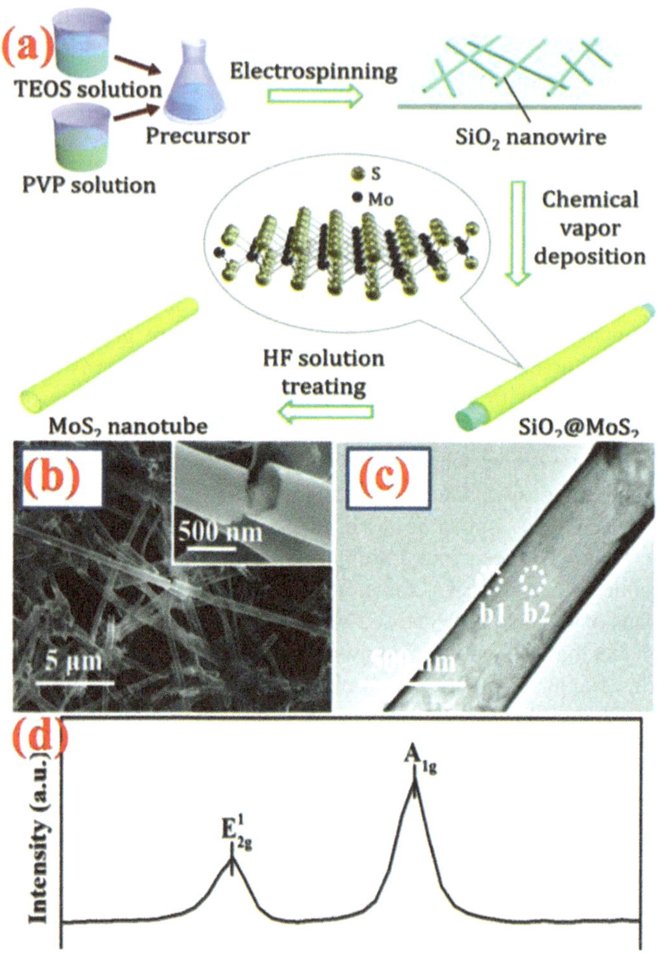

Fig. 8.2 Figure (**a**) shows the schematic diagram for synthesis process MoS_2 nanotubes. Figure (**b**) shows the HR-TEM image for prepared MoS_2 nanotubes, and figure (**c**) shows the flat edges of individual nanotube. Figure (**d**) represents the Raman spectra for prepared MoS_2 nanotubes. Images have been used with the permission of reference (Jia et al. 2020)

8.2.2 Liquid Phase Exfoliation Method

Liquid phase exfoliation technique is very useful to make nanoflakes and nanosheets of MoS_2. This technique has majorly three steps, first, bulk material is needed to disperse in ionic liquid, second for exfoliation of the bulk material in a few layers or monolayer form, ultrasonication is required, and in the final step, centrifugation followed by a filtration process needs to be done. Due to exceptional acoustic cavitation, high-power ultrasonication gives a facile synthetic approach for various nanomaterials.

Subitha et al. fabricated the hexagonal structures of MoS_2 nanosheets by liquid phase exfoliation technique. To fabricate MoS_2 nanostructure, bulk MoS_2 was ultrasonicated by a probe ultrasonicator in the ionic liquid mixture (ratio = 1:1) of dimethylformamide (DMF) and benzyl benzoate (BB) for 1 h. The resultant solution was centrifuged and filtered followed by overnight dried at 60 °C. The resultant sample was analysed by XRD as well as by Raman spectroscopy, and it is observed that as-grown 2D-MoS_2 nanostructures were well crystalline and consist of only a few layers. The lateral dimension of as-grown 2D-MoS_2 is observed as 242.5 nm (Subitha et al. 2019). For synthesizing big size as well as transparent thin nanosheets of MoS_2, Wang et al. reported the modified liquid phase exfoliation method. In this method, MoS_2 powder was ultrasonicated by a horn tip ultrasonicator in N-methyl-2-pyrrolidone solvent for 60 min. To maintain the constant temperature, an ice bath was utilized. The lateral dimension of as-fabricated MoS_2 is found to be around 3600 nm, which is much more than the previous case (Wang et al. 2019).

8.2.3 Hydrothermal Method

Hydrothermal is a conventional bottom-up technique to fabricate the quantum dots (QDs), nanodiscs as well as nanosheets of MoS_2. During synthesis of MoS_2 nanostructures by a hydrothermal technique, commonly molybdenum compounds such as sodium molybdate, ammonium orthomolybdate and ammonium tetrathiomolybdate are utilized as molybdenum source, while most common precursors used for sulphur are thiourea, dibenzyldisulphides and cysteine. During the hydrothermal reaction, first the sulphur is generated by decomposition of sulphur precursors. In the next step, the nucleation sites are produced and then epitaxial growth of MoS_2 nanostructures has been taken place.

Zhang et al. synthesized the 2D-MoS_2 nanosheets by utilizing the hydrothermal method for electromagnetic shielding application. For the same purpose, MoO_3, thioacetamide and urea have been mixed in deionized water and loaded in autoclave. The mixture-filled autoclave was heated at 300 °C for a different duration. In the final step, the prepared solution has been washed with deionized water as well as with ethanol. It has been observed that the sample prepared with 6-h reaction time is found in agglomerated form. It indicates that the reaction was not completed in 6-h duration. When the same experiment repeated for 10- as well as 12-hour reaction time, then the hexagonal structure of 2D-MoS_2 nanosheets has been observed. They also observed that the 2D-MoS_2 nanosheets grown by taking long reaction time (12 hours) show the large surface area, more transparent as well as thinner nanosheets as compares to 6-h reaction time. Interestingly, it has been observed that the number of layers in 2D-MoS_2 nanosheets is reduced with long reaction time. The sample prepared with 12-h reaction time is highly pure and consists of the lowest number of layers, i.e. 2–3 layers in one nanosheet (Zhang et al. 2019a).

Sangeetha et al. deposited the spherical shape nanosheets of hexagonal type MoS_2. As-prepared MoS_2 nanostructures were aggregated with each other, and the dimension of individual nanoparticles falls in the range of 30–40 nm. Figure 8.3

Fig. 8.3 Figure (**a**) and (**b**) shows the FE-SEM as well as HR-TEM image for synthesized spherical shape nanosheets of MoS_2, respectively. Images have been used with the permission of reference (Sangeetha and Madhan 2020)

shows the SEM and TEM images of prepared MoS_2 nanostructures. It is clearly visualized in TEM image that prepared nanostructures have spherical disc-type structure. To fabricate such type of structures, they utilized thiourea as sulphur precursor with sodium molybdate. In this case, first the mixture of both materials in deionized water with 7.5 pH value has been prepared, and then the standard hydrothermal reaction for making MoS_2 nanostructure has been processed for 12 h. Here, the reaction temperature maintained lower than previous cases; the hydrothermal reaction was performed at 200 °C instead of 300 °C (Sangeetha and Madhan 2020).

By utilizing the different fabrication process, Naz et al. fabricated the flower-shaped 1 T-phase of MoS_2 nanostructures on reduced graphene oxide (rGO). As-prepared MoS_2 nanostructures have high purity, stability as well as active sites, which make MoS_2 nanostructure suitable for supercapacitor application. For depositing mentioned nanostructures of MoS_2 as well as for reducing the toxicity

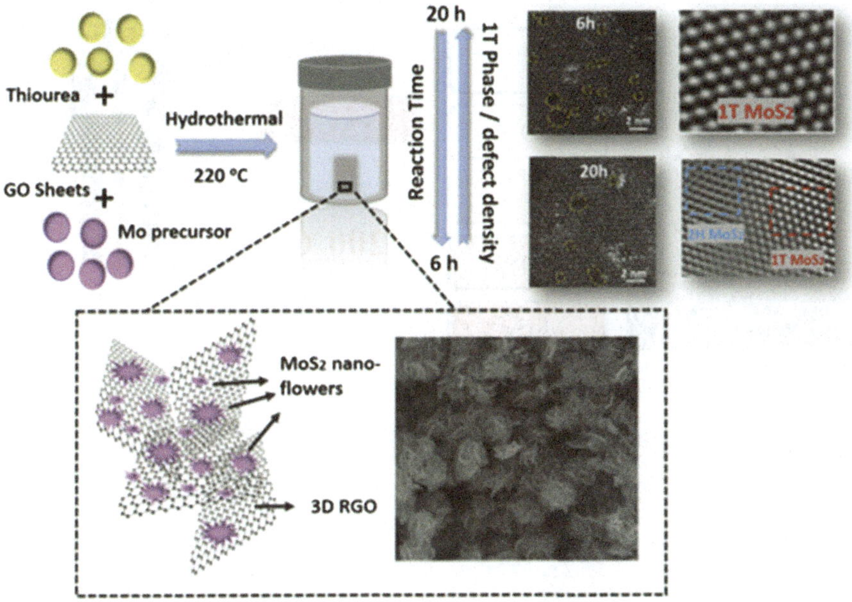

Fig. 8.4 Figure shows the fabrication steps for growing 1 T-MoS_2 nanoflowers on rGO. The figure also included the SEM and HR-TEM images for prepared nanostructures for different reaction time. Images have been used with the permission of reference (Naz et al. 2019)

of hydrothermal process, they used a one-pot hydrothermal process. To prepare the mentioned MoS_2 nanostructures on rGO, they mixed the rGO, ammonium molybdatetetrahydrate and thiourea and were dissolved in deionized water by the stirring process followed by ultrasonication. The prepared mixture was heated at 220 °C for 6 h in the autoclave. In the final step, the resultant solution was washed with deionized water as well as with absolute ethanol. The same experiment was repeated for a different heating time in the range of 7–20 h. Also, the experiment has been performed without taking rGO. Figure 8.4 shows the MoS_2 synthesis process as well SEM and HR-TEM images to verify the prepared 1 T phase of pure MoS_2 and flowerlike nanostructure with rGO. It has been observed that reaction time directly made an impact on thickness as well as size of MoS_2 sheets. The sizes of nanosheets for 6- and 20-h reaction time have been found to be around 400 and 680 nm, respectively. In the case of thickness, they observed contradictory results as reported by Zhang et al. Here the thickness is increased from 4.8 to 15 nm, when the reaction time increased from 6 to 12 h. These findings also are confirmed by Raman spectroscopic analyses. It has been also observed that defect density increased with decreasing the reaction time. The highest defect density is found in the sample, which is produced by keeping 6-h reaction time. The 1 T phase of MoS_2 is confirmed by XRD analysis. Another important finding has been reported that the sample produced in lesser time shows the highest specific surface area (Naz et al. 2019).

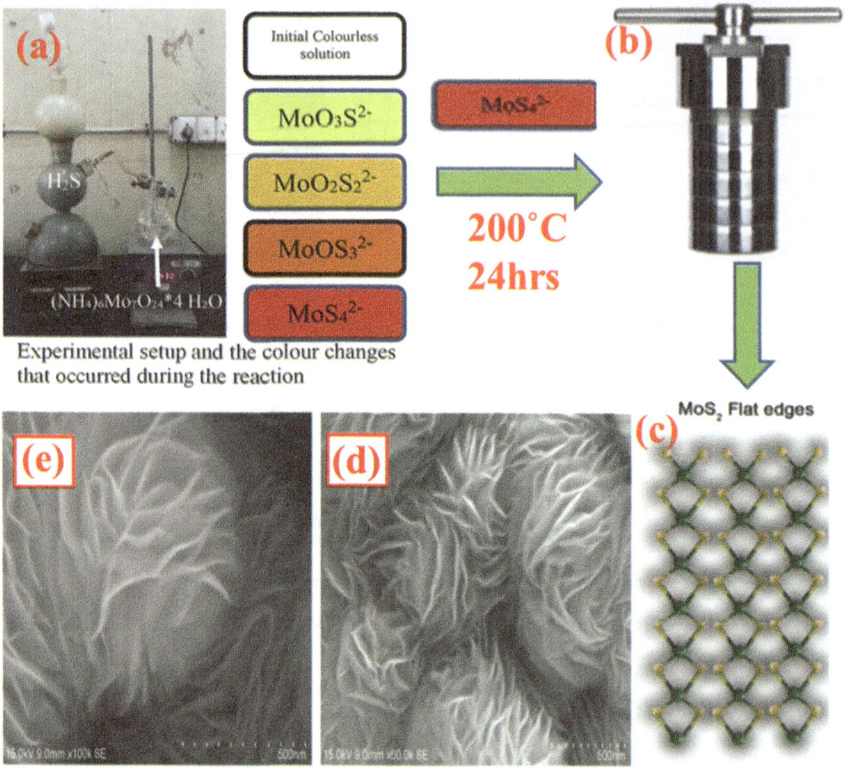

Fig. 8.5 Figure (**a**) and (**b**) shows the fabrication steps of flower-shaped MoS_2 nanostructures. Figure (**c**) shows the schematic diagram for prepared flat edge MoS_2 nanostructures. Figure (**d, e**) shows the surface morphology of prepared MoS_2 nanostructures. Images have been used with the permission of reference Mathialagan et al. (2020)

To enhance the energy storage properties of lithium-ion batteries, Mathialagan et al. developed the 3D nanoflower-type MoS_2 nanostructures with flat edge sites. For developing the mentioned structure, they utilized the hydrothermal technique with some modifications. Figure 8.5 shows the MoS_2 nanoflower fabrication steps with resultant material FE-SEM images. For synthesizing the mentioned nanostructures, first they produced the H_2S gas by utilizing ferrous sulphide and hydrochloric acid from Kipp's apparatus, and the produced H_2S gas had been supplied from a colourless solution of ammonium heptamolybdatetetrahydrate with deionized water. The red-colour resultant solution was heated into an autoclave at 200 °C for 24 h. The final product was washed with deionized water as well as with ethanol. The prepared 3D MoS_2 nanoflowers were analysed by FE-SEM and HR-TEM, and it has been observed that the nanoflowers are made with vertically aligned few layers of nanosheets. The flat edges of individual nanosheets have been verified by HR-TEM analysis (Meng et al. 2019).

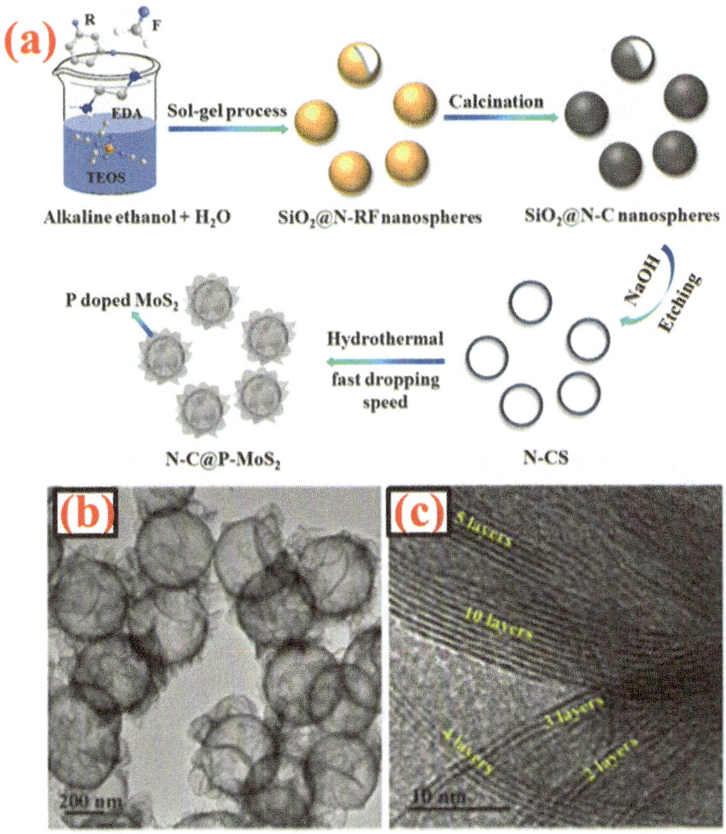

Fig. 8.6 Figure (**a**) shows the schematic diagram for synthesis steps for development N-C@P-MoS_2. Figure (**b**, **c**) shows the HR-TEM images on different magnification for as-prepared N-C@P-MoS_2 nanostructures. Images have been used with the permission of reference (Wei et al. 2019)

For utilization of MoS_2 nanostructures in hydrogen evolution reaction, Wei et al. fabricated the highly conductive as well as with high active site MoS_2 hybrid nanostructures. By utilizing calcination process and followed by hydrothermal process, the hierarchical P-doped MoS_2 nanopetals (P-MoS_2) decorated N-doped hollow carbon spheres (N-C@P-MoS_2) core-shell structures have been developed. It has been observed that N and P doping manipulates the electronic structure of MoS_2 and hence enhanced electronic conductivity. One can see the synthesis process in Fig. 8.6. The N-C was fabricated by the modified Stober method on SiO_2. The prepared sample was dissolved into NaOH for removing SiO_2 templates. During the making of the hollow carbon spheres, ethylenediamine has been absent in processing steps; the rest of the process remains the same. In the final step, three types of MoS_2 hybrid nanostructures have been developed, i.e. MoS_2 on carbon sphere (C@MoS_2), MoS_2 on N-doped carbon sphere (N-C@MoS_2) and P-doped MoS_2 on N-doped carbon sphere (N-C@P-MoS_2). For synthesizing the C@MoS_2-type nanostructure, the mixture of sodium molybdate, TAA and hollow carbon spheres was heated in an autoclave at 220 °C for 24 h. To get C@MoS_2, as-prepared solution was washed with deionized water as well as with ethanol followed by drying in a vacuum at 60 °C. For making N-C@P-MoS_2, in the previous procedure, ammonium dihydrogen phosphate was added, and the hollow carbon spheres were replaced by N-C; the remaining synthesis process followed the same synthesis of C@MoS_2. The N-C@MoS_2 has been synthesized by employing the same process which is used for making C@MoS_2; the only the difference is that here N-C has been used instead of hollow carbon spheres. Figure 8.5b, c shows the NC@P-MoS_2 structure at different magnifications. The size of as-prepared N-C has been found as 360 nm with shell thickness around 10 nm. The porous structure of N-Cs provides the nucleation sites for the fabrication of MoS_2 nanoplates. The size of as-prepared MoS_2 nanoplates on N-C has been estimated around 200 nm, and these MoS_2 nanostructures strongly attached with N-Cs, which provides the stability to the structure. The basic structure of MoS_2 nanoplates remains the same even after doping. The large surface area of prepared MoS_2 nanostructures (109.9 m^2 g^1) and its mesoporous structure enhanced the active sites and electron as well as photon transportation (Wei et al. 2019).

Shi et al. have fabricated the MoS_2 QDs for a tumour imaging application. To fabricate QDs with different optical properties, the sodium molybdate is utilized as a molybdenum source with a number of different sulphur precursors such as cysteine, ammonium sulphide, thiourea, thioacetamide, sodium thiosulfate hydrate, mercaptoacetic acid, mercaptopropionic acid, glutathione, etc. For developing MoS_2 QDs, in a certain ratio, all sulphur precursors with sodium molybdate are added in water, and well mixed by ultrasonication process. The resultant solution was heated in an autoclave for 36 h at 200 °C. In the final step, the resultant product was filtered and cleaned by water. It has been observed that MoS_2 QD fabrication highly depends on the used sulphur source; the sulphur source can change the structure of QDs and hence the fluorescence property. The best intensity of fluorescence has been detected when glutathione is used as sulphur source. It is also reported that the fluorescence intensity of grown MoS_2 QDs also depends on the ratio of molybdenum source to sulphur precursor. The best results have been observed for the ratio of glutathione to sodium molybdate kept around 20:1 (Shi et al. 2019).

Leng et al. developed the small 2D-MoS_2 flakes by adding ethylene glycol precursor in hydrothermal reaction. It has been observed that the ethylene glycol shows its capability in manipulating the size of MoS_2 nanoflakes and it also creates oxygen doping sites on as-grown MoS_2 nanoflakes. To make a more suitable material for the humidity sensors, the oxygen doping in 2D-MoS_2 has been further manipulated by the reaction temperature. They also reported that the MoS_2 nanostructures were grown in a temperature range from 180 to 200 °C show the perfect 2D structure. The MoS_2 nanostructures grown at 220 °C were found as a non-2D bulky shape. The traces of oxygen were verified by EDS as well as by Raman spectroscopic analyses; the MoS_2 nanostructures grown at 180 °C are found to be rich in oxygen (Nethravathi et al. 2013).

Among all techniques, the hydrothermal technique is more suitable to make composite materials of MoS_2 with other materials. For making composite materials, one needs to add the other materials during the hydrothermal reaction. By utilizing the mentioned facility of hydrothermal technique, researchers made the MoS_2-SWNTs, MoS_2-amorphous CNT, hybrid materials, etc. (Bera and Banerjee 2019b; Gan et al. 2020; Hwang et al. 2019a; Bera and Banerjee 2019c; Smagulova et al. 2019; Ma et al. 2019; Cao et al. 2019; Nethravathi et al. 2013).

8.2.4 Reactive Magnetron Sputtering Technique

All techniques mentioned above for fabrication of MoS_2 nanostructure used or its by-products are toxic and corrosive elements. Apart from the toxic process, the fabrication of MoS_2 nanostructures with a large area by utilizing the above-discussed methods is very hard. These problems can be solved by the sputtering technique. The sputtering technique is very useful for depositing uniform as well as a highly pure thin film of MoS_2 nanostructures. By utilizing magnetron sputtering equipment, Tyagi et al. deposited the thin film of MoS_2 on SiO_2 substrate (2.5 cm 1.5 cm) for the purpose of thin-film transistor. The MoS_2 thin film has been deposited in the duration of 40 S with 5 mTorr chamber pressure by taking MoS_2 as a sputtering target. Fig. 8.7 shows the FE-SEM as well as AFM image of as-grown film. It is confirmed

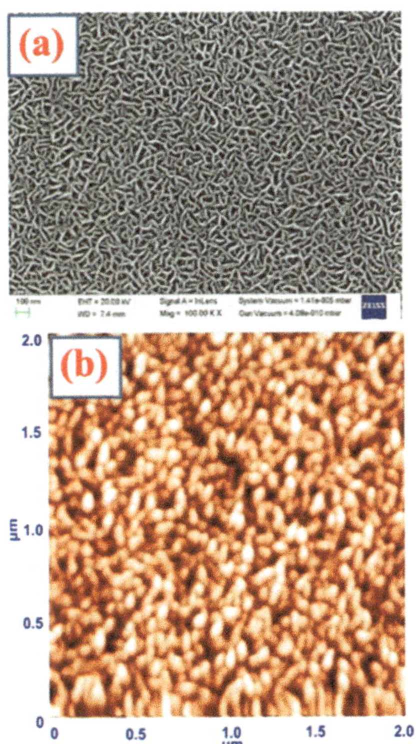

Fig. 8.7 Figure (**a**) and (**b**) shows FE-SEM and AFM images for thin-film MoS_2 deposited by sputtering technique. Images have been used with the permission of reference (Tyagi et al. 2019)

by XRD and Raman spectroscopic analysis that synthesized film is a multilayer form of a hexagonal phase of MoS_2. The MoS_2 layers have been deposited uniformly with a thickness of 30 nm in a vertically aligned manner. The thin-film transistor based on as-grown MoS_2 film shows excellent mobility, i.e. 24.17 cm^2 V^1 s^1 as well as a very high *ON/OFF* ratio (~10^6) (Tyagi et al. 2019).

8.3 MoS_2 Nanostructure-Based Photodetectors

The photodetectors are very useful devices for light detection, imaging science, spectroscopies as well as optical communication. Although researchers tried many materials for developing photodetectors, TMDs are established as the most suitable materials for photodetection applications. As per photodetection applications, TMDs show extraordinary properties such as optical transparency, high breakdown voltage, excitonic effects, tunable optoelectronics properties, mechanical flexibility, etc., which made them ideal materials for photodetection applications. As discussed in the introduction section, the layered nanostructures of MoS_2 show the direct bandgap, high electron mobility as well as high mobility among all TMD materials. For photodetection applications, the bandgap of material is the most important required property. MoS_2 nanostructures have potential to tune the bandgap in an easy way. In the present section, we are going to discuss the development of photodetectors based on MoS_2 nanostructures. Based on the above suitable properties of MoS_2 nanostructure, Zhou et al. developed the broadband (λ = 375–808 nm) photodetector with very high detectivity (10^{10} Jones). For developing photodetector, the Au/Ti electrodes were fabricated on CVD-deposited MoS_2 nanoflakes on sapphire substrates. It has been observed that the MoS_2 nanoflakes developed the Schottky contacts with electrodes, which also played a crucial role in photoreaction applications. As-developed Schottky barrier reduced the recombination rate of photogenerated charge carriers, and hence enhanced the separation of charge carriers. In the equilibrium state of a photodetector, the electronic bands of electrodes (Au/Ti) and photoactive material (MoS_2 nanoflakes) at interface bent to make equalize the Fermi energy level (see Fig. 8.8a). When the developed photodetector comes under the illumination of suitable light, the photocarrier generation phenomenon takes place in the valance band of active materials. When the photogenerated electrons get sufficient energy, then they jump from valance band to conduction band. The bias voltage helps in segregation of photogenerated carriers. And, hence the photocurrent starts to flow in circuit. The photodetection properties have been analysed for the different power densities of light sources. The photocurrent increased almost linearly with increment in power density (see Fig. 8.8b). The increment in photocurrent with power density has followed the power law equation ($I_{ph}\sim P^{\alpha}$). It has been found, by the curve fitting method, that the values of α were found around 0.865, which is less than its ideal valve ($\sim$1). The deviation in the value of α shows the partial recombination in photogenerated charge carriers. The responsivity of the prepared device decreased

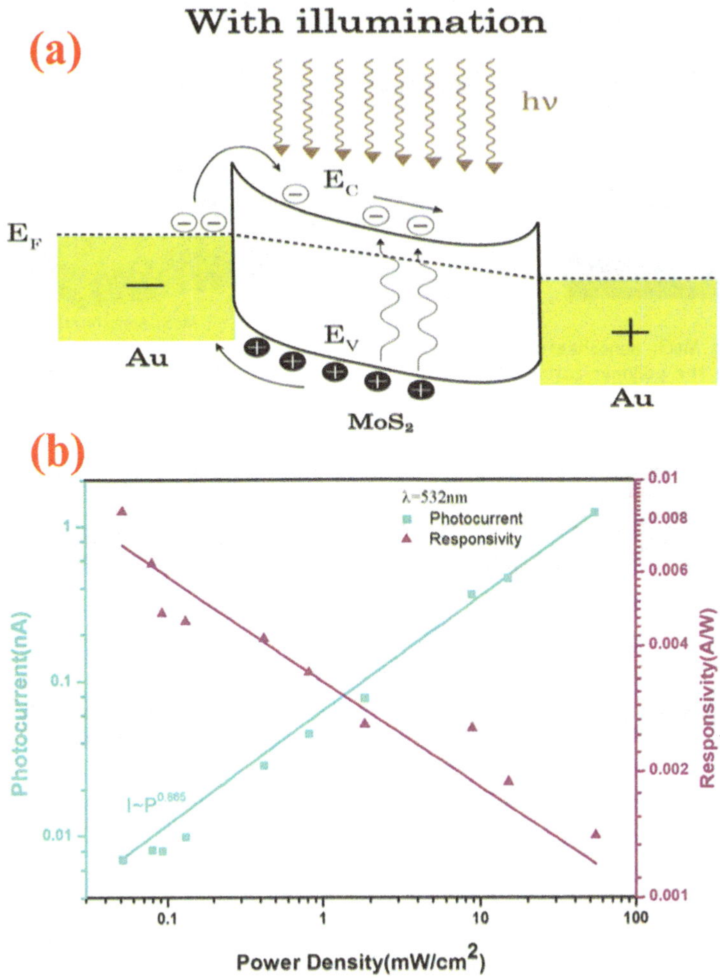

Fig. 8.8 Figure (**a**) shows the light sensing mechanism for MoS_2-based photodetector. Figure (**b**) shows the plots for photocurrent and responsivity with respect to power density of light. Images have been used with the permission of reference (Zhou et al. 2019)

with increment in power density. The proposed reason for this phenomenon is the trap centres, available in the device structure. The values of responsivity, external quantum efficiency and detectivity of prepared photodetector for 532 nm wavelength light source have been reported as 0.0084 A/W, 2% and 1.74×10^{10} Jones, respectively. The mentioned values calculated for the light source power density are around 51 μW/cm^2. The prepared photodetector gave satisfactory results for the wavelength range from 375 to 808 nm (Zhou et al. 2019).

To solve the problem of charge carrier trapping as well as for enhanced the responsivity of MoS_2-based photodetectors (~25A/W), Chaudhary et al. developed

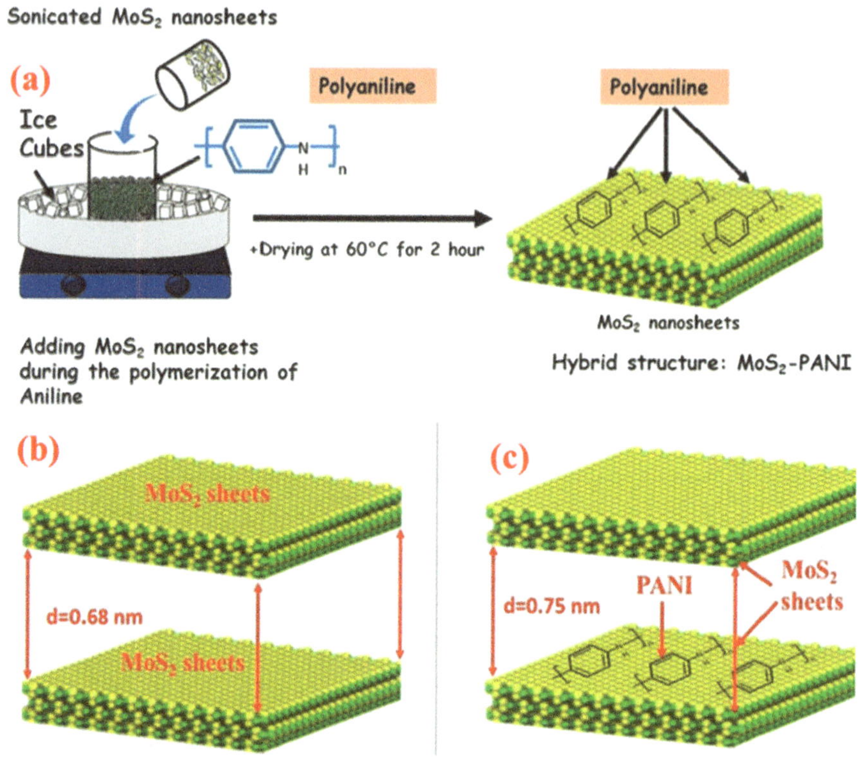

Fig. 8.9 Figure (**a**) shows the 3D-MoS_2-PANI hybrid material synthesis steps. Figure (**b**) and (**c**) represents the schematic diagram of 2D-MoS_2 nanosheets before and after composition of PANI. Images have been used with the permission of reference (Chaudhary et al. 2019)

the 3D MoS_2-PANI hybrid material for photodetection application. For the same purpose, 2D-MoS_2 nanosheets have been fabricated by hydrothermal technique, and 3D hybrid structure has been constructed with mixing of 1D polymer polyaniline (PANI) with 2D-MoS_2 nanosheets. The photodetection properties of prepared sensor were analysed for different wavelengths ($\lambda = 635$ nm, 785 nm and 1064 nm) of the light source as a function of power density. Figure 8.9a shows the 3D-MoS_2-PANI hybrid material synthesis steps. Figure 8.9b, c represents the schematic diagram of 2D-MoS_2 nanosheets before and after the composition of PANI. It has been observed by the HR-TEM analysis that the spacing between the adjacent layers increased after treatment with PANI. They observed an important finding of hybrid material by absorbance spectra that the hybrid material shows the different peaks as compared to pristine MoS_2. The pristine MoS_2 shows two absorbance peaks, one is at 438 nm and the second is on 638 nm, while the PANI shows the absorbance peaks at 430 and 765 nm. But the peaks in absorbance spectra of MoS_2-PANI hybrid material have been found at 455 and 779 nm. The proposed reason behind the different peak location in hybrid material is the transition phenomenon that occurred

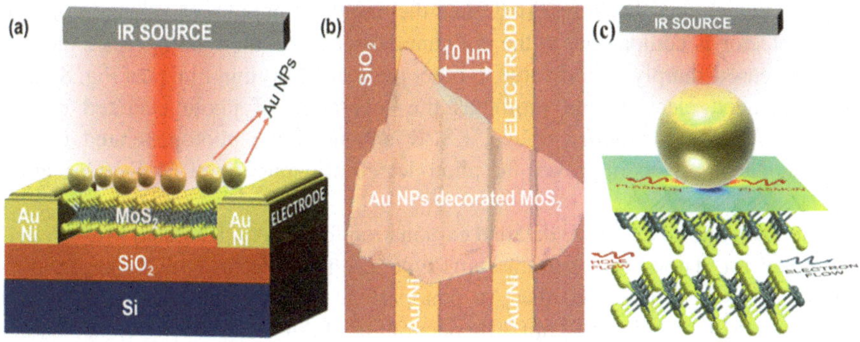

Fig. 8.10 Figure (**a**) and (**b**) shows the schematic diagram and optical image of prepared photodetector. Figure (**c**) shows the light sensing mechanism with plasmonic effects. Images have been used with the permission of reference (Guo et al. 2019)

in the polarization zone of PANI. As discussed above, the photogenerated charge carriers might be trapped in trap centres/defect states, and this trapping results in a reduction of responsivity. But in a hybrid structure, the photogenerated charge carriers can be de-trapped and can return to their respecting bands without recombination. The basic photodetection mechanism is the same as explained by Zhou et al. The advantage of using PANI is that the Fermi energy level MoS_2-PANI system is closer to valance band of PANI, so during the light illumination, the PANI supplied the extra electrons to valance band of MoS_2, which enhance the responsivity of the prepared photodetector. It has been observed that prepared 3D-MoS_2-PANI photodetector gave the best photodetection parameters for 785 nm light source; the proposed reason behind this is that the peak available at 779 nm in absorbance spectra of MoS_2-PANI hybrid material is well matched with 785 nm (Chaudhary et al. 2019).

Guo et al. proposed a method to enhance the photodetection property of MoS_2 nanostructures especially for near-infrared region (NIR). In this case, the photoresponsivity was enhanced by decorating the MoS_2 nanostructures with plasmonic nanoparticles. The plasmonic nanoparticles increased the interaction between incident photons and MoS_2 nanostructures; the enhanced light-matter interaction results in enhanced responsivity of the photodetector. For developing the mentioned photodetector, first, the thin layer of MoS_2 was fabricated by the sputtering system on SiO_2/Si substrate; then, the prepared layered was decorated by Au nanoparticles. For making electrical connections, the Au/Ni electrodes were fabricated on Au-decorated MoS_2 film. Figure 8.10a, b shows the schematic diagram and optical image of the prepared photodetector. The purpose of decoration with Au nanoparticle is an excitation of strong localized surface plasmon resonance (LSPR) on the surface of MoS_2 nanostructures. It has been observed that the decoration of Au nanoparticles enhanced the light-matter interaction for the NIR region around three times. By this technique, they enhance the photocurrent as well as the responsibility of photodetector for the 980 nm light source up to 480 nA and 64 mA/W, respectively. It has been reported that the incident photon wavelength on the photodetector is comparable to the Au nanoparticles and then the interface electric field increased sharply as compared to other parts of MoS_2 thin film (see Fig. 8.10c). The enhanced electric field promotes the separation of photogenerated charge carriers. And as mentioned above, LSPR enhanced the photogenerated charge carriers. These effects improved the photodetection parameter of MoS_2 (Guo et al. 2019).

Rahmati et al. gave a more detailed explanation for the Au-decorated MoS_2-based photodetector. They developed the same type of photodetector with further modification. It has been observed that Au nanoparticles represent the surface plasmon resonance (SPR) band at 532 nm. The decoration of Au nanoparticles enhanced the Raman count in Raman spectra of MoS_2, which shows the surface-enhanced Raman scattering (SERS) effect. It happens due to an increment in the light-induced electric field of Au nanoparticle's surface. When the incident light makes the resonance with the SPR band of Au nanoparticles, the hot electrons could generate on the Au nanoparticles, this process known as LSPR. The LSPR is responsible for the enhancement in SERS as well as light-absorbing capacity of MoS_2. As-generated hot electrons can jump into the conduction band of MoS_2 by crossing the interface barrier of Au and MoS_2, which create n-type doping in MoS_2. And this action of the hot electron can be responsible for enhancement of photocurrent. They utilized vertically aligned nanosheets (V-MoS_2) instead of the thin film of MoS_2. The V-MoS_2 provides the large surface area, catalytic active sites, extensively exposed edges as well as prominent photoluminescence (PL) response; all mentioned properties of V-MoS_2 provide the enhanced optical absorption as well as fast interlayer transportation. The mentioned properties of V-MoS_2 make it a more suitable material for photodetection applications. Another advantage of V-MoS_2nanosheets is that the Au nanoparticles may be trapped between the vertical aligned sheets of MoS_2, which can further enhance the light-absorbing capacity of MoS_2. Due to this phenomenon, the population of photogenerated carriers can be increased sharply and hence the photocurrent (Rahmati et al. 2019).

To enhance the further detectivity and gain of MoS_2-based photodetector, Zhang et al. proposed a similar kind of approach with different decoration materials. They decorate the thin film of MoS_2 by core-shell zinc cadmium selenide/zinc sulphide (ZnCdSe/ZnS) colloidal quantum dots (QDs). The proposed photodetector shows the responsivity (10^4A/W) around three times larger than the pristine MoS_2-based photodetector. With hybrid photoactive material, the value of detectivity (10^{12} Jones) as well as gain (10^5) is also found much larger than the pristine TMD-based photodetectors. The role of QDs in the present work is almost the same as the Au nanoparticle reported by Rahmati et al. The QD decoration enhanced the light absorption capacity of MoS_2, which enhances the generation of light-induced charge carriers and hence the photocurrent. The extra advantage of the decoration of QDs is that it can harvest the broad range (UV-NIR) of the solar spectrum. The light absorption capacity of QDs is highly dependent on their size; so, the bandgap energy of QDs can be easily tuned by the size of QDs. And, one can make the MoS_2 matched selective QDs for enhancing the energy transfer. Also, the

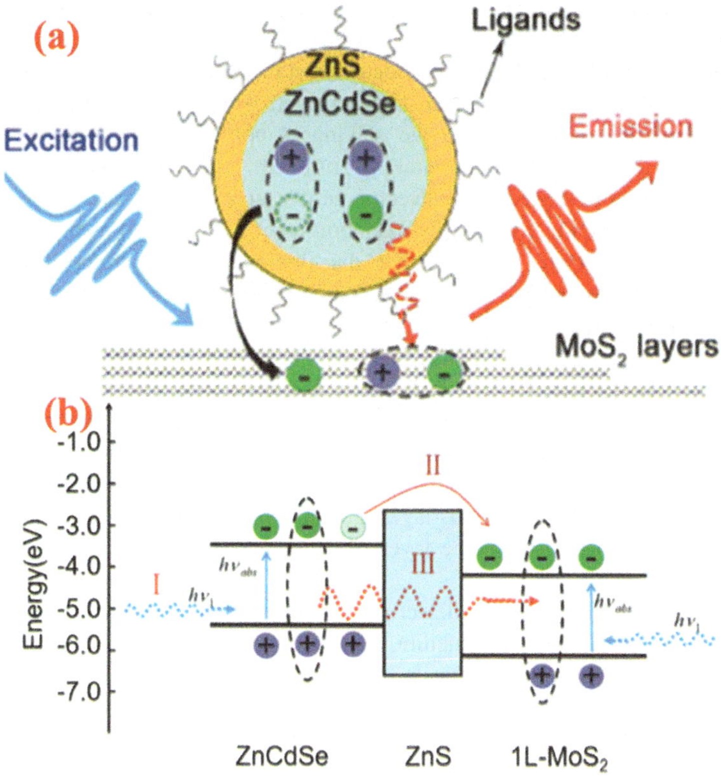

Fig. 8.11 Figure (**a**) shows the schematic diagram of the developed hybrid photodetector. Figure (**b**) shows the band structure of MoS_2-QDs hybrid system after alignment and electron-hole generation phenomenon under light illumination. Images have been used with the permission of reference Zhang et al. (2019c)

combination of MoS_2 and QDs provides the higher mobility of charge carriers, which is good for photodetection applications. In the present work, they used ZnCdSe QDs (diameter = 5.5 nm) as core and ZnS (thickness = 1.5 nm) as shell. Figure 8.11a shows the schematic diagram of the developed hybrid photodetector. It is clearly observable in the schematic that the QD's role is mainly as an energy harvester and MoS_2 works as the transport channel for charge carriers. The photocurrent in the developed light detector can be enhanced by carrier transport (black arrow of Fig. 8.11a) or by nonradiative energy transfer (red dashed arrow) or by both. As shown in Fig. 8.10b, the shell can act as a barrier; in core-shell structure, the injection can be possible by charge carrier tunnelling. So, in the carrier transport phenomenon, the photocurrent generation takes place by tunnelling of photogenerated electrons from QDs to MoS_2. In nonradiative energy transfer phenomenon, the excitons are generated on QDs and then excitons' energy transferred to the thin film of MoS_2; after absorbing energy the electron-hole pair generation phenomenon takes place on MoS_2. Fig. 8.11b shows the band structure of MoS_2-QD hybrid system after alignment and electron-hole generation phenomenon under light illumination (Zhang et al. 2019c).

In the case of V-MoS_2-based photodetector, it has been observed by Zhang et al. that the nanosheets of V-MoS_2 can stack together, which can change the optoelectronics properties of MoS_2. So, the staking is not suitable for photodetection applications. To minimize the staking problem, they suggested the MoS_2/MoO_2heterostructure-based photodetector. In the mentioned device, MoS_2 worked as an n-type semiconductor, and its oxidation layer MoO_3 acted as a p-type semiconductor. It has been reported that the photodetection parameters such as responsivity, detectivity and power conversion efficiency increased significantly as compared to pristine MoS_2-based photodetector. The values of responsivity, detectivity and power conversion efficiency have been observed as 670 mA/W, 4.77×10^{10} Jones and 3.5% at 0 V bias voltage, respectively. For preparing the proposed heterostructure, first, the mechanically exfoliated MoS_2 flakes were deposited on SiO_2/Si substrate. For making an electrical measurement, two electrodes (Cr/Au) were fabricated on the prepared MoS_2 sample, and in the next step, the entire sample has been coated by MoO_3 layer. For the purpose of the third electrode, the mechanically exfoliated graphene layer was transferred on the MoO_3 film (Zhang et al. 2019d).

For developing a very high detectivity photodetector, Liu et al. developed the lateral graphene-MoS_2 heterostructure. For making the mentioned device, first, the CVD grown graphene film was transferred on a SiO_2/Si substrate. In the next step, the graphene sheet was etched from the middle portion of the sample, and the empty area, which is created by graphene film etching, was filled by deposition of MoS_2 film using the CVD method (see Fig. 8.12). Two electrodes have been fabricated on graphene sheets available on the edges of the sample. It has been reported that two Schottky barriers developed on both graphene/MoS_2 interface are the main prime part of the prepared device. The light-induced electron-hole pair generation phenomenon took part at MoS_2. The main role of developed Schottky junctions was to immediately separate the photogenerated charge carriers and rapidly transfer the photogenerated electrons into graphene. Due to the fast and efficiently separation of electron-hole pairs by Schottky junction, the prepared device shows the extraordinary ON/OFF ratio ($\tilde{}10^6$) as well as ultra-high photoresponsivity (2×10^3 mA/W) and photodetectivity (10^{13} Jones). The photodetection parameters of the prepared device can be tuned by adjusting the Fermi energy level of graphene and hence the height of Schottky barriers by adjusting the gate voltage. It has been observed that as-prepared photodetector is highly sensitive with incident photon density; the photocurrent increased with photon density as well as for gate voltage (Liu et al. 2019).

A light wave always contained lots of information such as polarization, amplitude, frequency, as well as phase. Till now, in this section, the discussion made only amplitude-based light detection. The polarization is a very crucial parameter of light. When the light is interacted with any object, the reflected/transmitted light from the object can store the information about the object. So, the development of polarization-sensitive photodetector is a very important task. The polarization-

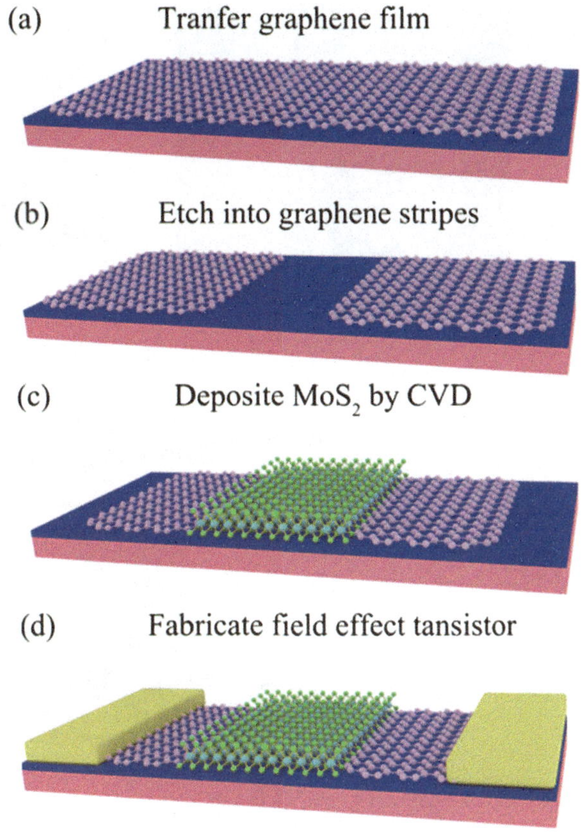

Fig. 8.12 The fabrication steps for the development of graphene-MoS_2 heterostructure-based photodetector. Figure (**a**) shows the schematic diagram for transfer of graphene on substrate while figure (**b**) shows the patterning of graphene on substrate. Figure (**c**) shows the fabrication of MoS_2 on as-prepared sample and figure (**d**) shows the device fabrication steps. Images have been used with the permission of reference Liu et al. (2019)

sensitive photodetectors show the integral applications in the fields of environmental sensing, defence equipment and remote sensing imaging as well as in medical detection. It has been observed that pristine MoS_2 nanostructures are insensitive towards polarization sensitivity. Recently, Deng et al. proposed the microtubular 3D-MoS2 nanostructure-based FET for polarization-sensitive photodetector. One can see the proposed structure in Fig. 8.13(a). It has been reported that the mentioned structure works as an optical resonate microcavity. Such type of nanostructure provides a high area for light-material interaction. Also, it enhanced the internal light field. They also reported that the prepared 3D-MoS_2 nanostructure is capable of detecting linear-polarized light. For 395 nm wavelength's light, the responsivity and the polarization ratio of the prepared device are estimated to be around 2.8A/W and 1.64, respectively. For fabrication of the proposed device, first, the silicon nitride (SiNx) layer was coated on a sacrificial layer of aluminium. After the fabrication of Cr/Au gate electrode, the prepared sample was coated with a dielectric layer (SiO_2). In the next step, the single layer of photoactive material (MoS_2) has been transferred over the prepared sample. The monolayer of MoS_2 was prepared by CVD method. On the top of MoS_2 monolayer, two electrodes were fabricated for the purpose of

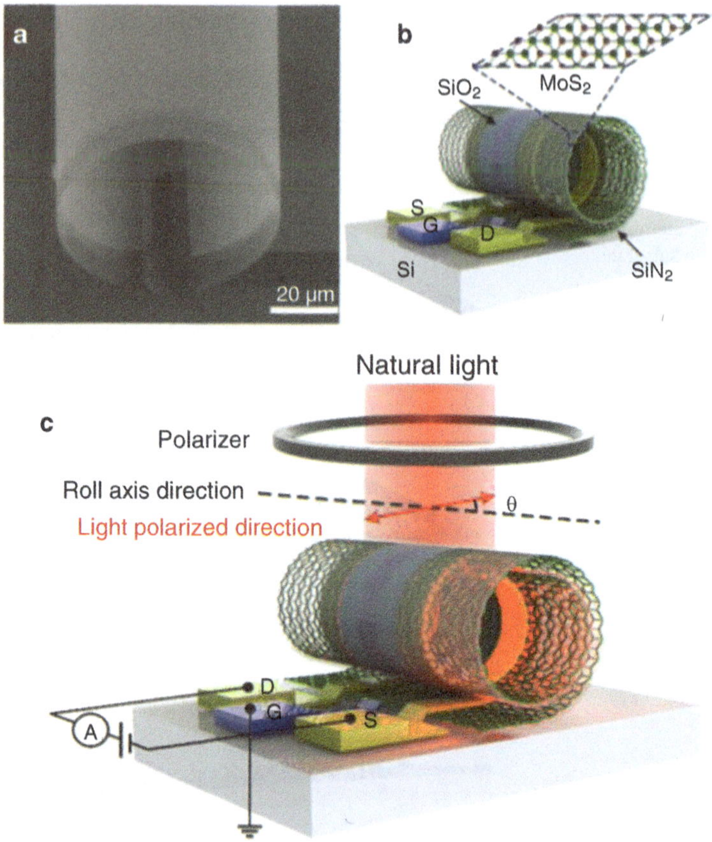

Fig. 8.13 Figure (**a**) shows the rolled microcavity structure of MoS_2 . Figure (**b**) and (**c**) shows the linear and polarized-light detector based on as-prepared microcavity. Images have been used with the permission of reference Deng et al. (2020)

source and drain connections. The prepared layer was rolled into the shape of a 3D tubular structure (see Fig. 8.13b). First, the prepared photodetector was analysed for power-dependent light detection property. The prepared photodetector was illuminated by a 635 nm LED light source with taking a power density of 14.7 mW/cm^2. The response and recovery time of prepared photodetector were found to be around 690 and 110 ms, respectively. The photocurrent was observed with taking the value of V_{ds} and V_{gs} as 1 V and 0 V, respectively. It has been observed that the response and recovery time are highly dependent on light wavelength. The fastest response of photodetector was observed for 395 nm light wavelength. For analysing polarization-sensitive photodetection property of prepared photodetector, the 635 nm laser light was allowed to pass through a polarizer. It has been observed that as-prepared sample was very highly sensitive to the polarization angle (angle between the light polarization direction and the axis of 3D-MoS_2

microtube) of light. They reported that the photocurrent of the prepared device varied periodically with respect to the variation in polarization angle (from 0° to 360°). The highest value of photocurrent has been observed at 90° and 270° and its lowest value found at 0° and 180° (Deng et al. 2020).

8.4 Conclusion

In summary, we have shown the advancement in the field of TMD-based nanomaterials and nanostructures. MoS_2 nanostructures are the latest and emerging material, which show better properties, ease of synthesis, and thermal as well as chemical stability and are cost-effective as compared to other nanomaterials. The various techniques for synthesizing the MoS_2 nanostructures such as CVD, hydrothermal, liquid exfoliation and sputtering have been discussed. It has been found that for making thin film and nanosheets of MoS_2, the CVD technique is more suitable, while the hydrothermal technique is more appropriate for synthesizing nanoflakes, nanoflowers as well as composite materials of MoS_2. The detailed analysis of comparative properties of synthesized MoS_2 nanostructures is, also, part of our chapter. From an application point of view, we made a detailed analysis of MoS_2 nanostructure-based photodetectors. Different types of photodetectors (linear and polarized-light) employed using MoS_2 nanostructures working at low temperature and room temperature. Effects of various parameters such as biasing and doping on photodetection have been studied. The various photodetector structures, heterojunctions and composites were proposed for improvements in the responsivity, detectivity, stability, repeatability and uniformity of detectors. The best outcomes for photodetection have been observed for plasmonic nanoparticle-decorated V-MoS_2 nanosheets. It has been observed that the pristine form of MoS_2 is not suitable for developing polarization-sensitive photodetectors. The hybrid MoS_2 nanostructures rolled in the cavity form show their capability for developing polarization-sensitive light detectors.

Manipulation in optical properties and device engineering in MoS_2-based photodetectors will lead to the development of future optoelectronics devices.

Acknowledgments This work was supported by the grant of project funded by Russian Science Foundation (Grant No. 21-79-00272).

References

Bera B (2019) Synthesis, properties and applications of amorphous carbon nanotube and MoS2 Nanosheets: a review. Nano Trends: J Nanotechnol Appl 21:36–52

Bera B, Banerjee D (2019a) Two-dimensional MoS2 nanosheets: preparation and characterization. Int J Appl Nanotechnol 5:1–26

Bera B, Banerjee D (2019b) Synthesis and characterization of amorphous carbon nanotube-molybdenum disulfide (ACNT-MoS2) nanohybrids using sodium molybdate dihydrate and LCysteine as precursor. Nano Trends J Nanotechnol Appl 21:41–47

Bera B, Banerjee D (2019c) Preparation and characterization of amorphous carbon nanotube-MoS2 Nanohybrid. Int J Res Eng Sci Manage 2:5–8

Bhakhar SA, Patel NF, Zankat CK, Tannarana M, Solanki GK, Patel KD, Pathak VM, Pataniya P (2019) Sonochemical exfoliation and photodetection properties of MoS2 Nanosheets. Mater Sci Semicond Process 98:13–18

Cao D, Ye K, Moses OA, Xu W, Liu D, Song P, Wu C, Wang C, Ding S, Chen S, Ge B, Jiang J, Song L (2019) Engineering the in-plane structure of metallic phase molybdenum disulfide via Co and O dopants toward efficient alkaline hydrogen evolution. ACS Nano 13:11733–11740

Chaudhary N, Khanuja M, Islam SS (2019) Broadband photodetector based on 3D architect of MoS2-PANI hybrid structure for high photoresponsive properties. Polymer 165:168–173

De Fazio D, Goykhman I, Yoon D, Bruna M, Eiden A, Milana S, Sassi U, Barbone M, Dumcenco D, Marinov K, Kis A, Ferrar AC (2016) High responsivity, large-area graphene/MoS2 flexible photodetectors. ACS Nano 10:8252–8262

Deng T, Li S, Li Y, Zhang Y, Sun J, Yin W, Weidong W, Zhu M, Wang Y, Liu Z (2020) Polarization-sensitive photodetectors based on three-dimensional molybdenum disulfide (MoS2) field-effect transistors. Nano 9:4719–4728

Duan Y, Liu Y, Chen Z, Liu D, Yu E, Zhang X, Fu H, Fu J, Zhang J, Du H (2020) Amorphous molybdenum sulfide nanocatalysts simultaneously realizing efficient upgrading of residue and synergistic synthesis of 2D MoS2 nanosheets/carbon hierarchical structures. Green Chem 22:44

Gan F, Dong N, Liu Z, Jia H, Wang J, Chen Y (2020) Organic small molecule covalently functionalized molybdenum disulfide hybrid material for optical limiting. 93:26–31

Guo J, Li S, He Z, Li Y, Lei Z, Liu Y, Huang W, Gong T, Ai Q, Mao L, He Y, Ke Y, Zhou S, Yu B (2019) Near-infrared photodetector based on few-layer MoS2 with sensitivity enhanced by localized surface plasmon resonance. Appl Surf Sci 483:1037–1043

Guoa J, Li S, Ke Y, Lei Z, Liu Y, Mao L, Gong T, Cheng T, Huang W, Zhang X (2020) Broadband photodetector based on vertically stage-liked MoS2/Si heterostructure with ultra-high sensitivity and fast response speed. Scr Mater 176:1–6

Gustavsson F, Svahn F, Bexell U, Jacobson S (2013) Nanoparticle based and sputtered WS_2 low-friction coatings—Differences and similarities with respect to friction mechanisms and tribofilm formation. Surf Coat Technol 232:616–626

Huang W-M, Liao W-S, Lai Y-M, Peter Chen I-W (2020) Tuning the surface charge density of exfoliated thin molybdenum disulfide sheets via noncovalent functionalization for promoting hydrogen evolution reaction. J Mater Chem C 8:510

Hwang T, Lee J, Jiseop O, Kim JM, Jeon Y, Park S-K, Piao Y (2019b) Facile synthesis of crumpled nitrogen-doped carbon/molybdenum disulfide hybrid sheets as high-rate anodes for lithium-ion batteries. 319:596–605

Hwang T, Lee J, Jiseop O, Kim JM, YoungmooJeon S-KP, Piao Y (2019a) Facile synthesis of crumpled nitrogen-doped carbon/molybdenum disulfide hybrid sheets as high-rate anodes for lithium-ion batteries. Electrochim Acta 319:596–605

Jia X, Zhu X, Tian W, Ding Y, Tian X, Cheng B, Cheng L, Bai S, Qin Y (2020) Nanowire templated CVD synthesis and morphological control of MoS2 nanotubes. J Mater Chem C 8: 4133

Kumar S, Pavelyev V, Mishra P, Tripathi N, Platonov V, Sharma P (2020) A review on 2D transition metal di-chalcogenides (TMDs) and metal oxide nanostructures based NO_2 gas sensor. Mater Sci Semicond Process 107:104865

Leng X, Wang Y, Wang F (2019) Ethylene glycol assisted hydrothermal synthesis of molybdenum disulfide for MEMS humidity sensor. MEMS 2019. IEEE:327–330

Li X, Zhu H (2015) Two-dimensional MoS2: properties, preparation, and applications. J Mater 1: 33–44

Liu B, Chen Y, You C, Liu Y, Kong X, Li J, Li S, Deng W, Li Y, Yan H, Zhang Y (2019) High performance photodetector based on graphene/MoS2/graphene lateral heterostructure with Schottky junctions. J Alloys Compd 779:140–146

Ma K, Sinha A, Dang X, Zhao H (2019) Electrochemical preparation of gold nanoparticles-polypyrrole co-decorated 2D MoS2 nanocomposite sensor for sensitive detection of glucose. J Electrochem Soc 166:147–154

Mathialagan S, Gomathi P-P (2020) Mo_2CeMoS_2 embedded reduced graphene oxide nanohybrid: Epitaxial synthesis of Mo_2C to augment the lithium storage properties of MoS_2. Carbon 158: 756–765

Meng C, Lin M-C, Xi-Wen D, Zhou Y (2019) Molybdenum disulfide modified by laser irradiation for catalyzing hydrogen evolution. ACS Sustain Chem Eng 7:6999–7003

Naz R, Imtiaz M, Liu Q, Yao L, Abbas W, Li T, Zada I, Yuan Y, Chen W, Gu J (2019) Highly defective 1T-MoS2 nanosheets on 3D reduced graphene oxide networks for supercapacitors. Carbon 152:697–703

Nethravathi C, Anto Jeffery A, Rajamathi M, Kawamoto N, Tenne R, Golberg D, Bando Y (2013) Chemical unzipping of WS2 nanotubes. ACSNano 7:7311

Pumera M (2013) Electrochemistry of graphene, graphene oxide and other graphenoids: review. Electrochem Commun 36:14–18

Rahmati B, Hajzadeh I, Taheri M, Karimzadeh R, Mohajerzadeh S, Mohseni SM (2019) Plasmonic improvement photoresponse of vertical-MoS2 nanostructure photodetector by Au nanoparticles. Appl Surf Sci 490:165–171

Sangeetha M, Madhan D (2020) Ultra sensitive molybdenum disulfide (MoS2)/graphene based hybrid sensor for the detection of NO2 and formaldehyde gases by fiber optic clad modified method. Opt Laser Technol 127:106193

Sharma P, Gupta N (2016) Model for threshold voltage instability in top-gated nanocrystalline silicon thin film transistor. J Comput Electron 15:666–671

Sharma P, Tripathi N, Gupta N (2017) Nanocrystalline silicon thin film prepared by e-beam evaporation for display application. J Mater Sci Mater Electron 28:3891–3896

Shi M, Dong L, Zheng S, Hou P, Cai L, Zhao M, Zhang X, Wang Q, Li J, Xu K (2019) "Bottom-up" preparation of MoS2 quantum dots for tumor imaging and their in vivo behavior study. Biochem Biophys Res Commun 516:1090–1096

Singh J, Tripathi N, Mohapatra S (2019) Synthesis of ag-TiO2 hybrid nanoparticles with enhanced photocatalytic activity by a facile wet chemical method. Nano-Struct Nano-Objects 18:100266

Smagulova S, Semenova A, Zakharkina E, Vinokurov P (2019) Investigation of the properties of two-dimensional molybdenum disulfide films synthesized by the CVD method. IOP Conf. Series: Mater Sci Eng 693:012030

Subitha M, Sasikanth SM, Bindhu B (2019) Ionic liquid assisted exfoliation and dispersion of molybdenum disulphide: synthesis and characterization. AIP Conf Proc 2100:020098

Suna B, Xi S, Liu Z, Liu X, Wang Z, Tan X, Shi T, Zhou J, Lia G (2019) Sensitive, fast, and stable photodetector based on perovskite/MoS2 hybrid film. 493:389–395

Tripathi N, Islam SS (2017) A new approach for orientation controlled growth of CNTs: an in-depth analysis on the role of oxygen plasma treatment to catalyst. Appl Nanosci 7:125–129

Tripathi N, Pavelyev V, Islam SS (2018) Tunable growth of single-wall CNTs by monitoring temperature increasing rate. Int Nano Lett 8:1–9

Tyagi S, Kumar A, Kumar M, Singh BP (2019) Large area vertical aligned MoS2 layers toward the application of thin film transistor. Mater Lett 250:64–67

Wang G, Zhou F, Ziwei L, Ma Y, Li X, Tong Y, Dong X (2019) Controlled synthesis of CoFe2O4/MoS2 nanocomposites with excellent sedimentation stability for magnetorheological fluid. J Ind Eng Chem 17:439–446

Wei Y, Zhang X, Zhao Z, Chen H-S, Matras-Postolek K, Wang B, Yang P (2019) Controllable synthesis of P-doped MoS2 nanopetals decorated N doped hollow carbon spheres towards enhanced hydrogen evolution. Electrochim Acta 297:553–563

Zhang D, Liang S, Chai J, Liu T, Yang X, Wang H, Chengc J, Zheng G, Cao M (2019a) Highly effective shielding of electromagnetic waves in MoS2 nanosheets synthesized by a hydrothermal method. J Phys Chem Solids 134:77–82
Zhang D, Pang M, Junfeng W, Cao Y (2019b) Experimental and density functional theory investigation of Pt-loaded titanium dioxide/molybdenum disulfide nanohybrid for SO2 gas sensing. New J Chem 43:4900
Zhang R, Ma X, An C, Zhang D, Sun D, Hu X, Liu J (2019d) Self-powered photodetector based on vertical MoO3/MoS2 hetero-structure with gate tunable photo-response. 2D Mater 6:035033
Zhang S, Wang X, Chen Y, Guangjian W, Tang Y, Zhu L, Wang H, Jiang W, Sun L, Lin T, Shen H, Weida H, Ge J, Wang J, Meng X, Chu J (2019c) Ultrasensitive hybrid MoS2–ZnCdSe quantum dot photodetectors with high gain. ACS Appl Mater Interfaces 11:23667–23672
Zhou YH, An HN, Gao C, Zheng ZQ, Wang B (2019) UV–vis-NIR photodetector based on monolayer MoS2. Mater Lett 237:298–302

Magnetic Characterization of Nanomaterials

9

Yassine Slimani, Sadik Guner, Munirah A. Almessiere, Essia Hannachi, Ayyar Manikandan, and Abdulhadi Baykal

Abstract

Nanomaterials display unique magnetic features dissimilar from those of the counterpart bulk material. To be efficient in practical technological applications, the best knowledge of these magnetic properties should be well investigated. In the present chapter, we propose an overview on the following interesting issues: magnetic character of magnetic nanomaterials, saturation magnetization, magnetic anisotropy, and surface effects, all with respect to their dimension and magnetic configurations. The magnetic traits of the nanomaterials are discussed

Y. Slimani (✉)
Department of Biophysics, Institute for Research and Medical Consultations (IRMC), Imam Abdulrahman Bin Faisal University, Dammam, Saudi Arabia
e-mail: yaslimani@iau.edu.sa

S. Guner
Institute of Inorganic Chemistry, RWTH Aachen University, Aachen, Germany

M. A. Almessiere
Department of Biophysics, Institute for Research and Medical Consultations (IRMC), Imam Abdulrahman Bin Faisal University, Dammam, Saudi Arabia

Department of Physics, College of Science, Imam Abdulrahman Bin Faisal University, Dammam, Saudi Arabia

E. Hannachi
Department of Nuclear Medicine Research, Institute for Research and Medical Consultations (IRMC), Imam Abdulrahman Bin Faisal University, Dammam, Saudi Arabia

A. Manikandan
Department of Chemistry & Center for Catalysis and Renewable Energy, Bharath Institute of Higher Education and Research (BIHER), Bharath University, Chennai, Tamil Nadu, India

A. Baykal
Department of Nanomedicine, Institute for Research and Medical Consultations (IRMC), Imam Abdulrahman Bin Faisal University, Dammam, Saudi Arabia

A. Thakur et al. (eds.), *Synthesis and Applications of Nanoparticles*,
https://doi.org/10.1007/978-981-16-6819-7_9

and compared to those of the counterpart bulk material. The magnetic dynamics in nanomaterials, the superparamagnetic relaxation, and magnetic interactions among nanoparticles as well as the utilization of dc magnetization and ac susceptibility were also discussed.

Keywords

Nanomaterials · Magnetic behavior · Magnetization · Magnetic anisotropy · Magnetic domains · Superparamagnetism · Magnetic interactions · Coercive field

9.1 Introduction

During the current epoch, the area of nanotechnology is prospering to a great extent, and nowadays numerous kinds of research are directly or indirectly correlated to the nanotechnology. Nanotechnology can be asserted as the capability to develop, prepare, characterize, and apply/manipulate the materials by altering their shape and size in nano-scale level. The nanomaterials (NMs) are the matters that display, at least in one dimension, a less than 100 nm size ones (Slimani et al. 2019a). Amazingly, NMs exhibit special, unique, and different physicochemical properties (chemical stability, electrical, magnetic, optical, mechanical, etc.) compared to bulk materials, which essentially are depending on their shapes and sizes that lead principally to greater surface-to-volume ratio (Slimani et al. 2019a; Almessiere et al. 2019a; Vinosha et al. 2020a). NMs could be of dissimilar morphologies such as nanowires, nanofibers, nanosheets, nanoparticles, nanowhiskers, nanorods, nanobelts, thin films, etc. that could be categorized on the basis of their dimensionality (Saleh 2020; Slimani and Hannachi 2020). NMs can be categorized as zero-dimensional (0D) NMs that are largely nanoparticles (NPs); one-dimensional NMs like nanowires, nanofibers, nanorods, etc.; and two-dimensional NMs that are mostly thin films. These categorize principally single isolated NMs. The particles with different interacted constituents, such as nanocomposites, are designated as bulk or three-dimensional NMs (Manikandan et al. 2020). Due to their specific and unique characteristics, NMs are employed in various areas like electronics (Slimani et al. 2020a; Vinosha et al. 2021; Slimani et al. 2019b; Slimani et al. 2019c), energy production/batteries (Ullah et al. 2020; Nadeem et al. 2021), energy transmission (Slimani et al. 2019d; Slimani et al. 2018a; Hannachi et al. 2018; Ben Salem et al. 2014; Slimani et al. 2018b; Hannachi et al. 2019a), energy storage (Slimani and Hannachi 2021; Yasin et al. 2020a; Seevakan et al. 2019; Seevakan et al. 2018), hard magnets (Almessiere et al. 2018a; Algarou et al. 2020a), microwave absorbers (Almessiere et al. 2020a; Almessiere et al. 2019b; Almessiere et al. 2019c; Slimani et al. 2021a), medicine (Nawaz et al. 2018; Akhtar et al. 2019; Almessiere et al. 2020b; Rehman et al. 2020; Rehman et al. 2019; Algarou et al. 2020b), environment (Vinosha et al. 2020b; Tombuloglu et al. 2018; Tombuloglu et al. 2019a; Tombuloglu et al. 2019b; Al-Amri et al. 2020; Tombuloglu et al. 2019c), catalysis (Nawaz et al. 2019; Kumar et al. 2020; Ajeesha et al. 2020), cosmetics (Nanda et al. 2020), corrosion (Yasin et al. 2020b), etc.

Table 9.1. Magnetization behaviors of NPs with respect to their dimensions and magnetic structures

Size	Magnetic structures	State of magnetization	Magnetization behavior
$D > D_c$	MMD	Non-uniform Stable	Like bulk sample Large hysteresis loops
$D_t \ll D < D_c$ In the vicinity of D_c	SMD	Uniform Stable	Occurrence of hysteresis (from rectangular to linear loop)
$D_t < D \ll D_c$ In the vicinity of D_t	SMD	Transition state (relaxation)	Small hysteresis loops to absence of hysteresis
$D < D_t$	SMD	Fluctuation	SPM

In accordance with their sizes (taking into consideration that "*D*" is the particles diameter), the nanomaterials (NMs) are in the dimensional interval ranging from a few nanometers (nm) up to few hundreds of nm. The impact of the size of nanoparticles (NPs) is greatly exposed on the magnetic structure of NPs and, as consequence, on the magnetic behaviors of NPs as a function of temperature and/or under an externally applied magnetic field. We consider in the following sentences that the notations D_c and D_t refer to the critical diameter and the threshold diameter of the nanoparticles, respectively. As mentioned in previous sections, the magnetic structure of NPs changes with diminishing the size of NPs (Table 9.1.): from a structure that display multi-magnetic domains (MMD) at $D > D_c$ wherein the magnetization is non-uniform (Weiss domains) and stable, to a structure single-magnetic domain (SMD) (i.e., with no magnetic domains) at $D_t < D < D_c$ wherein the magnetization is uniform and stable closer to D_c or non-stable (transition state or relaxation) closer to D_t, to also a structure of single-magnetic domain but with magnetization that fluctuate along the easy-axis direction in the crystal at very tiny sizes $D < D_t$ (superparamagnetic (SPM) interval). The circumstance of minimizing the crystal's free energy will result of these magnetic structures, leading to its stable configuration. At very tiny sizes (generally few nm), the magnetization of nanoparticles with single-magnetic domain is no longer stable, and below the impact of the thermal activations, it could reverse by π-angle. The transition state (relaxation) could be deliberated as the transition region among the stabile magnetic state and the SPM state. The behavior of the magnetization of these nanostructured materials involve the occurrence of hysteresis at $D > D_t$, and the absence of hysteresis at $D < D_t$, more accurately superparamagnetic character that is also greatly affected by the magnetic anisotropy of nanoparticles at a certain specific temperature. A summary of these information is listed in the Table 9.1..

Frequently, the magnetization of NPs is smaller than that of bulk counterpart. The achieved findings revealed that, for example, the saturation magnetization (M_s) of nanoparticles depends on their sizes; M_s reduces typically with reducing the NPs size due to the spins disorder at the surface of NPs that conduce to a governing effect at very tiny sizes. The effects of spins disorder at the surface of NPs play a great role in the evolution of the magnetization with respect to the temperature, which is uncommon in numerous cases. The magnetocrystalline anisotropy has also a great influence on the magnetization. Besides the nature of the product, the dimensions of the NPs are an essential parameter for several practical applications. Moreover, the size distribution and morphology (shape) of nanostructured materials, besides the magnetostatic interactions among NPs, could alter the magnetic properties of the samples. Furthermore, dispersing the NPs within a carrier liquid (nanofluids) or implanting the NPs into diverse crystalline or non-crystalline matrices (nanocomposites) or coating the NPs with a layer or utilizing surfactants, all these could alter the magnetic behavior of the nano-sized products. The knowledge of all these features and their influences on the magnetic properties of NPs in the presence of an applied magnetic field is a vital issue for forthcoming promising nanotechnological applications. In coming subsections and sections, we will consider these features in the circumstance of NPs with ferri- and ferromagnetic ordering of their internal magnetic moments. This is because of the exchange interactions (direct) and super-exchange interactions (indirect, via the ions of oxygen) of the magnetic moments.

9.2 Magnetic Anisotropy

The term "magnetic anisotropy" means that determined magnetic properties of a substance depends on the direction in which the measurement is conducted. Magnetically anisotropic nanoparticles (NPs) have at least two energetically preferential directions (also known as two directions of an easy axis) for their spontaneous magnetization vector (M) in the absence of an externally applied magnetic field, H. The number of preferential directions of magnetization depends on structure and present types of anisotropies in the nanomaterial. The frequently observed kinds of anisotropy in the nanomaterials are as follows:

- Magnetocrystalline anisotropy, also known as crystal anisotropy
- Shape anisotropy
- Stress anisotropy
- Exchange anisotropy

9.2.1 Magnetocrystalline Anisotropy in Cubic Crystals

Ferromagnetic materials are made of the elements like Fe (body-centered cubic, *bcc*) and nickel (face-centered cubic, *fcc*) or ferrimagnetic compounds like spinel ferrites having the common formula MFe_2O_4 where M divalent metal ions like Mn, Ni, Zn, Cu, Fe, Co, or Mg have the cubic crystal structure. For simple cubic crystals, the lattice vectors are orthogonal and of equal length. Miller indices in the forms [hkl] and (hkl) simply denote directions and planes in cubic crystals, respectively. Researchers who conduct angular-dependent magnetization measurements on ferromagnetic bulk crystal or film samples specify the easy and hard axes of magnetization as initial step of investigation. Figure 9.1a shows four kinds of domains

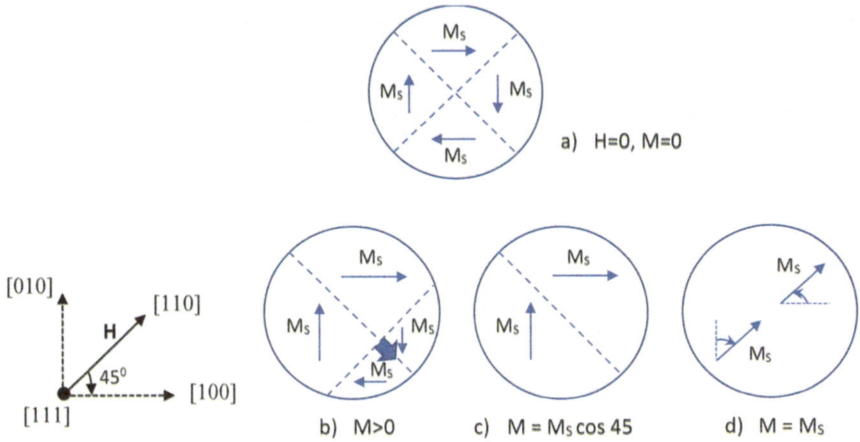

Fig. 9.1 (**a**) Domain structure of single Fe crystal (schematic); (**b**, **c**) magnetization due to domain wall motion, ($H_1 < H_2$) // (Hubert and Schäfer 1998); and (**d**) net magnetization due to domain rotation, ($H_2 < H_3$) // (Hubert and Schäfer 1998)

schematically. Magnetocrystalline (or crystal) anisotropy is responsible to hold the magnetic domains as aligned in certain stable directions. Figure 9.1b, c show the magnetization processes of Fe crystal disk when external static fields H_1 and H_2 are applied in Hubert and Schäfer (1998) direction, respectively. These two steps indicate that magnetization is realized by domain wall motion until just two domains are left. A fairly higher field H_3 causes the domain rotation by doing work against the crystal anisotropy, and magnetization process is exactly completed as M_S // (Hubert and Schäfer 1998) // H_3, as seen in Fig. 9.1d. Work done by fairly high field against the crystal anisotropy stores a potential energy in the crystal and called as crystal anisotropy energy, E_a. In a cubic crystal, let M_s vector make angles θ_1, θ_2, and θ_3 with the crystal axes, and let α_1, α_2, and α_3 be the cosines of these angles, which are also called as direction cosines. Then, E_a is expressed as (Cullity and Graham 2008):

$$E_a = K_0 + K_1\left(\alpha_1^2\alpha_2^2 + \alpha_2^2\alpha_3^2 + \alpha_3^2\alpha_1^2\right) + K_2\left(\alpha_1^2\alpha_2^2\alpha_3^2\right) + \ldots \tag{9.1}$$

where K_0, K_1, K_2, etc. are anisotropy energy constants for probed material under a particular temperature condition. Isotropic first term, K_0, is usually ignored since we are interested only with ΔE when the M_s vector rotates from one direction to another. Further, K_2 sometimes and mostly higher constants are so small that the terms involving them can be neglected. For instance, we can calculate the value of E when the M_s vector lies in a particular direction. All angles are $\theta_1 = \theta_2 = \theta_3 = 54.7^\circ$ for (Almessiere et al. 20191) direction in cubic crystal, and the corresponding direction cosines are $\alpha_1 = \alpha_2 = \alpha_3 = \frac{\sqrt{3}}{3}$, so anisotropy energy is $E_a = K_0 + \frac{K_1}{3} +. \frac{K_2}{27}$ (Erg/cm^3 (CGS) or J/m^3 (SI)).

9.2.2 Magnetocrystalline Anisotropy in Hexagonal Crystals

Ferromagnetic cobalt crystals or magnetically hard ferrimagnetics like M-type barium and strontium ferrites have hexagonal-closed packet structure (Almessiere et al. 2019d; Almessiere et al. 2019e; Almessiere et al. 2019f; Almessiere et al. 2019g; Unal et al. 2019; Algarou et al. 2020c; Algarou et al. 2020d; Slimani et al. 2020b). The hexagonal *c* axis is the single easy axis of magnetization, while all orientations in the basal plane are equally hard directions. That is why, anisotropy energy depends on a single angle θ between M_s vector and *c* axis. This type of anisotropy is described as uniaxial anisotropy. The uniaxial anisotropy energy is commonly written in powers of $sin\theta$ as below:

$$E_a = K_0 + K_1\, sin^2\theta + K_2\, sin^4\theta + \ldots \tag{9.2}$$

When isotropic K_0 and higher terms including K_2 are negligible, Eq. (9.2) converts to the simplest form for uniaxial symmetry,

$$E_a = K_1\, sin^2\theta \tag{9.3}$$

In general, the physical origin of magnetocrystalline anisotropy is the spin-orbit coupling. In the last step of magnetization process (as described in Fig. 9.1d), fairly high enough external field tries to reorient the spin of an electron; simultaneously the orbit of that electron also tends to be reoriented. But the orbit is strongly coupled to the crystal lattice and therefore resists to rotation of the spin axis. That is why, work to rotate the spin system of a domain away from the easy direction is just the work required to overcome the spin-orbit coupling. The magnitude of magnetocrystalline anisotropy generally decreases with temperature more rapidly with respect to magnetization and disappears at the T_c point of specimen. Since the coercive field (H_c) strongly originates from anisotropy, it generally goes to zero together with disappearing anisotropy.

9.2.3 Anisotropy Measurement

There are four different methods to determine the magnitude of anisotropy constants: (a) torque curve, (b) two different ways of calculation from magnetization curves, (c) torsion pendulum, and (d) magnetic resonance. We will briefly inform about three of them.

9.2.3.1 Torque Curve

Torque method is the most fundamental one. A high-field torque curve measurement involves only the rotation of M_s relative to the axes of a single-domain crystal; no wall motion is included. Torque curves are registered by torque magnetometers. For example, a hexagonal crystal ferrite having uniaxial anisotropy is cut in the form of a thin disk having the easy *c* axis parallel to the plane of disk. It is located in a saturating field that is parallel with the disk plane, too. The disk is rotated until a certain angle about its central axis from easy direction and torque curve is plotted as a function of the angle. The torque (L) magnitude exerted on crystal by M_s:

$$L = -\frac{dE}{d\theta} \tag{9.4}$$

If the angle derivative of E_a expression (Eq. (9.3)) belonging to a hexagonal crystal is taken, the expression of L became:

$$L = -2K_1 \sin\theta\cos\theta = -2K_1 \sin 2\theta \tag{9.5}$$

9.2.3.2 Torsion Pendulum

This method also uses a circular disk. Like at torque magnetometers, uniaxially anisotropic disk is rotated away from easy c axis direction (initially parallel to applied H again), released, and allowed to oscillate back and forth about the field direction, and the oscillation frequency is measured in a particular time interval as:

$$\frac{1}{f} = T = 2\pi\sqrt{\frac{I}{k_w + k_s}} \tag{9.6}$$

where f is oscillation frequency, I is the moment of inertia of the disk, k_w is the torsion constant of the wire, k_s is the torsional stiffness of the disk and correlated with the rate of change of torque with angle, $k_s = \frac{dL}{d\theta} = \frac{d^2E}{d\theta^2}$. Once the oscillation frequency is measured, I and k_w are already known, and then k_s and anisotropy constant might be determined. Determination of anisotropy constants by torsion pendulum is a seldomly used method.

9.2.3.3 Anisotropy Constant from Fitted Magnetization Curves

We can calculate the K_1 and K_2 anisotropy constants of a hexagonal Co crystal having the uniaxial symmetry. We can start by rewriting the anisotropy energy equation (Eq. (9.2)) including just three terms:

$$E_a = K_0 + K_1 \sin^2\theta + K_2 \sin^4\theta \tag{9.7}$$

The magnetic potential energy (when H is applied at the right angles to easy axis and is strong enough to rotate crystal's M_s vector by an angle θ):

$$E_p = -M_s \cdot H = -M_s H \cos\left(90^0 - \theta\right) \tag{9.8}$$

The condition for minimum total energy of crystal:

$$2K_1 \sin\theta \cos\theta + 4K_2 \sin^3\theta \cos\theta - M_s H \cos\theta = 0 \tag{9.9}$$

where $M = M_s \cos(90^0 - \theta) = M_s \sin\theta$; if $\sin\theta = \frac{M}{M_s}$ is substituted in Eq. (9.9), an angle independent field equation is obtained:

$$H = \frac{2K_1}{M_s}\left(\frac{M}{M_s}\right) + \frac{4K_2}{M_s}\left(\frac{M}{M_s}\right)^3 \tag{9.10}$$

when H is strong enough to saturate the sample, $M = M_s$, so saturating field equation becomes:

$$H = \frac{2K_1}{M_s} + \frac{4K_2}{M_s} \tag{9.11}$$

If K_2 is zero, then saturating field becomes:

$$H = \frac{2K_1}{M_s} \tag{9.12}$$

The second method related with the magnetization curves is the area method. Sample is magnetized by applying a fairly enough field along a non-easy direction. Anisotropy constants are determined utilizing from the equality of the stored energy and work expressions. Area between M-H curve and M-axis gives the work done by field during magnetization process.

9.2.3.4 Shape Anisotropy

Spherically symmetric magnetic objects or nanomaterials do not have anisotropy since the applied field magnetizes them at the same magnitude in any direction. If the sample is non-spherical, then magnetizing it along a long axis is easier than magnetizing along a short axis. Because demagnetizing field is stronger along the short axis, one can conclude that the shape of an object is the source of anisotropy under these circumstances. If a non-spherical object or nanomaterial is magnetized until some extent, M and H are removed later, and its magnetization drops to an order smaller than remnant magnetization level due to effect of demagnetizing field (H_d) of sample. The magnetization loop is denoted in Fig. 9.2. In the figure, OC is the demagnetizing field line with a slope of $-1/N_d$, where N_d is the demagnetizing factor of sample and the stored energy in the sample is equal to the shaded triangular area

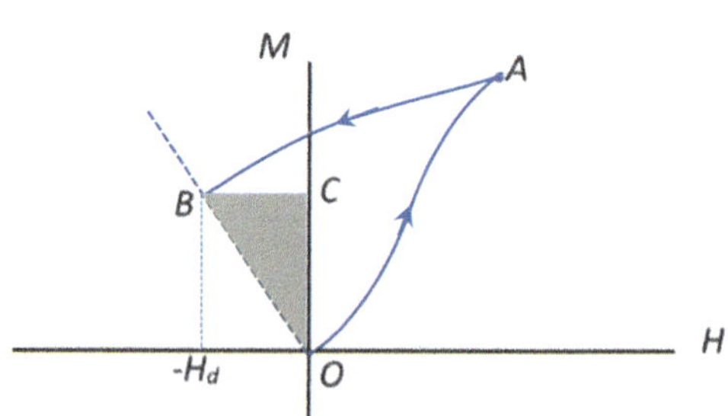

Fig. 9.2 Determination of magnetostatic energy from M-H curve

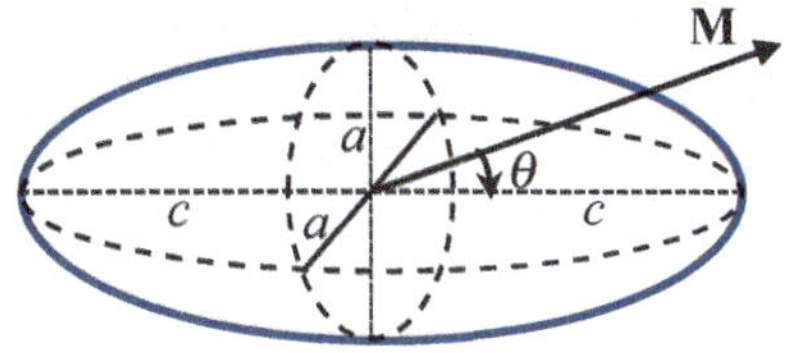

Fig. 9.3 Prolate ellipsoid

between field line and M-axis. The self-energy of magnetic sample correlated with the demagnetizing field is called as magnetostatic energy, E_{MS}:

$$E_{MS} = Area\ (BCO) = -\frac{1}{2} H_d \cdot M = \frac{1}{2} H_d M \tag{9.13}$$

Magnetostatic energy can be written in terms of demagnetizing factor by substitution, $H_d = N_d M$:

$$E_{MS} = \frac{1}{2} N_d M^2\ \mathrm{Erg/cm^3}\ (\mathrm{CGS})\ \text{or}\ E_{MS} = \frac{1}{2} \mu_0 N_d M^2\ \mathrm{J/m^3}\ (\mathrm{SI}) \tag{9.14}$$

We now consider a sample having the shape of a prolate ellipsoid with the semi-minor axis a and semi-major axis c as depicted in Fig. 9.3. Magnetization vector makes an angle θ with c axis. E_{MS} equation for this ellipsoid is:

$$E_{MS} = \frac{1}{2} \left[N_c (M\ cos\ \theta)^2 + N_a (M\ sin\ \theta)^2 \right] \tag{9.15}$$

where N_c and N_a are the demagnetizing factors along c and a axes, respectively. Substituting $\cos^2\theta = 1 - \sin^2\theta$, we obtain:

$$E_{MS} = \frac{1}{2} N_c M^2 + \frac{1}{2} (N_a - N_c) M^2\ sin^2\theta \tag{9.16}$$

The first term can be neglected since N_c is small along easy c axis, so we get an equation in the same form of uniaxial crystal anisotropy energy (Eq. (9.3)) as below:

$$E_{MS} = \left[\frac{1}{2} (N_a - N_c) M^2 \right] sin^2\theta \tag{9.17}$$

Term in square bracket of Eq. (9.17) must be shape anisotropy constant, K_S.

$$K_S = \frac{1}{2} (N_a - N_c) M^2\ \mathrm{Erg/cm^3}\ (\mathrm{CGS})\ \text{or}$$
$$K_S = \frac{1}{2} \mu_0 (N_a - N_c) M^2\ \mathrm{J/m^3}\ (\mathrm{SI}) \tag{9.18}$$

Saturating field equation for uniaxial shape anisotropy can be written utilizing Eq. (9.18) as:

$$H = \frac{2K_s}{M_s} = (N_a - N_c)M_s \tag{9.19}$$

In 1948, E.C. Stoner and E.P. Wohlfarth calculated the shape anisotropy constant of uniaxial Co as equal to $K_s = 4.5 \times 10^6$ Erg/cm^3 (CGS) or $K_s = 4.5 \times 10^5$ J/m^3 (SI) for the axial ratio $c/a = 3.5$ (Stoner and Wohlfarth 1948). Those values are the same with determined crystal anisotropy constant K_1 as reported in Sect. 2.3.3. A striking feature of dipolar interaction is that it decreases slowly as a function of the distance r_{ij} (like r_{ij}^{-3}); thus the summation over the magnetic moment pairs (m_i, m_j) converges very slowly. As a consequence, the dipolar field $H_{dip}(i)$ experienced by a given moment m_i depends significantly on the moments located at the boundary of the sample resulting in the shape anisotropy (Bruno 1993).

9.2.3.5 Stress Anisotropy

Stress anisotropy is counterpart to magnetostriction, and anisotropy equations are expressed in terms of magnetostriction parameters. That is why one should understand the phenomena of magnetostriction initially. The physical origin of the magnetostriction is the spin-orbit coupling like crystal anisotropy. Magnetostriction is a change of material's physical dimensions as a result of the change of the orientation of magnetization. The direction of magnetization changes under the influence of externally applied field H or temperature T. The fractional change in length is simply defined as strain. The common strain (ε) caused by applied stress on objects differs here from our interest the magnetically induced strain, λ:

$$\lambda = \frac{\Delta l}{l} \tag{9.20}$$

The magnitude of λ at the magnetic saturation level is called as saturation magnetostriction, λ_s. The value of λ_s can be positive, negative, or, in some alloys at some temperature, zero. Supposing that a cubic crystal, like Fe, is exposed to a saturating field in easy direction, the cubic dimension will change from l to $l + \Delta l$ in direction (Mazo-Zuluaga et al. 2008). The is cubic symmetry will convert to tetragonal symmetry where two dimensions are equal and the third one is longer. The degree of tetragonality can be determined by XRD measurements. However, direction-dependent λ_s is usually small for many substances, in the order of 10^{-6} to 10^{-5} (Chikazumi 1964). If the sample is exposed to a field much higher than saturating field, an additional strain at small magnitude is named as forced magnetostriction. Figure 9.4 indicates the dependence of λ on H schematically.

In most practical applications, such as for a cubic crystal, the λ_s can be described by expression with small number of constants:

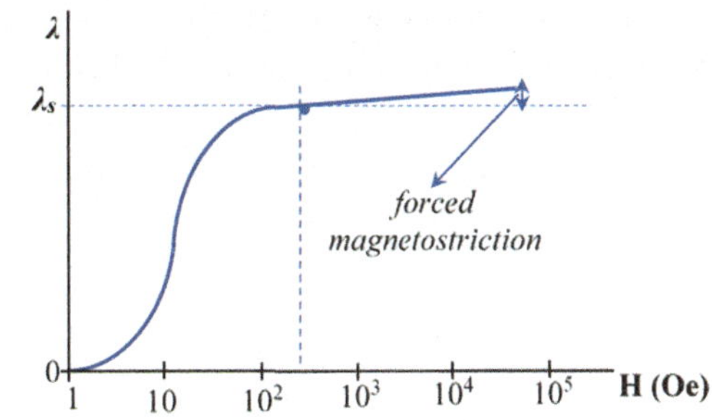

Fig. 9.4 Dependence of magnetostriction on applied magnetic field (H) is in logarithmic scale

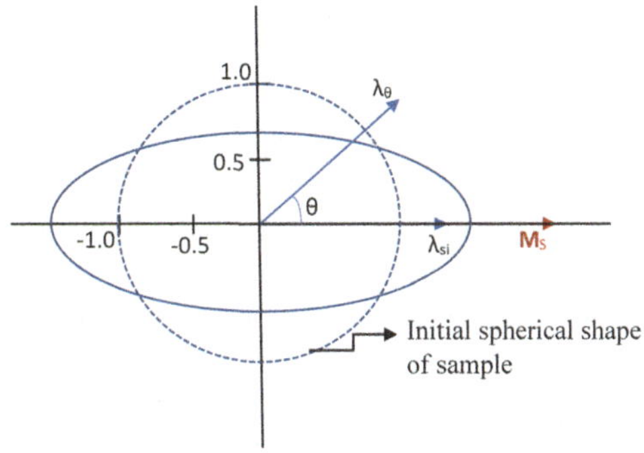

Fig. 9.5 Spherical material with isotropic magnetostriction

$$\lambda_s = \frac{3}{2}\lambda_{100}\left(\alpha_1^2\beta_1^2 + \alpha_2^2\beta_2^2 + \alpha_3^2\beta_3^2 - \frac{1}{3}\right) + 3\lambda_{111}(\alpha_1\alpha_2\beta_1\beta_2 + \alpha_2\alpha_3\beta_2\beta_3 + \alpha_3\alpha_1\beta_3\beta_1) \tag{9.21}$$

where α_1, α_2, and α_3 are the direction cosines in direction of M_s vector and β_1, β_2, and β_3 are the direction cosines along which the strain is measured. If magnetic sample is spherical, then magnetostriction is isotropic, and Eq. (9.21) gets a new form by substitution of $\lambda_{100} = \lambda_{111} = \lambda_{si}$ with the introduction of a new symbol:

$$\lambda_\theta = \lambda_{si}\left[(\alpha_1\beta_1 + \alpha_2\beta_2 + \alpha_3\beta_3)^2 - \frac{1}{3}\right] \tag{9.22}$$

where λ_{si} is the isotropic magnetostriction and refers the saturation magnetostriction measured from an ideal demagnetized state and λ_θ is the saturation magnetostriction at an angle θ to the direction of M_s (// λ_{si}). According to definition of θ, one can put instead of three terms in the parenthesis above $\alpha_1\beta_1 + \alpha_2\beta_2 + \alpha_3\beta_3 = cos\theta$:

$$\lambda_\theta = \lambda_{si}\left[(cos\theta)^2 - \frac{1}{3}\right] \tag{9.23}$$

The magnetostriction effect on a spherical object which was distorted from an ideal demagnetized sphere into an ellipsoid shape is illustrated in Fig. 9.5.

An equivalence of Eq. (9.21) for magnetostriction of the hexagonal crystals can be written having the four constants as below:

$$\lambda_{si} = \lambda_A\left[(\alpha_1\beta_1 + \alpha_2\beta_2)^2 - (\alpha_1\beta_1 + \alpha_2\beta_2)\alpha_3\beta_3\right]$$
$$+ \lambda_B\left[\left(1 - \alpha_3^2\right)\left(1 - \beta_3^2\right) - (\alpha_1\beta_1 + \alpha_2\beta_2)^2\right]$$
$$+ \lambda_C\left[\left(1 - \alpha_3^2\right)\beta_3^2 - (\alpha_1\beta_1 + \alpha_2\beta_2)\alpha_3\beta_3\right] + 4\lambda_D(\alpha_1\beta_1 + \alpha_2\beta_2)\alpha_3\beta_3 \quad (9.24)$$

When magnetostriction is measured along easy c axis direction like magnetization, since direction cosines α_1, α_2, $\alpha_3 = \beta_1$, β_2, β_3 and $\alpha_1^2 + \alpha_2^2 + \alpha_3^2 = 1$, this common equation converts to a two constants expression:

$$\lambda_{si} = \lambda_A\left[\left(1 - \alpha_3^2\right)^2 - \left(1 - \alpha_3^2\right)\alpha_3^2\right] + 4\lambda_D\left(1 - \alpha_3^2\right)\alpha_3^2 \quad (9.25)$$

The saturation magnetostriction of polycrystalline samples which is parallel to the magnetization is characterized by a single magnetostriction constant λ_p. Its value depends on the magnetostrictive properties of the individual crystals and on the way in which they are arranged, i.e., on the presence or absence of preferred domain or grain orientation. If the grain orientations are completely random, the saturation magnetostriction of the polycrystal is given as average over these orientations. In 1965, H.E. Callen and N. Goldberg claimed ideally averaged magnetostriction expression which has best fitting to experimental data (Callen and Goldberg 1965):

$$\bar{\lambda} si = \lambda_{111} + \left(\frac{2}{5} - \frac{lnc}{8}\right)(\lambda_{100} - \lambda_{111}) \quad (9.26)$$

where $c = \frac{2c_{44}}{c_{11} - c_{12}}$, c_{11}, c_{12} and c_{44} are the single crystal elastic constants. If a crystal is elastically isotropic, then $c = 1$. When the single crystal data belonging to Fe are submitted in Eq. (9.25), $\bar{\lambda} si = -9 \times 10^{-6}$ is obtained. One can find the saturation magnetostriction at an angle θ just by substituting λ_p instead of λ_{si} in Eq. (9.23):

$$\lambda_\theta = \lambda_p\left[(cos\ \theta)^2 - \frac{1}{3}\right] \quad (9.27)$$

It was mentioned that magnetostriction is counterpart to stress anisotropy. This means that an applied mechanical stress (δ) can alter the domain structure and create a new source of magnetic anisotropy. The applied stress might change the magnetization reversal characteristics. Low-field properties like magnetic permeability (μ) and remanence (M_r) are also affected significantly. Here are some examples, a polycrystalline Ni has (−) magnetostriction coefficient. When it is exposed to a 10^4 lb./in^2 of compressive stress under the 10 Oe of field, μ doubles, but same magnitude of tensile stress decreases μ to about one tenth of its zero stress value. A 68 permalloy (68% Ni, 32% Fe) has (+) magnetostriction coefficient. A reverse effect is observed, that is, tensile stress increases the permeability of 68 permalloy.

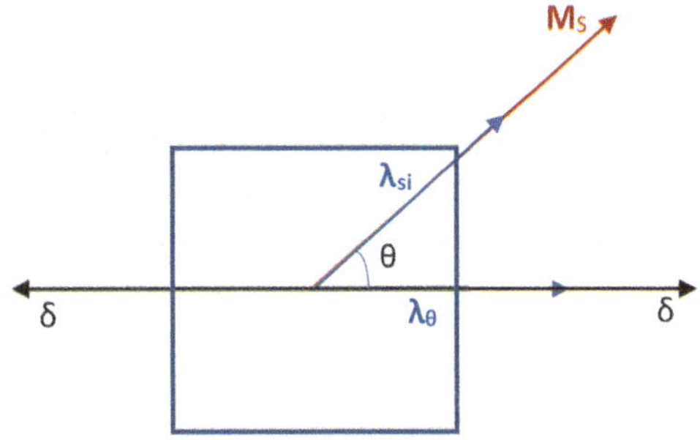

Fig. 9.6 Cubic crystal material with isotropic magnetostriction

The effect of stress on magnetization is referred as magneto-mechanical effect. If a substance has (+) magnetostriction, applied tensile stress elongates it and increases the magnetization. A compressive stress has reverse effects on both dimensions and magnetization. Stress applied to a ferromagnetic body may also affect the orientation of magnetization through magnetostriction. This means M_s and δ do not always have to be parallel. In presence of an elastic mechanical stress, the direction of M_s vector is determined by both δ and leading anisotropy (K) in magnetic sample. A common energy relation includes λ, δ, and K_1 constants for cubic crystal:

$$E = K_1\left(\alpha_1^2\alpha_2^2 + \alpha_2^2\alpha_3^2 + \alpha_3^2\alpha_1^2\right) - \frac{3}{2}\lambda_{100}\delta\left(\alpha_1^2\gamma_1^2 + \alpha_2^2\gamma_2^2 + \alpha_3^2\gamma_3^2\right)$$
$$- 3\lambda_{111}\delta(\alpha_1\alpha_2\gamma_1\gamma_2 + \alpha_2\alpha_3\gamma_2\gamma_3 + \alpha_3\alpha_1\gamma_3\gamma_1)\ (\mathrm{Erg/cm^3}) \quad (9.28)$$

where α_1, α_2, and α_3 are the direction cosines in direction of M_S vector and γ_1, γ_2, and γ_3 are the direction cosines of elastic mechanical stress δ. First term of Eq. (9.28) is crystal anisotropy energy; next two terms include λ and δ constants and are usually called the magnetoelastic energy, E_{ME}. These two terms reduce to a simple form for the conditions of isotropic magnetostriction and uniaxial anisotropy,

$$E_{ME} = \frac{3}{2}\lambda_{si}\delta\, sin^2\theta \quad (9.29)$$

Equation (9.29) physically means the stored magnetoelastic energy due to work on cubic material in the presence of elastic tensile stress by rotation of M_s until θ (Fig. 9.6). $E_{MS} = 0$ if M_s // δ, $E_{MS} = \frac{3}{2}\lambda_{si}\delta$ is maximum if $M_s \perp \delta$, and $\frac{3}{2}\lambda_{si}\delta$ is positive and $E_{MS} = -\frac{3}{2}\lambda_{si}\delta$ is minimum if $M_s \perp \delta$ and $\frac{3}{2}\lambda_{si}\delta$ are negative. For uniaxial anisotropy, the governing relation is $E_a = K_u\sin^2\theta$, and for uniaxial magnetoelastic energy, it is $E_{ME} = K_\delta\sin^2\theta$; this means uniaxial stress anisotropy is constant:

$$K_\delta = \frac{3}{2}\lambda_{si}\delta\ (\mathrm{CGS})\ \text{or}\ K_\delta = \frac{3}{2}\mu_0\lambda_{si}\delta\ (\mathrm{SI}) \quad (9.30)$$

Saturating field equation for uniaxial stress anisotropy can be written utilizing from Eq. (9.30) as:

$$H = \frac{2K_\delta}{M_s} = \frac{3\lambda_{si}\delta}{M_s} \tag{9.31}$$

Magnetostrictive transducers which convert electrical energy into mechanical energy are the main application of magnetostriction effect like at sonar (sound navigation and ranging) and ultrasonic sound generator for cleaning purposes.

9.2.3.6 Exchange Anisotropy

Exchange anisotropy defines the magnetic coupling of two different types of spin systems across the interface between them. So far, this type of magnetic coupling was observed between ferro-antiferromagnetic, ferri-antiferromagnetic, and ferri-ferromagnetic phases. A shifted hysteresis loop also known as exchange bias (H_E) and rotational hysteresis are expected results in magnetic field greater than $2K/M_s$ and in temperature below T_N, if one of the interacting phases is antiferromagnetic. In 1956 and 1957, first reports about the exchange anisotropy came from W. H. Meiklejohn and C. P. Bean (Meiklejohn and Bean 1956; Meiklejohn and Bean 1957). They performed low-temperature magnetization measurement on 20 nm single-domain size ferromagnetic core Co particles covered by antiferromagnetic CoO nanolayer. Fine particles were cooled down to 77 K in a strong applied field, and hysteresis curve was recorded between $\mp 10\ kOe$. The recorded M-H curve is shown in Fig. 9.7, is not symmetrical about the origin, shifted toward left ($H_E = 1600$ Oe), and has higher coercivity (Meiklejohn and Bean 1956; Meiklejohn and Bean 1957). However, a ZFC magnetization measurement provides a symmetric hysteresis loop obtained from the fine particles. A shifted hysteresis loop assigns a ferro-antiferromagnetic coupling as a system of $sin\theta$ torque function. The $sin\theta$ torque function indicates that anisotropy is unidirectional instead of expected uniaxial

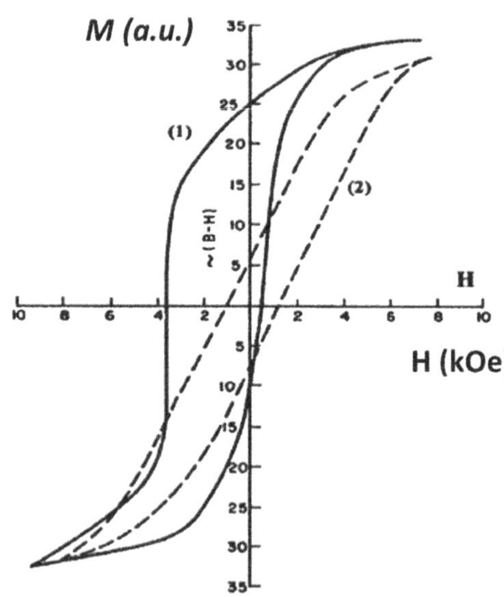

Fig. 9.7 Hysteresis loops at 77 K of partially oxidized Co particles. Solid curve (1) is the recorded loop after cooling the fine particle assembly in a 10 kOe field. Dashed curve (2) shows the recorded loop when assembly was cooled in zero field. *Reprinted with permission from* Meiklejohn and Bean (1956), Meiklejohn and Bean (1957)

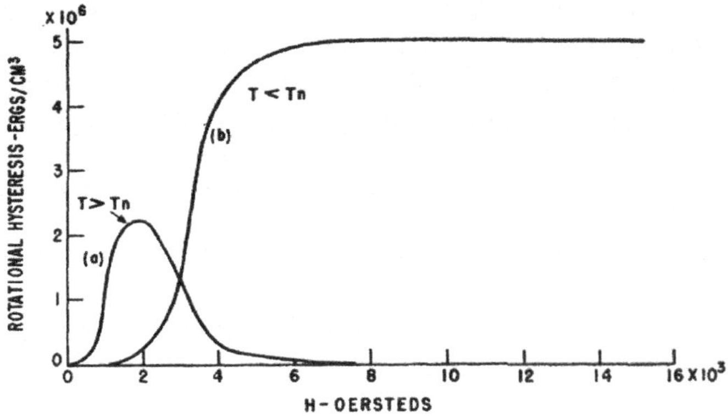

Fig. 9.8 Rotational hysteresis of 20 nm Co NPs covered by CoO shell. Curves (a) and (b) were recorded at 300 and 77 K, respectively

anisotropy represented by a torque function $L = -2K_1 \sin 2\theta$ (Meiklejohn and Bean 1956; Meiklejohn and Bean 1957). Total free energy expression for core Co particles is covered by CoO shell when they are exposed to an applied field anti-parallel to M_s (Berkowitz and Takano 1999):

$$E_F = M_s H cos\theta - K_{ud} cos\theta + K_1 \, sin^2\theta \tag{9.32}$$

where θ is the angle between the easy direction and M_s the direction of saturation magnetization and K_{ud} and K_1 are the unidirectional and uniaxial anisotropy energy constants, respectively. Unidirectional anisotropy energy is proportional to first power, rather than square of cosine function. Solution of Eq. (9.32) is expressed in terms of effective field H_{eff}:

$$H_{eff} = H - \frac{K_{ud}}{M_S} \tag{9.33}$$

where $\frac{K_{ud}}{M_S} = H_E$ specifies the displacement magnitude of hysteresis loop.

The sin θ torque curve is displaced from the zero L axis due to a rotational hysteresis loss, W_r. This loss at high magnetic fields is the second unusual feature of exchange anisotropy and is probably a more general characteristic feature than a shifted M-H curve (Meiklejohn 1962). Rotational hysteresis loss occurs in a pure ferromagnetic particle with uniaxial anisotropy for magnetic fields greater than K_1/M_S but less than $2K_1/M_S$ at temperature $T < (T_{\text{N,CoO}} \sim 290$ K). This type of loss is represented in Co-CoO fine particles by curve (*a*) of Fig. 9.8, where the temperature of the antiferromagnet is $T > T_{\text{N,CoO}}$, antiferromagnetic moments are disordered (paramagnetic state), and therefore the exchange anisotropy is zero. Curve (*b*) of Fig. 9.8 illustrates the rotational hysteresis when $T < T_{\text{N,CoO}}$ and the properly ordered antiferromagnetic moments are coupled to the ferromagnetic moments. The rotational hysteresis is finite for H up to 15 kOe. The observed nonvanishing rotational hysteresis in magnetic fields greater than $2K_1/M_S$ is an indication of an exchange coupling. However, a pure unidirectional anisotropy which accounts for *sin* θ torque function and shifted loop is definitely not the physical origin of rotational hysteresis. In magnetic fields much greater than $2K_1/M_S$, a finite W_r exits depending on a criteria which is $K_{ud} > K_{1,AFM}$. The magnitude of W_r depends on both thickness of antiferromagnetic material (AFM) and ratio $K_{ud}/K_{1,AFM}$ (Berkowitz et al. 1999).

A schematic diagram of the magnetic moment configuration can explain the concepts of unidirectional anisotropy and exchange bias by assuming the exchange interaction exits at an FM-AFM interface. Let's interpret each step by accepting upward direction of field as positive direction in Fig. 9.9. (a) When $T_N < T < T_C$, applied field H_1 magnetizes only the FM layer upward while AFM layer stays disordered; (b) temperature is cooling to $T < T_N < T_C$ for this and for all the rest steps. Both layers are properly ordered; (c) H_1 is reversed, and spins of FM layer could not align reversely due to the exchange coupling with spins of AFM layer. Antiferromagnetic interaction is so strong that just a partial rotation of a line of spin exits at interface. (d) A fairly high reverse field H_2 is applied; the spins of FM layer could coherently rotate and get aligned in reverse direction. In AFM layer, just a few lines of spins at interface partially rotate due exchange interaction. The present reverse direction of spins of FM layer is not stable, since the strong exchange coupling with spins of AFM layer still tries to reverse them in positive direction. Therefore, there is a single stable orientation for the spins of FM layer, i.e., the anisotropy is unidirectional; (e) H_3, which has smaller magnitude with respect to H_1, is applied upward again. The spins of FM layer can easily rotate via the extra torque contribution originates from the exchange coupling with AFM layer. Briefly, FM layer behaves as if it has an extra (internal) biasing field, i.e., exchange bias (Nogués and Schuller 1999). The extensive developments on both thin film growth technology and on the quantum theory of solids led a wide usage of exchange coupling at high technological devices like reading heads of hard disk drives of computer or the spin valves at spintronic devices.

9.3 Magnetic Domains

Ferromagnetic materials have so strong internal molecular magnetic field which can magnetize them until saturation at temperatures below their Curie temperature (T_C) as well as above. That is why ferromagnetics are accepted to be self-saturated or spontaneously magnetized substances even in absence of an externally applied magnetic field. The strong ferromagnetics like Fe, Co, and Ni have high T_C magnitudes of 1043, 1400, and 627 K, respectively. Here a question instantly comes to mind that how one can get a metal piece of Fe at room temperature in the demagnetized state-like paramagnets. In 1906, Pierre Weiss made a brilliant assumption to answer this question: a ferromagnet in the demagnetized state is divided into a number of small and magnetized multi-regions called "domains" (Weiss 1906). Each domain has self-saturated of magnetization, M_s, and the

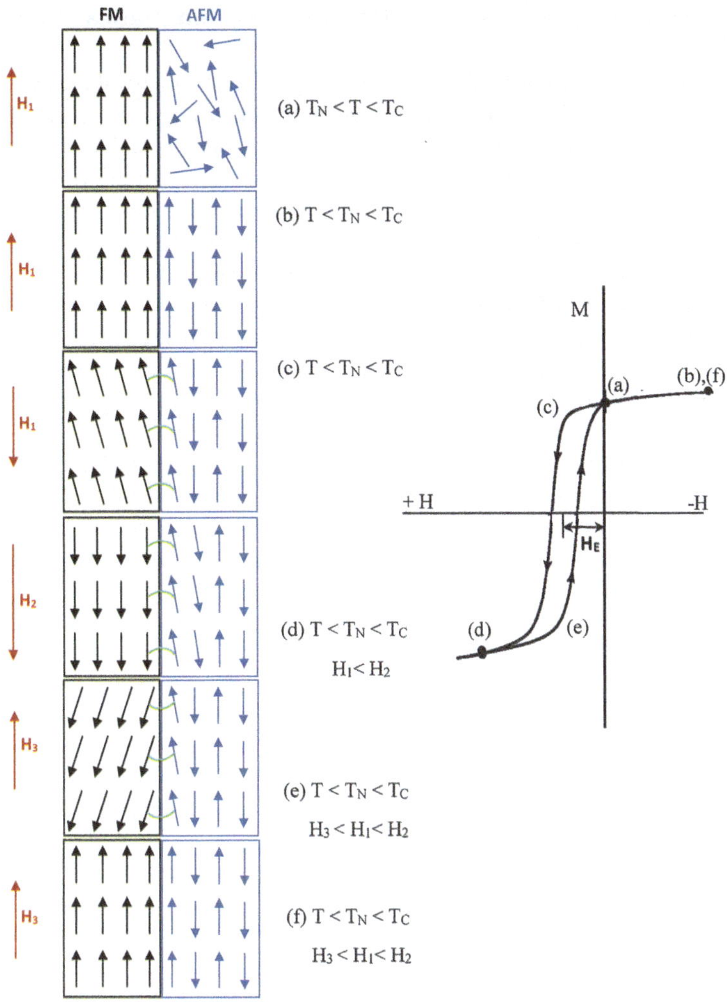

Fig. 9.9 Schematic representation of magnetic moments of a FM-AFM bilayer. (**a**) AFM layer is at paramagnetic state since $T > T_N$; (**b–f**) different stages of an exchange biased hysteresis loop

directions of M_s vectors of the nearby domains are such that metal Fe piece as a whole has no net magnetization. However, first experimental evidences on magnetic domain patterns in single crystals of silicon-iron were published by H. J. Williams, R. M. Bozorth, and W. Shockley in 1949 at the Bell Telephone Laboratories (Williams et al. 1949). Since then, domain theory has become central to any discussion of magnetization process.

9.3.1 Multi-magnetic Domain Structure and Magnetization Process

Quantum mechanical exchange forces were assigned as the physical origin of strong molecular field in ferro- and ferrimagnetics by Heisenberg in 1928. Type of exchange interaction retains the spins (or magnetic moments) parallel or antiparallel. However, there is a limit distance for spins to maintain their collinearity throughout the material. This distance depends on the type of material, but it is usually 100 nm or smaller. If we consider a circular disk made of a ferromagnetic material with 1 cm radius, it is very normal to have multi-domain structure or to have spontaneously magnetized magnetic regions more than 100. Even nanoparticles (NPs) which have a size larger than a critical diameter must have multi-domain structured nature.

Magnetization is a process of converting the ferromagnetic specimen from a multi-domain state into one in which it is a single domain magnetized in the same direction as the applied field. Let's imagine the simplest case for our disk-shaped ferromagnetic material in the demagnetized state that is provided by only two domains. Figure 9.10a shows the schematic representation of this simplest multi-domain structured material and its magnetization process in the Fig. 9.10b, c, d (Cullity and Graham 2008). Two domains are spontaneously magnetized at equal magnitude and separated by a boundary (illustrated with dashed line in the middle of disk) called as domain wall. Magnetization direction of domains is specified by the crystal structure of disk's material. In the absence of an external field, resultant magnetization (M) is zero due to opposite direction of spontaneous magnetizations (M_s) in the domains. In Fig. 9.10b, a static field H_1 is applied by making an angle θ with M_s vector of upper field, causing the upper domain to grow at the expense of the lower one by downward motion of the domain wall. In Fig. 9.10c, the wall moved right out of the region due to H_2 which has higher magnitude than H_1. Finally, at much higher fields like H_3, magnetic moment of each atom rotates until all are parallel with the applied field, that is, M_S vector parallel to H_3 as in Fig. 9.10d. This last process is called as domain rotation; a fairly large field H_3 does work against the

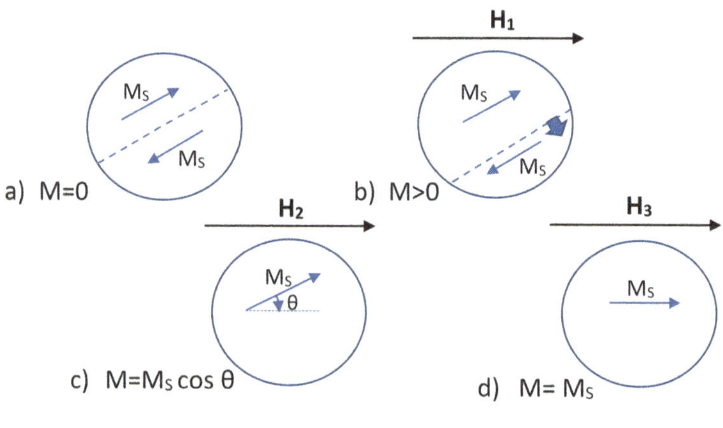

Fig. 9.10 Schematic representation of magnetization process for a ferromagnet ($H_3 >> H_2 > H_1$), where: (**a**) M = 0, (**b**) M > 0, (**c**) $M = M_s\cos\theta$, and (**d**) $M = M_s$

crystal anisotropy which is regarded as a force that tends to hold the magnetization in certain equivalent crystallographic directions of crystal structure. During the whole process, there were no changes in the magnitude of M_S of any domain, only in the direction of M_s.

9.3.2 Single-Domain Structure and Superparamagnetism

A fine particle is accepted to have single-domain structure when all spins (or magnetic moments) retain their parallel position, both in the zero applied field and during reversal process in an applied field. This type of reversal is called coherent rotation. A maximum critical size, mostly in the order of a few tens of diameter, is foreseen to maintain this nature. Such as critical size for spherical Fe particles is around 17 nm. Larger critical sizes are possible with the larger anisotropy. We learned the rotation of magnetic moments (or domain rotation) as last step of magnetization process of magnetic samples in the previous section. In single-domain structured particles, magnetization realizes only by rotation of magnetic moments, and there is no any domain wall in the structure. This case was investigated by E. C. Stoner and E. P. Wohlfarth in detail (Stoner and Wohlfarth 1948). When an applied field H rotates the M_s of a single-domain particle out of easy axis at angle θ, work is done against any type of uniaxial crystal, shape, or stress anisotropies. Analyses will continue using a prolate spheroid particle having uniaxial shape anisotropy as seen in Fig. 9.11.

Uniaxial anisotropy energy in general $E_a = K_u \sin^2\theta$, θ, is the angle between easy c axis and M_s. H makes angle α between easy c axis. Then, potential energy is:

$$E_p = -M_s.H = -M_sH\ cos\ (\alpha - \theta) \tag{9.34}$$

Total energy equals to summation of potential and uniaxial anisotropy energies:

$$E = E_a + E_p = K_u\, sin^2\theta - M_sH\ cos\ (\alpha - \theta) \tag{9.35}$$

Equilibrium condition for M_S is determined by $\frac{dE}{d\theta} = 0$:

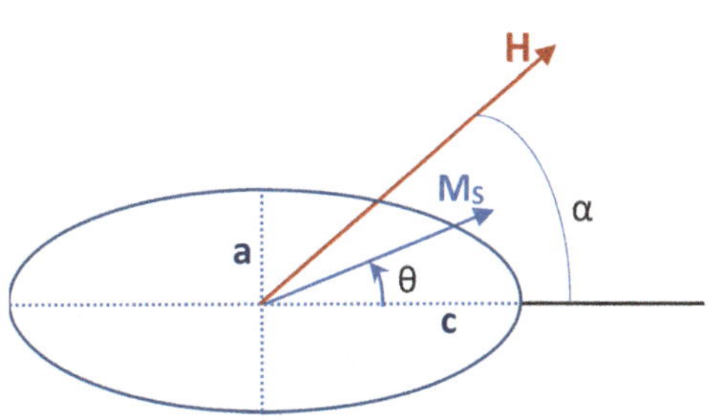

Fig. 9.11 Prolate spheroid nanoparticle having uniaxial shape anisotropy

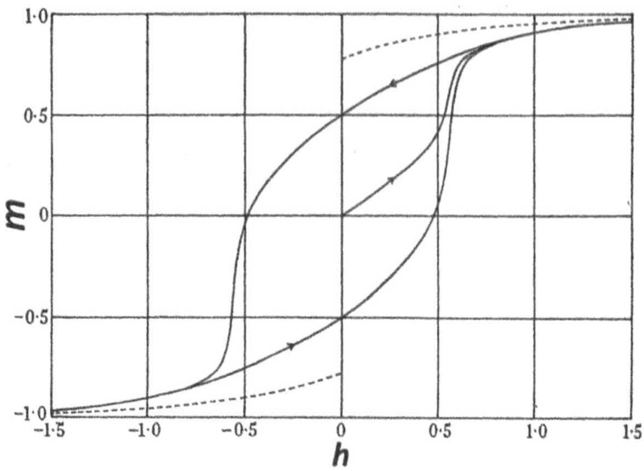

Fig. 9.12 Magnetization curves for prolate (solid line) and oblate (dashed line) spheroid non-interacting particles. *Reprinted with permission from* Stoner and Wohlfarth (1948)

$$\frac{dE}{d\theta} = 2K_u \sin\,\theta\,\cos\,\theta - M_s H\,\sin\,(\alpha - \theta) = 0 \tag{9.36}$$

Component of magnetization in the H direction:

$$M = M_s \cos\,(\alpha - \theta) \tag{9.37}$$

If H is applied along hard a-axis, in this case $\alpha = 90^0$, then from Eq. (9.36):

$$2K_u \sin\,\theta\,\cos\,\theta = M_s H\,\cos\,\theta \text{ and } M = M_s \sin\,\theta \tag{9.38}$$

Therefore,

$$2K_u \frac{M}{M_s} = M_s H \tag{9.39}$$

where $\frac{M}{M_s} = m$ is called as normalized magnetization, then:

$$m = H\frac{M_s}{2K_u} \tag{9.40}$$

This expression proves that normalized magnetization is a linear function of H. If the applied field is enough to saturate the particle, then its magnitude equals to anisotropy field H_K (also called as intrinsic coercive field, H_{ci}), that is, $H = H_K = \frac{2K_u}{M_s}$. The ratio $\frac{H}{H_K} = H\frac{M_s}{2K_u} = h$ is called normalized field. Then $m = h$ when $\alpha = 90^\circ$.

Stoner and Wohlfarth published some of the calculated hysteresis loops corresponding to various angle of α for prolate spheroid (Stoner and Wohlfarth 1948). Stoner and Wohlfarth have also plotted the hysteresis loops of non-interacting single-domain prolate and oblate spheroid non-interacting fine particles (Fig. 9.12).

The easy axes of assembly of fine particles are randomly oriented in space that is magnetically isotropic. Some characteristic data for randomly oriented prolate ellipsoids is tabulated as below (Stoner and Wohlfarth 1948):

Coercivity: $H_c = h(N_a - N_c)M_s = 0.479(N_a - N_c)M_s$, that is, $h \cong 0.48$

Initial susceptibility: $\chi_0 = \frac{2}{3}\frac{1}{(N_a - N_c)}$

Remanent magnetization: $M_r = mM_s = 0.5M_s$, that is, $m = 0.5$

The ratio M_r/M_S is also called squareness ratio (*SQR*), and it measures how square is the hysteresis loop. According to Stoner-Wohlfarth model, $SQR = m = 0.5$ is given as decisive value for single-domain NPs (Sadaqat et al. 2019; Almessiere et al. 2020c). We will now reconsider all three types of uniaxial anisotropies by defining a new governing equation, $H = h\left(\frac{2K_u}{M_s}\right)$.

$$Crystal: \quad H = h\left(\frac{2K_1}{M_s}\right) \tag{9.41}$$

$$Shape: H = h\,(N_a - N_c)M_s \tag{9.42}$$

$$Stress: H = h\left(\frac{3\lambda_{si}\delta}{M_s}\right) \tag{9.43}$$

The magnitude of H_K (or intrinsic coercivities, H_{ci}) can be calculated by using above equations. Since h is at the same side of each equation, anisotropy field is directly proportional with M_S for shape anisotropy and indirectly proportional with M_S for crystal and stress anisotropies. There is great influence of crystal and shape anisotropies on coercivity, in other words on magnetic hardness of fine particles.

All ferromagnetic or ferrimagnetic fine particles do not exist with single-domain structure. However, ones that are smaller than a critical size are expected to have single-domain structure in the absence of an externally applying field. All magnetic moments align in a preferential direction determined by the leading type of anisotropy in the structure. That is, a single-domain structured nanoparticle behaves like a single giant magnetic moment having a net spontaneous magnetization M_S. This type of magnetism exhibited by single-domain NPs is called the superparamagnetism (SP). Deeper descriptions will be given in the next sections.

9.3.3 Domain Wall Structure

A domain wall is a separating region or interface between two domains that are self-magnetized in different easy directions. All magnetic materials which are in partial or complete demagnetized state contain domain walls. Inside of wall region, spins reorient from one crystallographic easy direction to another. If we imagine a very thin domain wall for a ferromagnetic sample, exchange coupling between the antiparallel spins at both sides of wall would be very large. That is why wider domain walls are needed to retain the minimum energy equilibrium condition in the sample. This case requires the minimum magnitude of angle (∅) between adjacent spins of wall, too. Imagine that angle between two neighboring domain regions is 90° and total number of reorienting spins in the wall region is N. This means exchange coupling angle between two interacting spins in wall region is $\varnothing = \pi/2N$. On the other hand, crystal anisotropy energy is correlated with the angle between wall spin and easy axis. Crystal anisotropy tries to reduce the number of non-easy directional spins to minimize the energy. Hence, the wall thickness is determined according to relativistic magnitudes of two competing types of interactions in the ferromagnetic sample. Additionally, domain wall has energy per unit area of its surface, originating from the summation of exchange energy (E_{ex}) and crystal anisotropy energy (larger with respect to adjoining domain due to orienting spins in non-easy directions) within the wall.

9.3.3.1 Bloch Wall

First theoretical investigations on domain wall were conducted by Felix Bloch in 1932 (Bloch 1930). After his published study, domain walls are often called as Bloch walls. However, here we will use the term of "Bloch wall" for bulk materials that have usually much larger size with respect to thickness of their domain wall. Total domain wall energy includes both E_{ex} and E_K:

$$E_{wall} = E_{ex} + E_K = -2Acos\frac{d\varnothing}{dx} + g(\varnothing) \tag{9.44}$$

where A is the exchange stiffness constant in terms of Erg/cm, $\frac{d\varnothing}{dx}$ is the displacement rate of angular position of local magnetization in the wall, and $g(\varnothing)$ is a general function defining any type of crystal anisotropy. For uniaxial anisotropy $g(\varnothing) = K_u\sin^2\varnothing$, domain wall is called as *180°* domain wall, and its structure is schematically shown in Fig. 9.13. Considering the cubic anisotropy, general function is $g(\varnothing) = K_1\sin^2\varnothing\cos^2\varnothing$. Utilizing from the series expansion of cosine function, $\cos\varnothing \cong 1 - \frac{\varnothing^2}{2}$, we get exchange energy equation as $E_{ex} = -2A + A\left(\frac{d\varnothing}{dx}\right)^2$; here the first term can be dropped since it is independent of angle. So, for *total wall energy* ($E_{T,\ wall}$), Eq. (9.44) should be integrated over x:

$$E_{T,wall} = \int_{-\infty}^{\infty}\left[A\left(\frac{d\varnothing}{dx}\right)^2 + g(\varnothing)\right]dx \tag{9.45}$$

Considering again the simplest case where the 180° domain wall having uniaxial anisotropy, the following relations are obtained, respectively: a net domain wall energy expression and a relation between x and ∅ through the domain wall:

$$E_{T,uniaxial} = 4\sqrt{AK_1} \tag{9.46}$$

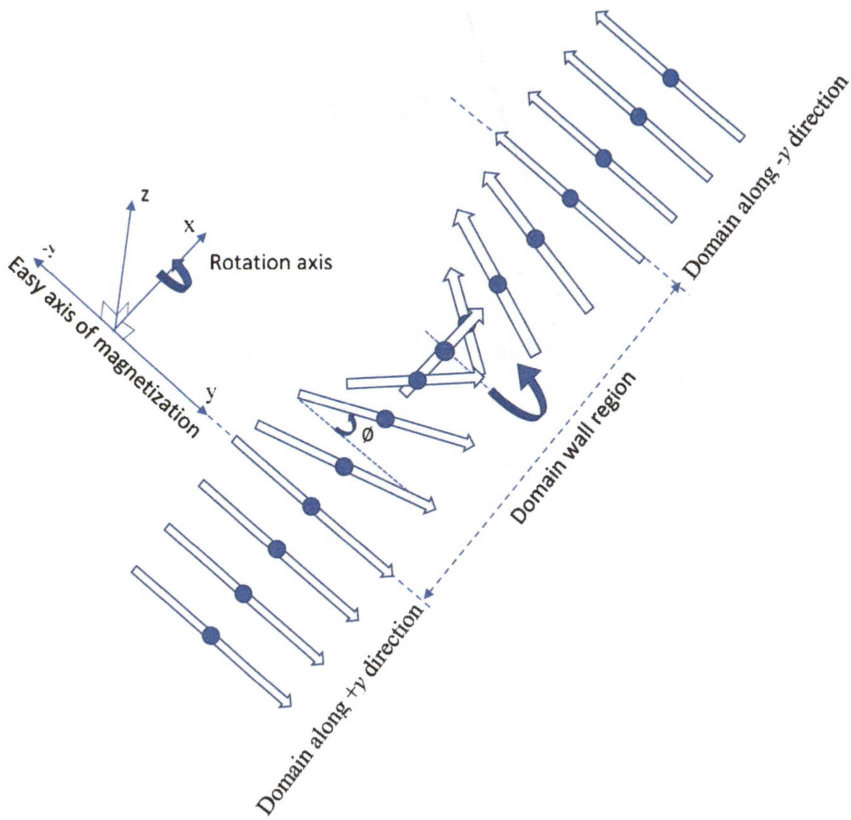

Fig. 9.13 Schematic structure of 180° domain wall for bulk materials. The spins of domains which have the $\pm y$ easy axis rotate around x-axis

$$x = \sqrt{\frac{A}{K_u}} ln \left(tan \frac{\varnothing}{2} \right) \tag{9.47}$$

For cubic anisotropy, exactly same expression with Eq. (9.46) is obtained. According to Eq. (9.47), the wall thickness is infinitely thick formally. However, an effective wall thickness (Δ) is defined for the constant value of $\frac{d\varnothing}{dx}$ at the center of wall thickness. For uniaxial anisotropy, the effective wall thickness expression is:

$$\Delta = \pi \sqrt{\frac{A}{K_u}} \tag{9.48}$$

The exchange stiffness constant is directly proportional to exchange constant, J as, $A = \frac{nJS^2}{a}$ where n is the number of atoms in unit cell, S is the spin of the adjacent identical atoms, and a is the lattice parameter. The magnitude of J for cubic iron can also be estimated from a direct relation $J \approx 3k_BT_C \approx 4 \times 10^{-14}$ Erg. Then

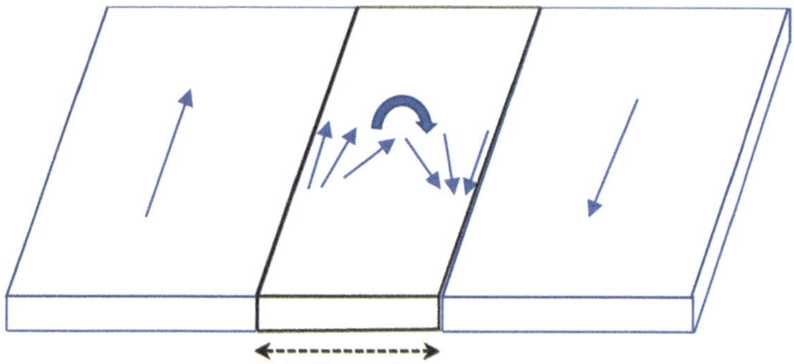

Fig. 9.14 Néel type of domain wall in thin film and rotation of spins in wall region

$A \approx 7 \times 10^{-7}$ Erg/cm is obtained for $n = 2$, $S = 1/2$ and $a = 2.87 \times 10^{-8}$ cm. Total domain wall energy and effective wall thickness are found to be $E_{T,\text{cubic}} = 4\sqrt{AK_1} \cong 2.32 \frac{\text{Erg}}{\text{cm}^2}$ and $\Delta = \pi\sqrt{\frac{A}{K_1}} = 3.8 \times 10^{-6}$ cm $= 38$ nm for $K_1 = 4.8x10^5 \frac{Erg}{cm^3}$ for cubic Fe. There exit two types of 90° wall (Cullity and Graham 2008). Magnetic samples like Nickel have $K_1 < 0$, and its domain wall separates magnetization directions of adjoining domains in two opposite directions of (Almessiere et al. 2019l) easy axis, that is, 180° type wall is expected. However, Ni has 70° and 109° walls instead of two 90° walls.

9.3.3.2 Néel Wall

Néel type of domain wall is usually observed in thin films. Basic difference according to Bloch wall is that the spins in wall region change their magnetization direction in a plane that is parallel to film plane. Rotation of spins in the film plane is shown in Fig. 9.14.

9.4 Magnetization Measurements

9.4.1 Influence of Surface Spins Disorder in Nanomaterials on the Magnetization

Most of findings, either theoretical or experimental, have showed that the magnetization of NPs is smaller compared to the equivalent bulk sample. Largely, the magnetization reduces quickly by reducing the dimension of very small NPs (Slimani et al. 2019e). The dissimilarity among the magnetization of NPs and bulk systems has been well explained experimentally by J.M.D. Coey (Coey 1971). It was found that at the surface of tiny NPs, the spins are disordered and randomly oriented, not like those in the core (interior). Actually, numerous models have been suggested for various nanostructured materials concerning the spins disorder at the surface of

NPs. J.M.D. Coey (Coey 1971) proposed the core-shell model, wherein the spins are ordered and aligned along the same direction in the core, whereas they are oriented along different direction at the surface layer. In two different studies, Berkowitz and co-workers (Berkowitz et al. 1975; Berkowitz et al. 1980) proposed the "spin canting" model for the spin disorder at the surface of Ni spinel ferrite NPs and the "spin pinning" model when the NPs are coated with an organic surfactant. In another study, R.H. Kodama and co-workers (Kodama et al. 1996) suggested the model where the spins in the core are ferrimagnetically aligned and the surface layer is of spin glass-like structure wherein the canted spins are frozen in such a state. The surface layer of NPs, where the spins are disordered, was experimentally verified through transmission electron microscopy (TEM) observations for NiZn spinel ferrite NPs dispersed in amorphous silica oxide matrix (Kodama et al. 1996); magnetization versus field (M-H) measurements for colloidal magnetite NPs (Hrianca et al. 2002); magnetic resonance (like electron spin resonance (ESR) and ferromagnetic resonance (FMR)) for NiZn spinel ferrite NPs (Caizer 2008), for nanocomposites of Fe_2O_3 dispersed over SiO_2 matrix (Cannas et al. 1998), and for Mn spinel ferrite NPs with oleic acid as surfactant (Upadhyay et al. 2000); polarized neutron powder diffraction technique for Co spinel ferrite NPs uncovered and covered with oleic acid (Lin et al. 1995); as well as via Mössbauer spectroscopy technique for Ni spinel ferrite and γ-Fe_2O_3 NPs (Tronc et al. 2000; Morr and Haneda 1981). The spins disorder at the surface is caused by the change of the exchange interactions among superficial magnetic ions in non-complete coordination (Berkowitz et al. 1999). In the event of ferrimagnetic NPs, the orientation of magnetic moments at the surface could be further adjusted since the exchange interactions are performed via the ions of oxygen O^{2-} (super-exchange interactions). Consequently, the non-existence of ions at the surface or the existence of other atoms (ions) as impurities leads to break the exchange interactions (broken exchange bonds) among the magnetic cations, which will induce the spins disorder at the surface (Kodama et al. 1996).

The spins disorder could alter the magnetic characteristics of NPs, particularly once the surface-to-volume ratio is high (Caizer 2015; Slimani et al. 2019f). Because of the surface effects stated above and to the core-shell structure, the magnetization of NPs will be significantly smaller compared to the bulk counterpart system.

Considering now the surface layer of NPs with no magnetic ordering (by taking into consideration the core-shell morphology wherein the spins are aligned in the core because of the ferri- and ferromagnetic exchange interactions, whereas they are disoriented at the shell), the magnetization formula of a NP should consider the magnetic volume ($V_{m,\ NP}$) of the NP describing the core volume of NPs wherein the spins are ordered and aligned, which is smaller compared to the physical volume of the NP (V_{NP}), i.e., $V_{m,\ NP} < V_{NP}$. For NPs with spherical shape (Fig. 9.15), one could measure the volume of the superficial layer of NPs that involves the disordered spins:

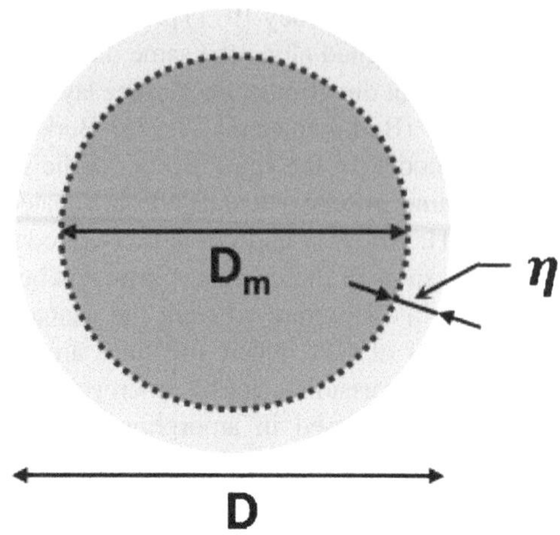

Fig. 9.15 Core-shell structure of a spherical NP

$$\Delta V = V_{NP} - V_{m,NP} = \frac{\pi}{6}\left(D^3 - D_m^3\right) \tag{9.49}$$

and the thickness of the superficial layer:

$$\eta = \frac{1}{2}(D - D_m) \tag{9.50}$$

The disparity among the saturation magnetization of the bulk and that of the NPs belonging to the similar compound (ΔM_{sat}) rises as the NPs are smaller. The difference reduces as the dimensions of the NPs increase, and it is being unimportant ($\Delta M_{sat} \rightarrow 0$) for very large nanoparticles (hundreds of nanometers or more). Such result could be simply understood when the surface-to-volume ratio of the NPs (S_{NP}/V_{NP}) is considered that could be determined in the approximation of spherical NPs as:

$$\frac{S_{NP}}{V_{NP}} = \frac{6}{D_m} \tag{9.51}$$

When the magnetic volume of NPs reduces (i.e., with reducing the diameter D_m), the S_{NP}/V_{NP} ratio would rise, which reach great values in the nm domain. For instance, for NPs having average diameter of 10 nm, the S_{NP}/V_{NP} ratio will be around $6 \times 10^8 m^{-1}$. For instance, if $\eta = 0.8$ and evaluating the surface layer contribution regarding the core of the NP, $\Delta V/V_m = (V - V_m)/V_m = (D/D_m)^3 - 1$, one will get the values of 0.05, 0.69, and 2.18 for 100, 10, and 5 nm, respectively. These results show that, for smaller NPs, the contribution of the surface spins of NPs is being very significant and will not be neglected and, consequently, an important decrease of the saturation magnetization will happen. The effect is reversed for larger NPs. There will be a rise in the contribution of the core spins, leading to an improvement of the magnetization. Once the contribution of the surface spins is being unimportant, it will conduce to the magnetization value of the bulk.

Considering the thickness of the nanoparticles surface layer, M_{sat} is estimated as:

$$M_{sat} = M_s(V_{m,NP}/V_{NP}) \tag{9.52}$$

By using Eqs. (9.49) and (9.50), one can write:

$$M_{sat} = M_s\left(\frac{V_{NP} - \Delta V}{V_{NP}}\right) = M_s\left(1 - \frac{2\eta}{D}\right)^2 \tag{9.53}$$

By knowing the experimental saturation magnetization and the diameter D (determined for example via TEM), this expression will lead to estimate more precisely the thickness of the surface layer of NPs through the experimental data, than by using the equation $M'_{sat} = M_s\left(1 - \frac{6}{D}\right)$ (Chen et al. 1996). The later expression is appropriate only for smallest thickness values, generally below 0.3 nm. At greater thickness, the important errors happen once utilizing the later expression.

The condition is being more complexed for the situation of surfacted NPs or those implanted in various non-magnetic solid matrices (non-crystalline or crystalline). After the synthesis procedure (via various chemical and physical techniques), in addition to the effect reported above, another layer at the surface of NPs (with no magnetic ordering and with thickness ranging between 1 and 2 up to tens of nm) could also appear as a consequence of various procedures (chemisorption, adsorption, creation of bonds with the magnetic ions at the surface of nanoparticles, etc.) (Rosensweig 1985). The matters are not yet elucidated for these conditions. For instance, whether one could take into account two layers resulted from the surface effect, or only of one layer decided by their collective effects. Numerous investigations have been performed on this subject (Caizer and Hrianca 2003a; Caizer and Hrianca 2003b); nevertheless a justifiable universal magnetic pattern of NP does not occur.

9.4.2 Influence of Temperature on the Magnetization of Nanomaterials

The dependence of the magnetization of nanomaterials on temperature is a widely debated subject; nonetheless it is not yet a well-elucidated matter, particularly at lower temperatures. There exist numerous investigations related to this issue (Almessiere et al. 2019h; Almessiere et al. 2019i; Almessiere et al. 2019j; Almessiere et al. 2019k; Slimani et al. 2019g). The surface effects for the situation of smaller NPs, or the interfacial effects for the situation of surfactants or those implanted in diverse matrices of NPs, could affect the evolution of the magnetization against the temperature of nanomaterials (Almessiere et al. 2020d).

Currently, this issue is being more comprehensible. Indeed, at higher temperatures, the temperature wherein M_s attains zero and a transition from

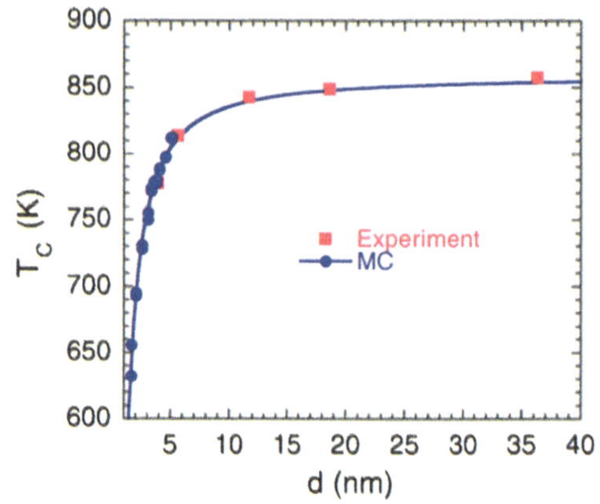

Fig. 9.16 Variation of T_C against the mean diameter d of Fe_3O_4 NPs. *Reprinted with permission from* Lu et al. (2008)

ferrimagnetic (or ferromagnetic) state to paramagnetic state happens (noted as Curie temperature, T_C) is influenced by the dimensions of NPs. As demonstrated by both experiment and theory, T_C of NPs diminishes when the diameter of NPs reduces (Lu et al. 2008; He and Shi 2012; Mayama and Naito 2009). It has demonstrated that T_C could be connected to the NP diameter (D) by a finite-size scaling law (in accordance with the simulation of Monte Carlo (MC)) (Fisher and Ferdinand 1967):

$$T_C(D) = T_{C,bulk}\left[1 - \left({}^{d_0}/_{D}\right)^{\frac{1}{\nu}}\right] \tag{9.54}$$

In this expression, $T_{C,\ bulk}$ is the Curie temperature of bulk material, d_0 is the microscopic length scale near the lattice constant, and ν is an exponent that takes the value of about 0.705 assessed theoretically from the Heisenberg model (Chen et al. 1993). Fig. 9.16 illustrates the evolution of T_C against the mean diameter d ($\equiv D$) of SiO_2-covering Fe_3O_4 NPs (Wang et al. 2011). It is obvious that the MC simulation (circles and solid line) (Mazo-Zuluaga et al. 2008) fits appropriately the experimental data (square symbols), which gives $d_0 = 0.51$ nm and $\nu = 0.82$.

Nevertheless, at lower temperatures, the attained findings have revealed that there exists a great impact of the surface layer of NPs (or NP dimension) on the evolution of magnetization against temperature. For bulk material and at lower temperatures, the evolution of M_s value versus temperature (T) obeyed a law in $T^{3/2}$ (Bloch law), determined from the model of spin wave (Bloch 1930; Caizer 2005):

$$M_s(T) = M_s(0)\left(1 - jT^{3/2}\right) \tag{9.55}$$

In this expression, $M_s(0)$ is M_s value at $T = 0K$ and the constant j is depending on the exchange integral J as $j \sim 1/J^{3/2}$. For example, the correlation reported in Eq. (9.55) was proved by experiment up to ambient temperature for some spinel ferrite NPs like Mn ferrites as well as for bulk systems (Caizer 2005; Aldred 1975).

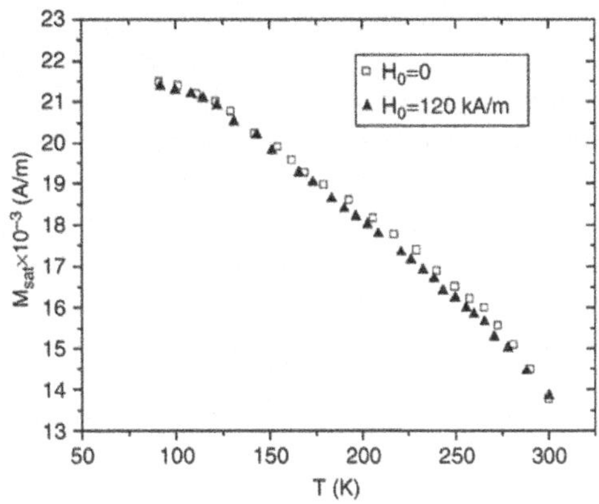

Fig. 9.17 Evolution of M_{sat} versus T for Mn spinel ferrite NPs coated with oleic acid in the absence ($H = 0$ *kA/m*) and the presence of an applied magnetic field ($H = 120$ *kA/m*). *Reproduced with permission from* Caizer (2005)

Certain disparities could only be noticed for the exponent value of temperature, such as for the case of Fe_3O_4 where this exponent takes a rough value of 2.

For clusters and fine particles, certain experimental findings along with theoretical calculations have revealed that the exponent of the temperature is greater than 3/2 (Hendriksen et al. 1992; Linderoth et al. 1993; Hendriksen et al. 1993). Nevertheless, Martinez and co-workers have proved by experiment that for maghemite NPs with size ranging between 10 and 15 nm in diameter, $M_s(T)$ do not obey the $T^{3/2}$ law up to ambient temperature (Kodama 1999). These findings indicated that $M_s(T)$ that was established for the bulk systems is not always applied to materials comprised of clusters and fine particles and there exist numerous explanations for such behavior. Moreover, once the NPs are coated by an organic surfactant (like oleic acid), the acid is greatly absorbed on the surface, and hence it creates a surface layer, which will alter the evolution of magnetization against temperature (Caizer 2008; Cannas et al. 1998).

Figure 9.17 shows a typical example for the evolution of M_{sat} versus T (from 300 down to 90 K) for Mn spinel ferrite NPs coated with oleic acid in the absence and the presence of an applied magnetic field (H) (Caizer 2005). Two significant parts could be noticed in this figure:

(i) A quick rise of the M_{sat} value by decreasing the temperature with a relative evolution of $\frac{\Delta M_{sat}}{M_{sat}(300K)}\tilde{5}6\%$, which is highly greater compared to that of corresponding bulk material (~20% (Caizer 2015)).
(ii) The rise in M_{sat} is independent to the fact that the NPs, and with their easy axes magnetization, were oriented due to the application of magnetic field along the subsequent measuring direction. This finding illustrates an irregular rise of M_{sat} of the coated NPs, and this rise is a fundamental characteristic of the particle.

In the same study (Caizer 2005), the authors showed the existence of a great deviation in M_{sat} versus T for coated NPs and that of bulk system. This difference is governed by the growth of the magnetic diameter attributed to the core of NPs in which the spins were aligned because of the super-exchange interactions. This explanation is built on their preceding findings that showed that the magnetic diameter of Mn spinel ferrite NPs coated by oleic acid rises with decreasing temperature (Caizer and Hrianca 2003a). The obtained findings conducted that also the magnetic packing fraction (f_m) rises with temperature, and this feature must be considered once the evolution of magnetization versus temperature for nanomaterials is established. $M_{sat} = M_s(0)$ for bulk ferrites, whereas M_{sat} vanishes entirely for ferrofluids case because $M_{sat} = f_m M_s(0)$. Below these conditions, f_m value for nanoparticles at a certain temperature T is written as:

$$f_m(T) = \frac{M_{sat}(T)}{M_s(T)} \tag{9.56}$$

and at 0 K, it could be written as:

$$f_m(0) = \frac{M_{sat}(0)}{M_s(0)} \tag{9.57}$$

It is not, anymore, a constant, but it rises when the temperature decreases. Moreover, if one replaces the $M_{sat}(T)$ and $M_{sat}(0)$ extracted from Eqs. (9.56) and (9.57), respectively, within the Bloch law (i.e., Eq. (9.55)), one can attain the following expression for coated NPs:

$$M_{sat}(T) = M_{sat}(0)\frac{f_m(T)}{f_m(0)}\left(1 - jT^{3/2}\right) = M_s(0)f_m(T)\left(1 - jT^{3/2}\right) \tag{9.58}$$

For further limited form, M_{sat} versus T for coated NPs could be written as:

$$M_{sat}(T) = M_s(T)\left(1 - jT^{3/2}\right) \tag{9.59}$$

in which here M_s is not a constant but depends on temperature as $M_s(T) = M_s(0)$ $f_m(T)$, not like for case of bulk ferrites.

Given that f_m is independent to temperature (i.e., constant like the one at ambient temperature), the Eq. (9.58) will be downgraded to the Eq. (9.55) (i.e., Bloch law for bulk materials) wherein $M_s(0) = constant$, and the rise of M_{sat} of a material will be the consequence of the evolution of M_s with respect to temperature.

If f_m is dependent to temperature (i.e., not constant), an additional term must be added to the expression to reflect this feature. Indeed, with the decreasing temperature, the magnetic diameter and the magnetic moment of NPs will increase. This means that the expression must be deliberated as it was expressed in Eq. (9.58).

Since $M_{sat}(T)$ of NPs is expressed as:

$$M_{sat}(T) = n\, m_{m,NP}(T) = nV_m(T)M_s(T) \tag{9.60}$$

where the concentration n is the number of NPs in the system volume (V_s), and by analogy with Eq. (9.56), it will result as:

$$f_m(T) = \frac{n\pi}{6} \langle D_m(T) \rangle^3 \tag{9.61}$$

where $\langle D_m(T) \rangle$ is the mean magnetic diameter of NPs versus temperature.

By introducing Eq. (9.61) within Eq. (9.58), and in the spherical NPs approximation, one gets the following expression:

$$M_{sat}(T) = \frac{n\pi}{6} M_s(0) \langle D_m(T) \rangle^3 \left(1 - jT^{3/2}\right) \tag{9.62}$$

According to this equation and because n is constant, it could be determined that the significant rise of M_{sat} value for coated NPs, in comparison to that for corresponding bulk ferrites, is resulted from the growth of $\langle D_m \rangle$ of NPs (where the spins are ordered by the effect of super-exchange interactions) and the rise of $M_s(T)$ with the decreasing temperature will be much lower. This interpretation is ascribed to the modified super-exchange energy (W_{exchange}) at the superficial layer of NPs because of the existence of surfactant molecules (Caizer and Hrianca 2003a). Therefore, the Néel temperature ($T_N \simeq W_{\text{exchange}}/k_B$, where k_B is the Boltzmann constant) at the surface layer will be changed, and it would be smaller than the ambient temperature. With decreasing T, T_N of the sub-layers that are adjoining the magnetic core of NPs will be surpassed progressively, and as a consequence, these sub-layers will be consecutively ferrimagnetically ordered. With the decreasing temperature, the outcome will be a growth of $\langle D_m \rangle$ associated to the cores of NPs wherein the spins are aligned, which will conduce to a rise of f_m, and indirectly, the saturation magnetization of NPs will enhance.

According to another route, and from Eq. (9.12) at $T < 300$ K and at ambient temperature ($T = 300$ K), one can get:

$$\langle D_m(T) \rangle^3 = \langle D_m \rangle^3_{T=300} \left(\frac{M_s}{M_{sat}} \right)_{T=300K} \left(\frac{M_{sat}(T)}{M_s(T)} \right) \tag{9.63}$$

$\langle D_m \rangle_{T=300K}$ is the mean magnetic diameter at $T = 300$ K, which can be deduced from magnetization experiment as described in the following study (Caizer 2002).

Furthermore, one can determine the thickness of the surface layer:

$$\langle \eta \rangle_{T=300K} = \frac{1}{2} \left(\langle D \rangle - \langle D_m \rangle_{T=300K} \right) \tag{9.64}$$

From Eq. (9.64), and when the D_m increases, it can be concluded that the superficial layer is reducing with decreasing the temperature. Taking into account that $\langle D_m \rangle$ is always lower than $\langle D \rangle$, the obtained findings reflect that the superficial layer of NPs is paramagnetic at ambient temperature. With the decreasing temperature, the superficial layer is being progressively magnetically ordered, beginning from the core side to the shell. By utilizing the ESR tool, the occurrence of a paramagnetic shell for ferrofluids of Mn spinel ferrite NPs coated by oleic acid has been exposed (Upadhyay et al. 2000; Sastry et al. 1995). The researchers have emphasized the presence of two absorption lines in the electron spin resonance spectra; the first one appeared because of the ferrimagnetic cores, and the second one is ascribed to Fe^{3+} ions in the complex structure comprised of molecules of oleic acid (in which the spectroscopic splitting factor ε is equal to 4). The line in which ε is equal to 4 vanishes at lower temperatures. Likewise, by using Mössbauer spectroscopy at lower temperatures, the researchers showed the presence of a paramagnetic layer at the superficial layer of phosphated Fe_2O_3 NPs (Tronc et al. 2000).

9.4.3 Magnetization of Nanomaterials with Applied Magnetic Field

From both practically and theoretically viewpoints, the understanding of the behavior of systems made of magnetic nanoparticles once they are magnetized by an externally applied magnetic field is very important. It has been well established that the magnetic comportment of NPs is extremely different to that of the bulk material counterpart since it is greatly affected by the dimensions of NPs, which decides a particular behavior. All these matters are described below, beginning from larger NPs that display multi-magnetic domains or single-magnetic structures until attaining NPs with no magnetic domains but with a fluctuated magnetization (superparamagnetic state).

9.4.3.1 Magnetic Hysteresis Behavior of Nanomaterials with Multi-magnetic Domains

The NPs with large dimensions, having a size greater than a certain critical value D_c, display a structure of magnetic domains, and their magnetization behavior in the presence of an external magnetic field shows a hysteresis, likewise the magnetic bulk material counterpart (Kneller et al. 1962; Hubert and Schäfer 1998). Magnetically stable structures are depending on the symmetry of crystal:

(i) For uniaxial symmetry (Fig. 9.18a), the magnetic structures are with free magnetic poles.
(ii) For cubic symmetry (Fig. 9.18b), the magnetic structures do not involve free magnetic poles (i.e., with magnetic flux closing domains). In this case, the spontaneous magnetization is oriented at a $\pi/4$ angle regarding the wall surfaces of the closing domains; thus no further magnetostatic energy (E_m) will appear. $\vec{M}_s$.

In stable magnetic structures, the number of magnetic domains could be determined by minimizing the crystal energy (equilibrium case).

The magnetization of NPs having domain structures is taking place via procedures of reversible and irreversible displacement of magnetic domain walls at lower fields and via procedures of reversible and irreversible rotation of the

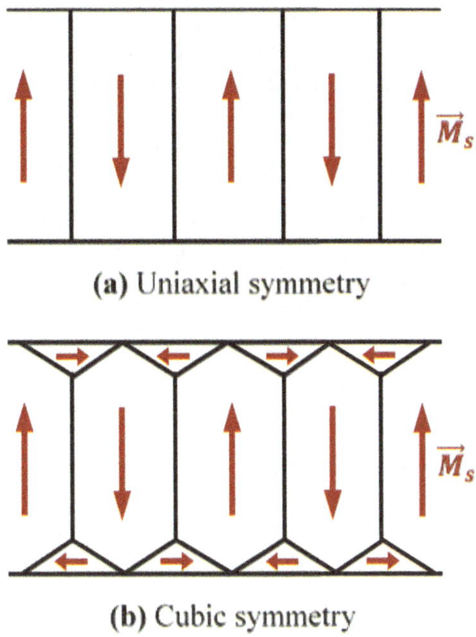

Fig. 9.18 Configurations of magnetic domains in large NPs: (**a**) uniaxial symmetry and (**b**) cubic symmetry

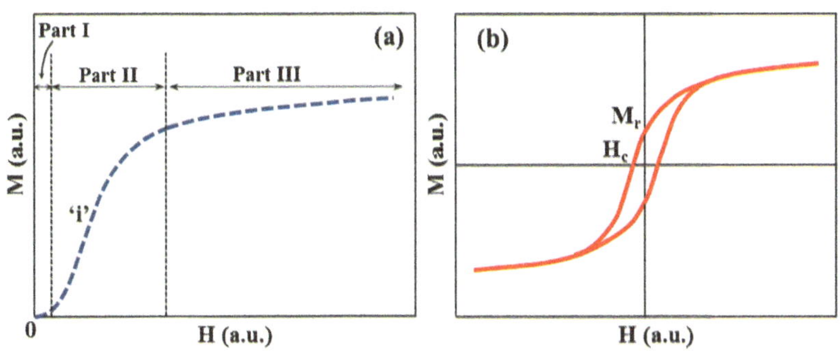

Fig. 9.19 (**a**) Curve of first magnetization and (**b**) M-H hysteresis loop for NPs with configuration of magnetic domains

spontaneous magnetization vector at higher fields once the walls are absent (it become a single-domain structure). The first magnetization curve is fundamentally known as a general feature of ferri- or ferromagnetic bulk materials (Fig. 9.19a):

(i) A nearly linear part at lower fields (part I).
(ii) A transition region at inflection point "*i*" for moderate fields (part II).
(iii) A region of saturated magnetization at higher fields (part III).

At lower magnetic fields (up to approximately $H_c/10$, where H_c is the coercive field), the law of Rayleigh is applied, giving the variation of magnetization once operating a variation of the externally applied magnetic field (ΔH):

$$\Delta M = \chi_i \Delta H + \frac{\alpha}{2}(\Delta H)^2 \tag{9.65}$$

In this expression, α is a constant and χ_i is the initial magnetic susceptibility.

At higher fields, the magnetization is appropriately described by the law of approaching to saturation (LAS) (Almessiere et al. 2019l; Almessiere et al. 2019m; Trukhanov et al. 2020):

$$M = M_s\left(1 - \frac{a}{H} - \frac{b}{H^2} - \frac{c}{H^3} - \ldots\right) + \chi_0 H \tag{9.66}$$

wherein a, b, and c are constants. The $\chi_0 H$ term is established by the existence of the χ_0 para-process contribution, which is not dependent to field. The $\chi_0 H$ term is being more important once the material is approaching to T_C. Nevertheless, several experiments revealed that the term in $1/H^3$ and the $\chi_0 H$ term could be neglected (Almessiere et al. 2020e). Moreover, the existence of the term in $1/H$ is not yet validated, excepting some particular conditions where its existence will involve an infinite magnetization that could not be the consequence of rotation of the magnetization processes. Accordingly, the Eq. (9.66) for NPs could be reduced to be as follows (Korkmaz et al. 2019; Almessiere et al. 2019n):

$$M \simeq M_s\left(1 - \frac{b}{H^2}\right) \tag{9.67}$$

The constant b is depending on the magnetic anisotropy (noted K_u or K_1 for uniaxial and cubic anisotropy constant, respectively, and sometimes it is simply noted as K without using indices):

(i) $b = \frac{4}{15}\left(\frac{K_u}{\mu_0 M_s}\right)^2$
(Eq. 9.68), for uniaxial symmetry (Kojima and Wohlfarth 1982)

(ii) $b = \frac{8}{105}\left(\frac{K_1}{\mu_0 M_s}\right)^2$
(Eq. 9.69), for cubic symmetry (Becker and Polley 1940).

For magnetic nanomaterials, the Eqs. (9.68) and (9.69) permit to determine experimentally the magnetic anisotropy constants by knowing M_s values from measurements of magnetization hysteresis plots (Almessiere et al. 2019b; Almessiere et al. 2019o; Almessiere et al. 2018b; Almessiere et al. 2019p; Almessiere et al. 2020f).

For moderate magnetic fields (i.e., transition region), there is no expression that describes properly the magnetization of nanomaterials. The attained experimental magnetization plots are the characteristic of each kind of sample.

For the magnetization and demagnetization of NPs, a magnetic hysteresis loop is attained (Fig. 9.19b) with a saturation magnetization (M_{sat}) that depends on the surface effects, a remanent magnetization (M_r), and the coercivity (H_c) that greatly depends on the dimensions of NPs.

The critical volume (V_C), which is associated to the D_C of NPs that corresponds to the transformation from multi-magnetic domains structure (where the magnetization is non-uniform) to single-magnetic domain structure (where the magnetization is stable and uniform), could be deduced from the circumstance of minimization of the crystal energy by utilizing the model of single-magnetic domain particles (Kittel 1946). With diminishing the volume of NPs that have a structure of magnetic domains (tens of nanometers), a particular critical size will be achieved wherein the NP does not display anymore a multi-domain structure; however it consists of a single-magnetic domain structure where the magnetization is uniform. C. Kittel (Kittel 1946) is the first scientist who highlighted that specific geometries could conduce to a uniform magnetization, hence with no multi-magnetic domains. For this condition, the difficulty that gets up is at which dimensions a structure with no magnetic domains is attained; hence the energy of single-magnetic domain structure is lesser compared to NPs with multi-magnetic domains configuration. The elucidation of this issue could be acquired on the basis of the case where the free energy of a NP having a uniform magnetization is equivalent, to the limit, to the free energy of a NP having a domain configuration (Kittel 1946; Kittel 1949; Néel 1947). Here, below, we take into consideration two states of domain structures.

9.4.3.1.1 Cubic Symmetry

For single-domain configuration (Fig. 9.20a), the energy is governed only by E_m:

$$E_1 = E_m = \frac{\mu_0}{2} NVM_s^2 \tag{9.70}$$

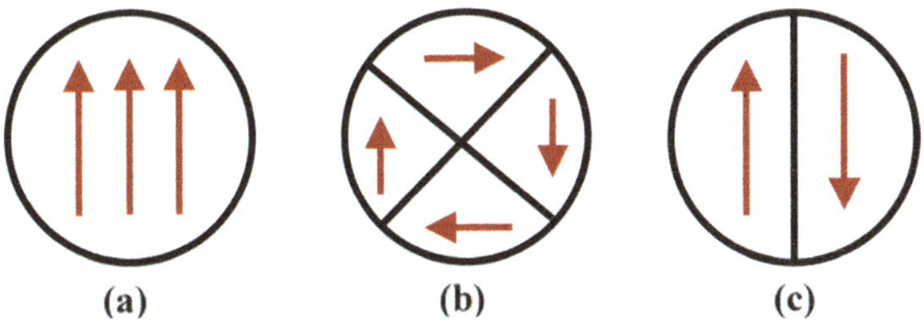

Fig. 9.20 Magnetic configurations in the model of spherical nanoparticles: (**a**) uniform magnetization, (**b**) non-uniform magnetization with cubic symmetry, and (**c**) non-uniform magnetization with uniaxial symmetry

here N is a demagnetizing factor. For cubic symmetry configuration (Fig. 9.20b), the Bloch walls mainly contribute to the energy. Hence, the energy could be written as:

$$E_2 = 2E_w = 2\varepsilon_p S \tag{9.71}$$

where E_w is the magnetic domain walls energy and S is the wall surface. Inserting the surface and the volume of NP (counted as sphere) and by making equal the Eqs. (9.70) and (9.71), one gets the following expression for critical diameter:

$$D_C = \frac{18\varepsilon_p}{\mu_0 M_s^2} \tag{9.72}$$

where ε_p is the energy density of domain walls and the V_C is linked to D_C as for spherical NPs as:

$$V_C = \frac{\pi}{6} D_C^3 \tag{9.73}$$

According to these results, the NPs are in single-magnetic domain and the magnetization is uniform once $D < D_C$. Contrarily, the NPs have multi-magnetic domains configuration, and the magnetization becomes non-uniform when $D > D_C$.

9.4.3.1.2 Uniaxial Symmetry

We consider here that the configuration of domains is as shown in Fig. 9.20c. The energy of NPs will promote the magnetostatic energy E_m (configuration with no closing poles) as well as the energy of the existing domains walls. Because of the existence of two domains magnetized at π angle, E_m will be lowered to 1/2 in comparison with the situations in which there exists a single domain (Fig. 9.20a). Therefore, at that situation, the energy of the configuration could be expressed as:

$$E_2' = \frac{E_m}{2} + E_w \tag{9.74}$$

By imposing the equality of Eqs. (9.70) and (9.71), one gets:

$$E_2' = \frac{E_m}{2} + \varepsilon_p S \tag{9.75}$$

And hence, the expression of D_C for uniaxial symmetry will be as follows:

$$D_C = \frac{18\varepsilon_p}{\mu_0 M_s^2} \tag{9.76}$$

The NPs will display a configuration with single-magnetic domain wherein the magnetization is uniform below D_C. By looking at Eq. (9.72) and Eq. (9.76), with no matter of what model is selected for the domain configuration, the expression for the critical diameter D_C stays unchanged.

L. Landau and E. Lifshitz determined ε_p on the basis of the magnetocrystalline anisotropy and the exchange energy of the walls comprised in the orientation of spins (Landau and Lifshitz 1935). Thus, the next expression has been attained:

$$\varepsilon_p = \left(\frac{2k_B T_c K_V}{a}\right)^{\frac{1}{2}} \tag{9.77}$$

where K_V is the bulk magnetocrystalline anisotropy, k_B is the Boltzmann constant, and a is the lattice constant.

At the end of this subsection, one should indicate here that for particular examples, one should take into consideration both the type and the size of NPs; hence, the observed M_{sat}, K_{eff} (effective magnetocrystalline anisotropy), and T_C are adjusted. With respect to this, one should take into consideration the experimentally deduced values, such as M_{sat}, K_{eff}, and T_C, for the situation of nanostructured materials, which display dissimilar magnitudes compared to their counterpart bulk materials.

9.4.3.2 Magnetic Hysteresis Behavior of Nanomaterials with Single-Magnetic Domain

As discussed in the previous parts, NPs that display a volume $V < V_c$ exhibit a single-magnetic domain configuration wherein the magnetization is stable and uniform. The magnetization under an externally applied field will be formed by the rotation procedures of the vector of spontaneous magnetization. The magnetic behaviors of a nanomaterial with SMD configuration and in which the magnetization is stable could be attained through the model of Stoner-Wohlfarth (S-W) (Stoner and Wohlfarth 1948). Therefore, by taking into consideration an isolated SMD NP system with uniaxial anisotropy, without thermal agitation (or with extremely low thermal agitation; hence it will not inverse the magnetization of the NPs), and under the application of magnetic field (H), the energy density will be expressed as:

$$w = K_u \sin^2\varphi - \mu_0\left(\vec{M}_s - \vec{H}\right) \tag{9.78}$$

We considered in this case the plane (xOy); hence the Ox axis accords with the easy magnetization axis (EMA) of NP, and the 0y axis corresponds to the hard magnetization axis (HMA; $\perp$ to 0x axis).

The vectors $\vec{H}$ and $\vec{M}_s$ could be expressed as:

$$\vec{H} = H_x\,\vec{i} + H_y\,\vec{j} \tag{9.79}$$

$$\vec{M}_s = M_x\,\vec{i} + M_y\,\vec{j} = (M_s \cos\,\varphi)\,\vec{i} + (M_s \sin\,\varphi)\,\vec{j} \tag{9.80}$$

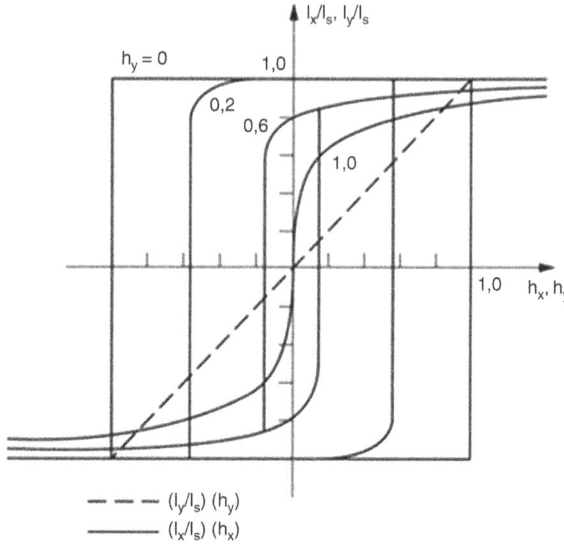

Fig. 9.21 Theoretical curves of longitudinal magnetization $(I_x/I_s) = (h_x)_{hy\ =\ constant}$ for a uniaxial thin layer. The transversal curve $(I_y/I_s) = (h_y)_{hx\ -\ 0}$ is presented by dashed line. *Reproduced with permission from* Caizer (2015)

By inserting the expressions of the vectors $\vec{H}$ and $\vec{M}_s$ in the Eq. (9.78), one can get (Kneller et al. 1962):

$$w = K_u \sin^2\varphi - \mu_0 H_x M_s \cos\varphi - \mu_0 H_y M_s \sin\varphi \tag{9.81}$$

In this equation, φ is the angle among $\vec{M}_s$ and the easy magnetization axis.

We consider in the following parts that the reduced fields are $h_x = H_x/H_{V,\ u}$ and $h_y = H_y/H_{V,\ u}$ in which $H_{V,\ u} = 2K_u/(\mu_0 M_s)$.

Stoner and Wohlfarth determined the plots of reduced magnetization, at magnetization-demagnetization, for diverse angles among the magnetic field and the EMA. The plots of $M_x/M_s = f(h_x)$ for various constant values of h_x are presented in Fig. 9.21 (where $I \equiv M$) (Smith 1958). In this case, a thin layer system consisted of non-interaction SMD nanoparticles with uniaxial anisotropy alongside the layer is considered.

As exposed, the coercive field for coherent rotations, $h_{xr} = H_{xr}/H_{V,\ u}$, is moving from the value 1 to 0 once h_y rises from 0 to 1. The $h_{xr} = f(h_y)$ function is similar to the critical curve and is presented by the asteroid equation in which h_x should be changed by h_{xr}. For the $M_y/M_s = f(h_y)$ plot, once $h_x = 0$, one obtains a line as presented by the dashed line in Fig. 9.21. When that magnetic field is applied along x (i.e., $h_y = 0$), the magnetization along EMA obeys a hysteresis loop with rectangular shape. This circumstance is crucial in terms of practical investigations for the reason that, (a) firstly, for these conditions, one obtains a reversing of magnetization by rotation among two stable states (very quick switch to π) and, (b) secondly, it permits to determine the uniaxial anisotropy field attained by experimental investigations. The rapid magnetization reversing gets real applications in the manufacture of magnetic memories like hard drives in computer.

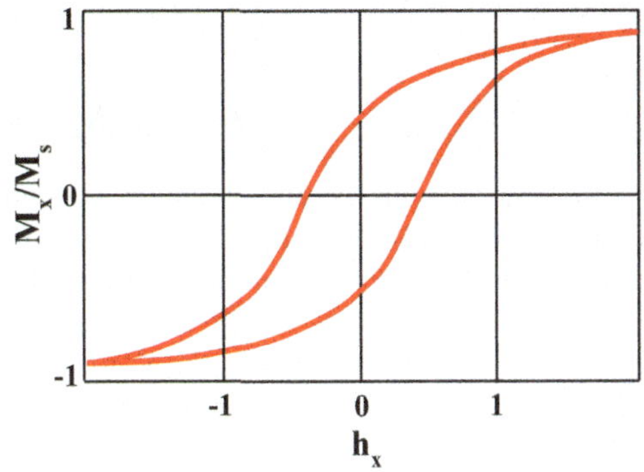

Fig. 9.22 Hysteresis loop for a system of NPs with arbitrarily oriented EMA

With the aim to determine the anisotropy field, supposing the circumstance $h_x = 1$, one obtains the expression:

$$H_{V,u} = \frac{2K_u}{\mu_0 M_s} = H_{xr} \approx H_c \tag{9.82}$$

Thus, K_u could be determined by the experimental measurement of H_c that could be extracted by the approximation with the critical field wherein the switching by jump of the magnetization takes place (Almessiere et al. 2020g).

For NPs with SMD configuration, uniaxial anisotropy, no interactions among them, and arbitrarily anisotropy axes, the hysteresis loop is attained as presented in Fig. 9.22. For this circumstance, the reduced field (h_c) in the magnetization direction is about 1/2, the mean coercive field is given as (Stoner and Wohlfarth 1948):

$$\langle H_c \rangle \approx \frac{H_{V,u}}{2} \approx \frac{K_u}{\mu_0 M_s} \tag{9.83}$$

and the theoretical value of reduced magnetization is given as:

$$\frac{M_r}{M_s} = 0.5 \tag{9.84}$$

The attained findings approve the validity of the model, and only little disparities happen in the values $H_{V,\ u}/2$ and $M_r/M_s = 0.5$ that are little smaller because of the effects of SPM relaxation or magnetization quantum tunneling that happens for smaller NPs (Caizer 2015).

For the case where the magnetocrystalline anisotropy is dissimilar to the uniaxial one, L. Néel and C. R. Acad determined the H_c value for arbitrarily oriented spherical NPs as $0.64K_u/\mu_0 M_s$ (Néel 1947). We consider the following expressions of the shape anisotropy energy (E_{sh}) and the shape anisotropy constant (K_{sh}) for the situation of a single crystal of ellipsoidal shape with the axes $a > > b = c$ (Kneller et al. 1962):

$$E_{sh} = \frac{\mu_0}{2}(N_a - N_b)M_s^2 \sin^2\theta \tag{9.85}$$

$$K_{sh} = \mu_0(N_a - N_b)M_s^2/2 \tag{9.86}$$

In these relations, N_a and N_b are the demagnetization factors of the ellipsoid axes a and b, respectively, and θ is the angle in which the M_s takes with the major ellipsoid a-axis. For a system of spherical NP, N_a becomes equal to N_b; hence, there will be no anisotropy due to the shape (i.e., $K_{sh} = 0$).

Accordingly, once the anisotropy due to the shape is governing, and by taking into consideration the expressions of E_{sh} and K_{sh}, one obtains:

$$H_c \approx (N_b - N_a) = N(r) \tag{9.87}$$

For a group of NPs, an expanded range of values could occur; thus it will be reflected as $H_c \approx \langle N_r \rangle$ or a function of volume distribution. Other reports disclosed that H_c of a group of NPs is inferior to the mean value, according to the size of particles (Campbell 1957; Bean 1955). Evidently, this problem is more complicated (Barbara 2001).

9.4.3.3 Superparamagnetic Relaxation

According to Néel (Neel 1949), in the existence of thermal activation ($T > 0$), the magnetization of NP fluctuates along EMA (switching at π) once $V < V_t$ (V_t is the threshold volume and is linked to D_t). For $H = 0$, to reverse the spontaneous magnetization, by taking into consideration magnetic NP with uniaxial anisotropy,

$$E_{V,u} = K_u V \sin^2\varphi \tag{9.88}$$

the energy barrier should be surpassed ($\varphi = \pi/2$):

$$E_b = K_u V \tag{9.89}$$

At this situation, there exists some probability to cross the potential barrier that is larger as lowering the volume of NPs and increasing the temperature (Brown 1959; Néel 1955). This mechanism is recognized as the Néel magnetic relaxation. Néel relaxation time is the one when this process occurs (Neel 1949; Almessiere et al. 2018c):

$$\tau_N = \tau_0 \exp\left(\frac{K_u V}{k_B T}\right) \tag{9.90}$$

In this expression, τ_0 is a time constant of the order of $10^{-9} - 10^{-13}$ s (Almessiere et al. 2018d). Experimentally, the time interval where the relaxation mechanism is recorded is termed as measurement time (τ_m). It is about 100 s for the condition of static measurements (Néel 1955). If $\tau_m = \tau_N$, one could deduce the volume of NPs that corresponds to the transition from a state where the magnetization is not relaxed (over time, the magnetization is stable) to a state where the magnetization is relaxed (in a certain time, it switches by π-angle). Hence, by using Eq. (9.90), one gets:

$$K_u V_t \approx 25 k_B T \tag{9.91}$$

and

$$V_t = \frac{25 k_B T}{K_u} \tag{9.92}$$

Furthermore, it describes a temperature that corresponds to the blocking temperature (magnetic moments are blocked) (Néel 1955):

$$T_B = \frac{K_u V}{25 k_B} \tag{9.93}$$

At $T > T_B$, the magnetic moments of NPs fluctuate along the EMA (they relax), whereas at $T < T_B$, the magnetic moments of NPs are blocked.

For the case of dynamic measurements, usually utilized in present practice (such as ac susceptibility and Mössbauer spectroscopy), by taking into consideration the experimental time of measurement (t_m), the Eqs. (9.91) and (9.92) could be more general (Bean and Livingston 1959):

$$V_t = \frac{k_B T}{K_u} \, ln\left(\frac{t_m}{\tau_0}\right) \tag{9.94}$$

$$T_B = \frac{K_u V}{k_B} \times \frac{1}{ln\left(\frac{t_m}{\tau_0}\right)} \tag{9.95}$$

By applying a magnetic field along the EMA of NPs, the energy barrier will be changed (E_{bH}). It could increase or decrease according to the direction of magnetic field. Indeed, it reduces once H is applied in the opposite direction of M_s and rises in reverse case. For the case of applying H in the reverse direction of M_s, beginning from the particle energy in the field, one gets the following expression (Dennis et al. 2002):

$$E_{bH} = K_u V_m \left(1 - \left(\frac{\mu_0 m_{m,NP}}{2 K_u V_m} \times H\right)\right)^2 \tag{9.96}$$

wherein the magnetic moment of the NP has been changed. At this condition, consistent with the expression of Eq. (9.91), it conduces to the existence of certain critical field (or threshold field, H_{ct}) wherein the magnetic moments will be reversed by π-angle. The expression of H_{ct} and its discussion with respect to the temperature and the particle (magnetic) size will be reported in the below sections.

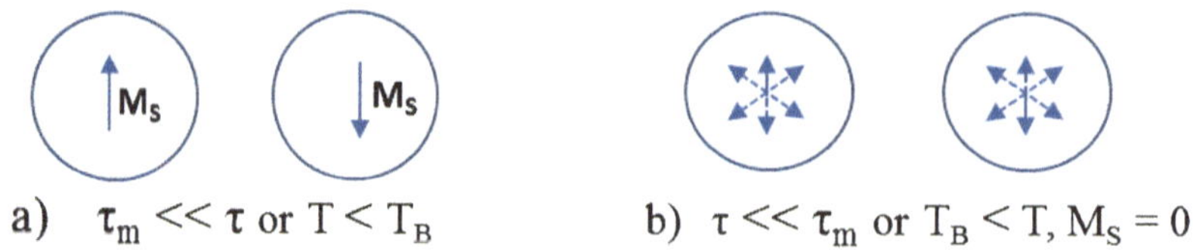

Fig. 9.23 (**a**) One of the two stable states is observed, *blocked* state; (**b**) due to the fluctuation of magnetization, net spontaneous magnetization is detected as zero, superparamagnetic state

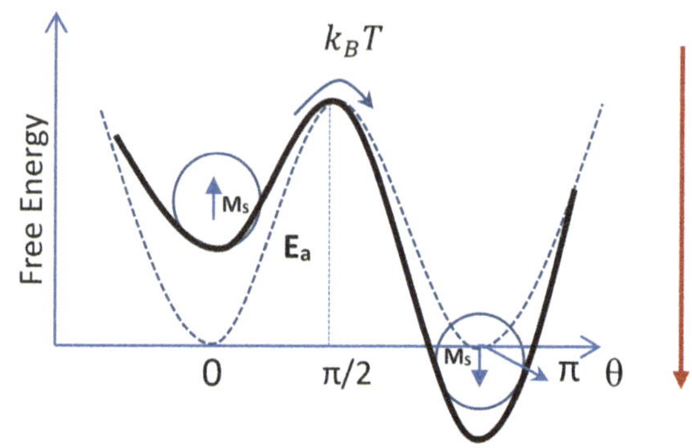

Fig. 9.24 Schematic picture of the free energy of a single-domain particle with uniaxial anisotropy as a function of magnetization direction

According to above discussions, there are two possible scenarios by considering the measurement temperature conditions, as shown in Fig. 9.23:

(a) $\tau_m < < \tau$ or $T < T_B$: When measurement time is much smaller than the relaxation time of the sample, then the magnitude of M_s is easily specified while it locates at one of the two stable orientations. This case is possible for SP NPs when the stimulant effect of temperature on spin flips is excluded and usually referred as blocked state. T_B is the maximum temperature for the sample to be in the blocked state and is called as blocking temperature.
(b) $\tau < < \tau_m$ or $T_B < T$: When measurement time is much greater than the relaxation time of SP sample as mentioned above, the device that measures the magnetization could not detect a stable state and record the magnetization value of sample as zero. This means that the sample is in the superparamagnetic state in the absence of an external H applied during measurement.

A further discussion explains how the anisotropic energy barrier E_a affects the spin flips and the corresponding relaxation time. A superparamagnetic particle with uniaxial anisotropy has $E_a = KV$, where K is the uniaxial anisotropy constant and V is the volume of nanoparticle. Two directions of an easy magnetization axis are stable directions for M_s. There are two energy minima for those two preferred directions and maxima for least preferred directions. Dashed curve simply identifies these conditions in the Fig. 9.24. When an external H applies, especially one direction which is parallel to field is more preferential as schematically demonstrated with the solid line in Fig. 9.24. M-H curve of single-domain ferromagnetic and ferrimagnetic particles is the form of sigmoidal shape but, without any coercivity or hysteresis loop, is generally known as S-shape (Almessiere et al. 2020h).

9.4.3.4 Magnetic Interactions into Nanomaterials

By using the experimental results of zero-field cooling (ZFC) magnetization of products with a recognized distribution of particles size, one could determine the average value of T_B. Nevertheless, it is impossible to determine the values of relaxation time (τ_m) and magnetic anisotropy constant (K) uniquely from a ZFC magnetization measurements. But this could be accomplished via utilizing measurements of ac susceptibility attained at various frequencies. For these experiments, a magnetic field $h(t) = h_0 \sin(\omega t)$ is applied, and hence the magnetization $M(\omega) = \chi_{ac}(\omega, t)h_0$ is recorded. The complex susceptibility is presented as:

$$\chi_{ac}(\omega, t) = \chi'(\omega, t) \pm i\chi''(\omega, t) \tag{9.97}$$

In this expression, $\chi'(\omega, t)$ and $\chi''(\omega, t)$ are the real and imaginary parts of the ac susceptibility. They are determined as (Gittleman et al. 1974):

$$\chi'(\omega, t) = \frac{\mu_0 M^2(V)}{3K}\left[\left(\frac{KV}{k_B T} \times \frac{1}{1 + (\omega\tau_m)^2}\right) + \left(\frac{(\omega\tau_m)^2}{1 + (\omega\tau_m)^2}\right)\right] \tag{9.98}$$

$$\chi''(\omega, t) = \frac{\mu_0 M^2(V)}{3K}\left[\left(\frac{\omega\tau_m}{1 + (\omega\tau_m)^2}\right) - \left(\frac{KV}{k_B T} \times \frac{\omega\tau_m}{1 + (\omega\tau_m)^2}\right)\right] \tag{9.99}$$

wherein $\tau_m = \tau_N$ is already reported in Eq. (9.90). If the product contains particle size distribution, the Eqs. (9.98) and (9.99) should be gathered with the function of volume-weighted distribution. $\chi'(\omega, t)$ displays a maximum at a certain temperature that is associated with T_B in an analogous manner as the peak temperature in a ZFC plot. $\chi''(\omega, t) \neq 0$ once τ_N of an important portion of the particles is comparable to the observation time. For the case of a lognormal size distribution of ferromagnetic particles, the peak in $\chi''(\omega, t)$ accords with T_B. Frequently, one performs the measurements of ac susceptibility versus temperature at various frequencies.

By considering the Eq. (9.90), one could find that:

$$ln(\tau_m) = ln(\tau_N) = ln(\tau_0) + \frac{KV}{k_B T_B} \tag{9.100}$$

Accordingly, by plotting $\ln(\tau_m)$ against $1/T_B$, a straight line with KV/k_B as slope and $\ln(\tau_0)$ as intercept is observed.

The SPM relaxation is very sensitive to the magnetic interactions among the NPs. Ferri- and ferromagnetic NPs could enter in interaction through dipole interactions of long range. Mössbauer investigations showed that weak dipolar interactions could cause a quicker relaxation (Prené et al. 1993; Mørup and Tronc 1994). This could be elucidated by the reduction of the energy barriers that separate the minimum of magnetic energy. Nevertheless, strong interactions cause a slower relaxation and manifestation of a collective state having resemblances to the spin glass at lower temperatures (Mørup 1994; Fiorani et al. 1999; Slimani et al. 2020c). The magnetic investigations of ferromagnetic NPs in liquid suspensions have revealed that τ_m could diverge at a certain temperature, noted as T_{VF}. Accordingly, the Eq. (9.90) can be changed by the following Vogel-Fulcher law (Zhang et al. 1996; Almessiere et al. 2018e; Almessiere et al. 2020i):

$$\tau_m = \tau_0 \, exp\left(\frac{KV}{k_B(T - T_{VF})}\right) \tag{9.101}$$

where T_{VF} temperature could be expressed as:

$$T_{VF} \cong \frac{\mu_0 \mu^2}{4\pi k_B d_p^3} \tag{9.102}$$

where d_p represents the average distance among the nanoparticles. The majority of systems of NPs display a broad particle size distribution such that freezing of the dynamics of the moments of the tiniest nanoparticles because of the interactions existing among them is governed by the thermal blocking of SPM relaxation of biggest nanoparticles. At this situation, the interaction effects are merged with the effects of individual particle (Hansen et al. 2002). Nevertheless, for the case of very narrow distribution of particles size, one could possibly separate the effects of interactions from the dynamics of individual particle, i.e., the effects of interactions happen on timescales, which are separated from those caused by the relaxation of single particles (Djurberg et al. 1997).

Figure 9.25 shows a typical illustration of the variation of magnetization versus temperature under ZFC and FC modes, which illuminate the impact of effects of interactions. Generally, the maxima in ZFC plot shift to elevated temperatures when the interactions between particles increase. The FC curve of particles with weak interactions (Fig. 9.25a) rises with the decrease in the temperature well below the maxima in ZFC plot owing to the blocking of the tiniest particles. Nevertheless, in systems of stronger interacting particles (Fig. 9.25b, c), the FC plot is relatively flat or plateau-like at lower temperatures, which indicate that the directions of magnetization of tiniest particles are frozen in a spin glass-like configuration.

Figure 9.26 presents the real and imaginary parts of ac susceptibility versus temperature curves for diluted and concentrated suspensions of magnetic particles. One should recall that the imaginary part of ac susceptibility is different to zero once an important portion of the particles displays relaxation times comparable to observation time. As shown in Fig. 9.26b, the area of $\chi'' \neq 0$ is considerably widened and moved in the direction of elevated temperatures for the concentrated product. For $T > 40\ K$, χ'' data for the diluted product is near to "0," and thus effectively the different particles display SPM character; however the maxima in χ'' for the concentrated product are fairly higher than 40 K. Therefore, the variation in the dynamics of relaxation in the concentrated product is caused by the collective

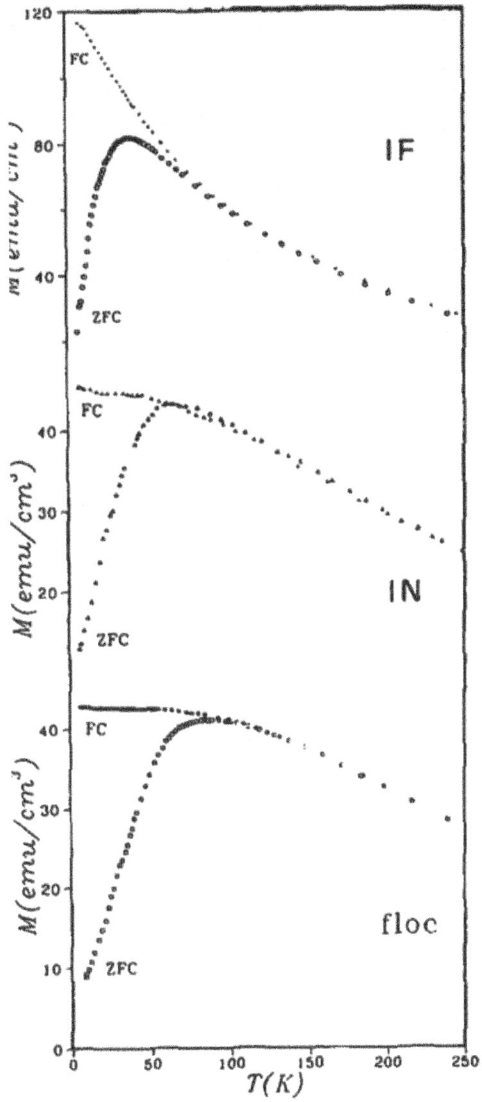

Fig. 9.25 Curves of ZFC-FC magnetization for Fe_2O_3 NPs dispersed in polyvinyl alcohol with $\langle D \rangle = 7.1\ nm$. "IF," "IN," and "FLOC" refer to inter-nanoparticles spacing of $5\langle D \rangle$, $1.4\langle D \rangle$, and strong aggregation. Reproduced with permission from (Tronc et al. 1995)

interaction effects instead of the combination of the effects of interactions and the slow dynamics of the biggest particles. Furthermore, the dynamics of relaxation are noticed to vary at around 40 K in the case of concentrated product that could be discerned as the existence of a bump in the data that is being asymmetric at lower frequency. In the vicinity of and under 40 K, the aging phenomenon was noticed that is a feature of collective dynamics in frustrated systems (Djurberg et al. 1997). Thus, the dynamics were examined with models that are usually utilized for spin glasses. From Fig. 9.26, one determines the variations of τ_m (or τ in the figure) versus the freezing temperature T_s (or T in the figure) as shown in Fig. 9.27. The inset in this

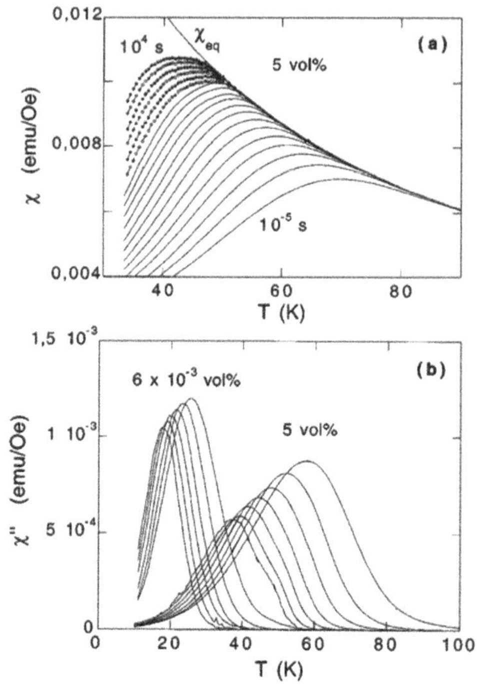

Fig. 9.26 (*a*) χ' and (**b**) χ'' parts versus temperature curves for diluted and concentrated suspensions of magnetic $Fe_{1-x}C_x$ nanoparticles. *Reproduced with permission from* Djurberg et al. (1997)

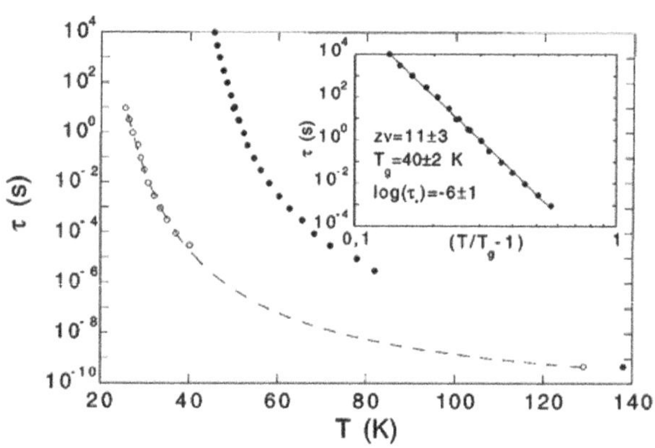

Fig. 9.27 Relaxation time versus temperature for diluted (non-filled circles) and concentrated (filled circles) suspensions of Fe-C nanoparticles. The inset represents the V-F fitting for the concentrated product. *Reproduced with permission from* Djurberg et al. (1997)

figure illustrates an investigation of the experimental data through the critical slowing down law that is usually utilized for the theory of spin glass (Almessiere et al. 2019q):

$$\tau_m = \tau^* \left(\frac{T_s}{T_g} - 1 \right)^{-z\nu} \tag{9.103}$$

where T_g is the is the spin glass freezing temperature and τ^* is the relaxation time of non-interacting particles. The critical exponent "$z\nu$" reflects the interaction strength and is varying between 4 and 12, where the critical exponent "z" is connecting τ and the correlation length (ξ) is $\tau \propto \xi^z$ and the critical exponent is of $\xi = (T_s/T_g - 1)^{-\nu}$ (Almessiere et al. 2019r; Slimani et al. 2021b). By fitting data in Fig. 9.26 via Eq. (9.103), one could estimate the values of τ^*, T_g, and "$z\nu$." In general, the magnetic interactions strengthen with the decrease in "$z\nu$" exponent.

Some examples of ac susceptibility measurements for different magnetic nanomaterials can be found in the following studies (Slimani et al. 2018c; Almessiere et al. 2018f; Hannachi et al. 2020; Slimani et al. 2018d), where the researchers analyzed the experimental curves using the above Néel-Arrhenius, Vogel-Fulcher, and critical slowing down models, and the obtained results were deeply discussed. Furthermore, one could look to the following investigations (Slimani et al. 2019h; Slimani et al. 2019i; Almessiere et al. 2019s; Hannachi et al. 2019b), which report the analyses of ZFC-FC magnetizations of diverse magnetic nanomaterials with respect to the above discussions (Algarou et al. 2020e; Slimani et al. 2019j; Almessiere et al. 2020j). In addition to these characterization techniques, the Mössbauer spectroscopy plays a significant role to elucidate the magnetic properties of nanomaterials (Almessiere et al. 2019t). It can deliver structural information, determine quantitatively the valence state of Fe, determine the coordination number of iron atoms, and identify the various Fe oxides.

9.4.3.5 Coercive Field of Nanomaterials and Its Size Dependence

In previous parts, we reported that there exists a certain critical field (or threshold field, H_{ct}) at which the magnetic moments will be reversed by π-angle. The expression of H_{ct} could be deduced by applying the condition imposed by Eq. (9.91):

$$K_u V_m \left(1 - \left(\frac{\mu_0 m_{m,NP}}{2K_u V_m} \times H_{ct}\right)\right)^2 = 25 k_B T \tag{9.104}$$

This will conduce to:

$$H_{ct} = \frac{2K_u V_m}{\mu_0 m_{m,NP}}\left[1 - \left(\frac{25 k_B T}{K_u V_m}\right)^{\frac{1}{2}}\right] = H_{V,u}\left[1 - \left(\frac{25 k_B T}{K_u V_m}\right)^{\frac{1}{2}}\right] \tag{9.105}$$

For $T = 0$ (i.e., without thermal activation), the previous equation leads to $H_{ct} = H_{V,\,u}$. This means that the critical field is becoming of the order of anisotropy field (as reported in Eq. (9.82)), an expression that is in line with S-W model.

By taking the expressions of V_t (Eq. (9.92)) and H_{ct} (Eq. (9.105)), one obtains the spherical NPs approximation:

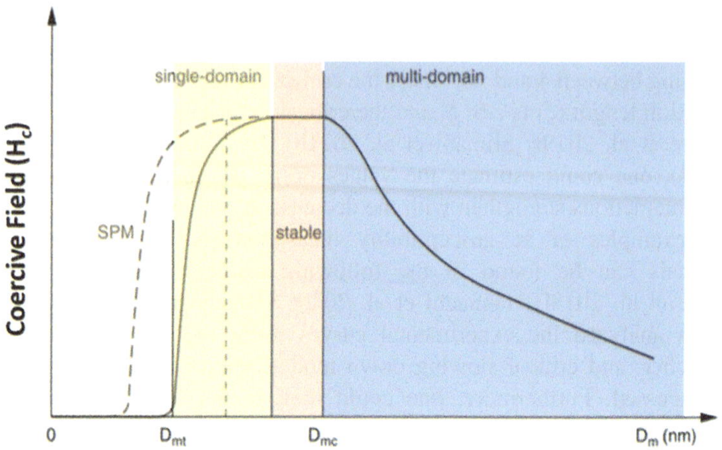

Fig. 9.28 Evolution of H_c as a function of magnetic diameter for a system of NPs

$$H_{ct} = H_{V,u}\left[1 - \left(\frac{V_{mt}}{V_m}\right)^{1/2}\right] = H_{V,u}\left[1 - \left(\frac{D_{mt}}{D_m}\right)^{3/2}\right] \tag{9.106}$$

where D_{mt} and V_{mt} designate the threshold magnetic volume and volume, respectively.

Employing the Eq. (9.93), the Eq. (9.106) could be expressed by inserting T_B as (Almessiere et al. 2019u):

$$H_{ct} = H_{V,u}\left[1 - \left(\frac{T}{T_B}\right)^{\frac{1}{2}}\right] \tag{9.107}$$

The Eq. (9.106) indicates that the critical field (the coercive field in this situation) reduces with reducing the D_m of NPs and is being nil at the D_{mt}. The findings obtained from experiments confirmed the variation law of Eq. (9.106) in the interval $D_m = (1\ to\ 5) \times D_{mt}$. F. Kneller and F.E. Luborsky (Kneller and Luborsky 1963) observed a well agreement among the experimental and calculated values for CoFe spherical NPs. For $D_m = (5\text{ to }6) \times D_{mt}$ values, divergences from these patterns happen wherein H_c decreases in this part, which approaches to the H_c value characteristic to the bulk system.

In the situation of NPs with arbitrary orientation of uniaxial anisotropy axis, the Eq. (9.83) will be utilized for the anisotropy field $H_{V,\ u}$.

Figure 9.28 presents the evolution of H_c of NPs with respect to their magnetic diameter, beginning from the region of MMD configuration (very large NPs ranging from tens to hundreds of nanometers) up to SPM region (very small NPs (few nanometers)). One must specify here that this variation is attained for NPs displaying low anisotropy and at temperatures at the thermal fluctuations of M_s (magnetic moments) of SMD NPs. In the region of NPs with MMD configuration ($D_m > D_{mc}$),

H_c increases with reducing D_m. The zone of SMD in which the magnetization is stable and relatively limited, at the vicinity of the critical diameter, for $D_m < D_{mc}$, could be expanded to lower D_m values as shown by dashed line in the Fig. 9.28. This is possibly occurred by the subsequent two ways:

(i) Decreasing the temperature ($T \rightarrow 0$) according to the prediction of S-W model.
(ii) At ambient temperature by rising the magnetic anisotropy. This situation could be attained for NPs exhibiting huge anisotropy value (hard), like the case of cobalt ferrites NPs.

According to Eq. (9.106), when $\frac{K_u V_m}{k_B T} \gg 1$, τ_N is extremely high and the magnetic moments of NPs are "frozen" on the uniaxial anisotropy axes. At that condition and under standard circumstances, the magnetization is stable. In the vicinity of D_{mt} (i.e., $D_{mt} < D_m \ll D_{mc}$), the H_c value diminishes with reducing D_m.

9.4.3.6 Superparamagnetic Behavior of Nanomaterials

In the region of very smaller dimensions of nanoparticles, wherein $D_m < D_{mt}$ (largely smaller than 10 nm for NPs with moderate anisotropy), and at ambient temperature, when the circumstance $(K_u V_m / k_B T) > 1$ occurred and $K_u V_m / k_B T$ is little superior than 1, the time of relaxation of the magnetic moments of NPs (oriented or not) is very small (of the order of 10^{-9} s). At these circumstances, the magnetization of NPs with SMD configuration is fluctuating quickly alongside the EMA and occurring continuously in thermodynamic equilibrium, and once an externally magnetic field is applied, it tracks (nearly instantly) the variations of field (Neel 1949). From a magnetic viewpoint, this system performs as a system of paramagnetic atoms (Langevin) with no interactions, in which the magnetic moments for atoms exist in place of the magnetic moments of nanoparticles (Langevin 1905). Taking into account this fundamental characteristic among these two paramagnetic systems, atoms or NPs (that comprises $>10^5$ atoms) with their magnetic moments, in this case the NP system is superparamagnetic (SPM), and it is said that it displays SPM behavior under an applied magnetic field (Bean and Livingston 1959). The magnetic moments for NP and that for paramagnetic atom will be noted by $\vec{m}_{n,NP}$ and $\vec{\mu}_a$, respectively. In these circumstances, one would apply the Langevin atomic paramagnetism theory (Langevin 1905). Therefore, the magnetization of superparamagnetic NPs (with a concentration of NPs "n" in the system) is done by the following expression (Jacobs and Bean 1963):

$$M_{SPM}(H, T) = n m_{m,NP} \left[coth\left(\frac{\mu_0 m_p H}{k_B T}\right) - \frac{k_B T}{\mu_0 m_p H} \right] \tag{9.108}$$

In this equation, the brackets contain the function of Langevin (Almessiere et al. 2019v; Almessiere et al. 2020k):

$$L(H,T) = coth\left(\frac{\mu_0 m_{m,NP} H}{k_B T}\right) - \frac{k_B T}{\mu_0 m_{m,NP} H} \tag{9.109}$$

$M_{SPM}(H,T) = f(H,T)$ curve is with no hysteresis loop (coercivity is null), and the first magnetization curve ($H > 0$) does not display an inflection point, which is different from that of other NPs, ferri- or ferromagnetic with SMD or MMD configuration in which there is permanently a narrower or wider hysteresis loop and with an inflection point in their curve of first magnetization under an applied magnetic field. In the case of NPs system displaying SPM behavior (with no interactions) under an applied magnetic field, two states should be occurred: (i) M-H curves registered at diverse temperatures should be with no hysteresis and obey the Langevin function, and (ii) the same M-H curves in the $M/M_{sat} = f(H,T)$ representation must be overlapped.

Usually, there exists a size distribution of NPs in a material. Thus, for a precise methodology, their distribution function must as well be taken into consideration. In the majority of cases, it has been observed that the distribution of NP is lognormal (Caizer 2015):

$$f(D) = \frac{1}{\sqrt{2\pi}\lambda D} exp\left(-\frac{[ln D - ln D_0]^2}{2\lambda^2}\right) \tag{9.110}$$

In this expression, λ and D_0 are distribution parameters. Accordingly, the magnetization for SPM NPs could be expressed as:

$$M_{SPM}(H,T) = M_{sat} \int_0^{\infty} \mathcal{L}\{\xi(H,T,D_m)\} f(D_m) d(D_m) \tag{9.111}$$

in which the argument of the Langevin function (in the case of spherical NPs approximation) is expressed as:

$$\xi(H,T,D_m) = \frac{\mu_0 \pi}{6} \times \frac{D_m^3 M_s H}{k_B T} \tag{9.112}$$

In the Eq. (9.111), the expression of $M_{sat,\ \mathrm{NP}} = nm_{\mathrm{NP}}$ and of $f_m(T) = \frac{n\pi}{6} \langle D_m(T) \rangle^3$ (Eq. (9.61)) was considered.

Y. Slimani's group performed numerous analyses of M-H curves for, as examples, rare earth substituted spinel ferrites, such as Eu-substituted NiCuZn ferrite, Tm-substituted CoZm ferrite, Eu-substituted Co ferrite, Dy-substituted MnZn ferrite, etc. (Almessiere et al. 2019v; Almessiere et al. 2020k; Slimani et al. 2020d; Almessiere et al. 2019w). These M-H curves revealed SPM behavior at ambient temperature, and they are well fitted by the Langevin function. The investigation of M-H plots attained experimentally allows to determine λ and D_0 and subsequently the $\langle D_m \rangle$ of the NPs in the system:

$$\langle D_m \rangle = D_0 \, exp \left(\frac{\lambda^2}{2} \right) \tag{9.113}$$

This is an interesting matter for the magnetic NPs since it permits to evaluate the thickness of the surface layer of NPs (SPM), considering that $D_m < D$ for diverse conditions (NPs implanted in different matrices, surfacted NPs, tinier NPs, etc.), utilizing TEM or HR-TEM or other tools to estimate the physical diameter (D). For more precise investigation, specifically for the condition of broad distribution of NP sizes, to determine $\langle D_m \rangle$ value, one should consider that the magnetic moments of NPs depend also on the D_m value:

$$m_{m,NP}(D_m) = \frac{\pi}{6} M_s D_m^3 \tag{9.114}$$

At this condition, the following function for the magnetization will be employed (Pshenichnikov et al. 1996):

$$M_{SPM} = n \int_0^{\infty} m_p(D_m) \mathcal{L}\{\xi(H, T, D_m)\} f(D_m) d(D_m) \tag{9.115}$$

in place of the one provided by Eq. (9.111).

9.5 Conclusions

The dimensions of NPs represent a critical parameter in which the change in size conduces to a specific magnetic structure: multi-magnetic domain where the magnetization is non-uniform or single-magnetic domain where the magnetization is stable and uniform or where the magnetization is fluctuating. As a consequence, NPs will display specific magnetic character in the presence of an applied magnetic field, from ferri- or ferromagnetic where M-H loop is wide for very large NPs, like the counterpart bulk material, to M-H loop with narrow hysteresis for smaller dimensions, to SPM character (no hysteresis) for very tiny NPs.

The dimension of NPs reflects also on the spins' configuration at the superficial layer of NPs, where the spins are no longer ordered by the exchange or super-exchange interactions. The spins in the NPs core are ferri- or ferromagnetically aligned. This configuration is being important for smaller NPs, provoking a significant reduction in the M_s of NPs. Accordingly, one should consider the structure of core-shell NPs in which the magnetic moments are ordered within the core and disordered at the shell. The non-collinearity of spins at the superficial layer of NPs reflects on the reduction of the M_s of NPs in comparison to that of counterpart bulk system. This effect becomes more important for very tiny NPs. Moreover, the reduction in M_s is greater for ferrimagnetic NPs in which the exchange interactions arise via the oxygen ions (super-exchange interaction).

The reduction in the dimension of NPs influences on the evolution of M_s against temperature that is dissimilar to that of the counterpart bulk system. In addition, T_C

of NPs reduces when their dimensions diminish. This reduction becomes more accentuated for very tiny NPs.

The magnetic anisotropy is altered for NPs (being remarkably great) in comparison to the counterpart bulk system. The imperative contribution to the magnetic anisotropy is owing to the surface anisotropy part that could be important for the situation of smaller NPs in comparison to the shape anisotropy or to the magnetocrystalline anisotropy. In addition to the nature of the system, the strength of this contribution is depending on the NPs being covered or not or implanted in diverse amorphous or crystalline matrices. Immense magnetic anisotropy values could happen for the case of NPs implanted in diverse matrices.

These different features should be considered for a precise fundamental investigation of the magnetic traits of NPs and their practical upcoming nanotechnological applications.

References

Ajeesha T, George AAM, Manikandan A, Mary JA, Slimani Y, Almessiere MA, Baykal A (2020) Nickel substituted MgFe2O4 nanoparticles via co-precipitation method for photocatalytic applications. Phys B Condens Matter:412660. https://doi.org/10.1016/j.physb.2020.412660

Akhtar S, Rehman S, Almessiere MA, Khan FA, Slimani Y, Baykal A (2019) Synthesis of Mn0.5Zn0.5SmxEuxFe1.8−2xO4 Nanoparticles via the Hydrothermal Approach Induced Anti-Cancer and Anti-Bacterial Activities. Nano 9:1635. https://doi.org/10.3390/nano9111635

Al-Amri N, Tombuloglu H, Slimani Y, Akhtar S, Barghouthi M, Almessiere M, Alshammari T, Baykal A, Sabit H, Ercan I, Ozcelik S (2020) Size effect of iron (III) oxide nanomaterials on the growth, and their uptake and translocation in common wheat (Triticum aestivum L.). Ecotoxicol Environ Saf 194:110377. https://doi.org/10.1016/j.ecoenv.2020.110377

Aldred AT (1975) Temperature dependence of the magnetization of nickel. Phys Rev B 11:2597–2601. https://doi.org/10.1103/PhysRevB.11.2597

Algarou NA, Slimani Y, Almessiere M, Baykal A (2020c) Exchange-coupling behavior in SrTb0.01Tm0.01Fe11.98O19/(CoFe2O4)x hard/soft nanocomposites. New J Chem 44:5800–5808. https://doi.org/10.1039/d0nj00109k

Algarou NA, Slimani Y, Almessiere MA, Alahmari FS, Vakhitov MG, Klygach DS, Trukhanov SV, Trukhanov AV, Baykal A (2020d) Magnetic and microwave properties of SrFe12O19/MCe0.04Fe1.96O4 (M = Cu, Ni, Mn, Co and Zn) hard/soft nanocomposites. J Mater Res Technol 9:5858–5870. https://doi.org/10.1016/j.jmrt.2020.03.113

Algarou NA, Slimani Y, Almessiere MA, Baykal A, Guner S, Manikandan A, Ercan I (2020a) Enhancement on the exchange coupling behavior of SrCo0.02Zr0.02Fe11.96O19/MFe2O4 (M = Co, Ni, Cu, Mn and Zn) as hard/soft magnetic nanocomposites. J Magn Magn Mater 499:166308. https://doi.org/10.1016/j.jmmm.2019.166308

Algarou NA, Slimani Y, Almessiere MA, Güner S, Baykal A, Ercan I, Kögerler P (2020e) Exchange-coupling effect in hard/soft SrTb0.01Tm0.01Fe11.98O19/AFe2O4 (where A = Co, Ni, Zn, Cu and Mn) composites. Ceram Int 46:7089–7098. https://doi.org/10.1016/j.ceramint.2019.11.201

Algarou NA, Slimani Y, Almessiere MA, Rehman S, Younas M, Unal B, Korkmaz AD, Gondal MA, Trukhanov AV, Baykal A, Nahvi I (2020b) Developing the magnetic, dielectric and anticandidal characteristics of SrFe12O19/(Mg0.5Cd0.5Dy0.03Fe1.97O4)x hard/soft ferrite nanocomposites. J Taiwan Inst Chem Eng 113:344–362. https://doi.org/10.1016/j.jtice.2020.07.022

Almessiere M, Slimani Y, Sertkol M, Nawaz M, Sadaqat A, Baykal A, Ercan I, Ozçelik B (2019o) Effect of Nb3+ substitution on the structural, magnetic, and optical properties of Co0.5Ni0.5Fe2O4 nanoparticles. Nano 9:430. https://doi.org/10.3390/nano9030430

Almessiere MA, Dabagh S, Slimani Y, Chaudhary K, Ali J, Baykal A (2018f) Investigation of structural and magnetic properties on Mg1−xZnxFe2−xAlxO4 (0.0 ≤ x ≤ 0.8) nanoparticles. J Inorg Organomet Polym Mater 28:942–953. https://doi.org/10.1007/s10904-017-0764-9

Almessiere MA, Slimani Y, Ali S, Baykal A, Ercan I, Sozeri H (2019b) Nd 3+ ion-substituted Co 1−2x Ni x Mn x Fe 2−y Nd y O 4 nanoparticles: structural, morphological, and magnetic investigations. J Inorg Organomet Polym Mater 29:783–791. https://doi.org/10.1007/s10904-018-1052-z

Almessiere MA, Slimani Y, Auwal İA, Shirsath SE, Manikandan A, Baykal A, Özçelik B, Ercan İ, Trukhanov SV, Vinnik DA, Trukhanov AV (2020d) Impact of Tm3+ and Tb3+ rare earth cations substitution on the structure and magnetic parameters of Co-Ni nanospinel ferrite. Nano 10:2384. https://doi.org/10.3390/nano10122384

Almessiere MA, Slimani Y, Baykal A (2018a) Structural, morphological and magnetic properties of hard/soft SrFe12-xVxO19/(Ni0.5Mn0.5Fe2O4)y nanocomposites: Effect of vanadium substitution. J Alloys Compd 767:966–975. https://doi.org/10.1016/j.jallcom.2018.07.212

Almessiere MA, Slimani Y, Baykal A (2019e) Impact of Nd-Zn co-substitution on microstructure and magnetic properties of SrFe12O19 nanohexaferrite. Ceram Int 45:963–969. https://doi.org/10.1016/j.ceramint.2018.09.272

Almessiere MA, Slimani Y, Baykal A (2020f) Synthesis and characterization of Co1–2-xNixMnxCeyFe2–yO4 nanoparticles. J Rare Earths 38:188–194. https://doi.org/10.1016/j.jre.2019.07.005

Almessiere MA, Slimani Y, Demir Korkmaz A, Güner S, Baykal A, Shirsath SE, Ercan I, Kögerler P (2020k) Sonochemical synthesis of Dy3+ substituted Mn0.5Zn0.5Fe2−xO4 nanoparticles: Structural, magnetic and optical characterizations. Ultrason Sonochem 61:104836. https://doi.org/10.1016/j.ultsonch.2019.104836

Almessiere MA, Slimani Y, El Sayed HS, Baykal A (2018b) Ca2+ and Mg2+ incorporated barium hexaferrites: structural and magnetic properties. J Sol-Gel Sci Technol 88:628–638. https://doi.org/10.1007/s10971-018-4853-1

Almessiere MA, Slimani Y, El Sayed HS, Baykal A (2018c) Ce-Y co-substituted strontium nanohexaferrites: AC susceptibility and Mossbauer studies. Ceram Int 44:12520–12527. https://doi.org/10.1016/j.ceramint.2018.04.046

Almessiere MA, Slimani Y, El Sayed HS, Baykal A, Ali S, Ercan I (2019p) Investigation of microstructural and magnetic properties of $BaV_xFe_{12-x}O_{19}$ nanohexaferrites. J Supercond Nov Magn 32. https://doi.org/10.1007/s10948-018-4856-8

Almessiere MA, Slimani Y, Guner S, Aldakhil S, Korkmaz AD, Sertkol M, Gungunes H, Yasin G, Baykal A (2020e) Ultrasonic synthesis, magnetic and optical characterization of Tm3+ and Tb3+ ions co-doped barium nanohexaferrites. J Solid State Chem 286:121310. https://doi.org/10.1016/j.jssc.2020.121310

Almessiere MA, Slimani Y, Güner S, Baykal A, Ercan I (2019u) Effect of dysprosium substitution on magnetic and structural properties of NiFe2O4 nanoparticles. J Rare Earths 37:871–878. https://doi.org/10.1016/j.jre.2018.10.009

Almessiere MA, Slimani Y, Güner S, Nawaz M, Baykal A, Aldakheel F, Akhtar S, Ercan I, Belenli BO (2019l) Magnetic and structural characterization of Nb 3+ −substituted CoFe 2 O 4 nanoparticles. Ceram Int 45:8222–8232. https://doi.org/10.1016/j.ceramint.2019.01.125

Almessiere MA, Slimani Y, Guner S, Nawaz M, Baykal A, Aldakheel F, Sadaqat A, Ercan I (2019h) Effect of Nb substitution on magneto-optical properties of Co0.5Mn0.5Fe2O4 nanoparticles. J Mol Struct 1195:269–279. https://doi.org/10.1016/j.molstruc.2019.05.075

Almessiere MA, Slimani Y, Guner S, Sertkol M, Demir Korkmaz A, Shirsath SE, Baykal A (2019i) Sonochemical synthesis and physical properties of Co0.3Ni0.5Mn0.2EuxFe2−xO4 nano-spinel ferrites. Ultrason Sonochem 58:104654. https://doi.org/10.1016/j.ultsonch.2019.104654

Almessiere MA, Slimani Y, Güner S, van Leusen J, Baykal A, Kögerler P (2019s) Effect of Nb^{3+} ion substitution on the magnetic properties of $SrFe_{12}O_{19}$ hexaferrites. J Mater Sci Mater Electron 30. https://doi.org/10.1007/s10854-019-01464-0

Almessiere MA, Slimani Y, Güngüneş H, Ali S, Baykal A, Ercan I (2019r) AC susceptibility and hyperfine interactions of Mg-Ca ions co-substituted $BaFe_{12}O_{19}$ nanohexaferrites. Ceram Int 45. https://doi.org/10.1016/j.ceramint.2019.02.050

Almessiere MA, Slimani Y, Güngüneş H, Ali S, Manikandan A, Ercan I, Baykal A, Trukhanov AV (2019q) Magnetic attributes of NiFe2O4 nanoparticles: influence of dysprosium ions (Dy3+) substitution. Nano 9:820. https://doi.org/10.3390/nano9060820

Almessiere MA, Slimani Y, Güngüneş H, Baykal A, Alhamed NA, Trukhanov AV, Trukhanov SV (2019t) Structure, Mössbauer and AC susceptibility of strontium nanohexaferrites: effect of vanadium ions doping. Ceram Int 45:11615–11624. https://doi.org/10.1016/j.ceramint.2019.03.033

Almessiere MA, Slimani Y, Güngüneş H, El Sayed HS, Baykal A (2018d) AC susceptibility and hyperfine interactions of vanadium substituted barium nanohexaferrites. Ceram Int 44:17749–17758. https://doi.org/10.1016/j.ceramint.2018.06.242

Almessiere MA, Slimani Y, Güngüneş H, El Sayed HS, Baykal A (2018e) AC susceptibility and Mossbauer study of Ce3+ ion substituted SrFe12O19 nanohexaferrites. Ceram Int 44:10470–10477. https://doi.org/10.1016/j.ceramint.2018.03.064

Almessiere MA, Slimani Y, Güngüneş H, Korkmaz AD, Baykal A, Trukhanov AV, Yasin G (2020i) SrCoxZrxFe12−2xO19 and SrNixZrxFe12−2xO19 hexaferrites: a comparison study of AC susceptibility, FC-ZFC and hyperfine interactions. Chin J Phys 66:596–605. https://doi.org/10.1016/j.cjph.2020.05.012

Almessiere MA, Slimani Y, Güngüneş H, Kostishyn VG, Trukhanov SV, Trukhanov AV, Baykal A (2020a) Impact of Eu3+ ion substitution on structural, magnetic and microwave traits of Ni–Cu–Zn spinel ferrites. Ceram Int 46:11124–11131. https://doi.org/10.1016/j.ceramint.2020.01.132

Almessiere MA, Slimani Y, Gungunes H, Manikandan A, Baykal A (2019d) Investigation of the effects of Tm 3+ on the structural, microstructural, optical, and magnetic properties of Sr hexaferrites. Results Phys 13:102166. https://doi.org/10.1016/j.rinp.2019.102166

Almessiere MA, Slimani Y, Korkmaz AD, Güner S, Maarouf AA, Baykal A (2020g) Comparative study of sonochemically synthesized Co-Zr and Ni-Zr substituted Sr-hexaferrites: Magnetic and structural investigations. J Magn Magn Mater 497:165996. https://doi.org/10.1016/j.jmmm.2019.165996

Almessiere MA, Slimani Y, Kurtan U, Guner S, Sertkol M, Shirsath SE, Akhtar S, Baykal A, Ercan I (2019w) Structural, magnetic, optical properties and cation distribution of nanosized Co0.7Zn0.3TmxFe2−xO4 ($0.0 \leq x \leq 0.04$) spinel ferrites synthesized by ultrasonic irradiation. Ultrason Sonochem 58:104638. https://doi.org/10.1016/j.ultsonch.2019.104638

Almessiere MA, Slimani Y, Rehman S, Khan FA, Polat EG, Sadaqat A, Shirsath SE, Baykal A (2020c) Synthesis of Dy-Y co-substituted manganese-zinc spinel nanoferrites induced anti-bacterial and anti-cancer activities: comparison between sonochemical and sol-gel auto-combustion methods. Mater Sci Eng C 116:111186. https://doi.org/10.1016/j.msec.2020.111186

Almessiere MA, Slimani Y, Sayed HSE, Baykal A, Ercan I (2019n) Microstructural and magnetic investigation of vanadium-substituted Sr-nanohexaferrite. J Magn Magn Mater 471. https://doi.org/10.1016/j.jmmm.2018.09.054

Almessiere MA, Slimani Y, Sertkol M, Nawaz M, Baykal A, Ercan I (2019k) The impact of Zr substituted Sr hexaferrite: Investigation on structure, optic and magnetic properties. Results Phys 13:102244. https://doi.org/10.1016/j.rinp.2019.102244

Almessiere MA, Slimani Y, Shirsath SE, Wudil YS, Baykal A, Ercan I (2020h) Customized magnetic properties of (Mn0.5Zn0.5)[EuxNdxFe2-2x]O4 nanospinel ferrites synthesized via ultrasonic irradiation approach. Results Phys 19:103350. https://doi.org/10.1016/j.rinp.2020.103350

Almessiere MA, Slimani Y, Tashkandi NA, Güngüneş H, Sertkol M, Nawaz M, Ali S, Baykal A, Ercan I (2019f) Tailored microstructures, optical and magnetic qualities of strontium hexaferrites: consequence of Tm^{3+} and Tb^{3+} ions Co-substitution. Ceram Int 45. https://doi.org/10.1016/j.ceramint.2019.07.126

Almessiere MA, Slimani Y, Trukhanov AV, Demir Korkmaz A, Guner S, Akhtar S, Shirsath SESE, Baykal A, Ercan I (2020j) Effect of Nd-Y co-substitution on structural, magnetic, optical and microwave properties of NiCuZn nanospinel ferrites. J Mater Res Technol 9:11278–11290. https://doi.org/10.1016/j.jmrt.2020.08.027

Almessiere MA, Sözeri H, Trukhanov AV, Slimani Y, Ali S, Acar E, Baykal A (2019c) Tb3+ ion substituted Sr-hexaferrites as high quality microwave absorbers. J Magn Magn Mater 491: 165595. https://doi.org/10.1016/j.jmmm.2019.165595

Almessiere MA, Trukhanov AV, Khan FA, Slimani Y, Tashkandi N, Turchenko VA, Zubar TI, Tishkevich DI, Trukhanov SV, Panina LV, Baykal A (2020b) Correlation between microstructure parameters and anti-cancer activity of the [Mn0.5Zn0.5](EuxNdxFe2-2x)O4 nanoferrites produced by modified sol-gel and ultrasonic methods. Ceram Int 46:7346–7354. https://doi.org/10.1016/j.ceramint.2019.11.230

Almessiere MA, Unal B, Slimani Y, Demir Korkmaz A, Algarou NA, Baykal A (2019g) Electrical and dielectric properties of Nb^{3+} ions substituted Ba-hexaferrites. Results Phys 14. https://doi.org/10.1016/j.rinp.2019.102468

Almessiere MA, Ünal B, Slimani Y, Korkmaz AD, Baykal A, Ercan I (2019a) Electrical properties of La3+ and Y3+ ions substituted Ni0.3Cu0.3Zn0.4Fe2O4 nanospinel ferrites. Results Phys. 15: 102755. https://doi.org/10.1016/j.rinp.2019.102755

Almessiere MA, Slimani Y, El Sayed HS, Baykal A (2019m) Morphology and magnetic traits of strontium nanohexaferrites: effects of manganese/yttrium co-substitution. J Rare Earths 37:732–740. https://doi.org/10.1016/j.jre.2018.09.014

Almessiere MA, Slimani Y, Korkmaz AD, Taskhandi N, Sertkol M, Baykal A, Shirsath SE, Ercan İ, Ozçelik B (2019v) Sonochemical synthesis of Eu^{3+}substituted $CoFe_2O_4$ nanoparticles and their structural, optical and magnetic properties. Ultrason Sonochem 58:104621. https://doi.org/10.1016/j.ultsonch.2019.104621

Almessiere MA, Slimani Y, Sertkol M, Khan FA, Nawaz M, Tombuloglu H, Al-Suhaimi EA, Baykal A (2019j) Ce–Nd Co-substituted nanospinel cobalt ferrites: an investigation of their structural, magnetic, optical, and apoptotic properties. Ceram Int 45:16147–16156. https://doi.org/10.1016/j.ceramint.2019.05.133

Barbara B (2001) Magnetization reversal of nano-particles. Springer, Berlin, Heidelberg, pp 157–208. https://doi.org/10.1007/3-540-44954-x_7

Bean CP (1955) Hysteresis loops of mixtures of ferromagnetic micropowders. J Appl Phys 26: 1381–1383. https://doi.org/10.1063/1.1721912

Bean CP, Livingston JD (1959) Superparamagnetism. J Appl Phys 30:S120–S129. https://doi.org/10.1063/1.2185850

Becker R, Polley H (1940) Influence of internal strain on the law of approach to saturation for nickel. Ann Phys 37:534–540

Ben Salem MK, Hannachi E, Slimani Y, Hamrita A, Zouaoui M, Bessais L, Ben Salem M, Ben Azzouz F (2014) SiO_2 nanoparticles addition effect on microstructure and pinning properties in $YBa_2Cu_3O_y$. Ceram Int 40:4953–4962. https://doi.org/10.1016/j.ceramint.2013.10.103

Berkowitz AE, Kodama RH, Makhloul SA, Parker FT, Spada FE, McNiff EJ, Foner S (1999) Anomalous properties of magnetic nanoparticles. J Magn Magn Mater 196:591–594. https://doi.org/10.1016/S0304-8853(98)00845-2

Berkowitz AE, Lahut JA, Jacobs IS, Levinson LM, Forester DW (1975) Spin pinning at ferrite-organic interfaces. Phys Rev Lett 34:594–597. https://doi.org/10.1103/PhysRevLett.34.594

Berkowitz AE, Lahut JA, VanBuren CE (1980) Properties of magnetic fluid particles. IEEE Trans Magn 16:184–190. https://doi.org/10.1109/TMAG.1980.1060632

Berkowitz AE, Takano K (1999) Exchange anisotropy—a review. J Magn Magn Mater 200:552–570. https://doi.org/10.1016/S0304-8853(99)00453-9

Bloch F (1930) Zur Theorie des Ferromagnetismus. Zeitschrift Für Phys 61:206–219. https://doi.org/10.1007/BF01339661

Brown WF (1959) Relaxational behavior of fine magnetic particles. J Appl Phys 30:S130–S132. https://doi.org/10.1063/1.2185851

Bruno P (1993) Physical origins and theoretical models of magnetic anisotropy. Magn von Festkörpern Und Grenzflächen 24:1–28

Caizer C (2002) Magnetic behaviour of $Mn_{0.6}Fe_{0.4}Fe_2O_4$ nanoparticles in ferrofluid at low temperatures. J Magn Magn Mater 251:304–315. https://doi.org/10.1016/S0304-8853(02)00701-1

Caizer C (2005) Deviations from Bloch law in the case of surfacted nanoparticles. Appl Phys A Mater Sci Process 80:1745–1751. https://doi.org/10.1007/s00339-003-2471-3

Caizer C (2008) Magnetic properties of the novel nanocomposite ($Zn_{0.15}Ni_{0.85}Fe_2O_4$)0.15/($SiO_2$)0.85 at room temperature. J Magn Magn Mater 320:1056–1062. https://doi.org/10.1016/j.jmmm.2007.10.012

Caizer C (2015) Nanoparticle size effect on some magnetic properties. In: Handb. Nanoparticles. Springer International Publishing, pp 475–519. https://doi.org/10.1007/978-3-319-15338-4_24

Caizer C, Hrianca I (2003a) The temperature dependence of saturation magnetization of γ-Fe2O3/SiO2 magnetic nanocomposite. Ann Phys 12:115–122. https://doi.org/10.1002/andp.200310008

Caizer C, Hrianca I (2003b) Dynamic magnetization of γ-Fe2O3 nanoparticles isolated in an SiO2 amorphous matrix. Eur Phys J B 31:391–400. https://doi.org/10.1140/epjb/e2003-00047-1

Callen HB, Goldberg N (1965) Magnetostriction of polycrystalline aggregates. J Appl Phys 36:976–977. https://doi.org/10.1063/1.1714285

Campbell RB (1957) Determination of the orientation of crystals in elongated particles. J Appl Phys 28:381–382. https://doi.org/10.1063/1.1722758

Cannas C, Gatteschi D, Musinu A, Piccaluga G, Sangregorio C (1998) Structural and magnetic properties of Fe2O2 nanoparticles dispersed over a silica matrix. J Phys Chem B 102:7721–7726. https://doi.org/10.1021/jp981355w

Chen J, Sorensen C, Klabunde K (1996) Size-dependent magnetic properties o fine particles synthesized by coprecipitation. Phys Rev B Condens Matter Mater Phys 54:9288–9296. https://doi.org/10.1103/PhysRevB.54.9288

Chen K, Ferrenberg AM, Landau DP (1993) Static critical behavior of three-dimensional classical Heisenberg models: a high-resolution Monte Carlo study. Phys Rev B 48:3249–3256. https://doi.org/10.1103/PhysRevB.48.3249

Chikazumi S (1964) Physics of magnetism. John Wiley, New York

Coey JMD (1971) Noncollinear spin arrangement in ultrafine ferrimagnetic crystallites. Phys Rev Lett 27:1140–1142. https://doi.org/10.1103/PhysRevLett.27.1140

Cullity BD, Graham CD (2008) Introduction to Magnetic Materials. John Wiley & Sons, Inc., Hoboken, NJ. https://doi.org/10.1002/9780470386323

Dennis CL, Borges RP, Buda LD, Ebels U, Gregg JF, Hehn M, Jouguelet E, Ounadjela K, Petej I, Prejbeanu IL, Thornton MJ (2002) The defining length scales of mesomagnetism: a review. J Phys Condens Matter 14:R1175. https://doi.org/10.1088/0953-8984/14/49/201

Djurberg C, Svedlindh P, Nordblad P (1997) Dynamics of an interacting particle system: evidence of critical slowing down. Phys Rev Lett 79:5154–5157. https://doi.org/10.1103/PhysRevLett.79.5154

Fiorani D, Dormann JL, Cherkaoui R, Tronc E, Lucari F, D'Orazio F, Spinu L, Nogues M, Garcia A, Testa AM (1999) Collective magnetic state in nanoparticles systems. J Magn Magn Mater 196:143–147. https://doi.org/10.1016/S0304-8853(98)00694-5

Fisher ME, Ferdinand AE (1967) Interfacial, boundary, and size effects at critical points. Phys Rev Lett 19:169–172. https://doi.org/10.1103/PhysRevLett.19.169

Gittleman JI, Abeles B, Bozowski S (1974) Superparamagnetism and relaxation effects in granular Ni-SiO2 and Ni-Al2O3 films. Phys Rev B 9:3891–3897. https://doi.org/10.1103/PhysRevB.9.3891

Hannachi E, Almessiere MA, Slimani Y, Baykal A, Ben Azzouz F (2020) AC susceptibility investigation of YBCO superconductor added by carbon nanotubes. J Alloys Compd 812:152150. https://doi.org/10.1016/j.jallcom.2019.152150

Hannachi E, Slimani Y, Ben Azzouz F, Ekicibil A (2018) Higher intra-granular and inter-granular performances of YBCO superconductor with TiO nano-sized particles addition. Ceram Int 44:18836–18843. https://doi.org/10.1016/j.ceramint.2018.07.118

Hannachi E, Slimani Y, Ekicibil A, Manikandan A, Ben Azzouz F (2019a) Excess conductivity and AC susceptibility studies of Y-123 superconductor added with TiO2 nano-wires. Mater Chem Phys 235:121721. https://doi.org/10.1016/j.matchemphys.2019.121721

Hannachi E, Slimani Y, Ekicibil A, Manikandan A, Ben Azzouz F (2019b) Magneto-resistivity and magnetization investigations of YBCO superconductor added by nano-wires and nanoparticles of titanium oxide. J Mater Sci Mater Electron 30:8805–8813. https://doi.org/10.1007/s10854-019-01205-3

Hansen MF, Jönsson PE, Nordblad P, Svedlindh P (2002) Critical dynamics of an interacting magnetic nanoparticle system. J Phys Condens Matter 14:4901–4914. https://doi.org/10.1088/0953-8984/14/19/314

He X, Shi H (2012) Size and shape effects on magnetic properties of Ni nanoparticles. Particuology 10:497–502. https://doi.org/10.1016/j.partic.2011.11.011

Hendriksen PV, Linderoth S, Lindgård PA (1992) Finite-size effects in the magnetic properties of ferromagnetic clusters. J Magn Magn Mater 104–107:1577–1579. https://doi.org/10.1016/0304-8853(92)91461-2

Hendriksen PV, Linderoth S, Lindgård PA (1993) Finite-size modifications of the magnetic properties of clusters. Phys Rev B 48:7259–7273. https://doi.org/10.1103/PhysRevB.48.7259

Hrianca I, Caizer C, Schlett Z (2002) Dynamic magnetic behavior of Fe_3O_4 colloidal nanoparticles. J Appl Phys 92:2125–2132. https://doi.org/10.1063/1.1493654

Hubert A, Schäfer R (1998) Magnetic domains. Springer, Berlin Heidelberg. https://doi.org/10.1007/978-3-540-85054-0

Jacobs IS, Bean CP (1963) In: Rado GT, Suhl H (eds) Magnetism III. Academic Press, New York

Kittel C (1946) Theory of the structure of ferromagnetic domains in films and small particles. Phys Rev 70:965–971. https://doi.org/10.1103/PhysRev.70.965

Kittel C (1949) Physical theory of ferromagnetic domains. Rev Mod Phys 21:541–583. https://doi.org/10.1103/RevModPhys.21.541

Kneller E, Seeger A, Kronmüller H, Kneller E, Seeger A, Kronmüller H (1962) Ferromagnetismus. In: Ferromagnetismus. Springer, Berlin Heidelberg, pp 30–37. https://doi.org/10.1007/978-3-642-86695-1_4

Kneller EF, Luborsky FE (1963) Particle size dependence of coercivity and remanence of single-domain particles. J Appl Phys 34:656–658. https://doi.org/10.1063/1.1729324

Kodama RH (1999) Magnetic nanoparticles. J Magn Magn Mater 200:359–372. https://doi.org/10.1016/S0304-8853(99)00347-9

Kodama RH, Berkowitz AE, Mc Niff EJ, Foner S (1996) Surface spin disorder in NiFe2O4 nanoparticles. Phys Rev Lett 77:394–397. https://doi.org/10.1103/PhysRevLett.77.394

Korkmaz ADD, Güner S, Slimani Y, Gungunes H, Amir MM, Manikandan A, Baykal A (2019) Microstructural, optical, and magnetic properties of Vanadium-substituted nickel spinel nanoferrites. J Supercond Nov Magn 32:1057–1065. https://doi.org/10.1007/s10948-018-4793-6

Kumar A, Yasin G, Korai RM, Slimani Y, Ali MF, Tabish M, Tariq Nazir M, Nguyen TA (2020) Boosting oxygen reduction reaction activity by incorporating the iron phthalocyanine nanoparticles on carbon nanotubes network. Inorg Chem Commun 120:108160. https://doi.org/10.1016/j.inoche.2020.108160

Landau LD, Lifshitz EM (1935) To the theory of magnetic permeability dispersion in ferromagnetic solids. Sov Phys 8:153–166

Langevin P (1905) Magnétisme et théorie des électrons. Ann Chim et Phys 8:203

Lin D, Nunes AC, Majkrzak CF, Berkowitz AE (1995) Polarized neutron study of the magnetization density distribution within a CoFe2O4 colloidal particle II. J Magn Magn Mater 145:343–348. https://doi.org/10.1016/0304-8853(94)01627-5

Linderoth S, Balcells L, Labarta A, Tejada J, Hendriksen PV, Sethi SA (1993) Magnetization and Mössbauer studies of ultrafine Fe-C particles. J Magn Magn Mater 124:269–276. https://doi.org/10.1016/0304-8853(93)90125-L

Lu HM, Cao ZH, Zhao CL, Li PY, Meng XK (2008) Size-dependent ordering and Curie temperatures of FePt nanoparticles. J Appl Phys 103:123526. https://doi.org/10.1063/1.2946724

Manikandan A, Yogasundari M, Thanrasu K, Dinesh A, Raja KK, Slimani Y, Jaganathan SK, Srinivasan R, Baykal A (2020) Structural, morphological and optical properties of multifunctional magnetic-luminescent ZnO@Fe3O4 nanocomposite. Phys E Low-Dimensional Syst Nanostruct 124:114291. https://doi.org/10.1016/j.physe.2020.114291

Mayama H, Naito T (2009) Correlation between Curie temperature and system dimension. Phys E Low-Dimension Syst Nanostruct 41:1878–1881. https://doi.org/10.1016/j.physe.2009.08.013

Mazo-Zuluaga J, Restrepo J, Mejía-López J (2008) Influence of non-stoichiometry on the magnetic properties of magnetite nanoparticles. J Phys Condens Matter 20:195213. https://doi.org/10.1088/0953-8984/20/19/195213

Meiklejohn WH (1962) Exchange anisotropy—a review. J Appl Phys 33:1328–1335. https://doi.org/10.1063/1.1728716

Meiklejohn WH, Bean CP (1956) New magnetic anisotropy. Am Phys Soc. https://doi.org/10.1103/PhysRev.102.1413

Meiklejohn WH, Bean CP (1957) New magnetic anisotropy. Phys Rev 105:904–913. https://doi.org/10.1103/PhysRev.105.904

Morr AH, Haneda K (1981) Magnetic structure of small NiFe2O4 particles. J Appl Phys 52:2496–2498. https://doi.org/10.1063/1.328979

Mørup S (1994) Superparamagnetism and spin glass ordering in magnetic nanocomposites. EPL 28:671–676. https://doi.org/10.1209/0295-5075/28/9/010

Mørup S, Tronc E (1994) Superparamagnetic relaxation of weakly interacting particles. Phys Rev Lett 72:3278–3281. https://doi.org/10.1103/PhysRevLett.72.3278

Nadeem M, Yasin G, Arif M, Tabassum H, Bhatti MH, Mehmood M, Yunus U, Iqbal R, Nguyen TA, Slimani Y, Song H, Zhao W (2021) Highly active sites of Pt/Er dispersed N-doped hierarchical porous carbon for trifunctional electrocatalyst. Chem Eng J 409:128205. https://doi.org/10.1016/j.cej.2020.128205

Nanda A, Nanda S, Nguyen TA, Rajendran S, Slimani Y (2020) Nanocosmetics: fundamentals, applications and toxicity, Elsevier. doi:https://doi.org/10.1016/c2019-0-00468-2.

Nawaz M, Almessiere MA, Almofty SA, Gungunes CD, Slimani Y, Baykal A (2019) Exploration of catalytic and cytotoxicity activities of CaxMgxNi1-2xFe2O4 nanoparticles. J Photochem Photobiol B Biol 196:111506. https://doi.org/10.1016/j.jphotobiol.2019.05.003

Nawaz M, Sliman Y, Ercan I, Lima-Tenório MK, Tenório-Neto ET, Kaewsaneha C, Elaissari A (2018) Magnetic and pH-responsive magnetic nanocarriers. In: Stimuli responsive Polym. Nanocarriers Drug Deliv. Appl. Vol. 2 Adv. Nanocarriers Ther. Elsevier, pp 37–85. https://doi.org/10.1016/B978-0-08-101995-5.00002-7

Néel L (1947) Propriétés d'un ferromagnétique cubique en grains fins. CR Acad Sci Paris 224:1488. https://hal.archives-ouvertes.fr/hal-02878474/document

Neel L (1949) Theorie du trainage magnetique des ferromagnetiques en grains fins avec applications aux terres cuites. Ann Geophys 5:99–136. https://ci.nii.ac.jp/naid/10017473358

Néel L (1955) Some theoretical aspects of rock-magnetism. Adv Phys 4:191–243. https://doi.org/10.1080/00018735500101204

Nogués J, Schuller IK (1999) Exchange bias. J Magn Magn Mater 192:203–232. https://doi.org/10.1016/S0304-8853(98)00266-2

Prené P, Tronc E, Jolivet JP, Livage J, Cherkaoui R, Noguès M, Dormann JL, Fiorani D (1993) Magnetic properties of isolated γ-Fe2O3 particles. IEEE Trans Magn 29:2658–2660. https://doi.org/10.1109/20.280834

Pshenichnikov AF, Mekhonoshin VV, Lebedev AV (1996) Magneto-granulometric analysis of concentrated ferrocolloids. J Magn Magn Mater 161:94–102. https://doi.org/10.1016/S0304-8853(96)00067-4

Rehman S, Almessiere MA, Khan FA, Korkmaz AD, Tashkandi N, Slimani Y, Baykal A (2020) Synthesis and biological characterization of Mn0.5Zn0.5EuxDyxFe1.8-2xO4 nanoparticles by sonochemical approach. Mater Sci Eng C 109:110534. https://doi.org/10.1016/j.msec.2019.110534

Rehman S, Almessiere MA, Tashkandi N, Baykal A, Slimani Y, Jermy R, Ravinayagam V, Yaman C (2019) Fabrication of Spinel Cobalt Ferrite ($CoFe_2O_4$) nanoparticles with unique earth element cerium and neodymium for anticandidal activities. Chem Select 4:14329–14334. https://doi.org/10.1002/slct.201901811

Rosensweig RE (1985) Ferrohydrodynamics. Cambridge University Press, Cambridge

Sadaqat A, Almessiere M, Slimani Y, Guner S, Sertkol M, Albetran H, Baykal A, Shirsath SE, Ozcelik B, Ercan I (2019) Structural, optical and magnetic properties of Tb3+ substituted Co nanoferrites prepared via sonochemical approach. Ceram Int 45:22538–22546. https://doi.org/10.1016/j.ceramint.2019.07.280

Saleh TA (2020) Nanomaterials: classification, properties, and environmental toxicities. Environ Technol Innov 20:101067. https://doi.org/10.1016/j.eti.2020.101067

Sastry MD, Babu Y, Goyal PS, Mehta RV, Upadhyay RV, Srinivas D (1995) Electron magnetic resonance of ferrofluids: Evidence for anisotropic resonance at 77 K in samples cooled in a magnetic field. J Magn Magn Mater 149:64–66. https://doi.org/10.1016/0304-8853(95)00339-8

Seevakan K, Manikandan A, Devendran P, Slimani Y, Baykal A, Alagesan T (2018) Structural, morphological and magneto-optical properties of $CuMoO_4$ electrochemical nanocatalyst as supercapacitor electrode. Ceram Int 44:20075–20083. https://doi.org/10.1016/j.ceramint.2018.07.282

Seevakan K, Manikandan A, Devendran P, Slimani Y, Baykal A, Alagesan T (2019) Structural, magnetic and electrochemical characterizations of $Bi_2Mo_2O_9$ nanoparticle for supercapacitor application. J Magn Magn Mater 486:165254. https://doi.org/10.1016/j.jmmm.2019.165254

Slimani Y, Hannachi E, Ben Azzouz F, Ben Salem M (2018b) Impact of planetary ball milling parameters on the microstructure and pinning properties of polycrystalline superconductor $Y_3Ba_5Cu_8O_y$. Cryogenics 92:5–12. https://doi.org/10.1016/j.cryogenics.2018.03.010

Slimani Y, Algarou NA, Almessiere MA, Sadaqat A, Vakhitov MG, Klygach DS, Tishkevich DI, Trukhanov AV, Güner S, Hakeem AS, Auwal IA, Baykal A, Manikandan A, Ercan I (2021a) Fabrication of exchanged coupled hard/soft magnetic nanocomposites: correlation between composition, magnetic, optical and microwave properties. Arab J Chem 14:102992. https://doi.org/10.1016/j.arabjc.2021.102992

Slimani Y, Almessiere MA, Baykal A (2018c) AC susceptibility study of Cu substituted BaFe12O19 nanohexaferrites. Ceram Int 44:13097–13105. https://doi.org/10.1016/j.ceramint.2018.04.130

Slimani Y, Almessiere MA, Guner S, Alahmari FS, Yasin G, Trukhanov AV, Baykal A (2020b) Influence of Tm–Tb substitution on magnetic and optical properties of Ba–Sr hexaferrites prepared by ultrasonic assisted citrate sol-gel approach. Mater Chem Phys 253:123324. https://doi.org/10.1016/j.matchemphys.2020.123324

Slimani Y, Almessiere MA, Güner S, Tashkandi NA, Baykal A, Sarac MF, Nawaz M, Ercan I (2019h) Calcination effect on the magneto-optical properties of vanadium substituted $NiFe_2O_4$ nanoferrites. J Mater Sci Mater Electron 30:9143–9154. https://doi.org/10.1007/s10854-019-01243-x

Slimani Y, Almessiere MA, Hannachi E, Al-qwairi FO, Manikandan A, Baykal A, Ben Azzouz F (2019j) AC susceptibility, DC magnetization and superconducting properties of tungsten oxide nanowires added $YBa_2Cu_3O_y$. Ceram Int 45. https://doi.org/10.1016/j.ceramint.2019.07.196

Slimani Y, Almessiere MA, Hannachi E, Mumtaz M, Manikandan A, Baykal A, Ben Azzouz F (2019i) Improvement of flux pinning ability by tungsten oxide nanoparticles added in $YBa_2Cu_3O_y$ superconductor. Ceram Int 45:6828–6835. https://doi.org/10.1016/j.ceramint.2018.12.176

Slimani Y, Almessiere MA, Korkmaz AD, Baykal A, Ercan I (2020c) AC susceptibility and FC-ZFC magnetic properties of SrTbxFe12−xO19 and SrTmxFe12−xO19 hexaferrites: comparative study. J Rare Earths. https://doi.org/10.1016/j.jre.2020.07.026

Slimani Y, Almessiere MA, Nawaz M, Baykal A, Akhtar S, Ercan I, Belenli I (2019f) Effect of bimetallic (Ca, Mg) substitution on magneto-optical properties of NiFe2O4 nanoparticles. Ceram Int 45:6021–6029. https://doi.org/10.1016/j.ceramint.2018.12.072

Slimani Y, Almessiere MA, Sertkol M, Shirsath SE, Baykal A, Nawaz M, Akhtar S, Ozcelik B, Ercan I (2019e) Structural, magnetic, optical properties and cation distribution of nanosized Ni0.3Cu0.3Zn0.4TmxFe2−xO4 ($0.0 \leq x \leq 0.10$) spinel ferrites synthesized by ultrasound irradiation. Ultrason Sonochem 57:203–211. https://doi.org/10.1016/j.ultsonch.2019.05.001

Slimani Y, Almessiere MA, Shirsath SE, Hannachi E, Yasin G, Baykal A, Ozçelik B, Ercan I (2020a) Investigation of structural, morphological, optical, magnetic and dielectric properties of (1-x)BaTiO3/xSr0.92Ca0.04Mg0.04Fe12O19 composites. J Magn Magn Mater 510:166933. https://doi.org/10.1016/j.jmmm.2020.166933

Slimani Y, Almessiere MA, Korkmaz AD, Guner S, Güngüneş H, Sertkol M, Manikandan A, Yildiz A, Akhtar S, Shirsath SE, Baykal A (2019g) $Ni_{0.2}Cu_{0.4}Zn_{0.4}TbxFe_{2-x}O_4$ nanospinel ferrites: ultrasonic synthesis and physical properties. Ultrason Sonochem:59:104757. https://doi.org/10.1016/j.ultsonch.2019.104757

Slimani Y, Hannachi E (2020) Magnetic nanosensors and their potential applications. In: Nanosensors for Smart Cities. Elsevier, pp 143–155. https://doi.org/10.1016/b978-0-12-819870-4.00009-8

Slimani Y, Hannachi E (2021) Ru-based perovskites/RGO composites for applications in high performance supercapacitors. In: Hybrid Perovskite Compos. Mater. Elsevier, pp 335–354. https://doi.org/10.1016/b978-0-12-819977-0.00016-0

Slimani Y, Hannachi E, Ben Salem MK, Ben Azzouz F, Ben Salem M (2018d) Comparative study of electrical transport and magnetic measurements of Y3Ba5Cu8O18±δ and YBa2Cu3O7−δ compounds: intragranular and intergranular superconducting properties. Appl Phys A Mater Sci Process. https://doi.org/10.1007/s00339-017-1547-4

Slimani Y, Hannachi E, Ekicibil A, Almessiere MA, Ben Azzouz F (2019d) Investigation of the impact of nano-sized wires and particles TiO_2 on Y-123 superconductor performance. J Alloys Compd 781:664–673. https://doi.org/10.1016/j.jallcom.2018.12.062

Slimani Y, Hannachi E, Hamrita A, Ben Salem MK, Ben Azzouz F, Manikandan A, Ben Salem M (2018a) Comparative investigation of the ball milling role against hand grinding on microstructure, transport and pinning properties of $Y_3Ba_5Cu_8O_{18\pm\delta}$and $YBa_2Cu_3O_{7-\delta}$. Ceram Int 44:19950–19957. https://doi.org/10.1016/j.ceramint.2018.07.261

Slimani Y, Hannachi E, Tombuloglu H, Güner S, Almessiere MA, Baykal A, Aljafary MA, Al-Suhaimi EA, Nawaz M, Ercan I (2019a) Magnetic nanoparticles based nanocontainers for biomedical application. Elsevier. https://doi.org/10.1016/B978-0-12-816770-0.00014-9

Slimani Y, Selmi A, Hannachi E, Almessiere MA, Baykal A, Ercan I (2019c) Impact of ZnO addition on structural, morphological, optical, dielectric and electrical performances of BaTiO3 ceramics. J Mater Sci Mater Electron 30:9520–9530. https://doi.org/10.1007/s10854-019-01284-2

Slimani Y, Unal B, Almessiere MA, Demir Korkmaz A, Baykal A (2021b) Investigation of AC susceptibility, dielectric and electrical properties of Tb–Tm co-substituted M-type Sr hexaferrites. Mater Chem Phys 260:124162. https://doi.org/10.1016/j.matchemphys.2020.124162

Slimani Y, Unal B, Almessiere MA, Korkmaz AD, Shirsath SE, Yasin G, Trukhanov AV, Baykal A (2020d) Investigation of structural and physical properties of Eu3+ ions substituted

Ni0.4Cu0.2Zn0.4Fe2O4 spinel ferrite nanoparticles prepared via sonochemical approach. Results Phys 17:103061. https://doi.org/10.1016/j.rinp.2020.103061

Slimani Y, Unal B, Hannachi E, Selmi A, Almessiere MA, Nawaz M, Baykal A, Ercan I, Yildiz M (2019b) Frequency and dc bias voltage dependent dielectric properties and electrical conductivity of BaTiO3-SrTiO3/(SiO2) x nanocomposites. Ceram Int 45:11989–12000. https://doi.org/10.1016/j.ceramint.2019.03.092

Smith DO (1958) Static and dynamic behavior of thin permalloy films. J Appl Phys 29:264–273. https://doi.org/10.1063/1.1723097

Stoner EC, Wohlfarth EP (1948) A mechanism of magnetic hysteresis in heterogeneous alloys. Philos. Trans. R. Soc. London. Ser. A, Math. Phys. Sci 240:599–642. https://doi.org/10.1098/rsta.1948.0007

Tombuloglu H, Slimani Y, Tombuloglu G, Almessiere M, Baykal A (2019a) Uptake and translocation of magnetite (Fe3O4) nanoparticles and its impact on photosynthetic genes in barley (Hordeum vulgare L.). Chemosphere 226:110–122. https://doi.org/10.1016/j.chemosphere.2019.03.075

Tombuloglu H, Slimani Y, Tombuloglu G, Almessiere M, Sozeri H, Demir-Korkmaz A, AlShammari TM, Baykal A, Ercan I, Hakeem KR (2019c) Impact of calcium and magnesium substituted strontium nano-hexaferrite on mineral uptake, magnetic character, and physiology of barley (Hordeum vulgare L.). Ecotoxicol Environ Saf 186:109751. https://doi.org/10.1016/j.ecoenv.2019.109751

Tombuloglu H, Slimani Y, Tombuloglu G, Demir Korkmaz A, Baykal A, Almessiere M, Ercan I (2019b) Impact of superparamagnetic iron oxide nanoparticles (SPIONs) and ionic iron on physiology of summer squash (Cucurbita pepo): A comparative study, Plant Physiol. Biochemist 139. https://doi.org/10.1016/j.plaphy.2019.03.011

Tombuloglu H, Tombuloglu G, Slimani Y, Ercan I, Sozeri H, Baykal A (2018) Impact of manganese ferrite (MnFe2O4) nanoparticles on growth and magnetic character of barley (Hordeum vulgare L.). Environ Pollut 243:872–881. https://doi.org/10.1016/j.envpol.2018.08.096

Tronc E, Ezzir A, Cherkaoui R, Chanéac C, Noguès M, Kachkachi H, Fiorani D, Testa AM, Grenèche JM, Jolivet JP (2000) Surface-related properties of γ-Fe2O3 nanoparticles. J Magn Magn Mater 221:63–79. https://doi.org/10.1016/S0304-8853(00)00369-3

Tronc E, Prene P, Jolivet JP, d'Orazio F, Lucari F, Fiorani D, Godinho M, Cherkaoui R, Nogues M, Dormann JL (1995) Magnetic behaviour of γ-Fe2O3 nanoparticles by mössbauer spectroscopy and magnetic measurements. Hyperfine Interact 95:129–148. https://doi.org/10.1007/BF02146310

Trukhanov AV, Astapovich KA, Turchenko VA, Almessiere MA, Slimani Y, Baykal A, Sombra ASB, Zhou D, Jotania RB, Singh C, Zubar TI, Tishkevich DI, Trukhanov SV (2020) Influence of the dysprosium ions on structure, magnetic characteristics and origin of the reflection losses in the Ni–Co spinels. J Alloys Compd 841:155667. https://doi.org/10.1016/j.jallcom.2020.155667

Ullah S, Yasin G, Ahmad A, Qin L, Yuan Q, Khan AU, Khan UA, Rahman AU, Slimani Y (2020) Construction of well-designed 1D selenium-tellurium nanorods anchored on graphene sheets as a high storage capacity anode material for lithium-ion batteries. Inorg Chem Front 7:1750–1761. https://doi.org/10.1039/c9qi01701a

Unal B, Almessiere M, Slimani Y, Baykal A, Trukhanov AV, Ercan I (2019) The conductivity and dielectric properties of neobium substituted Sr-hexaferrites. Nano 9:1168. https://doi.org/10.3390/nano9081168

Upadhyay RV, Srinivas D, Mehta RV (2000) Magnetic resonance in nanoscopic particles of a ferrofluid. J Magn Magn Mater 214:105–111. https://doi.org/10.1016/S0304-8853(99)00811-2

Vinosha PA, Manikandan A, Judith Ceicilia AS, Dinesh A, Francisco Nirmala G, Preetha AC, Slimani Y, Almessiere MA, Baykal A, Xavier B (2021) Review on recent advances of zinc substituted cobalt ferrite nanoparticles: Synthesis characterization and diverse applications. Ceram Int. https://doi.org/10.1016/j.ceramint.2020.12.289

Vinosha PA, Manikandan A, Ragu R, Dinesh A, Paulraj P, Slimani Y, Almessiere MA, Baykal A, Madhavan J, Xavier B, Nirmala GF (2020b) Exploring the influence of varying pH on structural, electro-optical, magnetic and photo-Fenton properties of mesoporous ZnFe2O4 nanocrystals. Environ Pollut 272:115983. https://doi.org/10.1016/j.envpol.2020.115983

Vinosha PAA, Manikandan A, Ragu R, Dinesh A, Thanrasu K, Slimani Y, Baykal A, Xavier B (2020a) Impact of nickel substitution on structure, magneto-optical, electrical and acoustical properties of cobalt ferrite nanoparticles. J Alloys Compd:157517. https://doi.org/10.1016/j.jallcom.2020.157517

Wang J, Wu W, Zhao F, Zhao GM (2011) Curie temperature reduction in SiO2 -coated ultrafine Fe 3 O4 nanoparticles: quantitative agreement with a finite-size scaling law. Appl Phys Lett 98: 083107. https://doi.org/10.1063/1.3558918

Weiss P (1906) La variation du ferromagnetisme avec la temperature. Comptes Rendus 143:1136–1139. 1137 https://ci.nii.ac.jp/naid/10006418112 (accessed January 20, 2021)

Wohlfarth EP (1982) Handbook of ferromagnetic materials, vol 3. Elsevier

Williams HJ, Bozorth RM, Shockley W (1949) Magnetic domain patterns on single crystals of silicon iron. Phys Rev 75:155–178. https://doi.org/10.1103/PhysRev.75.155

Yasin G, Anjum MJ, Malik MU, Khan MA, Khan WQ, Arif M, Mehtab T, Nguyen TA, Slimani Y, Tabish M, Ali D, Zuo Y (2020b) Revealing the erosion-corrosion performance of sphere-shaped morphology of nickel matrix nanocomposite strengthened with reduced graphene oxide nanoplatelets. Diam Relat Mater 104:107763. https://doi.org/10.1016/j.diamond.2020.107763

Yasin G, Arif M, Mehtab T, Shakeel M, Mushtaq MA, Kumar A, Nguyen TA, Slimani Y, Nazir MT, Song H (2020a) A novel strategy for the synthesis of hard carbon spheres encapsulated with graphene networks as a low-cost and large-scalable anode material for fast sodium storage with an ultralong cycle life. Inorg Chem Front 7:402–410. https://doi.org/10.1039/c9qi01105f

Zhang J, Boyd C, Luo W (1996) Two mechanisms and a scaling relation for dynamics in ferrofluids. Phys Rev Lett 77:390–393. https://doi.org/10.1103/PhysRevLett.77.390

Electrical Characterization of Nanomaterials

10

Artem Kozlovskiy, Inesh Kenzhina, Maxim V. Zdorovets, Kayrat K. Kadyrzhanov, and Alex V. Trukhanov

Abstract

To date, a key area of research in the field of nanostructured materials science and methods of obtaining nanostructures is the study of the effect of physicochemical and structural properties on their conductive characteristics. In contrast to macro-objects for nanostructures, factors such as the size of grains and the concentration

A. Kozlovskiy (✉)
L.N. Gumilyov Eurasian National University, Nur-Sultan, Kazakhstan

The Institute of Nuclear Physics, Almaty, Kazakhstan

Research School of Chemistry and Applied Biomedical Sciences National Research Tomsk Polytechnic University, Tomsk, Russia
e-mail: kozlovskiy.a@inp.kz

I. Kenzhina
L.N. Gumilyov Eurasian National University, Nur-Sultan, Kazakhstan

The Institute of Nuclear Physics, Almaty, Kazakhstan
e-mail: kenzhina@physics.kz

M. V. Zdorovets
L.N. Gumilyov Eurasian National University, Nur-Sultan, Kazakhstan

The Institute of Nuclear Physics, Almaty, Kazakhstan

Ural Federal University, Yekaterinburg, Russia
e-mail: mzdorovets@inp.kz

K. K. Kadyrzhanov
L.N. Gumilyov Eurasian National University, Nur-Sultan, Kazakhstan

A. V. Trukhanov
National University of Science and Technology MISiS, Moscow, Russia

SSPA "Scientific and Practical Materials Research Centre of NAS of Belarus", Minsk, Belarus

South Ural State University, Chelyabinsk, Russia

A. Thakur et al. (eds.), *Synthesis and Applications of Nanoparticles*,
https://doi.org/10.1007/978-981-16-6819-7_10

of their boundaries, dislocation and defect densities, the degree of ordering of the structure, etc. have a tremendous effect on the conducting properties, as well as the magnitude of the resistance. In this case, in most cases, nanosize can lead to extraordinary changes in the conductivity of nanostructures. The purpose of this chapter is to present the results of the influence of various conditions for the synthesis of copper nanostructures, as well as various methods of modification using ionizing radiation on their conducting properties.

Keywords

Nanotubes · Conductivity · Electrochemical synthesis · Defects · Ionizing radiation

10.1 Introduction

In the past few years, great attention has been paid to methods for producing metallic nanostructures of various geometries, such as cubes, spherical particles, tubes, wires, dendrites, etc. (Sharma et al. 2018; Yang et al. 2015; Matschuk et al. 2010; Bhanushali et al. 2015; Qiu et al. 2009). Such an increased interest in nanostructured materials is primarily due to the possibility of obtaining new data on the nature of quantum forces at the nanoscale, when the size effects begin to play a major role, as well as the wide potential of application as a basis for miniature devices, the demand for which is increasing every year. The use of nanostructures in microelectronics has made a breakthrough in the field of information storage, sensorics, energy storage, etc. Special attention in this direction is paid to nanowires and nanotubes, which, due to their high flexibility in terms of length, diameter, and their aspect ratio, open up broad prospects in the study of the properties of nanomaterials, as well as their practical application (Alia et al. 2013; Jiang et al. 2002). The most promising material among all metals is copper and its oxide forms. Interest in these types of nanostructures is due to their high electrical conductivity and resistance to degradation. Copper nanostructures are widely used as a basis for emitter bases, anode materials, and semiconductor devices (Zang et al. 2018; Zhang et al. 2019).

Among the variety of synthesis methods, a special place is occupied by the method of galvanic or electrochemical synthesis, which allows not only to control the geometry of the obtained nanostructures with high accuracy through the use of template matrices of a given geometry but also to scale this technology on a semi-industrial scale. This synthesis method makes it possible to obtain nanostructures of various lengths, which can vary from 100 nm to 100 μm or more, as well as diameters from several nanometers to tens of microns. In this case, the geometry of the synthesized nanostructures is completely specified by the template matrix, which is used for synthesis. At the same time, varying the synthesis conditions makes it possible not only to control the geometry of the synthesized nanostructures but also to control their phase and elemental composition, the main crystallographic characteristics, such as the crystallite size, crystal lattice parameter, dislocation density, degree of crystal structure perfection, etc. A change in all these parameters primarily has a significant effect on the conducting characteristics, while it can be both positive, associated with an increase in conductivity and a decrease in resistance, and negative, due to a sharp deterioration in conducting properties. In this case, the size effect plays an important role in determining the conducting characteristics of nanomaterials, since electron transport in metallic nanostructures, the size of which is comparable to the electron path length, is of great interest, both from a fundamental point of view and for practical application in microelectronics in order to increase the conductivity of nanostructures.

Another important factor in determining the conducting characteristics of nanostructures is their resistance to external influences, degradation as a result of interaction with acids or alkaline solutions, natural aging and oxidation in air, partial amorphization, etc. At the same time, an important task for researchers in this field is to find the optimal conditions for modification and increase the resistance to degradation of nanomaterials. A distinctive feature between nanomaterials and bulk analogs is the presence of a large number of point defects, nonequilibrium phases, which can have a negative effect on the conductive properties of nanostructures, as well as the ballistic nature of charge transfer.

However, despite a large number of studies in this area, there are still many questions associated primarily with the possibility of directed modification of nanostructures, as well as an increase in resistance to degradation, which will further help open up new promising opportunities for the use of nanostructures. This chapter is devoted to the consideration of such issues as the effect of synthesis conditions and the geometry of nanostructures on the conducting properties, assessing the prospects of using various types of ionizing radiation for targeted modification and increasing the conductivity of nanostructures and assessing the potential use of copper nanostructures as a basis for amplifying an electrical signal.

10.2 Influence of Synthesis Conditions and Geometry on the Conducting Characteristics of Nanostructures

Figure 10.1 shows a schematic representation of the production of Cu nanotubes in the pores of a template matrix and the subsequent release of their polymer template. In most cases, polymer matrices based on polycarbonate or polyethylene terephthalate, as well as based on aluminum oxide, are used as templates. At the same time, the choice of polymer matrices makes it possible to obtain nanostructures with diameters in the range from 30 to 50 nm to several microns, while matrices based on aluminum

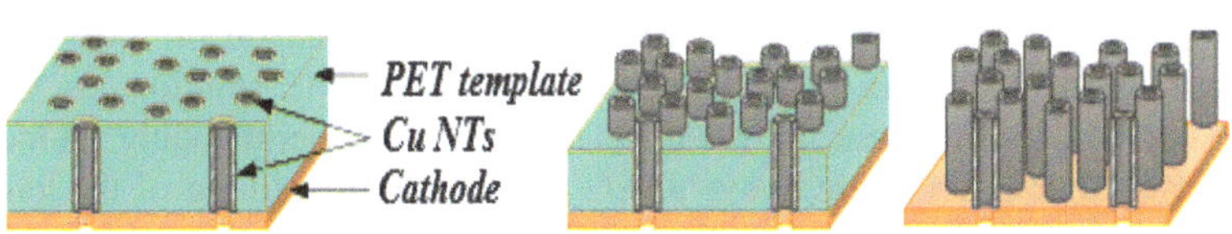

Fig. 10.1 Schematic representation of the template technology for producing Cu nanotubes

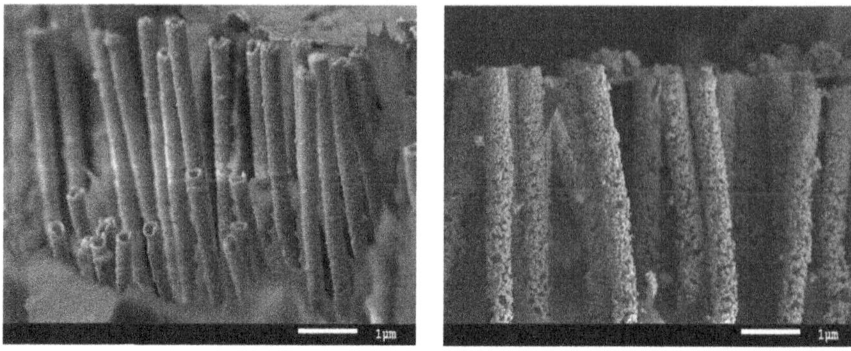

Nanotubes synthesized from sulfate electrolytes

Nanotubes synthesized from nitrate electrolytes

Fig. 10.2 Typical images of synthesized nanotubes from various electrolyte solutions

oxide limit the possibilities of obtaining ordered arrays of nanostructures with diameters greater than 300–500 nm, which is associated with the processes of anodizing matrices and etching of the matrix diameters.

In most cases, to obtain Cu nanostructures, copper sulfate salts ($CuSO_4 \cdot 5H_2O$) with the addition of sulfuric acid (H_2SO_4) are used to increase the acidity level. In this case, the range of applied potential differences in various works varies from 0.3 V to 2.0 V. The authors of various works associate such a wide range of applied potential differences with the possibility of obtaining both nanotubes and nanowires. Also, for the synthesis of Cu nanostructures, electrolyte solutions based on nitrates are used; however, in this case, the synthesized nanostructures have a low degree of crystallinity and order (see Fig. 10.2).

In this case, the presence of disordering regions, as well as the weak bonding of grains, has a negative effect on the conduction mechanisms.

Also, an important role, in addition to the salts used to prepare electrolyte solutions, is played by the synthesis conditions themselves, such as the difference in applied potentials, the temperature of the electrolyte, and the acidity of the solution. Thus, in (Kenzhina et al. 2018), the dependence of the change in the geometry of the obtained nanostructures on the synthesis conditions was shown. The authors varied not only the potential difference but also the temperature of the electrolyte. In this case, the choice of synthesis conditions was based on theoretical data on the deposition potential of copper structures by the electrochemical method, as well as on the conditions for creating an overvoltage gradient in the electrolyte solution. The range of deposition potentials for the search for optimal conditions was 0.5–1.5 V. The temperature range of 25–50 °C was based on the possibility of accelerating the reduction of metal ions in an electrolyte solution, as well as increasing the deposition rate. However, as it was found experimentally, an increase in the electrolyte temperature above 35 °C leads to uneven filling of the template pores with metal due to the uncontrolled growth of nanostructures, as well as a large number of amorphous inclusions resulting from the introduction of oxygen into the

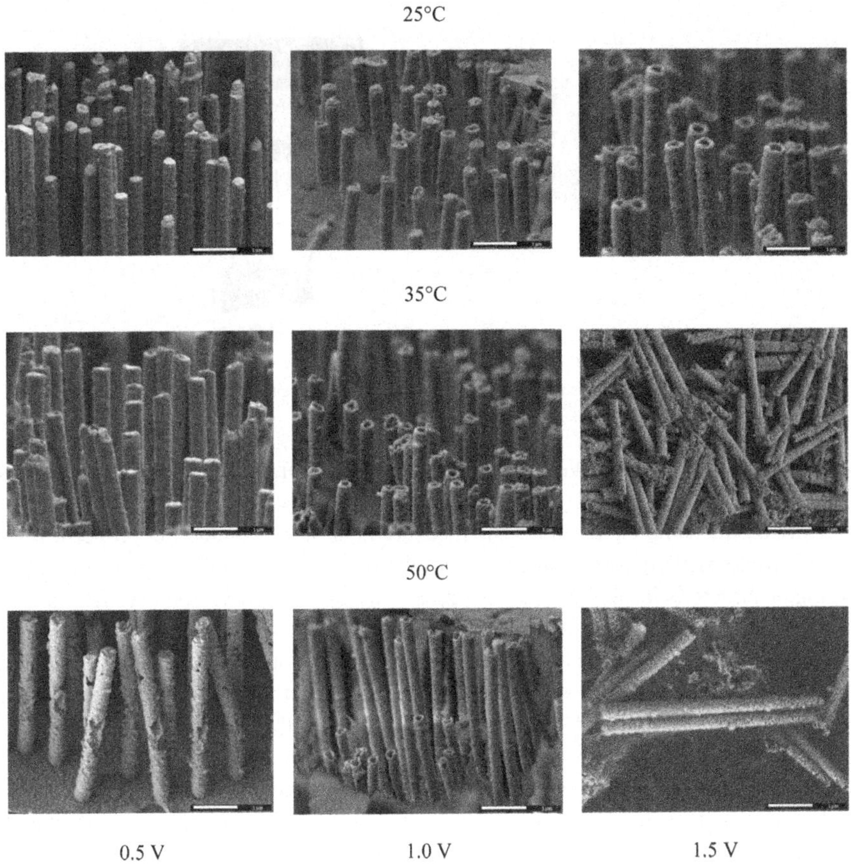

Fig. 10.3 Typical images of synthesized nanotubes obtained under different synthesis conditions (Kenzhina et al. 2018)

sites and interstices of the lattice during deposition. Amorphization or the formation of disordered regions as a result of the introduction of oxide impurities occurs due to the violent uncontrolled growth of nanostructures and a large amount of oxygen that is released during the deposition process. Also, at high electrolyte temperatures, air bubbles appear, which passivate the pores, thereby complicating the deposition process. In the case of high deposition potentials at high temperatures, the overvoltage gradient sharply increases, which leads to strong distortions and deformations of the structure, as well as the formation of regions of disorder. In this case, according to the authors, the optimal conditions, according to the experimental studies carried out, are an applied potential difference of 1.0–1.5 V at an electrolyte temperature of 25 °C and an applied potential difference of 1.0 V at an electrolyte temperature of 35 °C. These conditions make it possible to obtain nanotubes with a given geometry and wall thickness. The use of synthesis conditions of 0.5 V and an electrolyte

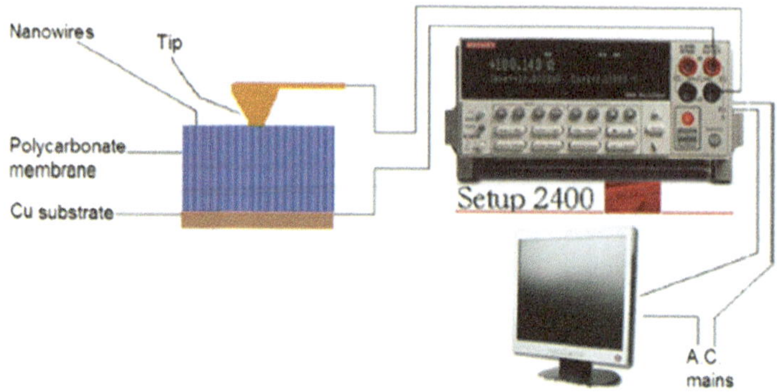

Fig. 10.4 Typical circuit for measuring current-voltage characteristics (Kaur and Chauhan 2014a)

temperature of 25 °C makes it possible to obtain ordered arrays of nanowires (Fig. 10.3).

Unlike bulk samples for which the measurement of the conducting properties is a fairly simple method, which includes various methods for measuring both current-voltage characteristics and directly the resistivity or conductivity, most methods are unacceptable for nanostructures, and methods for measuring conducting characteristics in most cases are difficult. Amandeep Kaur and R.P. Chauhan were among the first to propose measuring the current-voltage characteristics of nanostructures (Kaur and Chauhan 2014a). Their method is based on measuring the current-voltage characteristics of nanostructures located in a polymer matrix with a metal substrate deposited on one of the sides, which acts as an electrode. A probe with a diameter of several hundred microns is used as the second electrode, which allows simultaneous measurement of the current-voltage characteristics of several hundred nanostructures parallel to each other in a polymer matrix (see Fig. 10.4).

According to the proposed technique in (Kaur and Chauhan 2014a), the authors of (Kenzhina et al. 2018) measured the current-voltage characteristics of the synthesized nanostructures, and using expression (10.1), the values of the specific conductivity were calculated, which are shown in Fig. 10.5.

$$\sigma = \frac{dI}{dU}\frac{l}{A}, \tag{10.1}$$

where l is the length of the nanotubes, A is the area, and dI/dU is the tangent of the I-U slope.

As can be seen from the presented diagram of the change in the value of the specific conductivity depending on the synthesis conditions, in the case of an increase in the applied potential difference at an electrolyte temperature of 25 °C, it leads to an increase in the conductivity, which the authors of (Kenzhina et al. 2018) associate with an increase in the degree of structural ordering and, consequently, a decrease in concentration of obstacles to the ballistic movement of charge carriers. A decrease in resistance and an increase in conductivity are due to structural orderings,

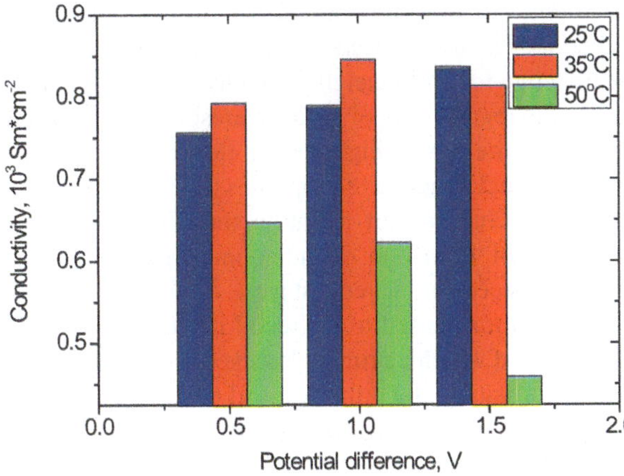

Fig. 10.5 Diagram of changes in the conductivity of nanostructures (Kenzhina et al. 2018)

as well as an increase in the size of crystallites, which leads to a decrease in the defect structure and a decrease in the density of dislocation defects. With the ballistic nature of the movement of electrons in the structure of nanotubes, a decrease in the dislocation density and concentration of defects with a change in the synthesis conditions leads to an increase in the conductivity of electrons by reducing the number of obstacles in their path. However, in the case of the formation of oxide nanostructures, which are obtained under synthesis conditions of 1.0–1.5 V and an electrolyte temperature of 50 °C, the value of the specific conductivity decreases sharply, which indicates a deterioration in the conducting properties.

10.3 Radiation Modification of Nanostructures to Increase Conductivity

10.3.1 Mechanisms of Interaction with Ionizing Radiation

One of the ways to change the physicochemical properties of nanostructured materials is radiation modification. One of the urgent problems of the radiation modification of nanomaterials is the controlled formation of defects in the crystal structure and their recombination in order to improve the strength and conductive properties of nanostructures (Empizo et al. 2015; Chauhan et al. 2016; Kumari et al. 2014). In this case, an important factor is the further migration of defects, which can significantly change the structure of materials. Meanwhile, it is these processes that determine the disturbances in the crystal lattice during irradiation or ion implantation of nanomaterials. In this case, point radiation defects can annihilate or flow down the grain boundaries. Radiation effects arising in nanomaterials under the action of ionizing radiation have a number of features that differ from similar effects in micro- and macro-sized objects. In this case, the presence of a large number of grain boundaries and joints, which serve as sinks for the removal of radiation defects, contributes to an increase in the stability of nanomaterials in comparison with bulk materials. The use of electronic and gamma radiation for modifying nanostructures is an effective tool for changing the physicochemical properties of nanostructures.

In this regard, it is of great interest to study the effect of ionizing radiation, in particular, the flux of high-energy electrons, gamma quanta, and heavy ions on the structural and conductive properties of Cu nanotubes.

The mechanisms of radiation modification are primarily associated with the processes of formation of radiation defects and their subsequent time evolution, which proceeds in several stages. At the first stage, the energy of the incident particles is transferred to the recoil nuclei for very short periods of time $\sim 10^{-18}$ s, as a result of which a primary knocked out atom can form. In this case, the condition for the formation of a primary knocked out atom is the excess of the threshold value of the binding energy upon collision and subsequent knockout of the atom from the site of the crystal lattice (Choudhary and Chauhan 2016).

The second stage consists in the transfer of energy from the initially knocked out atom to other atoms of the crystal lattice with the subsequent formation and development of branching cascades of atomic displacements in times not exceeding $\sim 10^{-13}$ s.

The third stage is characterized by the formation of local atomic rearrangements caused by the spontaneous recombination of defects in a time of $\sim 10^{-11}$ s. In this case, the formed point defects can migrate along the crystal lattice with subsequent recombination or the formation of clusters and regions of disorder. Energy transfer from incident particles to lattice atoms occurs through elastic scattering and energy losses on electrons and nuclei, as well as inelastic collisions with electrons, followed by ionization of lattice atoms and excitation of the electronic subsystem.

In the case of irradiation with heavy ions with energies above 1 MeV/nucleon, a colossal amount of energy is transferred into the crystal structure. At the same time, the intensity of the "injection" of energy into the electronic subsystem is 100–1000 times higher than into the nuclear one. As a result, the contribution of electronic excitations to the processes of defect formation is much greater. In this case, the density of energy losses along the trajectory of ion movement is 10–50 keV/nm. As a result, the formation of amorphous inclusions, phase transformations, and the appearance of shock waves is observed in the irradiated structures.

In turn, for copper nanostructures, the main processes of defect formation are elastic collisions under irradiation with heavy ions with energies up to 100 MeV. The main factor determining the mechanisms of defect formation upon irradiation with heavy ions is the relaxation of strong electronic excitations created as a result of elastic and inelastic collisions and the subsequent transfer of part of the energy into the crystal lattice.

Unlike ions, electrons have a significantly lower mass, and therefore structural defects such as atomic displacements can be observed only when irradiated with high-energy electrons. In this connection, the description of the interaction of electrons with matter and their motion is carried out within the framework of relativistic quantum mechanics. The relativistic kinetic energy of an incident electron can be calculated by the formula (10.2):

$$E = \frac{m_{0e}c^2}{(1 - v^2 / c^2)^{1/2}} - m_{0e}c^2, \tag{10.2}$$

where $m_{0e}c^2$ is the rest energy (0.51 MeV). For electrons with energies above 500 keV, which can lead to the formation of atomic displacements, the screening effects are neglected, and the Coulomb potential is used to describe the interaction processes. In this case, the loss of energy as a result of a collision with the nucleus can be written as:

$$T = \frac{2E(E + 2m_{0e}c^2)}{M_2c^2} \sin^2\left(\frac{\varphi}{2}\right), \tag{10.3}$$

where E is the energy of the incident electron. The maximum energy transferred to an atom in elastic collision is calculated using (10.4):

$$T_{\max} = \frac{2E(E + 2m_{0e}c^2)}{M_2c^2}, \tag{10.4}$$

According to the data presented, as a result of collisions of electrons with energies of 1–5 MeV, $T_{\max}$ = 180–250 eV, which leads to the formation of single defects, since the energy of the primary knocked out atom is insufficient to create cascades of secondary defects. This circumstance served as the basis for the use of electron radiation for the radiation annealing of defects in the crystal structure of materials.

In the case of irradiation with electrons with an energy of more than 10 MeV, which exceeds the energy of the threshold of nuclear reactions, it is necessary to take into account structural disturbances as a result of the inelastic effects of recoil nuclei, gamma quanta, and nucleons. In this case, a significant contribution to the energy losses is made by the deceleration of electrons due to radiation losses in the electric field of nuclei. The emerging bremsstrahlung X-ray radiation in this case can lead to ionization with the subsequent formation of electron-positron pairs and the formation of avalanche-like electron-photon showers in the structure.

In the case of irradiation of gamma quanta, the main feature is the small amount of absorption, as a result of which the depth of penetration and damage is much greater than that of electrons and heavy ions. In this case, the interaction of gamma quanta with energies up to 10 MeV is carried out through three mechanisms: the photoeffect, the Compton effect, and the effect of the formation of electron-positron pairs. The photoelectric effect is typical for the energy range from 10 to 100 keV, the Compton effect is a characteristic for the region from 100 keV to 5 MeV, and the formation of electron-positron pairs is observed at energies above 10 MeV. As a result of the absorption of gamma quanta in the process of the photoelectric effect and the ejection of an electron, the atom passes into an excited state, which is removed by internal electronic transitions. The Compton effect is the elastic scattering of gamma quanta by electrons.

In general, gamma quanta, when interacting with the crystal structure, do not create a large number of atomic displacements. The maximum energy that a gamma quantum is able to transmit as a result of collision can be written as:

$$T_{\max} = \frac{2E_\gamma^2}{M_2c^2 + 2E_\gamma} = E_\gamma\left(1 - \frac{M_2c^2}{2E_\gamma}\right)^{-1}, \quad (10.5)$$

On average, the value $T_{\max} = 10–50$ eV, which is significantly less than the energy required to create a large number of structural defects. In this case, the displacement of atoms can occur only in the case of elastic collisions of Compton electrons with target atoms. In this case, most of the gamma quanta are spent on ionization, followed by the release of heat in the form of thermal energy.

10.3.2 Examples of the Use of Ionizing Radiation for the Modification of Nanostructures

At the moment, the priority direction of development of the possibility of using ionizing radiation for peaceful purposes is the use of ionizing radiation to modify the properties of nanostructures. These processes are based on changes in the concentration of defects in the structure of nanomaterials during irradiation and the interaction of an ionizing substance with target atoms (Zhao et al. 2017; Tuboltsev and Räisänen 2009; Kaur and Chauhan 2014b). In this case, in the case of high-energy heavy ions, the energy losses, which can lead to the formation of displacement cascades, are not always applicable for modification. For example, in (Kozlovskiy et al. 2019), the effect of irradiation with C^{2+} and O^{2+} ions with an energy of 1.75 MeV/nucleon and doses of 1×10^9 to 5×10^{12} cm^{-2} on the change in the degree of structural ordering and conductive properties of Cu nanotubes obtained by the method of electrochemical deposition was considered. During the experiments, the authors established a three-stage nature of the change in conducting properties depending on the radiation dose and the type of ions (see Fig. 10.6).

The first stage is characterized by the retention of the specific conductivity, which is due to the insignificant contribution of the formed single defects to the change in the structural characteristics and conducting properties. An increase in the irradiation fluence from 5×10^{10} to 5×10^{11} ion/cm^2 leads to an increase in the specific conductivity and corresponds to the second stage. The change in the conductive properties is due to a decrease in the concentration of disordered regions and an increase in the degree of perfection of the crystal structure as a result of radiation annealing of defects and stresses arising in the course of synthesis, as well as a change in the density of dislocations and free charge carriers in the nanotube structure.

The first stage is characterized by the retention of the specific conductivity, which is due to the insignificant contribution of the formed single defects to the change in the structural characteristics and conducting properties. An increase in the irradiation

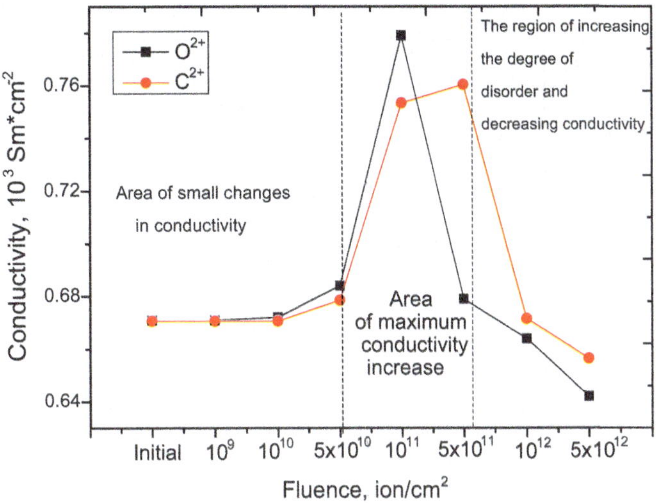

Fig. 10.6 Graph of changes in the conductivity of nanostructures (Kozlovskiy et al. 2019)

fluence from 5×10^{10} to 5×10^{11} ion/cm^2 leads to an increase in the specific conductivity and corresponds to the second stage. The change in the conductive properties is due to a decrease in the concentration of disordered regions and an increase in the degree of perfection of the crystal structure as a result of radiation annealing of defects and stresses arising in the course of synthesis, as well as a change in the density of dislocations and free charge carriers in the nanotube structure.

The most promising methods of irradiation are fluxes of gamma rays and electrons, which lead to the so-called processes of electron annealing of defects and an increase in structural ordering (Nagel and Balogh 1999). Below is a series of studies devoted to assessing the possibility of using ionizing radiation for directional modification.

The effect of gamma irradiation on electrical properties of Cu nanowires (Rana et al. 2014) considered the production of copper-based nanowires with diameters of 80, 100, and 200 nm, obtained by electrochemical deposition, as well as the effect of gamma radiation with different radiation doses (100 and 150 kGy) on the change in the current-voltage characteristics of nanostructures. Polycarbonate membranes were used as template matrices. After irradiation with gamma rays, there is a change in the linearity of the current-voltage characteristics shown in Fig. 10.7, which leads to an increase in resistance.

Size and irradiation effects on the structural and electrical properties of copper nanowires by *Pallavi Rana and R.P. Chauhan* (Rana and Chauhan 2014) considered the influence of size effects and gamma ray learning on changes in the structural and electrical properties of copper nanowires of various diameters (80, 100, and 200 nm). The authors found that after irradiation, the volt-ampere characteristics of nanowires

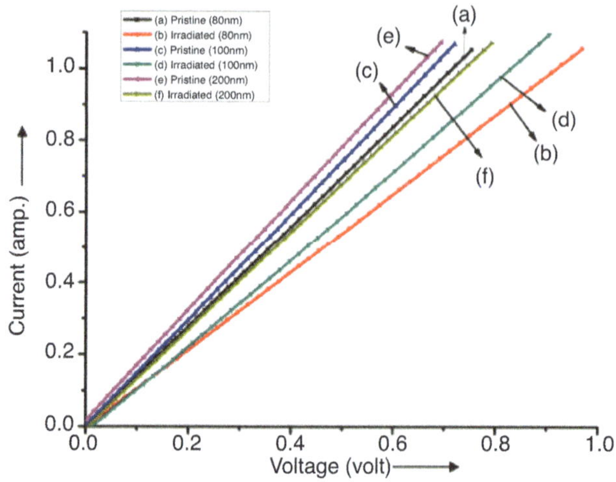

Fig. 10.7 Current-voltage characteristics of the samples under study before and after irradiation (Rana et al. 2014)

with a diameter of 80 nm change more strongly than that of nanostructures with a larger diameter.

Based on the data obtained, it can be concluded that the change in the conducting characteristics of Cu nanostructures is influenced not only by the geometry and synthesis conditions but also by ionizing radiation. At the same time, the use of irradiation with specified characteristics and doses can be considered as one of the promising methods for modifying nanostructures in order to increase their conducting characteristics.

10.4 Potential Applications of Nanostructures

In recent years, the use of the field emission effect for the generation of free electrons has limitations in comparison with thermionic emission, which is actively used in scanning electron microscopy. This effect is based on the creation of electric field gradients ($E \geq 10^7$ V/cm) in order to obtain high values of the emission current. Special attention is paid to determining the high electric field strength, which must be achieved at sufficiently low-voltage values. However, despite all these remarks, the field emission effect has a number of useful functions that can be used to create emitting devices with a current density of 10^6 A/cm^2, nanodiodes with a response time of no more than 10^{-13} s, etc.

In this case, the main requirements for micro and nanostructures used as elements of microelectronics are high values of the electric field strength obtained at low applied voltages (about 100–200 V). In this case, an important condition is a small interelectrode distance between the emitting tips and the surface of the receiver.

The phenomenon of cold field emission is based on the effect of quantum tunneling of electrons that are inside a grounded conductor and an external electric field. Using quantum mechanical calculation methods, it is possible to determine the dependence of the current density on the electric field strength E, which is called the Fowler-Nordheim relation (10.6):

$$J = C_1 E^2 \exp(-C_2 / E), \qquad (10.6)$$

where C_1 and C_2 are parameters expressed through the value of the work function of the electron $\varphi = 3.63$ eV, the charge and mass of the electron.

$$C_1 = \frac{e^3}{8\pi h t^2(y)\varphi}, \quad C_2 = \frac{8\pi\sqrt{2m}}{3he}\varphi^{3/2}\theta(y),$$

$$t(y) = 1, \quad \theta(y) = 1 - y^2, \quad y = \frac{e(eE)^{1/2}}{\varphi}$$

The Fowler-Nordheim relation describes the situation when the emitting surface is an infinite plane perpendicular to the direction of the external electric field. The value of E in the immediate vicinity of the apex of the nanostructure can be calculated using the analytical formula (10.7):

$$E = \frac{2U}{r\ln(\frac{4d}{r})}, \qquad (10.7)$$

where U is the applied voltage, r is the radius of the nanostructure, and d is the distance between the electrodes. The factor influencing the calculation of the Fowler-Nordheim dependence is the Schottky effect, based on the assumption of the interaction of an emitted electron from the emitter surface and its mirror image, leading to a decrease in the potential barrier, which is created by an external electric field. The correction taking into account this effect is contained in the function $\theta(y)$.

The most convenient way to analyze the obtained experimental data determined using (10.6) is based on the logarithmic representation of this equation, to which there is a proportional relationship between the ratio I/E^2 and the inverse value of the electric field strength $1/E$ (10.8):

$$\ln(\frac{I}{E^2}) = C_1 - \frac{C_2}{E}, \qquad (10.8)$$

The nature of the dependence indicates the mechanism of electron emission associated with field emission. The results obtained show that the emission properties of nanostructures are well described by the Fowler-Nordheim relation. Figure 10.8 shows the dependence of the change in the emission current for various Cu nanostructures irradiated with various types of ionizing radiation.

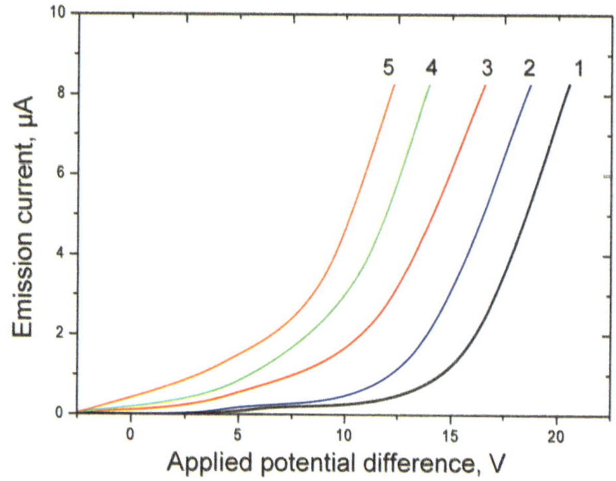

Fig. 10.8 Dependence of the change in the emission current on the difference in applied potentials for various nanostructures: (1) initial Cu nanostructures; (2) irradiated with a flux of gamma quanta with a dose of 100 kGy; (3) irradiated with an electron beam with an energy of 5 MeV with a dose of 100 kGy; (4) irradiated with C^{2+} ions with a fluence of 10^{11} ion/cm^2; (5) irradiated with O^{2+} ions with a fluence of 10^{11} ion/cm^2

As can be seen from the data presented, for modified nanostructures, the achievement of a given value of the emission current is observed at lower applied voltages, which indicates a more efficient enhancement of the emission current by modified nanostructures. An increase in the emission current for irradiated nanostructures is due to an increase in the concentration of charge carriers, as well as an improvement in the structural characteristics and a decrease in the amorphous regions leading to an increase in the conductive and emission properties of nanostructures.

One of the important properties of nanotubes used as emitters is a high aspect ratio (length to diameter ratio). Due to this feature of nanostructures, the value of the electric field E in the vicinity of the tip of the nanostructures is many times higher than the average value of E_0—the ratio of the applied voltage U to the distance between the tip of the nanostructure and the anode.

Expression (10.9) allows us to determine the ability of the emitter to amplify the electric field, which is characterized by the field gain β:

$$\beta = \frac{E}{E_0} = \frac{Ed}{U}, \quad (10.9)$$

Since the aspect ratio for nanostructures reaches ~10^2–10^3, the field emission of nanotubes is observed at lower applied voltages than in the case of traditional cold field emitters. Figure 10.9 shows a graph of the dependence of the change in the field gain for various types of nanostructures.

As can be seen from the presented graph, the modification of nanostructures leads to an increase in the value of the field gain, which is due to a change in the crystal structure as a result of irradiation. High values of the gain make it possible to study the Stark effect in order to determine the intramolecular forces and make it possible to explain the dielectric properties of molecules, etc.

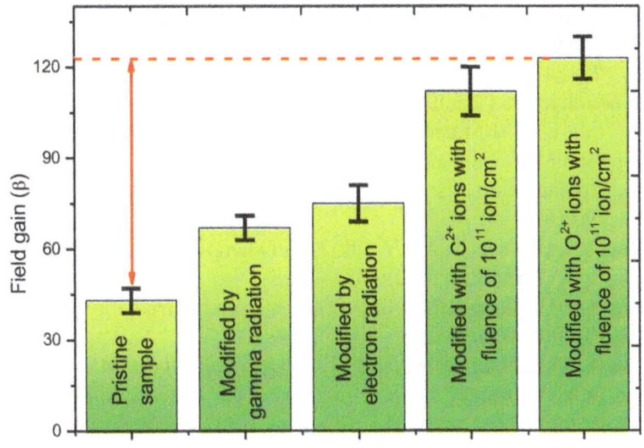

Fig. 10.9 Graph of field gain variation for different types of nanostructures

10.5 Conclusion

Despite a large number of works devoted to methods of obtaining and modifying nanostructures, there are still no effective methods for the large-scale production of ordered arrays of copper nanostructures with the same morphological and structural characteristics. The main difficulties in the synthesis are short length, nonlinear morphology, polydispersity, low crystallinity, and low content of nanostructures in matrices (no more than 5–7% of the matrix weight).

A promising method for modifying nanomaterials and their properties is the use of beams of heavy ions or gamma rays and electrons. In this case, in the case of irradiation with ionizing radiation, in contrast to thermal annealing, the modification of properties occurs locally within the structure. By controlling the energy and radiation dose, it is possible to modify the properties of nanomaterials at any depth, as well as significantly increase the resistance of nanostructures to external influences, by means of radiation hardening and changes in structural properties.

This chapter briefly shows the prospects of using various types of ionizing radiation for the targeted modification of copper nanostructures in order to increase the conducting characteristics, which have potential use as emitter bases and anode materials for lithium-ion batteries.

References

Alia SM, Jensen K, Contreras C, Garzon F, Pivovar B, Yan Y (2013) Platinum coated copper nanowires and platinum nanotubes as oxygen reduction electrocatalysts. ACS Catal 3(3): 358–362

Bhanushali S, Ghosh P, Ganesh A, Cheng W (2015) 1D copper nanostructures: progress, challenges and opportunities. Small 11(11):1232–1252

Chauhan RP, Rana P, Narula C, Panchal S, Choudhary R (2016) Variation in electrical properties of gamma irradiated cadmium selenate nanowires. Nucl Instrum Methods Phys Res Sect B 379: 78–84

Choudhary R, Chauhan RP (2016) Gamma irradiation induced modifications in spin coated CdSe thin films. J Mater Sci Mater Electron 27(11):11674–11681

Empizo MJF, Yamanoi K, Mori K, Arita R, Iwano K, Takabatake M, Abe Y (2015) Gamma-ray irradiation effects on the optical properties of bulk ZnO single crystals. Appl Phys Express 8(6): 061101

Jiang X, Herricks T, Xia Y (2002) CuO nanowires can be synthesized by heating copper substrates in air. Nano Lett 2(12):1333–1338

Kaur A, Chauhan RP (2014a) Effect of gamma irradiation on electrical and structural properties of Zn. Radiat Phys Chem 100:59–64

Kaur A, Chauhan RP (2014b) Carbon ion beam-induced variation in orientation of crystal planes of polycrystalline Zn nanowires. Radiat Eff Defects Solids 169(6):513–521

Kenzhina IE, Zdorovets MV, Kozlovskiy AL, Kadyrzhanov KK (2018) Synthesis and properties of Cu/CuO nanostructures obtained by electrochemical deposition. Mater Res Express 5(3): 035052

Kozlovskiy A, Kenzhina I, Zdorovets M, Petrov A, Kadyrzhanov K (2019) Effect of irradiation with C2+ and O2+ ions on the structural and conductive characteristics of copper nanostructures. Mater Res Express 6(7):075072

Kumari M, Rana P, Chauhan RP (2014) Modifications in structural and electrical properties of gamma irradiated CdSe nanowires. Nucl Instrum Methods Phys Res Sect A 753:116–120

Matschuk M, Bruus H, Larsen NB (2010) Nanostructures for all-polymer microfluidic systems. Microelectron Eng 87(5–8):1379–1382

Nagel R, Balogh AG (1999) Atomic transport in metal/ceramic interfaces under heavy ion irradiation. Nucl Instrum Methods Phys Res Sect B 156(1–4):135–142

Qiu R, Cha HG, Noh HB, Shim YB, Zhang XL, Qiao R, Kang YS (2009) Preparation of dendritic copper nanostructures and their characterization for electroreduction. J Phys Chem C 113(36): 15891–15896

Rana P, Chauhan RP (2014) Size and irradiation effects on the structural and electrical properties of copper nanowires. Phys B Condens Matter 451:26–33

Rana P, Gehlawat D, Chauhan RP (2014) Effect of gamma irradiation on electrical properties of Cu nanowires. AIP Conf Proc—Am Inst Phys 1591(1):265–266

Sharma A, Srivastava AK, Jeon Y, Ahn B (2018) Template-assisted fabrication of nanostructured tin (β-Sn) arrays for bulk microelectronic packaging devices. Metals 8(5):347

Tuboltsev V, Räisänen J (2009) Sculpturing nanowires with ion beams. Small 5(23):2687–2691

Yang D, Kim D, Ko SH, Pisano AP, Li Z, Park I (2015) Focused energy field method for the localized synthesis and direct integration of 1D nanomaterials on microelectronic devices. Adv Mater 27(7):1207–1215

Zang G, Hao W, Li X, Huang S, Gan J, Luo Z, Zhang Y (2018) Copper nanowires-MOFs-graphene oxide hybrid nanocomposite targeting glucose electro-oxidation in neutral medium. Electrochim Acta 277:176–184

Zhang L, Liang H, Ma X, Ye C, Zhao G (2019) A vertically aligned CuO nanosheet film prepared by electrochemical conversion on Cu-based metal-organic framework for non-enzymatic glucose sensors. Microchem J 146:479–485

Zhao Y, Sadat ME, Dunn A, Xu H, Chen CH, Nakasuga W et al (2017) Photothermal effect on Fe3O4 nanoparticles irradiated by white-light for energy-efficient window applications. Sol Energy Mater Sol Cells 161:247–254

Electromagnetic Characterizations of Nanomaterials

11

Atul Thakur, Preeti Thakur, Dinesh Kumar, and P. B. Sharma

Abstract

Nanoferrite materials of cobalt-substituted M-type barium were found suitable for defense applications and were prepared via a chemical method. The prepared samples were characterized by different techniques like X-ray diffractometer (XRD), field emission scanning electron microscopy (FESEM), Fourier transform infrared spectroscopy (FTIR), Raman spectroscopy (RS), and vector network analyzer (VNA). Pure single-phase M-type barium ferrite was confirmed by using XRD investigations with an average crystallite size in the range 65–85 nm. An average grain size in the range 250–325 nm was noticed by using FESEM. Strong Me–O bonds were confirmed by vibrational bands at 466 cm^{-1} and 580 cm^{-1} and Raman vibrational modes at 329 cm^{-1} and 676 cm^{-1}. Real, imaginary, and loss tangent parts of permittivity and permeability in the frequency range of 12–18 GHz (Ku band). $BaCoFe_{11}O_{19}$ materials were found to have a maximum reflection loss of −45 dB making suitable candidate for radar-absorbing material (RAM) in Ku band.

Keywords

Nanoferrite · Electromagnetic properties · Giga hertz frequency · Absorbing

A. Thakur (✉) · D. Kumar · P. B. Sharma
Centre of Nanotechnology, Amity University Haryana, Gurugram, India
e-mail: athakur1@ggn.amity.edu

P. Thakur
Department of Physics, Amity School of Applied Sciences, Amity University Haryana, Gurugram, India

A. Thakur et al. (eds.), *Synthesis and Applications of Nanoparticles*,
https://doi.org/10.1007/978-981-16-6819-7_11

11.1 Introduction

Due to the fantastic electromagnetic properties of nanomaterials; stealth technology has drawn a lot of attention of scientific community specially nano-scientists and nano-technologists. These nanomaterials find enormous applications in high-frequency devices, radar-absorbing materials, antenna miniaturization, etc. (Rana et al. 2018). Magnetic nanomaterials especially soft ferrites are quite useful and suitable for these applications, but these are restricted below 200 MHz frequencies (Thakur et al. 2011). Above 200 MHz, hard magnetic materials specially hexa-ferrites are found to be more useful in gigahertz frequency region (Li et al. 2013a). M-type barium ferrite is an important hard magnetic material with a space group of *P63/mmc* with magnetoplumbite structure (Batlle et al. 1991; Rashid and Ibarhim 2013). Substituted barium hexaferrites have been extensively investigated for improved electromagnetic properties (Ozgur et al. 2009; Cho and Kim 1999).

11.2 RF Network Analyzer

RF network analyzers are essential items of test instrumentation for RF design laboratories.

The RF network analyzer gives a stimulus for the network and then monitors the response. In this way, the operation and performance can be seen and assessed for its suitability.

RF network analyzers can be used for all RF and microwave frequencies. The network analyzers are basically of three types:

1. ***Scalar network analyzer (SNA):*** The scalar network analyzer, SNA is a form of RF network analyzer that only measures the amplitude properties of the device under test—i.e., its scalar properties. In view of this, it is the simpler of the various types of analyzer.
2. ***Vector network analyzer (VNA):*** The VNA is a more beneficial form of RF network analyzer than the SNA as it can measure more parameters of the device under test. It measures the amplitude response along with the phase. Therefore, VNA is also named as a gain-phase meter or an automatic network analyzer.
3. ***Large signal network analyzer (LSNA):*** The LSNA is a highly useful for investigating the characteristics of devices under large signal conditions. It is able to investigate the harmonics and nonlinearities of a network under these conditions and also provides a full analysis of its operation. Previously it was called as Microwave Transition Analyzer, MTA.

11.3 Difference Between RF Network Analyzers and Spectrum Analyzers

Although there are many similarities between RF network analyzers and spectrum analyzers, there are also several major differences in terms of parameter measurements.

A spectrum analyzer is intended for analyzing the nature of signals that are fed into them. On the other hand, a network analyzer generates a signal and uses this to analyze a network or device.

RF network analyzers are used to measure components, devices, circuits, and subassemblies. It contains both a source and multiple receivers then displays amplitude and often phase information. An RF network analyzer looks for a known signal and with vector-error correction; network analyzers provide much higher measurement accuracy than spectrum analyzers.

Whereas, spectrum analyzers are normally used to measure the characteristics of a signal rather than a device. It can measure the parameters like signal or carrier level, sidebands, harmonics, phase noise, etc. They are the most commonly configured as a single-channel receiver, without a source and they have a wide range of IF bandwidths available than most RF network analyzers.

11.4 Concept of Vector Network Analyzer

The vector network analyzer utilizes the concept of measuring the transmitted and reflected waves as a signal passes through a device under test.

Measuring the transmitted and reflected signals across the band of interest, and often beyond, enables the characteristics of a device to be determined. If both transmitted and reflected signals are used to characterize the input and also the output then the device can be fully characterized. This can form a key part of any design or test for an RF circuit. The basic concept of Vector Network Analyzer is shown in Fig. 11.1.

11.5 X-Ray Diffraction

Different characteristic peaks (110), (114), (107), (203), (205), (206), (217), (2011), and (220) were recorded on XRD observed for the cobalt-substituted M-type $BaCo_xFe_{12-x}O_{19}$ with $x = 0.4$, 0.8, and 1.0 (Fig. 11.2). Scherrer formula was used to calculate the average crystallite size on the most intense peaks along (107) and (114) planes (Li et al. 2014; Mathur et al. 2008):

$$D = \frac{0.9\lambda}{\beta \mathrm{Cos}\theta} \tag{11.1}$$

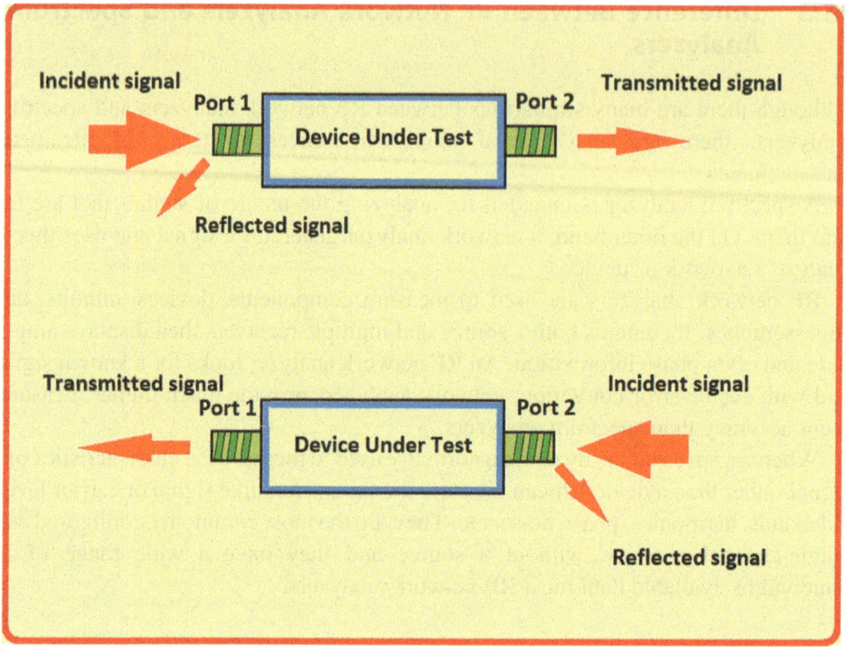

Fig. 11.1 Basic concept of Vector Network Analyzer

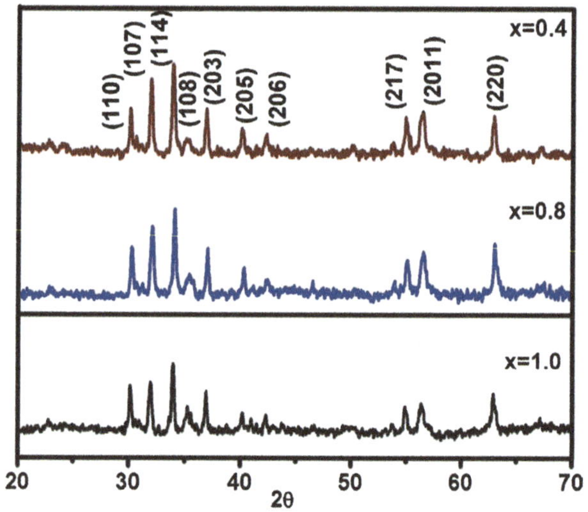

Fig. 11.2 XRD pattern of $BaCo_{0.4}Fe_{11.6}O_{19}$, $BaCo_{0.8}Fe_{11.2}O_{19}$, and $BaCoFe_{11}O_{19}$ sintered at 900 °C for 3 h. (With permission from Elsevier License Number 5133631119568)

where λ wavelength of Cu (K_{α}), β is full width at half maxima, and θ is Bragg's angle which the diffracted beam makes with diffracting planes.

Lattice parameters (a and c), volume (V), and density (ρ) of these materials were calculated by using Eqs. (11.2), (11.3), and (11.4), respectively:

$$\frac{1}{d}=\left(\frac{4}{3}\frac{h^2+hk+k^2}{a^2}+\frac{l^2}{c^2}\right)^{\frac{-1}{2}} \tag{11.2}$$

$$V=\frac{\sqrt{3}}{2}a^2c \tag{11.3}$$

$$\rho=\frac{2M}{N_A V} \tag{11.4}$$

where 2 is the number of barium atoms in unit cell, "M" is the mass, "V" is volume of the unit cell, and "N_A" is the Avogadro's number.

Average crystallite size for the samples was found in the range 65–85 nm and is comparable to the reported values (Din et al. 2014). The strain was calculated by using Williamson–Hall (W–H) plots (Sharma et al. 2014). The sign of the slope gives the nature of strain. Positive slope corresponds to tensile strain, whereas negative corresponds to compressive strain. It is clearly seen that the slope is positive that implies the nanoparticles experience a tensile strain (Rana et al. 2015).

This strain may be responsible for an increase in the average crystallite size because greater tensile strain leads to elongation of the particles (Wang et al. 2002).

11.6 Fourier Transform Infrared Spectroscopy (FTIR)

IR spectra of cobalt-substituted $BaCo_xFe_{12-x}O_{19}$ with $x = 0.4$, 0.8, and 1.0 is recorded in the range 2500–400 cm^{-1} as depicted in Fig. 11.3. The absorption bands in the range 1000–400 cm^{-1} imply the formation of inorganic or ferrites ions in the crystal lattice (Sertkol et al. 2009). In case of hexaferrite, atoms are distributed in three different sites, i.e., tetrahedral, octahedral, and trigonal bypyramidal. The observed bands at 466 cm^{-1}, 523 cm^{-1}, 580 cm^{-1}, 606 cm^{-1}, 768 cm^{-1}, 856 cm^{-1}, 1053 cm^{-1}, 1381 cm^{-1}, 1438 cm^{-1}, and 1631 cm^{-1} for different composition of cobalt-substituted M-type BaFe12O19 ($x = 0.4$, 0.8, and 1.0) are shown in Fig. 11.5. The position of different vibrational bands in cobalt-substituted M-type barium ferrite is shown in Table 11.1. The bands range from 580 to 560 cm^{-1} and 470–430 cm^{-1} correspond to the symmetric stretching of tetrahedral site (Fe–O) and octahedral sites (Fe–O) (Zhao et al. 2008). The vibrational bands at 763 cm^{-1} and 876 cm^{-1} correspond to N–O and Co–O stretching vibration (Kaur and Srivastava 2013). Similarly, the bands at 1381 cm^{-1} and 1438 cm^{-1}correspond to the symmetric stretching of COO group, whereas the vibrational band at 1641 cm^{-1} implies the antisymmetric stretching of COO (Asl et al. 2014; Chawla et al. 2014). With an increase in the cobalt content in M-type barium ferrite, the sharpness of the dips increases. This increase in the sharpness of dips corresponds to the strong Me–O bond formation, whereas N–O bond diminishes from the composition $x = 1.0$. The change in the sharpness of dips is due to the replacement of Fe^{3+} ion with larger ionic radii by Co^{2+} ion from 2b site.

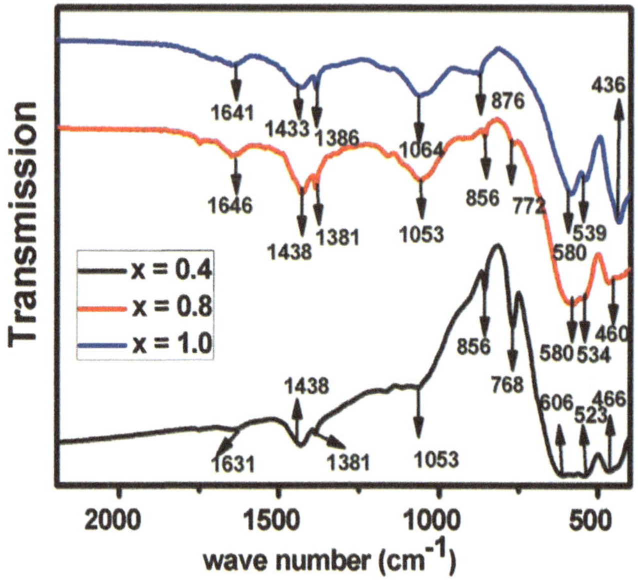

Fig. 11.3 FTIR spectra of $BaCo_{0.4}Fe_{11.6}O_{19}$, $BaCo_{0.8}Fe_{11.2}O_{19}$, and $BaCoFe_{11}O_{19}$ sintered at 900 °C for 3 h. (With permission from Elsevier License Number 5133631119568)

Table 11.1 FTIR vibrational bands of $BaCo_{0.4}Fe_{11.6}O_{19}$, $BaCo_{0.8}Fe_{11.2}O_{19}$, and $BaCoFe_{11}O_{19}$ sintered at 900 °C with composition

Composition (x)	Vibrational bands (ν) in cm^{-1}								
	ν_1	ν_2	ν_3	ν_4	ν_5	ν_6	ν_7	ν_8	ν_9
$BaCo_{0.4}Fe_{11.6}O_{19}$	466	523	606	768	856	1053	1381	1438	1631
$BaCo_{0.8}Fe_{11.2}O_{19}$	460	534	580	772	856	1053	1381	1438	1646
$BaCo_{1.0}Fe_{11}O_{19}$	436	539	580	-	876	1064	1386	1433	1641

11.7 Field Emission Scanning Electron Microscopy (FESEM)

FESEM images of samples ($BaCo_{0.4}Fe_{11.6}O_{19}$, $BaCo_{0.8}Fe_{11.2}O_{19}$, and $BaCoFe_{11}O_{19}$) are shown in Fig. 11.4a–c. A porous and layered structure of these samples was observed. The average grain size was found to increase with an increase in content of cobalt ions from 250 to 325 nm. The grains in Fig. 11.4c were densely packed as compared to other samples. Therefore, with an increase in the concentration of Co^{2+} ions, the average grain size and the density of these hexa particles are found to increase. The grains or the particles size appeared as elongated toward the horizontal axis. The layers of barium hexaferrites become thicker with an increase in the cobalt content. Moreover, the formation of hexa plates confirmed the formation of M-type barium hexaferrite.

Fig. 11.4 (a–c) FESEM micrograph of $BaCo_{0.4}Fe_{11.6}O_{19}$, $BaCo_{0.8}Fe_{11.2}O_{19}$, and $BaCoFe_{11}O_{19}$ sintered at 900 °C for 3 h. (With permission from Elsevier License Number 5133631119568)

11.8 Raman Spectroscopy

The lattice dynamics in ferrites Raman are examined with the Raman Spectroscope. The formation of gamma Fe_2O_3 phase during synthesis is confirmed by Raman. The scattering was excited using argon ion excimer with wavelength of 488 nm having power of 25 mW. According to the well-established group theory, a unit cell of M-type Barium hexaferrite includes 64 atoms which give rise to 189 optical modes ($k = 0$) which further revealed 42 Raman active mode, i.e., (11A1g + 14E1g + 17E2g), 30 IR active (13A2u + 17E1u) modes and the remaining modes (3A1u + 4A2g + 13B1g + 4B1g + 4B1u + 3B2g + 12B2u + 15E2u) are silent (Gazzali and Chandrasekaran 2014), where A, B, and E stand for the one-dimensional, two-dimensional, and three-dimensional representations of optical modes, "g" and "u" represent the symmetry and antisymmetry with respect to the center of inversion. For cobalt-substituted M-type barium ferrite, all Raman active modes and IR active modes are shown in Fig. 11.5a without fitting and Fig. 11.5b

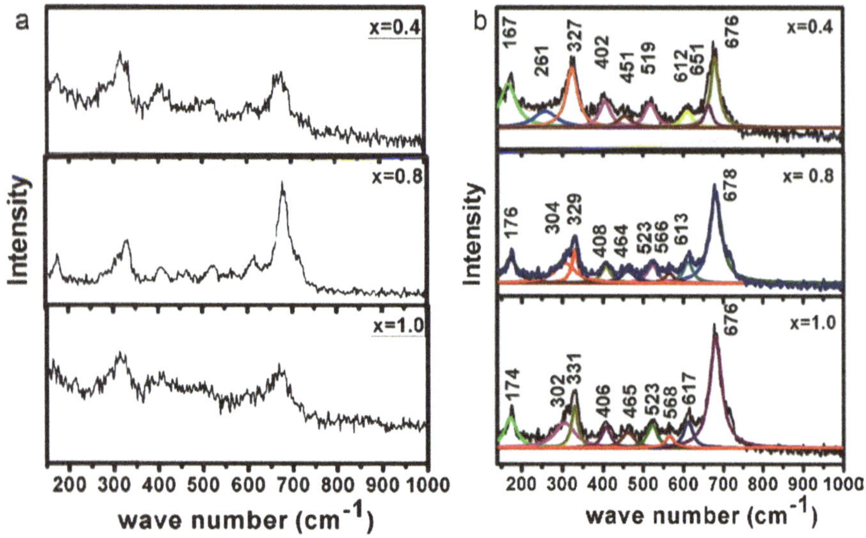

Fig. 11.5 (**a**) RAMAN spectra of $BaCo_{0.4}Fe_{11.6}O_{19}$, $BaCo_{0.8}Fe_{11.2}O_{19}$, and $BaCoFe_{11}O_{19}$ raw (**b**) Fitted RAMAN spectra of $BaCo_{0.4}Fe_{11.6}O_{19}$, $BaCo_{0.8}Fe_{11.2}O_{19}$, and $BaCoFe_{11}O_{19}$ sintered at 900 °C for 3 h. (With permission from Elsevier License Number 5133631119568)

with fitting. The vibrational bands for all compositions are found in good agreement with M-type barium hexaferrite (Ounnunkad 2006).

The real and imaginary part of permittivity (ε' and ε'') and real and imaginary part of permeability (μ', μ'') as a function of frequency for all the samples is shown in Fig. 11.6a, b, respectively. At 17 GHz, the observed electromagnetic parameters for $x = 0.4$ are $\varepsilon' = 3.90$, $\varepsilon'' = -1.19$ whereas $\mu' = 1.02$, $\mu'' = 0.66$. With an increase in the concentration of Co^{2+} from $x = 0.4$ to $x = 0.8$, the real part of permittivity (ε') has almost attained double value (Fig. 11.6).

This increase in the value permittivity may be due to the following reasons:

1. The higher positive magnetic crystalline anisotropy (MCA) of Co^{2+} and
2. Increase in space charge polarization at the grain boundaries as reported in the other literature (Li et al. 2013b; Tang and Hu 2007).

The electron exchange between Fe^{2+} and Fe^{3+} ions indicates the local displacement of charge carriers in applied field direction.

The space charge polarization arises only when the motion of charge carrier is restricted.

So the charges trapped within the boundary of the material and the motion of these charges can also be limited. This is how the space charge polarization is increased.

As the polarization of inducing dielectric moment cannot harmonize with the applied frequency and hence, the material attains a constant value of permittivity in

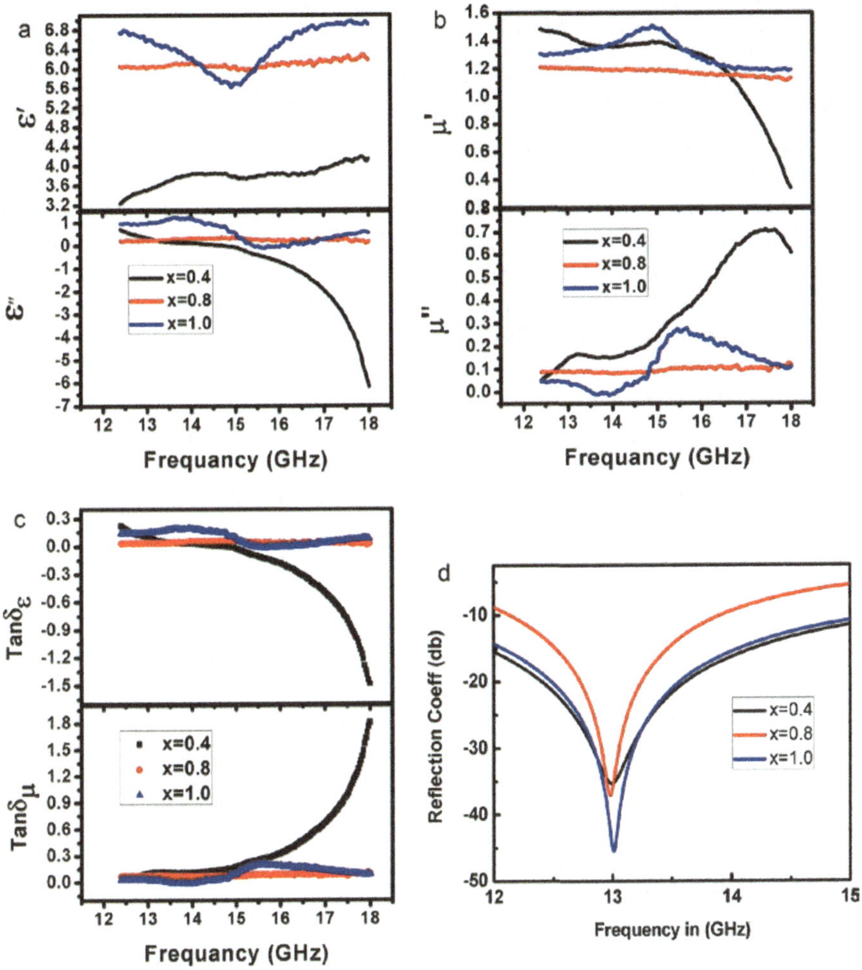

Fig. 11.6 (**a**) Real and imaginary part of permittivity, (**b**) real and imaginary part of permeability, (**c**) electrical loss (Tanδ_ε), magnetic loss (Tanδ_μ), (**d**) reflectance loss (dB) $BaCo_{0.4}Fe_{11.6}O_{19}$, $BaCo_{0.8}Fe_{11.2}O_{19}$, and $BaCoFe_{11}O_{19}$ sintered at 900 °C for 3 h. (With permission from Elsevier License Number 5133631119568)

high-frequency region (Mathur et al. 2010). The permittivity of nanoferrites is also affected by grain size density and the impurities present (Sharma et al. 2013). Figure 11.6c shows the electrical and magnetic loss tangents for cobalt-substituted M-type barium ferrite. The microwave absorbing property strongly relies on RL values. The RL values can be calculated by using Eq. (11.5)

$$RL(\text{dB}) = 20\log||\frac{Z_{in} - 1}{Z_{in} + 1}|$$

where $Z_{in} = \sqrt{(\frac{\mu^*}{\epsilon^*})}$ and $\tanh\,(\frac{j2\pi d}{\lambda})(\sqrt{(\frac{\mu^*}{\epsilon^*})})$ is the normalized input impendence in the free space. $\mu^* = \mu' - j\mu''$ and $\varepsilon^* = \varepsilon$ "$-j\varepsilon$" are the real and the imaginary parts respectively.

As the concentration of cobalt is increased ($x = 0.4$, 0.8, and 1.0), the values of reflection losses are also found to increase with values −35.07 dB, −36.98 dB, and −45.00 dB, respectively as seen in Fig. 11.6d.

For the sample of composition $BaCoFe_{11}O_{19,}$ the RL value is −45 dB which is in good agreement with Co–Mn-doped M-type barium ferrite (Wang et al. 2014). The increased value of RL may be attributed to increase in the aspect ratio that leads to increase in the size of the pores and ultimately leads to enhancement of the microwave penetration more effectively with less direct reflection on the surface of substituted barium ferrite composite (Mu et al. 2006).

Second probable reason may be the effective anisotropy which includes the magnetic crystalline anisotropy and the shape anisotropy as cobalt possesses very large value of positive magnetic crystalline anisotropy that balances the crystal structure. Furthermore, it has been observed earlier that the critical grain size of barium hexaferrite is 460 nm which is much higher than the value of 325 possessed by these materials (Rezlescu et al. 1999). Hence, the smaller grain size resulted in an increase in the surface area, dangled bonds, and unsaturated coordination on the surface of developed composite materials. All these parameters led to improvement of interface polarization and multiple scattering making $BaCoFe_{11}O_{19}$ ferrites an excellent absorbing material in Ku band.

11.9 Conclusion

The electromagnetic parameter is essential for the application of given materials at a particular frequency range. Cobalt-substituted M-type barium ferrites with an average crystallite size in the range 65–85 nm are potential candidates for very high-frequency applications FTIR and Raman spectroscopy confirmed the strong metal oxide bond formation. All the samples exhibited good microwave absorbing properties in the frequency region 12–18 GHz (Ku band). Enhanced interface polarization and multiple scattering with maximum reflectance loss of −45 dB for $BaCoFe_{11}O_{19}$ make this composition as an excellent absorbing material in Ku band.

References

Asl MJP, Ghasemi A, Gordani GR (2014) J Supercond Nov Magn 10948-014-2814-7
Batlle X, Obradors X, Carvajal JR, Pernet M, Cabanas MV, Vallet M (1991) J Appl Phys 70:1614
Chawla SK, Kaur P, Mudsainiyan RK, Meena SS, Yusuf SM (2014) J Supercond Nov Magn 10948-014-2893-5

Cho H-S, Kim S-S (1999) IEEE Trans Magn 35:3151–3153
Din MF, Ahmed I, Ahmed M, Farid MT, Iqbal MA, Murtaza G, Akhtar MN, Shakir I, Warsi MF, Khan MA (2014) J Alloy Compd 584:646–651
Gazzali PMM, Chandrasekaran G (2014) J Mater Sci Mater Electron 25:702–709
Kaur T, Srivastava AK (2013) Int J Res Mech Eng Technol 3:2249–5770
Li ZW, Yang ZH, Kong LB, Zhang YJ (2013a) J Magn Magn Mater 325:82–86
Li L, Chen K, Liu H, Tong G, Qian H, Hao B (2013b) J Alloy Compd 557:11–17
Li J, Zhang H, Li Y, Li Q, Yo G (2014) J Supercond Nov Magn 27:793–797
Mathur P, Thakur A, Singh M, Harris G (2008) Z Phys Chem 222:621–633
Mathur P, Thakur A, Lee JH, Singh M (2010) Mater Lett 64:2738–2741
Mu G, Shen H, Qiu J, Gu M (2006) Appl Surf Sci 253:2278–2281
Ounnunkad S (2006) Solid State Commun 138:472–475
Ozgur U, Alivov Y, Morkoc H (2009) J Mater Sci 20:789–834
Rana K, Thakur P, Sharma P, Tomar M, Gupta V, Thakur A (2015) Ceram Int 41:4492–4497
Rana K, Thakur P, Tomar M, Gupta V, Thakur A (2018) Ceram Int 44(6):6370–6375
Rashid MM, Ibarhim IA (2013) J Supercond Nov Magn 26:1639–1644
Rezlescu L, Rezlescu E, Popa PD, Rezlescu N (1999) J Magn Magn Mater 193:288
Sertkol M, Koseoglu Y, Baykal A, Kavas H, Bozkurt A, Toprak MS (2009) J Alloy Compd 486: 325–329
Sharma S, Daya KS, Sharma S, Singh M (2013) Appl Phys Lett 103:112402
Sharma M, Kashyap SC, Gupta HC, Dimri MC, Ashokan K (2014) AIP Adv 4:077129
Tang X, Hu KA (2007) Mater Sci Eng B 139:119–123
Thakur A, Thakur P, Hsu JH (2011) J Alloy Compd 505:5315–5319
Wang Y, Chen M, Zhou F, Ma E (2002) Nature 419:912–915
Wang L, Yu H, Ren X, Xu G (2014) J Alloy Compd 588:212–216
Zhao WY, Wei P, Wu XY, Wang W, Zhang QJ (2008) J Appl Phys 103:063902

Electromagnetic Characterization of Nanomaterials: Preliminary Study of 60 GHz Millimetre Wave Li-NGD Circuit in Microstrip Technology

12

Fayu Wan, Xiaoyu Huang, Samuel Ngoho,
Kishore Ajay Kumar Ayyala, Preeti Thakur, M. S. Prasad, Atul Thakur,
Sébastien Lalléchère, Wenceslas Rahajandraibe,
Nour Mohammad Murad, and Blaise Ravelo

F. Wan · X. Huang · B. Ravelo (✉)
School of Electronics and Information Engineering, NUIST, Nanjing, China
e-mail: fayu.wan@nuist.edu.cn

S. Ngoho
Association Française de Science des Systèmes (AFSCET), Paris, France

K. A. K. Ayyala · A. Thakur
Centre of Nanotechnology, Amity University Haryana, Gurugram, India
e-mail: athakur1@ggn.amity.edu

P. Thakur
Department of Physics, Amity School of Applied Sciences, Amity University Haryana, Gurugram, India
e-mail: pthakur@ggn.amity.edu

M. S. Prasad
Amity Institute of Space Science and Technology, Amity University Noida, Noida, India
e-mail: msprasad@amity.edu

S. Lalléchère
Université Clermont Auvergne (UCA), CNRS, SIGMA Clermont, Institut Pascal, Aubière, France
e-mail: sebastien.lallechere@uca.fr

W. Rahajandraibe
Aix-Marseille University, CNRS, University of Toulon, IM2NP UMR7334, Marseille, France
e-mail: wenceslas.rahajandraibe@im2np.fr

N. M. Murad
PIMENT Lab, Network and Telecom Lab, Institut Universitaire de Technologie, University of La Reunion, Saint Pierre, France
e-mail: nour.murad@univ-reunion.fr

A. Thakur et al. (eds.), *Synthesis and Applications of Nanoparticles*,
https://doi.org/10.1007/978-981-16-6819-7_12

Abstract

This book chapter describes a circuit investigation on millimetre wave negative group delay (NGD) function. The main specifications of unfamiliar bandpass (BP) NGD function are defined. The topological description of the li-topology to be exploited in the chapter is developed. The design method of millimetre wave NGD li-circuit is described. Proof of concept (PoC) of microstrip circuits designed on alumina dielectric substrate is introduced. Then, simulations with commercial tool which are presented to highlight the design feasibility of 60 GHz NGD centre frequency PoC are discussed. Then, parametric study with respect to the physical and geometrical parameters of li-circuit on the BP-NGD performance as NGD value, NGD centre frequency and NGD bandwidth is proposed.

Keywords

Negative group delay (NGD) · Li-topology · Millimetre wave circuit · Design method · Microstrip circuit · Bandpass NGD function · 60 GHz NGD circuit

12.1 Introduction

The traditional progress of the communication technology leads to the race around the 5G network (https://spectrum.ieee.org/static/the-race-to-5g; https://www.spirent.com/solutions/5g-network-testing?gclid=EAIaIQobChMI_IGt5pb77AIV2fhRCh37fA5TEAAYASABEgL0Q_D_BwE). Such a technology is expected to be the emerging and the most promising future wireless communication network (https://futurenetworks.ieee.org/; Yang et al. 2015). The exploration of such technology requires challenging research in particular on the implementation of the physical layer (Rappaport et al. 2013). For example, the massive amount of data communication leads to the development of higher frequency wireless communication devices. To face up this challenging aspect, design solutions were expected with the development of millimetre wave components (Hur et al. 2013; Dehos et al. 2014). Among the allocated frequency band, the 60 GHz millimetre wave was opened for the communication operator. However, more and more innovative electronic circuit design is necessary in this frequency band.

According to this technological need, we would like to bring a brick of solution against the signal delay issues with the innovative equalization by using the unfamiliar negative group delay (NGD) function as suggested in (Ravelo et al. 2016; Ravelo 2012). Moreover, some tentative applications of the NGD function as design of high-performance phase shifter, microwave amplifier linearization, antenna array beamforming and non-Foster microwave passive elements were proposed by some research groups in the area of microwave engineering (Ravelo 2014a; Keser and

Mojahedi 2010; Broomfield and Everard 2000; Choi et al. 2010a, b; Alomar and Mortazawi 2012; Mirzaei and Eleftheriades 2013, 2015; Zhu and Wu 2019; Zhang et al. 2017). In the last decade, considerable research work on progressive microwave designs of NGD circuits was performed (Wu and Itoh 2014; Choi et al. 2011; Shao et al. 2018; Chaudhary et al. 2011, 2013a, b; Wang et al. 2018; Junior et al. 2019; Ravelo 2014b, 2016a, b, c, 2017; Wan et al. 2020; Zhou et al. 2020; Ravelo et al. 2020; Ayyala et al. 2020). The main challenge in NGD engineering is the design of low attenuation loss, high operation frequency, miniature and wide band NGD circuit. The main solutions proposed in terms of design were:

- The design of maximally flat NGD active circuit by using a microwave transversal filter approach (Wu and Itoh 2014)
- The design of transmission line-based NGD microstrip structures (Choi et al. 2011; Shao et al. 2018) with high potential miniaturization
- The design of defected ground structure (DGS)-based NGD microwave compact circuits (Chaudhary et al. 2011, 2013a, b)
- Design of NGD microwave circuit based on signal interference techniques (Wang et al. 2018)
- The design of NGD structure with folded circuit with the introduction of distributed broadside parallel line (Junior et al. 2019)
- The design of NGD distributed circuit with parallel interconnect microstrip line (Ravelo 2016a)
- The design of NGD distributed circuits with coupled line (CL) which enable to proposed several planar passive structures (Ravelo 2016b; Wan et al. 2020; Zhou et al. 2020; Ravelo et al. 2020; Ayyala et al. 2020)

However, it should be underlined that the available research works (Ravelo et al. 2016, 2020; Ravelo 2012, 2014a, 2016a, b; Keser and Mojahedi 2010; Broomfield and Everard 2000; Choi et al. 2010a, b, 2011; Alomar and Mortazawi 2012; Mirzaei and Eleftheriades 2013, 2015; Zhu and Wu 2019; Zhang et al. 2017; Wu and Itoh 2014; Shao et al. 2018; Chaudhary et al. 2011, 2013a, b; Wang et al. 2018; Junior et al. 2019; Wan et al. 2020; Zhou et al. 2020; Ayyala et al. 2020) on the NGD circuit design were limited for the validation in the L-, S- and C-bands. As a matter of fact, huge effort must be urgently spent on the design, manufacturing and test methodologies of millimetre wave NGD circuits. For this reason, the present chapter which is opening a preliminary study of millimetre wave NGD circuit design was written. Because of the design simplicity and also the miniaturization constraint advantage, the proposed NGD circuit is based on the exploitation of the innovative "li" microstrip topology which was recently revealed in (Ravelo et al. 2020; Ayyala et al. 2020).

Before starting the NGD study, we need a democratic and easier approach for the readers to present and to describe the NGD function. It is important to remind that despite the variety of NGD circuit structures proposed so far, the NGD function is not familiar to non-specialist electronic, microwave and communication engineers. Undoubtedly because of its counterintuitive aspect, many engineers are still wondering about the meaning and the physical interpretation of the NGD phenomenon. Under such a situation, we can point out that one of the most pedagogical approaches to address such a question is the innovative theorization about the analogy between the NGD function and the filter behaviours (Ravelo 2014b, 2016c, 2017). Depending on the frequency bands were the group delay (GD) is negative, the concept of unfamiliar low-pass (LP), high-pass (HP) and bandpass (BP) NGD functions was introduced (Ravelo 2014b, 2016c, 2017). We can identify from these functions the categories of electronic circuits operating as LP, HP and BP NGD functions.

In the present study, the main originality of the research work is essentially focussed on the category of unfamiliar BP NGD function which is naturally adapted to most of microwave and millimetre wave circuits. The preliminary feasibility study of the NGD millimetre wave around the 60 GHz frequency will be performed with the NGD li-topology which was introduced in (Ravelo et al. 2020; Ayyala et al. 2020) for the case of the L- and S-band validations. Doing this, the proposed chapter is organized in five principal sections as follows:

- Section 12.2 is focussed on the basic definition of the li-topology and the basic specifications of the ideal unfamiliar BP NGD function. Based on the S-matrix modelling approach, the main parameters enabling to specify the BP NGD circuit will be defined.
- The feasibility study of the unfamiliar BP NGD concept should be illustrated with a proof of concept (POC). The magneto-dielectric substrate design feasibility with li-topology was investigated in (Ayyala et al. 2020; Saini et al. 2016; Thakur et al. 2015). Section 12.3 is presenting innovatively the design of the millimetre wave li-circuit. The main parameters of the li-circuit design are defined based on the microstrip TL theory (Hammerstad and Jensen 1980; Hammerstad 1975; Eudes et al. 2012).
- Section 12.4 discusses about the simulation results of the li-microstrip circuit POC by using the commercial tool ADS® from Keysight Technologies® (Advanced Design System 2020). Wide discussions about the influences of the different physical parameters of the li-microstrip circuits on the NGD performances will be elaborated. The study will be based on the parametric analyses versus material characteristics (dielectric substrate permittivity, metallization thickness and metallization conductivity) and the geometrical parameters (physical width, length and interspace) of the conductor line.
- Then, Sect. 12.5 is the conclusion of the chapter.

12.2 Theoretical Description on the NGD Specifications and Li-Topology Introduction

As the BP NGD aspect remains as an unfamiliar function to most of microwave and millimetre wave design and research engineers, it would be important to start this section with ideal descriptions. First, the topological description of the circuit under investigation will be introduced. Then, the basic elements allowing to define the NGD specifications will be defined based on the S-matrix modelling of the li-topology.

12.2.1 Ideal S-Parameter and NGD Specifications

The beginning of the millimetre wave NGD design should start with the definition of the desired specifications. However, the BP NGD specifications are not well known by all the microwave and millimetre wave engineers as the classical electronic functions. Therefore, a particular attention is paid in this subsection on how to define the NGD specifications based on the ideal S-parameter of the li-topology.

12.2.1.1 Description of Li-Topology

The li-topology equivalent diagram is depicted in Fig. 12.1. As illustrated in the figure, it acts as two-port passive circuit with reference port ① and port ② connected to the two access of circuit. The li-topology is constituted by two parallel transmission lines (TL) in which one of the lines (here the top one) is open ended. To feed the circuit, voltage source, U, is connected in left side of the diagram. Along the chapter, the modelling of our circuit is based on the source and load reference impedance, denoted, R_0, which is fixed to 50 Ω. The microwave investigation of the BP NGD circuit (Ravelo et al. 2020; Ayyala et al. 2020) is generally performed with S-matrix approach. The analytical way to model the li-topology will be described in the following paragraph.

12.2.1.2 Two-Port S-Matrix Description

First and foremost, our NGD circuit can be assumed as a two-port system. Therefore, the diagram introduced in Fig. 12.1 can be considered as equivalent to the black box system represented in Fig. 12.2.

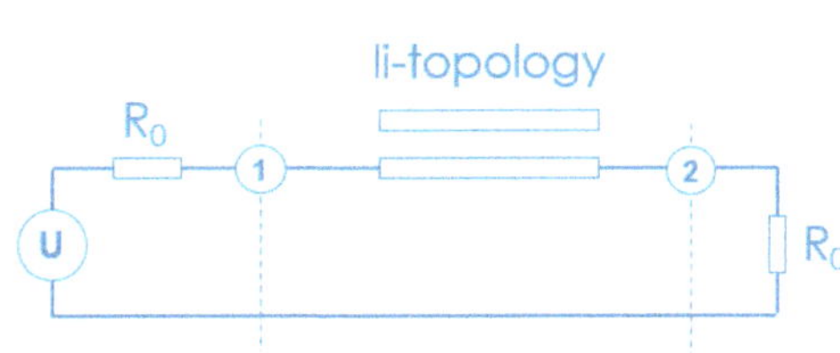

Fig. 12.1 Diagram of li-topology under study

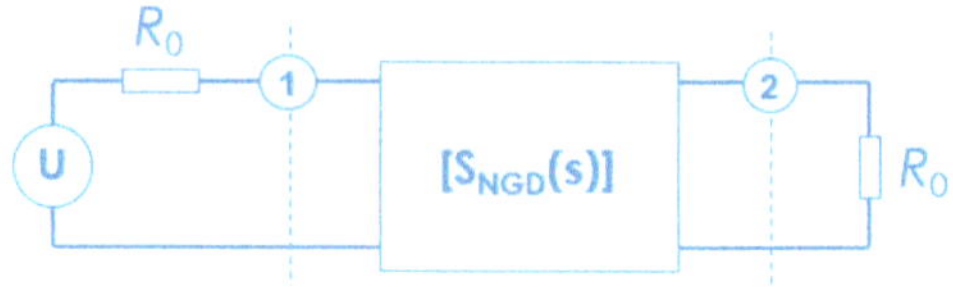

Fig. 12.2 Two-port black box microwave system

By denoting the Laplace variable, s, our two-port system can be analytically modelled by the 2D S-matrix:

$$[S_{\mathrm{NGD}}(s)] = \begin{bmatrix} S_{11}(s) & S_{12}(s) \\ S_{21}(s) & S_{22}(s) \end{bmatrix}. \tag{12.1}$$

As the li-topology is a passive and symmetrical circuit, the analytical investigation can be reduced to the two following components:

$$\begin{cases} S_{11}(s) = S_{22}(s) \\ S_{12}(s) = S_{21}(s) \end{cases}. \tag{12.2}$$

In difference to most of classical RF and microwave circuit theory, the NGD study requires intensive investigation on the GD response. Before the analysis of the GD, let us remind the main frequency responses in the following paragraph.

12.2.1.3 Magnitude and Phase Frequency Responses of Reflection and Transmission Coefficients

To explore the frequency response, let us denote the complex angular frequency variable:

$$s = j\omega = j2\pi\omega. \tag{12.3}$$

By definition, the magnitudes of the reflection and transmission coefficients of the S-matrix introduced in Eq. (12.1) are given by, respectively:

$$S_{11}(\omega) = |S_{11}(j\omega)| = \sqrt{\mathrm{Re}\left[S_{11}(j\omega)\right]^2 + \mathrm{Im}[S_{11}(j\omega)]^2} \tag{12.4}$$

$$S_{21}(\omega) = |S_{21}(j\omega)| = \sqrt{\mathrm{Re}\left[S_{21}(j\omega)\right]^2 + \mathrm{Im}[S_{21}(j\omega)]^2} \tag{12.5}$$

where Re(z) and Im(z) are the real and imaginary parts of z, respectively. The phase associated to the transmission coefficient of S-matrix introduced in Eq. (12.2) is equal to:

$$\varphi(\omega) = \arg[S_{21}(j\omega)] = \arctan\left[\frac{\mathrm{Im}[S_{21}(j\omega)]}{\mathrm{Re}\left[S_{21}(j\omega)\right]}\right]. \tag{12.6}$$

The GD depending on this phase expression will be defined in the next paragraph.

12.2.1.4 GD Model

As its description said, the BP NGD analysis is fundamentally depended to the GD analytical expression. For this reason, it would be important to recall the mathematical definition of this key parameter. According to the microwave theory, the analytical model of the associated GD is defined by:

$$GD(\omega) = \frac{-\partial\varphi(\omega)}{\partial\omega}.. \tag{12.7}$$

With this expression, we are going to build an easy way to understand and to explore the BP NGD concept. Knowing this analytical definition, we will propose in the next subsection the basic specifications of the unfamiliar BP NGD function to be exploited in the following sections of the chapter.

12.2.2 Bandpass (BP) NGD Specifications

Before all, it is important to emphasize that, in the present study, the term "bandpass (BP)" of the NGD (Ravelo 2014b, 2016c, 2017) is associated to the sign of the transmission coefficient GD but not related to the magnitude.

Figure 12.3 shows the three diagrams with frequency in *x*-axis and the key real parameters to be analysed in the present study represented in *y*-axis. Figure 12.3a represents the ideal GD response of BP NGD function. Figure 12.3b, c illustrate the ideal behaviours of the reflection and transmission coefficients, respectively. The analytical definitions of each parameters of these diagrams will be described in the following paragraphs of the subsection.

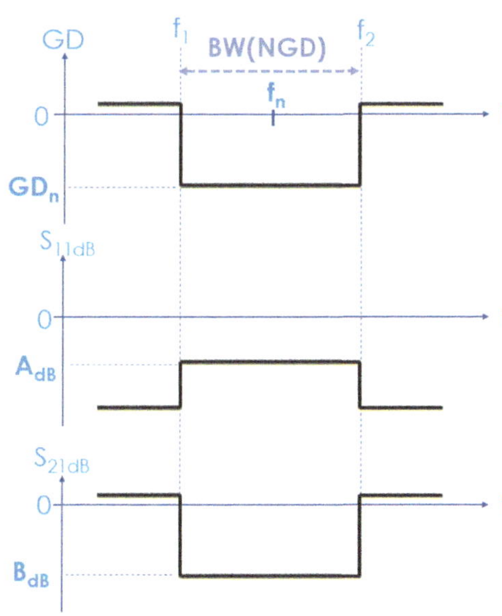

Fig. 12.3 (**a**) GD, (**b**) S_{21} and (**c**) S_{11} BP NGD ideal specifications

12.2.2.1 NGD Frequency Specifications

The BP NGD function must be defined with the NGD cut-off and centre frequencies graphically introduced in Fig. 12.3a, which are denoted, respectively:

$$\begin{cases} \omega_1 = 2\pi f_1 \\ \omega_2 = 2\pi f_2 \end{cases} \tag{12.8}$$

$$\omega_n = 2\pi f_n. \tag{12.9}$$

As illustrated in Fig. 12.3a, the BP NGD frequency band is belonging in the frequency interval:

$$\omega_1 \leq \omega \leq \omega_2 \tag{12.10}$$

The NGD bandwidth indicated in Fig. 12.3a is equal to:

$$\mathrm{BW(NGD)} = f_2 - f_1 = \frac{\omega_2 - \omega_1}{2\pi}. \tag{12.11}$$

The NGD cut-off frequencies, $\omega_{1,2}$, are defined by:

$$\mathrm{GD}(\omega_1) = \mathrm{GD}(\omega_2) = 0. \tag{12.12}$$

The specifications of the NGD value in the different frequency bands are introduced in the following paragraph.

12.2.2.2 NGD Value

The NGD value by negative real parameter, GD_n, indicated in Fig. 12.3a is analytically defined by:

$$\mathrm{GD}(\omega_n) = \mathrm{GD}_n \leq 0 \tag{12.13}$$

It is very important to keep in mind that the ideal response expected to BP NGD function illustrated in Fig. 12.3a is defined by the following ideal condition with the NGD frequency band defined in inequality (12.10):

$$\mathrm{GD}(\omega) \approx \mathrm{GD}_n. \tag{12.14}$$

It means that we should have the different values outside the NGD the frequency band:

$$\omega < \omega_1 \cup \omega_2 < \omega. \tag{12.15}$$

In addition, the GD must be positive under condition (12.15):

$$GD(\omega) > 0. \tag{12.16}$$

Knowing the GD specifications and the NGD frequency band, the insertion and reflection loss specifications need to be recalled also. The classical constraints associated to these parameters will be described in the following paragraph.

12.2.2.3 S-Parameter Specifications

As shown in Fig. 12.3b and in Fig. 12.3c, the main S-parameters to be considered for the li-topology NGD analysis are the reflection and transmission coefficients similar to the classical millimetre wave circuit constraints. In the present BL NGD function analysis, the main constraint limitations associated to the S-parameters are defined in function of the given positive real parameters A and B as follows:

- As seen in Fig. 12.3b, within the NGD frequency band situated in (12.10), given the li-topology access matching constraints quantified by the real positive value:

$$S_{11_{\max}} = A = 10^{A_{\mathrm{dB}}/20} \tag{12.17}$$

the reflection coefficient must respect the condition:

$$S_{11}(\omega) \leq A. \tag{12.18}$$

- Then, in the same NGD frequency band, as seen in Fig. 12.3c, the li-topology insertion loss constraints quantified by the real positive value:

$$S_{21_{\min}} = B = 10^{B_{\mathrm{dB}}/20} \tag{12.19}$$

the transmission coefficient must respect the classical constraint:

$$S_{21}(\omega) \geq B. \tag{12.20}$$

To be familiar with these ideal specifications, it would be necessary to study the feasibility of BP NGD circuit design with POC. As a matter of fact, a microstrip structure will be described in the following section.

12.3 Design of Millimetre Wave Li-Microstrip Circuit

The proposed millimetre wave li-circuit POC is designed in microstrip technology. The present section describes the geometrical implementation and the key parameters of the circuit necessary for the S-parameters and GD simulations.

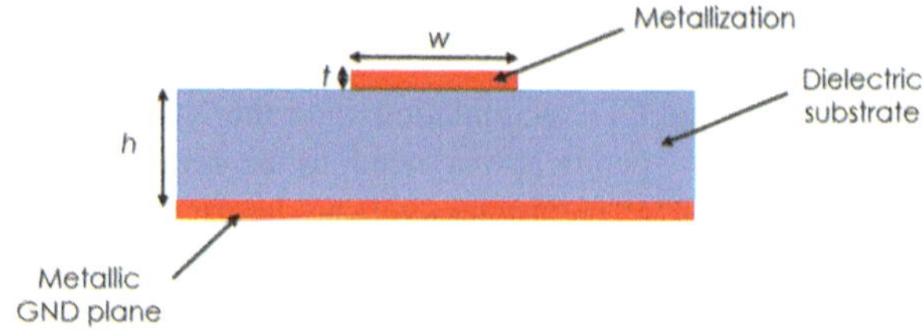

Fig. 12.4 Geometrical representation of microstrip TL

12.3.1 Microstrip Line Description

We remind that a microstrip TL is characterized by its geometrical length d, width w and metallization thickness t as depicted in Fig. 12.4. The metallizations are indicated in top and bottom planes of the structure. The conductor line is usually striped on the dielectric substrate characterized by its height h, relative permittivity ε_r and loss constant $\tan(\delta)$.

12.3.2 Microstrip Line Physical Parameters

The microstrip line is essentially parametrized by its frequency-dependent characteristic impedance, Z_c, and its propagation constant in function of the attenuation constant α and phase constant β defined by:

$$\gamma = \alpha + j \cdot \beta. \tag{12.21}$$

Giving the physical properties of the microstrip structure, Z_c can be deduced according to the formulations reported in (Hammerstad and Jensen 1980; Hammerstad 1975; Eudes et al. 2012) expressed as:

$$Z_c = \frac{Z_{air}}{2\pi\sqrt{\varepsilon_{eff}}} \ln\left[\frac{h \cdot \psi(w/h)}{w} + \sqrt{1 + \frac{4h^2}{w^2}}\right] \tag{12.22}$$

where the ψ-function is given by:

$$\psi(x) = 6 + (2\pi - 6) \cdot \exp\left[-\left(\frac{30.66}{x}\right)^{0.7528}\right] \tag{12.23}$$

where $Z_{air} \approx 377\ \Omega$ is the air impedance and ε_{eff} is the substrate effective relative permittivity.

12.3.3 NGD Centre Frequency of Li-Topology

As we are dealing with dielectric substrate, we can assume that the substrate relative permeability is equal to unity. By denoting the vacuum light speed, c, the expected NGD centre frequency of li-topology is expressed as (Ravelo et al. 2020; Ayyala et al. 2020):

$$f_n = \frac{d\sqrt{\varepsilon_{\text{eff}}}}{4c}. \tag{12.24}$$

By inverting this equation, we can determine the physical length of the li-microstrip circuit in function of the desired NGD centre frequency. Substantially, the synthesis formula is given by:

$$d = \frac{4f_n c}{\sqrt{\varepsilon_{\text{eff}}}}. \tag{12.25}$$

The other expressions of the NGD design equations can be found in (Ravelo et al. 2020). For the sake of the page limitation, they will not be explored in details in the present chapter. However, more details about the influences of the microstrip line parameters on the li-circuit BP NGD performances will be discussed in the next section.

12.4 Feasibility Study with Simulations of Millimetre Wave Li-Microstrip Circuit

The present section presents the ADS® (Advanced Design System 2020) simulation results of the millimetre wave li-microstrip circuit POC. Different to the NGD research work available in the literature (Wu and Itoh 2014; Choi et al. 2011; Shao et al. 2018; Chaudhary et al. 2011, 2013a, b; Wang et al. 2018; Junior et al. 2019; Ravelo 2014b, 2016a, b, c, 2017; Wan et al. 2020; Zhou et al. 2020; Ravelo et al. 2020; Ayyala et al. 2020; Saini et al. 2016; Murad et al. 2020), this feasibility study will be originally performed from 59 to 61 GHz.

12.4.1 Design Parameters

To illustrate the design feasibility of the BP NGD li-circuit, let us consider the POC of microstrip PCB with schematic shown in Fig. 12.5a. The microstrip circuit is designed on Cu-metallized (in top and in bottom ground plane) alumina dielectric substrate having height h, relative permittivity ε_r and loss constant $\tan(\delta)$. Figure 12.5b represents the 3D design of the considered POC millimetre wave circuit. The li-circuit is defined with geometrical length d, width w and metallization thickness t. The POC circuit is composed of two-straight parallel microstrip CL with

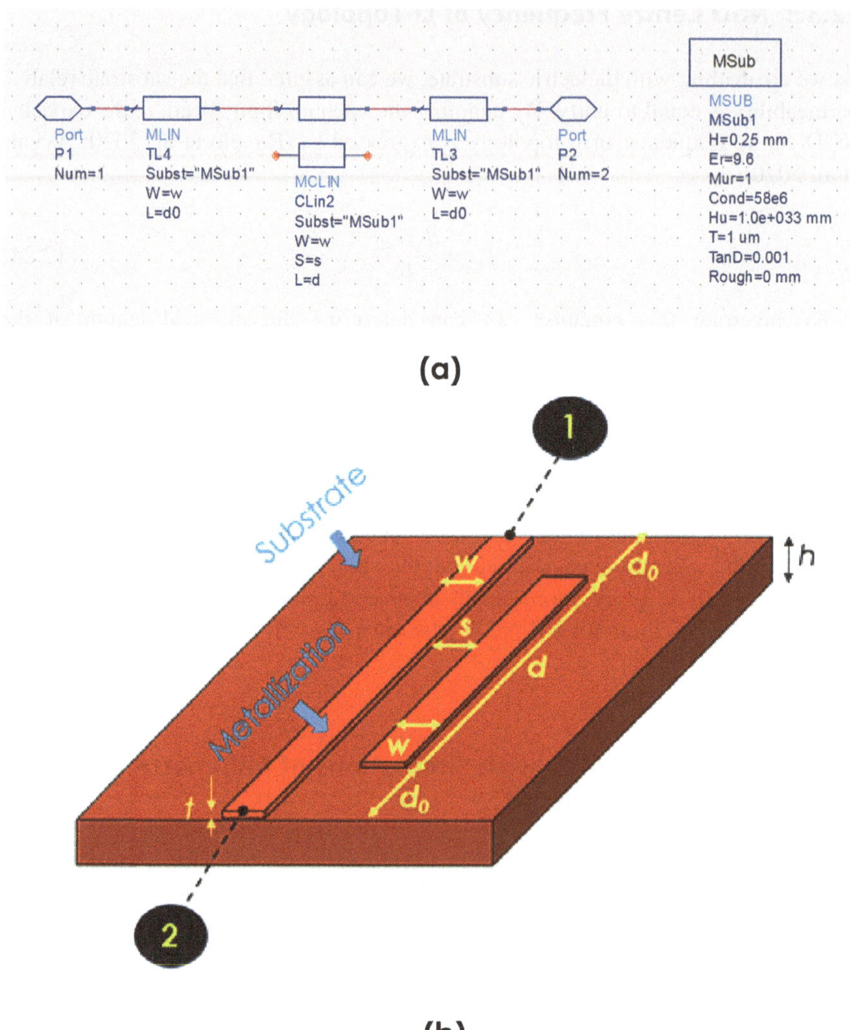

Fig. 12.5 (**a**) Schematic and (**b**) 3D design of millimetre wave microstrip li-circuit

same physical widths and same physical lengths. The circuit was designed with Eq. (12.25) under the desired NGD centre frequency, $f_n = 60$ GHz.

The considered parameters of the POC circuit as the substrate and geometrical parameters are summarized in Table 12.1.

With the introduced POC, S-parameter simulations were performed. The next subsection will discuss on the obtained results.

Table 12.1 Physical characteristics of the li-circuit POC

Description	Parameter	Value
Relative permittivity	ε_r	9.6
Loss tangent	$\tan(\delta)$	0.001
Thickness	h	0.25 mm
Material	Copper (Cu)	–
Thickness	t	1 μm
Conductivity	σ	58 MS/m
Physical width	w	625 μm
Physical length	d	873 μm
Physical access line length	d_0	1 mm
Physical interspace	s	0.5 mm

12.4.2 Simulation Results

To validate the NGD aspect, the previously described POC of microstrip li-circuit was simulated in schematic environment of standard microwave electronic tool ADS® from Keysight Technologies® (Advanced Design System 2020). The simulations were based on the frequency domain S-parameters in the frequency band defined from 59 to 61 GHz with 400 frequency samples.

12.4.2.1 NGD Response

Consequently, we obtain the GD response displayed in Fig. 12.6. The associated NGD specifications are shown in Table 12.2. It can be noticed that the circuit presents an NGD behaviour with NGD value, $GD_n \approx -1.84$ ns, at the NGD centre

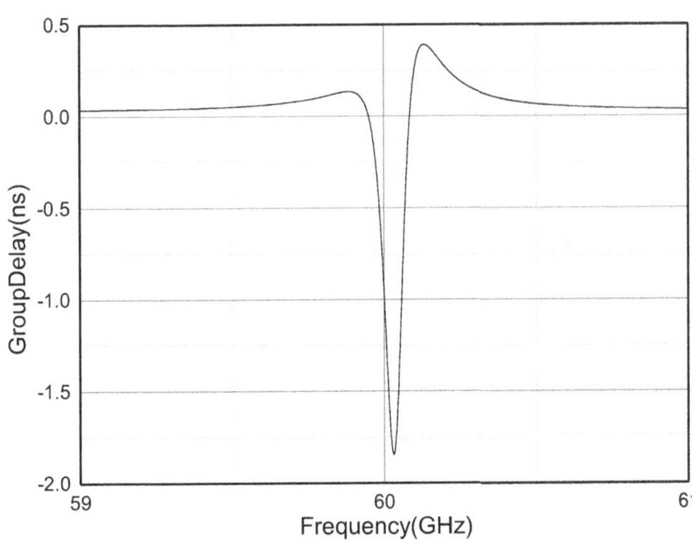

Fig. 12.6 GD response of ADS® design of the microstrip li-circuit shown in Fig. 12.5

Table 12.2 BP NGD specifications of the simulated circuit

Parameter	f_n (GHz)	GD_n (ns)	BW (MHz)	$S_{21}(f_n)$ (dB)	$S_{11}(f_n)$ (dB)
Value	60.032	−1.84	135	−7.23	−4.84

frequency, $f_n \approx 60.032$ GHz. The cut-off frequencies are approximately equal to $f_1 \approx 59.951$ GHz and $f_2 \approx 60.086$ GHz. These characteristics correspond to challenging NGD bandwidth, BW = 300 MHz. It means that the millimetre wave li-circuit can be assumed as a BP NGD function if the transmission and reflection coefficients satisfy the expected constraints in the NGD frequency band.

12.4.2.2 Reflection and Transmission Coefficients

The frequency responses of the reflection and transmission parameters of the NGD li-circuit POC are plotted in Fig. 12.7 and in Fig. 12.8, respectively. It can be seen that despite the NGD response, around the NGD centre frequency, the circuit is not perfectly matched if the required access matching is fixed to $A = -10$ dB. We found also the traditional similar behaviour between the NGD and transmission parameter curves in the NGD bandwidth. Moreover, the transmission coefficient is worse than −7 dB. In the continuation of the present research work, these NGD specification imperfections can be improved with further optimization on the geometrical parameters of the li-circuit.

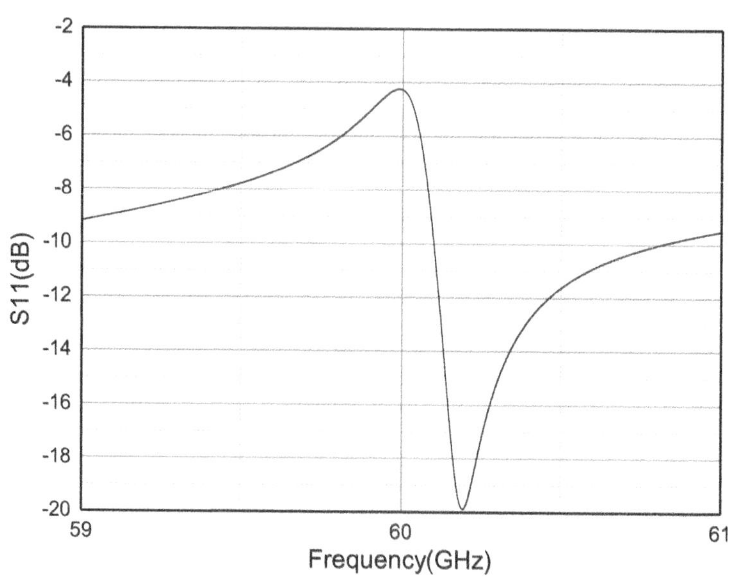

Fig. 12.7 $S_{11} \approx S_{22}$ response of ADS® design of the microstrip li-circuit shown in Fig. 12.5

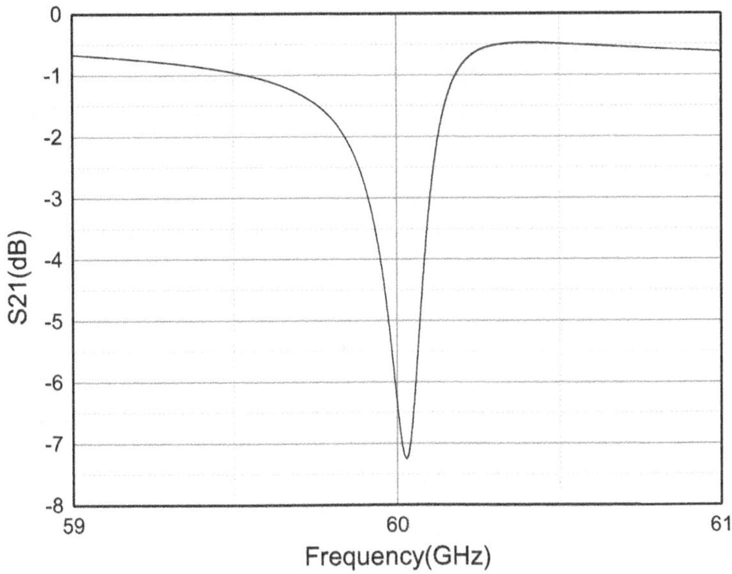

Fig. 12.8 S_{21} response of ADS® design of the microstrip li-circuit shown in Fig. 12.5

12.4.3 Comparisons Between the NGD Specifications of Microwave and 60 GHz Millimetre Wave Li-Circuits

Comparative study between the performances of the existing li-NGD microstrip circuits operating at the centre frequencies $f_0 = 0.92$ GHz and $f_0 = 2.56$ GHz proposed in (Ravelo et al. 2020) is discussed in the present paragraph. The existing li-NGD circuits operates in L- and S-band and implemented on Cu-metallized FR4-substrate presenting relative permittivity, $\varepsilon_r = 5$ and 1.6 mm thickness (Ravelo et al. 2020). Table 12.3 summarizes the NGD specifications of the considered

Table 12.3 Comparisons of NGD specifications of microwave and millimetre wave NGD li-microstrip circuits

NGD li-microstrip circuit		Microwave frequencies (Ravelo et al. 2020)		Millimetre wave frequency
NGD centre frequency	f_0	0.92 GHz	2.56 GHz	60.032 GHz
Size	Physical length	95 mm	40 mm	2.9 mm
	Physical width	10 mm	10 mm	2 mm
NGD value	GD_n	−3.7 ns	−0.9 ns	−1.84 ns
NGD bandwidth	BW	18 MHz	46 MHz	135 MHz
Transmission coefficient	$S_{21}(f_0)$	−3.18 dB	−2.49 dB	−7.23 dB
Reflection coefficient	$S_{11}(f_0) \approx S_{22}(f_0)$	−11.43 dB	−14.86 dB	−4.28 dB

microwave and millimetre wave circuits. The millimetre wave circuit presents less performing insertion and reflection losses. But these parameters can be reoptimized carefully in function of the available substrate and measurement technique. However, the millimetre wave li-circuit enables to achieve more significantly wider NGD bandwidth. It can be underlined that the millimetre wave NGD circuit presents a significant compactness compared to the microwave circuits.

In addition to this comparative study, we may wonder on the influences of the proposed millimetre wave li-circuit physical parameters on the NGD specifications. An overview on the li-circuit parametric analyses will be elaborated in the following section.

12.5 NGD Frequency Analyses with Respect to the Different Physical Design Parameters

For the further understanding about the influences of material and geometrical parameters of the li-circuit, wide parametric analysis is proposed in the present sections.

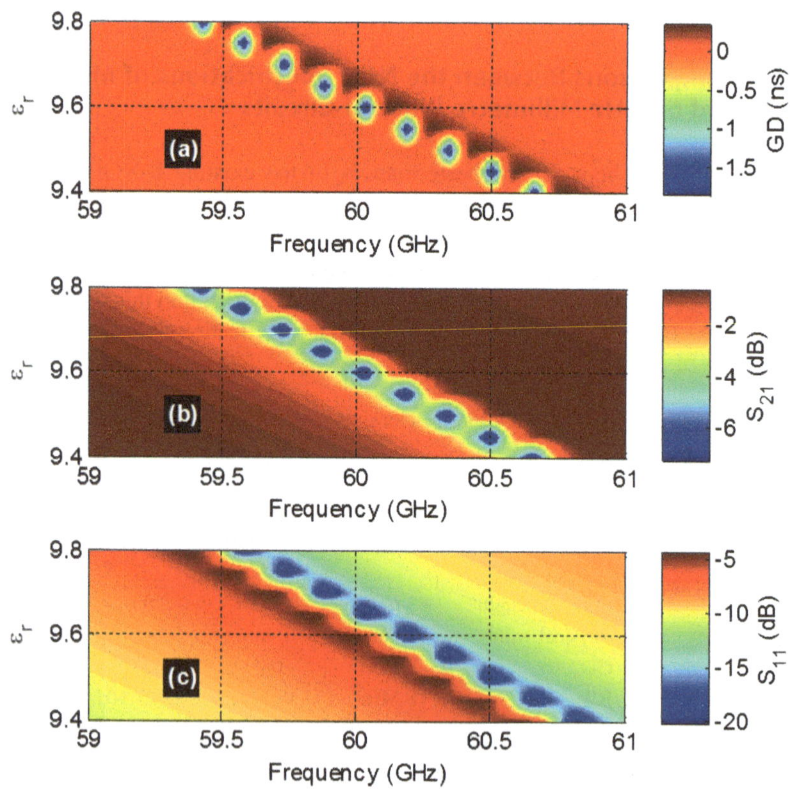

Fig. 12.9 GD, S_{21} and S_{11} mapping with respect to couple variable (ε_r, f)

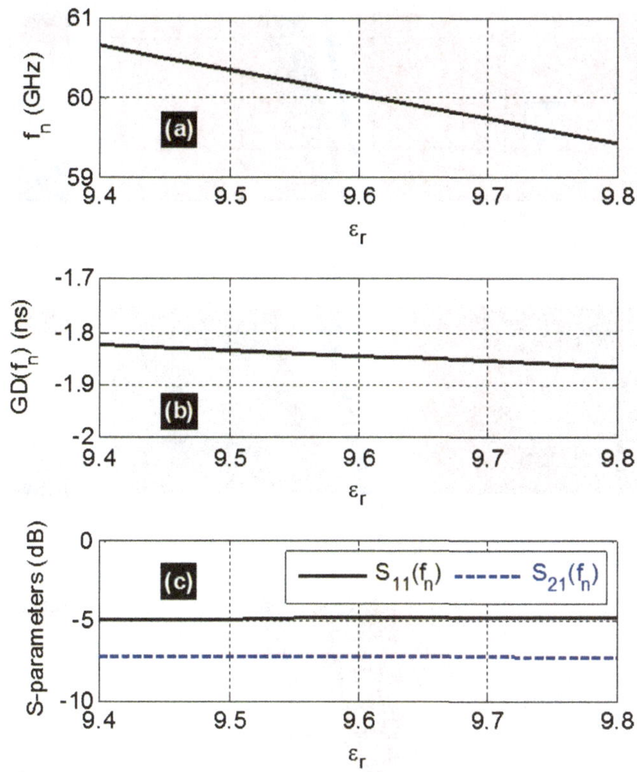

Fig. 12.10 NGD centre frequency, GD, S_{21} and S_{11} values versus ε_r

Table 12.4 BP NGD specifications of the simulated circuit in function of the relative permittivity

Relative permittivity	f_n (GHz)	GD_n (ns)	BW (MHz)	$S_{21}(f_n)$ (dB)	$S_{11}(f_n)$ (dB)
9.4	60.663	−1.825	136	−7.183	−4.886
9.5	60.348	−1.835	136	−7.205	−4.865
9.6	60.032	−1.845	135	−7.227	−4.844
9.7	59.731	−1.855	135	−7.245	−4.831
9.8	59.428	−1.865	133	−7.271	−4.797

12.5.1 Parametric Versus Material Characteristics

One of the open and regularly wondered questions for the NGD engineers is the influence of the material parameters on the NGD specifications. The main material parameters considered in this study are the relative permittivity of the substrate and the metallization conductivity and thicknesses.

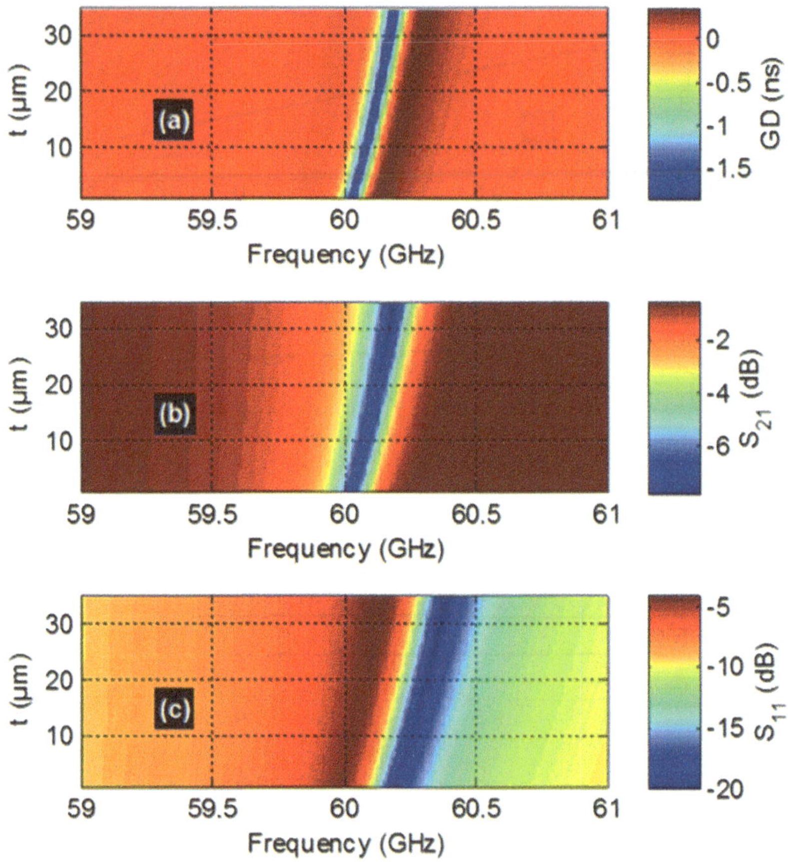

Fig. 12.11 GD, S_{21} and S_{11} mapping with respect to couple variable (t, f)

12.5.1.1 Influence of the Dielectric Substrate Permittivity

Figures 12.9 and 12.10 represent the li-circuit frequency responses with respect to relative permittivity, ε_r, varied from 9.4 to 9.8. It can be underlined that the most significant change is the NGD centre frequency. When ε_r is increased from 9.4 to 9.8, this centre frequency is decreased from 60.663 to 59.428 GHz. The associated NGD specification variations are tabulated in Table 12.4.

12.5.1.2 Influence of the Metallization Thickness

In this case, Figs. 12.11 and 12.12 highlight the influences of the li-microstrip circuit thickness of metal layer, t, which is varied from 1 mm to 35 µm on the NGD frequency responses. The associated NGD specifications are tabulated in Table 12.5. As illustrated in Fig. 12.10, the most significant changes in the responses are as follows:

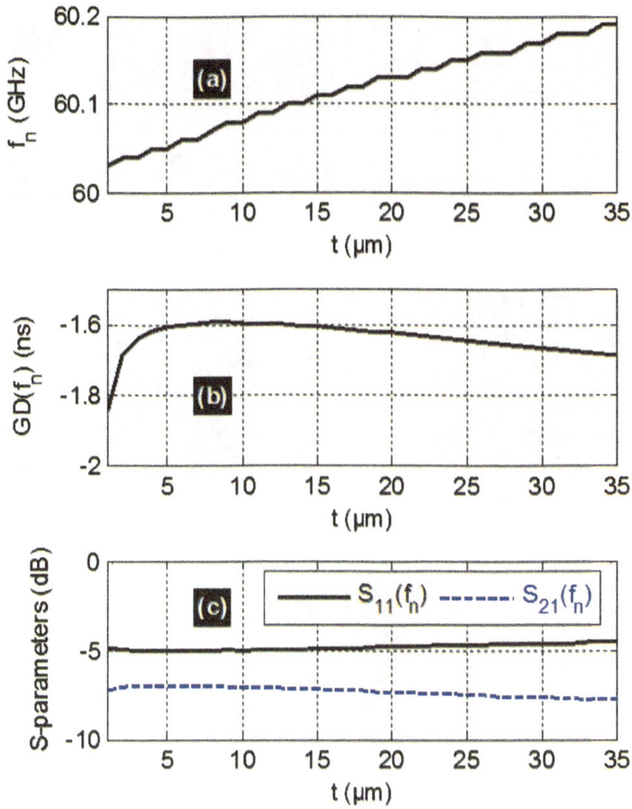

Fig. 12.12 NGD centre frequency, GD, S_{21} and S_{11} values versus t

Table 12.5 BP NGD specifications of the simulated circuit in function of the metallization thickness

t (μm)	f_n (GHz)	GD_n (ns)	BW (MHz)	$S_{21}(f_n)$ (dB)	$S_{11}(f_n)$ (dB)
1	60.037	−1.845	135	−7.227	−4.844
5	60.058	−1.608	147	−6.916	−5.053
10	60.086	−1.596	149	−7.005	−5.003
15	60.111	−1.608	151	−7.143	−4.905
20	60.135	−1.626	154	−7.292	−4.796
25	60.156	−1.647	155	−7.440	−4.700
30	60.178	−1.668	157	−7.586	−4.609
35	60.198	−1.689	157	−7.732	−4.519

- The GD value which increases from −1.845 to −1.596 ns but not linearly
- The centre frequency, which increases from 60.037 to 60.198 GHz

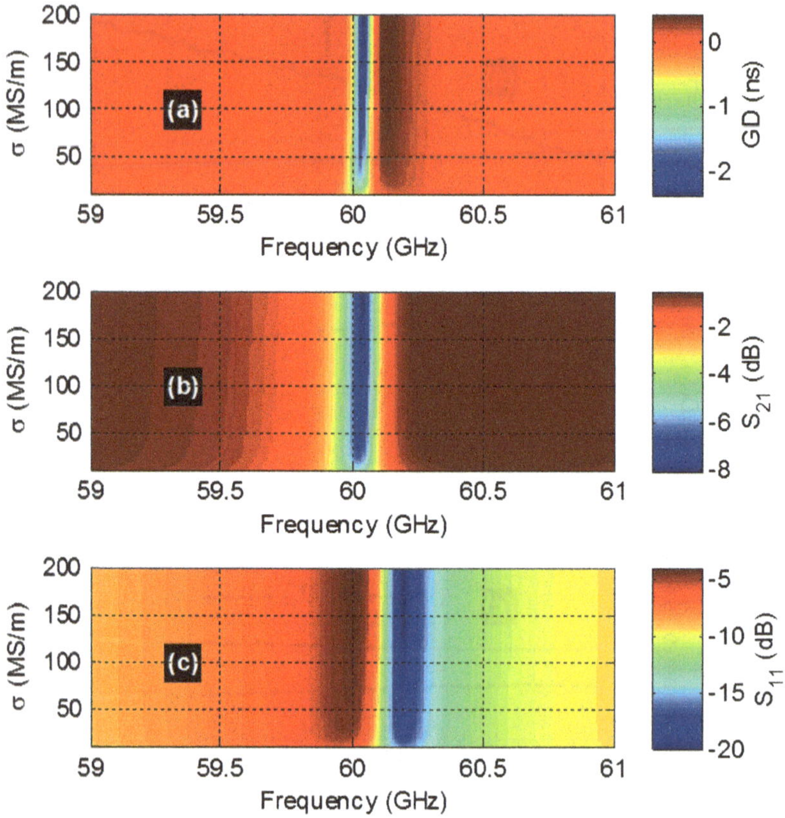

Fig. 12.13 GD, S_{21} and S_{11} mapping with respect to couple variable (σ, f)

12.5.1.3 Influence of the Metallization Conductivity

In the present case of study, we reported in Figs. 12.13 and 12.14 the influence of the li-microstrip circuit metallization conductivity, σ, on the represents the NGD frequency responses. The associated NGD specifications are tabulated in Table 12.6. This later is varied from 10 to 200 MS/m. We can point out that the most significant change is the NGD value. When σ is increased from 10 to 200 MS/m, the NGD value of the li-circuit is improved from −0.784 to −2.386 ns.

12.5.2 Influence of Li-Microstrip Geometrical Parameters

Different to the previous subsection, the present one is focussed on the parametric studies of the different geometrical parameters of the microstrip li-circuit.

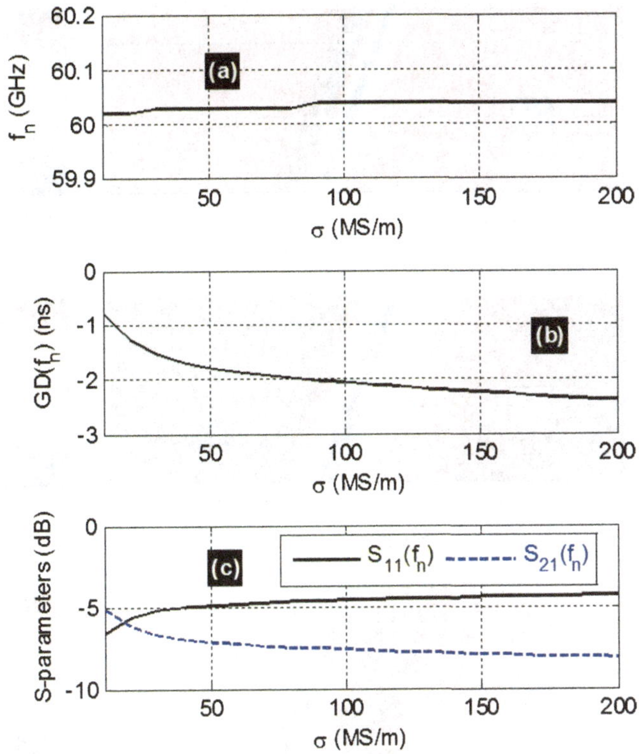

Fig. 12.14 NGD centre frequency, GD, S_{21} and S_{11} values versus σ

Table 12.6 BP NGD specifications of the simulated li-circuit in function of the metallization conductivity

σ (MS/m)	f_n (GHz)	GD_n (ns)	BW (MHz)	$S_{21}(f_n)$ (dB)	$S_{11}(f_n)$ (dB)
10	60.026	−0.784	197	−5.074	−6.590
50	60.036	−1.782	137	−7.121	−4.918
100	60.041	−2.068	128	−7.590	−4.599
150	60.044	−2.251	129	−7.870	−4.424
200	60.046	−2.386	119	−8.066	−4.039

12.5.2.1 Influence of the Width

Figures 12.15 and 12.16 show the li-circuit NGD frequency responses with respect the variation of width, w. The associated NGD specifications are given in Table 12.7. The microstrip line width is varied from 0.6 to 0.65 mm in the present case of study. We can see that the most significant change is the centre frequency, which decreases from 60.183 to 59.898 GHz.

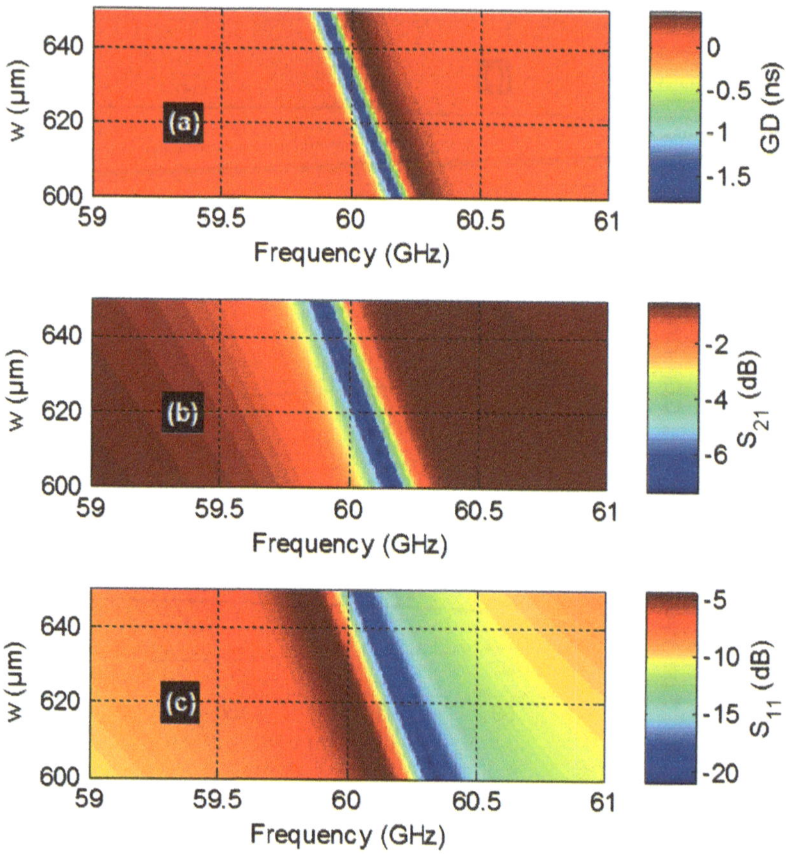

Fig. 12.15 GD, S_{21} and S_{11} mapping with respect to couple variable (w, f)

12.5.2.2 Influence of the Li-Circuit Constituting Microstrip CL Physical Length

In this paragraph, we study the influence of the li-circuit physical length. Figures 12.17 and 12.18 represent the NGD frequency responses versus the CL length, *d*, varied from 0.87 to 0.88 mm. The associated NGD specifications are tabulated in Table 12.8. We can see that the most significant changes on the NGD responses are the centre frequency and the GD value. This first one is decreasing from 60.230 to 59.585 GHz. And the other one is decreasing from −1.829 to −1.881 ns.

12.5.2.3 Influence of the Li-Circuit Microstrip CL Interspace

This last paragraph introduces the parametric study of the microstrip CL interspace. Figures 12.19 and 12.20 display the mapping and the curve showing the influence of

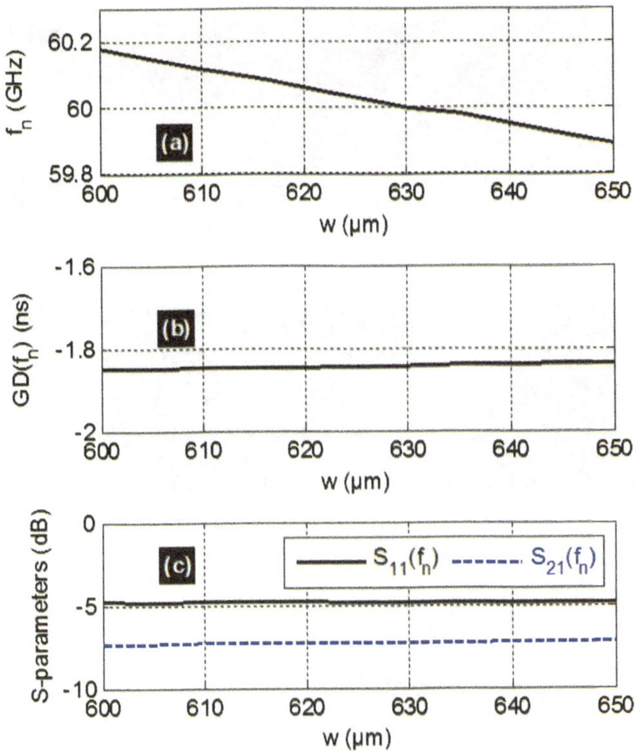

Fig. 12.16 NGD centre frequency, GD, S_{21} and S_{11} values versus w

Table 12.7 BP NGD specifications of the simulated li-microstrip circuit in function of the microstrip line physical width

w (µm)	f_n (GHz)	GD_n (ns)	BW (MHz)	$S_{21}(f_n)$ (dB)	$S_{11}(f_n)$ (dB)
600	60.183	−1.851	137	−7.259	−4.830
610	60.123	−1.849	136	−7.264	−4.845
620	60.064	−1.846	135	−7.247	−4.821
630	60.007	−1.843	135	−7.220	−4.825
640	59.953	−1.840	136	−7.184	−4.858
650	59.898	−1.836	135	−7.159	−4.854

the CL interspace, s. The associated NGD specifications are tabulated in Table 12.9. In this case, s is varied from 0.1 to 1 mm.

Based on Table 12.9, notable changes are observed around the following specifications around the NGD centre frequency:

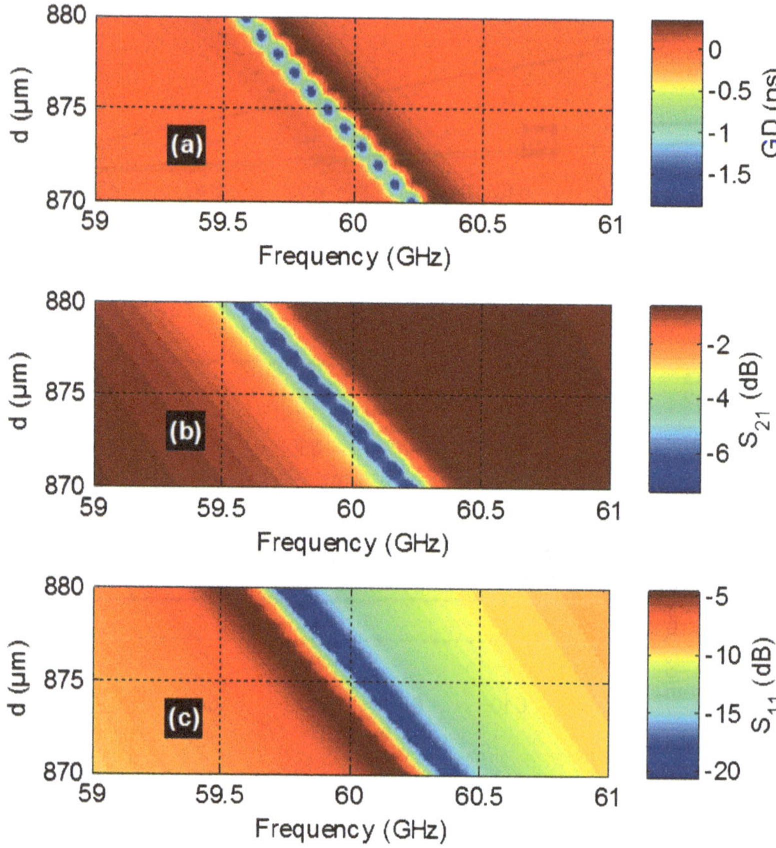

Fig. 12.17 GD, S_{21} and S_{11} mapping with respect to couple variable (d, f)

- The li-circuit NGD value which increases from −2.702 to −0.572 ns
- The li-circuit transmission coefficient which increases from −14.340 to −2.246 dB
- The li-circuit reflection coefficient which decreases from −1.905 to −9.113 dB

12.6 Conclusion

An original study of BP NGD behaviour of li-topology in millimetre waves is introduced. The basic specifications of unfamiliar NGD function are described. The analytical definition of the GD is expressed in function of the S-parameter approach applied to two-port system. The S-parameter is used to describe analytically the li-NGD topology.

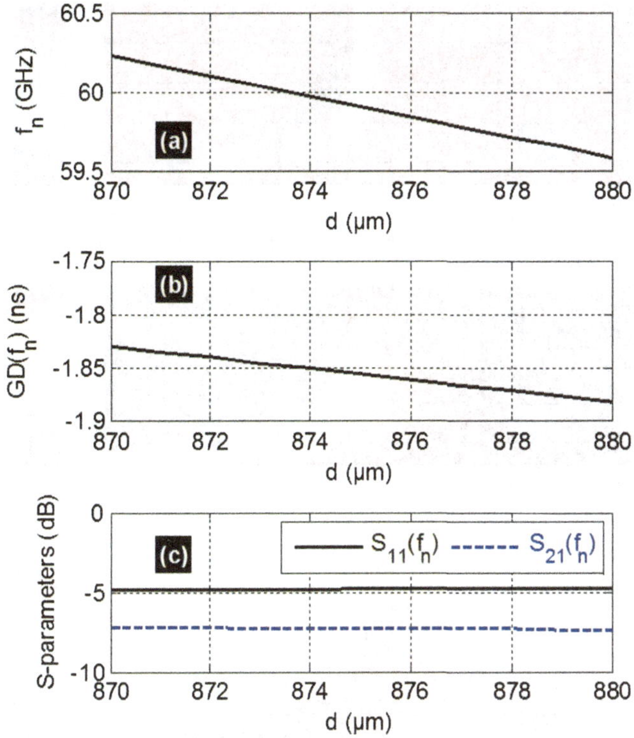

Fig. 12.18 NGD centre frequency, GD, S_{21} and S_{11} values versus d

Table 12.8 BP NGD specifications of the simulated li-circuit in function of the microstrip CL physical length

d (μm)	f_n (GHz)	GD_n (ns)	BW (MHz)	$S_{21}(f_n)$ (dB)	$S_{11}(f_n)$ (dB)
870	60.230	−1.829	135	−7.188	−4.841
872	60.101	−1.839	135	−7.219	−4.826
874	59.971	−1.850	134	−7.246	−4.825
876	59.843	−1.861	136	−7.278	−4.805
878	59.714	−1.871	134	−7.305	−4.799
880	59.585	−1.881	137	−7.338	−4.776

The design method of li-topology in microstrip technology is considered as a proof of concept of the present NGD study. The li-NGD circuit is innovatively investigated in mm wave. The variation of the NGD specifications in function of the substrate material and metallization, is added with geometrical parameter analyses.

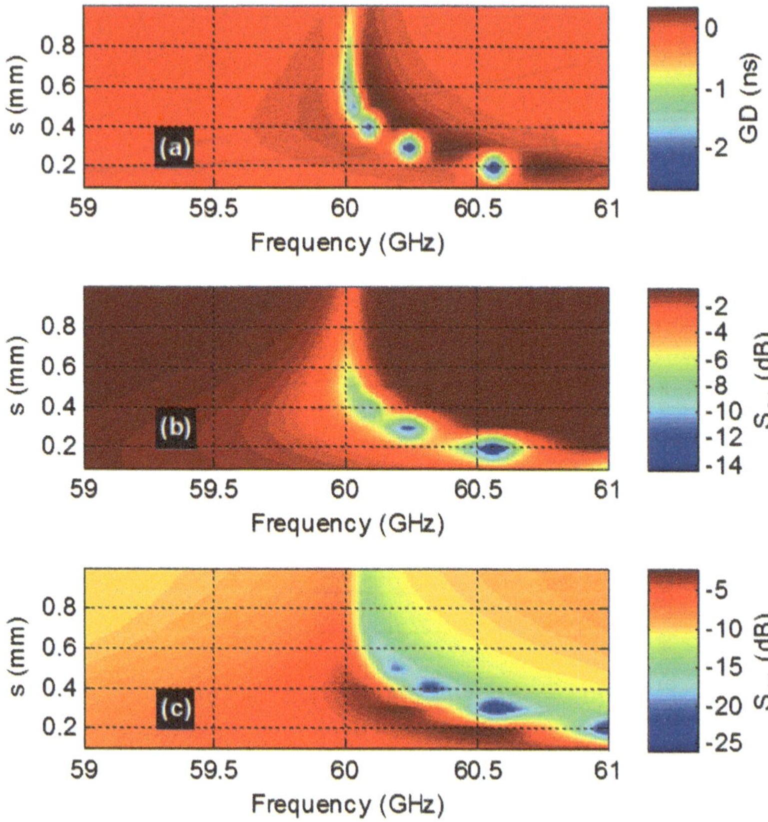

Fig. 12.19 GD, S_{21} and S_{11} mapping with respect to couple variable (s, f)

The proposed NGD study in this chapter opens the way of the design of future NGD structure in millimetre wave and also in terahertz wavelength.

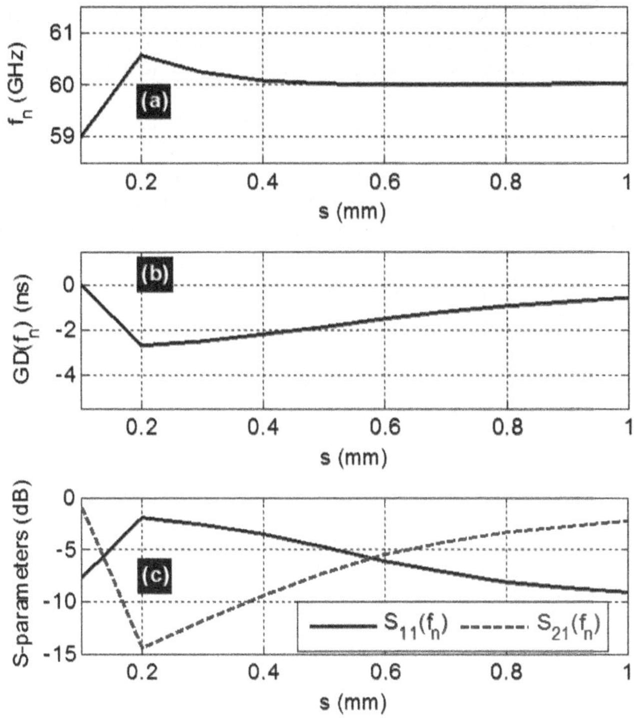

Fig. 12.20 NGD centre frequency, GD, S_{21} and S_{11} values versus s

Table 12.9 BP NGD specifications of the simulated li-circuit in function of the CL interspace

s (mm)	f_n (GHz)	GD_n (ns)	BW (MHz)	$S_{21}(f_n)$ (dB)	$S_{11}(f_n)$ (dB)
0.2	60.569	−2.702	206	−14.340	−1.905
0.4	60.096	−2.194	153	−9.352	−3.620
0.6	60.019	−1.490	121	−5.521	−6.140
0.8	60.018	−0.923	104	−3.355	−8.175
1	60.028	−0.572	95	−2.246	−9.113

References

[Online]. https://spectrum.ieee.org/static/the-race-to-5g. Accessed Nov 2020

[Online]. https://www.spirent.com/solutions/5g-network-testing?gclid=EAIaIQobChMI_IGt5pb77AIV2fhRCh37fA5TEAAYASABEgL0Q_D_BwE. Accessed Nov 2020

[Online]. https://futurenetworks.ieee.org/. Accessed Nov 2020

Advanced Design System (ADS) [Online]. https://www.keysight.com. Accessed Nov 2020

Alomar W, Mortazawi A (2012) Elimination of beam squint in uniformly excited serially fed antenna arrays using negative group delay circuits. In: Proc. of IEEE Int. Symp. Antennas Propag., Chicago, IL, USA, Jul 2012, pp 1–2

Ayyala KAK, Ngoho S, Thakur P, Thakur A, Prasad MS, Ravelo B (2020) Magneto-dielectric substrate effect on bandpass li-circuit NGD performance. In: Proc. of International Conference on Smart Electronics and Communication (ICOSEC 2020), 10–12 Sept 2020, Trichy, India, pp 984–987

Broomfield CD, Everard JKA (2000) Broadband negative group delay networks for compensation of oscillators, filters and communication systems. Electron Lett 36(23):1931–1933

Chaudhary G, Jeong Y, Lim J (2011) Miniaturized dual-band negative group delay circuit using dual-plane defected structures. IEEE Microw Wireless Compon Lett 21(1):19–21

Chaudhary G, Jeong Y, Lim J (2013a) Miniaturized negative group delay circuit using defected microstrip structure and lumped elements. In: Proc. of 2013 IEEE MTT-S International Microwave Symposium Digest (MTT), Seattle, WA, USA, 2–7 June 2013, pp 1–3

Chaudhary G, Jeong J, Kim P, Jeong Y, Lim J (2013b) Compact negative group delay circuit using defected ground structure. In: Proc. of 2013 Asia-Pacific Microwave Conference Proceedings (APMC), Seoul, South Korea, 5–8 Nov 2013, pp 22–24

Choi H, Jeong Y, Kim CD, Kenney JS (2010a) Efficiency enhancement of feedforward amplifiers by employing a negative group delay circuit. IEEE Trans Microw Theory Tech 58(5): 1116–1125

Choi H, Jeong Y, Kim CD, Kenney JS (2010b) Bandwidth enhancement of an analog feedback amplifier by employing a negative group delay circuit. Prog Electromagn Res 105:253–272

Choi H, Jeong Y, Lim J, Eom SY, Jung YB (2011) A novel design for a dual-band negative group delay circuit. IEEE Microw Wireless Compon Lett 21(1):19–21

Dehos C, Gonzalez JL, Dmoenico AD, Ktennas D, Dussopt L (2014) Millimeter-wave access and backhauling: the solution to the exponential data traffic increase in 5G mobile communications systems? IEEE Commun Mag 52(9):88–95

Eudes T, Ravelo B, Louis A (2012) Experimental validations of a simple PCB interconnect model for high-rate signal integrity. IEEE Tran EMC 54(2):397–404

Hammerstad E (1975) Equations for microstrip circuit design. In: Proc. of 5th EuMC, Sept 1975, pp 268–272

Hammerstad E, Jensen O (1980) Accurate models for microstrip computer aided design. IEEE Trans MTT 407–409

Hur S, Kim T, Love DJ, Krogmeier JV, Thomas TA, Ghosh A (2013) Millimeter wave beamforming for wireless Bachaul and access in small cell networks. IEEE Trans Wirel Commun 61(10):4391–4403

Junior RA, Fontgalland G, Rodrigues RAA, Oliveira AHS, Humberto D (2019) Negative group delay folded circuit with distributed broadside parallel line. In: Proc. of 2019 IEEE Int. Symp. Antennas and Propagation and USNC-URSI Radio Science Meeting, Atlanta, GA, USA, USA, 7–12 July 2019, pp 1943–1944

Keser S, Mojahedi M (2010) Removal of beam squint in series fed array antennas using abnormal group delay phase shifters. In: Proceedings of IEEE Antennas Propagation Society International Symposium (APSURSI), Toronto, ON, Canada, 11–17 July 2010, pp 1–4

Mirzaei H, Eleftheriades GV (2013) Squint-free beam forming in series fed antenna arrays using synthesized non-foster elements. In: Proc. of IEEE Antennas Propagation Society Int. Symp. (APSURSI), Orlando, FL, USA, pp 2209–2210

Mirzaei H, Eleftheriades GV (2015) Arbitrary-angle squint-free beamforming in series-fed antenna arrays using non-foster elements synthesized by negative-group-delay networks. IEEE Trans Antennas Propag 63(5):1997–2010

Murad NM, Rajaoarisoa L, Lalléchère S, Fontgalland G, Ravelo B (2020) Analysis of microstrip coupled line based data signal and energy hybrid receiver. J Electromagn Waves Appl 34 (18):2433–2454. https://doi.org/10.1080/09205071.2020.1819443

Rappaport TS, Shu S, Mayzus R et al (2013) Millimeter wave mobile communications for 5G cellular: it will work! IEEE Access 1:335–349

Ravelo B (2012) Theory and design of analogue and numerical elementary NGD circuits: theoretical characterization of analogue and numerical NGD circuits. LAP Lambert Academic Publishing, Germany, 352 p, ISBN-13: 978-8484-1558-8, ISBN-10: 3848415585, 14 Mar 2012

Ravelo B (2014a) Distributed NGD active circuit for RF-microwave communication. Int J Electron Commun 68(4):282–290

Ravelo B (2014b) Similitude between the NGD function and filter gain behaviours. Int J Circ Theory Appl 42(10):1016–1032

Ravelo B (2016a) Negative group-delay phenomenon analysis with distributed parallel interconnect line. IEEE Trans Electromagn Compat 58(2):573–580

Ravelo B (2016b) Theory on coupled line coupler-based negative group delay microwave circuit. IEEE Trans Microw Theory Tech 64(11):3604–3611

Ravelo B (2016c) First-order low-pass negative group delay passive topology. Electron Lett 52(2): 124–126

Ravelo B (2017) On the low-pass, high-pass, bandpass and stop-band NGD RF passive circuits. URSI Radio Sci Bull 2017(363):10–27

Ravelo B, Lalléchère S, Thakur A, Saini A, Thakur P (2016) Theory and circuit modelling of baseband and modulated signal delay compensations with low- and band-pass NGD effects. Int J Electron Commun 70(9):1122–1127

Ravelo B, Wu L, Wan F, Rahajandraibe W, Mourad NM (2020) Negative group delay theory on li topology. IEEE Access 8(1):47596–47606

Saini A, Ravelo B, Thakur A, Thakur P, Lalléchère S (2016) Magneto-dielectric properties of doped ferrite based nanosized ceramics over very high frequency range. Int J Eng Sci Technol 19(2): 911–916

Shao T, Fang S, Wang Z, Liu H (2018) A compact dual-band negative group delay microwave circuit. Radio Eng 27(4):1070–1076

Thakur A, Sharma P, Thakur P, Ravelo B, Hsu J-H (2015) Effect of high spin Mn2+/Mn3+ ions on microstructural, optical, magnetic & electrical properties of hydrothermally prepared Ni-Mg nano-ferrites. Int J Mod Phys B 29(24) (1550183):1–13

Wan F, Gu T, Ravelo B, Lalléchère S (2020) S-parameter model of IB-shape interconnect lines including crosstalk perturbation. In: IEEE Transactions on Electromagnetic Compatibility, Early Access, pp 1–9

Wang Z, Cao Y, Shao T, Fang S, Liu Y (2018) A negative group delay microwave circuit based on signal interference techniques. IEEE Microw Wireless Compon Lett 28(4):290–292

Wu C-T-M, Itoh T (2014) Maximally flat negative group-delay circuit: a microwave transversal filter approach. IEEE Trans Microw Theory Tech 62(6):1330–1342

Yang N, Wang L, Geraci G, Elkashlan M, Yuan J, Di Renzo M (2015) Security and privacy in emerging networks: safeguarding 5G wireless communication networks using physical layer security. IEEE Commun Mag 20–27

Zhang T, Xu R, Wu CM (2017) Unconditionally stable non-Foster element using active transversal-filter-based negative group delay circuit. IEEE Microw Wireless Compon Lett 27(10):921–923

Zhou X, Li B, Li N, Ravelo B, Hu X, Ji Q, Wan F, Fontgalland G (2020) Analytical design of dual-band negative group delay circuit with multi-coupled lines. IEEE Access 8(1):72749–72756

Zhu M, Wu C-TM (2019) Reconfigurable series feed network for squint-free antenna beamforming using distributed amplifier-based negative group delay circuit. In: Proc. of 2019 49th European Microwave Conference (EuMC), Paris, France, 1–3 Oct 2019, pp 256–259

Nanomaterials for Antenna Applications 13

Anwer S. Abd El-Hameed, Asmaa I. Afifi, Moustafa A. Darwish, and Trukhanov Alex

Abstract

This chapter discusses the contribution of the nanomaterial in evolving the front-end RF passive circuits of wireless communication devices. The antenna is addressed due to its indispensable role in any wireless system. Various antenna types like patch, slot, wearable, and dielectric resonators are explored. The nanomaterial impact on antenna performance like bandwidth, size, cost, applicability, and reconfigurability is presented. Graphene, carbon nanotubes (CNT), and nanowires are considered in this chapter as the main nanomaterials used by antenna engineers. The advantages and disadvantages of the discussed nanomaterials are also highlighted by providing practical examples.

A. S. Abd El-Hameed (✉)
Center for Northeast Asian Studies, Tohoku University, Sendai, Miyagi, Japan

Electronics Research Institute, Giza, Egypt
e-mail: anwer.sayed@eri.sci.eg

A. I. Afifi
Electronics Research Institute, Giza, Egypt
e-mail: asmaa.afifi@eri.sci.eg

M. A. Darwish
Physics Department, Faculty of Science, Tanta University, Tanta, Egypt

National University of Science and Technology MISiS, Moscow, Russia
e-mail: mostafa_ph@science.tanta.edu.eg

T. Alex
National University of Science and Technology MISiS, Moscow, Russia

SSPA “Scientific and Practical Materials Research Centre of NAS of Belarus”, Minsk, Belarus

South Ural State University, Chelyabinsk, Russia

A. Thakur et al. (eds.), *Synthesis and Applications of Nanoparticles*,
https://doi.org/10.1007/978-981-16-6819-7_13

Keywords

Antenna · Carbon nanotubes · Graphene · Nanowires · Nanomaterials

13.1 Introduction

Recently, significant progress in developing nanomaterials has been witnessed throughout fabrication technologies, characteristics, and applications (Lee et al. 2018; Cheng and Li 2016; Li et al. 2018, 2019a; Shen et al. 2017; Wang et al. 2017, 2018; Xu et al. 2016; Aravind et al. 2015; Liu et al. 2019; Xiao et al. 2018; Ayodhya and Veerabhadram 2019; Sahoo and Kale 2019; Kolavekar and Ayachit 2019; Kampylafka et al. 2019; Chen et al. 2019; Ge et al. 2019; Chauhan et al. 2019; Binas et al. 2019; Tian et al. 2015). Nanomaterials have obtained a considerable interest in numerous fields due to their unexpected properties. Consequently, many scientific research types have been reported related to nanomaterial scale; moreover, they are not experienced with macroscale materials (Dahl et al. 2014; Smith and Gambhir 2017; Chen and Selloni 2014; Fattakhova-Rohlfing et al. 2014; Aragay et al. 2012; Yang et al. 2015; Liu and Chen 2014; Bai and Zhou 2014; Cheng et al. 2014; Lee et al. 2014; Jing et al. 2016). Nanomaterials open the window toward various applications in optical, microwave, and RF systems (Mohamed et al. 2016). Nanomaterials can be developed for microwave absorption at the high-frequency spectrum (Green and Chen 2019). Electromagnetic absorption is considerably heightened in nanomaterials due to their interaction with the magnetic field, the electric field, or the electromagnetic fields (Liang et al. 2019). The electromagnetic behavior of nanomaterials is obtained by the permeability and the permittivity, correlated with the stored and dissipated energy. The amount of reflection loss at the interface between nanomaterial and air is mainly determined by the real part of permittivity and permeability. Most microwave energy is reflected by the substrate, while little energy is reflected from the air-nanomaterials interface (Park et al. 2003).

Alternatively, the nanomaterials contributed to developing the front-end RF passive circuits of wireless communication devices. The antenna is one of the most essential passive circuits in the front-end stage due to its function as an interface between transmitters and receivers (Abd El-Hameed et al. 2013a, b, c, 2018, 2019; El-Hameed et al. 2013, 2016; Afifi et al. 2016, 2020). The main nanostructures which received great concern in antenna development embrace graphene sheets, carbon nanotubes (CNT), nanocomposites, and nanowires. The use of graphene nanomaterial imported several features for antennas with the compact size, which operates at high frequencies (Yao et al. 2014; Mehta et al. 2016). The impedance of the graphene surface, which is the core point on antenna design, relies on the intrinsic conductivity, which depends on the electrical doping of the material, means, controlled by the biased electric field of graphene (Sharma et al. 2012; Gomez-Diaz and Perruisseau-Carrier 2012; Dragoman et al. 2010).

CNT-based antennas received great interest because they host low-power dissipation features, which increases the antenna radiation efficiency compared to the

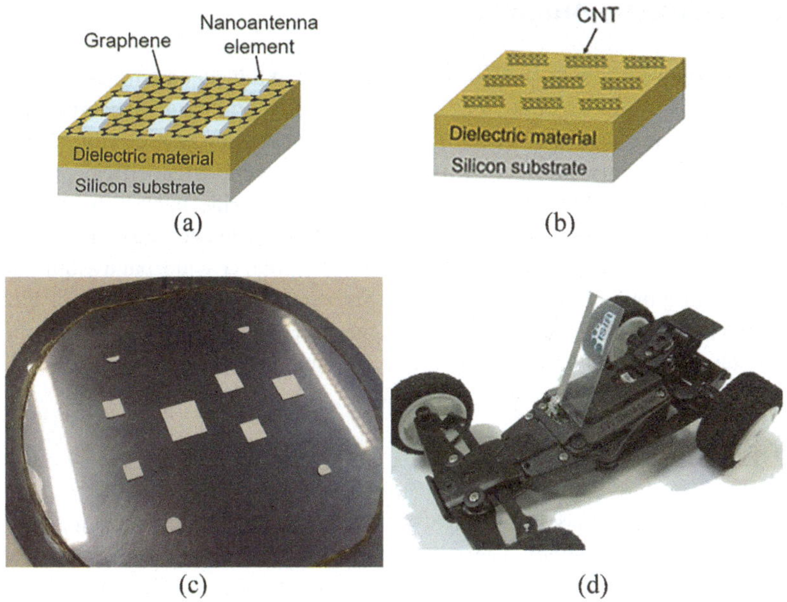

Fig. 13.1 (**a**) Graphene-based biosensor for detection of bacteria (Huang et al. 2011). (**b**) CNT-based antenna (Huang et al. 2011). (**c**) Nanocomposite antenna elements (Aligodarz et al. 2013). (**d**) Commercialize car with a printed silver nanowire monopole antenna (Komoda et al. 2012)

conventional antennas (Lee et al. 2011). When a CNT-based antenna accepts some micro-amperes of current and few volts across its input, it radiates the maximum of a few microwatts in the near field zone (Atakan and Akan 2010). As the demanded communication range increases, the amplification process is required to transmit power by utilizing a group of parallel nanotubes.

The nanocomposite of polyester-styrene polymer consists of high permittivity nano-ceramic powders for designing low-cost antennas by utilizing fabrication techniques of hot embossing and tape casting ceramic injection molding and screen printing (Fig. 13.1) (Aligodarz et al. 2013).

Silver nanowires also showed conductive and printable features and are thought to be an attractive material in the printed antennas. Printed silver nanowire lines have been used to design microstrip antennas with signal losses lower than those obtained in the conventional copper at high frequency (Komoda et al. 2012). Figure 13.1 presents a different antenna application for nanomaterials.

In conclusion, the potential for using nanomaterial-based antennas is a crucial point in the development of high-performance wireless architecture, leading to a new era of smart wireless nano-systems. This chapter discusses the advancement in the antenna by nanomaterials in terms of wireless applications.

13.2 Microstrip Antenna

Microstrip patch antennas are highly recommended in advanced wireless communication systems compactness, performance, and integration capabilities. The microstrip antenna is characterized by resonant frequency, directivity, gain, and efficiency. In general, the microstrip antenna has several geometrical profiles of metallic radiators printed on a dielectric substrate top side backed by a metallic ground plane, as described in Fig. 13.2 (Balanis 2016). The geometry of the metallic radiator determines the antenna features. For example, a wide bandwidth can be offered by an annular ring. Moreover, the substrate dielectric constant or electric permittivity is an essential parameter for the antenna design. Small permittivity thicker substrate is preferred for improving the antenna radiated power.

13.2.1 Substrate Material

The significant parameters that influence the antenna characteristic are the substrate permittivity and loss tangent. These two parameters affect the electric wave enactment of the antenna and consequently the radiation properties. Many commercially available materials are used for RF circuits and antennas specifically. For example, the glass-based FR4 gloss epoxy of 4.4 permittivity and 0.02 loss tangent has been widely used for antenna fabrication (Kanth et al. 2012). Polymer-based material can be another alternative, such as silicon and alumina, with permittivity of 11.9 and 9.8, respectively. Although these last mentioned substrates contribute to size miniaturization, they provide lower radiation efficiency due to their high permittivity.

13.3 Antenna Advancement by Nanomaterials

13.3.1 Patch Antenna

Recently, effective methods based on nanomaterials have been developed to improve antenna characteristics. These methods have focused on modifying the antenna patch or the substrate material properties. Various nanomaterials such as carbon nanotubes, graphene, conductive polymers, carbon nanowires, and graphite have been proposed to improve antenna performance, particularly at THz frequency

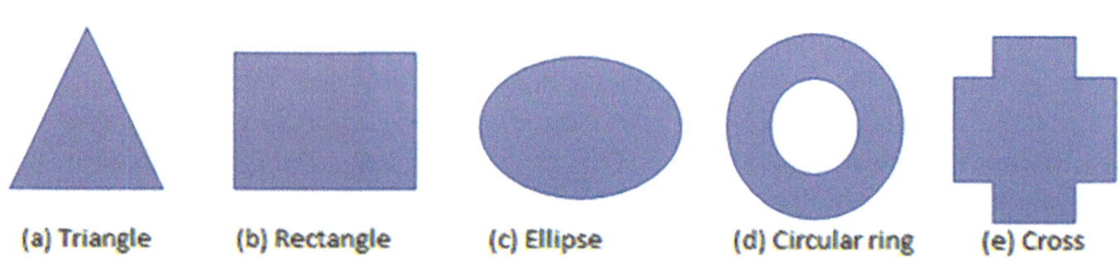

Fig. 13.2 Various shapes of the patch antenna

(Chaya Devi et al. 2017; Dashti and Carey 2018; Goyal and Vishwakarma 2018; Qu et al. 2018; Pan et al. 2014; Gan et al. 2019; Staaf et al. 2014; Gao et al. 2019; Jian et al. 2017; Salim et al. 2019; Li et al. 2019b; Wajahat et al. 2018; Wu et al. 2019).

13.3.1.1 Bandwidth Enhancement

The multiwalled carbon nanotube (MWCNT) was adopted as the rectangular patch material to increase the antenna bandwidth (Chaya Devi et al. 2017). The spin coating method was used to fabricate the MWCNT rectangular patch, as demonstrated in Fig. 13.3. The impedance bandwidth has been improved by 20% without affecting the radiation properties.

Mojtaba and Carey (Dashti and Carey 2018) introduced graphene as a nanomaterial instead of metal to increase the antenna's bandwidth to be hundreds of GHz at THz frequency.

13.3.1.2 Applicability Improvement

Since the traditional metallic films have problems at the terahertz frequency, such as the microcracks, and lose a lot of energy, Goyal and Vishwakarma in (Goyal and Vishwakarma 2018) have promoted the graphene nanomaterial for use at these frequencies. Figure 13.4 shows the graphene-based antenna geometry and its potential chemical effect on the antenna's reflection coefficient and resonance frequency. Moreover, Fig. 13.4 indicates that the graphene can be used for frequency reconfigurability by changing its chemical potential.

13.3.1.3 Frequency Reconfigurability

In Qu et al. (2018), the graphene was integrated with a frequency-selective surface (FSS) for reconfigurable applications at terahertz frequency as given in Fig. 13.5. Because of the configurable properties of the graphene strips embedded in the FSS by controlling the chemical potential, the reconfigurable features can be achieved for FSS as indicated by transmission and refection characteristics, as shown in Fig. 13.6. Equation (13.1) describes the graphene surface conductivity (Hanson 2008).

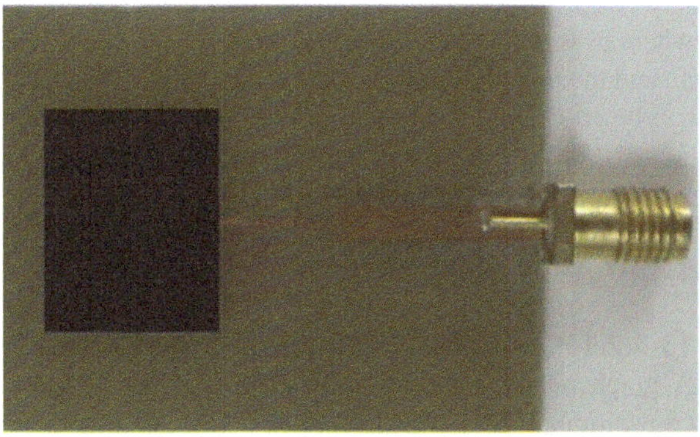

Fig. 13.3 Fabricated MWCNT patch antenna (Chaya Devi et al. 2017)

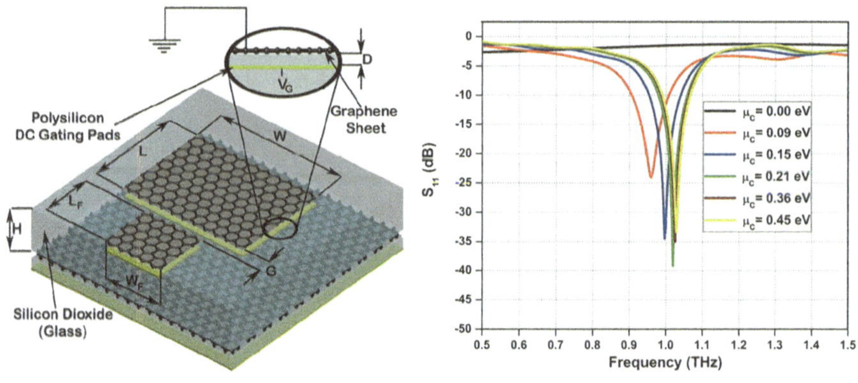

(a) (b)

Fig. 13.4 (**a**) Orthographic view of the proposed graphene-based patch antenna. (**b**) Variation in return loss of the graphene antenna with different chemical potential at a temperature of 300 K and transport relaxation time of 1 ps (Goyal and Vishwakarma 2018)

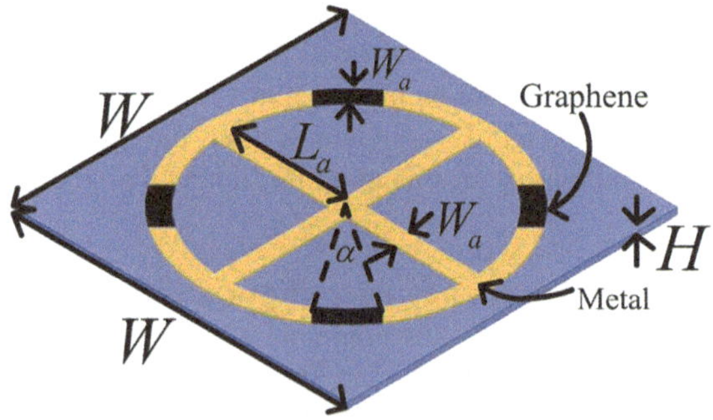

Fig. 13.5 Layout of the proposed graphene-based FSS with dimensions marked (Qu et al. 2018)

$$\sigma_{\mathrm{s}} = \frac{-je^2 k_{\mathrm{B}} T}{\pi \hbar^2 (\omega - j\tau^{-1})} \left(\frac{\mu_{\mathrm{c}}}{k_{\mathrm{B}} T} + 2 \ln \left(e^{-\frac{\mu_{\mathrm{c}}}{k_{\mathrm{B}} T}} + 1 \right) \right), \tag{13.1}$$

where σ_s is determined by temperature T, chemical potential μ_c, radiation frequency ω, and the relaxation time τ. It can be extracted that graphene's surface conductivity can be efficiently controlled by applying DC bias to change its chemical potential.

13.3.2 Dipole and Monopole Antennas

13.3.2.1 Cost-Effective and Eco-Friendly

A graphene nanomaterial is used in designing a dipole antenna instead of using conventional metal for UHF-RFID, as illustrated in (Akbari et al. 2016). The

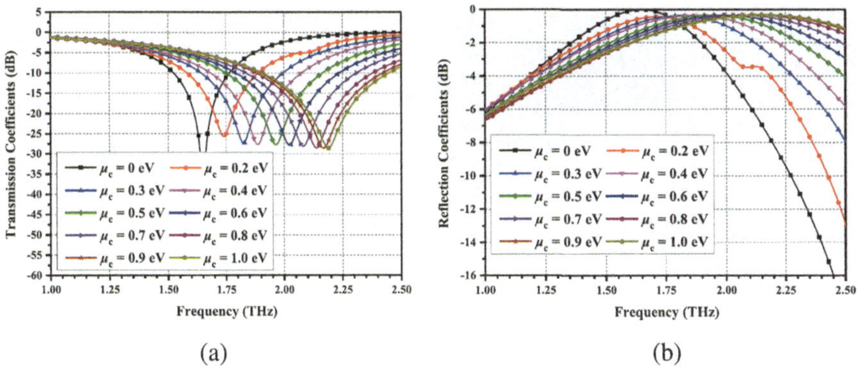

Fig. 13.6 The simulated results of the graphene-based FSS with different chemical potentials ($\mu_c = 0$, 0.2, 0.3, 0.4, 0.5, 0.6, 0.7, 0.8, 0.9, and 1.0 eV): (**a**) transmission coefficients, (**b**) reflection coefficients (Qu et al. 2018)

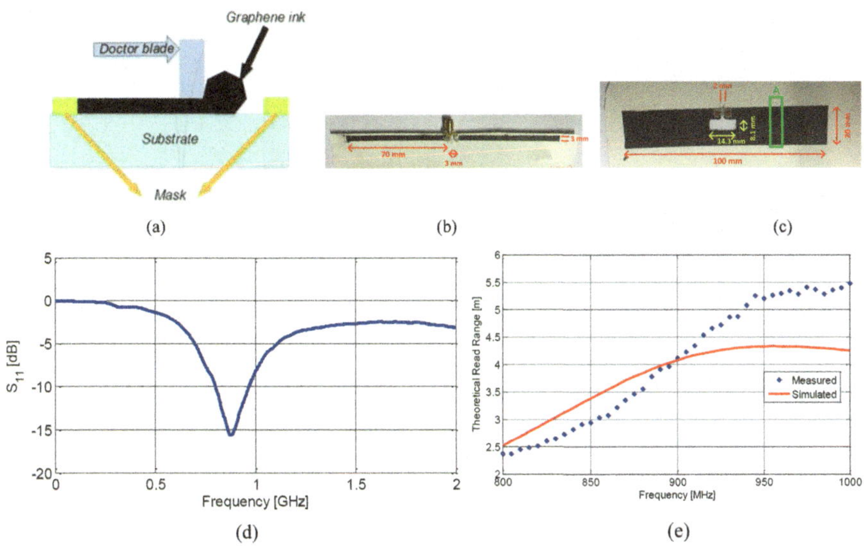

Fig. 13.7 Geometry and results of the graphene-based dipole antenna. (**a**) Schematic image of the doctor blade technique, (**b**) dipole antenna, (**c**) RFID tag, (**d**) measured S11 [dB] of dipole antenna, and (**e**) measured and simulated theoretical read range [m] of RFID tag (Akbari et al. 2016)

graphene-based dipole antenna achieves low cost and eco-friendly because it can be printed on cardboard as a substrate by using the process of doctor blading that is a fast, simple, and easy way. The geometry and results of the graphene-based dipole antenna in (Akbari et al. 2016) are shown in Fig. 13.7.

13.3.2.2 Miniaturization

In Inui et al. (2014), a silver nanowire was inserted inside the cellulose nano paper to minimize the dipole antenna's size. Adding the silver nanowires to the nano paper

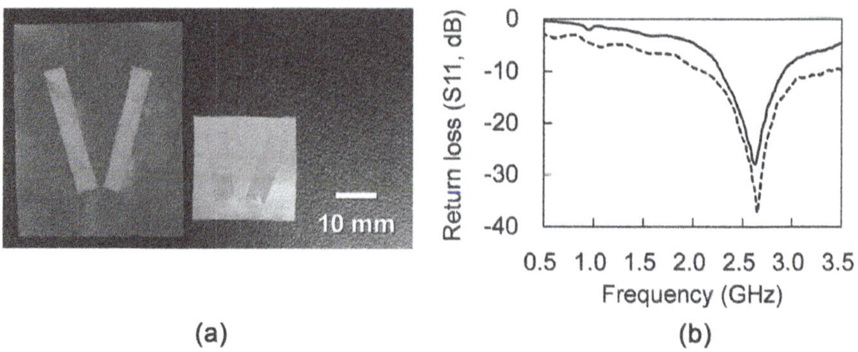

Fig. 13.8 (**a**) Dipole antenna on the nano paper (left), dipole antenna on the silver nanowire/nano paper composite (right). (**b**) Return losses of the dipole antenna on the nano paper (solid line) and antenna on the silver nanowire/nano paper composite (dotted line) (Inui et al. 2014)

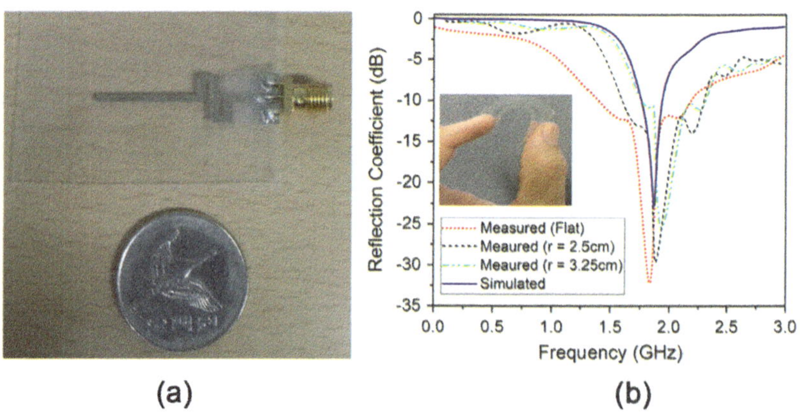

Fig. 13.9 (**a**) Fabricated antenna on PET (with 0.04 mm thick) a size. (**b**) Measured (over different bending radii) and simulated reflection coefficient (Hassan et al. 2016)

substrate increases the overall dielectric (k) of the substrate, reducing the antenna size. As shown in Figure 13.8, the antenna's size was downsized to half by using the modified substrate compared to the original one.

In Hassan et al. (2016), UV-treated polyethylene terephthalate (PET) substrate and silver nanoparticle radiator were more attractive as an alternative to the normal materials for designing a helical monopole antenna as shown in Fig. 13.9. Designing the antenna with these materials reduced the antenna's size to become more compact and controlled cost limits. The presented antenna exhibited 2.72 dBi gain and 93.33% efficiency at 1.8 GHz.

13.3.2.3 Isolation Improvement

Although the monopole antenna's semi-infinite ground plane helps isolate improvement between the elements, it requires a 3D structure, which will be more

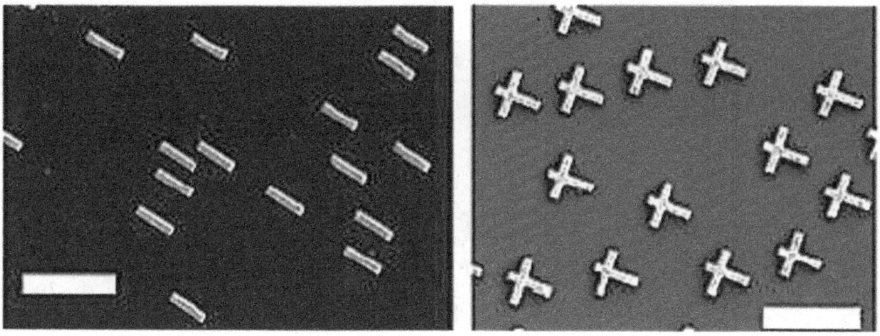

(a) (b)

Fig. 13.10 Representative scanning electron microscope (SEM) images of fabricated structures: (**a**) nanorod and (**b**) composite double monopole. In both, the white scale bar is 2 μm long (Adato et al. 2011)

complicated for chip-scale circuits. Consequently, the plasmonic nanoparticle was used to implement a wire nano reflector to isolate the nanoparticles on chip, as illustrated in (Adato et al. 2011). Nanomaterial hosting for the monopole antenna design shown in Fig. 13.10 helps this antenna be more compatible with the chip-scale fabrication.

13.3.3 Slot Antenna

13.3.3.1 Cost-Effective

High concentration silver nanowire contributes to designing simple slot antenna with low cost and ease of fabrication features for 2.4 GHz, the band of ISM applications (Fu et al. 2018). Unlike the conventional materials, using such nanomaterial helps the scientist print out the slot antenna on the common photo paper that acts as a substrate, as shown in Fig. 13.11.

13.3.3.2 Impedance Matching and Loss Reduction

The graphene film is also used to design a dual-port printed slot antenna to overlay in the microstrip feed line to improve the antenna impedance matching over broadband: 100% at the center of frequency 3.5 GHz. By configuring a graphene film into a small area outside the radiator, the antenna loss is minimized (Naishadham 2014). Figure 13.12 shows the geometry of the slot antenna and its reflection coefficient.

13.3.3.3 Frequency Reconfigurability

The property of graphene conductivity variation based on DC voltage biasing was exploited for designing novel frequency reconfigurable antennas. A short-circuiting stub was connected between a simple circular copper antenna and a graphene slot, as shown in Fig. 13.13 (Zhang et al. 2015). The graphene can work as the extension of the copper circular monopole antenna. When the graphene biasing voltage changed,

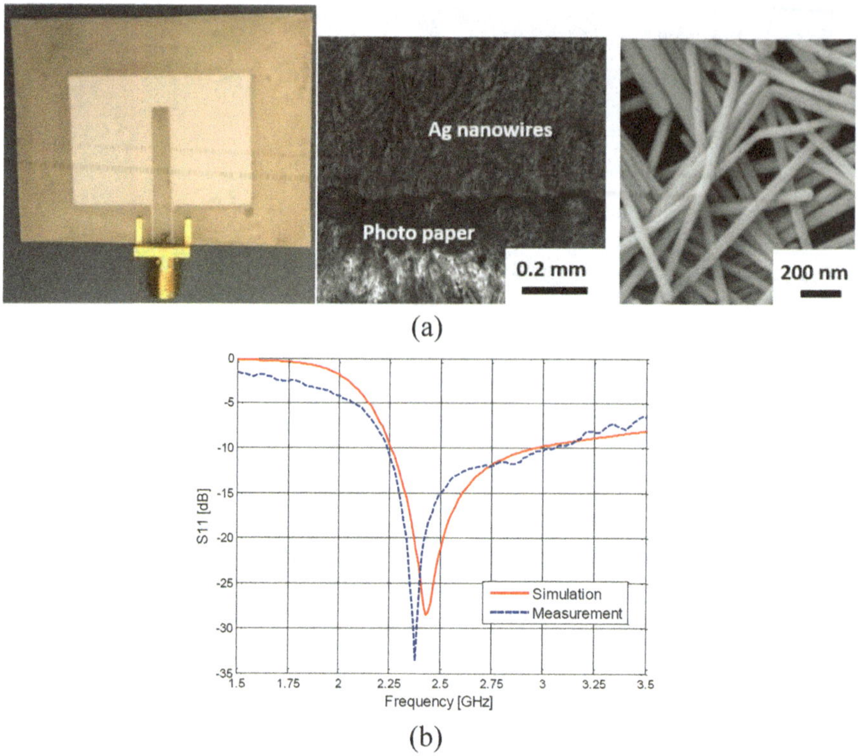

Fig. 13.11 Low-cost silver nanowire-based slot antenna (Fu et al. 2018). (**a**) The photo image and SEM images of the proposed antenna. (**b**) Reflection coefficient

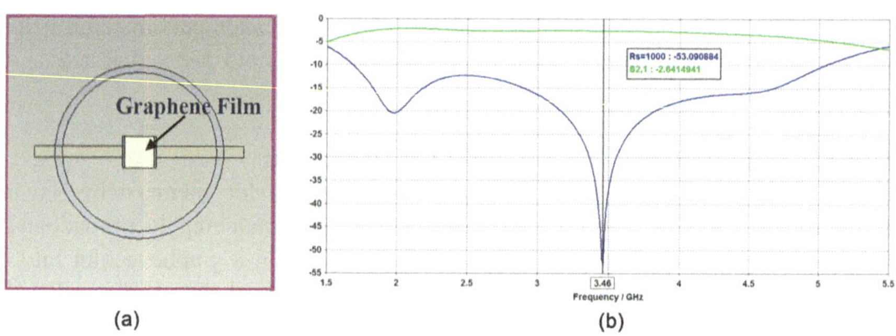

Fig. 13.12 Graphene-based broadband slot antenna (Naishadham 2014). (**a**) Design geometry. (**b**) Reflection coefficient when $Z_s = 100 - j150\ \Omega$

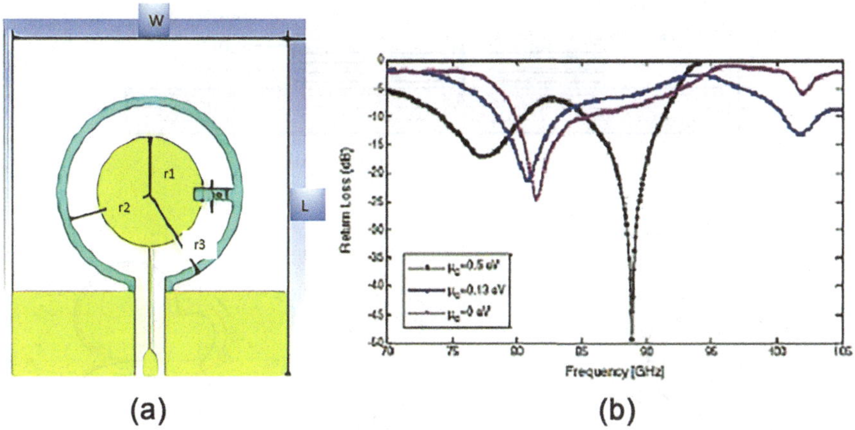

Fig. 13.13 Graphene-based frequency reconfigurable antenna (Hanson 2008). (**a**) Antenna structure. (**b**) Resonant frequency shift for the reconfigurable antenna under different chemical potential

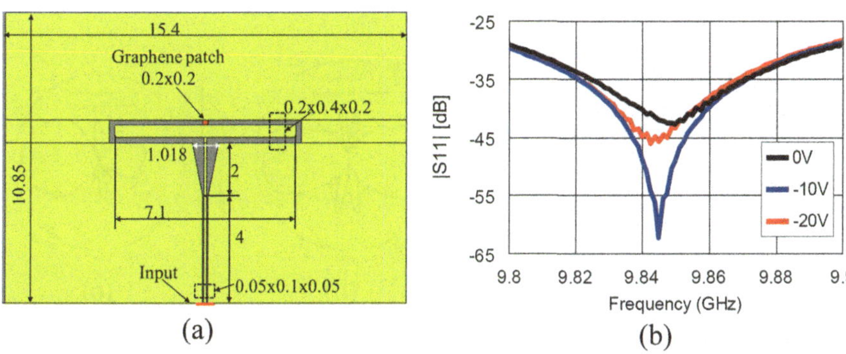

Fig. 13.14 (**a**) Electromagnetic model and general layout dimensions of the X band slot antenna with a graphene patch (all dimensions are given in millimeters) (Bunea et al. 2015). (**b**) Measured |S11| parameter for different biasing voltages

its impedance changed, and hence the slot effective electrical length changed. Consequently, the resonant frequency changed to different values.

Similarly, a graphene patch is loaded to the signal line and the ground of a CPW of X band slot antenna to control its position of the center frequency Fig. 13.14 (Bunea et al. 2015).

13.3.3.4 Radiation Pattern Reconfigurability

Radiation pattern configurability was also conducted by graphene/Al_2O_3 sheets, which impeded the slot ring antenna's substrate. The antenna bandwidth was also controlled based on the same phenomena of graphene conductivity variation corresponding to voltage biasing. Figure 13.15 describes the radiation pattern-controllable microstrip slot antenna's structure using alternating graphene sheet

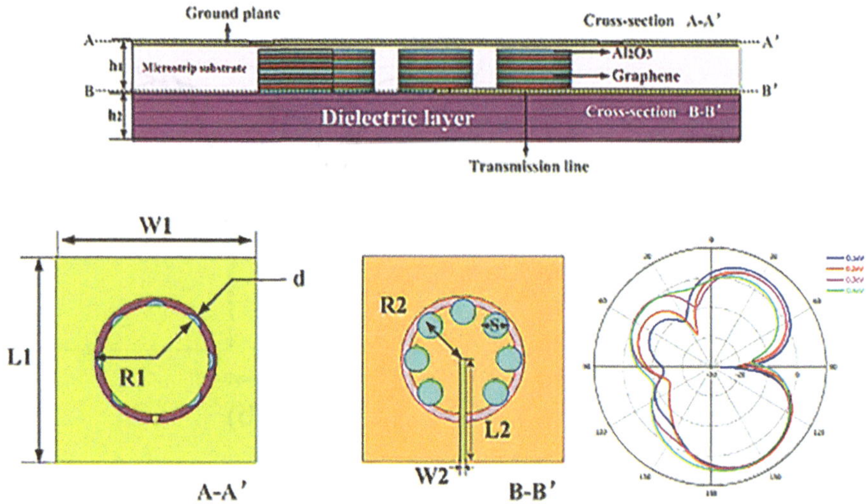

Fig. 13.15 Structure of the graphene/Al_2O_3-based microstrip slot antenna and radiation pattern with different voltage biasing (Liu et al. 2017)

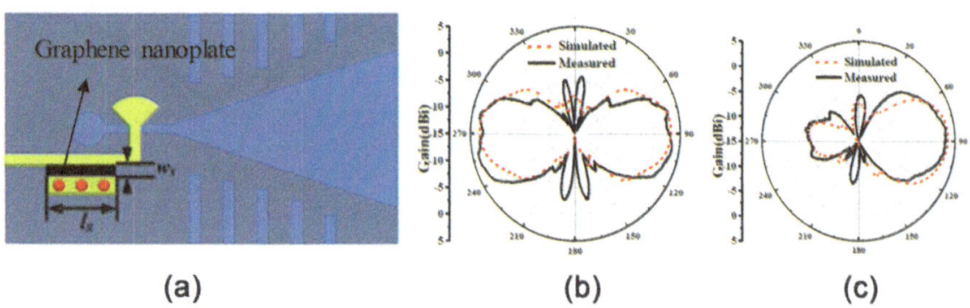

Fig. 13.16 Fabricated pattern reconfigurable Vivaldi antenna. (**a**) Modified tapered slot antenna. (**b**) Radiation pattern when graphene resistance equal to 200 Ω. (**c**) Radiation pattern when graphene resistance equal to 20 Ω (Fan et al. 2020)

and Al_2O_3 insulating film and radiation pattern performance (Liu et al. 2017). The antenna supports dual frequency bands at the THz frequency range. It was noticed that the higher band is more sensitive to the DC biasing than the lower band.

Similarly, a graphene-based tunable resistor was used to reconfigurable the tapered slot Vivaldi antenna in millimeter wave (MMW) (Fan et al. 2020). To achieve the radiation pattern's dynamical management, the graphene nano-plat pads were uniformly loaded onto the feed line. According to the biasing voltage between two graphene pads, the resistance will vary between them. Consequently, the Vivaldi antenna radiation pattern will be a single beam rather than two opposite beams, as shown in Fig. 13.16.

13.3.4 Wearable Antennas

One of the most important challenges faced by the conventional wearable antennas is the integration with flexible substrates like papers and textiles. Flexible substrates metal coating was the key solution. However, fabrication complexity, cost, and low efficiency are still obstacles. Another problem is the ease of damage with multiple uses. Nanomaterials such as silver nanowires, conductive polymers, and carbon nanotubes were adopted to overcome the abovementioned issues (Rai et al. 2013; Mannoor et al. 2012; Park et al. 2012; Yao and Zhu 2015; Kim et al. 2016, 2017; Song et al. 2018; Tang et al. 2018; Zhou et al. 2010).

For example, in Kim et al. (2017), the silver nanowires (Ag NWs) have been encouraged as conductors for wearable dives because of its transparency and biaxial stretchability. The wavy Ag NWs were produced by a floating compression process, which allows the Ag NWs to have a biaxial strain of 30%. Moreover, a comparison between wavy or straight Ag NWs for wearable devices has been explained in Kim et al. (2016). According to the previous study, the wavy Ag NWs is more attractive than the planar because the wavy Ag NWs provides a lower production cost, and it is more transparent because it contains a small amount of NWs. Furthermore, the wavy Ag NWs enhance the stretchability and present low resistance. The monopole patch antenna's procedure and results with wavy Ag NWs are shown in Fig. 13.17.

In Song et al. (2018), the flexible graphite films (FGF) were treated with high thermal to improve their properties for use in place of the rigid copper in wearable antennas. Compared to the identical copper, the FGF-based dipole antenna shows higher gain than the copper-based dipole antenna. Besides, the FGF-based dipole antenna can withstand bending a lot of time in different directions without affecting the antenna performance, as illustrated in Fig. 13.18.

The flexible multilayer graphene film-based antenna can pay attention to its use in a wearable device as it has higher strain sensitivity for both compressive and tensile bending than copper antenna (Tang et al. 2018). The FGF-based patch antenna with its return losses for compression and tensile bending is illustrated in Fig. 13.19.

In Zhou et al. (2010), it has been suggested that the polymer CNT is suitable for load-bearing antenna applications due to its acceptable mechanical performance. The CNTs patch antenna can endure stretching up to 13% and bending to 130° angle with a good performance, as shown in Fig. 13.20.

13.3.5 Dielectric Resonator Antennas (DRA)

13.3.5.1 Bandwidth Enhancement

Three stacked polyurethane (PU) materials with different CNT concentrations are used to design wideband DRA, as shown in Fig. 13.21a (Alrayes and Hussein 2017). The open-ended coaxial probe technique was used to measure the dielectric characteristics and loss tangent of the three composite materials. The control on the CNT concentration helped design DR with gradient dielectric constant and

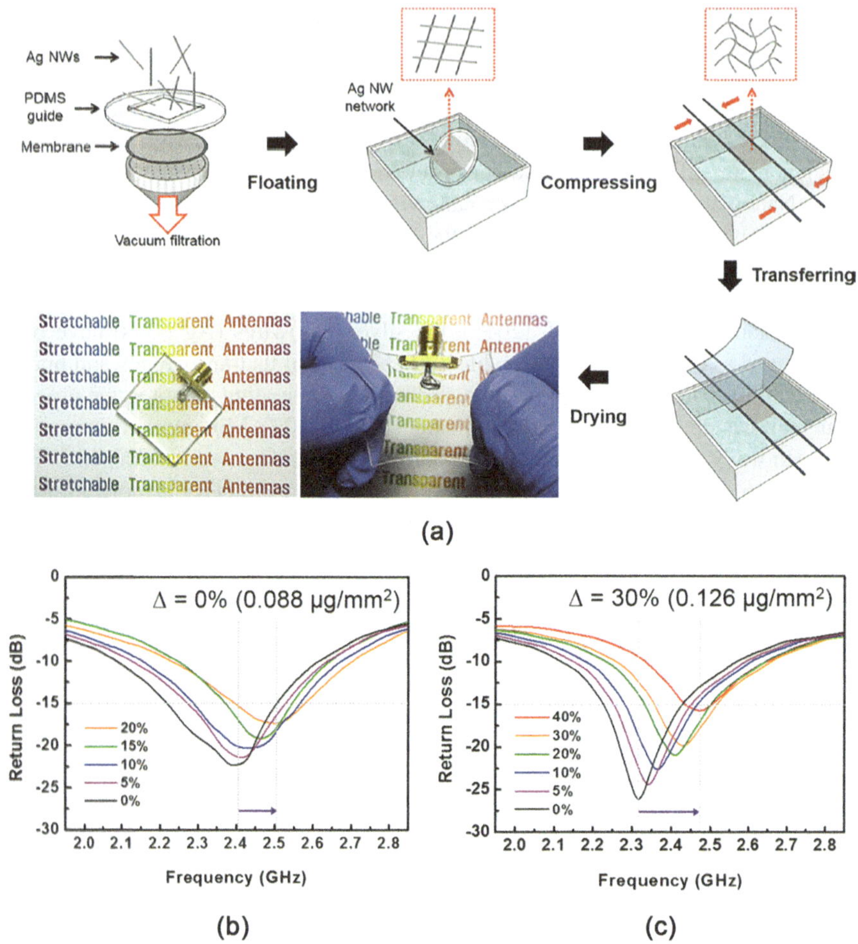

Fig. 13.17 Monopole patch antenna with Ag NWs, (**a**) diagram showing the fabrication process, (**b**) RL of the straight NW networks, (**c**) RL of the wavy NW networks (Kim et al. 2016)

improve the antenna bandwidth. Comparison between the DRA with stacked PU-CNA and non-stacked DRA is presented in Fig. 13.21b.

In Zhang et al. (2017), a thin film graphene is placed on the top surface of rectangle DRA to improve its impedance bandwidth. It was noticed that when the thin graphene film's conductivity was decreased, the impedance bandwidth improved and vice versa. Figure 13.22 presents the antenna geometry and S-parameters results.

13.3.5.2 Gain Enhancement

Hybrid nanomaterial has been used to design high gain DRA. According to the coupling between the plasmonic graphene mode and the dielectric wave mode, a

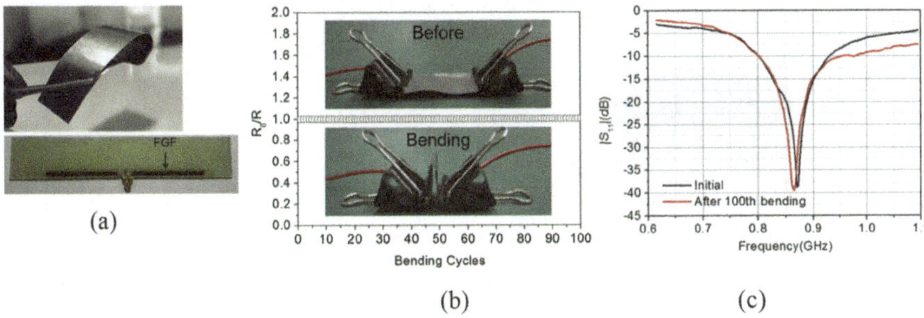

Fig. 13.18 (**a**) Photo of the FGF and FGF antenna, (**b**) resistance change of FGF with the bending test, (**c**) S11 of the FGF antenna (before and after bending) (Song et al. 2018)

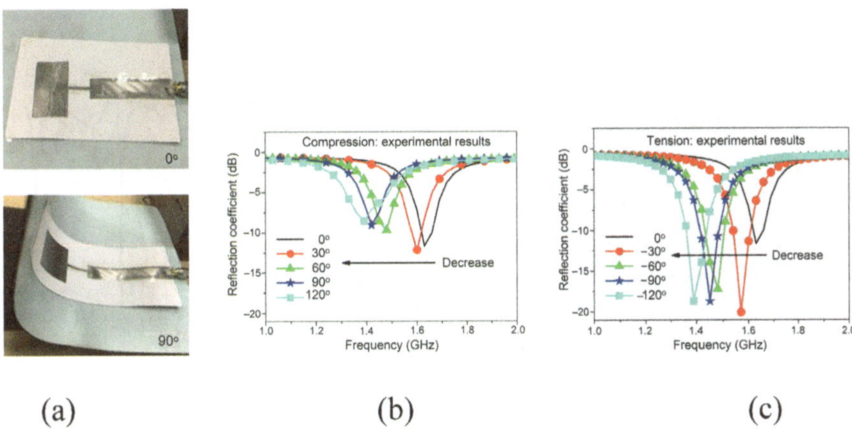

Fig. 13.19 Tension and compression strain measurement based on the proposed antenna. (**a**) Photographs of an FGF antenna sensor attached to a soft mouse pad and bent at different angles. (**b**) The reflection coefficient of experimental results under compression bending strain. (**c**) The reflection coefficient of experimental under tension bending strain (Tang et al. 2018)

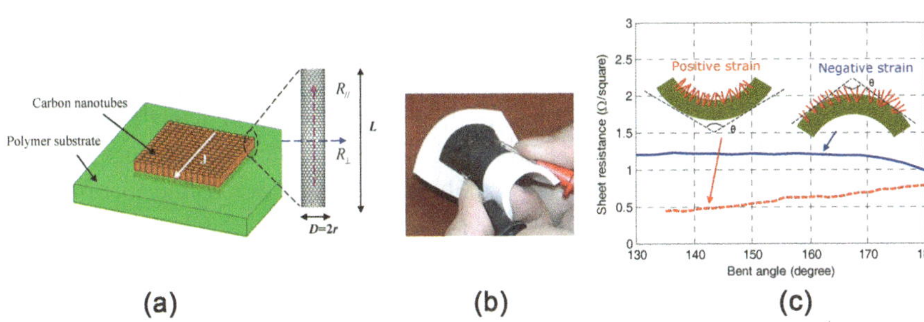

Fig. 13.20 Polymer-carbon nanotube sheets for conformal load-bearing antennas. (**a**) Illustrative model of vertically aligned CNTs, (**b**) fabricated polymer-CNT patch, (**c**) DC sheet resistances versus strain (Zhou et al. 2010)

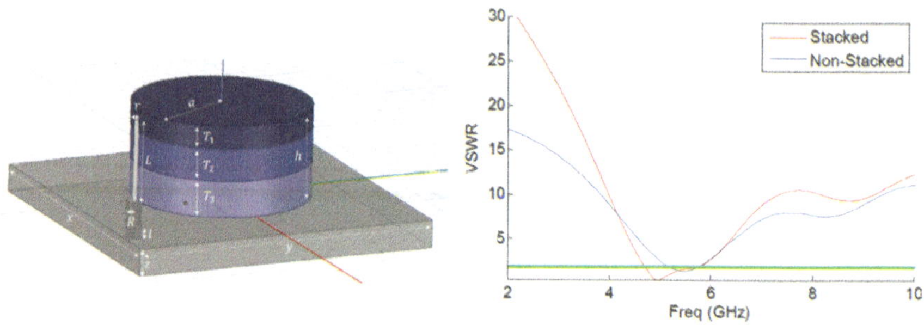

(a) (b)

Fig. 13.21 Stacked DRA antenna geometry (**a**); VSWR comparison between stacked and non-stacked DRA (**b**) (Alrayes and Hussein 2017)

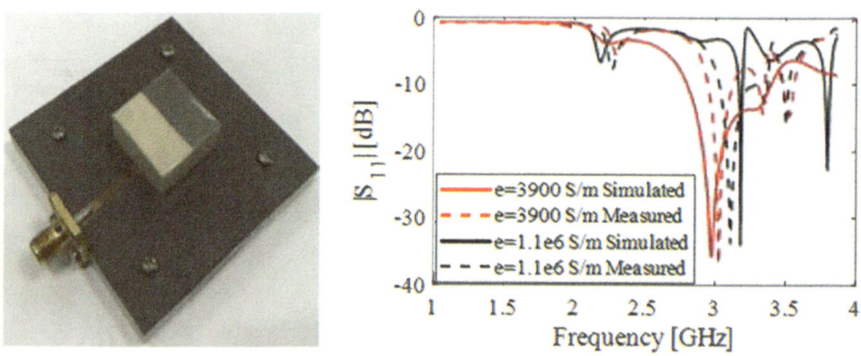

(a) (b)

Fig. 13.22 (**a**) Rectangle DRA with 50% thin graphene film coverage on the top surface. (**b**) The reflection coefficient of the DRA corresponding to different values of thin graphene film conductivity (Zhang et al. 2017)

novel THz antenna was proposed (Ehsan Hosseininejad et al. 2018). Few-layer graphene with five layers were considered as a plasmonic material, PMMA with 2.4 dielectric constant and thickness dL as the L-layer and GaAs with 12.9 dielectric constant and thickness d as the H-layer, Fig. 13.23. Results showed 6.5 dB gain improvement by stimulating the DRA with higher-order modes. Table 13.1 presents a comparison between the DRA antennas with different d_H.

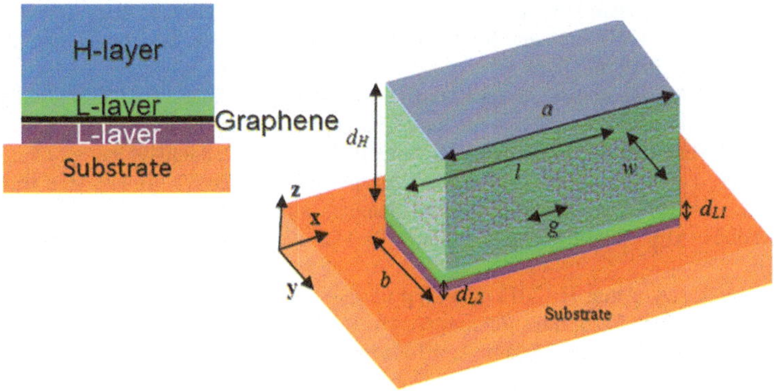

Fig. 13.23 Configuration of a dielectric resonator antenna coupled to a graphene plasmonic dipole (Ehsan Hosseininejad et al. 2018)

Table 13.1 Summary of antenna types

Structure	d_H (μm)	Radiation mechanism	e_r (%)	Gain (dBi)
Type 1	5	Purely plasmonic	65	0.5
Type 2	60	TE_y^{111} DRA and plasmonic	70	4
Type 3	120	TE_y^{112} DRA and plasmonic	70	7

13.4 Conclusion

The nanomaterials' influence has been considered for different types of the antenna: patch antenna, slot antenna, wearable antenna, DRA, etc. Graphene, CNT, and nanowires opened the window for the antenna engineers toward numerous innovations. An apparent enhancement in many antenna parameters has been investigated, such as bandwidth enhancement, gain improvement, and cost reduction. Ease of change in most nanomaterials' surface conductivity by applying different voltage values enabled building a lot of frequency and radiation pattern configurable antennas by very simple structures.

Acknowledgement It was supported by the Russian Science Foundation (Agreement No. 19-72-10071 from 06 Aug. 2019).

References

Abd El-Hameed AS, Salem DA, Abdallah EA, Hashish EA (2013a) Notched quasi self-complementary UWB microstrip antenna. In: 2013 IEEE Antennas Propag. Soc. Int. Symp. https://doi.org/10.1109/aps.2013.6710954

Abd El-Hameed AS, Salem DA, Abdallah EA, Hashish EA (2013b) Ultra Wide Band CPW-fed circularly polarized square slot antenna. In: 2013 IEEE Antennas Propag. Soc. Int. Symp. https://doi.org/10.1109/aps.2013.6710662

Abd El-Hameed AS, Salem DA, Abdallah EA, Hashish EA (2013c) Fractal quasi-self complimentary miniaturized UWB antenna. In: 2013 IEEE Antennas Propag. Soc. Int. Symp. https://doi.org/10.1109/aps.2013.6710668

Abd El-Hameed AS, Barakat A, Abdel-Rahman AB, Allam A, Pokharel RK, Yoshitomi K (2018) Broadband printed-dipole antenna for future 5G applications and wireless communication. In: 2018 IEEE Radio Wirel. Symp. https://doi.org/10.1109/rws.2018.8304959

Abd El-Hameed AS, Wahab MG, Elboushi A, Elpeltagy MS (2019) Miniaturized triple band-notched quasi-self complementary fractal antenna with improved characteristics for UWB applications, AEU. Int J Electron Commun 108:163–171. https://doi.org/10.1016/j.aeue.2019.06.016

Adato R, Yanik AA, Altug H (2011) On chip plasmonic monopole nano-antennas and circuits. Nano Lett 11:5219–5226. https://doi.org/10.1021/nl202528h

Afifi AI, Abdel-Rahman AB, Allam A, El-Hameed ASA (2016) A compact ultra-wideband monopole antenna for breast cancer detection. In: 2016 IEEE 59th Int. Midwest Symp. Circuits Syst. https://doi.org/10.1109/mwscas.2016.7870066

Afifi AI, Abdel-Rahman AB, El-Hameed ASA, Allam A, Ahmed SM (2020) Small frequency ratio multi-band dielectric resonator antenna utilizing vertical metallic strip pairs feeding structure. IEEE Access 8:112840–112845. https://doi.org/10.1109/access.2020.3002789

Akbari M, Khan MWA, Hasani M, Bjorninen T, Sydanheimo L, Ukkonen L (2016) Fabrication and characterization of graphene antenna for low-cost and environmentally friendly RFID tags. IEEE Antennas Wirel Propag Lett 15:1569–1572. https://doi.org/10.1109/lawp.2015.2498944

Aligodarz MT, Rashidian A, Klymyshyn DM, Schulz M, Boerner M, Hanemann T, Meyer P, Mohr J (2013) Polyester-styrene/ceramic nanocomposites for antenna applications. In: 2013 IEEE Antennas Propag. Soc. Int. Symp. https://doi.org/10.1109/aps.2013.6711611

Alrayes NM, Hussein MI (2017) Design and analysis of stacked CNT loaded polyurethane composite DRA antenna. In: 2017 Int. Conf. Electr. Comput. Technol. Appl. https://doi.org/10.1109/icecta.2017.8252046

Aragay G, Pino F, Merkoçi A (2012) Nanomaterials for sensing and destroying pesticides. Chem Rev 112:5317–5338. https://doi.org/10.1021/cr300020c

Aravind G, Raghasudha M, Ravinder D (2015) Electrical transport properties of nano crystalline Li–Ni ferrites. J Mater 1:348–356. https://doi.org/10.1016/j.jmat.2015.09.003

Atakan B, Akan O (2010) Carbon nanotube-based nanoscale ad hoc networks. IEEE Commun Mag 48:129–135. https://doi.org/10.1109/mcom.2010.5473874

Ayodhya D, Veerabhadram G (2019) Fabrication of Schiff base coordinated ZnS nanoparticles for enhanced photocatalytic degradation of chlorpyrifos pesticide and detection of heavy metal ions. J Mater 5:446–454. https://doi.org/10.1016/j.jmat.2019.02.002

Bai J, Zhou B (2014) Titanium dioxide nanomaterials for sensor applications. Chem Rev 114: 10131–10176. https://doi.org/10.1021/cr400625j

Balanis CA (2016) Antenna theory: analysis and design. Wiley

Binas V, Stefanopoulos V, Kiriakidis G, Papagiannakopoulos P (2019) Photocatalytic oxidation of gaseous benzene, toluene and xylene under UV and visible irradiation over Mn-doped TiO2 nanoparticles. J Mater 5:56–65. https://doi.org/10.1016/j.jmat.2018.12.003

Bunea A-C, Neculoiu D, Dragoman M, Konstantinidis G, Deligeorgis G (2015) X band tunable slot antenna with graphene patch. In: 2015 Eur. Microw. Conf. https://doi.org/10.1109/eumc.2015.7345838

Chauhan NS, Bathula S, Vishwakarma A, Bhardwaj R, Johari KK, Gahtori B, Dhar A (2019) Enhanced thermoelectric performance in p-type ZrCoSb based half-Heusler alloys employing nanostructuring and compositional modulation. J Mater 5:94–102. https://doi.org/10.1016/j.jmat.2018.11.003

Chaya Devi KS, Angadi B, Mahesh HM (2017) Multiwalled carbon nanotube-based patch antenna for bandwidth enhancement. Mater Sci Eng B 224:56–60. https://doi.org/10.1016/j.mseb.2017.07.005

Chen X, Selloni A (2014) Introduction: titanium dioxide (TiO2) nanomaterials. Chem Rev 114: 9281–9282. https://doi.org/10.1021/cr500422r

Chen F, Kong L, Song W, Jiang C, Tian S, Yu F, Qin L, Wang C, Zhao X (2019) The electromechanical features of LiNbO3 crystal for potential high temperature piezoelectric applications. J Mater 5:73–80. https://doi.org/10.1016/j.jmat.2018.10.001

Cheng L-Q, Li J-F (2016) A review on one dimensional perovskite nanocrystals for piezoelectric applications. J Mater 2:25–36. https://doi.org/10.1016/j.jmat.2016.02.003

Cheng L, Wang C, Feng L, Yang K, Liu Z (2014) Functional nanomaterials for phototherapies of cancer. Chem Rev 114:10869–10939. https://doi.org/10.1021/cr400532z

Dahl M, Liu Y, Yin Y (2014) Composite titanium dioxide nanomaterials. Chem Rev 114:9853–9889. https://doi.org/10.1021/cr400634p

Dashti M, Carey JD (2018) Graphene microstrip patch ultrawide band antennas for THz communications. Adv Funct Mater 28:1705925. https://doi.org/10.1002/adfm.201705925

Dragoman M, Muller AA, Dragoman D, Coccetti F, Plana R (2010) Terahertz antenna based on graphene. J Appl Phys 107:104313. https://doi.org/10.1063/1.3427536

Ehsan Hosseininejad S, Neshat M, Faraji-Dana R, Abadal S, Lemme MC, Haring Bolivar P, Alarcon E, Cabellos-Aparicio A (2018) Terahertz dielectric resonator antenna coupled to graphene plasmonic dipole. In: 12th Eur. Conf. Antennas Propag. (EuCAP 2018). https://doi.org/10.1049/cp.2018.1041

El-Hameed ASA, Salem DA, Abdallah EA-F, Hashish EA (2013) Quasi self-complementary UWB notched microstrip antenna for USB application. Prog Electromagn Res B 56:185–201. https://doi.org/10.2528/pierb13040807

El-Hameed ASA, Mahmoud N, Barakat A, Abdel-Rahman AB, Allam A, Pokharel RK (2016) A 60-GHz on-chip tapered slot Vivaldi antenna with improved radiation characteristics. In: 2016 10th Eur. Conf Antennas Propag. https://doi.org/10.1109/eucap.2016.7481426

Fan C, Wu B, Hu Y, Zhao Y, Su T (2020) Millimeter-wave pattern reconfigurable Vivaldi antenna using tunable resistor based on graphene. IEEE Trans Antennas Propag 68:4939–4943. https://doi.org/10.1109/tap.2019.2952639

Fattakhova-Rohlfing D, Zaleska A, Bein T (2014) Three-dimensional titanium dioxide nanomaterials. Chem Rev 114:9487–9558. https://doi.org/10.1021/cr500201c

Fu Q, Nie M, Wu Q, Yang J, Li R (2018) Direct-writing on paper of CPW-fed slot antenna with silver nanowire conductive ink. In: 2018 IEEE Int. Conf. Comput. Commun. Eng. Technol. https://doi.org/10.1109/ccet.2018.8542175

Gan Q, Qin N, Zhu Y, Huang Z, Zhang F, Gu S, Xie J, Zhang K, Lu L, Lu Z (2019) Polyvinylpyrrolidone-induced uniform surface-conductive polymer coating endows Ni-rich LiNi0.8Co0.1Mn0.1O2 with enhanced cyclability for lithium-ion batteries. ACS Appl Mater Interfaces 11:12594–12604. https://doi.org/10.1021/acsami.9b04050

Gao L, Wang Y, Hu X, Zhou W, Cao K, Wang Y, Wang W, Lu Y (2019) Cellular carbon-film-based flexible sensor and waterproof supercapacitors. ACS Appl Mater Interfaces 11:26288–26297. https://doi.org/10.1021/acsami.9b09438

Ge P-Z, Jian X-D, Lin X-W, Tang X-G, Zhu Z, Liu Q-X, Jiang Y-P, Zhang T-F, Lu S-G (2019) Composition dependence of giant electrocaloric effect in Pb Sr1-TiO3 ceramics for energy-related applications. J Mater 5:118–126. https://doi.org/10.1016/j.jmat.2018.10.002

Gomez-Diaz JS, Perruisseau-Carrier J (2012) Microwave to THz properties of graphene and potential antenna applications. In: 2012 Int. Symp. Antennas Propag., IEEE, pp 239–242

Goyal R, Vishwakarma DK (2018) Design of a graphene-based patch antenna on glass substrate for high-speed terahertz communications. Microw Opt Technol Lett 60:1594–1600. https://doi.org/10.1002/mop.31216

Green M, Chen X (2019) Recent progress of nanomaterials for microwave absorption. J Mater 5: 503–541. https://doi.org/10.1016/j.jmat.2019.07.003

Hanson GW (2008) Dyadic Green's functions and guided surface waves for a surface conductivity model of graphene. J Appl Phys 103:64302. https://doi.org/10.1063/1.2891452

Hassan A, Ali S, Bae J, Lee CH (2016) All printed antenna based on silver nanoparticles for 1.8 GHz applications. Appl Phys A Mater Sci Process 122. https://doi.org/10.1007/s00339-016-0286-2

Huang Y, Dong X, Liu Y, Li L-J, Chen P (2011) Graphene-based biosensors for detection of bacteria and their metabolic activities. J Mater Chem 21:12358. https://doi.org/10.1039/c1jm11436k

Inui T, Koga H, Nogi M, Komoda N, Suganuma K (2014) A miniaturized flexible antenna printed on a high dielectric constant nanopaper composite. Adv Mater 27:1112–1116. https://doi.org/10.1002/adma.201404555

Jian M, Wang C, Wang Q, Wang H, Xia K, Yin Z, Zhang M, Liang X, Zhang Y (2017) Advanced carbon materials for flexible and wearable sensors. Sci China Mater 60:1026–1062. https://doi.org/10.1007/s40843-017-9077-x

Jing L, Kershaw SV, Li Y, Huang X, Li Y, Rogach AL, Gao M (2016) Aqueous based semiconductor nanocrystals. Chem Rev 116:10623–10730. https://doi.org/10.1021/acs.chemrev.6b00041

Kampylafka V, Kostopoulos A, Modreanu M, Schmidt M, Gagaoudakis E, Tsagaraki K, Kontomitrou V, Konstantinidis G, Deligeorgis G, Kiriakidis G, Aperathitis E (2019) Long-term stability of transparent n/p ZnO homojunctions grown by rf-sputtering at room-temperature. J Mater 5:428–435. https://doi.org/10.1016/j.jmat.2019.02.006

Kanth RK, Liljeberg P, Tenhunen H, Chen Q, Zheng L, Kumar H (2012) Study on glass-epoxy-based low-cost and compact tip-truncated triangular printed antenna. Int J Antennas Propag 2012:1–8. https://doi.org/10.1155/2012/184537

Kim BS, Shin K-Y, Pyo JB, Lee J, Son JG, Lee S-S, Park JH (2016) Reversibly stretchable, optically transparent radio-frequency antennas based on wavy ag nanowire networks. ACS Appl Mater Interfaces 8:2582–2590. https://doi.org/10.1021/acsami.5b10317

Kim BS, Pyo JB, Son JG, Zi G, Lee S-S, Park JH, Lee J (2017) Biaxial stretchability and transparency of ag nanowire 2D mass-spring networks prepared by floating compression. ACS Appl Mater Interfaces 9:10865–10873. https://doi.org/10.1021/acsami.7b00449

Kolavekar SB, Ayachit NH (2019) Synthesis of praseodymium trioxide doped lead-boro-tellurite glasses and their optical and physical properties. J Mater 5:455–462. https://doi.org/10.1016/j.jmat.2019.01.010

Komoda N, Nogi M, Suganuma K, Kohno K, Akiyama Y, Otsuka K (2012) Printed silver nanowire antennas with low signal loss at high-frequency radio. Nanoscale 4:3148. https://doi.org/10.1039/c2nr30485f

Lee H, Shaker G, Naishadham K, Song X, McKinley M, Wagner B, Tentzeris M (2011) Carbon-nanotube loaded antenna-based ammonia gas sensor. IEEE Trans Microw Theory Tech 59: 2665–2673. https://doi.org/10.1109/tmtt.2011.2164093

Lee K, Mazare A, Schmuki P (2014) One-dimensional titanium dioxide nanomaterials: nanotubes. Chem Rev 114:9385–9454. https://doi.org/10.1021/cr500061m

Lee Y, Kim E, Park Y, Kim J, Ryu W, Rho J, Kim K (2018) Photodeposited metal-semiconductor nanocomposites and their applications. J Mater 4:83–94. https://doi.org/10.1016/j.jmat.2018.01.004

Li Q, Danilkin S, Deng G, Li Z, Withers RL, Xu Z, Liu Y (2018) Soft phonon modes and diffuse scattering in Pb(In1/2Nb1/2)O3-Pb(Mg1/3Nb2/3)O3-PbTiO3 relaxor ferroelectrics. J Mater 4: 345–352. https://doi.org/10.1016/j.jmat.2018.06.001

Li J, Zhang X, Duan B, Cui Y, Yang H, Wang H, Li J, Hu X, Chen G, Zhai P (2019a) Pressure induced convergence of conduction bands in Al doped Mg2Si: experiment and theory. J Mater 5:81–87. https://doi.org/10.1016/j.jmat.2018.11.002

Li J-C, Xiao F, Zhong H, Li T, Xu M, Ma L, Cheng M, Liu D, Feng S, Shi Q, Cheng H-M, Liu C, Du D, Beckman SP, Pan X, Lin Y, Shao M (2019b) Secondary-atom-assisted synthesis of single iron atoms anchored on N-doped carbon nanowires for oxygen reduction reaction. ACS Catal 9: 5929–5934. https://doi.org/10.1021/acscatal.9b00869

Liang L, Yang R, Han G, Feng Y, Zhao B, Zhang R, Wang Y, Liu C (2019) Enhanced electromagnetic wave-absorbing performance of magnetic nanoparticles-anchored 2D Ti3C2Tx MXene. ACS Appl Mater Interfaces 12:2644–2654. https://doi.org/10.1021/acsami.9b18504

Liu L, Chen X (2014) Titanium dioxide nanomaterials: self-structural modifications. Chem Rev 114:9890–9918. https://doi.org/10.1021/cr400624r

Liu H, Liu P, Yu D (2017) An adjustable microstrip slot antenna using grapheme stacks. In: 2017 Int. Appl. Comput. Electromagn. Soc. Symp., IEEE, pp 1–2

Liu R, Wu Z, He P, Fan H, Huang Z, Zhang L, Chang X, Liu H, Wang C, Li Y (2019) A self-standing, UV-cured semi-interpenetrating polymer network reinforced composite gel electrolytes for dendrite-suppressing lithium ion batteries. J Mater 5:185–194. https://doi.org/10.1016/j.jmat.2018.12.006

Mannoor MS, Tao H, Clayton JD, Sengupta A, Kaplan DL, Naik RR, Verma N, Omenetto FG, McAlpine MC (2012) Graphene-based wireless bacteria detection on tooth enamel. Nat Commun 3. https://doi.org/10.1038/ncomms1767

Mehta B, Benkstein KD, Semancik S, Zaghloul ME (2016) Gas sensing with bare and graphene-covered optical nano-antenna structures. Sci Rep 6:21287. https://doi.org/10.1038/srep21287

Mohamed AF, Amin Babiker A, Mustafa N (2016) Nanotechnology for 5G. Int J Sci Res 5:1044–1047. https://doi.org/10.21275/v5i2.nov152542

Naishadham K (2014) Design of a graphene loaded slot antenna with 100:1 bandwidth for wireless sensor applications. In: 2014 IEEE Antennas Propag. Soc. Int. Symp. https://doi.org/10.1109/aps.2014.6904775

Pan L, Chortos A, Yu G, Wang Y, Isaacson S, Allen R, Shi Y, Dauskardt R, Bao Z (2014) An ultra-sensitive resistive pressure sensor based on hollow-sphere microstructure induced elasticity in conducting polymer film. Nat Commun 5. https://doi.org/10.1038/ncomms4002

Park S-E, Ryoo R, Ahn W-S, Lee CW (2003) Preface, nanotechnol. mesostructured mater. In: Proc. 3rd Int. Mater. Symp., pp xix–xx. https://doi.org/10.1016/s0167-2991(03)80309-9

Park M, Im J, Shin M, Min Y, Park J, Cho H, Park S, Shim M-B, Jeon S, Chung D-Y, Bae J, Park J, Jeong U, Kim K (2012) Highly stretchable electric circuits from a composite material of silver nanoparticles and elastomeric fibres. Nat Nanotechnol 7:803–809. https://doi.org/10.1038/nnano.2012.206

Qu M, Song J, Yao L, Li S, Deng L, Yang Y (2018) Design of a graphene-based tunable frequency selective surface and its application for variable radiation pattern of a dipole at terahertz. Radio Sci 53:183–189. https://doi.org/10.1002/2017rs006401

Rai T, Dantes P, Bahreyni B, Kim WS (2013) A stretchable RF antenna with silver nanowires. IEEE Electron Device Lett 34:544–546. https://doi.org/10.1109/led.2013.2245626

Sahoo MK, Kale P (2019) Integration of silicon nanowires in solar cell structure for efficiency enhancement: a review. J Mater 5:34–48. https://doi.org/10.1016/j.jmat.2018.11.007

Salim NV, Mateti S, Cizek P, Hameed N, Parameswaranpillai J, Fox B (2019) Large, mesoporous carbon nanoparticles with tunable architectures for energy storage. ACS Appl Nano Mater 2:1727–1736. https://doi.org/10.1021/acsanm.9b00213

Sharma P, Gomez-Diaz JS, Ionescu AM, Perruisseau-Carrier J (2012) Determination of minimum conductivity of graphene from contactless microwaves measurements. In: 2012 12th IEEE Int. Conf Nanotechnol. https://doi.org/10.1109/nano.2012.6322060

Shen Z, Grüner D, Eriksson M, Belova LM, Nan C-W, Yan H (2017) Ordered coalescence of nano-crystals in alkaline niobate ceramics with high remanent polarization. J Mater 3:267–272. https://doi.org/10.1016/j.jmat.2017.08.004

Smith BR, Gambhir SS (2017) Nanomaterials for in vivo imaging. Chem Rev 117:901–986. https://doi.org/10.1021/acs.chemrev.6b00073

Song R, Wang Q, Mao B, Wang Z, Tang D, Zhang B, Zhang J, Liu C, He D, Wu Z, Mu S (2018) Flexible graphite films with high conductivity for radio-frequency antennas. Carbon, NY 130:164–169. https://doi.org/10.1016/j.carbon.2018.01.019

Staaf LGH, Lundgren P, Enoksson P (2014) Present and future supercapacitor carbon electrode materials for improved energy storage used in intelligent wireless sensor systems. Nano Energy 9:128–141. https://doi.org/10.1016/j.nanoen.2014.06.028

Tang D, Wang Q, Wang Z, Liu Q, Zhang B, He D, Wu Z, Mu S (2018) Highly sensitive wearable sensor based on a flexible multi-layer graphene film antenna. Sci Bull 63:574–579. https://doi.org/10.1016/j.scib.2018.03.014

Tian H, Xin F, Wang X, He W, Han W (2015) High capacity group-IV elements (Si, Ge, Sn) based anodes for lithium-ion batteries. J Mater 1:153–169. https://doi.org/10.1016/j.jmat.2015.06.002

Wajahat M, Lee S, Kim JH, Chang WS, Pyo J, Cho SH, Seol SK (2018) Flexible strain sensors fabricated by meniscus-guided printing of carbon nanotube–polymer composites. ACS Appl Mater Interfaces 10:19999–20005. https://doi.org/10.1021/acsami.8b04073

Wang L, Di Q, Sun M, Liu J, Cao C, Liu J, Xu M, Zhang J (2017) Assembly-promoted photocatalysis: three-dimensional assembly of CdS x Se $1-x$ ($x = 0$–1) quantum dots into nanospheres with enhanced photocatalytic performance. J Mater 3:63–70. https://doi.org/10.1016/j.jmat.2016.11.008

Wang J, Hu J, Yang L, Zhu K, Li B-W, Sun Q, Li Y, Qiu J (2018) High discharged energy density of polymer nanocomposites induced by Nd-doped BaTiO3 nanoparticles. J Mater 4:44–50. https://doi.org/10.1016/j.jmat.2018.01.001

Wu S, Lin Y, Xing L, Sun G, Zhou H, Xu K, Fan W, Yu L, Li W (2019) Stabilizing LiCoO2/graphite at high voltages with an electrolyte additive. ACS Appl Mater Interfaces 11:17940–17951. https://doi.org/10.1021/acsami.9b01053

Xiao E-C, Li J, Wang J, Xing C, Guo M, Qiao H, Wang Q, Qi Z-M, Dou G, Shi F (2018) Phonon characteristics and dielectric properties of BaMoO4 ceramic. J Mater 4:383–389. https://doi.org/10.1016/j.jmat.2018.08.004

Xu Y-H, Zhao S-X, Deng Y-F, Deng H, Nan C-W (2016) Improved electrochemical performance of 5 V spinel LiNi0.5Mn1.5O4 microspheres by F-doping and Li4SiO4 coating. J Mater 2:265–272. https://doi.org/10.1016/j.jmat.2016.04.005

Yang X, Yang M, Pang B, Vara M, Xia Y (2015) Gold nanomaterials at work in biomedicine. Chem Rev 115:10410–10488. https://doi.org/10.1021/acs.chemrev.5b00193

Yao S, Zhu Y (2015) Nanomaterial-enabled stretchable conductors: strategies, materials and devices. Adv Mater 27:1480–1511. https://doi.org/10.1002/adma.201404446

Yao Y, Shankar R, Rauter P, Song Y, Kong J, Loncar M, Capasso F (2014) High-responsivity mid-infrared graphene detectors with antenna-enhanced photocarrier generation and collection. Nano Lett 14:3749–3754. https://doi.org/10.1021/nl500602n

Zhang X, Huang X, Leng T, Hu Z, Auton G, Hill E, Ouslimani H, Abdalla MA (2015) Graphene reconfigurable coplanar waveguide (CPW)-fed circular slot antenna. In: 2015 IEEE Int. Symp. Antennas Propag. Usn. Natl. Radio Sci. Meet. https://doi.org/10.1109/aps.2015.7305535

Zhang B, Xia W, Zhang J, He D, Liu C, Wu Z (2017) Rectangular dielectric resonator antennas with a thin graphene film. In: 2017 Int. Symp. Antennas Propag. https://doi.org/10.1109/isanp.2017.8228789

Zhou Y, Bayram Y, Du F, Dai L, Volakis JL (2010) Polymer-carbon nanotube sheets for conformal load bearing antennas. IEEE Trans Antennas Propag 58:2169–2175. https://doi.org/10.1109/tap.2010.2048852

Advancement in Crops and Agriculture by Nanomaterials

14

Sandeep Kaur, Kajal Sharma, Rajat Singh, and Naveen Kumar

Abstract

Agriculture is an important sector that provides immense opportunities for development and livelihood for over half of the world's population. Globally, India is the second leading country in the production of agricultural commodities. The agricultural sector is confronted with huge issues such as rapid climatic change, a decline in soil fertility, nutrient deficiency, excessive use of chemicals and pesticides, and the presence of toxic metals in soil. However, the world population growth has subsequently increased the food demand. Nanomaterials have gotten a lot of attention in recent decades due to their multiple applications in industries like health, chemistry, energy, and textiles. Nanomaterials have recently been explored as an alternative approach to control plant pests, provide nutrients to soil, and help in the protection of the environment. Several nanosensors have been used for the detection and monitoring of plant illnesses, pesticide residues, pH, and soil fertility. Therefore, in this chapter, we highlight the role of nanomaterials in disease management, crop protection, and the development of sustainable agricultural practices.

S. Kaur
Department of Life Sciences, RIMT University, Mandi Gobindgarh, Punjab, India

K. Sharma
Chitkara School of Health Sciences, Chitkara University, Rajpura, Punjab, India

R. Singh
Department of Food Technology, Eternal University, Sirmour (HP), India

N. Kumar (✉)
Amity Institute of Biotechnology, Amity University Rajasthan, Jaipur, India
e-mail: nkumar2@jpr.amity.edu

A. Thakur et al. (eds.), *Synthesis and Applications of Nanoparticles*,
https://doi.org/10.1007/978-981-16-6819-7_14

Keywords

Nanomaterial · Agriculture · Soil · Pesticide · Crop protection

14.1 Introduction

Agriculture is the major sector of Indian economy as it shares the 17% of total GDP and offers employment to 60% of its population (Arjun 2013). Besides, agriculture industry also addresses the food security for the country (Deshpande 2017). In India, the West Bengal, Uttar Pradesh, Punjab, Gujarat, Haryana, Madhya Pradesh, Assam, Andhra Pradesh, Karnataka, and Chhattisgarh are the top ten leading states in agriculture sector (https://www.tractorjunction.com/blog/top-10-agriculture-states-in-india/). India is the world's second largest producer of rice, wheat, sugarcane, cotton, tea, vegetables, fruits, and cultured fish (Kumar 2019). Moreover, acceptance of modern farming techniques, high-yielding variety of seeds, use of fertilizers, increasing irrigation facility, and ease of use of electricity have brought the incredible advancement in agriculture sector of India (Prajapati and Datta 2014). Not only the modern technology but also the geographical position and physical factors like climate and soil are favorable for the cultivation of many crops (Rafie and Kumar 2020). According to Indian Agriculture and Allied Industries report (May 2021), the production of food grain was 296.65 million tonnes during the year 2019–2020 which is a recorded value (www.ibef.org/industry/agriculture-india.apex). New techniques and systems like genetically modified crops, organic farming, etc. are now become popular as they help in mitigating the agricultural challenges which is beneficial for farming (Alam et al. 2014). Nonetheless, better growth and stability of agriculture lead to the advancement of agro-based industries which further helps to raise the economy of the nation. However, agro-processing sector in India is relatively small as compared to the primary agriculture sector. So, export of agro-based stuff should be promoted to export in foreseeable future (Paramasivan and Pasupathi 2016).

Among the modern techniques, nanotechnology attracts the agriculturists from all over the world due to its tremendous benefits in agriculture sector. Nanotechnology is a Greek word where nano stands for dwarf. The term nanotechnology was coined by Norio Taniguchi in 1947 (EPA 2007). Nanotechnology has been defined as relating to material, system, and process which operate at a level of 100 nm or less, and the convergence of nanotechnology with biology at the nanoscale is called nanobiotechnology. Different nanostructures do not only help the plant to easily absorb the nutrients but also help to combat viruses and other pathogens by releasing their content at targeted site (Abd-Elrahman and Mostafa 2015). Nanoparticles of essential minerals and non-essential elements have an impact on plant growth, physiology, and development depending on their size, composition, concentration, and mode of application (Achari and Kowshik 2018). Nanotechnology has been used for developing nanofertilizers, herbicides, fungicides, pesticides, and nanosensors in agriculture sector. It can be helpful in enhancing the quality and yield of crop and alleviate chemical pollution and prevent the crops against environmental stress (Iavicoli et al. 2017; Liu and Lal 2015). Nanomaterials possess miraculous properties. For instance, surface area, cation exchange capacity, ion adsorption, complexation, and other functions of clay would improve if they were processed at nanoscale. Additionally, ago-nanotechnology is a blessing for poor countries as it effectively tackles the issues related to water scarcity and poor sanitary conditions. This is the reason that nanotechnology has become the future of any nation (Mukhopadhyay 2014). It is widely used in contemporary fields of agriculture, processing and prevention of food, packaging as well as transportation, and quality control of agricultural goods (Pandey 2018). It is to say that nanotechnology brings the revolution and transformation in food supply system in global scope (Mir et al. 2015). According to current statistics, approximately 90% of nano-based patents and products have been obtained from just seven countries: China, Germany, France, Japan, Switzerland, South Korea, and the United States, while India's investments and growth remain unsatisfactory (Elizabath et al. 2019).

14.2 Different Types of Nanoparticles and Their Use in Agriculture

Nanoparticles are widely used in increased production of agriculture. Figure 14.1 depicts a schematic illustration of nanoparticle's possible applications in sustainable agriculture production. Silver nanoparticles are widely studied for their use in bio-systems due to their strong inhibitory and antimicrobial effects. AgNPs play a vital role in the plant's development from growth to pest control, thus preventing them from the destruction caused by pathogenic organisms (Kale et al. 2021). Silver is most commonly used nanomaterial in the consumer products which are included in the revised consumer products inventary (CPI) released in October 2013 (Vance et al. 2015). Similarly, the nanomaterials based on carbon (carbon nanotubes,

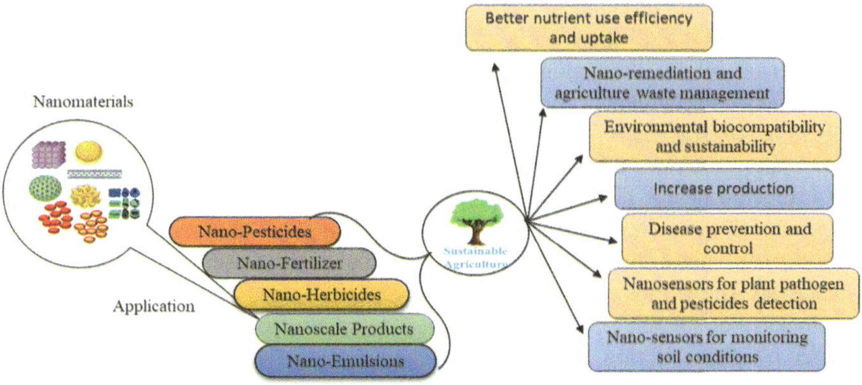

Fig. 14.1 Schematic representation of potential nanomaterials applications in sustainable agriculture production

fullerenes, and graphene) have been widely used in the scientific and industrial areas due to their wonder properties. These nanomaterials cannot be obtained from natural resources, so the basic component for their production is carbon vapor (Zaytseva and Neumann 2016). They are incorporated into polymers due to their superior mechanical properties and other materials to produce advanced structural and composite materials (Ahmad and Pan 2015). Carbon nanotubes can be used as a vehicle to deliver desired molecules into seeds during germination, such as nutrients or biocides (Qureshi et al. 2018). Additionally, other nanoparticles like titanium, zinc, iron, etc. possess numerous properties including large surface, stability, antimicrobial activity, and many more aspects which are discussed along with their functioning in the upcoming paragraphs.

14.3 Nanopesticides for the Enhancement of Crop Productivity

The compounds which control pests and pathogens are called pesticides. They have a wide range of chemicals like herbicides, insecticides, fungicides, bactericides, and antimicrobials to reduce the rate of environmental stress agents (Mali et al. 2020). However, they are not target specific in action, and they cause water, air, and soil pollution (Vivekanandhan and Duraisamy 2012; Mahmood et al. 2016). Moreover, the overuse of the same pesticides leads to pesticide resistance and the development of new pesticides which are more powerful and hazardous not only to pests but also to human and environmental health. Nanotechnology is important in the development of eco-friendly nanopesticides that aid in the development of disease-free crops in fields (Rastogi et al. 2019; Mali et al. 2020). Different forms of nanopesticides are described as below:

14.3.1 Nano-insecticides

Nano-insecticides have many advantages over chemical insecticides. AgNPs are widely studied due to their use as nanopesticides. Several kinds of nanoparticles such as nano-silica, silver, aluminum oxide, zinc oxide, titanium dioxide, and nano-emulsions can be efficiently used as nanocides in the management of rice weevil, *Sitophilus oryzae*; lesser grain borer, *Rhyzopertha dominica*; red flour beetle, *Tribolium castaneum*; tobacco caterpillar, *Spodoptera litura*; oleander aphid, *Aphis nerii*; bruchid beetle, *Callosobruchus maculatus*; and diamondback moth, *Plutella xylostella* (Kannan et al. 2020). They can be applied directly in the field as nanocarriers, such as mesoporous silica nanoparticles, and can effectively kill insects and larvae. The killing mechanisms involve the dysfunctioning of the digestive tract or blockage of spiracles and tracheas, or it may involve the breakage of lipid water barrier by physisorption of nano-silica, thus causing the death of the organism (Rastogi et al. 2019). Silver nanoparticles are also used to control the activity and viability of the land snail *Eobania vermiculata* which is a major pest in agriculture (www.nanoshel.com/silver-nanoparticles-agriculture). Moreover, mosquito larvae can be controlled by UV-irradiated silver nanoparticles effectively (Sap-lam et al. 2010).

14.3.2 Nanoherbicides

Nanoherbicides are safer and provide effective delivery as compared to conventional herbicides. Nanoherbicides include polymeric and metallic nanoparticles which allow the controlled release of agrochemicals and lower herbicide consumption rate which contributes to increasing crop productivity (Chidambaram and Abigail 2017). Nanoherbicide having features like effective concentration, time-controlled release in response to particular stimuli, better activity at the target site, and less toxicity for bionetwork can be a good substitute for chemicals. Consequently, they check the occurrence of troubles plus challenges arising from the use of herbicides in weed management. Moreover, nanoherbicides are designed to attack the seed coat of the weed and their germination (Dhillon and Mukhopadhyay 2015). Nanocapsule is a perfect way for slow-release herbicides to achieve weed control for a long season in an environment-friendly manner, without leaving any toxic material in soil and environment (Pérez de Luque and Rubiales 2009). Developing a target-specific herbicide molecule encapsulated with nanoparticles is intended for the receptor in the roots of target weeds, where it penetrates the root system and translocates to parts that inhibit glycolysis of food stores in the root system, causing food starvation and weed death (Chinnamuthu and Kokiladevi 2007). SiNPs are used as potential herbicides. Many herbicides, namely, chloroacetanilide, anilide, and benzimidazole, have been incorporated in a diatom fistula and in their active state and delivered to the field (Lodriche et al. 2012). In addition to this, alginate/chitosan nanoparticles can be used as herbicide carrier material especially for herbicide such as paraquat (Singh et al. 2017). Nanoparticles are also used in detoxification of herbicide residue such as application of silver tailored with nanoparticles of magnetite stabilized with carboxymethyl cellulose nanoparticles which recorded 88% degradation of herbicide atrazine residue under controlled ecological conditions (Yadav and Srivastava 2015).

14.3.3 Nano-emulsions

Emulsions are defined as the dispersion of two immiscible liquids. Nano-emulsion is made in a viscous liquid via the dispersion of polymer, droplet, or other solid material and is referred to as dispersed or discontinuous phase. Physical properties of these liquids, such as viscosity, phase behavior, and density, are influenced by oil phase components (Szöllösi et al. 2020). The use of nano-emulsion systems is becoming increasingly popular due to their practical properties such as increased physical stability, solubility, biodegradability, bioavailability, and environmental friendliness (Mou et al. 2008). Adamu et al. (2021) concluded that ginger essential oil-loaded nano-emulsions can control *Xanthomonas oryzae*-caused bacterial leaf blight in rice. This formulation is far better than synthetic antibiotics in suppressing *Xoo (Xanthomonas oryzae* pv. *oryzae)* growth, and it was also reported to increase the rice yield under a glasshouse trial. Equally, green tea essential oil nano-emulsion could be used as a potential candidate for the fungal disease of rice blast caused by *Magnaporthe oryzae*. Microscopic and molecular analysis revealed that the nano-emulsion destroyed the cell membrane significantly (Perumal et al. 2021). Additionally, nano-emulsions are also used to control pests due to their strong adhesion to surfaces, high penetration power, and the broad range of applicability (Feng et al. 2018). For example, when a nano-emulsion formulation of norcantharidin, a poor water-soluble pesticide, was used, it demonstrated effective insecticidal activity against *Plutella xylostella* (Zeng et al. 2019). Apart from this, thymol nano-emulsions have been reported to control *Xanthomonas axonopodis* which is responsible for bacterial pustule in soybean, and it significantly affects the plant growth and thus could be used as a plant growth-promoting agent for agriculture (Kumari et al. 2018). Furthermore, nano-emulsion-based pesticide delivery methods can be effective, safe, and inexpensive, with superior physicochemical constancy, functional properties such as spreading and penetration, and enhanced bioavailability (Feng et al. 2018).

14.4 Nanofertilizers

Nanofertilizers are important in agriculture because they reduce the use of harsh chemicals, improve soil quality, and prevent environmental pollution such as eutrophication of water bodies. They also reduce the cost of manufacturing and subsequently lead to agricultural sustainability (Bratovcic et al. 2021). Moreover, they are used in less quantity as compared to chemical fertilizers but results in better plant yield and food which is safe for health. In comparison to chemical fertilizers, nanofertilizers increase crop production by up to 30% (Kah et al. 2018). Based on the type of formulation, nanofertilizers are divided into three classes: (a) nanoscale fertilizer, which corresponds to the conventional fertilizer reduced in size typically in the form of nanoparticles; (b) nanoscale additive fertilizer, which is a traditional fertilizer containing a supplement nanomaterial; and (c) nanoscale coating fertilizer which refers to nutrients encapsulated by nanofilms or intercalated into nanoscale pores of a host material (Mastronardi et al. 2015). Discerning the different nanoparticles, silver possesses unique properties. It improved the uptake of nutrients from the soil than the bulk one. As a fertilizer, it boosts seed germination as well as prevents nutrient release by providing their slow and effective release (Anand and Bhagat 2019). The slow release of nanofertilizer is due to the strong holding of surface-coated nanomaterials on fertilizer particles (Banotra et al. 2017). Nevertheless, nano-iron and copper fertilizers helped to improve nutrient deficiencies of economically important plants. Due to their superior nutrient control, nanofertilizers can become a better alternative for sustainable and precise agriculture (Heinisch et al. 2019). Furthermore, foliar application of combination/individual manganese and iron nano oxides significantly increased squash fruit yield, while Fe nano oxide increased the content of organic matter, proteins, lipids, and energy levels in squash plant fruits (Shebl et al. 2019). Apart from this, nanofertilizers improve soybean seed germination and make the fruits more succulent and nutritious (Al-Juthery et al. 2021).

Almost every crop needs the nutrient for increasing soil fertility and healthy growth. Nitrogen, phosphorus, and potassium are the main macronutrients required to obtain higher agricultural yields. However, it is estimated that 40–70% of nitrogen, 80–90% of phosphorus, and 50–90% of potassium will be lost in the environment, preventing these fertilizers from reaching the plant (Ombodi and Saigusa 2000). The application of nanofertilizer increased the nutrient use efficiency (NUE) of several plants, reduced nutrient leaching into groundwater, and increased abiotic stress tolerance in certain crops (Zulfiqar et al. 2019). For example, when compared to traditional fertilizers, phosphatic nanofertilizer increased soybean growth rate and seed yield by 32 and 20%, respectively (Liu and Lal 2014), and the foliar application of nano-phosphorus enhanced the crop yield under arid conditions (Tarafdar et al. 2012). Similarly, SiO_2 and TiO_2 nanoparticles enhanced the action of nitrate reductase in soybeans and increased plant absorption capacity, allowing for more efficient water and fertilizer use (Lu et al. 2002). Besides, ZnNPs and their nanocapsule formulation provide a resourceful way to distribute pesticide and fertilizer with high site specificity and thus collateral damage (Nair and Durga 2021). Moreover, the foliar application of various nanofertilizers significantly improves the growth, productivity, biochemical content of yielded seeds, and anti-oxidant system of plants (Abd El-Azeim et al. 2020). Furthermore, nanofertilizer helps plants to resist diseases and provides superior strength through anti-bending and deeper rooting of the crops (Singh et al. 2017).

14.5 Nano-formulations in Bacterial and Fungal Plant Disease Management

Plant diseases are mainly caused by bacterial species and viruses. Nanotechnology gives rise to many approaches through which plant infections can be controlled. It offers eco-friendly choices for plant disease management. Undoubtedly, a lot of literature is present which describes the use of nanomaterial in disease control and management. The generation of reactive oxygen species (ROS), membrane damage, loss of enzyme activity, protein dysfunction, and a variety of other factors all contribute to nanoparticle antimicrobial activity (Lemire et al. 2013; Swarnkar et al. 2016). In this context, nanoparticles are used alone or in combination with some other compounds to combat plant diseases. Recently, a new nano-fungicide based on the halogen-substituted azomethines was reported to be the sheath blight disease caused by *Rhizoctonia solani* in rice (Siddhartha et al. 2020). Similarly, green AgNP biosynthesis demonstrated strong antifungal activity against *Bipolaris sorokiniana* and effectively controlled its infection in wheat plants (Mishra et al. 2014). Silver nanoparticles are also observed to control soft rot disease caused by *Pectobacterium carotovorum* in sugar beet (Ghazy et al. 2021). Mishra et al. (2017)

revealed that the biosynthesis of AgNPs using *Stenotrophomonas* sp. shows antifungal activity against foliar and soil-borne phytopathogens. It considerably inhibited the growth of conidia and sclerotia. Moreover, this biosynthesis successfully managed the chickpea collar rot disease caused by *S. rolfsii*. Copper-chitosan nanoparticles also possess maximum inhibition rate of spore germination of *A. alternata* (Abd-Elsalam and Alghuthaymi 2015). What's more, the green synthesis of ZnONPs using *Eucalyptus globulus* showed remarkable fungicidal activity against phytopathogenic fungi of apple orchards, namely, *Alternaria mali*, *Botryosphaeria dothidea*, and *Diplodia seriata*. Thus, it shows that fungal pest management can be done by using ZnNPs as they guarantee protection in fruit crops (Ahmad et al. 2020). Likewise, zinc-based nanoparticles were observed to inhibit the spore germination of *Peronospora tabacina*, *Botrytis cinerea*, and *Penicillium expansum* (Wagner et al. 2016; He et al. 2011). Nano-copper alone or in combination with other molecules is used to manage or control many bacterial and fungal causative agents which cause serious plant infections such as *Phytophthora infestans*, *Xanthomonas axonopodis* pv. *punicae*, *Fusarium graminearum*, *F. culmorum*, *F. oxysporum*, and *Fusarium equiseti* (Rai et al. 2018). It has been found that ZnONPs were more lethal to fungal species than $ZnSO_4$, while CuNPs were more fungitoxic to $CuSO_4$ excluding *B. cinerea*, *A. alternata*, and *M. fructicola* (Anastasios et al. 2019). Jo et al. (2009) demonstrated that silver ions and nanoparticles significantly inhibit the viability of spores and disease progress caused by plant pathogenic fungi, namely, *Bipolaris sorokiniana* and *Magnaporthe grisea*. The coffee fungi, namely, *Mycena citricolor* and *Colletotrichum* sp., have been observed for showing inhibition on applying ZnO-based nanoparticles, but the nanofungicide activity depends on the physicochemical characteristics of the cell wall and the evolutionary adaptation of fungus (Arciniegas-Grijalba et al. 2019). During an in vitro study, the nanoform of a commercial fungicide trifloxystrobin 25% + tebuconazole 50% showed antifungal activity against *Macrophomina phaseolina* where the nanoform causes the abnormality of hyphae and sclerotia along with hyphal lysis (Dileep Kumar et al. 2016). Tang et al. (2021) also developed carrier-free self-assembled nanoparticles based on fenhexamid and polyhexamethylene biguanide for long-term plant disease control. This combination possesses antimicrobial activity against *P. syringae* pv. *lachrymans*, *Botrytis cinerea*, and *Sclerotinia sclerotiorum*.

14.6 Nanotechnology in Seed Germination and Growth

Seed germination is the important step for the development of plant. Different nanoparticles are used in seed germination in different sizes and concentration. Some nanoparticles are used as nanocarriers while some are applied directly. Nanotechnology helps in the early germination of seed; thus, it contributes to end up the dormant state of seeds. The nanoparticles also reduce the chance of attack of any microbe or fungal disease to seeds and protect them from the environmental stress or unfavorable conditions. However, not all nanoparticles have impact on seed germination as suggested by recent studies. Kim et al. (2020) demonstrated that silver-graphene oxide improved the growth of roots and shoots in radish while inhibiting cucumber shoot and alfalfa root growth. Surprisingly, these Ag-GO nanostructures notably affect the early development of plants in species-specific manner. Besides, Ag nanoparticles are also found to have positive impact as compared to gold nanoparticles on leaf growth of seedlings of *Mimusops laurifolia* which is a slow-growing tree (Alshehddi and Bokhari 2020). Similarly, BSM (bovine submaxillary mucin)-AgNPs possess strong antimicrobial effect. This nano-complex protects the transmission of *Acidovorax citrulli* from melon seeds to seedlings which is a major pathogen of cucurbit production (Makarovsky et al. 2018). Apart from this, silver nanoparticles have been effective in enhancing salinity tolerance in the *Satureja hortensis* seedlings and thus help the plant to grow against salt toxicity (Nejatzadeh 2021). Likewise, it was concluded by Ivani et al. (2018) that nanosized SiO_2 improves the seedling growth attributes of fenugreek and alleviates the adverse effects of salt stress on shoot, root, and seedling length more efficiently as compared to bulk SiO_2. Furthermore, another essential nanoparticle in agriculture is cerium oxide. Lizzi et al. (2020) reported that cerium oxide nanoparticles ($nCeO_2$) did not show any negative effect on the development of seedlings of different plants, namely, *Holcus lanatus*, *Lychnis flos-cuculi*, and *Diplotaxis tenuifolia*, even at higher concentrations, and different sizes of $nCeO_2$ do not influence the germination and root development of these plants so powerfully. In comparison to bulk titanium oxide, appropriate concentration of nano-TiO_2 has been found to quicken the germination of aged seeds of spinach and wheat. It also has a positive effect on the dry weight of the plant, chlorophyll formation in spinach (Zheng et al. 2005), and shoot and seedling length in wheat (Feizi et al. 2012). Not only CeO_2 and TiO_2 nanoparticles but also the applications of $nSiO_2$ to tomato seed possess dramatically properties as it enhances the percent seed germination, mean germination time, seed vigor, and seed germination index as well as the fresh and dry weight of seedling (Siddiqui and Al-Whaibi 2014). Nonetheless, MgO nanoparticles also enhanced the seed germination and growth rate mechanism of peanut seeds due to their ability to penetrate seed coats and support the water uptake ability of seeds (Jhansi et al. 2017). Additionally, carbon nanotubes (CNTs) and graphene also show positive results when exposed to the seeds of fiber-producing crops (cotton, *Gossypium hirsutum*) and ornamental species (vinca, *Catharanthus roseus*). Also, the flower production of *Catharanthus* plants cultivated in CNTs and graphene-supplemented soil increases by 37 and 58%, respectively, and carbon-based nanomaterials increase the plant survival without any symptoms of leaf wilting (Pandey et al. 2019). On the contrary, sometimes nanoparticles such as AgNPs and nanoscale silica platelets reduced the seed germination, root elongation, and seedling growth of some plants (Chang et al. 2020).

14.7 Nanoparticle-Based Sensors/Nanosensors

Nano-phytopathology is a cutting-edge field that employs nanotechnology to detect, diagnose, and control plant disease and its pathogens early on, hence preventing epidemic disease in the crop. Early and efficient diagnosis of the disease is essential for proper disease control and management to lessen crop loss (Hussain 2017). According to the study conducted during the year 2015 by the Associated Chambers of Commerce and Industry of India, the annual crop loss of $500 billion was reported in the country due to pests and diseases (https://croplife.org/news/keeping-indias-pests-in-line/). Visual detection of morphological changes, serological and electron microscope, and polymerase chain reaction-based examinations of diseased plants are some traditional methods for identification of disease (Shang et al. 2011; Lobert et al. 1987; Hongyun et al. 2008), but these methods have limited characters. Surprisingly, the emergence of nanomaterials with various miraculous properties like optical, high electrical conductivity, better shock bearing, and ultrasensitive response mechanism results in the development of nanobiosensor that makes use of nano and biomolecule assemblies to cater to challenges of routine methods of diagnosis (Kumar and Arora 2020). Moreover, the synthesis of nanomaterial by utilizing functionalized metal nanoparticles as a sensing constituent presents numerous advantageous features necessary for pathogen detection (Kahyap et al. 2016).

The biosensor offers the early detection of plant disease with high sensitivity and specificity (Khater et al. 2017). Many nanoparticles are used in the development of these biosensors. To explain, silver nanorods are used for the detection of a variety of plant pathogens. These nanoparticles can identify toxins using surface-enhanced Raman spectroscopy (Li et al. 2020). Similarly, Pt nanosensors are used to sense bacterial disease in soil and vegetables (Ahmad et al. 2012). Fluorescent Si nanoparticles conjugated with a secondary antibody also help to detect plant pathogen such as *Xanthomonas axonopodis* pv. *vesicatoria* which is a causative agent of bacterial spot disease in the Solanaceae plant (Yao et al. 2009). Further, Au nanoparticles conjugated with a specific single-stranded DNA detect *Ralstonia solanacearum* which is responsible for bacterial wilt disease in potatoes (Khaledian et al. 2017). The unamplified genome of tomato yellow leaf curl virus (TYLCV) in infected plants can be detected by localized surface plasmon resonance (LSPR) of AuNPs (Razmi et al. 2019). Besides, Lau et al. (2017) described the gold nanoparticle-based electrochemical biosensor for quick and sensitive detection of plant pathogen DNA on disposable screen-printed carbon electrodes. They also suggested that this sensor was 10,000 times more sensitive as compared to the conventional PCR method. Plant infection caused by *P. syringae* can be detected by this nanosensor even before the appearance of symptoms of the disease. Nonetheless, plant signaling pathways and metabolism can be monitored with nano-engineered nanomaterial-based nanosensors that are non-destructive, are minimally invasive, and can enable real-time analysis of biotic and abiotic risks for improved plant health. These sensors can measure chemical flux even at the single-molecule level. Therefore, plant health could be monitored through nutrient management, disease management, plant hormone level, environmental pollution, etc. (Hasan Dad

Ansari et al. 2020). Likewise, many kinds of biosensors are developed using nanoparticles for the detection of pesticides. For instance, Bao et al. (2015) developed a plant esterase-chitosan/gold nanoparticles-graphene nanosheet-based biosensor for the detection of organophosphate pesticides in agro-foods which cause neurotoxicity in humans. Similarly, carbon nanotube sensors were productive utilized as stress indicators for in vivo monitoring of ROS formation in plant tissues (Ren et al. 2013). The large surface area of C-based nanomaterials provides enough space for interactions with the sensed molecules. Certain nanoparticles have this feature, which makes them a better matrix for immobilizing desirable biomolecules for signal amplification in biosensors for the detection of mycotoxins (Thipe et al. 2018). Zhang et al. (2019) also developed a highly sensitive SERS (surface enhanced Raman scattering) for the detection of carbaryl pesticides for various fruit surfaces, using in situ grown silver nanoparticles on non-woven fabric surfaces based on mussel-inspired polydopamine molecules. Moreover, deltamethrin and carbofuran residues in soil could be easily detected by SERS with the flexible AuNPs colloids (He et al. 2019; Singh et al. 2020a, b). Chemiluminescent sensor array based on the triple channel properties of the luminal-functionalized silver nanoparticles has been developed for the discrimination of organophosphate and carbamate pesticides (He et al. 2015; Singh et al. 2020a, b). Equally, a preliminary tool for the label-free colorimetric chlorpyrifos pesticide sensing in water and agriculture products is developed using gold nanoparticles-silk fibroin dispersion. The dispersion offers a swift and first-rate pesticide sensing response even at 10 ppb concentration (Mane et al. 2020). Apart from this, nanobiotechnology has the potential to enable smart plant sensors that communicate with and actuate electronic devices for improving plant productivity, optimize and automate water and agrochemical application, and enable highly efficient plant chemical phenotype. Nanomaterials enable the conversion of chemical signals from plants into digital data that can be monitored by standoff electronic devices (Giraldo et al. 2019). Levar (2007) designed and fabricated a plant monitoring system that connected a variety of sensors to a data acquisition system and used Bluetooth to send data from the data acquisition system to a remote computer monitoring station. The system was observed to perform different functions such as monitoring and recording of temperature, humidity, soil moisture, and ripeness of fruits and vegetables.

14.8 Conclusion

The present study highlights the role of nanomaterials in agriculture. Majority of the world's population relies on agriculture for their livelihood. The better growth and stability of the agriculture sector leads to the advancement of agro-based industries which further helps to raise the economy of the nation. The agricultural sector is facing numerous challenges which include climatic changes, a soil fertility reduction, macro and micronutrient deficiencies, excessive use of agrochemicals, and the presence of heavy metals in the soil. But, with the advent of nanotechnology, these challenges are significantly addressed using nanomaterial in agriculture. The use of nanopesticides and fertilizers provides better outcomes and results in environmental protection because of the exceptional properties of nanomaterials; a series of novel nanosensors have been developed which aid in the monitoring of pesticide residues as well as other soil characteristics like pH and moisture.

References

Abd El-Azeim MM, Sherif MA, Hussien MS, Tantawy IAA, Bashandy SO (2020) Impacts of nano- and non-nanofertilizers on potato quality and productivity. Acta Ecol Sin 40:388–397

Abd-Elrahman SH, Mostafa MAM (2015) Applications of nanotechnology in agriculture: an overview. Egypt J Soil Sci 55(2):197–214

Abd-Elsalam KA, Alghuthaymi MA (2015) Nanobiofungicides: is it the next-generation of fungicides? J Nanotechnol Mater Sci. https://doi.org/10.15436/2377-1372.15.0

Achari GA, Kowshik M (2018) Recent developments on nanotechnology in agriculture: plant mineral, nutrition, health and interactions with soil microflora. J Agric Food Chem 66:8647–8661

Adamu A, Ahmad K, Siddiqui Y, Ismail IS, Asib N, Kutawa AB, Adzmi F, Ismail MR, Berahim Z (2021) Ginger essential oils-loaded nanoemulsions: potential strategy to manage bacterial leaf blight disease and enhanced rice yield. Molecules 26(13):3902

Ahmad K, Pan W (2015) Microstructure-toughening relation in alumina based multiwall carbon nanotube ceramic composites. J Eur Ceram Soc 35:663–671

Ahmad F, Siddiqui MA, Babalola OO, Wu H-F (2012) Biofunctionalization of nanoparticles assisted mass spectrometry as biosensors for rapid detection of plant associated bacteria. Biosens Bioelectron 35(1):235–242

Ahmad H, Venugopal K, Rajagopal K, De Britto S, Nandini B, Pushpalatha HG, Konappa N, Udayashankar AC, Geetha N, Jogaiah S (2020) Green synthesis and characterization of zinc oxide nanoparticles using Eucalyptus globules and their fungicidal ability against pathogenic fungi of apple orchards. Biomolecules 10(3):425

Alam A, Rizvi AH, Verma K, Gautam C (2014) The changing scenario in India Agriculture: a review. Int J Sci Res Agric Sci 1(7):118–127

Al-Juthery HWA et al (2021) Intelligent, nano-fertilizers: a new technology for improvement nutrient use efficiency (article review). IOP Conf Ser Earth Environ Sci 735:012086

Alshehddi LAA, Bokhari N (2020) Influence of gold and silver nanoparticles on the germination and growth of *Mimusops laurifolia* seeds in the south-western regions in Saudi Arabia. Saudi J Biol Sci 27(1):574–580

Anand R, Bhagat M (2019) Silver nanoparticles (AgNPs): as nanopesticides and nanofertilizers. MOJ Biol Med 4(1):19–20. https://doi.org/10.15406/mojbm.2019.04.00107

Anastasios M, Nektarios K, Constantinos C (2019) Nano-fungicides against plant pathogens: copper, silver and zinc NPs. Geophys Res Abstr 21:1

Arciniegas-Grijalba PA, Patiño-Portela MC, Mosquera-Sánchez LP, Guerra Sierra BE, Muñoz-Florez JE, Erazo-Castillo LA, Rodríguez-Páez JE (2019) ZnO-based nanofungicides: synthesis, characterization and their effect on the coffee fungi *Mycena citricolor* and *Colletotrichum* sp. Mater Sci Eng C 98:808–825

Arjun KM (2013) Indian agriculture—status, importance, and role in Indian economy. Int J Agric Food Sci Technol 4(4):343–346

Banotra M, Kumar A, Sharma BC, Nandan B, Verma A, Kumar R, Gupta V, Bhagat S (2017) Prospectus of use of nanotechnology in agriculture—a review. Int J Curr Microbiol App Sci 6(12):1541–1551

Bao J, Hou C, Chen M, Li J, Huo D, Yang M, Luo X, Lei Y (2015) Plant esterase—chitosan/gold nanoparticles—graphene nanosheet composite-based biosensor for the detection of organophosphate pesticides. J Agric Food Chem 63(47):10319–10326

Bratovcic A, Hikal WM, Said-Al Ahl HAH, Tkachenoko KG et al (2021) Nanopesticides and nanofertilizers and agricultural development: scopes, advances and applications. Open J Ecol. https://doi.org/10.4236/oje.2021.114022

Chang PFL, Chang TH, Liu YW, Chen CC, Li WY, Chung WH, Lin JJ, Huang JW (2020) Effect of nanomaterials on seedling growth and disease control. Acta Hortic 1269:269–272. https://doi.org/10.17660/ActaHortic.2020.1269.36

Chidambaram R, Abigail EA (2017) In: Seehra MS (ed) Nanotechnology in herbicide resistance, nanostructured materials-fabrication to applications. IntechOpen. https://doi.org/10.5772/intechopen.355

Chinnamuthu CR, Kokiladevi E (2007) Weed management through nanoherbicide. In: Chinnamuthu CR, Chandrasekaran B, Ramasamy C (eds) Application of nanotechnology in agriculture. Tamil Nadu Agricultural University, Coimbatore

Deshpande T (2017) State of agriculture in India. PRS Legislative Research. https://prsindia.org/policy/analytical-reports/state-agriculture-india. Accessed 15 Dec 2021

Dhillon NK, Mukhopadhyay SS (2015) Nanotechnology and allelopathy: synergism in action. J Crop Weed 11(2):187–191

Dileep Kumar G, Natarajan N, Nakkeeran S (2016) Antifungal activity of nanofungicide trifloxystrobin 25% + tebuconazole 50% against *Macrophomina phaseolina*. Afr J Microbiol Res 10(4):100–105

Elizabath A, Babychan M, Mathew AM, Syriac GM (2019) Application of nanotechnology in agriculture. Int J Pure Appl Biosci 7(2):131–139

EPA (2007) Nanotechnology white paper. US Environmental Protection Agency Report. EPA100/B-07/001, Washington DC 20460, USA

Feizi H, Moghaddam PR, Shahtahmassebi N, Fotovat A (2012) Impact of bulk and nanosized titanium dioxide on wheat seed germination and seedling growth. Biol Trace Elem Res 146: 101–106

Feng J, Zhang Q, Liu Q, Zhu Z (2018) Applications of nanoemulsions in formulation of pesticides. In: Mahdi Jafari S, McClements DJ (eds) Nanoemulsions: formulation, applications, and characterization. Academic, Waltham, pp 379–413. https://doi.org/10.1016/B978-0-12-811838-2.00012-6

Ghazy NA, Abd El-Hafez OA et al (2021) Impact of silver nanoparticles and two biological treatments to control soft rot disease in sugar beet (*Beta vulgaris* L). Egypt J Biol Pest Control 31:3

Giraldo JP, Wu H, Newkirk GM, Kruss S (2019) Nanobiotechnology approaches for engineering smart plant sensors. Nat Nanotechnol 14:541–553

Hasan Dad Ansari M, Lavhale S, Kalunke RM, Srivastava PL, Pandit V, Gade S, Yadav S, Laux P, Luch A, Gemmati D, Zamboni P, Singh AV (2020) Recent advances in plant nanobionics and nanobiosensors for toxicological applications. Curr Nanosci 16(1):27–41

He L, Liu Y, Mustapha A, Lin M (2011) Antifungal activity of zinc oxide nanoparticles against Botrytis cinerea and Penicillium expansum. Microbiol Res 166:207–215

He Y, Xu B, Li W, Yu H (2015) Silver nanoparticles based chemiluminescent sensor array for pesticide discrimination. J Agric Food Chem 63(11):2930–2934

He Y, Xiao S, Dong T, Nie P (2019) Gold nanoparticles for qualitative detection of deltamethrin and carbofuran residues in soil by surface enhanced Raman scattering (SERS). Int J Mol Sci 20(7):1731

Heinisch M, Jácome J, Miricescu D (2019) Current experience with application of metal-based nanofertilizers. In: MATEC Web of Conferences, vol 290, p 03006

Hongyun C, Wenjun Z, Quinsheng G, Qing C, Shiming L, Shuifang Z (2008) Real time Taqman RT-PCR assay for the detection of Cucumber green mottle mosaic virus. J Virol Methods 149(2):326–329

https://croplife.org/news/keeping-indias-pests-in-line/. Accessed 10 Sept 2021

https://www.tractorjunction.com/blog/top-10-agriculture-states-in-india/. Accessed 7 July 2021

Hussain T (2017) Nanotechnology: diagnosis of plant diseases. Agric Res Technol: Open Access J 10(1):555777

Iavicoli I, Leso V, Beezhold DH, Shvedova AA (2017) Nanotechnology in agriculture: opportunities, toxicological implications and occupational risks. Toxicol Appl Pharmacol 329: 96–111

Ivani R, Nejad SHS, Ghahraman B, Astaraei AR, Feizi H (2018) Role of bulk and nanosized SiO2 to overcome salt stress during fenugreek germination (Trigonella foenum-graceum L.). Plant Signal Behav 13(7):e1044190

Jhansi K, Jayarambabu N, Paul Reddy K, Manohar Reddy N, Suvarna RP, Rao KV, Kumar VR, Rajendar V (2017) Biosynthesis of MgO nanoparticles using mushroom extract: effect on peanut (Arachis hypogaea) seed germination. 3 Biotech 7(4):263

Jo Y-K, Kim BH, Jung G (2009) Antifungal activity of silver ions and nanoparticles on phyopathogenic fungi. Plant Dis 93(10):1037–1043

Kah M, Kookana RS, Gogos A, Bucheli TD (2018) A critical evaluation of nanopesticides and nanofertilizers against their conventional analogues. Nat Nanotechnol 13:677–684

Kahyap PL, Rai P, Sharma S, Chakdar H, Kumar S, Pandiyan K, Srivastava AK (2016) Nanotechnology for the detection and diagnosis of plant pathogens. Nanosci Food Agric 2:253–276

Kale SK, Parishwad GV, Husainy ASN, Patil AS (2021) Emerging agriculture applications of silver nanoparticles. ES Food Agroforest 3:17–22

Kannan M, Elango K, Tamilnayagan T, Preetha S, Kasivelu G (2020) Impact of nanomaterials on beneficial insects in agricultural ecosystem. In: Nanotechnology for food, agriculture, and environment. pp 373–393

Khaledian S, Nikkhah M, Shams-bakhshn M et al (2017) A sensitive biosensor based on gold nanoparticles to detect Ralstonia solanacearum in soil. J Gen Plant Pathol 83:231–239

Khater M, de la Escosura-Muñiz A, Merkoçi A (2017) Biosensors for plant pathogen detection. Biosens Bioelectron 93:72–86

Kim MJ, Kim W, Chung H (2020) Effects of silver graphene oxide on seed germination and early growth of crop species. PeerJ 8:e8387

Kumar M (2019) Agriculture: status, challenges, policies, and strategies for India. Int J Eng Res Technol 8(12):1–5

Kumar V, Arora K (2020) Trends in nano-inspired biosensors for plants. Mater Sci Energy Technol 3:255–273

Kumari S, Kumaraswamy RV, Choudhary RC et al (2018) Thymol nanoemulsion exhibits potential antibacterial activity against bacterial pustule disease and growth promotory effect on soybean. Sci Rep 8:6650. https://doi.org/10.1038/s41598-018-24871-5

Lau HY, Wu H, Wee EJH, Trau M, Wang Y, Botella JR (2017) Specific and sensitive isothermal electrochemical biosensor for plant pathogen DNA detection with colloidal gold nanoparticles as probes. Sci Rep 7:38896

Lemire JA, Harrison JJ, Turnr RJ (2013) Antimicrobial activity of metals: mechanisms, molecular targets and applications. Nat Rev Microbiol 11:371–384

Levar O (2007) Effect of silver nanoparticles on tomato plants and development of a plant monitoring system (PMS). A thesis. http://hdl.handle.net/10415/109

Li Z, Yu T, Paul R, Fan J, Yang Y, Wei Q (2020) Agricultural nanodiagnostics for plant disease: recent advances and challenges. Nanoscale Adv 2:3083–3094

Liu R, Lal R (2014) Synthetic apatite nanoparticles as a phosphorus fertilizer for soybean (Glycine max). Sci Rep 4:5686. https://doi.org/10.1038/srep05686

Liu R, Lal R (2015) Potential of engineered nanoparticles as fertilizers for increasing agronomic productions. Sci Total Environ 514:131–139

Lizzi D, Mattiello A, Piani B, Fellet G, Adamiano A, Marchiol L (2020) Germination and early development of three spontaneous plant species exposed to nanoceria (nCeO2) with different concentrations and particle sizes. Nanomaterials (Basel) 10(12):2534

Lobert S, Heil PD, Namba K, Stubbs G (1987) Preliminary X-ray fiber diffraction studies of Cucumber green mottle mosaic virus, watermelon strain. J Mol Biol 196(4):935–938

Lodriche SS, Soltani S, Mirzazadeh R (2012) U.S. Patent application no. 13/406, 538
Lu CM, Zhang CY, Wen JQ, Wu GR, Tao MX (2002) Research of the effect of nanometer materials on germination and growth enhancement of Glycine Max and its mechanism. Soybean Sci 3: 168–172
Mahmood I, Imadi SR, Shazadi K, Gul A, Hakeem KR (2016) Effects of pesticides on environment, plant, soil and microbes. Springer, Cham, pp 253–269
Makarovsky D, Fadeev L, Salam BB, Zelinger E, Matan O, Inbar J, Jurkevitch E, Gozin M, Burdman S (2018) Silver nanoparticles complexed with bovine submaxillary mucin possess strong antibacterial activity and protect against seedling infection. Appl Environ Microbiol 84(4):e02212–e02217
Mali SC, Raj S, Trivedi R (2020) Nanotechnology a novel approach to enhance crop productivity. Biochem Biophys Rep 24:100821
Mane PC, Shinde MD, Varma S, Chaudhari BP, Fatehmulla A, Shahabuddin M, Amalnerkar DP, Aldhafiri AM, Chaudhari RD (2020) Highly sensitive label-free bio-interfacial colorimetric sensor based on silk fibroin-gold nanocomposite for facile detection of chlorpyrifos pesticide. Sci Rep 10:4198
Mastronardi E, Tsae P, Zhang X, Monreal C, Derosa M (2015) Strategic role of nanotechnology in fertilizers: potential and limitations. Springer, Berlin, pp 25–67. https://doi.org/10.1007/978-3-319-14024-7_2
Mir S, Sirousmehr A, Shirmohammadi E (2015) Effect of nano and biological fertilizers on carbohydrate and chlorophyll content of Forage sorghum (Speed feed hybrid). Int J Biosci 6(4):157–164
Mishra S, Singh BR, Singh A et al (2014) Biofabricated silver nanoparticles act as a strong fungicide against *Bipolaris* sorokiniana causing spot blotch disease in wheat. PLoS One 9(5): e97881
Mishra S, Singh B, Naqvi A et al (2017) Potential of biosynthesized silver nanoparticles using *Stenotrophomonas* sp. BHU-S7 (MTCC 5978) for management of soil-borne and foliar phytopathogens. Sci Rep 7:45154
Mou D, Chen H, Du D, Mao C, Wan J, Xu H, Yang X (2008) Hydrogel-thickened nanoemulsion system for topical delivery of lipophilic drugs. Int J Pharm 353(1–2):270–276
Mukhopadhyay SS (2014) Nanotechnology in agriculture: prospects and constraints. Nanotechnol Sci Appl 7:63–71
Nair AS, Durga C (2021) Introduction of nanoparticles in agriculture. Biotica Res Today 3(5):321–323
Nejatzadeh F (2021) Effect of silver nanoparticles on salt tolerance of Satureja hortensis I. during invitro and in vivo germination tests. Heliyon 7(2):e05981
Ombodi A, Saigusa M (2000) Broadcast application vs band application of polyolefin-coated fertilizer on green peppers grown on andisol. J Plant Nutr 23:1485–1493
Pandey G (2018) Challenges and future prospects of agri-nanotechnology for sustainable agriculture in India. Environ Technol Innov 11:299–307
Pandey K, Anas M, Hicks VK, Green MJ, Khodakovskaya MV (2019) Improvement of commercially valuable traits of industrial crops by application of carbon-based Nanomaterials. Sci Rep. 9:19358
Paramasivan C, Pasupathi R (2016) Performance of agro-based industries in India. Natl J Adv Res 2(6):25–28
Pérez de Luque A, Rubiales D (2009) Nanotechnology for parasitic plant control. Pest Manag Sci 65:540–545
Perumal AB, Li X, Su Z, He Y (2021) Preparation and characterization of a novel green tea essential oil nanoemulsion and its antifungal mechanism of action against *Magnaporthe oryzae*. Ultrason Sonochem 76:105649. https://doi.org/10.1016/j.ultsonch.2021.105649
Prajapati HR, Datta I (2014) Future of Indian agriculture: prospects and challenges. In: Agriculture situation in India

Qureshi A, Singh DK, Dwivedi S (2018) Nano-fertilizers: a novel way for enhancing nutrient use efficiency and crop productivity. Int J Curr Microbiol App Sci 7(2):3325–3335

Rafie J, Kumar R (2020) A review on scenario of agriculture in India and punjab1900-2019. Int J Curr Microbiol App Sci 9(6):4149–4170

Rai M, Ingle AP, Pandit R, Paralikar P, Shende S, Gupta I, Biswas JK, Silvério da Silva S (2018) Copper and copper nanoparticles: role in management of insect-pests and pathogenic microbes. De Gruyter. https://doi.org/10.1515/ntrev-2018-0031

Rastogi A, Tripathi DK, Yadav S, Chauhan DK, Živčák M, Ghorbanpour M, El-Sheery NI, Brestic M (2019) Application of silicon nanoparticles in agriculture. 3 Biotech 9(3):90

Razmi A, Golestanipour A, Nikkhah M, Bagheri A, Shamsbakhsh M, Malekzadeh-Shafaroudi S (2019) Localized surface Plasmon resonance biosensing of tomato yellow leaf curl virus. J Virol Methods 267:1–7

Ren Q, Yuan X, Huang X, Wen W, Zhao Y, Chen W (2013) In vivo monitoring of oxidative burst on aloe under salinity stress using haemoglobin and single walled carbon nanotubes modified carbon fiber ultramicroelectrode. Biosens Bioelectron 50:318–324

Sap-lam N, Homklinchan C, Larpudomlert R et al (2010) UV irradiation-induced silver nanoparticles as mosquito larvicides. J Appl Sci 10(23):3132–3136

Shang H, Xie Y, Zhou X, Qian Y, Wu J (2011) Monoclonal antibody-based serological methods for detection of Cucumber green mottle mosaic virus. Virol J 8:228

Shebl A, Hassan AA, Salama DM, Abd El-Aziz ME, Abd Elwahed MSA (2019) Green synthesis of nanofertilizers and their application as a foliar for Cucurbita pepo L. J Nanomater 2019: 3476347. https://doi.org/10.1155/2019/3476347

Siddhartha, Verma A, Bashyal BM, Gogoi R, Kumar R (2020) New nano-fungicide for the management of sheath blight (Rhizoctonia solani) in rice. Int J Pest Manag. https://doi.org/10.1080/09670874.2020.1818870

Siddiqui MH, Al-Whaibi MH (2014) Role of nano-SiO2 in germination of tomato (Lycopersicum esculentum seeds Mill.). Saudi J Biol Sci 21(1):13–17

Singh MD, Chirag G, Prakash PO, Mohan MH, Prakasha G, Vishwajith (2017) Nanofertilizer is a new way to increase nutrient efficiency in crop production. Int J Agric Sci 9(7):3831–3833

Singh R, Kumar N, Mehra R, Kumar H, Singh VP (2020a) Progress and challenges in the detection of residual pesticides using nanotechnology based colorimetric techniques. Trends Environ Anal Chem 26:e00086

Singh R, Thakur P, Thakur A, Kumar H, Chawla P, Rohit JV, Kaushik R, Kumar N (2020b) Colorimetric sensing approaches of surface-modified gold and silver nanoparticles for detection of residual pesticides: a review. Int J Environ Anal Chem 101(15):3006–3022

Swarnkar RK, Pandey JK, Soumya KK, Dwivedi P, Sundaram S, Prasad S, Gopal R (2016) Enhanced antibacterial activity of copper/copper oxide nanowires prepared by pulsed laser ablation in water medium. Appl Phys A 122:704

Szőllősi R, Molnár A, Kondak S, Kolbert Z (2020) Dual effect of nanomaterials on germination and seedling growth: stimulation vs. phytotoxicity. Plants (Basel) 9(12):1745

Tang G, Tian Y, Niu J, Tang J, Yang J, Gao Y, Chen X, Li X, Wang H, Cao Y (2021) Development of carrier-free self-assembled nanoparticles based on fenhexamid and polyhexamethylene biguanide for sustainable plant disease management. Green Chem 23:2531–2540

Tarafdar JC, Raliya R, Rathore I (2012) Microbial synthesis of phosphorus nanoparticles from Tri-calcium phosphate using *Aspergillus tubingensis* TFR-5. J Bionanosci 6:84–89

Thipe VC, Keyster M, Katti KV (2018) Sustainable nanotechnology: mycotoxin detection and protection. In: Nanobiotechnology applications in plant protection. pp 323–349

Vance ME, Kuiken T, Vejerano EP, McGinnis SP, Hochella MF Jr, Rejeski D, Hull MS (2015) Nanotechnology in the real world: redeveloping the nanomaterial consumer products inventory. Beilstein J Nanotechnol 6:1769–1780

Vivekanandhan N, Duraisamy A (2012) Ecological impact of pesticides principally organochlorine insecticide endosulfan: a review. Univers J Environ Res Technol 2(5):369–376

Wagner G, Korenkov V, Judy JD, Bertsch PM (2016) Nanoparticles composed of Zn and ZnO inhibit *Peronospora tabacina* spore germination in vitro and *P. tabacina* infectivity on tobacco leaves. Nanomaterials 6:50. https://doi.org/10.3390/nano6030050

www.ibef.org/industry/agriculture-india.apex. Accessed 7 July 2021

www.nanoshel.com/silver-nanoparticles-agriculture

Yadav AS, Srivastava DS (2015) Application of nanotechnology in weed management: a review. Res Rev: J Crop Sci Technol 4(2):21–23

Yao KS, Li SJ, Tzeng KC, Cheng TC, Chang CY, Chiu CY, Liao CY, Hsu JJ, Lin ZP (2009) Fluorescence silica nanoprobe as a biomarker for rapid detection of plant pathogens. pp 513–516

Zaytseva O, Neumann G (2016) Carbon nanomaterials: production, impact on plant development, agricultural and environmental applications. Chem Biol Technol Agric 3:17

Zeng L, Liu Y, Pan J, Liu X (2019) Formulation and evaluation of norcantharidin nanoemulsions against the *Plutella xylostella* (Lepidoptera: Plutellidae). BMC Biotechnol 19:16

Zhang Z, Si T, Liu J, Zhou G (2019) In-situ silver nanoparticles on nonwoven fabrics based on mussel-inspired polydopamine for highly sensitive SERS carbaryl pesticides detection. Nanomaterials (Basel) 9(3):384

Zheng L, Hong F, Lu S, Liu C (2005) Effect of nano-TiO2 on strength of naturally aged seeds and growth of spinach. Biol Trace Elem Res 104:83–91

Zulfiqar F, Navarro M, Ashraf M, Akram NA, Munné-Bosch S (2019) Nanofertilizer use for sustainable agriculture: advantages and limitations. Plant Sci 289:110270. https://doi.org/10.1016/j.plantsci.2019.110270

Micronutrient Nanoparticles: Synthesis, Properties and Application in Agriculture 15

Kartik Pal, S. M. Paul Khurana, Nitai Debnath, and Sumistha Das

Abstract

Nutrients which are required in trace amount for plant growth are known as micronutrients. Adequate amount of micronutrients in soil not only helps the overall development of plants but also increases their endurance against abiotic stresses. Long-term cropping, soil erosion and excessive usage of commercial fertilizers are some of the reasons for which micronutrients are continuously being depleted from soil. Though nowadays a balanced mixture of macro and micronutrients is used to replenish micronutrient deficiency, indiscriminate usage of this formulation again escalates the cost and hinders the balance of nutrients in the soil. Because of higher surface-area-to-volume ratio, better stability and dispersibility, micronutrients in their nano form will be required in much less amount to cater the nutritional requirement of the plants, and can make the agriculture practice more sustainable. It was observed that micronutrient nanoparticles have many other advantages like these can be utilized to deliver genetic materials required for better plant growth and may also be utilized to decontaminate water. But before wide-scale application of these nanoparticles, the detailed toxicity needs to be studied.

Keywords

Sustainable agriculture · Green synthesis · Precision farming · Nanofertilizers · Slow release

K. Pal · N. Debnath · S. Das (✉)
Amity Institute of Biotechnology, Amity University Haryana, Gurugram, India

S. M. P. Khurana
Science Instrumentation Centre, Amity University Haryana, Gurugram, Haryana, India

A. Thakur et al. (eds.), *Synthesis and Applications of Nanoparticles*,
https://doi.org/10.1007/978-981-16-6819-7_15

15.1 Introduction

Soil is the reservoir of many nutrients which help the plants to grow. These nutrients can be broadly classified into macronutrients and micronutrients. This classification is based on the requirement by the plants for their proper growth, which is necessary to achieve maximum agricultural yield. Nutrients which are required in large volumes are called macronutrients (carbon, hydrogen, oxygen, nitrogen, potassium, sulphur, etc.); on the contrary, nutrients which are required in small amounts are called micronutrients (copper, iron, boron, manganese, zinc, molybdenum). Although the micronutrients are required in trace amounts, they play a vital role in several aspects of plant development like reproductive growth, seed germination, fruiting, chlorophyll synthesis, increase in crop biomass production, enhancement of nitrogen uptake from soil, etc. They help the plants to grow endurance so that they can protect themselves from attacks of phytopathogens and insect pests. They also help the plants to overcome abiotic stresses like UV radiation, higher soil salinity, drought, etc. Micronutrient deficiency results in numerous diseases and developmental challenges in plants. As for example, the deficiency of the micronutrient like B causes necrotic spots on the tips and edges of the leaves (Brdar-Jokanović 2020). Similarly, Mn deficiency causes growth reduction (Ohki 1984), whereas Zn and Fe deficiency causes severe stunting (Tewari et al. 2008) and bronzing (Rout 2015), respectively. Because of these reasons, micronutrient supply should be adequate in the soil throughout.

Globally, NPK (nitrogen, phosphorus, potassium) fertilizers are used in excessive amount, and this led to micronutrient depletion and adversely affected the overall soil quality. Initially the importance of micronutrients was neglected, and fertilizers containing only macronutrients were used. Though, nowadays, balanced fertilizers containing both macro and micronutrients are being used for better crop growth, nano formulations of these micronutrients can increase their availability manifold. According to Dimkpa et al. (2019) and Tripathi et al. (2015), nanotechnology has a great potential to address the issues of agricultural stress related to micronutrient deficiency.

Nanotechnology deals with particles of at least one dimension less than 100 nm. The properties of the bulk materials tend to change when their size approaches nanoscale, and they start exhibiting novel properties. Due to higher surface-area-to-volume ratio, quantum confinement effect, dominance of electromagnetic force, etc., these materials show new optical, electrical, physical properties, unique chemical features, catalytic efficiency, etc. Because of these reasons, nanomaterials are now being used in myriads of fields like electronics, robotics, environmental monitoring and remediation, drug delivery, disease diagnosis and therapeutics, etc. (Khalid et al. 2020). Apart from these, people are also utilizing nanomaterials in agricultural sector for production of crops with higher yield and nutritional value. For example, to achieve a higher quality of mustard plants (*Brassica juncea*), nanocapsules containing herbicides are being applied in the agricultural field. These nanoscale carriers were long being used for drug delivery. The same concept can be utilized to make controlled release formulations of pesticides and fertilizers, and this not only reduces the cost of agrochemical input but also controls environmental pollution arising from the excessive use of these toxic chemicals. This type of on-demand input of agricultural resources is called as "precision farming", and this ultimately improves crop productivity without much polluting the soil and water table. This technique improves plants germination and its growth, too (He et al. 2018). Yang et al. (2007) used nano TiO_2 instead of bulk TiO_2 in naturally aged spinach seeds, and this led to higher chlorophyll content and higher photosynthetic rate. By using such potential nanomaterials, photosterilization and photogeneration of active oxygen were observed. Different nanopesticides are being used for protecting plants from insect pest infestation. Similarly, many nanomaterials (like nano ZnO, nano silver, etc.) show broad spectrum antimicrobial property. Nano formulations have many superior characteristics in comparison with commercial formulations like better dispersibility, higher stability, etc., and these in turn increase the crop yield (Qureshi et al. 2018; Khot et al. 2012; Duhan et al. 2017)

Despite so much advancement in the field on agri-nanotechnology, the potential of micronutrient nanoparticles (NPs) is not much explored. In this chapter, we tried to explore synthetic routes and various applications of micronutrient NPs in agriculture.

15.2 Micronutrient Nanoparticles: Properties and Methods of Synthesis

15.2.1 Zn Nanoparticles

It is one of the most common types of micronutrients obligatory for crop growth. Deficiency of Zn possesses huge threat to agriculture. This problem may arise due to the presence of $CaCO_3$ in soil which tends to decrease Zn solubility. The common strategy applied to address this issue is the utilization of fertilizers enriched with Zn oxides and Zn sulphates. Another approach to address this issue is the use of nanoparticulate form of Zn as the solubility of zinc increases rapidly at nano domain ensuring better availability of Zn in soil. Moreover, nano ZnO is well known for its antimicrobial property. For these reasons, nano ZnO is becoming very popular as micronutrient NP.

There are many ways of ZnO NP synthesis. In chemical route, one of the most common ways of ZnO NP synthesis is called polyol method. In this method, multiple reflux reaction of zinc acetate (0.1 M) is carried out in presence of diethylene glycol at 180 °C, followed by further reflux reaction in presence of triethylene glycol at 220 °C. Initially before reflux, the solution is kept on a magnetic stirrer at 80 °C for 1.5 h. Then finally the sample is centrifuged at 8000 rpm for 15 min which is followed by washing with distilled water and ethanol for two to three times. Finally, it is dried at 80 °C, and ZnO NPs are obtained (Mahamuni et al. 2019). Sol-gel method is another approach commonly practiced for synthesis of ZnO NPs. Here, initially 2 g of zinc acetate dihydrate and 8 g of sodium hydroxide are measured. Then two solutions are prepared: 2 g of zinc acetate is dissolved in 15 mL, and 8 g of sodium hydroxide is dissolved in 10 mL of distilled water. This is followed by mixing of these two aqueous solutions of salts with magnetic stirring for 5 min. Then this solution is titrated using a burette which is filled with 100 mL of ethanol, and a white precipitate is formed which indicates the formation of ZnO NPs (Hasnidawani et al. 2016). Pulsed laser ablation is a mechanical method adopted for ZnO NPs preparation. In this method, zinc target is used (99.9% pure) at the base of the vessel containing 1 mL double distilled water. Then the zinc target is subjected to a pulsed Nd:YAG (neodymium-doped yttrium aluminium garnet) laser with a repetition rate of 10 Hz and pulse width of 10 ns operating at 1064 and 534 nm. Commonly a convex lens is used to focus the beam with focal length of 11 cm. At the last a dark yellow colloidal solution of ZnO NPs is obtained after laser irradiation of nearly 50 pulses (Ismail et al. 2011).

Because of huge toxicity issues related to conventional fertilizers and pesticide in agriculture sector, researchers are now very much tuned towards biological or green method of NP synthesis, specifically for agricultural use. Generally, biological method is preferred over chemical one because chemical synthesis requires toxic chemicals, whereas biological method is eco-friendly. In green approach, the plant extract of *Moringa oleifera* is used along with zinc sulphate heptahydrate ($ZnSO_4.H_2O$) or zinc acetate dihydrate ($Zn\ (CH_3COO)_2.2H_2O$) as salt source of Zn dissolved in water. The plant extract is prepared by dissolving the plant leaves in distilled water or in methanol. Later, a mixture of plant leaf extract with zinc sulphate heptahydrate or zinc acetate dihydrate solution is prepared at desired pH until formation of ZnO NP precipitate is reached.

15.2.2 Cu Nanoparticles

Cu plays an important role in different physiological processes. Cu binds with proteins to form metalloprotein which is essential for electron transport system (ETS) in chloroplasts and mitochondria. It is also required for biochemical processes like photosynthesis and pigment synthesis, etc. and growth and overall development of plant (Yruela 2009). Deficiency of Cu causes stunted growth, chlorosis of plant leaves, curling of the leaf margins, decrease in fruit formation, etc. (Burkhead et al. 2009). Apart from the low dosage requirement of copper at nano domain, it has additional benefits like bactericidal efficacies due to which nano Cu formulations are preferred to bulk Cu-based fertilizers.

In chemical method of nano Cu synthesis, copper sulphate may be used as precursor and starch as capping agent (Khan et al. 2016). In this method, copper sulphate and starch solution are mixed with ascorbic acid solution (reducing agent). This is followed by addition of sodium hydroxide, and this results in Cu NP precipitate. Pulsed wire discharge method is a mechanical method which is also used for synthesizing Cu NPs. In this method a copper wire with diameter of 125 μm and length of 6.1 cm is subjected to 100,000 Pa, 50,000,000 Pa, 10,000,000 Pa and 5,000,000 Pa ambient chamber pressures. The current density of 106 A cm^{-2} from a 1.85 micro F capacitor is applied which charges the wire up to 10 kV and evaporates the wire. Then the vapours are cooled by an ambient gas to produce Cu NPs (Tamilvanan et al. 2014). In green synthesis method, *Citrus medica* fruit extract was reported to be used in $CuSO_4.5H_2O$ solution under continuous stirring at 50 °C for 15 min. The facile synthesis process is carried out in an aluminium vessel. Powdered Cu NPs are obtained following this method. These Cu NPs were further studied for their efficacy on plant cell division (Nagaonkar et al. 2015). In another green synthesis approach, Cu NP is synthesized using plant leaf extract. For this, *Magnolia kobus* leaves are initially dried for 2 days. The plant broth solution is created by washing leaves (25 g) with distilled water which is then followed by heating this mixture for 5 min. Then, it is stored at 4 °C in refrigerator until used.

15.2.3 Fe Nanoparticles

Fe plays an important role in various plant developmental processes. It is the third most limiting element for plant growth, DNA synthesis, respiration and photosynthesis. It also plays a vital role in biochemical pathways like ETS (Rout and Sahoo 2015). The uptake of Fe by plant is done in the form of insoluble hydroxides, oxides and carbonates-bicarbonates complexes. The deficiency of Fe could occur due to increase in soil salinity, low temperature, high pH, low water drainage, etc. (Boamponsem et al. 2017).

Chemically, Fe NPs can be synthesized by using 0.1 M ferric chloride along with 2.5 M sodium borohydride as reducing agent. Initially, 1.89 g of sodium borohydride is dissolved in 20 mL of distilled water. In this method, trisodium citrate dihydrate is used as a stabilizing agent. 0.1 M ferric chloride is mixed with 2.205 g of trisodium citrate, to which 20 mL of sodium borohydride is added slowly under constant stirring condition. This induces the transition in colour from light orange to dark black. Then the solution is incubated at 30 °C for 24 h. Finally, the Fe NPs are obtained (Thyagarajan et al. 2019). Fe NPs can also be produced by various green methods using leaf extract, amino acids, bacteria, etc. In nano form Fe possesses good antimicrobial properties as well. One very easy green method of synthesizing Fe NPs is by using tea leaves (*Camellia sinensis*) which are rich sources of different polyphenols. These polyphenols act both as reducing and surface capping agent imparting stability to the particles. This method is cost effective, and tea leaves are abundantly available. 20 g L^{-1} green tea leaves are used for preparing the leaf extract. Then, a solution of 0.1 M $FeCl_3$ is added to the leaf extract in 2% volume ratio. This finally helps in the synthesis of Fe NPs with a diameter of 5–10 nm (Saif et al. 2016).

15.2.4 B Nanoparticles

B is another important micronutrient for plants. Various studies have shown that B is a very much important candidate for plant cell division. In most cases, this micronutrient is required for reproductive growth for plant. Boron has huge role in development of seed and fruits. N_2 metabolism of plant is very much dependent on appropriate level of B. Phosphate and nitrate level are also regulated by this micronutrient. Deficiency of B in plants results in impaired development of nascent growing tips like root and shoot tips. Fruit and flower development also gets impeded due to B deficiency.

There are many ways of B NP synthesis. Kocakuşak et al. (1996) explained physical method of boron oxide synthesis by heating boric acid at 300 °C. In this process metaboric acid is produced at 170 °C. But gradual heating at 300 °C releases further steam, and boron trioxide NP is synthesized. Plant extract-mediated facile synthesis of boron nanosheet was reported by Deshmukh et al. (2019). Here, they used a green approach to synthesize nano B using by exfoliation method. This novel method is simple, eco-friendly and does not require high temperature for nano B synthesis. In this method, mixture of plant extract is filtered with a 0.45 μm filter paper and mixed with boron nitride. This mixture is kept in a sonication bath for 24 h at 30 °C. The temperature of the bath sonicator is maintained by surrounding cold water. This is followed by collection of the supernatant from the solution settled overnight. Supernatant is then centrifuged at 15,000 rpm for an hour to rescue the pellet. The pellet containing B nanosheet is freeze-dried to obtain nano powder.

15.2.5 Mn Nanoparticles

Mn plays a vital role in the growth and development of the plant. It is an important element which helps in the formation of oxygen-evolving complex (OEC) in photosystem II (PSII). Other than photosynthesis, Mn is also required in respiration, hormone signalling and ensuring defence against plant pathogens, etc. Mn deficiency is seen in the plant where the soil possesses high pH and elevated partial pressure of O_2. The conventional source of Mn fertilizer is $MnSO_4$. Mn deficiency imparts toxic effect on growth of plant that includes lower number of chloroplasts, lower net photosynthetic rate, higher susceptibility to pathogen infections, decreased tolerance to low temperature, etc. Such symptoms become evident from appearance of chlorotic leaves and necrotic spots on the leaves (Tanoi and Kobayashi 2015). Mn in its nano form helps in increased electron transportation within ETS. It also helps in increased photosynthetic capability which in turn improves the crop yield.

In inorganic synthesis method, apoferritin and manganese(II) tetrahydrate is added to deionised water. Sodium hydroxide is added to bring pH 9, and reaction is continued under stirring condition for 6 h in a test tube, and finally the MnO NPs are obtained. Hot injection is another physicochemical method (Duong et al. 2019), where the thermal decomposition of magnesium oleate is carried out to achieve monodisperse MnO nanocrystals (Liu et al. 2013). Soxhlet solvent method is used for green synthesis approach of Mn NPs. In this technique, Mn NPs are synthesized using lemon and turmeric extracts. Initially the lemon extract is filtered using a Whatman filter paper and is stored for further use. Curcumin is extracted from turmeric using Soxhlet solvent method where 95% ethanol is added for the extraction and then it is evaporated. Already prepared lemon extract is added to the aqueous solution of manganese acetate while under constant stirring. Then it is heated at 50–60 °C for an hour until colour changes from pale green to pale yellow which shows the metal reduction. Now, previously prepared curcumin extract is added to it at constant stirring condition for an hour. This process stabilizes Mn NPs which is confirmed from the yellowish-brown coloration (Vergheese and Kiran Vishal 2018). In another green synthesis approach, Mn NPs were synthesized by Mohammed et al. using lemon extract as reducing agent.

15.2.6 Mo Nanoparticles

Mo is another important micronutrient required for plant growth. It is required for normal activity of several enzymes like nitrate reductase, xanthine dehydrogenase, aldehyde oxidase and sulphite oxidase. Mo is essential for conversion of nitride from nitrate and then to ammonia. It can eventually be used to form amino acids by the plant. Mo deficiency takes place due to increased acidity of soil, excess intake of anions of soil oxides, etc. The deficiency reported to cause moulting, leaf coupling, grey tinting and appearance of flaccid leaves (Kaiser et al. 2005). Mo deficiency is connected to N_2, and therefore Mo deficiency sometimes is mistaken as N_2 deficiency.

In chemical method, 0.025 mM $(NH_4)_6Mo_7O_{24}{\cdot}4H_2O$ (molybdic acid ammonium salt tetrahydrate) is dispersed in 5 mM $[Emim][N(CN)_2]$ ionic liquid. The solution is sonicated for 10 min. This step is then followed by addition of 1 mM $NaBH_4$ into previously sonicated solution. Now this colourless solution 5 $(NH_4)_6Mo_7O_{24}{\cdot}4H_2O$: $2NaBH_4$ is heated at 180 °C for 24 h. This results in formation of dark red colloidal precipitate which is isolated by centrifugation. Finally, the Mo NPs are obtained (Ayi et al. 2015). In green approach NPs are synthesized using a fungi *Aspergillus tubingensis* TFR-29 which is initially grown in 250 mL of potato dextrose broth. Before the onset of reaction, pH is maintained at 5.8 and temperature at 28 °C for 72 h. After the incubation period, the fungal mycelia are separated using a Whatman filter paper, and the process is followed by vigorous washing with sterile double distilled water. After some time, these mycelia are reintroduced in sterile Milli-Q water and are again subjected to an incubator for the next 48 h at 28 °C. After incubating two times, cell-free filtrate is obtained by separating out the fungal biomass. Finally, 0.5 mM of aqueous solution of ammonium molybdenum is added to synthesize Mo NPs (Thomas et al. 2017).

15.3 Applications of Micronutrient Nanoparticle

15.3.1 Micronutrient Nanoparticles as Fertilizers and Other Beneficial Applications

15.3.1.1 Zn Nanoparticles

Zn-based micronutrient NPs (MNPs) have a huge potential as fertilizers because it supplies the nutrient to the soil and helps the soil to regain its vitality. Due to their higher surface area, these fertilizers are required in much less amount in comparison with the conventional Zn-based fertilizer formulations. These NPs have shown antimicrobial activities against *Staphylococcus aureus*, *Escherichia coli* and a broad spectrum fungal strain (De la Rosa-García et al. 2018). The antimicrobial property of ZnO NPs mostly arises from generation of reactive oxygen species. They inhibit the growth of microorganisms by directly interacting with the cell surfaces which hinders the cell permeability. Later, they enter the cell and further induce oxidative stress within the microbial cells. This leads to the inhibition of microbial cell growth, and finally the cell dies. It also helps in improving the yield and growth of the crop. Along with their antimicrobial properties, ZnO NPs promote seed germination, seedling vigour, and stem and root growth (Sabir et al. 2014). So far most of the Zn-based nanofertilizers are basically ZnO nanofertilizers. According to Lin and Xing (2007), ZnO NP can elongate root length in radish (*Raphanus sativus*) and rape (*Brassica napus*) seedling, and metallic Zn NPs treatment has shown to promote growth of ryegrass (*Lolium perenne*) seedlings.

15.3.1.2 Cu Nanoparticles

Cu NPs as fertilizers enhance fruit quality, as it increases firmness of the plant which occurs due to root lignification resulting in an increase in the pericarp cell wall. The treated fruits are found to be less acidic than control. This confers that there is an increase in the pH of fruits because of NPs. Cu NPs also promote vitamin C, lycopene, total phenol and flavonoid content in fruits. The antioxidant capacity is also enhanced. So, the yield and quality of the fruits get enhanced due to Cu NP treatment (López-Vargas et al. 2018).

According to the report by Nekrasova et al. (2011), CuO NPs have shown high assimilation of Cu in plant. The treatment also improves antioxidant enzymatic activity, enhanced lipids peroxidation and stimulates rate of photosynthesis (by 35%) in *Elodea densa* (Planch). Although phytotoxicity is a concern at high concentration, plant growth improvement can be assured with very low concentration Cu, and it can be as low as 0.02 mg L^{-1}. Cu NPs show antimicrobial activity as well. CuO NPs eliminate a number of microbes including *E. coli*, *Bacillus subtilis*, *Vibrio cholerae*, *Pseudomonas aeruginosa*, *Syphilis typhus*, *S. aureus*, etc. This antimicrobial activity is achieved by the release of Cu ions which interact with the microbial membrane. Due to its ability of changing its oxidation states back and forth from cuprous (Cu^{1+}) to cupric (Cu^{2+}), it results in the production of hydroxyl radicals that binds with the DNA molecules. This ultimately hinders the helical structure of DNA because of cross-linking within and between the nucleic acid strands. Cu NP can also denature the essential microbial proteins by binding to the sulfhydryl, amino and carboxyl groups. The denatured proteins result in ineffective enzymes that inactivate the surface proteins which is necessary for transport of the essential metabolites. This affects the membrane lipids and overall membrane integrity (Shobha et al. 2014). This is why, along Cu supplementation for crop yield, these NPs can also protect the plants from unwanted microbial infection.

15.3.1.3 Mn Nanoparticles

These NPs enhance the photosynthetic ability of plants and eventually increase the crop productivity. Moreover, they also accelerate the assimilation of N_2. Table 15.1 shows advantageous effect of Mn NPs on different plant growth parameters.

Liu et al. (2016) also reported the beneficial effect of Mn NPs in its oxide form in promotion of lettuce growth, and hence claimed it to be a potential candidate as fertilizer for agricultural products. Their study reported seedling growth promotion to an extent of 12–54% under Mn NP treatment. Another study by Tombuloglu et al. (2018) emphasized the effect of manganese ferrite ($MnFe_2O_4$) NPs on growth of barley (*Hordeum vulgare* L.). This treatment resulted in 7–95% and 4–7% increase in Mn and Fe contents, respectively in the plant. Plant primary growth parameters like fresh weight, total chlorophyll and carotenoid contents were also significantly improved. These observations together with better translocation of Mn in its nano form through root to leaf make this NP a potential fertilizer in promoting overall plant growth. Yuvaraj and Subramanian (2014) showed a novel nano core/shell structure made up of manganese nitrate can encapsulate zinc sulphate from which Zn is released in a sustained way (Lin et al. 2018).

15.3.1.4 Fe Nanoparticles

These NPs help in enhancement of root length, plant growth and overall plant biomass. The promotion of plant growth is majorly regulated by the phytohormone content which is again quite sensitive to concentration of the nutrients. Fe is very much important for photosynthesis, specifically for the electron transportation during this process. Delfani et al. (2014) showed that foliar spray of Fe NPs has many beneficial effects on black-eyed pea. Here, not only the iron content but also the chlorophyll content was elevated, and plasma membrane stability was also increased due to application of Fe NPs. Ghafariyan et al. (2013) showed that treatment with SPIONs (superparamagnetic iron oxide NPs) can significantly increase chlorophyll content in soybean leaf. The treatment did not cause any toxic effect on plant growth and development. The treatment had beneficial effects on different biochemical and enzymatic reactions related to photosynthesis. Zhu et al. (2008) reported translocation of magnetite (Fe_2O_4) NPs in different plant tissues of pumpkin (*Cucurbita maxima*). Kim et al. (2014) showed beneficial effect of zero valent Fe NPs on root elongation of *Arabidopsis thaliana* via cell wall loosening mechanism of released OH. Iron oxide NPs were also evaluated for their efficiency in growth promotion of peanut. Peanut plant requires good amount of Fe which is mostly inadequately

Table 15.1 Enhancement of plant growth parameters after treatment with Mn nanoparticles (Liu and Lal 2015; Pradhan et al. 2014)

Growth parameters	Growth enhancement (%)
Root length	52
Shoot length	38
No. of rootlets	71
Free biomass	38
Dry biomass	100

present in commercial Fe fertilizers. In this study Fe NPs showed to promote root and shoot length and biomass of peanut. This growth promotion was attributed to better antioxidant enzymatic activity and phytohormone regulation (Rui et al. 2016).

15.3.1.5 Mo Nanoparticles

Increased concentration of Mo in plant causes the "growth inhibition effect". This adverse effect can be overcome by application of Mo nanofertilizer. Mo NPs not only promote plant growth by lowering the germination energy but also help in increase of chlorophyll a, b and carotenoid levels. Like Fe NPs, this NP also helps in increase of the length of sprouts, roots and chlorophyll content. A very interesting study by Taran et al. (2014) revealed treatment of colloidal Mo NP caused growth promotion of agronomically valuable microorganisms. Seed treatment with Mo NPs along with microorganisms had significant beneficial effect on root nodule formation in chickpea, and this was four times better than control condition.

15.3.1.6 B Nanoparticles

B as a micronutrient plays a pivotal role in major cell development processes. Deficiency of this micronutrient accounts for multiple plant disorders related to nitrogen and phosphorous metabolism, tissue development, fruit and flower generation, etc. Overall yield of plant is immensely dependent on appropriate availability of B from soil. To address the issue of B deficiency, researchers are now concentrating on application of boron nitrite NPs for plant growth development. As, for example, Genaidy et al. (2020) experimented on application of nano Zn and nano B together to improve yield of olives. Field application of these two nanofertilizers was organized in picual olive orchards for two seasons spanning year 2017 and 2018, and growth and yield of cultivar were studied in detail during this time span. Various parameters like pigment and mineral content of leaves, appearance of fruit-set, fruit yield and drop of fruit set were considered as olive yield parameters. It was noticed that application of these micronutrient fertilizers in their nano form was very useful for improvement of picual olive tree cultivar.

Another study by Ibrahim and Al Farttoosi (2019) also showed efficiency of B NP spray on growth of *Vigna radiata*. This study was carried out on three growth stages with varied dosages of B NPs. Three growth stages selected were initial stage of bud appearance, 50% flowering stage and 100% flowering stage. The data showed that B NP can help in enhancement of plant growth. Ibrahim and Al Farttoosi (2019) categorically proved that in comparison with metallic B, nano B is more beneficial for mung plant growth in terms of increased height, number of pods and seed yield.

Abiotic stress in the form of high salinity hinders crop growth to a great extent. To address this issue, Mahmoud et al. (2020) treated soil synergistically with B, Zn, Si and zeolite NPs. This treatment resulted in growth enhancement of plant cultivar (*Solanum tuberosum*). Even field application of these micronutrients in soil with more salinity caused improvement of almost all the possible growth parameters. Root length, shoot length, leaf pigment content, mineral content and rate of photosynthesis were found to be increased. Other growth factors like protein, carbohydrate, lipid content, hormone level, etc. were also reported to be better. This study

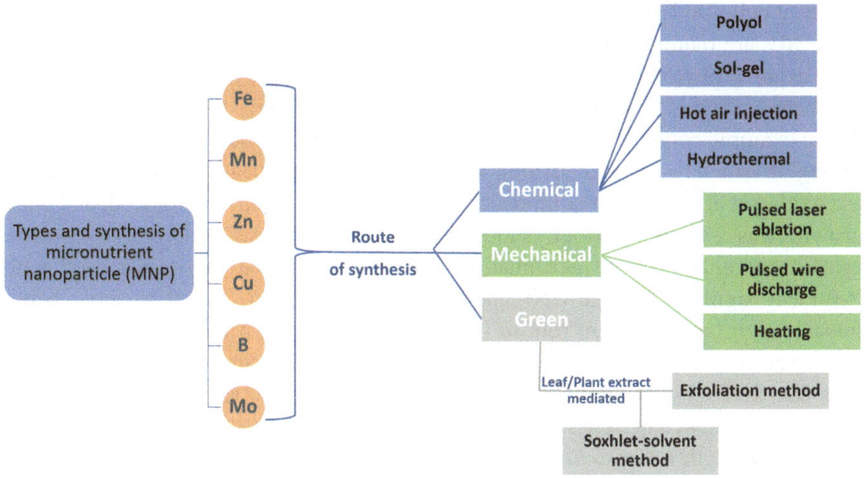

Fig. 15.1 Schematic representation of possible synthetic routes of micronutrient nanoparticles

finally showed high hope on micronutrient-based nanofertilizer as potent fertilizers for the agricultural field.

The synthesis techniques of micronutrient NPs can be summarized in Fig. 15.1.

15.3.2 Micronutrient Nanoparticles as Agrochemical Delivery System

The delivery of fertilizers at an adequate amount is very important in agriculture because excessive amounts of fertilizers not only cause environmental pollution but also affect the soil microflora. There are some chemical fertilizers which are used to compensate the deficiency of micronutrients in the soil for proper development of the plant. The conventional fertilizers have a low crop nutrient use efficiency (NUE). Due to its low NUE, a high supply of micronutrient fertilizers is needed, and this causes toxicity of the metal elements like Cu, Fe and Zn in the soil. Moreover, due to leaching, most of these elements cause eutrophication of the water bodies. These metals may also cause soil structure disturbances and increase the risk of heavy metal contamination in the soil, too. So, a precise amount of nutrients should be delivered in the field.

One of the ways to address this issue is to encapsulate the nutrients within a nanocarrier system. It promotes slow release of fertilizers because NPs hold the fertilizers with higher surface tension. It prevents the nutrient loss, and also unwanted interaction with the microorganisms can be avoided. Many of the micronutrient NPs also exhibit broad-spectrum antimicrobial properties. These nanomatrerials are easily absorbed by the bacterial cell where precipitation occurs and leads to cell death. However, they can also affect the cell exterior by binding to the cilia present on the surface of the bacteria imparting changes in the electrostatic properties of the bacterial cell, hindering the vital functions that lead to the apoptosis of the cell. As these NPs are required in less quantity, they are comparatively safe for the soil. But their detailed phytotoxic effect is yet to be studied.

There are many studies on nanocarrier-mediated delivery of micronutrient NPs. As for example colloidal solution of Cu NPs is grown on carbon nanofiber (Cu-CNF) through chemical vapour deposition method for efficient release of Cu micronutrient as (Ashfaq et al.). The nanocomposite was applied to *Cicer arietinum* seed, and plant growth was examined. Root, shoot growth, water intake, chlorophyll content, protein content and seed germination rate were observed to be increased with treatment significantly. Cu-CNF nanocomposite was also found to have better translocation capacity from root to the shoot. Similarly, Zn deficiency in soil was addressed by Deshpande et al. (2017) by developing Zn complexed with chitosan NPs (Zn-CNP). Foliar application of this nanocomposite was carried out in wheat, grown in Zn-deficient states. The treatment resulted in increased uptake of Zn in plant, and growth was also improved because of the micronutrient treatment with the nanocarrier systems.

Patel et al. (2017) reported the microemulsion-based method of Fe nano reservoir synthesis to address Fe deficiency in plant. In this study, scientists developed a facile synthesis method to produce iron-loaded calcium alginate nanocarrier system. The study exhibited beneficial effect on plants as the nanocarrier imparted stable release of Fe leading to growth augmentation in comparison with traditional Fe fertilizers. Kumar et al. (2018) exhibited facile synthesis method of polymer nanoconjugate of polyvinyl alcohol (PVA)-starch carrying Cu, Zn and carbon nanofiber for slow release of these micronutrients. The treatment resulted in growth promotion and improved reactive oxidative stress management.

15.3.3 Micronutrient Nanoparticles for Water Quality Management

Among other benefits these micronutrients in their nano form can also be utilized for water quality management. Decontamination of water is very important for sustainable agriculture practice, as polluted water may lead to accumulation of toxic materials in the crops. The NPs used for wastewater remediation may either be in zero valent or in their oxide form (Lu et al. 2016).

15.3.3.1 Iron Nanoparticles for Water Pollution Remediation

Nanoparticulate forms of metals like Fe, Zn, etc. are used for disinfecting water. Specifically, Fe in its zero valent form is used for degradation of water contaminants as Fe has high adsorption, precipitation and oxidation capability, and moreover, it is not costly. These NPs are useful for removal of contaminants like halogenated and aromatic compounds, organic dyes, phenols, heavy metals, inorganic dyes, metalloids, etc. There are some disadvantages of using Fe NP like as this NP tends to aggregate, and it is difficult to separate the NP from the treatment setup. These limitations can be overcome by surface coating, encapsulation in matrix, emulsification, etc. (Table 15.2).

Table 15.2 Methods to overcome the limitations of zero valent nano Zn for water treatment

Methods	Effects
Doping	Enhances the reactivity
Surface coating	Prevents aggregation and enhances the dispersibility
Conjugation	Prevents aggregation and enhances the dispersibility
Encapsulation	Makes separation of zero valent nano Zn easier
Emulsification	Enhances fertilizer delivery

15.3.3.2 Zinc Nanoparticles for Management of Water Pollution

Zn NPs is also considered as a potent nanomaterial for degradation of halogenated organic contaminants like CCl_4, etc. from polluted water system. They can be used for removal of dye pollutant from effluents of textile industry. This micronutrient NP serves as a good electron donor, whereas the dye molecules are excellent electron acceptor. Zero valent Zn NP converts Fe^{2+} to Fe^{3+} which in turn leads to the formation of reaction intermediates like hydroxyl and hydrogen ions due to reduction of ferrite ions. These ions interact with the dye pollutant and breaks the chromophore bond (–N=N–). Along with it, the zero valent Zn NP also decolorizes the dye molecules and forms intermediate organic compounds which undergo further mineralization. After mineralization, the intermediate compounds are converted into CO_2, H_2O and inorganic ions to achieve complete degradation of the dye. The efficacy of this NP can be further enhanced by metal doping. This in turn increases the stability and reactivity of zero valent Zn NP towards the dye pollutant present in the water. However, there are more materials like biochar, kaolin, rectorite and clinoptilolite which are used as the supports for zero valent Zn NP. They simply adsorb the dye pollutant and then transfer them by adsorption process rather than degrading the dye pollutants.

Apart from this, ZnO NPs can also adsorb other toxicants as smaller size and higher surface area of nano ZnO ensures higher catalytic activity and increased interaction with microbes. ZnO NP easily gets attached with the microbial cell membrane, and generation of reactive oxygen species ensures removal of microbial contaminants from wastewater (Dimapilis et al. 2018). Besides, these NPs are also used for purification of water due to their wide band gap in the near UV spectrum, strong oxidation capability and photocatalytic activity. Moreover, ZnO NPs are found to be environment friendly. Though ZnO NPs have much resemblance with TiO_2 NPs so far catalytic activity is concerned, nano ZnO can absorb a wider range of solar species and are relatively less costly than TiO_2 NPs. To increase the photodegradation efficiency of the ZnO NPs, mostly metal doping is done using anionic and cationic dopants, rare earth dopants, etc. (Vega-Vásquez et al. 2020).

15.3.4 Micronutrient Nanoparticles as Delivery Vehicle of Genetic Material

Biofortification is a technique used for improving the micronutrient content of the crops. However, this can be achieved through two ways. One way is the conventional breeding in which a crop with less nutrient content is bred with a crop that has high micronutrient content. This latter crop is initially selected, purified and then multiplied so that they can be bred with the crop having a less micronutrient content. Eventually the crop is obtained with desired trait. But there are not always the crops present in our environment which we need for the conventional breeding. So these traits are integrated in the crops genetically. It also enables us to infuse the genes from the wild which was difficult to achieve using the conventional breeding practices. By increasing the nutritional value of the crop, one can solve the problem of hidden hunger which arises due to the less nutritional value of our food. This technique is costlier in comparison to the conventional breeding. But once the crop is obtained, then there are no further investments and remains more beneficial than the crops obtained after the conventional breeding (Raman and Kanmani 2016).

NP-mediated gene delivery in plant system is still in its infancy. Several works have been reported with engineered NPs for gene delivery in animal cells. From these studies it is very much understood that nanomaterial-based delivery vehicle for genes should consist of a few important attributes. First, the material should be very much biocompatible in nature. Various metal (gold, silver), metal oxides (iron oxides) and metal alloys (FePt, FeCO) are most common candidates for genetic material. Second, the material should carry positive charge at surface to facilitate interaction with negatively charged nucleic acid. In this context, surface functionalization of engineered nanostructures with positively charged group can very easily be conducted. Surface tunability of several nanomaterials makes them very good as carrier molecule for precise loading and target-specific delivery of genetic material. Micronutrient NPs can also have huge prospect in gene delivery mechanism because many of them (Fe, Co, Cu) can easily be surface modified with highly positively charged group with the help of PEI (polyethylene imine), etc. Such conjugate can easily be used to deliver genetic material of interest to plants as well.

15.3.5 Nanocomposite of Micronutrient Nanoparticles for Crop Growth Promotion

It is already evident that micronutrient-based NPs promote plant growth by increasing plant metabolic processes like photosynthesis, etc. Due to increase in such processes, the meristematic activity is increased that eventually promotes the apical growth like growth of leaves, increase in shoot length, increase in root length and increase in root area (Elemike et al. 2019). Nowadays, micronutrient NPs are also being used in composite form to promote plant growth. As for example Sabir et al. (2014) showed application of nanocalcite with nano SiO_2 (4%), MgO (1%) and Fe_2O_3 (1%) improved availability of both micro and macronutrients to crops.

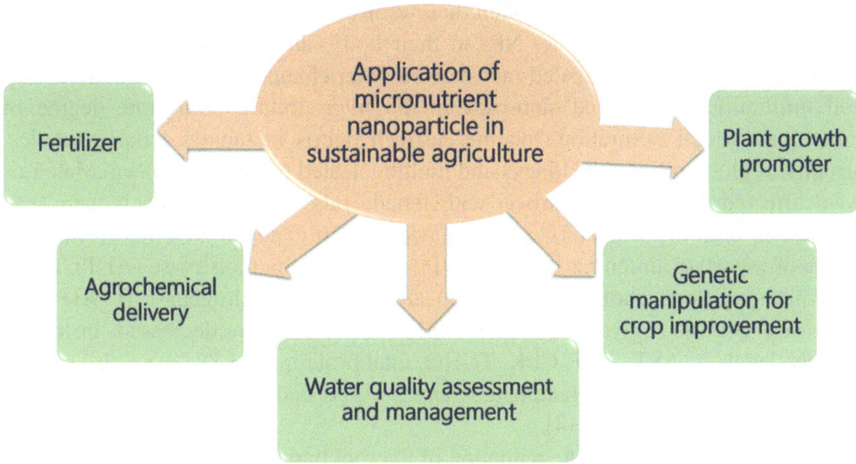

Fig. 15.2 Different applications of micronutrient nanoparticles in agriculture system

Similarly, it is also evident from the work of Disfani et al. (2017) that Fe-SiO_2 nanoconjugate can effectively promote rate of seed germination. Srivastava et al. (2014a, b) also reported that soil mixture with CuO NP can promote photosynthesis and biomass production in *Spinacia oleracea*. Even seed treatment with FeS_2 nanoconjugate shows enhancement of plant biomass (Srivastava et al. 2014a, b). Researchers have also shown the efficacious application of Cu NP in watermelon plant growth.

Different applications of micronutrient NPs in agriculture are summarized in Fig. 15.2.

15.4 Toxicity of Micronutrient Nanoparticles

Though NPs have various advantages, they can cause phytotoxicity when its application dosage is more than the threshold. Excessive exposure to these nano formulations may lead to inhibition of plant growth and changes in physiological properties of plants like decline in biomass, reduction in leaf number and lessening in root elongation and seed germination, and sometimes it may also lead to plant death. Along with these changes are the following: they also alter the subcellular metabolism, oxidative damage to biological membranes, decreased photosynthetic rate, chromosomal abnormalities, disturbance in water transport, decrease in plant growth hormones, and changes in transcription profile of genes. However, this toxicity occurs due to various characteristics of NPs like particle size, dosage of application, soil type and soil moisture (Lin and Xing 2007).

Zn NP toxicity leads to reduced number of roots and length of seedlings, and it also inhibits photosynthesis (Raman and Kanmani 2016). Toxicity of Cu NP leads to stunted growth of plants, cell death and loss of coloration. Even they may cause adverse effect in human beings when these crops are consumed. Experiments were conducted on rats having CuO NPs in their body, and loss of body weight was observed. These Cu NPs-treated rats showed other changes like mild inflammatory cell infiltration and dilated sinusoids in the liver, mild to moderate degree of inflammatory cell infiltration, hyaline cast/cell debris in tubules, dilated tubules, atrophy of glomeruli in the kidneys and multinucleated cells in the spleen. Male rats were affected more in comparison with female rats. The serum levels were also checked in both the sexes of Cu NPs-treated rats. In male Cu NPs-treated rats, the levels of aspartate aminotransferase (AST), alanine aminotransferase (ALT), blood urea nitrogen, total cholesterol (TCHO) and lactate dehydrogenase (LDH) were increased. But the triglycerides (TG), Na and Cl levels were decreased. In female rats, the levels of AST, ALT, CPK, TCHO, total protein and LDH were also seen to be increased due to Cu NP treatment, whereas TG, K and Cl levels were found to be decreased (Delfani et al. 2014).

Fe NPs toxicity leads to the reduction of the root hydraulic conductivity (Lo) and the nutrient uptake. Concentrations of the macronutrients like Ca, K, Mg and S are decreased due to the phytotoxicity of Fe NPs. The reduction in the concentration of the macronutrients causes the reduction of the chlorophyll pigments (Ghafariyan et al. 2013). Mn NPs show low in vivo toxicity when consumed by the humans. Experiments were conducted using murine model system, and various adverse changes were observed in the skin, fur and eyes. Some minor brain anomalies were also seen in Mn NP treatment group at much higher dosage. It was found to affect the activity of Complex II–III and/or Complex IV resulting in reduction in energy conversion through electron transport chain (ETC) (Pradhan et al. 2014). Mg NPs toxicity is harmful for the membrane of the crops and causes severe deterioration in the plant growth. Liu et al. (2016) carried out an extensive study of different micronutrient NPs on lettuce seed germination to study their toxicity profile at extremely low concentrations. This study revealed that all the oxide forms of micronutrient NPs (ZnO, CuO) are slightly more toxic than Zn and Cu NPs. Although it was also reported that it is not necessary that engineered micronutrient nanostructures are more toxic than commercial fertilizers.

15.5 Conclusions and Future Perspective

Globally a large percentage of agricultural soil is deficient in micronutrients. This deficiency does not only affect the crop yield, but also human beings dependent on these crops become deficient in these nutrients. Hence optimization of micronutrient availability is of paramount importance throughout the world. Though there are some commercial micronutrient fertilizers to address this issue, most of the time they are used indiscriminately in the field, and majority of these do not reach the target site. Nano formulations of micronutrients can be utilized to overcome many of the limitations of commercial formulations like these (nano formulation) are required in much less amount, these have higher dispersibility, their delivery is much more target oriented, and all these make the agricultural practices more sustainable. Other than catering the nutrient requirements, these elements in their nano form can be utilized to deliver other agrochemicals and genetic materials to crops and can be utilized for wastewater remediation. Despite having so many advantages, the over-exposure to nano micronutrients may cause phytotoxicity and mammalian toxicity as well. Hence, before wide-scale application of these NPs, detailed toxicity of these materials need to be evaluated.

References

Ayi AA, Anyama CA, Khare V (2015) On the synthesis of molybdenum nanoparticles under reducing conditions in ionic liquids. J Mater 2015:372716. 7 p

Boamponsem GA, Leung DWM, Lister C (2017) Insights into resistance to Fe deficiency stress from a comparative study of in vitro-selected novel Fe-efficient and Fe-inefficient potato plants. Front Plant Sci 8:1581

Brdar-Jokanović M (2020) Boron toxicity and deficiency in agricultural plants. Int J Mol Sci 21(4): 1424

Burkhead JL, Gogolin Reynolds KA, Abdel-Ghany SE, Cohu CM, Pilon M (2009) Copper homeostasis. New Phytol 182(4):799–816

De la Rosa-García SC, Martínez-Torres P, Gómez-Cornelio S, Corral-Aguado MA, Quintana P, Gómez-Ortíz NM (2018) Antifungal activity of ZnO and MgO nanomaterials and their mixtures against Colletotrichum gloeosporioides strains from tropical fruit. J Nanomater 2018:3498527

Delfani M, Firouzabadi MB, Farrokhi N, Makarian H (2014) Some physiological responses of black-eyed pea to iron and magnesium nanofertilizers. Commun Soil Sci Plant Anal 45:530–540

Deshmukh AR, Jeong JW, Lee SJ, Park GU, Kim BS (2019) Ultrasound-assisted facile green synthesis of hexagonal boron nitride nanosheets and their applications. ACS Sustain Chem Eng 7:17114–17125

Deshpande P, Dapkekar A, Oak M, Paknikar K, Rajwade J (2017) Zinc complexed chitosan/TPP nanoparticles: a promising micronutrient nanocarrier suited for foliar application. Carbohydr Polym 165:394–401. https://doi.org/10.1016/j.carbpol.2017.02.061

Dimapilis EAS, Mendoza RMO, Lu M-C (2018) Zinc oxide nanoparticles for water disinfection. Sustain Environ Res 28(2):47–56

Dimkpa CO, Singh U, Bindraban PS, Adisa IO, Elmer WH, Gardea-Torresdey JL, White JC (2019) Micronutrients and their diverse role in agricultural crops: advances and future prospective. Acta Physiol Plant 37(7):1–14

Disfani MN, Mikhak A, Kassaee MZ, Maghari A (2017) Effects of nano Fe/SiO2 fertilizers on germination and growth of barley and maize. Arch Agron Soil Sci 63:817–826

Duhan JS, Kumar R, Kumar N, Kaur P, Nehra K, Duhanc S (2017) Nanotechnology: the new perspective in precision agriculture. Biotechnol Rep (Amst) 15:11–23

Duong THY, Nguyen TN, Oanh HT, Thi TAD, Giang LNT, Phuong HT, Anh NT, Nguyen BM, Quang VT, Le GT, Van Nguyen T (2019) Synthesis of magnesium oxide nanoplates and their application in nitrogen dioxide and sulfur dioxide adsorption. J Chem 2019:4376429. 9 p

Elemike EE, Uzoh IM, Onwudiwe DC, Babalola OO (2019) The role of nanotechnology in the fortification of plant nutrients and improvement of crop production. Appl Sci 9:499

Genaidy AE, Abd-Alhamid N, Hassan HSA, Hassan AM, Hagagg LF (2020) Effect of foliar application of boron trioxide and zinc oxide nanoparticles on leaves chemical composition, yield and fruit quality of Olea europaea L. cv. Picual. Esmail. Bull Natl Res Cent 44:106

Ghafariyan MH, Malakouti MJ, Dadpour MR, Stroeve P, Mahmoudi M (2013) Effects of magnetite nanoparticles on soybean chlorophyll. Environ Sci Technol 47:10645–10652

Hasnidawani JN, Azlina HN, Norita H, Bonnia NN, Ratim S, Ali ES (2016) Synthesis of ZnO nanostructures using sol-gel method. Proc Chem 19:211–216

He X, Deng H, Hwang H-M (2018) The current application of nanotechnology in food and agriculture. J Food Drug Anal 27(1):1–21

Ibrahim NK, Al Farttoosi HAK (2019) Response of mung bean to boron nanoparticles and spraying stages (Vigna radiata l.). Plant Arch 19(Suppl 2):712–715

Ismail RA, Ali AK, Ismail MM, Khaleel I (2011) Hassoon preparation, and characterization of colloidal ZnO nanoparticles using nanosecond laser ablation in water. Appl Nanosci 1:45–49

Kaiser BN, Gridley KL, Brady JN, Phillips T, Tyerman SD (2005) The role of molybdenum in agricultural plant production. Ann Bot 96(5):745–754

Khalid K, Tan X, Zaid HFM, Tao Y, Chew CL, Chu D-T, Lam MK, Ho Y-C, Icon JWL, Wei LC (2020) Advanced in developmental organic and inorganic nanomaterial: a review. Bioengineered 11(1):328–355

Khan A, Rashid A, Younas R et al (2016) A chemical reduction approach to the synthesis of copper nanoparticles. Int Nano Lett 6:21–26

Khot LR, Sankaran S, Maja JM, Ehsani R, Schuster EW (2012) Applications of nanomaterials in agricultural production and crop protection: a review. Crop Prot 35:64–70

Kim J-H, Lee Y, Kim E-J, Gu S, Sohn EJ, Seo YS, An HJ, Chang Y-S (2014) Exposure of iron nanoparticles to Arabidopsis thaliana enhances root elongation by triggering cell wall loosening. Environ Sci Technol 48(6):3477–3485

Kocakuşak S, Akçay K, Ayok T, Koöroğlu HJ, Koral M, Savaşçi ÖT, Tolun R (1996) Production of anhydrous, crystalline boron oxide in fluidized bed reactor. Chem Eng Process 35(4):311–317

Kumar R, Ashfaq M, Verma N (2018) Synthesis of novel PVA–starch formulation-supported Cu–Zn nanoparticle carrying carbon nanofibers as a nanofertilizer: controlled release of micronutrients. J Mater Sci 53(10):7150–7164

Lin D, Xing B (2007) Phytotoxicity of nanoparticles: inhibition of seed germination and root growth. Environ Pollut 150:243–250

Lin C, Juanni C, Zhongwei L, Hancheng W, Huikuan Y, Wei D (2018) Magnesium oxide nanoparticles: effective agricultural antibacterial agent against Ralstonia solanacearum. Front Microbiol 98:790

Liu R, Lal R (2015) Potentials of engineered nanoparticles as fertilizers for increasing agronomic productions. Sci Total Environ 514:131–139

Liu X, Chen C, Zhao Y, Jia B (2013) A review on the synthesis of manganese oxide nanomaterials and their applications on lithium-ion batteries. J Nanomater 2013:736375. 7 p

Liu R, Zhang H, Lal R (2016) Effects of stabilized nanoparticles of copper, zinc, manganese, and iron oxides in low concentrations on lettuce (Lactuca sativa) seed germination: nanotoxicants or nanonutrients? Water Air Soil Pollut 227:42

López-Vargas ER, Ortega-Ortíz H, Cadenas-Pliego G, de Alba Romenus K, de la Fuente MC, Benavides-Mendoza A, Juárez-Maldonado A (2018) Foliar application of copper nanoparticles increases the fruit quality and the content of bioactive compounds in tomatoes. Appl Sci 8:1020

Lu H, Wang J, Stoller M, Wang T, Bao Y, Hao H (2016) Hindawi. An overview of nanomaterials for water and wastewater treatment. Publishing Corporation. Adv Mater Sci Eng 2016:4964828. 10 p

Mahamuni PP, Patil PM, Dhanavade MJ, Badiger MV, Shadija PG, Lokhande AC, Boharaa RA (2019) Synthesis and characterization of zinc oxide nanoparticles by using polyol chemistry for their antimicrobial and antibiofilm activity. Biochem Biophys Rep 17:71–80

Mahmoud AWM, Abdeldaym EA, Abdelaziz SM, El-Sawy MBI, Mottaleb SA (2020) Synergetic effects of zinc, boron, silicon, and zeolite nanoparticles on confer tolerance in potato plants subjected to salinity. Agronomy 10:19

Nagaonkar D, Shende S, Rai M, Kim BS (2015) Biosynthesis of copper nanoparticles and its effect on actively dividing cells of mitosis in Allium cepa. Biotechnol Prog 31(2):557–565

Nekrasova GF, Ushakova OS, Ermakov AE et al (2011) Effects of copper(II) ions and copper oxide nanoparticles on Elodea densa Planch. Russ J Ecol 42:458

Ohki K (1984) Manganese deficiency and toxicity effects on growth, development, and nutrient composition in wheat. Agron J 6(2):213–218

Patel S, Bajpai AK, Bajpai J, Saini RK, Acharya S (2017) Facile preparation of iron loaded calcium alginate nanocarriers and study of controlled release of iron. J Environ Chem Eng 5(6): 5337–5346

Pradhan S, Patra P, Mitra S, Dey KK, Jain S, Sarkar S, Roy S, Palit P, Goswami A (2014) Manganese nanoparticles: impact on non-nodulated plant as a potent enhancer in nitrogen metabolism and toxicity study both in vivo and in vitro. J Agric Food Chem 62(35):8777–8785

Qureshi A, Singh DK, Dwivedi S (2018) Nano-fertilizers: a novel way for enhancing nutrient use efficiency and crop productivity. Int J Curr Microbiol Appl Sci 7:3325–3335. https://doi.org/10.20546/ijcmas.2018.702.398

Raman CD, Kanmani S (2016) Textile dye degradation using nano zero valent iron: a review. J Environ Manag 177:341–355

Rout GR (2015) Role of iron in plant growth and metabolism, Gyana, Role of iron in plant growth and metabolism. Rev Agric Sci 3:1–2

Rout GR, Sahoo S (2015) Role of iron in plant growth and metabolism. Rev Agric Sci 3:1–24

Rui M, Ma C, Hao Y et al (2016) Iron oxide nanoparticles as a potential iron fertilizer for peanut (Arachis hypogaea). Front Plant Sci 7:815

Sabir S, Arshad M, Chaudhari SK (2014) Zinc oxide nanoparticles for revolutionizing agriculture: synthesis and applications. ScientificWorldJournal 2014:925494. 8 p

Saif S, Tahir A, Chen Y (2016) Green synthesis of iron nanoparticles and their environmental applications and implications. Nanomaterials 6(11):209

Shobha G, Moses V, Ananda S (2014) Biological synthesis of copper nanoparticles, and its impact—a review. Int J Pharm Sci Invent 3(8):06-28-38

Srivastava G, Das A, Kusurkar TS, Roy M, Airan S, Sharma RK, Singh SK, Sarkar S, Das M (2014a) Iron pyrite, a potential photovoltaic material, increases plant biomass upon seed pretreatment. Mater Express 4:23–31

Srivastava G, Das CK, Das A, Singh SK, Roy M, Kim H, Sethy N, Kumar A, Sharma RK, Singh SK (2014b) Seed treatment with iron pyrite (FeS2) nanoparticles increases the production of spinach. RSC Adv 4:58495–58504

Tamilvanan A, Balamurugan K, Ponappa K, Madhan Kumar B (2014) Copper nanoparticles: synthetic strategies, properties and multifunctional application. Int J Nanosci 13(02):1430001

Tanoi K, Kobayashi NI (2015) Leaf senescence by magnesium deficiency. Plants (Basel) 4(4): 756–772

Taran NY, Gonchar OM, Lopatko KG, Batsmanova LM, Patyka MV, Volkogon MV (2014) The effect of colloidal solution of molybdenum nanoparticles on the microbial composition in rhizosphere of Cicer arietinum L. Nanoscale Res Lett 9(1):289

Tewari RK, Kumar P, Sharma PN (2008) Morphology and physiology of zinc-stressed mulberry plants. J Plant Nutr Soil Sci 171(2):286–294

Thomas E, Rathore I, Tarafdar J (2017) Bioinspired production of molybdenum nanoparticles and its effect on chickpea (Cicer arietinum L). J Bionanosci 11:1–7

Thyagarajan R, Kumar S, Ramesh Kumar V, Narendrakumar G (2019) Synthesis of Fe nanoparticles using biological and. Chem Methods Appl 16:567–584

Tombuloglu H, Kekec G, Slimani Y, Ercan I, Sözeri H, Baykal A (2018) Impact of manganese ferrite (MnFe2O4) nanoparticles on growth and magnetic character of barley (Hordeum vulgare L.). Environ Pollut 243:872–881

Tripathi DK, Singh S, Singh S, Mishra S, Chauhan DK, Dubey NK (2015) Micronutrients and their diverse role in agricultural crops: advances and future prospective. Acta Physiol Plant 37(7): 1–14

Vega-Vásquez P, Mosier NS, Irudayaraj J (2020) Nanoscale drug delivery systems: from medicine to agriculture. Front Bioeng Biotechnol 8:79

Vergheese M, Kiran Vishal S (2018) Green synthesis of magnesium oxide nanoparticles using Trigonella foenum-graecum leaf extract and its antibacterial activity. J Pharmacogn Phytochem 2018(3):7

Yang F, Liu C, Gao F, Su M, Wu X, Zheng L, Hong F, Yang P (2007) The improvement of spinach growth by nano-anatase TiO2 treatment is related to nitrogen photoreduction. Biol Trace Elem Res 119(1):77–88

Yruela I (2009) Copper in plants: acquisition, transport and interactions. Funct Plant Biol 36:409–430

Yuvaraj M, Subramanian KS (2014) Controlled release fertilizer of zinc encapsulated by a manganese hollow core shell. Soil Sci Plant Nutr 2014:1–8

Zhu H, Han J, Xiao JQ, Jin Y (2008) Uptake, translocation, and accumulation of manufactured iron oxide nanoparticles by pumpkin plants. J Environ Monit 10(6):713–717

Phytosynthesis of Silver Nanoparticles and Their Role as Antimicrobials

16

Alka Yadav and Mahendra Rai

Abstract

Bionanotechnology is an interdisciplinary field involving biotechnology, nanotechnology, chemistry, physics, and material science. Nanoparticles exhibit novel properties depending on the size, shape, and morphology which enable them to interact with plants, animals, and microorganisms. Some researchers have utilized the chemical and physical methods for the synthesis of metal nanoparticles; however, environment-friendly synthesis of metal nanoparticles is becoming popular in recent times. Also, the biological synthesis method is cost effective, and the raw material is available in abundance. Phytosynthesis of nanoparticles is also explained as the green chemistry approach for the synthesis of metal nanoparticles. It has emerged as a promising field of research in the field of bionanotechnology. Various studies have showcased diverse aspects of phytosynthesis of metal nanoparticles that include methods of synthesis, mechanism, and the applications offered by the biosynthesized particles. Among the nanoparticles researched till date, silver nanoparticles have gained significant position owing to their inherent characteristic as an antimicrobial agent. Although the antimicrobial potential of silver metal has been recognized since centuries, after the discovery of silver in the nanoparticle form, it has gained tremendous impetus due to the exceptional rise in its antimicrobial property. Phytosynthesized silver nanoparticles show efficient antimicrobial activity and are used in different biomedical applications. In the present chapter, we highlight the biosynthesis of nanoparticles, plants as a system for synthesis process,

A. Yadav
Department of Biotechnology, Sant Gadge Baba Amravati University, Amravati, India

M. Rai (✉)
Department of Biotechnology, Sant Gadge Baba Amravati University, Amravati, India

Department of Microbiology, Nicolaus Copernicus University, Torun, Poland

A. Thakur et al. (eds.), *Synthesis and Applications of Nanoparticles*,
https://doi.org/10.1007/978-981-16-6819-7_16

biomaterials used, and mechanism of formation of nanoparticles and silver nanoparticles as an efficient antimicrobial agent against bacteria, fungi, and viruses.

Keywords

Bionanotechnology · Nanoparticles · Plants · Phytosynthesis · Silver · Antimicrobial agent

16.1 Introduction

The term nanotechnology is related to the science and engineering involved in the synthesis, design, and characterization of materials at the nanoscale level (1–100 nm) (Rai et al. 2008). Nanotechnology offers application in different fields of science and technology ranging from devices to materials with at least one dimension at the nanometer level (De Morais et al. 2014). Bionanotechnology is a multidisciplinary field of science that involves intersection of biotechnology, nanotechnology, physics, chemistry, and material science for the development of different applications in the field of science and technology (Rai and Yadav 2013; Soman and Ray 2017). Bionanotechnology includes the synthesis of nanoparticles based on biologically available sources like plants, microorganisms, algae, and planktons (Siddiqui et al. 2018). Nanoparticles are of extreme scientific interest as they fill the gap between bulk materials and the atomic materials (Rauwel et al. 2015). The shape, size, and morphology of nanoparticles facilitate their application in medicine, drug delivery, sensors, and catalyst (Santoshkumar et al. 2017). There are number of synthesis methodologies developed for the fabrication of nanoparticles involving physical, chemical, and biological methods (Rai et al. 2008). However, the biological process is the most sought-after synthesis methodology applied for the synthesis of nanoparticles (Rai and Yadav 2013). As for the chemical synthesis process, the chemicals that are employed are hazardous in nature, while for the physical synthesis process, high energy is required which is not economic in nature (Saravanan et al. 2021). For the biological synthesis process, a variety of plant species and microorganisms like bacteria, fungi, yeast, and actinomycetes are harnessed for the synthesis of silver, gold, copper, platinum, palladium, and silica nanoparticles (Prasad et al. 2011; Santoshkumar et al. 2017). Among the wide variety of nanoparticles studied, silver nanoparticles are known for their potential antimicrobial activity (Fierascu et al. 2020). Silver in its ion form also depicted efficient antibacterial property, but silver in the form of nanoparticles shows exponential properties against a large number of microorganisms (Tang and Zheng 2018). With the major increase in the number of drug-resistant bacteria, fungi, and viruses due to the high usage of antibiotics, rising pollution, and changes in the environmental conditions, multiple infectious diseases are rising (Chang et al. 2019). To overcome this situation, scientists are working tirelessly to develop new drugs and treatment measures to fight rising cases of infection (Soman and Ray 2017). Silver nanoparticles function as efficient growth inhibitors and work by disrupting the cell wall and cell organelles of the microbial cell (Siddiqui et al. 2018). These properties of silver nanoparticles gained insight for its use in a number of commercial applications (Soman and Ray 2017). Silver nanoparticles are extensively used in applications like food packaging, antimicrobial topical agent, pharmaceutical products, and biomedical materials (Singh et al. 2018). The present book chapter also reviews biological synthesis of silver nanoparticles using plant as an efficient fabrication system. Also, the biological synthesis of silver nanoparticles and its antimicrobial application as antibacterial, antifungal, and antiviral agent will be discussed. Toxicity studies of biologically synthesized silver nanoparticles will also be explained in the chapter.

16.2 Methods of Synthesis

There are a number of methods employed for the synthesis of metal nanoparticles including physical, chemical, biological, and enzymatic method (Singh et al. 2018). A number of physical and chemical methods are employed for the synthesis of metal nanoparticles; however, these methods prove to be quite expensive and potentially harmful to the environment (Siddiqui et al. 2018). Utilization of a number of toxic chemicals and synthesis and elevated temperature and pressure conditions often result in hazardous by-products which in turn affect the micro and bio-flora of the environment (Saravanan et al. 2021). Biologically inspired synthesis methodologies have evolved with a new branch of nanotechnology, i.e., bionanotechnology which includes synthesis of nanoparticles using microorganisms like bacteria, fungi, viruses, yeast, actinomycetes, plant parts (leaf, stem, roots, fruits, seeds, peel, and biomass), and enzymes (Bilal et al. 2017; Saravanan et al. 2021). Majorly two approaches are in practice for the synthesis process; top-down and bottom-up approaches are involved for the synthesis of metal nanoparticles (Fig. 16.1).

16.3 Biological Synthesis Methods

Biological synthesis of nanoparticles is gaining great interest in research and development due to its application in different fields of science and technology (Priyadarshini and Mahalingam 2017). Owing to their environment-friendly approach, regulation and control of size and shape, use of safer solvents, and easy clean up, researchers are giving impetus on synthesis of nanoparticles using biological entities like bacteria, fungi, yeast, virus, algae, plants, enzymes, and bioactive compounds (Rai et al. 2008; Chand et al. 2020). Compared to the numerous physical and chemical methods used for the synthesis process, biological agents offer a single-step reduction method which needs comparatively low energy for the initiation of the synthesis process. Thus, biological synthesis methods are a cost-effective process (Rai and Yadav 2013).

Phytonanotechnology and phytosynthesis process has acquired considerable attention as a potential technique for synthesis of metal nanoparticles with multi-functional properties (Bilal et al. 2017). The single-step synthesis methodology using plant extracts has gained tremendous attention due to the eco-friendly

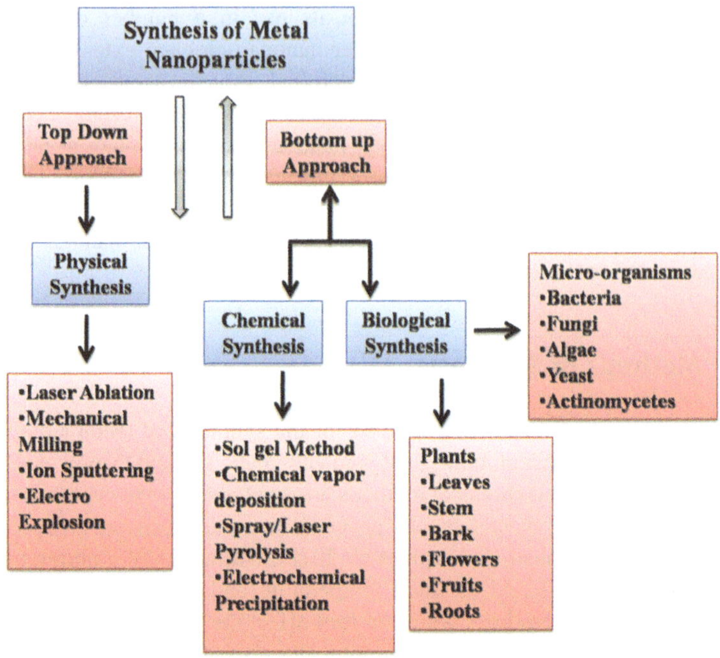

Fig. 16.1 Methods of synthesis of nanoparticles

approach, cost-efficient technology, and rapid results (Bilal et al. 2017). Although the potential of plant extracts to reduce metal ions has been recognized since the 1900s, the knowledge has been applied since the last 35 years for the reduction of metal salts; in the recent decade, the use of plant extracts for the synthesis of nanoparticles has been introduced (Mittal et al. 2014). Plant contain numerous bioactive compounds including phenols, steroids, flavonoids, poly phenols, ascorbic acid, terpenoids, and reductases which function as reducing agents for the synthesis of metal nanoparticles (Rai and Yadav 2013; Santoshkumar et al. 2017). Plant-induced synthesis of metal nanoparticles proves as an important biosynthesis procedure as the plant extract acts as both reducing and capping agent for the synthesis of metal nanoparticles. Synthesis of nanoparticles using the plant system can be followed by both the intracellular and extracellular methods (Bilal et al. 2017). The intracellular synthesis of nanoparticles can be performed by growing the plants in metal-rich organic media, metal-rich soil, or metal-rich hydroponic system, while the extracellular synthesis of nanoparticles is executed by preparing the extract of plant parts (leaves, stem, bark, seeds) through boiling or crushing the plant sample (Rauwel et al. 2015; Kuppusamy et al. 2016).

Incorporation of green synthesis procedures for nanofabrication of nanoparticles has decreased the use of lethal chemical agents and has also ensured noteworthy measures to assure social and environmental well-being. Plant-mediated synthesis protocols depend on various reaction constraints like temperature, pH, pressure, and

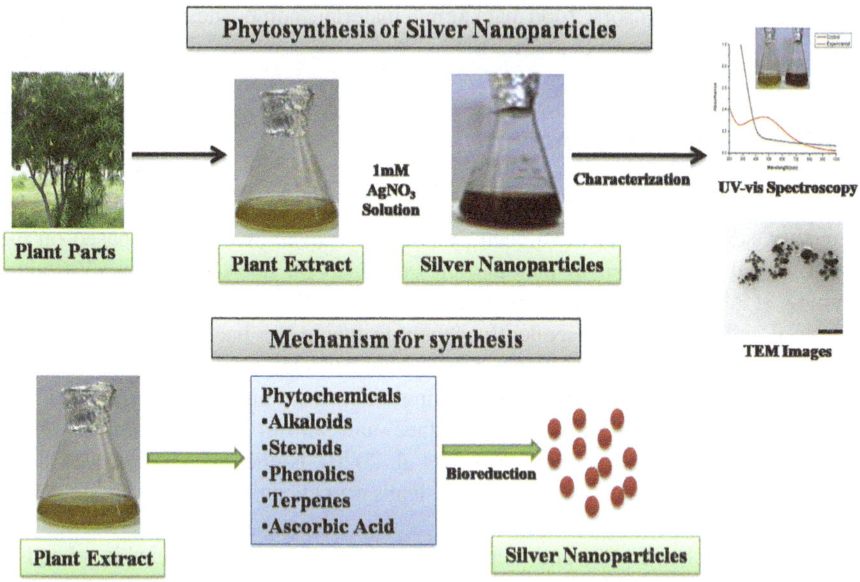

Fig. 16.2 Phytosynthesis of silver nanoparticles and mechanism of synthesis

solvent concentration (Chand et al. 2020). Due to the availability of a variety of phytochemicals like aldehydes, ketones, flavonoids, terpenoids, and phenolic compounds, a wide array of plant material is available for the synthesis process (Rajeshkumar and Bharath 2017) (Fig. 16.2). Thus, eco-friendly synthesis of metal nanoparticles using plant extracts serves as a novel substrate for bulk synthesis of nanoparticles. Nanoparticles synthesized using the plant system offer different applications in healthcare and commercial products.

16.4 Silver Nanoparticles

Silver metal ions have been known in the nineteenth century for its medicinal property and antimicrobial applications (Rai et al. 2009). Silver ions and silver-based compounds are recognized to depict antimicrobial activity against 16 major bacterial organisms (Dakal et al. 2016). Silver ions are mostly utilized in their nitrate form for the induction of antimicrobial activity. However, due to rising increase of antibiotics, the use of silver ion solution as a topical antimicrobial agent decreased (Rai et al. 2009). Silver-based compounds in the form of silver nanoparticles are employed for a number of physical, biological, and medicinal applications (Prabhu and Poulose 2012). Silver nanoparticles can be synthesized through the physical and chemical methods; however, due to the feasibility and environment-friendly nature of the process, biological methods are preferred for the fabrication of silver nanoparticles (Rai et al. 2009). Silver nanoparticles offer applications in diverse fields as catalyst, sensors, food packaging material, composites, bactericidal, and therapeutic agent (Chand et al. 2020).

16.5 Mechanism of Action

The exact mechanism of action of silver nanoparticles has not been postulated yet; however, several theories have been proposed by several researchers based on the microbicidal effect of silver nanoparticles on different microorganisms (Vasquez-Munoz et al. 2019). Silver nanoparticles possess the ability to get adhered to bacterial cell wall and subsequently penetrate inside the bacterial cell membrane (Pulit-Prociak and Banach 2016). Silver nanoparticles cause structural changes inside the cell membrane thereby increasing its permeability which in turn leads to the formation of pits on the cell surface and allows accumulation of silver nanoparticles on the cell surface (Singh et al. 2020). Higher accumulation of silver nanoparticles on the cell surface finally leads to cell death (Prabhu and Poulose 2012).

Another mechanism for antimicrobial action of silver nanoparticles is related to the interaction of silver nanoparticles with the thiol groups. It is supposed that silver nanoparticles release silver ions inside the cells, and these silver ions react with the thiol groups of the essential bacterial enzymes and inactivate them. The bacterial cells uptake silver ions from silver nanoparticles which slow down functioning of bacterial cells resulting in cell damage (Feng et al. 2008). Generation of reactive oxygen species is also a possible hypothesis for antimicrobial activity of silver nanoparticles (Morones et al. 2005). The silver ions react with the respiratory enzymes and inhibit their activity thereby resulting in the generation of reactive oxygen species that itself attack and damage the cells (Fig. 16.3). Liao et al. (2019) also the supported the theory of generation of reactive oxygen species; for the study, the authors chose multidrug-resistant bacteria *Pseudomonas aeruginosa* and checked the effect of silver nanoparticles on its growth. It was observed that silver nanoparticles depicted significant effect on the growth of bacteria in a time- and concentration-dependent manner. The bacterial cell wall became fragile and ruptured after interaction with silver ions resulting in spilling of bacterial cell constituents. Also, due to the rise in oxidative stress, response levels of catalase, peroxidase, and superoxide dismutase increased, while silver nanoparticles inhibited the activity of catalase and peroxidase leading to excessive generation of reactive oxygen species which conferred in impairment of DNA and ribosomes and decline in synthesis of macromolecules following cell death.

Bacterial DNA is majorly composed of sulfur and phosphorus groups; sulfur and phosphorus are soft bases, while silver is a soft acid; hence the interaction of silver ions with bacterial DNA causes hurdles in DNA replication thus resulting in cell death (Shrivastava et al. 2007). Ahmad et al. (2020) also suggested that positively charged silver ions play an important role in the exhibition of antimicrobial activity of silver nanoparticles. Silver ions interact with the negatively charged nucleosides; this electrostatic interaction of negatively charged bacterial cells and positively

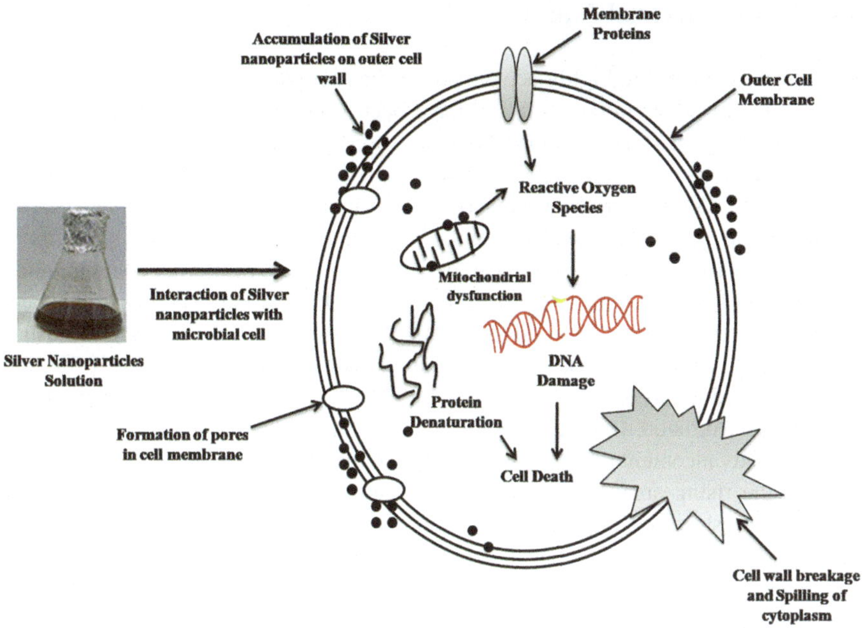

Fig. 16.3 Mechanism of action of silver nanoparticles on bacterial cell

charged silver ions leads to the disruption of cell wall and increase in the permeability of cell membrane, thus leading to cell disruption and cell death.

Signal transduction in microbial cells is affected by silver nanoparticles via the alternation of tyrosine phosphorylation. Silver nanoparticles also inhibit biofilm formation by affecting its growth; increased membrane permeability results in interaction of silver ions with cellular biomolecules causing damage to the microbial cells. Silver ions also bind with bacterial DNA thereby affecting DNA replication and cell division which leads to cell apoptosis and inhibition of cell growth (Singh et al. 2020). Similarly, silver nanoparticles while interacting with viruses get attached to the outer surface of the virus coat leading to disruption of glycoproteins, and also silver nanoparticles bind with viral DNA/RNA and trigger cell apoptosis, thereby blocking replication process (Singh et al. 2020).

16.6 Applications of Silver Nanoparticles as an Antimicrobial Agent

Infectious diseases are a serious threat to global health. Microbial infections often lead to serious medical conditions. Silver has been recognized for its microbicidal properties since centuries. However, silver nanoparticles have emerged as the new generation of antimicrobial agent with diverse applications (Chang et al. 2019).

16.6.1 Antibacterial Agent

Silver nanoparticles are highly acclaimed for their antibacterial activity. Silver ions in the form of silver nitrate solution have been used since centuries as a topical antibacterial agent. Silver nanoparticles are used in burn and wound infections, wound dressings, and antibacterial fabrics and for coating surgical instruments, catheters, etc. (Verma and Maheshwari 2019; Ahmad et al. 2020).

One of the practical applications of the food industry is to ensure the shelf life of the packaged food products to prevent them from spoilage. Microbial contamination of food products is a major issue for food industry. Prevention of food spoilage and maintaining the quality of food products are essential parts of food industry. Traditionally used food packaging materials do not provide efficient shelf life, but with the introduction of silver nanoparticles, development of nanocomposite material has been initiated. Impregnation of silver nanoparticles on biopolymer material has led to advancement of food packaging systems (Simbine et al. 2019).

With the rising areas of research, silver nanoparticles are used in dentistry for performing restorative dentistry. Silver nanoparticles along with polymers are used which act as adhesive and act against biofilm-forming bacteria. Silver nanoparticles due to their antibacterial properties are also used in root canal irrigation and denture implants and as root canal sealers or cement (Niska et al. 2016; Salas-Orozco et al. 2019). Chang et al. (2019) developed nanohybrids of albumin-conjugated silver-diamond and analyzed its antibacterial activity. The authors reported that the silver nanohybrids showed efficient antibacterial activity against gram-positive and gram-negative bacteria.

16.6.2 Antifungal Agent

Skin infections require use of antifungal agents like ketoconazole, itraconazole, Nystatin, and amphotericin. Mallmann et al. (2015) tested the antifungal activity of biologically synthesized silver nanoparticles against *Candida albicans* and *Candida tropicalis*. Silver nanoparticles showed substantial activity against *C. albicans* and *C. tropicalis* which was found to be equal to the activity showed by amphotericin B. Thus, it was suggested that silver nanoparticles can be efficiently used as a topical antifungal agent. Mussin et al. (2019) reported antifungal activity of silver nanoparticles against human skin microbiota *Malassezia furfur*. The authors checked synergistic activity of silver nanoparticles and ketoconazole and also prepared an antimicrobial gel with combination of silver nanoparticles and ketoconazole. The results of the study depicted that silver nanoparticles enhanced the antifungal activity of ketoconazole against *M. furfur*.

Apart from being used as a topical antifungal agent, silver nanoparticles can also be used for the management of plant diseases. Elgorban et al. (2016) used different concentrations of silver nanoparticles solution against six different *Rhizoctonia solani* anastomosis groups that infect cotton plants. The results of the study showed that silver nanoparticles strongly inhibit the activity of all six groups of *R. solani*.

Thus, silver nanoparticles can be used as a topical antifungal agent in diverse applications.

16.6.3 Antiviral Agent

The antiviral potential of biosynthesized silver nanoparticles is a field which has not been much investigated. Haggag et al. (2019) biosynthesized silver nanoparticles using the aerial part extracts of *Lampranthus coccineus* and *Malephora lutea*. The biosynthesized silver nanoparticles were checked for their antiviral activity against HAV-10 (hepatitis A) virus, HSV-1 (herpes simplex) virus, and CoxB4 (Coxsackie) virus. The results of the study showed that nano hexane extract of *L. coccineus* depicted potential activity against HAV-10, HSV-1, and CoxB4 virus, while extract of *M. lutea* showed significant activity against HAV-10 and CoxB4 virus thus confirming that biosynthesized silver nanoparticles can also be used as an efficient antiviral agent. Dung et al. (2020) reported the antiviral activity of silver nanoparticles against swine fever virus which a highly contagious disease. It was suggested that silver nanoparticles inhibited the activity of the virus and can be used as a disinfectant against the viral infection.

With the rise of the global pandemic COVID-19, scientist across the world started searching of probable treatment strategies that could be used to deal with the rising infections of coronavirus. Jeremiah et al. (2020) studied the effect of different concentrations of silver nanoparticles to inhibit the activity of SARS-CoV-2. It was observed that silver nanoparticles show antiviral activity in a concentration-dependent manner. Silver nanoparticles with the size range of 1–10 nm were found be highly effective. Silver nanoparticle-coated fabric masks, air filters, and polycotton fabrics can be efficiently used for protection against coronavirus. Silver nanoparticles bind with the spike glycoproteins of coronavirus and inhibit its binding with the cells, and also silver ions decrease the environmental pH of respiratory epithelium making it more acidic and difficult for the virus to reside (Jeremiah et al. 2020).

16.7 Toxicity of Silver Nanoparticles

Researchers throughout the world have centered their attention on the synthesis and application of silver nanoparticles focusing on their antibacterial and antifungal activities and optoelectronic properties. However, the large-scale use of metal nanoparticles in several applications has ensured their release into the environment raising concerns about the negative effect of nanoparticles on plants-soil and the aquatic bio-life and also on human health. Based on several studies, it is confirmed that silver nanoparticles have the tendency to get accumulated. Plants and soil can absorb metal nanoparticles thus paving way for their entrance into the food chain (Dobrucka et al. 2019). Hence, extensive research is needed for nano-toxicological studies, and their potential health impact cannot be neglected.

Sehnal et al. (2019) performed a study on the effect of silver nanoparticles on the growth of *Zea mays* (maize) plants. Silver nanoparticles and silver ions in the form of silver nitrate solution were tested for growth inhibition of aboveground plant parts. In the study it was observed that the effect of silver nanoparticles depends not only on the concentration of nanoparticles but also on the size of nanoparticles; smaller size leads to higher toxic effects. Plant cell walls allow entrance of nanoparticles through cell membranes and epidermal layers through the roots to reach the stem and leaves. As the pore size of cell wall is a few nanometers in size, smaller nanoparticles depict higher effect of plants. Also, the silver ions released from silver nanoparticles generate reactive oxygen species leading to cellular toxicity. The phytotoxic effect of silver nanoparticles was observed on the nucleus, mitochondria, and chloroplast of maize plants.

Dobrucka et al. (2019) studied the phytotoxic effect of biologically synthesized silver nanoparticles using plant extract of *Veronica officinalis*. For the study, biologically synthesized silver nanoparticles were checked for their phytotoxic effect on different stages of plant (*Linum flavum* and *Lepidium sativum*) development. The authors proposed that cell walls have small pore size of 5–20 nm and nanoparticles after entering the cell wall increase the pore size and form larger pores. After entering the cell wall, silver nanoparticles interact with the cytoplasmic membrane which becomes convex and traps nanoparticles in bubbles and allows their entry into inside the cell. However, in the present study, the biologically synthesized silver nanoparticles were around 40 nm in size, spherical, and agglomerated; hence, in the phytotoxic study, initially the silver nanoparticles stimulated the growth of seedlings (*L. sativum*), then inhibited the growth, and finally in later stages again stimulated the growth. For *L. flavum* seedlings, silver nanoparticles stimulated the growth of plants. Hence, it can be concluded that silver nanoparticles at a smaller size ($\leq$10 nm) have higher phytotoxic effects.

Ferdous and Nemmar (2020) studied the effect of silver nanoparticles on human health and its bio-distribution to local organs. The authors concluded that toxic effect of silver nanoparticles highly depends on the shape, size, and concentration of nanoparticles. Human exposure to nanoparticles comes through inhalation, oral, dermal, or intravenous implantation which assures its translocation and accumulation to various organs thereby depicting toxic effects. Toxic effects on nanoparticles on local and remote organs of the body are based on the duration of exposure, particle size of nanoparticles, route of administration, dose value, and the end concentration of the nanoparticles. The mechanism of action of silver nanoparticles on the human body is still a matter of research before concluding with the results.

Thus, with the increasing use of silver nanoparticles in diverse fields of medicine and technology, there is still great scope for the study of toxic effects of silver nanoparticles on the environment and the human health. As silver nanoparticles possess the tendency of accumulation, it is very likely that they enter the food chain and can adversely affect plant system, aquatic animals, and humans.

16.8 Conclusion

Silver nanoparticles provide a satisfactory answer to the growing concern of antibiotic resistance in microorganisms. As due to the growing number of infections, the human population is facing the number of effective antibiotics are depleting. Silver nanoparticles can be easily synthesized using the physical, chemical, and biological synthesis methods; however, much attention is given towards the synthesis of nanoparticles using the biological system as it offers clean, eco-friendly, and cost-effective synthesis process. Silver nanoparticles depict efficient antimicrobial activity and can be used as an effective antibacterial, antifungal, and antiviral agent both individually and in combination with antibiotics. With the rising global concern towards COVID-19, we should harness the multiple applications offered by silver nanoparticles. The present review also supports the use of silver nanoparticles as an effective antimicrobial agent.

Acknowledgments M.R. would like to thank the Polish National Agency for Academic Exchange (NAWA) for financial support under the grant PPN/ULM/2019/1/00117/DEC/1 2019- 10-02.

References

Ahmad SA, Das SS, Khatoon A, Ansari MT, Afzal M, Hasnain MS, Nayak AK (2020) Bactericidal activity of silver nanoparticles: a mechanistic review. Mat Sci Energy Technol 3:756–769

Bilal M, Rasheed T, Iqbal HMN, Hu H, Zhang X (2017) Silver nanoparticles: biosynthesis and antimicrobial potentialities. Int J Pharmacol 13(7):832–845

Chand K, Cao D, Fouad DE, Shah AH, Dayo AQ, Zhu K, Lakhan MN, Mehdi G, Dong S (2020) Green synthesis, characterization and photocatalytic application of silver nanoparticles synthesized by various plant extracts. Arab J Chem 13:8248–8261

Chang BM, Pan L, Lin HH, Chang HC (2019) Nanodiamond-supported silver nanoparticles as potent and safe antibacterial agents. Sci Rep 9:13164

Dakal TC, Kumar A, Majumdar RS, Yadav V (2016) Mechanistic basis of antimicrobial actions of silver nanoparticles. Front Microbiol 7:1831

De Morais MG, Martins VG, Steffens D, Pranke P, da Costa JAV (2014) Biological applications of nanobiotechnology. J Nanosci Nanotechnol 14:1007–1017

Dobrucka R, Szymanski M, Przekop R (2019) The study of toxicity effects of biosynthesized silver nanoparticles using *Veronica officinalis* extract. Int J Environ Sci Technol 16:8517–8526

Dung TTN, Nam VN, Nhan TT, Ngoc TTB, Minh LQ, Nga BTT, Le VP, Quang DV (2020) Mater Res Express 6(12):1250

Elgorban AM, El-Samawaty AM, Yassin MA, Sayed SR, Adil SF, Elhindi KM, Bakri M, Khan M (2016) Antifungal Silver Nanoparticles: synthesis, characterization and biological evaluation. Biotechnol Biotechnol Equip 30(1):56–62

Feng QL, Wu J, Chen GQ, Cui FZ, Kim TN, Kim JO (2008) A mechanistic study of the antibacterial effect of silver ions on *Escherichia coli* and *Staphylococcus aureus*. J Biomed Mater Res 52:662–668

Ferdous Z, Nemmar A (2020) Health impact of silver nanoparticles: a review of the bio-distribution and toxicity following various routes of exposure. Int J Mol Sci 21:2375

Fierascu I, Fierascu IC, Brazdis RI, Baroi AM, Fistos T, Fierascu RC (2020) Phytosynthesized metallic nanoparticles—between nanomedicine and toxicology. A brief review of 2019's findings. Materials 13:574

Haggag EG, Elshamy AM, Rabeh MA, Gabr NM, Salem M, Youssif KA, Samir A, Muhsinah AB, Alsayari A, Abdelmohsen UR (2019) Antiviral potential of green synthesized silver nanoparticles of *Lampranthus coccineus* and *Malephora lutea*. Int J Nanomedicine 14:6217–6229

Jeremiah SS, Miyakawa K, Morita T, Yamaoka Y, Ryo A (2020) Potent antiviral effect of silver nanoparticles on SARS-CoV-2. Biochem Biophys Res Commun 533(1):195–200

Kuppusamy P, Yusoff MM, Maniam GP, Govindan N (2016) Biosynthesis of metallic nanoparticles using plant derivatives and their new avenues in pharmacological applications—an updated report. Saudi Pharm J 24:473–484

Liao S, Zhang Y, Pan X, Zhu F, Jiang C, Liu Q, Cheng Z, Dai G, Wu G, Wang L, Chen L (2019) Antibacterial activity and mechanism of silver nanoparticles against multidrug-resistant *Pseudomonas aeruginosa*. Int J Nanomed 14:1469–1487

Mallmann EJJ, Cunha FA, Castro BNMF, Maciel AM, Menezes EA, Fechine PBA (2015) Antifungal activity of silver nanoparticles obtained by green synthesis. Rev Inst Med Trop Sao Paulo 57(2):165–167

Mittal J, Batra A, Singh A, Sharma MM (2014) Phytofabrication of nanoparticles through plants as nanofactories. Adv Nat Sci Nanosci Nanotechnol 5:043002

Morones JR, Elechiguerra JL, Camacho A, Holt K, Kouri JB, Ramirez JT, Yacaman MJ (2005) The bactericidal effect of silver nanoparticles. Nanotechnology 16:2346–2353

Mussin JE, Roldan MV, Rojas F, Sosa MA, Pellergi N, Giusiano G (2019) Antifungal activity of silver nanoparticles in combination with ketoconazole against *Malassezia furfur*. AMB Express 9:131

Niska K, Knap N, Kedzia A, Jaskiewicz M, Kamysz W, Inkielewicz-Stepniak I (2016) Capping agent-dependent toxicity and antimicrobial activity of silver nanoparticles: an *in vitro* study. Concerns about potential applications in dental practice. Int J Med Sci 13:772–782

Prabhu S, Poulose EK (2012) Silver nanoparticles: mechanism of antimicrobial action, synthesis, medical applications and toxicity effects. Int Nano Lett 2:32

Prasad TNVKV, Kambala VSR, Naidu R (2011) A critical review on biogenic silver nanoparticles and their antimicrobial activity. Curr Nanosci 7:531–544

Priyadarshini KC, Mahalingam PU (2017) Antimicrobial and anticancer activity of silver nanoparticles from edible mushroom: a review. Asian J Pharm Clin Res 10(3):37–40

Pulit-Prociak J, Banach M (2016) Silver nanoparticles—a material of the future. . . .? Open Chem 76–91

Rai M, Yadav A (2013) Plants as potential synthesizers of metal nanoparticles. IET Nanobiotechnol 7(3):117–124

Rai M, Yadav A, Gade A (2008) Current trends in phytosynthesis of metal nanoparticles. Crit Rev Biotechnol 28(4):277–284

Rai M, Yadav A, Gade A (2009) Silver nanoparticles as a new generation of antimicrobial. Biotechnol Adv 27(1):76–83

Rajeshkumar S, Bharath LV (2017) Mechanism of plant-mediated synthesis of silver nanoparticles—a review on biomolecules involved, characterization and antibacterial activity. Chem Biol Interact 273:219–227. https://doi.org/10.1016/j.cbi.2017.06.019

Rauwel P, Kuunal S, Ferdov S, Rauwel E (2015) A review on the green synthesis of silver nanoparticles and their morphologies studied via TEM. Adv Mat Sci Eng 2015:682749

Salas-Orozco M, Nino-Martinez N, Matinez-Castanon GA, Mendez FT, Jasso MEC, Ruiz F (2019) Mechanisms of resistance to silver nanoparticles in endodontic bacteria: a literature review. J Nanomater 2019:7630316

Santoshkumar J, Rajeshkumar S, Kumar SV (2017) Phyto-assisted synthesis, characterization and applications of gold nanoparticles—a review. Biochem Biophys Rep 11:46–57

Saravanan A, Kumar PS, Karishma S, Vo DN, Jeevanantham S, Yaashika PR, George CS (2021) A review on biosynthesis of metal nanoparticles and its environmental applications. Chemosphere 264:128580

Sehnal K, Hosnedlova B, Docekalova M, Stankova M, Uhlirova D, Tothova Z, Kepinska M, Milnerowicz H, Fernandez C, Ruttkay-Nedecky B, Nguyen HV, Ofomaja A, Sochor J, Kizek R (2019) An assessment of the effect of green synthesized silver nanoparticles using Sage leaves (*Salvia officinalis* L.) on germinated plants of Maize (*Zea mays* L). Nanomaterials 9:1550

Shrivastava S, Bera T, Roy A, Singh G, Ramachandrarao P, Dash D (2007) Characterization of enhanced antibacterial effects of novel silver nanoparticles. Nanotechnology 18:1–9

Siddiqui KS, Husen A, Rao RAK (2018) A review on biosynthesis of silver nanoparticles and their biocidal properties. J Nanobiotechnol 16:14

Simbine EO, Rodrigues LC, Lapa-Guimaraes J, Kamimura ES, Corassin CH, Oliveira CAF (2019) Application of silver nanoparticles in food packages: a review. Food Sci Technol 39(4):793–802

Singh J, Dutta T, Kim KH, Rawat M, Samddar P, Kumar P (2018) Green synthesis of metals and their oxide nanoparticles: applications for environmental remediation. J Nanobiotechnol 16:84

Singh A, Gautam PK, Verma A, Singh V, Shivapriya PM, Shivalkar S, Sahoo AK, Samanta SK (2020) Green synthesis of metallic nanoparticles as effective alternatives to treat antibiotics resistant bacterial infections: a review. Biotechnol Rep 25:e00427

Soman S, Ray G (2017) Chapter 10: Phytosynthesis of silver nanoparticles and its potent antimicrobial efficacy. In: Thomas J, Thomas S, Jose J, Kalarikkal N (eds) Recent trends in nanomedicine and tissue engineering. River Publisher, pp 249–289

Tang S, Zheng J (2018) Antibacterial activity of silver nanoparticles: structural effects. Adv Healthc Mater 7(13):e1701503

Vasquez-Munoz R, Meza-Villezcas A, Fournier PGJ, Soria-Castro E, Juarez-Moreno K, Gallego-Hernandez AL, Bogdanchikova N, Vazquez-Duhalt R, Huerto-Saquero A (2019) Enhancement of antibiotics antimicrobial activity due to the silver nanoparticles impact on the cell membrane. PLoS One 14(11):e0224904

Verma P, Maheshwari SK (2019) Application of silver nanoparticles in diverse sectors. Int J Nano Dimens 10(1):18–36

Microbicidal Nanoparticles

17

Nishant Srivastava and Shailendra K. Saxena

Abstract

Nanomaterials exhibit tremendous exciting properties and transform every aspect of science and technology. Microbiology has also not remained untouched from nanotechnology applications and utilization. Antimicrobial materials are one of the major concerns of microbiology research due to increasing challenges of microbial resistance for antibiotics and continuous mutation in microorganisms. Nanoparticles are found to exhibit excellent antimicrobial properties. Nanoparticles of gold, silver, selenium, tin oxide, zinc oxide, nickel, platinum, etc. were found to have excellent antimicrobial properties. These nanoparticles can be utilized for a wide range of antimicrobial applications. Antimicrobial coating, targeted delivery of antibiotics, nanoparticle conjugate with antibiotics, and antimicrobials are a few of the applications of nanoparticles as antimicrobials. Researchers utilized nanoparticles for coating on medical devices to keep them infection-free. The traditional antibiotics are found ineffective on several multidrug-resistant bacteria, and nanoparticles in conjugation or alone may provide the solution for such challenges. The nanoparticles are found to have great antioxidant activity, and nanoparticles in the size range of 5–100 nm can directly forage free radicals in vitro. Researchers also reported excellent antimicrobial properties of nanoparticles when capped or conjugated with antibiotics/ antibodies. The agar well diffusion method and minimum inhibition

N. Srivastava
Department of Biotechnology, Meerut Institute of Engineering and Technology, Meerut, India

S. K. Saxena (✉)
Centre for Advanced Research (CFAR), Faculty of Medicine, King George's Medical University (KGMU), Lucknow, India
e-mail: shailen@kgmcindia.edu

A. Thakur et al. (eds.), *Synthesis and Applications of Nanoparticles*,
https://doi.org/10.1007/978-981-16-6819-7_17

concentration (MIC) method are widely used by most of the researchers for determination of antimicrobial properties of nanoparticles.

Keywords

Antimicrobial · Nanoparticles · Bactericidal · Fungicidal · MIC

17.1 Introduction

Nanotechnology emerged as a novel and efficient tool for the treatment, diagnosis, and monitoring of diseases. These applications of nanotechnology in medicine help in the speedy growth of the most advanced branch in medicine known as nanomedicine. Nanoparticles as antimicrobial agents are one of the important areas of study in nanomedicine. Eclectic research is going on throughout the world to understand the effect, toxicity, and mechanism of nanoparticles as antimicrobials. Metallic nanoparticles, metal oxides, quantum dots, etc. are found to have excellent antimicrobial properties (Dizaj et al. 2014). From the beginning of nanotechnology, a variety of nanomaterials, namely, nanoparticles, nanowires, nanorods, nanotubes, and nanocomposites, have been reported useful as antimicrobial agents (Geoffrion et al. 2020).

In ancient time and in Indian native medicine system of Ayurveda, the use of Bhasma (metal/metallic preparation of fine metal powder containing particles of size less than 100 nm treated with herbal juices) can be traced back since its inception (Sharma and Prajapati 2016). The procedure of Bhasma preparation, purification, and utilization is well scripted and certainly comparable with present-day available modern technologies of nanoparticle synthesis. The ash or Bhasma of gold containing globular gold particles of size 56–57 nm is reported for treatment of mycobacterium, ischemia in global and focal model, and pristine-induced arthritis and for enhancement of apoptosis in B-chronic lymphocytic leukemia (Mukherjee et al. 2007; Srivastava and Mukhopadhyay 2015a, b). Similarly, ash/or Bhasma of silver is reported for its use in treatment of bacterial, fungal, and other forms of microbial infections, acidity, and irritable bowel syndrome. Swarnmakshik Bhasma (copper pyrite, mercury, and sulfur) utilizes for treatment of anemia, jaundice, and chronic fever, whereas Loknath Ras (mercury, sulfur, conch shell) is reported to be used for treatment of cancer, ovarian cyst, and diarrhea. Kumar kalyan ras (gold, iron, mica, copper pyrite, red sulfide of mercury) is used in Ayurveda for treatment of respiratory tract infections and fever in children (Pal et al. 2014). Recently, the world is passing through a medical emergency situation due to the outbreak of COVID-19 (Srivastava et al. 2020). Due to lack of proper treatment and vaccination, many lives are already lost and suffering is also huge (Srivastava and Saxena 2020). In this emergency situation, Ayurvedic nanoparticles emerged as savior because of their excellent biological activity, stability in solid state, and safety. Many Bhasma preparations containing Swarna Bhasma (gold nanoparticles), Rajata Bhasma (silver nanoparticles), Tamra Bhasma (copper nanoparticles), Lauha Bhasma

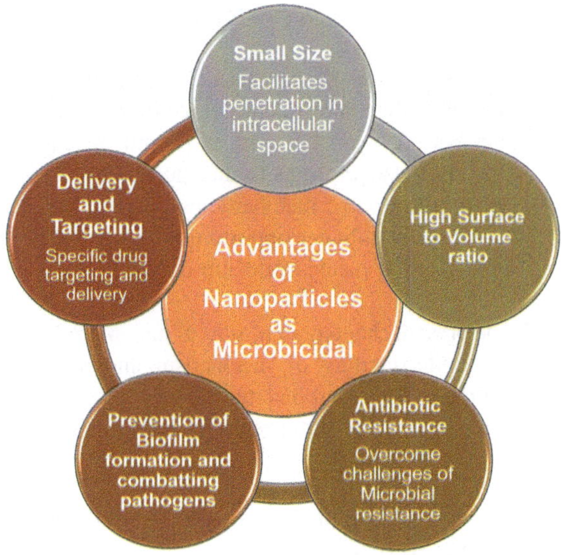

Fig. 17.1 Advantages of using nanoparticles as microbicidal

(iron nanoparticles), and Vanga Bhasma (tin nanoparticles) are proposed for preventive and curative preparation of COVID-19. Oral administration of Siddha Makardhwaja (nano-HgS particles) and Swarna Bhasma (nanogold particles) is revealed to be effective with black pepper powder and ghee for treatment of moderate to severe COVID-19 patients. These preparations are found to be better if administered through oral or sublingual route because of better bioavailability and absorption achieved in presence of nanoparticles (Sarkar and Das Mukhopadhyay 2021). Nanoparticles not only are known for their microbicidal properties but also can be used as the carrier for antimicrobial drugs. Nanoparticles of silver, gold, iron oxide, selenium, copper, copper oxide, tin oxide, quantum dots, bimetallic nanomaterials, etc. are reported for their antimicrobial efficiency, whereas nanoparticles like liposomal, magnetic, and solid lipid, quantum dots, mesoporous silica, carbon, polymer micelles, dendrimers, and many more can be utilized as a carrier for antibiotics, antifungal, and other forms of drugs (Fernando et al. 2018). Furthermore, nanoparticle conjugates with existing antibiotics also exhibit exciting results against pathogens. The important advantages of using nanoparticles as antimicrobials in comparison with existing traditional antibiotics are as follows (Fig. 17.1) (Srivastava and Mukhopadhyay 2013, 2015a, b; Fernando et al. 2018):

(a) Size: Very small size of nanoparticles facilitates easy escape from cell wall and penetration to intracellular space and destruction of microbial genetic material and other important cellular components. Additionally, a high surface to volume ratio provides determined reacting sites.
(b) Surface to volume ratio: High surface to volume ratio provides more exposure to reactive sites, and it is beneficial for conjugation and drug delivery.

(c) Metal oxide nanoparticles form a toxic ion-rich environment inside the cell and lead to cell lysis.
(d) The application of nanoparticles provides an opportunity to overcome the present challenge of microbial antibiotic resistance.
(e) Nanoparticles can be used as prominent carriers of antimicrobial drugs and efficiently deliver drugs to the specific target site.
(f) The application of nanoparticles helps to prevent microbial biofilm formation and provides a better opportunity to combat pathogenic microorganisms via a wide range of mechanistic approaches.

In the present chapter, an insight of nanoparticles as antimicrobials is provided for a better understanding of mechanism and advancements achieved till now by application in the microbial world with the special outlook of nanomaterials as microbiocidal on multidrug-resistant microorganisms, superbugs, and viruses.

17.2 Antimicrobial Properties of Nanoparticles and Action Mechanism

Nanotechnology plays a very decisive role in the advancement of science and technology. The successful application of nanotechnology in life sciences especially in medicine has opened up several doors of opportunity in diagnostics and treatment. Antibiotic resistance pathogens are becoming a serious challenge to scientists and medical professionals. The continuous consumption and exploitation of antibiotics deteriorates the effectiveness which gradually enhances the cases of antimicrobial resistance (Geoffrion et al. 2020). Nanoparticles have a wide spectrum of antimicrobial activity and can act via the various mechanisms. Though the exact mechanism of nanoparticle-based-antimicrobial activity is not completely known or/ understood the nanoparticles exhibit negligible chance for the development of antimicrobial resistance.

Engineered nanoparticles of size 1–100 nm are found to be more efficient microbicidal as smaller size facilitates unrestricted movement of nanoparticles inside the microbial cell and intracellular spaces. Nanoparticles of various metals and metal oxides like silver, gold, zinc oxide, iron, tin oxide, selenium, iron oxide, copper oxide, tungsten carbide, etc. are found to have very promising antibacterial, antifungal, and virucidal properties. These nanoparticles also show excellent result in combination with available antibiotics. Nanoparticles act against microbes via several mechanisms or in combination of these mechanisms which are as follows (Srivastava and Mukhopadhyay 2015a, b; Gakiya-Teruya et al. 2020; Geoffrion et al. 2020):

(a) Attachment and penetration into cell wall
(b) Unsettling of cell wall
(c) Formation of outer membrane vesicles (OMV)
(d) DNA condensation

(e) DNA inhibition and DNA damage
(f) Generation of reactive oxygen species (ROS)
(g) Restricting cellular respiration

The mechanism of nanomaterial-mediated microbicidal activity is initiated by deposition and internalization of nanomaterial on the surface of cell wall, cell membrane, and peptidoglycan layer to cytoplasm. Nanomaterial penetrates cells and is attached with cellular components. The small size and high surface to volume ratio facilitate easy diffusion of nanomaterial across the cell wall and in intracellular space. The infiltration of nanomaterial across the cell membrane disrupts the integrity of microbial cell membrane which further leads to leaching of intracellular material and cell lysis.

The infiltration of nanomaterial into microbial cell leads to the formation of outer membrane vesicles (OMV). The cytoplasm and other intracellular organelles were leaked through a disrupted portion of cell membrane (formed due to OMV). The severe pressure of cytoplasmic material on OMV causes disruption of OMV and leads to leakage of a high amount of cellular material. The high loss of cytoplasmic material due to disrupted cell membrane and cell wall through OMV further leads to cell lysis (Fig. 17.2).

Additionally, the infiltrated nanomaterials attacked DNA molecules of the cell. Researchers observed by the help of transmission electron microscopy that there were a condensed DNA molecule and electron light region showing presence of nanomaterials. DNA is storage of all the necessary information required for any cell growth and replication. Any damage to DNA may lead to mutation of the cell and cell death. The DNA agglomeration is part of cellular response for preventing DNA damage. Small molecular weight proteins are produced by cells to prevent DNA damage. Infiltration of nanomaterial initiates cellular defense response and smaller molecular weight proteins produced by the microbial cell to be conglomerated around the cell for its defense. The electron light region in cells is observed due to the conglomerated DNA and presence of nanomaterial. The high concentration of nanomaterial inside the cell inhibits all types of vital response for cellular protection which further leads to cell wall breakdown and cell death (Feng et al. 2000; Srivastava and Mukhopadhyay 2015a, b).

Another widely reported mechanism for microbicidal nanoparticles is formation of reactive oxygen species (ROS). ROS is a group of oxygen containing radical and non-radical compounds such as superoxides (O_2^-) and hydrogen peroxide, respectively. The generation of ROS occurs due to infiltration and interaction of nanomaterial with microbial cells affecting the survival of cells. The ROS species are capable of damaging lipid membrane of bacterial cell and producing different types of responses (free radicals) which leads to microbicidal activity (Gakiya-Teruya et al. 2020; Geoffrion et al. 2020). Researchers across the globe reported several studies related to microbicidal nanoparticles and their course of action. The nanomaterial as microbicidal has huge advantages and may become an important tool to fight acute to chronic infections more efficiently in future.

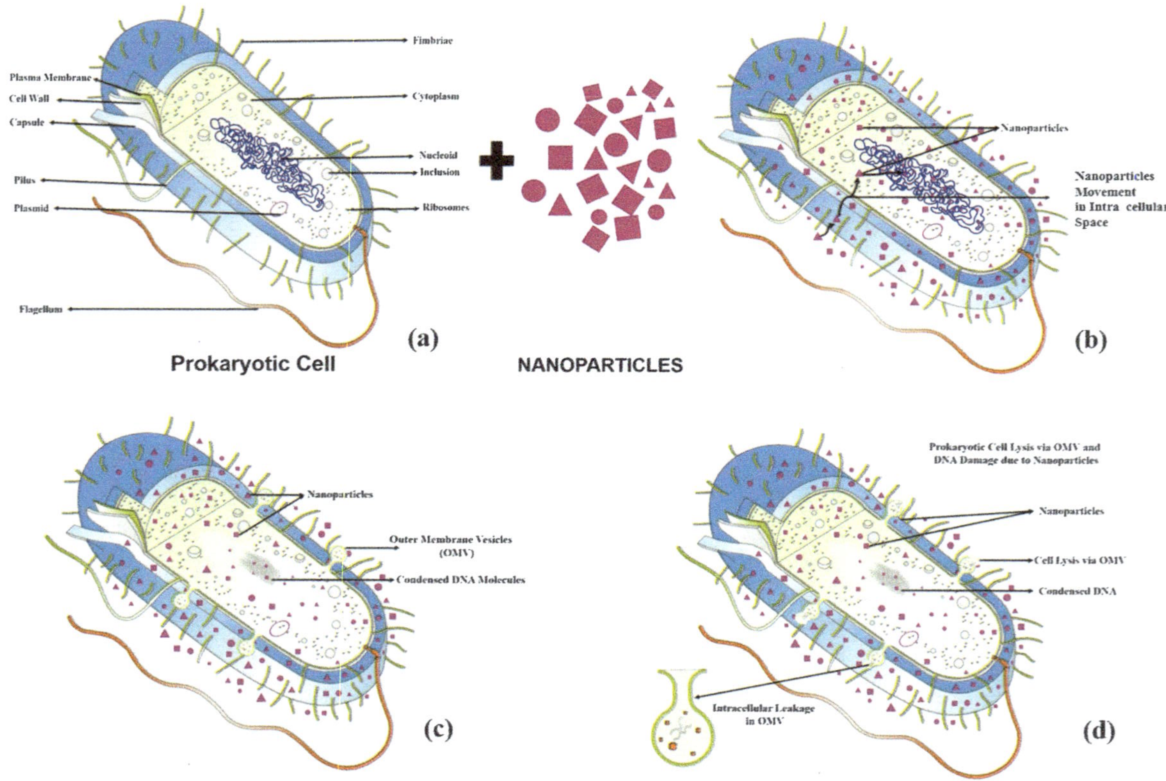

Fig. 17.2 Microbicidal mechanism of nanoparticles; (**a**) bacterial cell and nanoparticles; (**b**) infiltration of nanoparticles into bacterial cell through cell wall and plasma membrane; (**c**) formation of OMV and condensed DNA due to activity of nanoparticles; (**d**) cell lysis via OMV and cell death

17.3 Bactericidal Nanoparticles

Living organisms either plant or animals are eternally challenged with bacterial infection, and sometimes these infections appear into life intimidating ailment. Nanotechnology came up with the promising solution for diseases ensued due to bacterial infection. The conventional method of fighting bacterial infections gradually becomes ineffective due to mutation and antibacterial drug-resistant strains. Multidrug-resistant (MDR) bacteria are becoming a serious threat to life. As per the World Health Organization (WHO), MDR-tuberculosis (TB) is a type of TB that does not respond to two most powerful TB drugs rifampicin and isoniazid. Additionally, another form of MDR-TB is reported from 117 countries known as extensively drug-resistant TB (XDR-TB) which is a form of TB which does not respond to most of the TB medicines (WHO 2018). These MDR and XDR forms of TB become very critical challenges for microbiology and medicine scientists.

Another MDR strain posing a great challenge is methicillin-resistant *Staphylococcus aureus* (MRSA) also known as superbug, able to cause severe infection in hospitals as well as community settings with a high mortality rate. The bacteria form biofilm in combination with drug resistance which makes it more difficult to kill from any conventional antibiotic (Vanamala et al. 2021).

Researchers of nanobiotechnology and nanomedicine reported promising responses of nanomaterials as antibacterial against MDR and XDR microorganisms (Natan and Banin 2017; Yousefi et al. 2017; Vanamala et al. 2021). Many nanoparticles are found to have potential to inhibit bacterial growth. Researchers reported promising bactericidal property of gold nanoparticles in one of the studies conducted upon Gram-positive *Staphylococcus aureus*, *Streptococcus pyogenes*, Gram-negative *Pseudomonas aeruginosa*, and *Escherichia coli* and against *Mycobacterium tuberculosis* H37RV bacterial pathogens.

With the help of TEM micrograph and minimum inhibition concentration (MIC) analysis, the study reveals formation of OMV and DNA conglomeration occur due to presence of gold nanoparticles and further cell lysis (Srivastava and Mukhopadhyay 2015a, b). Nanoparticles of silver are extensively studied for antibacterial activity followed with gold, zinc oxide, and copper. Selenium nanoparticles also exhibit prominent response as antibacterial. Several studies reported exciting bactericidal activity of selenium against various pathogens (Srivastava and Mukhopadhyay 2013, 2015a, b; Stevanović et al. 2015; Nguyen et al. 2017; Shoeibi and Mashreghi 2017; Geoffrion et al. 2020; Ahmed et al. 2021; Salem et al. 2021).

The antibacterial activity of nanoparticles against various bacterial species is summarized in Table 17.1 for better understanding. Nanoparticles can be utilized as potential antibacterial substances, as independent material, in combination with conventional drugs, as drug carrier, for targeted delivery, for antibacterial protection film, for antifouling, and for antibacterial coating over tools and instruments in hospitals, laboratories, and other places of high contamination.

Table 17.1 Bactericidal nanoparticles

Nanomaterial	Size of nanomaterial	Bactericidal activity on bacterial pathogen(s)	References
Selenium	~80 nm and ~10 nm	*Multidrug-resistant E. coli (MDR-EC)* *Pseudomonas aeruginosa (PA)* *Methicillin-resistant S. aureus (MRSA)* *Staphylococcus epidermidis (SE)*	Geoffrion et al. (2020)
Selenium	40–120 nm	*Pseudomonas aeruginosa* *Staphylococcus aureus* *Escherichia coli* *Streptococcus pyogenes*	Srivastava and Mukhopadhyay (2015a, b)
Silver	25–50 nm	*Staphylococcus aureus* *Acinetobacter baumannii* *Escherichia coli* *Escherichia coli O157:H7* *Klebsiella pneumoniae* *Pseudomonas aeruginosa*	Ontong et al. (2020)
Silver	10–50 nm	*Methicillin-resistant Staphylococcus aureus* *Proteus mirabilis* *Salmonella typhi* *Acinetobacter baumannii* *Escherichia coli* *Salmonella typhimurium*	Mishra et al. (2020)
Silver	Avg 37.9 nm	*Staphylococcus aureus*	Mbae and Umesha (2020)
Gold	4–16 nm	*Pseudomonas aeruginosa* *Staphylococcus aureus* *Escherichia coli* *Streptococcus pyogenes* *Mycobacterium tuberculosis H37Rv*	Srivastava and Mukhopadhyay (2015a, b)
Silver	10–20 nm	*Streptococcus haemolyticus* *Corynebacterium diphtheriae* *Klebsiella ornithinolytica* *Salmonella paratyphi*	Jose and Kuriakose (2019)
Gold	25 nm	*Staphylococcus aureus* *Enterococcus faecalis* *Klebsiella pneumonia* *Clostridium sporogenes*	Folorunso et al. (2019)
ZnO	238 nm	*Escherichia coli* *Salmonella choleraesuis* *Staphylococcus aureus* *No activity against Pseudomonas aeruginosa, Lactobacillus plantarum, and Listeria monocytogenes*	Espitia et al. (2013)
Silver	~13.4 nm	*Staphylococcus aureus* *Escherichia coli*	IllanesTormena et al. (2020)
Gold	80 nm	*Serratia marcescens, Streptococcus mutans* *Candida albicans, Proteus* sp., *Pseudomonas fluorescens*	Rajasekar et al. (2020)

(continued)

Table 17.1 (continued)

Nanomaterial	Size of nanomaterial	Bactericidal activity on bacterial pathogen(s)	References
		Micrococcus luteus, Salmonella typhi, Escherichia coli, and Staphylococcus aureus	
Selenium Selenium-lysozyme hybrid	35.6 nm	*Escherichia coli and Staphylococcus aureus*	Vahdati and Tohidi Moghadam (2020)
Selenium	29.1–48.9 nm	*Staphylococcus aureus* *Bacillus subtilis* *Escherichia coli* *Pseudomonas aeruginosa*	Salem et al. (2021)
Curcumin	170–230	*Escherichia coli*	Shlar et al. (2015)
CdO nanoparticle	20 nm	*Staphylococcus aureus, Bacillus subtilis, Escherichia coli, and Salmonella paratyphi*	Somasundaram and Rajan (2019)
Cu	40–50 nm	*Escherichia coli*	Viswadevarayalu et al. (2016)
Graphene oxide	NA	MDR	Yousefi et al. (2017)

17.4 Fungicidal Nanoparticles

Nanomaterials not only act as great bactericidal but also are known for its excellent fungicidal activity. Diseases associated with fungi are another form of challenge in medical science research and practice. There are several fungal diseases of plants, animals, and humans which are able to cause huge loss in forms of food crop, economy, vegetation, farm, and life. Fungal diseases in humans associated with fungus like *Aspergillus clavatus*, *Aspergillus niger*, *Candida auris*, and *Cryptococcal meningitis* are few from the list which can cause serious illness leading to death (Srivastava and Mukhopadhyay 2015a, b; CDC 2021). As per the Centers for Disease Control and Prevention (CDC), opportunistic pathogens such as cryptococcosis and aspergillosis affect immune-compromised weak people like cancer patients, organ transplant recipients, and people with HIV/AIDS, whereas candidemia infection in healthcare workers and *coccidioidomycosis* (valley fever) and histoplasmosis diseases from fungi that occur due to change in environment are some of the serious fungal infections that need serious attention of researchers (CDC 2021). Table 17.2 summarizes various types of fungal pathogens and nanoparticles found to be effective against those fungi.

Table 17.2 Fungicidal nanoparticles

Nanomaterial	Size of nanomaterial	Fungicidal activity on fungal pathogen(s)	References
Silver	25–50 nm	*Candida albicans*	Ontong et al. (2020)
Selenium	40–120 nm	*Aspergillus clavatus*	Srivastava and Mukhopadhyay (2015a, b)
Silver	50 nm	*Aspergillus fumigatus* *Alternaria alternata* *Trichoderma parceramosum* *Penicillium citrinum* *Paecilomyces variotii* *Candida albicans* *Candida glabrata* *Trichophyton mentagrophytes* *Microsporum gypseum*	Gholami-Shabani et al. (2014)
Gold	25 nm	*Aspergillus flaws* *Candida albicans* *Fusarium oxysporum* *Penicillium camemberti*	Folorunso et al. (2019)
ZnO	238 nm	*Saccharomyces cerevisiae* *Aspergillus niger*	Espitia et al. (2013)
Silver	~12 nm	*Candida albicans* *Aspergillus fumigatus* *Aspergillus flavus* *Cryptococcus neoformans*	Singh et al. (2013)
CdO nanoparticle	20 nm	*Candida albicans* *Aspergillus niger* *Aspergillus fumigates*	Somasundaram and Rajan (2019)
Cu	40–50 nm	*Candida tropicalis*	Viswadevarayalu et al. (2016)

17.5 Virucidal Nanoparticles

Viruses are always a remaining pathogen of high concern. Since the inception of mankind, viruses alone claimed millions of human lives at various intervals. Recent outbreak of SARS-CoV-2 to outbreak of Ebola in Africa, SARS-CoV, MERS, H1N1 Spanish flu, Zika virus, H1N5, encephalitis, and HIV are few names from the deadliest list that claimed millions of human lives till date. Additionally, every year a large number of plants and crops are lost due to viral infection. The incessant challenge of viruses needs to be addressed and a proper solution for the same needs to be devised. Taking all the severity and challenges associated with viral diseases, it is important to find and adopt cost-effective, efficient, target-specific, safe, nontoxic, environmentally friendly, and biocompatible approach for treatment and prevention. Nanomaterials are one of the potential candidates which may fit in all of these parameters due to their exciting properties, small size, high surface to volume

Table 17.3 Virucidal nanoparticles

Nanomaterial	Size of nanomaterial	Virucidal activity on viruses	References
Silver	>100 nm	*Banana bunchy top virus* (BBTV)	Mahfouze et al. (2020)
Gold	7–15 nm	Measles virus (Edmonston strain)	Meléndez-Villanueva et al. (2019)
Gold, silver, copper, zinc and iron oxide	>100 nm	SARS-CoV-2	Sarkar and Das Mukhopadhyay (2021)
Gold	~2 nm	HIV Influenza virus (H1N1, H3N2, H5N1)	Di Gianvincenzo et al. (2010)
Silver	20 nm	Transmissible gastroenteritis virus (TGEV)-coronavirus	Lv et al. (2014)
Silver	1–10 nm	HIV-1	Elechiguerra et al. (2005)
Iron oxide		Influenza virus	Qin et al. (2019)

ratio, high surface plasmon resonance, and supramagnetic, paramagnetic, biocompatibility, and adjustable surface potential. Nanoparticles of silver, gold, ZnO, quantum dots, graphene oxide, and liposome were reported for their virucidal activity against HIV-1, human papillomavirus, herpes simplex virus, SARS, MERS, human parainfluenza virus, H1N1, H3N2, hepatitis B, dengue virus, and foot and mouth disease. Researchers continuously work to develop nanotechnology-driven sustainable solutions for viral diseases. Table 17.3 depicts the virucidal nanoparticles and their activity against viruses.

17.6 Challenges and Future Directions

Nanotechnology being a very advanced branch of materials science is very less known, and several aspects of it need to be unfolded. The mechanism of action of nanomaterials as antimicrobial is still unclear and demands more focused and specified research. Microbial pathogens are always a topic of major concern for the scientific fraternity and need to be dealt with in a sophisticated manner. There are various reports stating the emergence of microbial resistance toward conventional antibiotics. The antibiotic resistance is rising as a strong challenge and demands an urgent solution. Nanotechnology undoubtedly has the potential to become a good option for existing antibiotics or can be utilized in combination with existing antibiotics; still the dark side of technology needs to be addressed. Nanotoxicity, effect of nano-system on plants and animals, cytotoxicity of nanomaterials, effect on environment, and economical aspect need to be addressed before incorporating it as microbicidal. Moreover, more extensive and detailed research is required to be carried out to explore nanomaterials' potential as independent microbiocidal, in combination with one or more nanomaterials, in combination with antibiotics, for drug delivery, target-specific delivery, and therapeutic purpose. Furthermore, pre-clinical and clinical trials at several levels are required for better understanding of nanomaterials' antimicrobial efficiency in light of all pros and cons. In contrast to conventional antibiotics, nanomaterials have multiple mechanisms to initiate a variety of cellular responses making them suitable candidates for fighting with multidrug-resistant microbes and complex microbes like parasites and viruses.

17.7 Conclusions

Nanomaterials have a variety of applications in each and every branch of science and technology. Life sciences, biotechnology, bioinstrumentation, and medicine have wide applications of nanotechnology. Antimicrobial application of nanomaterials for inhibition and control of microbial growth is one of the major areas of research in microbiology, biotechnology, materials science, and medicine. Nanomaterials are found to be very useful in surface coating to prevent rusting and microbial infections, surface coating of nanomaterial on hospital instruments and surgical tools can prevent variety of microbial contamination, and combination of nanomaterials with conventional antibiotics is found to be very effective against pathogenic microbes and can be a target-specific vehicle for delivery of drugs. The rising microbial resistance can be ruled out by application of various types of available nanomaterials. Moreover, life-threatening virus-borne diseases can be tackled more accurately by application of nanomaterials. Nanomaterials display their antimicrobial potential via complex mechanisms. Nanoparticles' adhesion to microbial cells, infiltration inside the microbial cell, free radical and ROS generation, and variation of microbial signal transduction pathways have been recognized as the most prominent means of antimicrobial activity (Dakal et al. 2016). Nanomaterials are found to have promising virucidal properties and it can be of high use in control of viral pandemics in future. In contrast to medical science, the research on nanoparticles as plant microbicidal is at the very early stage and requires more focused research. The exciting physicochemical properties of nanomaterials make its application in various fields, and microbicidal is one of the applications of these tiny smart particles. In the near future, these small materials are going to be part of our life and also prevent us from various infections as microbicides are sustainable microbicidal.

Acknowledgments The authors are grateful to the Vice Chancellor, King George's Medical University (KGMU), Lucknow, India, for the encouragement of this work. The authors have no other relevant affiliations or financial involvement with any organization or entity with a financial interest in or financial conflict with the subject matter or materials discussed in the manuscript apart from those disclosed.

References

Ahmed MK, Moydeen AM, Ismail AM, El-Naggar ME, Menazea AA, El-Newehy MH (2021) Wound dressing properties of functionalized environmentally biopolymer loaded with selenium nanoparticles. J Mol Struct 1225:129138

CDC (2021) CDC at work: mycotic diseases branch. https://www.cdc.gov/fungal/cdc-and-fungal.html?CDC_AA_refVal=https%3A%2F%2Fwww.cdc.gov%2Ffungal%2Fglobal%2Findex.html

Dakal TC, Kumar A, Majumdar RS, Yadav V (2016) Mechanistic basis of antimicrobial actions of silver nanoparticles. Front Microbiol 7:1831

Di Gianvincenzo P, Marradi M, Martínez-Ávila OM, Bedoya LM, Alcamí J, Penadés S (2010) Gold nanoparticles capped with sulfate-ended ligands as anti-HIV agents. Bioorg Med Chem Lett 20(9):2718–2721

Dizaj SM, Lotfipour F, Barzegar-Jalali M, Zarrintan MH, Adibkia K (2014) Antimicrobial activity of the metals and metal oxide nanoparticles. Mater Sci Eng C 44:278–284

Elechiguerra JL, Burt JL, Morones JR, Camacho-Bragado A, Gao X, Lara HH, Yacaman MJ (2005) Interaction of silver nanoparticles with HIV-1. J Nanobiotechnol 3:6–6

Espitia PJP, Soares N d FF, Teófilo RF, Vitor DM, Coimbra JS d R, de Andrade NJ, de Sousa FB, Sinisterra RD, Medeiros EAA (2013) Optimized dispersion of ZnO nanoparticles and antimicrobial activity against foodborne pathogens and spoilage microorganisms. J Nanopart Res 15(1):1324

Feng QL, Wu J, Chen GQ, Cui FZ, Kim TN, Kim JO (2000) A mechanistic study of the antibacterial effect of silver ions on Escherichia coli and Staphylococcus aureus. J Biomed Mater Res 52(4):662–668

Fernando S, Gunasekara T, Holton J (2018) Antimicrobial nanoparticles: applications and mechanisms of action. Sri Lankan J Infect Dis 8(1):2–11

Folorunso A, Akintelu S, Oyebamiji AK, Ajayi S, Abiola B, Abdusalam I, Morakinyo A (2019) Biosynthesis, characterization and antimicrobial activity of gold nanoparticles from leaf extracts of Annona muricata. J Nanostruct Chem 9(2):111–117

Gakiya-Teruya M, Palomino-Marcelo L, Pierce S, Angeles-Boza AM, Krishna V, Rodriguez-Reyes JCF (2020) Enhanced antimicrobial activity of silver nanoparticles conjugated with synthetic peptide by click chemistry. J Nanopart Res 22(4):90

Geoffrion LD, Hesabizadeh T, Medina-Cruz D, Kusper M, Taylor P, Vernet-Crua A, Chen J, Ajo A, Webster TJ, Guisbiers G (2020) Naked selenium nanoparticles for antibacterial and anticancer treatments. ACS Omega 5(6):2660–2669

Gholami-Shabani M, Akbarzadeh A, Norouzian D, Amini A, Gholami-Shabani Z, Imani A, Chiani M, Riazi G, Shams-Ghahfarokhi M, Razzaghi-Abyaneh M (2014) Antimicrobial activity and physical characterization of silver nanoparticles green synthesized using nitrate reductase from Fusarium oxysporum. Appl Biochem Biotechnol 172(8):4084–4098

IllanesTormena RP, Rosa EV, Oliveira Mota B d F, Chaker JA, Fagg CW, Freire DO, Martins PM, Rodrigues da Silva IC, Sousa MH (2020) Evaluation of the antimicrobial activity of silver nanoparticles obtained by microwave-assisted green synthesis using Handroanthus impetiginosus (Mart. ex DC.) Mattos underbark extract. RSC Adv 10(35):20676–20681

Jose LM, Kuriakose S (2019) Spectroscopic and thermal investigation of silver nanoparticle dispersed biopolymer matrix bovine serum albumin: a promising antimicrobial agent against the pathogenic bacterial strains. Macromol Res 27(7):670–678

Lv X, Wang P, Bai R, Cong Y, Suo S, Ren X, Chen C (2014) Inhibitory effect of silver nanomaterials on transmissible virus-induced host cell infections. Biomaterials 35(13): 4195–4203

Mahfouze HA, El-Dougdoug NK, Mahfouze SA (2020) Virucidal activity of silver nanoparticles against Banana bunchy top virus (BBTV) in banana plants. Bull Natl Res Cent 44(1):199

Mbae KM, Umesha S (2020) Physicochemical and antimicrobial properties of post-synthesis betanin and chitosan oligosaccharide functionalized silver nanoparticles. J Nanopart Res 22(11):346

Meléndez-Villanueva MA, Morán-Santibañez K, Martínez-Sanmiguel JJ, Rangel-López R, Garza-Navarro MA, Rodríguez-Padilla C, Zarate-Triviño DG, Trejo-Ávila LM (2019) Virucidal activity of gold nanoparticles synthesized by green chemistry using garlic extract. Viruses 11(12):1111

Mishra AK, Tiwari KN, Saini R, Kumar P, Mishra SK, Yadav VB, Nath G (2020) Green synthesis of silver nanoparticles from leaf extract of Nyctanthes arbor-tristis L. and assessment of its antioxidant, antimicrobial response. J Inorg Organomet Polym Mater 30(6):2266–2278

Mukherjee P, Bhattacharya R, Bone N, Lee YK, Patra CR, Wang S, Lu L, Secreto C, Banerjee PC, Yaszemski MJ, Kay NE, Mukhopadhyay D (2007) Potential therapeutic application of gold nanoparticles in B-chronic lymphocytic leukemia (BCLL): enhancing apoptosis. J Nanobiotechnol 5:4–4

Natan M, Banin E (2017) From Nano to Micro: using nanotechnology to combat microorganisms and their multidrug resistance. FEMS Microbiol Rev 41(3):302–322

Nguyen THD, Vardhanabhuti B, Lin M, Mustapha A (2017) Antibacterial properties of selenium nanoparticles and their toxicity to Caco-2 cells. Food Control 77:17–24

Ontong JC, Singh S, Nwabor OF, Chusri S, Voravuthikunchai SP (2020) Potential of antimicrobial topical gel with synthesized biogenic silver nanoparticle using Rhodomyrtus tomentosa leaf extract and silk sericin. Biotechnol Lett 42(12):2653–2664

Pal D, Sahu C, Haldar A (2014) Bhasma: the ancient Indian nanomedicine. J Adv Pharm Technol Res 5(1):4–12

Qin T, Ma R, Yin Y, Miao X, Chen S, Fan K, Xi J, Liu Q, Gu Y, Yin Y, Hu J, Liu X, Peng D, Gao L (2019) Catalytic inactivation of influenza virus by iron oxide nanozyme. Theranostics 9(23): 6920–6935

Rajasekar T, Karthika K, Muralitharan G, Maryshamya A, Sabarika S, Anbarasu S, Revathy K, Prasannabalaji N, Kumaran S (2020) Green synthesis of gold nanoparticles using extracellular metabolites of fish gut microbes and their antimicrobial properties. Braz J Microbiol 51(3): 957–967

Salem SS, Fouda MMG, Fouda A, Awad MA, Al-Olayan EM, Allam AA, Shaheen TI (2021) Antibacterial, cytotoxicity and larvicidal activity of green synthesized selenium nanoparticles using Penicillium corylophilum. J Clust Sci 32(2):351–361

Sarkar PK, Das Mukhopadhyay C (2021) Ayurvedic metal nanoparticles could be novel antiviral agents against SARS-CoV-2. Int Nano Lett 6:1–7. https://doi.org/10.1007/s40089-020-00323-9

Sharma R, Prajapati P (2016) Nanotechnology in medicine: leads from Ayurveda. J Pharm Bioallied Sci 8(1):80–81

Shlar I, Poverenov E, Vinokur Y, Horev B, Droby S, Rodov V (2015) High-throughput screening of nanoparticle-stabilizing ligands: application to preparing antimicrobial curcumin nanoparticles by antisolvent precipitation. Nano-Micro Lett 7(1):68–79

Shoeibi S, Mashreghi M (2017) Biosynthesis of selenium nanoparticles using Enterococcus faecalis and evaluation of their antibacterial activities. J Trace Elem Med Biol 39:135–139

Singh M, Kumar M, Kalaivani R, Manikandan S, Kumaraguru AK (2013) Metallic silver nanoparticle: a therapeutic agent in combination with antifungal drug against human fungal pathogen. Bioprocess Biosyst Eng 36(4):407–415

Somasundaram G, Rajan J (2019) Ascendancy of Polianthes tuberosa, Nerium oleander, Hibiscus rosa sinensis and Dalia flower extracts on CdO nanoparticle morphologies and their effectiveness in photocatalytic and antimicrobial activities. J Inorg Organomet Polym Mater 29(6): 2145–2160

Srivastava N, Mukhopadhyay M (2013) Biosynthesis and structural characterization of selenium nanoparticles mediated by Zooglearamigera. Powder Technol 244:26–29

Srivastava N, Mukhopadhyay M (2015a) Biosynthesis and characterization of gold nanoparticles using Zooglea ramigera and assessment of its antibacterial property. J Clust Sci 26(3):675–692

Srivastava N, Mukhopadhyay M (2015b) Green synthesis and structural characterization of selenium nanoparticles and assessment of their antimicrobial property. Bioprocess Biosyst Eng 38(9):1723–1730

Srivastava N, Saxena SK (2020) Prevention and control strategies for SARS-CoV-2 infection. In: Saxena SK (ed) Coronavirus Disease 2019 (COVID-19): epidemiology, pathogenesis, diagnosis, and therapeutics. Springer, Singapore, pp 127–140

Srivastava N, Baxi P, Ratho RK, Saxena SK (2020) Global trends in epidemiology of Coronavirus Disease 2019 (COVID-19). In: Saxena SK (ed) Coronavirus Disease 2019 (COVID-19): epidemiology, pathogenesis, diagnosis, and therapeutics. Springer, Singapore, pp 9–21

Stevanović M, Filipović N, Djurdjević J, Lukić M, Milenković M, Boccaccini A (2015) 45S5Bioglass®-based scaffolds coated with selenium nanoparticles or with poly(lactide-co-glycolide)/selenium particles: processing, evaluation and antibacterial activity. Colloids Surf B: Biointerfaces 132:208–215

Vahdati M, Tohidi Moghadam T (2020) Synthesis and characterization of selenium nanoparticles-lysozyme nanohybrid system with synergistic antibacterial properties. Sci Rep 10(1):510

Vanamala K, Tatiparti K, Bhise K, Sau S, Scheetz MH, Rybak MJ, Andes D, Iyer AK (2021) Novel approaches for the treatment of methicillin-resistant Staphylococcus aureus: using nanoparticles to overcome multidrug resistance. Drug Discov Today 26(1):31–43

Viswadevarayalu A, Venkata Ramana P, Sreenivasa Kumar G, Rathna Sylvia L, Sumalatha J, Adinarayana Reddy S (2016) Fine Ultrasmall Copper Nanoparticle (UCuNPs) synthesis by using Terminalia bellirica fruit extract and its antimicrobial activity. J Clust Sci 27(1):155–168

WHO (2018) Tuberculosis: multidrug-resistant tuberculosis (MDR-TB). Global Tuberculosis Programme. https://www.who.int/news-room/q-a-detail/tuberculosis-multidrug-resistant-tuberculosis-(mdr-tb)

Yousefi M, Dadashpour M, Hejazi M, Hasanzadeh M, Behnam B, de la Guardia M, Shadjou N, Mokhtarzadeh A (2017) Anti-bacterial activity of graphene oxide as a new weapon nanomaterial to combat multidrug-resistance bacteria. Mater Sci Eng C 74:568–581

Advancement in Water Purification by Nanomaterials

18

Andrei Ivanets, Vladimir Prozorovich, and Tatyana Kouznetsova

Abstract

Nanomaterials are widely used for practical application in environmental chemical engineering and water purification technologies. The non-reagent methods (adsorption, catalysis, membrane technology, etc.) attracts the attention as the most prospect approaches for liquid radioactive waste treatment and waste water purification. Nanostructured metal oxides materials showed the enhanced characteristics for preparation of mesoporous adsorbents for selective radionuclides removal, heterogeneous catalysts for harmful organic pollutants degradation Fenton-like processes, as well as ceramic membranes for microfiltration water treatment. The chapter described the novel results for preparation, characterization and practical application of nanomaterials based on various metal oxides for water purification. The main factors affected on regularities between synthesis conditions, structure and functional properties were studied. The obtained results have significant impact for advancement in water purification by nanomaterials.

Keywords

Nanostructured metal oxides · Mesoporous adsorbents · Fenton-like catalysts · Microfiltration membrane · Water treatment

A. Ivanets (✉) · V. Prozorovich · T. Kouznetsova
Institute of General and Inorganic Chemistry, National Academy of Sciences of Belarus, Minsk, Belarus
e-mail: ivanets@igic.bas-net.by

A. Thakur et al. (eds.), *Synthesis and Applications of Nanoparticles*,
https://doi.org/10.1007/978-981-16-6819-7_18

18.1 Introduction

Intensive operation of nuclear power facilities is certainly accompanied by the formation and accumulation of large volumes of liquid radioactive waste (LRW). In accordance with the IAEA regulatory documents on the safe management of LRW, this waste is subject to mandatory processing before subsequent disposal. LRW most often have a complex chemical and radionuclide composition; in particular, it contains long-lived isotopes ^{137}Cs ($\tau_{1/2} = 30.2$ years), ^{60}Co ($\tau_{1/2} = 5.3$ years), and ^{90}Sr ($\tau_{1/2} = 28.8$ years). According to the literature data, ^{90}Sr radionuclide poses a serious hazard to humans and the environment (Pathak and Gupta 2020).

One of the main disadvantages of classical methods of LRW treatment, including ion exchange, membrane technology, and precipitation, is the lack of the possibility of selective removal of the ^{90}Sr radionuclide from multicomponent aqueous solutions containing competing ions of alkali and alkaline earth metals (Okoshi and Momma 2015). At the same time, sorption technology makes it possible to use efficient sorption materials with unique selectivity. Such sorption materials include manganese oxides, which exhibit exceptional selectivity toward ^{90}Sr radionuclide even in the presence of Na^+, K^+, Ca^{2+}, and Mg^{2+} ions (Pakarinen et al. 2010; Ivanets et al. 2019a, b). High sorption characteristics, such as unique selectivity, high sorption capacity, and fast sorption kinetics, can be assumed to be due to a number of factors. First, selectivity is related to the presence and availability of micropores and is due to the layered (interlayer space ≤ 10 Å) or channel structure (channel size 4.6–7.0 Å). Second, the sorption capacity and sorption kinetics depend on the availability of active sites for the ^{90}Sr radionuclide in micropores and their good communication with mesopores, which play the role of transport channels and improve electrolyte diffusion toward active sites.

Currently, pharmacologically active compounds (PhACs) and their metabolites are found everywhere in wastewater, groundwater, and surface and even drinking water, posing a great threat to human health and the aquatic ecosystem (Majumder et al. 2019). The high stability, low volatility, accumulative and mutagenic properties, contamination with antibiotics, hormonal and nonsteroidal anti-inflammatory drugs leads to irreversible negative consequences for humans and the environment. The main sources of environmental pollution by these pollutants are agriculture, wastewater from pharmaceutical enterprises, garbage disposal, etc. Also, many PhACs are not completely destroyed in the process of metabolism by humans and animals and are found in domestic and agricultural wastewater (Michael et al. 2014). The Fenton reaction is one of the most studied varieties of advanced oxidation processes (AOPs), which produces OH radicals. Of particular interest as heterogeneous Fenton catalysts are metal ferrites with the general formula $M^{2+}Fe^{3+}{}_2O_4$, where M: Mg^{2+}, Mn^{2+}, Co^{2+}, Ni^{2+}, Cu^{2+}, characterized by high stability, developed surface, small particle size, and high catalytic activity (Kefeni and Mamba 2020; Amiri et al. 2019; Ivanets et al. 2020).

Due to their high chemical and thermal stability, ceramic membranes are widely used in wastewater treatment processes (Goh and Ismail 2018). The use of ceramic membranes is limited by their low productivity compared to polymer membranes and the high sintering temperature of traditional metal oxides (Al_2O_3, ZrO_2, TiO_2) used for their production (Zuo et al. 2020). This is why researchers are interested in finding more affordable materials for creating ceramic membranes. The use of natural minerals and industrial waste is most attractive (Abdullayev et al. 2019), which reduces the cost of ceramic precursors and the temperature of heat treatment of membranes. Previously, the possibility of obtaining ceramic microfiltration membranes based on natural silicon oxide was shown. The main factors that determine the mechanical and transport characteristics are the nature and concentration of the binder and pore-forming additives, as well as the chemical and phase composition of natural quartz sand, which significantly varies depending on the deposit (Ivanets et al. 2016a; Ivanets and Agabekov 2017).

18.2 Mesoporous Manganese Oxides for ^{85}Sr Radionuclide Adsorption

Manganese oxides are widely used as effective sorption materials for heavy metal ions and radionuclides ($^{85,90}Sr$, ^{137}Cs, etc.) (Aminul Islam et al. 2018; Voronina et al. 2020). This is due to the presence of more than 14 polymorphic modifications and, accordingly, the characteristic features of its crystalline and chemical structure, which largely determine high sorption and selective properties (in the presence of competing ions Na^+, Ca^{2+}, etc.) toward ^{90}Sr radionuclide. In a negatively charged manganese-oxygen framework, MnO_6 octahedra can form tunnel (cryptomelane, todorokite, etc.) or layered (birnessite, etc.) structures that are capable of accommodating positively charged cations and/or water molecules, stabilizing their structure. In this work, mesoporous manganese oxides were obtained using a template-free solgel synthesis method. An aqueous solution of potassium permanganate was used as a precursor; hydrogen peroxide and a manganese (II) salt were used as reducing agents. In addition, the adsorption properties (removal efficiency, distribution coefficient) of the obtained oxides toward ^{85}Sr radionuclide were studied (Ivanets et al. 2015, 2016b).

The gas adsorption-textural properties of obtained manganese oxides were determined by using the low-temperature adsorption/desorption of N_2 analysis (Fig. 18.1). The type IV of sorption isotherms was determined by the presence of the loops of the capillary-condensation hysteresis for obtained manganese oxides (samples 1–8), which also indicates the presence of a mesoporous structure by the IUPAC classification (Thommes et al. 2015).Different shapes of hysteresis loops indicate the formation of mesopores of various shapes. So, on the isotherms of samples 1–3 (reducing agent (RA), H_2O_2; processing temperature (PT), 80, 200, and 400 °C), there is H2 hysteresis, which is characteristic of bottle-like and wide-neck mesopores, for which the size of the cylindrical necks is greater than the half-width of spherical voids (Fig. 18.1a, b). The pores of sample 4 (RA, H_2O_2; PT, 600 °C) were equivalent to the cylindrical pores according to H1 hysteresis (Fig. 18.1b). The hybrid shape of hysteresis loops (Fig. 18.1a, c) consists of H2 (at $p/p_0 \leq 0.60$), and H3 (at $p/p_0 \geq 0.60$) is typical for obtained samples 5, 6, and 8 (RA, $MnCl_2$; PT,

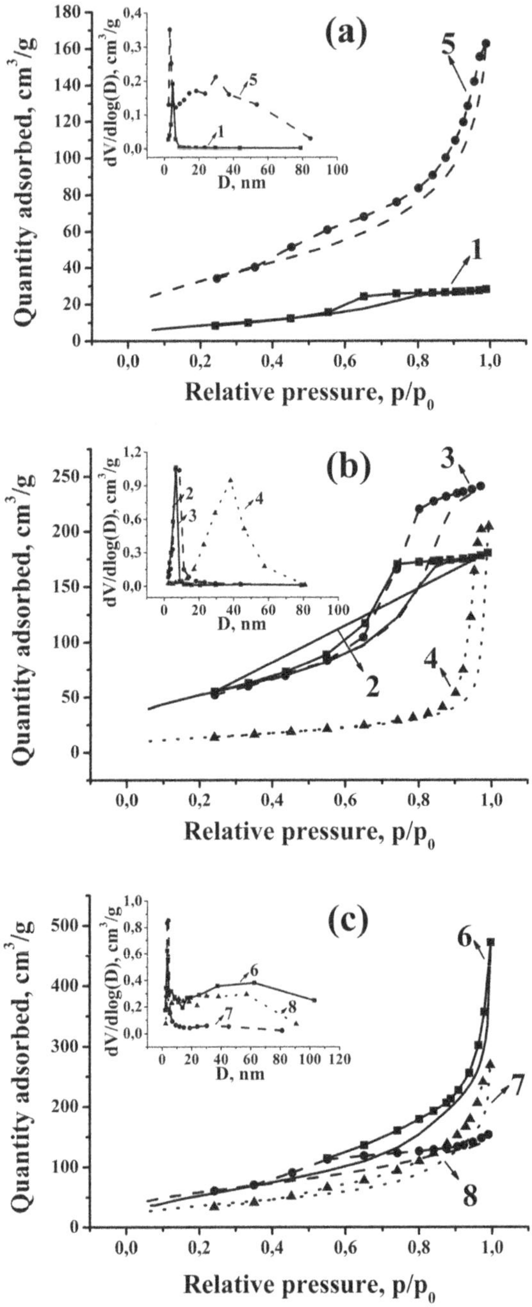

Fig. 18.1 Adsorption/desorption isotherms of N_2 and pore distribution curves of manganese oxide sorption materials (The samples were marked according to Table 18.1)

80, 200, and 600 °C) and indicates the presence of slit-shaped mesopores, observed in plate-like structures. For sample 7 (RA, $MnCl_2$; PT, 400 °C), the mesopores are mainly bottle-like according to the H2 hysteresis (Fig. 18.1c). Based on an analysis of the differential curves of the pore distribution size (PDS) for obtained samples 1–3, quite homogeneous structure was determined (Fig. 18.1a, b). As the temperature increased to 600°C, the intensity of *dV/dlogD* on the distribution curves decreased. The distribution of mesopores in sample 4 (PT—600 °C) was characterized by the presence of hysteresis in the range p/p_0 0.42–0.96 (Fig. 18.1b).

For obtained samples 1–3 and sample 4, the dominant bottleneck size was measured as the abscissa of the maximum on the PSD curve and amounted to ~5–8 nm and ~39 nm, respectively (D_{pr}; Fig. 18.1, Table 18.1). Close values (D_{BJHads} and D_{pr}) only confirm that the structure of the obtained manganese oxides (samples 1–3) is homogeneous. The PSD curves of samples 5, 6, and 8 (RA, $MnCl_2$; PT, 80, 200, and 600 °C) had two segments: (1) with well-discernible mode at D_{pr} ~ 3–4 nm and (2) with indistinguishable, diffuse peak maxima in the range of D_{pr} ~ 5–100 nm (at p/p_0 0.58–0.96).

This configuration of the distribution curves reflected the morphological properties of mesopores that differ in the shape, size, and volume (Table 18.1). Thus, the difference in mesopore shapes (from bottle-like pores with a homogeneous neck to slit-like pores with a wide scatter in thicknesses) indicates a significant effect of the processing temperature on phase transformations during obtaining.

When comparing the specific surface areas (Table 18.1), it was shown that for crystalline samples (2, 3 and 6, 7, respectively), it is significantly higher than for X-ray amorphous samples (1 and 5) and decreased at 600 °C due to the sintering process (samples 4 and 8). Analysis of XRD data indicates the formation of mixed phases with different compositions at the processing temperature range of 200–600 °C. The specific pore volumes changed in a more complex way.

During the synthesis of manganese oxides at the processing temperature (80 °C), particles of a dispersed phase are formed with an almost spherical shape diameter (200 nm), regardless of the choice of the reducing agent (Fig. 18.2).

Thus, the adsorption properties and morphology of manganese oxides obtained by the solgel method using reducing agents hydrogen peroxide and manganese chloride depend on the chemical processes at the stages of the conversion of sol into gel and subsequent heat treatment at the temperature range 80–600 °C. The factors responsible for the formation of crystalline nuclei in amorphous xerogel particles as the dominant of its internal rearrangement serve as the mechanism that triggers the subsequent phase transformations of xerogel. The X-ray amorphous samples (PT—80 °C) had a dense pore structure and low specific surface areas (32–120 m^2/g). At higher temperatures (200–600 °C), obtained xerogels crystallized into mixed phases with different compositions (excluding sample 4). The transformation of manganese oxide is important for its formation; variation of the treatment temperature allows to control over its phase composition and mesoporous structure.

The second stage of the work was the study of the sorption-selective properties of the obtained sorption materials based on manganese oxides toward ^{85}Sr radionuclide in static conditions at *V*/*m* ratio = 250 cm^3/g. The activity of initial model solutions (A_0, kBq·cm^3) and after sorption (A_e, kBq·cm^3) was carried out on a gamma beta spectrometer of MKS-AT1315 (Belarus). It was also investigated the effect of competing ions (Na^+, Ca^{2+}) presented in the model solution (with addition of

Table 18.1 The texture properties and phase composition of manganese oxide sorption materials

No	PT, °C	Hysteresis loops	A_{BET}, m^2/g	V_{ads}, cm^3/g	V_{BJHads}, cm^3/g	D_{ads}, nm	D_{BJHads}, nm	Phase composition
RA—H_2O_2								
1	80	H2	32	0.04	0.05	5.4	5.5	Amorphous
2	200	H2	188	0.28	0.28	6.1	6.3	α-MnO_2 + γ-MnO_2
3	400	H2	189	0.38	0.39	7.9	8.1	β-MnO_2 + γ-MnO_2
4	600	H1	49	0.26	0.32	21.1	28.5	β-MnO_2
RA—$MnCl_2$								
5	80	H2 + H3	120	0.23	0.25	7.7	9.6	Amorphous
6	200	H2 + H3	204	0.50	0.74	10.9	13.0	MnO + α-MnO_2
7	400	H2	212	0.23	0.20	4.5	6.1	α-MnO_2 + Mn_2O_3
8	600	H2 + H3	131	0.31	0.41	9.8	14.2	α-MnO_2 + Mn_5O_8

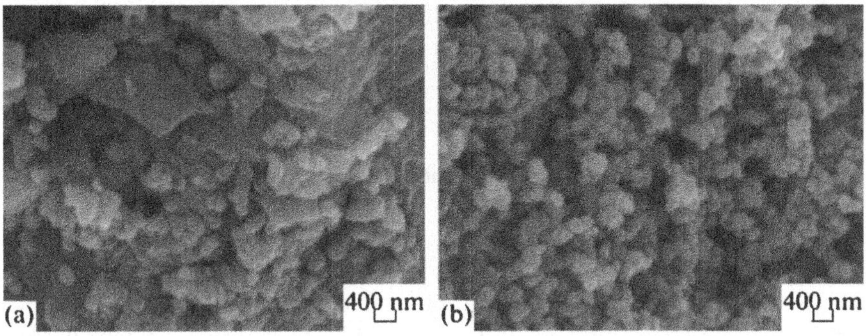

Fig. 18.2 SEM images of obtained manganese oxides synthesized with the reducing agent (**a**) H_2O_2 (×30 k) and (**b**) $MnCl_2$ (×40 k) and dried at 80 °C

Table 18.2 Adsorption of ^{85}Sr by mesoporous manganese oxides

	Chemical composition of the model solution					
	H_2O		0.1 MNaCl		0.05 M $CaCl_2$	
Sample	*S*, %	$K_d \cdot 10^{-2}$, cm^3/g	*S*, %	$K_d \cdot 10^{-2}$, cm^3/g	*S*, %	$K_d \cdot 10^{-2}$, cm^3/g
RA—H_2O_2						
1	94	96.7	17	1.1	7	0.4
2	95	106	19	1.2	10	0.6
3	100	3670	99	474	22	1.5
4	98	273	95	107	47	4.6
RA—$MnCl_2$						
5	71	12.8	41	3.5	11	0.6
6	86	30.6	45	4.1	12	0.7
7	96	211	63	1.2	19	8.7
8	99	725	75	15.6	18	1.1

0.1 M NaCl, 0.05 M $CaCl_2$) onto ^{85}Sr radionuclide removal efficiency (Ivanets et al. 2016b).

To assess the efficiency of the obtained sorption materials based on manganese oxides, the removal efficiency (*S*) and the distribution coefficient (K_d) of ^{85}Sr radionuclide in the presence of NaCl and $CaCl_2$ electrolytes (Table 18.2) were calculated, which amounted to 70–100% and (0.72–3.67) × 10^5 cm^3/g, respectively. Manganese oxide samples 1 and 5 demonstrated the lowest removal efficiency of ^{85}Sr radionuclide regardless of properties of initial sols (Table 18.2). Apparently this is due to small degree of samples' crystallinity and absence of the regular layer or channel structure forming micropores, as well as the lowest specific surface area of mesopores among the studied sorption materials. Introduction of the competing Na^+ ions into the model solution by adding 0.1 M NaCl led to significant decrease of the removal efficiency and distribution coefficient for all samples (Table 18.2). The highest removal efficiency of ^{85}Sr radionuclide in the presence of Na^+ ions was shown by samples 3–4 (RA, H_2O_2; PT, 400 and 600 °C). For these samples the K_d reached (1.06–4.74) × 10^4 cm^3/g. For all other samples the K_d value didn't exceed 1.56 × 10^3 cm^3/g.

In the case of using a model solution, containing Ca^{2+} ions, there was a further decrease in the efficiency of removing ^{85}Sr radionuclide by obtained sorption materials based on manganese oxides (Table 18.2).

The highest distribution coefficients of ^{85}Sr radionuclide in the presence of Ca^{2+} ions were shown by samples 3–4 (RA, H_2O_2; PT, 400 and 600 °C). For these samples the K_d reached 1.5–4.6 × 10^2 cm^3/g and 1.1–8.7 × 10^2 cm^3/g, respectively. The obtained results demonstrated that sorption-selective properties of the synthesized materials were compared with the widely using aluminasilicates, titanosilicates, and silicaantimonate adsorbents.

Thus, it was established that obtained manganese oxides using different types of the reducing agents (H_2O_2, $MnCl_2$) are effective sorption materials toward ^{85}Sr radionuclide and its K_d reached (0.72–3.67) × 10^5 cm^3/g. Presence of 0.1 M NaCl in model solution leads to decrease of K_d up to (4.74–1.06) × 10^4 cm^3/g. For comparison, the introduction of 0.05 M $CaCl_2$ to model solution of ^{85}Sr radionuclide was accompanied by an even stronger decrease of K_d up to (1.1–8.7) × 10^2 cm^3/g. It was found the influence of a reducing agent onto sorption properties, which are much higher when using H_2O_2 than $MnCl_2$. Samples of manganese oxides obtained at processing temperature of 400–600 °C demonstrated the highest sorption-selective properties in the presence of 0.1 M NaCl and 0.05 M $CaCl_2$.

18.3 Ibuprofen Degradation onto Nanostructured Magnesium Ferrite

Despite toxicological studies of the side effects of drugs on human and animal health, the potential environmental consequences of pharmaceutical production only became a topic of scientific interest in the 2000s. The presence of PhACs in wastewater and surface water was reported as early as the 1970s in the United States, which did not cause much concern at the time. Later, it was discovered that even small amounts of PhACs metabolites affect the environment and living organisms. The continuous increase in the population over the past few decades, its marked aging in many countries, globalization, and increased investment in health care have led to a significant increase in the use of pharmaceuticals. It is important to note that most medicines are not removed during the treatment of wastewater and domestic water, due to the low efficiency of treatment facilities in the removal of PhACs and their metabolites. This makes it necessary to develop new materials and technologies for natural and wastewater treatment (Rodriguez-Narvaez et al. 2017).

Catalytic degradation of organic compounds that are resistant to chemical and biological degradation, including PhACs, is one of the most promising methods for their removal. In this regard, in the last two decades, active research has been conducted on the development and application of AOPs, which involve the use of chemically stable, nontoxic, and highly active to a wide range of organic pollutants

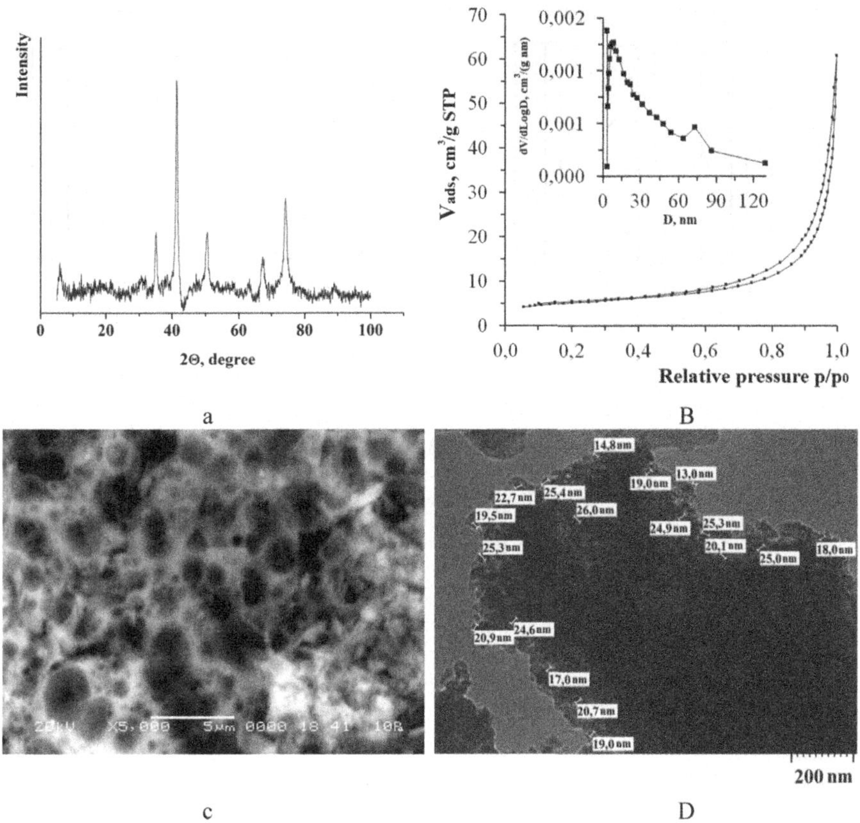

Fig. 18.3 XRD spectra (**a**), adsorption/desorption isotherms of N_2 and pore distribution curves (**b**), SEM image × 5 k (**c**), and TEM image × 50 k (**d**) of the $MgFe_2O_4$ sample

catalytic systems based on nanostructured metal oxides. Nanoparticles of magnesium ferrite were selected as the object of research due to its low toxicity and safety for living organisms in comparison with transition metal ferrites, which allows it to be used for solving a wide range of environmental problems (Ivanets et al. 2019c).

Figure 18.3a shows that the catalyst is a crystalline magnesium ferrite ($MgFe_2O_4$). The parameter a of the crystal lattice was 8.393 Å, which slightly differs from the reference value of 8.370 Å and indirectly indicated the presence of defects in the crystal structure of the studied catalyst. The crystallite size calculated by the Scherrer formula was 8.2 nm. It is obvious that the real structure was formed from agglomerates of the abovementioned crystallites, which lead to the formation of a mesoporous structure. This fact was confirmed by the presence of a hysteresis loop, which allows the isotherm, shown in Fig. 18.3b, to be classified as type IV according to the IUPAC classification (Thommes et al. 2015). The H1 hysteresis is typical for cylindrical pores formed by agglomerates of spherical particles-globules that are uniformly packed and close in size. The specific surface area calculated by the

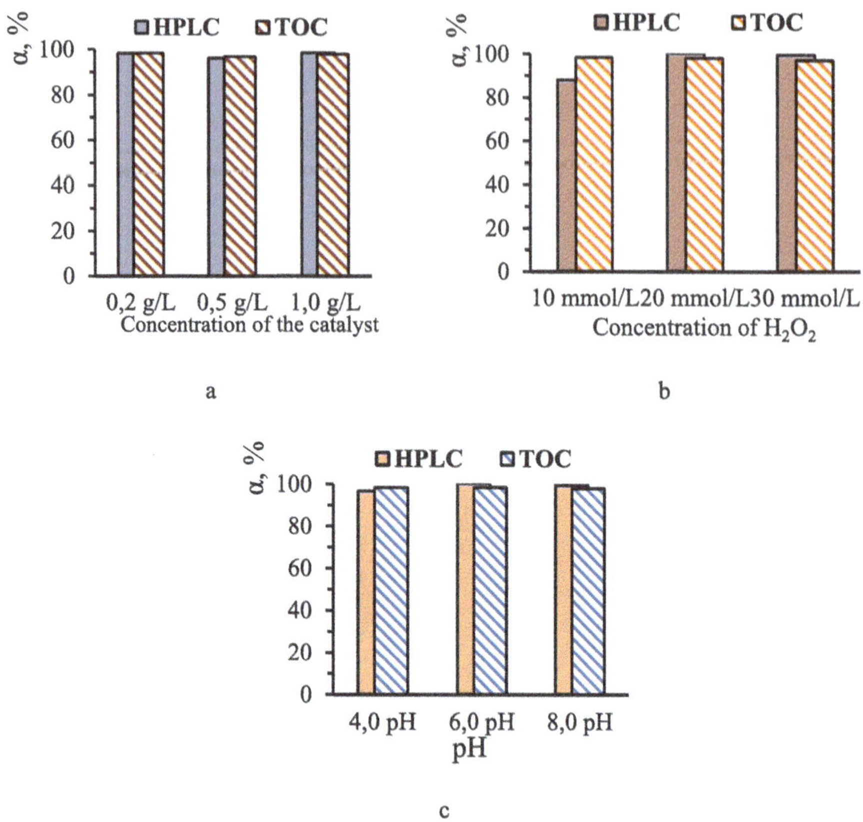

Fig. 18.4 Effect of (**a**) catalyst dose, (**b**) H_2O_2 concentration, (**c**) solution pH on the degree of ibuprofen catalytic degradation (α_{HPLC}, %) and mineralization (α_{TOC}, %)

single-point BET method was 14 m^2/g, the pore volume was 0.03 cm^3/g, and the average pore size was 8 nm. The $MgFe_2O_4$ sample consists of spherical agglomerates <1 μm in size (Fig. 18.3c), consisting of primary particles (crystallites) of 16–26 nm in size (Fig. 18.3d).

According to Fig. 18.4, the conditions of the catalytic process (the concentration of the catalyst and H_2O_2, pH of the solution) in the studied ranges did not significantly affect the efficiency of ibuprofen degradation. Thus, the degree of catalytic degradation of ibuprofen according to HPLC data for 40 min reached 98–100%, while the minimum residual concentration of ibuprofen under optimal conditions was below the detection limit. Despite the high efficiency of the process (Fig. 18.4a), reducing the concentration of catalyst of less than 0.2 g/L is impractical due to the complexity of the real aim of the amount of material in a given volume of treated solution, which corresponds to the ratio *V*(solution)/*m*(catalyst) more than 5000 L/g.

To ensure the highest efficiency of ibuprofen degradation under the specified conditions, the concentration of H_2O_2 should be at least 20.0 mmol/L (Fig. 18.4b).

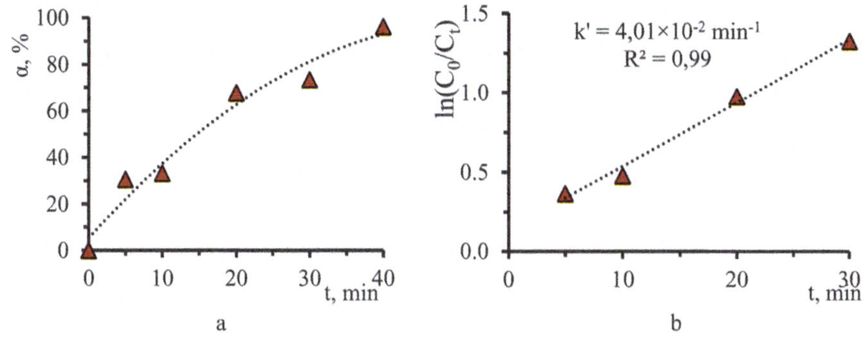

Fig. 18.5 Kinetics of ibuprofen degradation (**a**) and linear plot of first-order model (**b**). Reaction conditions: C($MgFe_2O_4$) 0.5 g/L, C(H_2O_2) 20.0 mmol/L, and pH 6.0

The pH range of the model solution varies from acidic (4.0) to slightly alkaline (8.0), which corresponds to conditions close to real wastewaters (Fig. 18.4c). It is important to note that the degree of ibuprofen degradation practically corresponded to the degree of mineralization calculated from the values of the residual total organic carbon. This indicates that it is possible to treat wastewater from pharmaceutically active compounds by completely mineralizing them without forming by-products and toxic products using a Fenton-like catalyst based on $MgFe_2O_4$ nanoparticles.

According to Fig. 18.5a, complete oxidative degradation of ibuprofen (100%) was achieved within 40 min of interaction. The dependence of the degree of destruction on time was monotonous, and the kinetics of the catalytic process was described with a high degree of confidence (R^2 0.99) by a first-order model (Fig. 18.5b). The calculated value of the apparent rate constant was 4.01×10^{-2} min^{-1}, which corresponded to the most effective catalysts described in the literature (Davarnejad et al. 2018).

Thus, it was shown the efficiency (degree of mineralization reached 100%) of catalytically active materials based on $MgFe_2O_4$ nanoparticles in the catalytic destruction of nonsteroidal anti-inflammatory drug ibuprofen. The conducted research showed the prospects of practical application of the developed Fenton-like heterogeneous catalyst for wastewater treatment from pharmaceutically active compounds.

18.4 Microfiltration Membranes Based on Silicate Ceramic

Ceramic membranes usually have an asymmetric structure: the top thin layer determines the separation properties and a large-pore base, which provides mechanical strength and increased permeability. In this case, as a rule, multistage sintering is used, which negatively affects the mechanical and transport properties and complicates the process of manufacturing membranes. The use of powders with the same chemical composition and different dispersion to obtain a porous base and

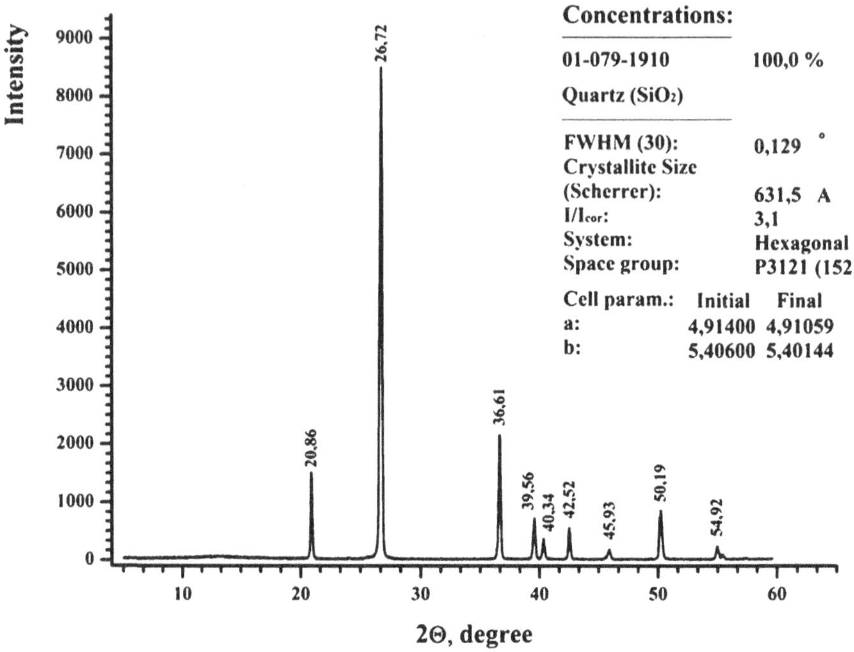

Fig. 18.6 XRD spectra of natural quartz sand

Table 18.3 Fractional composition of quartz sand

Fraction, μm	Content, wt.%
400–630	30.0
315–400	24.3
200–315	43.0
100–200	2.5
<100	0.2

membrane layers allows to ensure high adhesion and the necessary performance characteristics of ceramic membranes (Ivanets et al. 2016a).

The initial raw material for the production of ceramic membranes was natural quartz sand from the Henan deposit (China), which, according to X-ray analysis, consisted of quartz with a high degree of crystallinity, as evidenced by the calculated values of the quartz unit cell, and did not contain other impurities of crystalline compounds (Fig. 18.6). When creating porous materials from quasi-spherical particles, the size of the initial particles should exceed the pore size by four to five times. Therefore, to obtain a ceramic substrate with a pore size of 50–100 μm, it is advisable to use a fraction of 200–630 μm, which is more than 97 wt.% of natural quartz raw (Table 18.3).

Microfiltration membrane layers were applied using a suspension of fine crystalline silicon oxide in an aqueous solution of an aluminosilicate binder (5.0–15.0 wt.

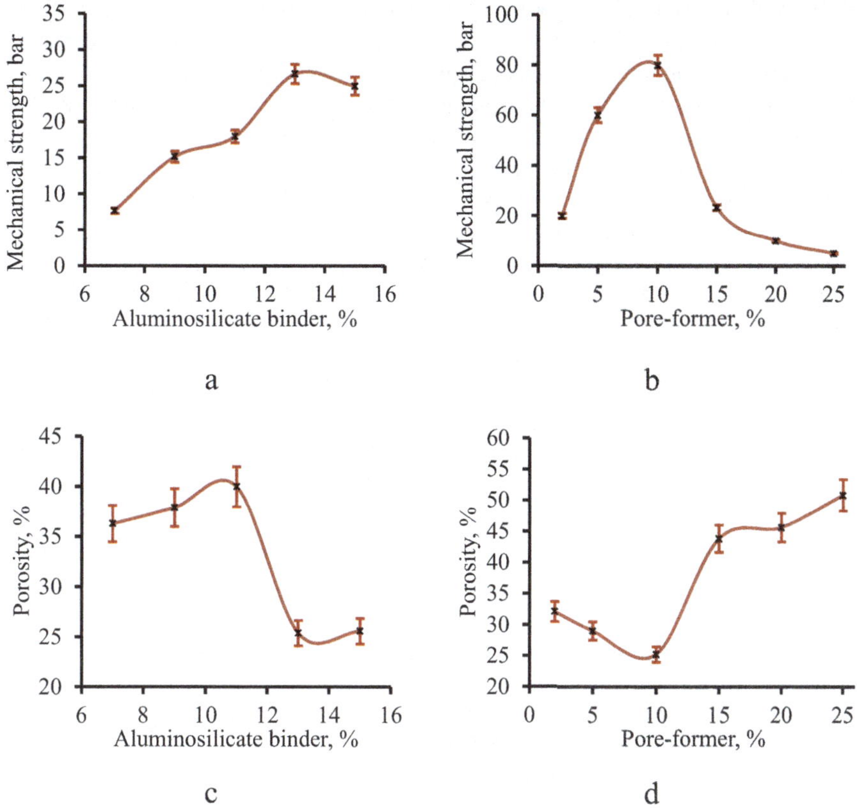

Fig. 18.7 Effect of aluminosilicate binder (**a**, **c**) and pore-former (**b**, **d**) contents on mechanical strength (**a**, **b**) and porosity (**c**, **d**) of ceramic samples

%): the intermediate layer was formed using particles of 10–40 μm and the membrane layer 0.5–2.0 μm.

The nature and content of the binder and pore-forming agent have a significant impact on the physical and chemical properties of silicate ceramics. To determine the effect of the binder and pore-forming agent concentration on the porosity and mechanical strength of silicate ceramics, a series of samples with a content of sodium aluminosilicate of 7.0–15.0 wt.% and burning additives of 2.0–25.0 wt.% were prepared (Fig. 18.7).

Figure 18.7a shows that with an increase in the content of the aluminosilicate binder from 7.0 to 13.0 wt.%, if there was a linear increase in mechanical strength, a further increase in concentration leads to a slight decrease in strength. This was due to the formation of a complete and optimal thickness of the binder film on the surface of the frame-forming silicon oxide particles, which provided high mechanical characteristics of ceramic samples. The mechanical strength decreased with increasing content of the pore-forming agent, which was due to a decrease in the number of interparticle contacts of the formed porous solid (Fig. 18.7b). However, only the samples containing 5.0–10.0 wt.% of the burning additive met the mechanical characteristics required for porous ceramics for filtration purposes.

Porosity is the most important parameter that determines the permeability and transport characteristics of porous ceramics. In this case, an increase in porosity is usually accompanied by a decrease in mechanical characteristics. Therefore, the choice of the optimal composition should be carried out when comparing the specified mechanical and transport characteristics of the resulting materials (Fig. 18.7c, d). For example, an increase in the binder content from 7.0 to 11.0 wt. % practically did not affect the porosity and was 35–40%, which is sufficient for their practical use in the filtration of liquid media. A further increase in the binder concentration leads to a sharp decrease in porosity, which was due to the filling of the binder in the porous space of the ceramic (Fig. 18.7c). The porosity of ceramics was expected to increase with an increase in the content of the pore-forming agent. In this case, samples containing a burning additive of 10–15 wt.% were characterized by porosity of 25–45% (Fig. 18.7d).

SEM images of ceramic samples with different binder and burning additive contents were shown in Fig. 18.8. The surface of samples with a binder content of 15.0 wt.%, as well as with the lowest pore-forming agent content of 2.0 wt.% had an

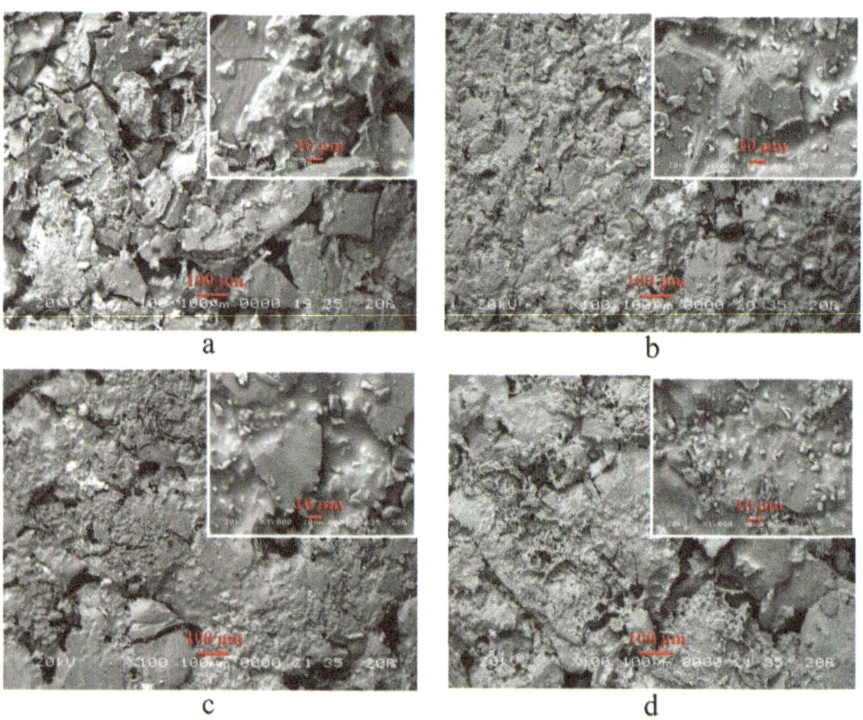

Fig. 18.8 SEM images of ceramic samples (× 100 and × 1 k): content of aluminosilicate binder of 7.0 (**a**) and 15.0 (**b**) wt.%; content of pore-former of 2.0 (**c**) and 25.0 (**d**) wt.%

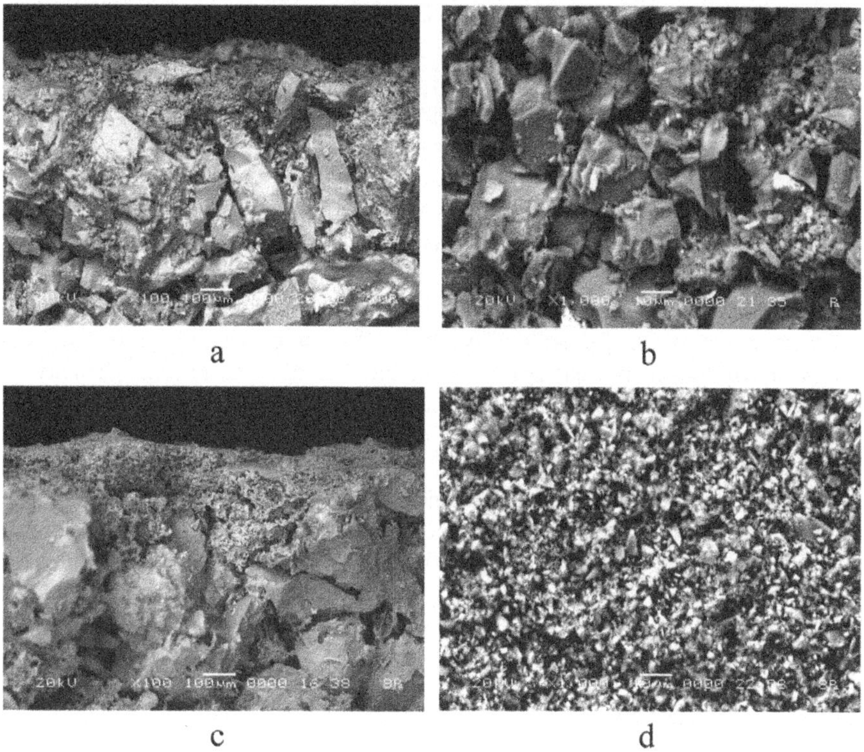

Fig. 18.9 SEM images of the cleavage (**a**, **c**) (× 100) and surface (**b**, **d**) (× 1 k) of ceramic samples with an intermediate (**a**, **b**) and microfiltration (**c**, **d**) layers

undeveloped porous structure, which was fully consistent with the data on the effect of the ceramic composition on the porosity of ceramic samples.

Preliminary studies have shown that during the formation of intermediate and microfiltration layers on the surface of a large-pore substrate, optimal mechanical and transport characteristics are achieved when using a suspension with a concentration of 15.0 wt.% of an aluminosilicate binder and the content of silicon oxide of 30.0 wt.%. Figure 18.9a shows that the thickness of the intermediate layer was 50–100 μm, while during the deposition of a suspension of silicon oxide, it partially penetrated into the porous space. The surface of the sample with an intermediate layer was not completely covered with fine particles of ground quartz sand, while the SEM image clearly identified the presence of a significant number of particles with size of ~40 μm, which formed the largest pores during sintering. Smaller particles were located in the interparticle space between large particles (Fig. 18.9b). The formations of a microfiltration layer lead to a slight increase in the thickness of the membrane, which did not exceed 150–200 μm (Fig. 18.9c). The surface of the microfiltration layer was uniform and defect-free (Fig. 18.9d).

Table 18.4 Characteristics of porous support and ceramic membranes

Sample	r_{av}, μm	r_{max}, μm	Q, $m^3/m^2 \times h \times bar$	Tensile strength, bar
Support	22 ± 3.0	29 ± 4.0	54 ± 5.0	9.0 ± 0.6
Intermediate layer	9.2 ± 0.5	18.6 ± 2.0	38 ± 2.0	7.8 ± 0.4
Microfiltration layer	2.3 ± 0.2	5.0 ± 0.3	26 ± 1.0	6.5 ± 0.3

Membrane transport characteristics and pore size determine its selectivity and performance. The performance of membranes is an integral characteristic that depends on the porosity, size, and morphology of the pores (tortuosity, shape, closed or open pores, etc.). According to Table 18.4, a coarse-pore substrate was characterized by an average pore size of 22 ± 3.02 μm, a water capacity of 54 ± 5.0 $m^3/(h \times m^2 \times bar)$, and a tensile strength of 9.0 ± 0.6 bar.

At the same time, the formation of an intermediate layer leads to a natural decrease in the average pore size to 9.2 ± 0.5 μm, water productivity to 38 ± 2.0 $m^3/m^2 \times h \times bar$, and mechanical tensile strength to 7.8 ± 0.4 bar. The prepared microfiltration ceramic membranes with an average pore size of 2.3 ± 0.2 μm, a water capacity of 26 ± 1.0 $m^3/(h \times m^2 \times bar)$, and a tensile strength of 6.5 ± 0.3 bar correspond to the most effective analogues.

18.5 Conclusions

In this chapter we present the results of systematic investigations of the main factors which influence on the physicochemical properties of mesoporous manganese oxide obtained via solgel method and demonstrate its effectiveness for sorption of ^{85}Sr radionuclide from model solutions with different chemical composition. This allowed obtaining the efficient adsorbents for the removal of ^{85}Sr radionuclide from aqueous media, to find out the optimal conditions of their production and to establish the relationship between the chemical and phase compositions, the parameters of porous structure, and sorption activity of developed sorbents.

A glycine-nitrate method was used to synthesize a Fenton-like catalyst consisting of $MgFe_2O_4$ nanoparticles with a parameter a 8.393 Å and crystallite size 8.2 nm, characterized by a mesoporous structure (A_{BET} 14 m^2/g, V_{ads} 0.03 cm^3/g, and D_{des} 8 nm), and consisting of agglomerates less than 1 μm in size formed by particles of 16–26 nm. The prepared catalysts have considerable interest for wastewater treatment from pharmaceutically active compounds.

The main factors affecting on the physicochemical properties of microfiltration ceramic membranes based on natural quartz sand were studied. Large-porous support with a content of aluminosilicate binder of 11.0 wt.% and burning additive of 10.0 wt.% was characterized by average pore size of 22 ± 3.02 μm, water capacity of 54 ± 5.0 $m^3/(h \times m^2 \times bar)$, and tensile strength of 9.0 ± 0.6 bar. The optimal conditions for membrane layers coating were determined that allowed to obtain microfiltration ceramic membranes with average pore size of 2.3 ± 0.2 μm, water capacity of 26 ± 1.0 $m^3/(h \times m^2 \times bar)$, and tensile strength of 6.5 ± 0.3 bar. The developed membranes are promising material for wastewater treatment.

Acknowledgments The authors are grateful to Artsiom Radkevich (Joint Institute for Power and Nuclear Research—Sosny NAS of Belarus) for conducting experiments with ^{85}Sr radionuclides.

References

Abdullayev A, Bekheet MF et al (2019) Materials and applications for low-cost ceramic membranes. Membranes 9(9):105–136. https://doi.org/10.3390/membranes9090105

Aminul Islam M, Morton DW, Johnson BB et al (2018) Manganese oxides and their application to metal ion and contaminant removal from wastewater. J Water Process Eng 26:264–280. https://doi.org/10.1016/j.jwpe.2018.10.018

Amiri M, Eskandari K, Salavati-Niasari M (2019) Magnetically retrievable ferrite nanoparticles in the catalysis application. Adv Colloid Interf Sci 271:101982. https://doi.org/10.1016/j.cis.2019.07.003

Davarnejad R, Soofi B, Farghadani F et al (2018) Ibuprofen removal from a medicinal effluent: a review on the various techniques for medicinal effluents treatment. Environ Technol Innov 11: 308–320. https://doi.org/10.1016/j.eti.2018.06.011

Goh PS, Ismail AF (2018) A review on inorganic membranes for desalination and wastewater treatment. Desalination 434:60–80. https://doi.org/10.1016/j.desal.2017.07.023

Ivanets AI, Agabekov VE (2017) Ceramic microfiltration membranes based on natural silica. Pet Chem 57:117–126. https://doi.org/10.1134/S0965544117020037

Ivanets AI, Kouznetsova TF, Prozorovich VG (2015) Sol-gel synthesis and adsorption properties of mesoporous manganese oxide. Russ J Phys Chem A 89:481–486. https://doi.org/10.1134/S0036024415030140

Ivanets AI, Azarova TA, Agabekov VE et al (2016a) Effect of phase composition of natural quartz raw material on characterization of microfiltration ceramic membranes. Ceram Int 42(15): 16571–16578. https://doi.org/10.1016/j.ceramint.2016.07.077

Ivanets AI, Prozorovich VG, Kouznetsova TF et al (2016b) Mesoporous manganese oxides prepared by sol-gel method: synthesis, characterization and sorption properties towards strontium ions. Environ Nanotechnol Monit Manag 6:261–269. https://doi.org/10.1016/j.enmm.2016.11.004

Ivanets AI, Milutin VV, Prozorovich VG et al (2019a) Adsorption properties of manganese oxides prepared in aqueous-ethanol medium toward Sr(II) ions. J Radioanal Nucl Chem 321(1): 243–253. https://doi.org/10.1007/s10967-019-06557-y

Ivanets AI, Milutin VV, Prozorovich VG et al (2019b) Sorption of ^{90}Sr onto manganese oxides prepared in aqueous-ethanol media. Radiochemistry 61:707–713. https://doi.org/10.1134/S1066362219060110

Ivanets A, Roshchina M, Srivastava V et al (2019c) Effect of metal ions adsorption on the efficiency of methylene blue degradation onto $MgFe_2O_4$ as Fenton-like catalysts. Colloids Surf A 571:17–26. https://doi.org/10.1016/j.colsurfa.2019.03.071

Ivanets A, Prozorovich V, Roshchina M et al (2020) Heterogeneous Fenton oxidation using magnesium ferrite nanoparticles for Ibuprofen removal from wastewater: optimization and kinetics studies. J Nanomater 2020:8159628. https://doi.org/10.1155/2020/8159628

Kefeni KK, Mamba BB (2020) Photocatalytic application of spinel ferrite nanoparticles and nanocomposites in wastewater treatment: review. Sustain Mater Technol 23:e00140. https://doi.org/10.1016/j.susmat.2019.e00140

Majumder A, Gupta B, Gupta AK (2019) Pharmaceutically active compounds in aqueous environment: a status, toxicity and insights of remediation. Environ Res 176:108542. https://doi.org/10.1016/j.envres.2019.108542

Michael I, Vasquez MI, Hapeshi E et al (2014) Metabolites and transformation products of pharmaceuticals in the aquatic environment as contaminants of emerging concern. In: Lambropoulou DA, Nollet LML (eds) Transformation products of emerging contaminants in the environment. Wiley, pp 425–470. https://doi.org/10.1002/9781118339558.ch14
Okoshi M, Momma T (2015) Radioactive waste treatment technologies. In: Nagasaki S, Nakayama S (eds) Radioactive waste engineering and management. Springer, Japan, pp 119–151. https://doi.org/10.1007/978-4-431-55417-2
Pakarinen J, Koivula R, Laatikainen M et al (2010) Nanoporous manganese oxides as environmental protective materials—effect of Ca and Mg on metals sorption. J Hazard Mater 180(1–3): 234–240. https://doi.org/10.1016/j.jhazmat.2010.04.019
Pathak P, Gupta DK (eds) (2020) Strontium contamination in the environment. Springer International Publishing. https://doi.org/10.1007/978-3-030-15314-4
Rodriguez-Narvaez OM, Peralta-Hernandez JM et al (2017) Treatment technologies for emerging contaminants in water: a review. Chem Eng J 323:361–380. https://doi.org/10.1016/j.cej.2017.04.106
Thommes M, Kaneko K et al (2015) Physisorption of gases, with special reference to the evaluation of surface area and pore size distribution (IUPAC Technical Report). Pure Appl Chem 87:1051–1069. https://doi.org/10.1515/pac-2014-1117
Voronina AV, Noskova AY, Semenishchev VS et al (2020) Decontamination of seawater from ^{137}Cs and ^{90}Sr radionuclides using inorganic sorbents. J Environ Radioact 217:106210. https://doi.org/10.1016/j.jenvrad.2020.106210
Zuo H-R, Shi P, Duan M (2020) A review on thermally stable membranes for water treatment: material, fabrication, and application. Sep Purif Technol 236:116223. https://doi.org/10.1016/j.seppur.2019.116223

19 Nanotechnology in Healthcare

Deepak Kala, Shagun Gupta, and Ankur Kaushal

Abstract

Nanotechnology is a versatile multidimensional science that covers different aspects and areas of biology, chemistry, physics, as well as engineering and gains worldwide attention in research day by day. Nanomaterials possess an excellent optical, electromagnetic property at the nanoscale level and can open new frontiers in the field of medical and health sciences that include diagnostics, drug delivery, robotics, tissue engineering, regenerative medicine, and vaccine development. Nanotechnology also played an important role in the diagnosis and treatment of the COVID-19 infections during corona pandemic time using different technologies like lateral flow assays, microfluidics technology, and nano-targeted drug delivery systems. Medical nanorobotics technology also offers new tools for the treatment of various human diseases and allows us a personalized treatment, hence achieving high efficacy against many diseases. So, the present chapter provides a brief overview about different nanomaterials and their clinical applications in the medical sciences and healthcare.

Keywords

Nanotechnology · Nanomedicine · Nano diagnostics · Nanorobotics

D. Kala
Department of Biotech Engineering and Food Technology, Chandigarh University, Mohali, Punjab, India

S. Gupta · A. Kaushal (✉)
Department of Biotechnology, Maharishi Markandeshwar University, Mullana, Ambala, Haryana, India

A. Thakur et al. (eds.), *Synthesis and Applications of Nanoparticles*,
https://doi.org/10.1007/978-981-16-6819-7_19

19.1 Introduction

Nanotechnology is an interdisciplinary subject that's infused different branches of science such as biology, chemistry, physics, and engineering to manipulate the system at the molecular and atomic level (Porter and Youties 2009). It combines components of molecular chemistry and physics with engineering to gain an advantage over the unique changes to the properties of materials that occur at a nanoscale (between 1 and 100 nm at least at one dimension) (McNeil 2005). A material at nanoscale shows tremendous changes in its properties due to the higher surface area to volume ratio. Nanomaterials possess an excellent optical, electromagnetic property at the nanoscale that has huge applications in the field of electronics and medicine (Sim and Wong 2021). The applications of nanomaterials in healthcare are opening new frontiers in the field of medical sciences. It has the potential to revolutionize the different aspects of medical industries (Fig. 19.1) including diagnostics, drug delivery, tissue engineering, regenerative medicine, and vaccine development. Nanotechnology brings advancements in healthcare, pharmaceuticals, and biomedical sciences in terms of improved drug delivery vehicles to develop effective therapeutic modalities, biocompatible scaffolds at nanoscale for tissue engineering, wearable biosensors, nanorobots, nanomedicine, etc. The application of nanoparticles as a drug carrier and delivery vehicles is getting the attention of researchers across the globe due to their advantages like the ability to cross the

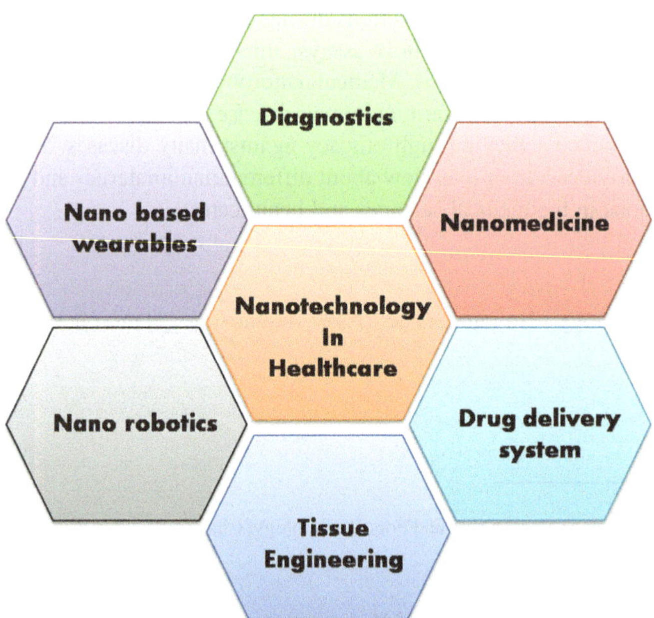

Fig. 19.1 Nanotechnology and its applications in healthcare

blood-brain barrier (Wang et al. 2021; Meng et al. 2021; Estelrich and Busquets 2021) and multifunctional mode of action that includes identifying the target site and carrying the drug for treatment (Fernandes et al. 2021; Allawadhi et al. 2021). Smart pills are the next-generation medicines that resemble morphologically pharmaceutical pills but are nanoscale electronic devices with the ability of sensing, imaging, and drug delivery (Romo et al. 2021; Suguna et al. 2021). The nanomaterials such as nanofibers are also being used in the development of smart bandages that can carry antibiotics for wound healing and also able to absorb into the tissues for complete healing.

In the future, smart bandages will facilitate the real-time monitoring of wound parameters with the help of integrated sensors, which will reduce the effort and time required to monitor and examine the wound (Almeida et al. 2021). Such nanomaterials are also being used in the development of nano-patch vaccines (Wan et al. 2021) that use nanoparticles to deliver the vaccine to the immune cells present below the skin. Nanotechnology is taking tissue engineering to the next level by making things possible that cannot be imagined earlier. Nanotechnology enables the designing and fabrication of biocompatible scaffolds similar to the extracellular matrix that helps in the development of implantable tissues. These scaffolds are under extensive research due to their ability to regenerate various tissues like bones, skeletal muscles, blood vessels, etc. (Edwards et al. 2009; Shi et al. 2010).

Nanotechnology is also being used in the field of diagnostics for the development of electrochemical and optical biosensors to improve the limitation of current technology such as lower sensitivity, higher cost, and requirement of sophisticated instrument facilities. Nanotechnology is becoming a pioneer in the field of diagnostics by introducing lab-on-a-chip technology that enables disease diagnosis with higher precession and short turnaround time. Nanoparticles such as AuNPs-, AgNPs-, quantum dots-based bioconjugate, etc. are being used in the development of a wide range of biosensing devices including SERS and SPR-based lateral flow assays to the microfluidic-based electrochemical devices (Liu et al. 2021; Yadav et al. 2021; Li et al. 2021; Shirshahi and Liu 2021; Huang et al. 2021).

Nanotechnology is opening new frontiers in the field of research and development with a promising future in the field of medical sciences. The discipline is not limited to a particular area as nanotechnology is already being used in different industries such as textile, chemicals, automobile, electrical, and electronics (Patra and Gouda 2013; Zhao et al. 2003; Venkatesan et al. 2017; Contreras et al. 2017; Rosenman et al. 2011). According to a new report published by Global Industry Analysts Inc. (GIA), the global market value of nanotechnology despite the COVID-19 crisis is estimated at 42.2 billion US$ in the year 2020 that is projected to touch the value of 42.2 billion US$ with a compound annual growth rate of 9.2% by 2026. Nanotechnology is playing an important role during this public health crisis by nano-interventions with regard to designing new ways to diagnose, treat, and eliminate the spread of the COVID-19 infections. Nanotechnology at the current situation is facing a mixed experience with several challenges and growth opportunities as the COVID-19 pandemic continues to take over the globe.

The chapter summarizes in detail the applications of nanotechnology in different areas of healthcare with a focus on different nano-interventions developed amid the COVID-19 spread across the globe.

19.2 Nanotechnology in Diagnostics

The point of care devices with the ability to detect infections at early stages with higher accuracy are the need of the diagnostics industry. Nanotechnology has revolutionized this area by miniaturization of the technology with rapid and better analytical abilities. Various nanomaterials such as metallic nanoparticles, GO, GQDs, CNTs, and CNFs have been used to develop rapid diagnostic tests (Verma et al. 2021; Kala et al. 2020; Kaushal et al. 2016). The nanomaterials have been used in the development of biosensors that offer better sensitivity, selectivity, and robustness. The nanomaterials are either used to enhance the sensitivity of the sensing matrix or to tag a biomolecule for monitoring the analyte receptor interaction by generating the detectable signals in the form of fluorescence, chemiluminescence, or redox reactions. Several electrochemical and optical biosensors have been designed for the detection of infectious diseases using various nanomaterials. The electrochemical sensors consist of three electrode systems including working (WE), counter (CE), and reference electrodes (RE) that measure changes in the current response on the WE as a result of applied voltage. The sensing matrix (WE) was modified with nanomaterials to enhance its conductivity and provide a larger surface area to immobilize the biomolecule (receptor) specific to the analyte (Ag, Abs, cDNA, protein) in the samples. Nanomaterials also have been used to tag the reporter biomolecule to monitor the analyte in the samples and to provide measurable signals.

19.2.1 Microfluidics Technology

Incorporation of the microfluidic technology with biosensors is a great initiative toward the development of lab-on-a-chip and point of care devices. The microfluidic system consists of micrometer-sized channels that can process small quantities of fluids by using tiny channels having dimensions at the microscale – typically tens to hundreds of micrometers. The microfluidic systems are developed using photolithography (Fig. 19.2), a method used to transfer geometric shapes from a mask to the surface of a substrate. Photolithography is an advanced micro fabrication technique used to create microstructures on the surfaces like designing computer chips, integrated circuits, etc. Several microfluidic-integrated DNA (Alsabbagh et al. 2021; Feng et al. 2021; Papamatthaiou et al. 2021; Arshavsky et al. 2021) and immuno-based (Arshavsky et al. 2021; Qi et al. 2021) sensors have been developed so far for the detection of a range of disease markers (protein, DNA, PSA, CA125, etc.). These microfluidic systems have several advantages like required lower sample volume, higher resolution, and sensitivity in the detection and separation of

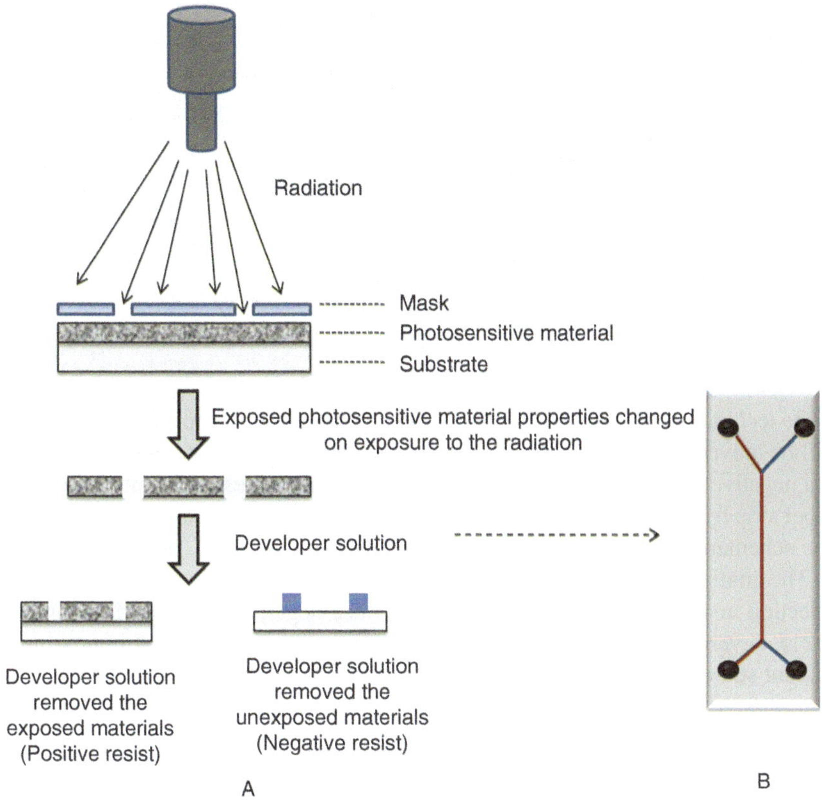

Fig. 19.2 Schematic illustration of (**a**) steps involved in photolithography to developed (**b**) microfluidic platform

molecules, greater control of experimental parameters at the microscale, and the ability to run multiple reactions simultaneously.

19.2.2 Lateral Flow Assays

Another POC system that is in trend is lateral flow assays (LFAs) due to their features such as user-friendliness, rapid response, portability, and low cost. They didn't require sophisticated instrument facilities and experts to operate and can be used even outside the laboratory settings. The LFAs are made of the paper strip and contain four major parts, i.e., sample pad, conjugate pad, biorecognition pad (contains test and control line), and absorbent pad (Ratajczak and Stobiecka 2020). The sample to be tested is loaded in sample pad which then moved to the conjugate pad which contains the analyte-specific antibodies tagged with nanoparticles or fluorescent particles as a tracer molecule. The analyte conjugate complex will bind

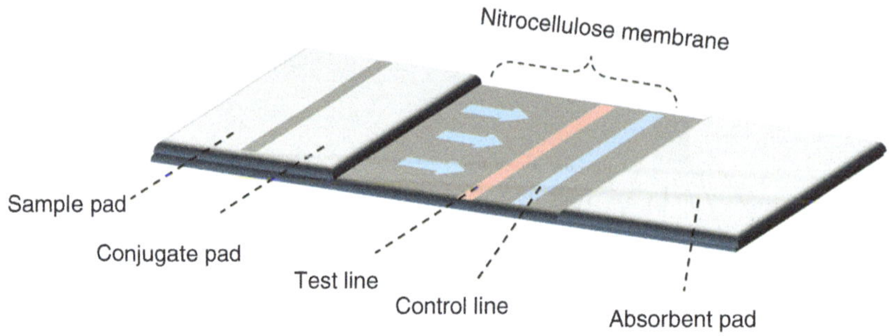

Fig. 19.3 Different components of lateral flow assay

to the test line (contains Abs specific to the analyte); it will be positive; otherwise it will be moved to the control line (contains Abs against conjugate Abs). The positive and negative samples can be confirmed by observing the test and control line on the paper strip by the naked eye or using a dedicated reader (Koczula and Gallotta 2016). The schematic of LFAs is shown in Fig. 19.3.

The traditional colorimetric lateral flow assays do not show a lower limit of detection and often result in a high rate of false-negative output (Frimpong et al. 2021). Several efforts have been made to overcome the limitation by improving the readout strategies to enhance the sensitivity.

19.2.2.1 SERS and LSPR

Moderate success has been achieved to overcome the limitations using the enzymes, surface-enhanced Raman spectroscopy (SERS), chemiluminescence, and localized surface plasmon resonance (LSPR) detection strategies (Frimpong et al. 2021; Pal and Dhar 2004; Rundstrom et al. 2007; Choi et al. 2004; Oh et al. 2005; Zaytseva et al. 2004; Hwang et al. 2016; Rong et al. 2018). The LSPR and SERS are the most advanced scheme used to improve the sensitivity and other limitations of current LFAs. Both methods are based on label-free, optical readout-based monitoring of association and dissociation rates of antibodies (Zhang et al. 2018). These assays must be integrated with some emerging detection techniques to fully address the limitation of current LFAs. The LFAs are the most successful point of care devices used commercially for the detection of several infectious diseases, and efforts are continuously made to overcome the limitations with the help of nanotechnology interventions.

19.3 Nanorobotics

Nanorobotics is an emerging field of nanotechnology which deals with the design and construction of devices at an atomic, molecular, or cellular level. In 1986, K. Eric Drexler introduced the idea of injecting small robots into the human body (Ross et al. 2018). "Nanorobots" will be the nanomachines that will repair the damage which accumulates as a result of metabolism (being alive) by performing nanorobotic therapeutic procedures. These hypothetical nanorobots will be extremely small and would transverse inside the human blood. As these nanorobots would have special sensors to detect the target molecules, they can be programmed to diagnose and treat various viral diseases (Drexler 2006). Medical nanorobotics (Manjunath and Kishore 2014; Freitas and Nanomedicine 1999) offers the prospect of powerful new tools for the treatment of various human diseases like cancer, heart attack, diabetes, arteriosclerosis, kidney stones, etc. and the improvement of human biological system. The nanorobots can allow us a personalized treatment, hence achieving high efficacy against many diseases.

Nanoscale robotics emerges as a novel tool in recent years to fulfill unmet clinical needs in the field of medicine, biosensing, and imaging. This nanoscale machine has the potential to convert diverse energy sources into movement and force of action. Moreover, it has multivalent functionalities such as large cargo-towing force for directional and long-distance transport, easy surface functionalization for precise capture and isolation of target subjects, and excellent biocompatibility for in vivo operation. These attractive functionalities and capabilities of micro/nanorobots have facilitated biomedical applications, ranging from targeted delivery of payloads and precise surgery on a cellular level to ultrasensitive detection of biological molecules and rapid removal of toxic compounds. Medical nanorobotics is used in the early stages of addressing healthcare issues. By bridging knowledge gaps in nanorobotics, different medical fields could be impacted profoundly.

19.4 Drug Delivery Systems

The importance of nanomaterials in the development of drug delivery systems is raising day by day due to the major challenges posed by the use of bulk materials including problems with targeted delivery, in vivo instability, poor bioavailability, adverse effects of drugs, poor solubility, and absorption in the body (Patra et al. 2018). The development of new drug delivery systems with the ability to target specific body parts might be the solution to the existing issues (Martinho et al. 2011; Jahangirian et al. 2017). Nanotechnology is playing a key role in the development of advanced medicine, controlled drug release, and targeted drug delivery systems with immense success. The smaller size of nanoparticles enables them to move more freely in the human body as compared to the bulk materials. They can be used to deliver the therapeutic drug to the targeted tissues with more accuracy and controlled release as their nanostructures can encapsulate drugs and have a larger surface area for their attachment.

19.4.1 Nano Drug Delivery Systems

Nano drug delivery systems (NDDSs) are a category of nanomaterials with the ability (Fig. 19.4) to improve the safety and effectiveness of drugs by enhancing

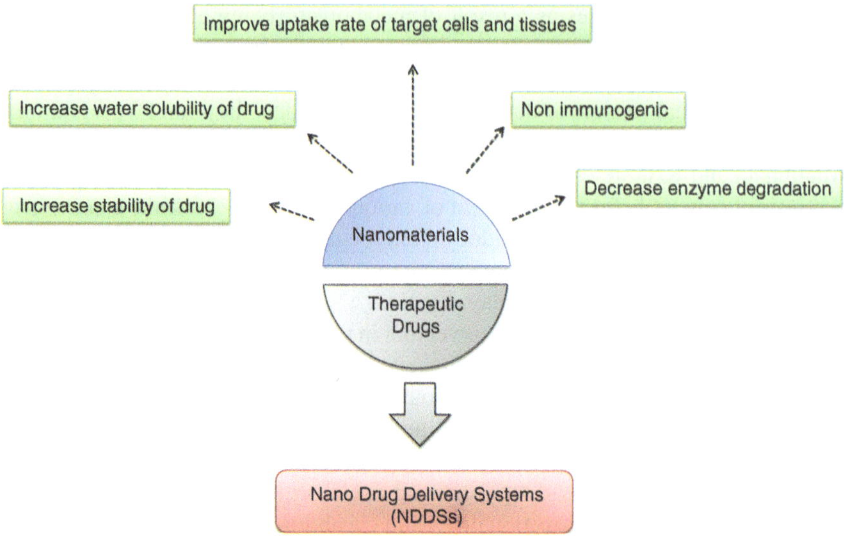

Fig. 19.4 Characteristics of nanomaterial required for the application of NDDSs

their stability and water solubility with increased uptake rate of target cells and tissues (Gupta et al. 2019; Quan et al. 2015). There are various routes from which the NDDSs can be administered in the body including oral, inhalation, and intravenous. Toxicity is one of the major problems exhibited by nanomaterials when considering it for in vivo applications (Lam et al. 2017).

As the application of nanomaterials is expanding in the field of medicine, it is also raising concern about its impact on human health. However, at present researchers are using nanoparticles in combination with natural products to minimize their toxicity. Green chemistry has been used to synthesize nanomaterials as it reduces the use of hazardous materials in biosynthetic processes. The use of such green nanoparticles for drug delivery can minimize the adverse effects of the medications (Lam et al. 2017).

19.4.1.1 Materials Used in NDDSs

The nanomaterials used for the NDDSs can be categorized as organic, inorganic, and composite materials (Deng et al. 2020). The organic nanomaterials used for NDDSs include liposomes, synthetic polymers, and dendrimers. Liposomes are spherical vesicles made up of one or more phospholipid bilayers (Akbarzadeh et al. 2013). The liposome can entrap both the hydrophobic and hydrophilic compounds and prevent them from degradation and release it at target sites (Atrooz 2011). The drug can be loaded in liposomes in two ways active and passive. The drug is entrapped during the liposome formation in passive and after liposome formation in an active method. The liposomes have several properties that make them an ideal material for drug delivery systems including low toxicity, biocompatibility, biodegradability, non-immunogenicity, and ability to entrap a range of compounds of both hydrophilic and hydrophobic nature. One of the major challenges with the use of liposomes as a drug delivery system is the charge on it. The cationic liposomes contain a positive charge on their surface that increases the probability of nonspecific interaction with negatively charged serum proteins (Deng et al. 2020). The uses of neutral and pH-sensitive liposomes are one of the possible solutions for these issues.

The most often used metallic nanomaterials are gold and silver nanomaterials, which exist in different structures such as nanoparticles, nanowires, nanorods, nanocuboids, and nanocapsules (Baeza et al. 2017). The gold nanoparticles are also used in the photothermal treatment of tumors and rheumatoid arthritis. The silver nanomaterial possesses antibacterial and antitumor properties (Di Pietro et al. 2016). These metallic nanomaterials can be used as a delivery vehicle by loading the drug physically in hollow gold and silver nanostructures or can be bonded to their surface by chemical methods. However, the slow removal of metallic nanoparticles from the human body and in vivo toxicity issues make it an unsuitable carrier for the treatment of chronic disease.

The inorganic nonmetallic nanomaterials include quantum dots, graphene, silicon, etc. Nonmetallic nanomaterials like mesoporous silicon gain the attention of researchers for the therapy of diseases due to their large surface area and porous structure (Wang et al. 2016). They have the potential to improve the transport efficiency of drugs and genes in target cells by integrating them with different functional groups. However, the biosafety of the nonmetallic nanomaterials may be an obstacle to its clinical applications. The composite nanomaterials are also in limelight due to their multifunctional abilities. The composite nanomaterials such as the combination of inorganic nonmetallic nanomaterials with lipid or polymers have multifunctional NDDSs' characteristics as they contain therapeutic drugs and contrast agents for medical imaging. The metallic and inorganic nanomaterials have been used with organic materials to improve their chemical and physical properties, biocompatibility, and pharmacokinetics.

19.5 Conclusion

The advancements in the field of nanotechnology offer the ability to develop next-generation tools for diagnostics and medicine. Nanotechnology is almost touching every aspect of healthcare and making a major contribution to the development of a large number of products that are incredibly powerful. Nanotechnology in diagnostics is playing a major role in the development of point of care (POC) devices for instant diagnosis of life-threatening diseases at early stages with more precision and less time. Nanotechnology-based products have the potential to overcome the limitations of traditional methods. Nanotechnology in the near future will revolutionize the field of diagnostics by providing cost-effective nanodevices that will offer real-time monitoring ability of patient's health and the therapy according to their genetic and disease profile. The major challenges in the application of nanotechnology-based products are their toxicity, environmental hazards, higher production cost, and accessibility to the unreachable and far-off areas.

References

Akbarzadeh A, Rezaei-Sadabady R, Davaran S, Joo SW, Zarghami N, Hanifehpour Y, Samiei M, Kouhi M, Nejati-Koshki K (2013) Liposome: classification, preparation, and applications. Nanoscale Res Lett 8(1):1–9

Allawadhi P, Singh V, Khurana A, Khurana I, Allwadhi S, Kumar P, Banothu AK, Thalugula S, Barani PJ, Naik RR, Bharani KK (2021) Silver nanoparticle based multifunctional approach for combating COVID-19. Sens Int:100,101

Almeida IB, Teixeira LGB, de Carvalho FO, Silva ÉR, Nunes PS, Dos Santos MRV, de Souza Araújo AA (2021) Smart dressings for wound healing: a review. Adv Skin Wound Care 34(2): 1–8

Alsabbagh K, Hornung T, Voigt A, Sadir S, Rajabi T, Länge K (2021) Microfluidic impedance biosensor chips using sensing layers based on dna-based self-assembled monolayers for label-free detection of proteins. Biosensors 11(3):80

Arshavsky-Graham S, Enders A, Ackerman S, Bahnemann J, Segal E (2021) 3D-printed microfluidics integrated with optical nanostructured porous aptasensors for protein detection. Microchim Acta 188(3):1–12

Atrooz OM (2011) Effects of alkylresorcinolic lipids obtained from acetonic extract of Jordanian wheat grains on liposome properties. Int J Biol Chem 5(5):314–321

Baeza A, Ruiz-Molina D, Vallet-Regí M (2017) Recent advances in porous nanoparticles for drug delivery in antitumoral applications: inorganic nanoparticles and nanoscale metal-organic frameworks. Expert Opin Drug Deliv 14(6):783–796

Choi S, Choi EY, Kim DJ, Kim JH, Kim TS, Oh SW (2004) A rapid, simple measurement of human albumin in whole blood using a fluorescence immunoassay (I). Clin Chim Acta 339(1-2): 147–156

Contreras JE, Rodriguez EA, Taha-Tijerina J (2017) Nanotechnology applications for electrical transformers—a review. Electr Pow Syst Res 143:573–584

Deng Y, Zhang X, Shen H, He Q, Wu Z, Liao W, Yuan M (2020) Application of the nano-drug delivery system in treatment of cardiovascular diseases. Front Bioeng Biotechnol 7:489

Di Pietro P, Strano G, Zuccarello L, Satriano C (2016) Gold and silver nanoparticles for applications in theranostics. Curr Top Med Chem 16(27):3069–3102

Drexler, K.E., 2006. Engines of Creation 2.0: The Coming Era of Nanotechnology.

Edwards SL, Werkmeister JA, Ramshaw JA (2009) Carbon nanotubes in scaffolds for tissue engineering. Expert Rev Med Devices 6(5):499–505

Estelrich J, Busquets MA (2021) Magnetic nanoparticles as delivery systems to penetrate the blood-brain barrier. In: Nanomedicines for brain drug delivery. Humana, New York, NY, pp 173–208

Feng D, Su J, Xu Y, He G, Wang C, Wang X, Pan T, Ding X, Mi X (2021) DNA tetrahedron-mediated immune-sandwich assay for rapid and sensitive detection of PSA through a microfluidic electrochemical detection system. Microsyst Nanoeng 7(1):1–10

Fernandes DA, Fernandes DD, Malik A, Gomes GNW, Appak-Baskoy S, Berndl E, Gradinaru CC, Kolios MC (2021) Multifunctional nanoparticles as theranostic agents for therapy and imaging of breast cancer. J Photochem Photobiol B Biol 218:112,110

Freitas RA, Nanomedicine VI (1999) Basic capabilities. Landes Bioscience, Georgetown, TX

Frimpong R, Jang W, Kim JH, Driskell JD (2021) Rapid vertical flow immunoassay on AuNP plasmonic paper for SERS-based point of need diagnostics. Talanta 223:121739

Gupta P, Garcia E, Sarkar A, Kapoor S, Rafiq K, Chand HS, Jayant RD (2019) Nanoparticle based treatment for cardiovascular diseases. Cardiovasc Haematol Disorders -Drug Targets (Formerly Current Drug Targets-Cardiovasc Hematol Disorders) 19(1):33–44

Huang T, Fu Q, Sun LP, Liu P, Wu Z, Li K, Xiao R, Yang X, Huang Y, Lin W, Lu H (2021) Photothermal lateral flow immunoassay using microfiber long-period grating. Sens Actuators B:130,283

Hwang J, Lee S, Choo J (2016) Application of a SERS-based lateral flow immunoassay strip for the rapid and sensitive detection of staphylococcal enterotoxin B. Nanoscale 8(22):11418–11425

Jahangirian H, Lemraski EG, Webster TJ, Rafiee-Moghaddam R, Abdollahi Y (2017) A review of drug delivery systems based on nanotechnology and green chemistry: green nanomedicine. Int J Nanomedicine 12:2957

Kala D, Sharma TK, Gupta S, Nagraik R, Verma V, Thakur A, Kaushal A (2020) AuNPs/CNF-modified DNA biosensor for early and quick detection of O. tsutsugamushi in patients suffering from scrub typhus. 3Biotech 10(10):1–13

Kaushal A, Singh S, Kala D, Kumar D, Kumar A (2016) speB genosensor for rapid detection of Streptococcus pyogenes causing damage of heart valves in human. Cell Mol Biol 62:140

Koczula KM, Gallotta A (2016) Lateral flow assays. Essays Biochem 60(1):111–120

Lam PL, Wong WY, Bian Z, Chui CH, Gambari R (2017) Recent advances in green nanoparticulate systems for drug delivery: efficient delivery and safety concern. Nanomedicine 12(4):357–385

Li Y, Liu X, Guo J, Zhang Y, Guo J, Wu X, Wang B, Ma X (2021) Simultaneous detection of inflammatory biomarkers by SERS nanotag-based lateral flow assay with portable cloud raman spectrometer. Nanomaterials 11(6):1496

Liu H, Dai E, Xiao R, Zhou Z, Zhang M, Bai Z, Shao Y, Qi K, Tu J, Wang C, Wang S (2021) Development of a SERS-based lateral flow immunoassay for rapid and ultra-sensitive detection of anti-SARS-CoV-2 IgM/IgG in clinical samples. Sens Actuators B 329:129196

Manjunath A, Kishore V (2014) The promising future in medicine: nanorobots. Biomed Sci Eng 2(2):42–47

Martinho, N., Damgé, C. and Reis, C.P., 2011. Recent advances in drug delivery systems. Journal of biomaterials and nanobiotechnology, 2(05), p.510.

McNeil SE (2005) Nanotechnology for the biologist. J Leukoc Biol 78(3):585–594

Meng L, Wang C, Lu Y, Sheng G, Yang L, Wu Z, Xu H, Han C, Lu Y, Han F (2021) Targeted regulation of blood–brain barrier for enhanced therapeutic efficiency of hypoxia-modifier nanoparticles and immune checkpoint blockade antibodies for glioblastoma. ACS Appl Mater Interfaces 13(10):11,657–11,671

Oh SW, Moon JD, Park SY, Jang HJ, Kim JH, Nahm KB, Choi EY (2005) Evaluation of fluorescence hs-CRP immunoassay for point-of-care testing. Clin Chim Acta 356(1-2):172–177

Pal A, Dhar TK (2004) An analytical device for on-site immunoassay. Demonstration of its applicability in semiquantitative detection of aflatoxin B1 in a batch of samples with ultrahigh sensitivity. Anal Chem 76(1):98–104

Papamatthaiou S, Estrela P, Moschou D (2021) Printable graphene BioFETs for DNA quantification in Lab-on-PCB microsystems. Sci Rep 11(1):1–9

Patra JK, Gouda S (2013) Application of nanotechnology in textile engineering: an overview. J Eng Technol Res 5(5):104–111

Patra JK, Das G, Fraceto LF, Campos EVR, del Pilar Rodriguez-Torres M, Acosta-Torres LS, Diaz-Torres LA, Grillo R, Swamy MK, Sharma S, Habtemariam S (2018) Nano based drug delivery systems: recent developments and future prospects. J Nanobiotechnol 16(1):1–33

Porter AL, Youties J (2009) How interdisciplinary is nanotechnology? J Nanopart Res 11(5): 1023–1041

Qi W, Zheng L, Wang S, Huang F, Liu Y, Jiang H, Lin J (2021) A microfluidic biosensor for rapid and automatic detection of Salmonella using metal-organic framework and Raspberry Pi. Biosens Bioelectron 178:113020

Quan, X.Q., Kang, L., Yin, X.Z., Jin, Z.H. and Gao, Z.G., 2015. Synthesis of PEGylated hyaluronic acid for loading dichloro(1,2-diaminocyclohexane)platinum(II) (DACHPt) in nanoparticles for cancer treatment. Chin Chem Lett, 26(6), pp.695-699.

Ratajczak K, Stobiecka M (2020) High-performance modified cellulose paper-based biosensors for medical diagnostics and early cancer screening: A concise review. Carbohydr Polym 229: 115463

Romo JE, Gutiérrez S, Rodrigo PM, Cardona M, Solanki VK (2021) Smarter pills: low-cost embedded device to elders. In: Research in intelligent and computing in engineering. Springer, Singapore, pp 663–671

Rong Z, Xiao R, Xing S, Xiong G, Yu Z, Wang L, Jia X, Wang K, Cong Y, Wang S (2018) SERS-based lateral flow assay for quantitative detection of C-reactive protein as an early bio-indicator of a radiation-induced inflammatory response in nonhuman primates. Analyst 143(9): 2115–2121

Rosenman G, Beker P, Koren I, Yevnin M, Bank-Srour B, Mishina E, Semin S (2011) Bioinspired peptide nanotubes: deposition technology, basic physics and nanotechnology applications. J Pept Sci 17(2):75–87

Ross, G., Bremer, M.G., Wichers, J.H., Van Amerongen, A. and Nielen, M.W., 2018. Rapid antibody selection using surface plasmon resonance for high-speed and sensitive hazelnut lateral flow prototypes. Biosensors, 8(4), p.130.

Rundstrom G, Jonsson A, Martensson O, Mendel-Hartvig I, Venge P (2007) Lateral flow immunoassay using europium (III) chelate microparticles and time-resolved fluorescence for eosinophils and neutrophils in whole blood. Clin Chem 53(2):342–348

Shi J, Votruba AR, Farokhzad OC, Langer R (2010) Nanotechnology in drug delivery and tissue engineering: from discovery to applications. Nano Lett 10(9):3223–3230

Shirshahi V, Liu G (2021) Enhancing the analytical performance of paper lateral flow assays: from chemistry to engineering. TrAC Trends Anal Chem:116,200

Sim S, Wong NK (2021) Nanotechnology and its use in imaging and drug delivery. Biomed Rep 14(5):1–9

Suguna R, Shanmugapriya M, Nishanthi S, Saranya V, Sowmiya J (2021) Swallowable glass pill for digestive motility & toxin detection using cell-based biosensor. J Phys Conf Ser 1916(1): 012150. IOP Publishing.

Venkatesan H, Sivamani S, Sampath S, Gopi V, Kumar D (2017) A comprehensive review on the effect of nano metallic additives on fuel properties, engine performance and emission characteristics. Int J Renewable Energy Res (IJRER) 7(2):825–843

Verma V, Kala D, Gupta S, Kumar H, Kaushal A, Kuča K, Cruz-Martins N, Kumar D (2021) Leptospira interrogans outer membrane protein-based nanohybrid sensor for the diagnosis of leptospirosis. Sensors 21(7):2552

Wan Y, Gupta V, Bird C, Pullagurla SR, Fahey P, Forster A, Volkin DB, Joshi SB (2021) Formulation development and improved stability of a combination measles and rubella live-viral vaccine dried for use in the nanopatch tm microneedle delivery system. Human Vacc Immunother:1–16

Wang W, Sun X, Zhang H, Yang C, Liu Y, Yang W, Guo C, Wang C (2016) Controlled release hydrogen sulfide delivery system based on mesoporous silica nanoparticles protects graft endothelium from ischemia–reperfusion injury. Int J Nanomedicine 11:3255

Wang Z, Zhang C, Huang F, Liu X, Wang Z, Yan B (2021) Breakthrough of ZrO2 nanoparticles into fetal brains depends on developmental stage of maternal placental barrier and fetal blood-brain-barrier. J Hazard Mater 402:123,563

Yadav S, Sadique MA, Ranjan P, Kumar N, Singhal A, Srivastava AK, Khan R (2021) SERS based lateral flow immunoassay for point-of-care detection of SARS-CoV-2 in clinical samples. ACS Appl Bio Mater 4(4):2974–2995

Zaytseva NV, Montagna RA, Lee EM, Baeumner AJ (2004) Multi-analyte single-membrane biosensor for the serotype-specific detection of Dengue virus. Anal Bioanal Chem 380(1):46–53

Zhang D, Huang L, Liu B, Ni H, Sun L, Su E, Chen H, Gu Z, Zhao X (2018) Quantitative and ultrasensitive detection of multiplex cardiac biomarkers in lateral flow assay with core-shell SERS nanotags. Biosens Bioelectron 106:204–211

Zhao QQ, Boxman A, Chowdhry U (2003) Nanotechnology in the chemical industry–opportunities and challenges. J Nanopart Res 5(5):567–572

Graphene Quantum Dots and Their Hybrid Hydrogels: A Multifaceted Platform for Theranostic Applications

20

Sujata Sangam, Piyush Garg, Trinanjana Sanyal, Siddhartha Pahari, S. M. Paul Khurana, and Monalisa Mukherjee

Abstract

Advances in materials science are an outcome of novel technologies that emerge from amalgamation of known components. In this arena, hydrogels are a promising material for a series of biomedical and biological applications and can simulate human's soft tissues, due to their unique 3D matrix, flexibility and high water content. Interestingly, graphene quantum dots can be used as reinforcing agents to alter the mechanical properties of hydrogels. Hybrid hydrogels are used because of flexibility, biocompatible nature and the ease of fabrication, as well as their wide range of mechanical, chemical, thermal and electrical behaviours, when combined with different materials as composites. The structural combination of a polymer network with graphene quantum dots, highly biocompatible carbon-based nanomaterial, can be utilized for biomedical applications like wound healing, drug delivery, antibacterial, bioimaging and tissue engineering. Furthermore, this mixing may result in synergistic property enhancement of each component. The combination of properties imparted by

S. Sangam · P. Garg · M. Mukherjee (✉)
Amity Institute of Biotechnology, Amity University, Noida, Uttar Pradesh, India

Amity Institute of Click Chemistry Research and Studies, Amity University, Noida, Uttar Pradesh, India
e-mail: mmukherjee@amity.edu

T. Sanyal
Amity Institute of Biotechnology, Amity University, Noida, Uttar Pradesh, India

S. Pahari
Amity Institute of Click Chemistry Research and Studies, Amity University, Noida, Uttar Pradesh, India

S. M. P. Khurana
Science Instrumentation Centre, Amity University Haryana, Gurugram, Haryana, India

A. Thakur et al. (eds.), *Synthesis and Applications of Nanoparticles*,
https://doi.org/10.1007/978-981-16-6819-7_20

both hydrogels and their nanocomposites can pave new horizons in therapeutics. The unique electronic structure of GQDs confers advantages such as tunable photoluminescence properties for bioimaging, enhancing the loading capacity of aromatic drug molecules for drug delivery and supporting stem cell differentiation for tissue engineering applications. In this chapter, we systematically delineate the recent advances in GQD hydrogel composites with a focus on their synthesis, applications and future perspectives.

Keywords

Hydrogels · Graphene quantum dots · Nanocomposites · Hybrid

20.1 Introduction

Virtually, a plethora of polymeric materials are prevalent in various aspects of life and biomedical applications. Interestingly, living systems are mainly poised with hydrogel-like polymer matrices infiltrated with water (Lin et al. 2016). Hydrogels are basically three-dimensional, physically or chemically cross-linked polymeric networks entrapping water in intermolecular space, with tunable mechanical properties (Massoumi et al. 2018). Most hydrogels are mechanically flexible yet robust, and they can accommodate large amount of solvents including water and therapeutic agents. They motivate various life value-added substances, endowing living bodies with exquisite functions such as vehicles for drug delivery; model extracellular matrices for biological studies; scaffolds for tissue engineering and self-healing; and actuators for optics, fluidics, sensing, responding, self-reinforcing and self-regulating (Lin et al. 2016; Liu et al. 2017). To harness hydrogels' unique properties and functions, new technologies devoted to the development of various biomimetic structures and devices based on the innovative combination of nanomaterials, particularly graphene quantum dots (GQDs), hold the promise of conferring superior functionality to the nanocomposite hydrogel with applications in diverse fields (Geng 2018). This results in an upshot of synergistic enhancement of several factors like mechanical strength of the hydrogel, rheological properties, etc. Hybrid hydrogels are used because of flexibility, biocompatible nature and the ease of fabrication, as well as their wide range of mechanical, chemical, thermal and electrical behaviours, when combined with different materials as composites. Recent progresses in GQDs and their hybrid hydrogel nanocomposites are herein delineated with a focus on potential of biomedical applications and their future prospects (Fig. 20.1).

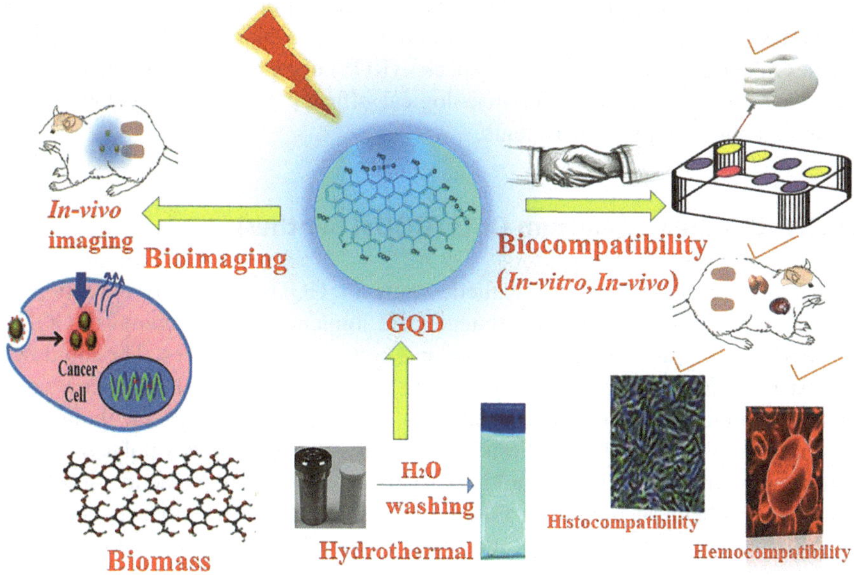

Fig. 20.1 Schematic representation of the synthetic route, bioimaging, biocompatibility of GQD

20.2 Graphene Quantum Dots (GQDs)

GQDs have emerged as a brilliant candidate for controlled release of drug molecules, immobilization, delivery, bioimaging, biosensing, etc. Furthermore, GQDs are striking due to their excellent water dispersibility, biodegradability, renewability and biocompatibility (Joshi et al. 2016). GQDs are zero-dimensional, sp^2-hybridized, atomic layer-thickened and fluorescent nanocrystals, typically below 20 nm, coupling the peculiar features of both graphene and quantum dots whose excitons are confined in all three spatial dimensions. Although quantum dots (QDs) are semiconductor in nature and also attractive labelling mediator for various biomedical applications, however, their compromised biocompatibility in comparison to GQDs limits their pharmaceutical applications. GQDs have unique electronic, photoelectric, spin and optical properties such as resonance light scattering and surface plasma resonance absorption (Suzuki et al. 2016). GQDs can act as an effective signal amplifier due to wide range of excitation and stable optical properties, size-tunable narrow and symmetric emission spectra, high photostability and fluorescence quantum yields. Furthermore, surface chemistry controls stability and bigger surface area, suitable for labelling biomolecules, and single molecule analysis, resistant to chemical and thermal effects. The fluorescence in GQDs is due to quantum confinement effect, surface defect and edge effect. Their high surface-to-volume ratio and solubility in aqueous media provides a better driving force for diffusion in living system (Joshi et al. 2016).

GQDs have also been recognized to exhibit special properties to act as oxidase or peroxidase via electron transportation to adopt certain biomolecules from normal species to cytotoxic reactive oxygen species (ROS) upon light irradiation. This also has the potential to deliver pharmacological efficacy in biomedical domains as photodynamic therapy (PDT) agents.

20.2.1 Graphene Quantum Dots (GQDs) Synthesis

Remarkable advancements in synthetic route of GQDs incline its execution in clinical operations. Two distinct strategies are implemented to accomplish GQDs' successful synthesis, namely, top-down and bottom-up approaches. The top-down approach mainly focuses on the exfoliation and cleavage of bulk graphene-based materials under harsh reaction conditions. On the other hand, the bottom-up approach employed in the synthesis route of this nanodot utilizes polycyclic aromatic compounds based on the multistep oxidative condensation reaction in organic solutions or chemical rupture of fullerenes (Joshi et al. 2016). Both the synthetic strategies exhibit their advantages and disadvantages. Moreover, the choice to trace a particular strategy relies on the type of application. The lack of morphological control, harsh conditions, high cost and environmental unfriendliness are the major hindrances of the top-down approach. However, the advantage of precise control on morphology, size and shape provides GQDs for various applications. GQDs synthesized from the bottom-up approach have the disadvantage of utilizing expensive precursors, and acquire a stronger tendency to aggregate and many complex synthetic steps. In addition to the above properties, GQDs synthesized via bottom-up approach possess remarkable optical properties that contribute to in vitro and in vivo imaging (Iannazzo et al. 2017).

20.2.1.1 Top-Down Approaches

This approach mainly depends on physical or chemical techniques for GQD synthesis. Based on the distinct chemical mechanism, top-down approaches are commonly described as "defect-mediated fragmentation processes" because of oxygen-containing reactive groups, i.e. epoxy, hydroxyl and carboxyl groups produced on oxidation process generate defects on graphene sheets which consecutively serve as a chemical reactive site to cleave graphene into smaller fragments. The distinctive top-down approach includes chemical vapour deposition method, electrochemical processes, pulse laser ablation and microwave-assisted hydrothermal and solvothermal process (Joshi et al. 2016; Iannazzo et al. 2017).

20.2.1.1.1 Hydrothermal/Solvothermal Synthesis

To synthesize GQDs, hydrothermal/solvothermal process is useful for good crystal morphology with atom-economical reactions. Water and oxidizing agents such as strong acid and alkali are commonly utilized to cleave carbon precursors to GQDs (Iannazzo et al. 2017). Hydrothermal route is a good example for top-down approach, and it is a suitable, easy and eco-friendly way to produce GQDs from the carbonaceous precursor. In this process the parameters like temperature and autogenous pressure are playing an important role to cut the large sheets of few μm into small dimensions of few nanometres.

20.2.1.1.2 Electrochemical Approaches

In this approach exploitation of electrochemical exfoliation and intercalation is a major step to obtain product by utilizing hydroxyl and oxygen as electrochemical "scissors" in the oxidative cleavage reaction (Iannazzo et al. 2017).

20.2.1.1.3 Laser Ablation

This method presents fast, flexible and environmentally friendly technique for nanostructure synthesis that strongly influences its scalability and reproducibility. The laser ablation focuses on the femtosecond laser pulse exposure to graphite and induces its plane to colloid which can be removed via the graphite exfoliation process (Iannazzo et al. 2017).

20.2.1.1.4 Chemical Vapour Deposition (CVD)

Chemical vapour deposition method was used to generate GQDs of uniform size with enhanced physical and chemical properties from self-assembled block copolymers (BCP) as an etch mask on graphene films (Iannazzo et al. 2017).

20.2.1.2 Bottom-Up Approaches

Contrary to the top-down strategy, this approach exploits the small precursors including citric acid, glutamic acid, glucose, small aromatic molecules, etc. for GQD synthesis, dominating it with controlled morphology and evenly distributed sizes. Solution chemistry approach limits the GQD synthesis procedure with harsh reaction conditions, long treatment duration and the necessity for separation processes (Iannazzo et al. 2017).

20.3 Hydrogels

Hydrogels can simulate human's soft tissues, due to their unique 3D matrix, flexibility and high water content, and have been pursued as promising materials for a series of biomedical and biological applications, such as tissue engineering, minimally invasive drug release systems, biological sensors, microarrays, bioimaging and actuators, cancer research, cellular therapies, stem cell bioengineering and implants (Teo et al. 2016; Liu et al. 2017). The hydrogel's resistance to dissolution increases due to the fabrication of cross-links between matrix chains, while the knack to integrate water increases with hydrophilic functional groups attached to the hybrid biopolymeric chains. Many endeavours of hydrogels have been exerted in several therapeutic applications that necessitate contact with living systems because of their closeness to native extracellular matrix. Additionally, various kinds of organic/inorganic nanoparticle types have been used to achieve comprehensive design of hydrogel architecture with organized physical and biological properties which can be implemented to curb tissue morphogenesis and cellular functionality. The development of advanced hydrogel nanocomposites with GQDs confers tunable physiochemical properties and has improved mechanical, electrical and biological properties (Annabi et al. 2014).

The strategies to combine functional nano-constructs with hydrogel towards physiological variability mimicking bioactive molecules, cells, tissues, etc. play a remarkable role in endorsing the regeneration of traumatic tissues or therapeutic systems impacting the whole spectrum of medicine and health care (Mantha et al. 2019). Nevertheless, these methodologies are restricted by short half-life, poor binding efficiency of the bioactive molecules or drugs and failure to induce long stretched therapeutic effects. GQD hybrid hydrogels can serve as a platform to support physiochemical interactions including non-viral, viral and graphene-based gene delivery systems. Encapsulating such therapeutic agents or delivery vectors inside hydrogels allows their meticulous delivery at the target site, protecting undesired loss during the transit to the specific site and facilitating proficient localized therapeutic therapy with reduced non-specific dispersal to other tissues. Thus, the GQD hybrid hydrogels increase the long retention time of the delivery vector at the target site, enhancing their chances of getting available long by the tissues, which is presently a major challenge of nanoparticle-based delivery systems (Paul et al. 2014). Based on maximizing the efficacy and patient compliance, GQD hybrid hydrogels can be the best choice of therapeutic applications in many branches of medicine, including oncology, immunology, pain management, wound healing and cardiology. Importantly, Liu et al. affirmed a drug-self-gated hydrogel system that has leveraged the drug discharge capacity compared to free drug molecules. Controlled drug release property and pH sensitivity have been awarded by benzoic-imine covalent bond designed between doxorubicin (DOX) and benzaldehyde group. The GQD hybrid hydrogels with fluorescence property extended the horizons of bioimaging arena and encouraging the expansion of theragnostic applications such as cell labelling, disease treatment and disease diagnosis (Liu et al. 2019).

Hydrogel nanoparticles hold versatility and suitable properties as carriers for efficient drug delivery vehicles, tissue engineering, proteins, peptides, oligosaccharides, vaccines and nucleic acid composite biomaterials and as injectable fillers in minimally invasive surgeries (Unterman et al. 2017). The successful application of GQD hybrid hydrogel in the body is due to their biocompatibility, biodegradability, injectability, mucoadhesiveness, relatively low cytotoxicity and tunable mechanical, bioadhesive and other functional properties by chemical modification (Chai et al. 2017).

Broad applicability of the GQD hybrid hydrogel, which has the advantage of precise control of molar mass, can be tailored to have a wide range of properties such as mechanical, flow, etc. and low immunogenicity with minimum risk of biological pathogens or contaminants. Balance of components in GQD hybrid hydrogels allows for considerably altering macromer solution flow, expediting injection and operating in a comprehensive range of implantation environments augmenting compressive modulus of hydrogels at low loading. Remarkably, mechanical and rheological effects of GQD hybrid hydrogel were dependent on aspect ratio (Unterman et al.

2017). The GQDs-nanostructured hybrid hydrogels may show very significant characteristics, i.e. the dynamic chemistry to develop possible self-adapting solid materials, and can spontaneously answer the ever-changing inner environment of living bodies (Li et al. 2018). Hydrogels are categorized as natural (e.g. alginate, dextran (Dex), gelatin, chitosan, collagen, agar-agar, etc.) and synthetic (e.g. poly (ethylene glycol)-poly(lactone)-poly(ethylene glycol) (PEG-PL-PEG), poly(vinyl alcohol) (PVA)) established on their source (Ahmed 2015). Hydrogels may also be classified as multipolymer, homopolymer and copolymer networks, interpenetrating networks, semi-interpenetrating networks and double networks based on the nature of the network, as homogeneous hydrogels and microporous and macroporous hydrogels based on pores, as permanent or chemical gels when they feature covalently cross-linked networks and as reversible or physical gel when the networks are held together by molecular entanglements and/or secondary forces such as ionic, hydrogen bonding or hydrophobic interactions (Subramanian and Vijayakumar 2015).

20.3.1 Hydrogel Synthesis

Herein, an economical, simple and environmentally friendly method has been deployed for the synthesis of dynamic GQD-based hydrogels using GQDs as nanofillers and monomers as main component. Hydrogels are primarily synthesized through either non-covalent cross-linking or covalent cross-linking between hydrogel network chains. Non-covalent cross-linking is more preferable over covalent cross-linking since covalently cross-linked hydrogels are non-injectable and not self-healable, restricting their therapeutic applications. Non-covalently cross-linked hydrogels are produced through transitory and feeble interactions between hybrid hydrogel polymer chains, subsequently resulting in the reversible construction of the hydrogels. Hence, non-covalently cross-linked hybrid hydrogels are anticipated for many therapeutic applications (Zhang et al. 2011).

Hydrogels may be fabricated in a number of "classical" chemical ways, namely, one-step procedures like polymerization and parallel cross-linking of multifunctional monomers or multiple step procedures concerning synthesis of polymer molecules having reactive groups and their subsequent suitable cross-linking (Ahmed 2015). Herein, the polymerization techniques have been designated as follows.

20.3.1.1 Bulk Polymerization

Preparation of hydrogels by bulk polymerization includes the intimate interaction of one or more types of monomers like vinyl monomers. Usually this reaction requires the addition of a small amount of cross-linking agent along with the help of radiation or catalyst as initiator. The selection of initiator solely depends upon the types of the monomers that are being used and also the solvents. The rate of reaction is directly proportional to the concentration of monomers present in the solvent. The higher the concentration of the monomer, the higher is the reaction rate. Thus, the homogenous hydrogel that is produced by this method is usually hard, but when in contact with water, it softens and becomes more flexible. There are different varieties of polymerized hydrogels that are produced which includes rods, particles, emulsions, films and even membranes (Panchal et al. 2017).

20.3.1.2 Solution Polymerization

The process of hydrogel synthesis by solution polymerization includes the mixing of monomers that are either neutral or ionic in nature along with multifunctional cross-linking agents. Thus, the hydrogels produced by this method are heterogenous in nature. The initiation of this polymerization is usually done thermally with the help of UV irradiation or by a redox initiator system. The solvent used in this reaction acts as a heat sink which serves as a major advantage over the bulk polymerization. The solvents that are generally used in this case comprises ethanol mixtures, benzyl alcohol, water, etc. The removal of unreacted mixture, initiator, cross-linking agents and other impurities has been carried out by washing with distilled water (Panchal et al. 2017; Singh et al. 2017).

20.3.1.3 Emulsion/Suspension Polymerization

Emulsion/suspension polymerization is employed for the synthesis, when we are aiming for particulate hydrogels or hydrogel beads as products. Here, a variety of hydrophilic monomers are used to produce hydrogel networks by the use of multi-functional cross-linkers and initiators in a hydrocarbon phase. The reaction mass is dispersed as small droplets in continuous aqueous (organic) phase where each droplet acts as a small bulk reactor. High-energy gamma radiation is used, and the transfer of heat occurs from the tiny droplets to water having large heat capacity and low viscosity. Thus, to facilitate the heat removal and to govern the particle size, shape and other specific properties, the viscosity of the monomer solution, agitation speed, rotor design and dispersant type are taken into consideration (Argenta et al. 2019).

20.3.1.4 Polymerization by High-Energy Radiation

Hydrogels that are produced by high-energy polymerization using gamma radiations and electron beam are pure and initiator-free. The energy of these electron beam and gamma rays is so high that it can ionize simple molecules which are even present in air or water. During these processes, many reactive sites are produced alongside the polymer strands, and their combination leads to higher cross-linking. The irradiation of the solution results in radical formation on the polymer chain along with radiolysis of water. Further, attack on the polymer chain leads to macroradicals, which gives the characteristic cross-linked structure of the hydrogel (Argenta et al. 2019; Ranganathan et al. 2019).

20.3.1.5 Free Radical Polymerization

In this process, the free radicals are produced on the surface of a strong support material by direct polymerization of monomers. This involves the attachment of a monomer chain covalently on the support followed by three particular steps of characterization:

- **Initiation:** This step involves the generation of active free radical species. This chain initiation process is one of the distinguishable features of the polymerization process.
- **Propagation:** This is a rapid and successive step in which the monomers are accumulated in the chain that grows with a steady active centre.
- **Termination:** This step deals with the formation of dead polymers via coupling and/or disproportionation (Argenta et al. 2019).

20.4 GQD Hydrogel Nanocomposites

Considering GQD hydrogel nanocomposites as a revolutionary avenue offers an enabling technology that presents the possibility of fine-tuning the physicochemical properties of nanomaterials and has contributed in the spectrum of medicine and health care. Colloidal nanostructures using functional macromolecules in hydrogels are an approach through which multifunctional hydrogels with additive properties can be achieved. Colloidal assemblies with utilization of GQDs lead to high level of modulation of structural parameters over other inorganic nanoparticles that are difficult to attain via conventional approaches. This procedure of colloidal synthesis of the nanocomposite can exploit the functions of GQDs within the hydrogel and can open up new horizons to multifunctional and monodisperse nanostructures consisting of discrete chemistries with complementary properties. GQDs can significantly help to manipulate the order and design of responsive functional nanocomposite systems with precise control over mechanics that has numerous impending clinical uses. GQD fillers are reported to affect mechanical and rheological properties of the hydrogels in a modest manner that possess the ability to be implanted and tuned readily in wet, flowing and hostile conditions. With GQDs, one can selectively tune the interfacial properties of the copolymers, breaking symmetry due to surface energy incongruity, producing exotic nanostructure to provide a useful platform for targeted drug delivery (Shakeel et al. 2019).

Due to the presence of hydroxyl, epoxy and carboxyl at the edges of GQDs, they can be easily combined with hydrophilic drugs. Stimulated by advances in bioengineering, GQD-based hydrogels possess excellent biocompatibility and good mechanical properties. Concurrently, the sp^2 conjugate region on the surface of GQDs can also aid in the loading of hydrophobic drugs, thus engendering them to be used as a drug carrier. Moreover, GQDs and their derivatives hold the ability of photothermal conversion and can be used in the photothermal therapy.

Owing to the tunable chemical, physical and biological properties, biocompatibility, versatility in fabrication and mimicking potency of the extracellular matrix, hydrogel emerged as a powerful tool to mend the human machinery defects. Although naturally derived hydrogels like chitosan, heparin, alginate, elastin, gelatin, collagen and hyaluronic acid due to their biodegradability, cell-interactive and cell signalling properties are employed in biological purposes, however, their low mechanical properties, potential immunogenicity, uncontrollable degradation and structural intricacy limit its practice in the promotion of health, while synthetic hydrogel imparts great mechanical strength and controlled degradation but it lacks biological moieties. The combination of properties imparted by both hydrogel classes and their nanocomposites can pave new horizon towards therapeutics. The physical interaction including non-covalent bond and chemical interaction including covalent bond formation in monomeric units harnesses the 3D polymeric networks of the hydrogel (Annabi et al. 2014).

Fabrication of nanostructured hydrogels using activated nanogels as cross-linker has proven to be gifted with rapid stimuli-responsive characteristics along with concomitant presence of properties such as withstanding high level of deformation due to twisting, bending and extensive stretching (Xia et al. 2013). Endowed with promising features including reduced immunogenicity and superior biosafety profile, GQD-based hydrogel nanocomposites can be employed to achieve bioactive hydrogels with tissue-specific functionalities in a controlled and localized manner. According to a report, graphene oxide (GO) nanosheets can help to deliver genes in an efficient manner when they are ionically bonded to cationic polymers such as polyethylenimine (PEI) and show low cytotoxicity (Paul et al. 2014). A careful examination of hydrogel-based GQD gene delivery system approach can be advantageous for enabling a range of therapeutic applications due to their innate unique features encompassing high surface area, water dispersibility, aspect ratio, effective cell internalization properties and efficient biomolecule loading.

GQDs have shown to hold promise in various areas such as delivery, immobilization and controlled release of drug molecules. To study their interaction with drugs, the following factors can be taken into account: (i) physically, their increased surface area; (ii) chemically, the aromatic structure that results in the π-π stacking interactions; (iii) electrostatic interactions that occur between drug molecules of opposite charges and GO derivatives; (iv) hydrophobic interactions attributing to their distinct hydrophilicities; and (v) the intermolecular hydrogen bonding. Delineating the mechanism of GQD-based nanocomposites for biomedical applications demonstrates that their interactions with hydrogel is the potential root for the improving therapeutic efficacies, such as highly sensitive electrical properties, enhanced photothermal property, a superior loading capacity for anticancer drugs and outstanding fluorescence quenching ability. However, with the development of personalized cancer treatments, the controlled release of drugs into the specific sites in the cancerous organs is of practical interest (Liu et al. 2019). In fact, significant strategies in the synthesis of hydrogel nanocomposites with GQDs can improve the restoration, retention and revitalization of lost tissues and organs, impacting the whole spectrum of medicine and health care. Dimensional and morphology-controlled graphene-based nanomaterials' fabrication in hydrogel through in situ and ex situ polymerization techniques suggests its broader application, for instance, polymer fabricated with GQD imparts greater mechanical strength to the polymer. The attribute of GQDs as gelator in GQD-based hydrogel assists its self-assembly. Intriguingly, GQD acts as a nanofiller in hydrogel assembly promoting its multifunctionality.

20.4.1 GQD Hydrogel Nanocomposite Synthesis

Generally, two major processes have been employed in the synthesis of nanocomposite: in situ method and ex situ method, elucidated as follows:

20.4.1.1 In Situ Method

This procedure, also referred to as one-step fabrication process, focuses on the addition of GQD suspension during the polymerization of monomer units under suitable conditions. Good spatial distribution and prevention of nanoparticle aggregation are the peculiar advantages of this route. Unreacted adduct formation in this reaction is the major drawback which influences the properties of the final product (Guo et al. 2014).

20.4.1.2 Ex Situ Method

In contrast to the in situ method, this synthetic route employs the addition of GQD nanoparticles externally in a polymeric matrix instead of growing it from inside the matrix. On account of successful nanoparticle formation by heeding on its dispersible properties and its distribution in polymer, a widely applicable hydrogel could be formulated (Guo et al. 2014). This technique delineates the successful formation of GQD hydrogel nanocomposites, namely, Kagome lattice (KL) and Janus nanomicelles (JNM). GQDs employed as a filler with concentration of 0.05% and 0.3% have been used for copolymerization with poly(AAC-*co*-DEAEMA) resulting into KL and JNM hybrid hydrogel, respectively (Shakeel et al. 2019).

20.5 Physicochemical Characterization of GQDs and Hydrogel Nanocomposites

Employing a suite of characterization methods, the intrinsic structural, morphological, chemical and optical properties of nanofiller, hydrogels and hybrid constructs are analysed by scanning electron microscopy (SEM), confocal microscopy, optical microscopy, high-resolution transmission electron microscopy (HRTEM), selected area electron diffraction (SAED) pattern, atomic force microscopy (AFM), X-ray diffraction (XRD), particle size, porosity, Fourier transform infrared (FTIR) spectroscopy, X-ray photoelectron spectroscopy (XPS), Raman spectroscopy and photoluminescence (PL) spectroscopy.

Through imaging ellipsometry, micro-patterned thermoresponsive hydrogel films can be characterized to distinguish the different regions of interest. The spatial heterogeneity in GQD hydrogel nanocomposites arising due to irregular spreading of cross-link density leads to reduced optical clarity and strength of hydrogels as examined by light scattering, small-angle neutron scattering (SANS) and small-angle X-ray scattering. The origin of the toughness and the mechanisms of deformation of hybrid hydrogels can be elucidated by SANS.

20.5.1 Morphology

Morphological studies provide a dais for the analysis of various information such as surface topography (texture and thickness), size, shape, pore presence, nanofiller or drug distribution in hybrid hydrogels or nanocomposites.

20.5.1.1 Transmission Electron Microscopy (TEM)

Nanostructure of the as-synthesized hybrid hydrogels can be investigated by employing TEM and HRTEM. The morphology of the drug carrier is one of the factors that influence the release pattern of the drug. Previous studies indicate that the surface of chitosan-GQD composite had many wrinkles and cavities which were attributed to the cross-linking of chitosan with GQDs (Javanbakht and Namazi 2018). Increase in the surface roughness was observed with rise in GQD concentration. Furthermore, the increase in the nanoparticle concentration witnesses aggregation phenomenon on the nanocomposite's surface. The H-bonding between the nanoparticle and polymer matrix could possibly explain the better dispersion of GQDs at low concentration (Yue et al. 2019). The GQDs, due to their ultrasmall particle size, homogeneous distribution and spherical morphology, can be employed as suitable drug carrier. When GQDs are added to the hydrogel network, an increase in number of pores with decreased pore size is seen, causing accretion in the water holding capacity of the nanocomposite. In swollen state, the crystallization of the water molecules facilitates pore formation. Furthermore, increase in porosity with increase in GQD content suggests that they can act as template for pore formation (Havanur and JagadeeshBabu 2018).

20.5.1.2 Scanning Electron Microscopy (SEM)

SEM can be used to confer information about surface topography, porosity, composition, electrical conductivity and the characteristic structure of a material. The larger pores are more easily detectable by SEM. Porosity of the hydrogel is dependent on pH of the medium. The samples are swollen to equilibrium in buffer solution and then frozen at -80 °C followed by lyophilization. Subsequently, freeze-dried samples are kept in vacuum condition until further use to spot their inner morphology. Distinctive nano- to macroscale structural topographies can be understood from SEM micrographs. The interior of hydrogels revealed that the free radical copolymerization results in the formation of heterogeneous polymer network in the presence of an inert diluent.

20.5.1.3 Atomic Force Microscopy (AFM)

AFM can be used to observe 3D surface topology at nanometre resolution with minimal sample preparation under ambient environment. The gel modulus is measured by AFM force-distance curves. GQDs were analysed by tapping mode.

Depending on the concentration, the GQDs have a tendency to self-aggregate in an aqueous medium. These self-assembled GQDs may be J-type aggregate or end-to-end stacked aggregate. Well-dispersed and uniform dots with a particle height distribution ranging from 0.9 to 1.1 nm and a topographic height of ~1 nm were observed in the very dilute sample (Sangam et al. 2018).

20.5.2 Porosity Measurement

The increase in the number of pores in the hydrogel nanocomposites or matrices facilitates the higher drug loading and consequently its release rate. In this framework, the pore size can be leveraged for chosen application. Porosity of swollen hydrogels was determined from the equilibrium water content assuming that the hydrogels prepared without NaCl in polymerization medium will have only a molecular porosity and that prepared with NaCl will have a macroporous structure. The porosity and mean pore size of the dehydrated gels can also be determined by mercury porosimetry.

20.5.3 Swelling Behaviour

Nanocomposite hydrogels have been employed for the delivery of myriad of therapeutic agents including small molecular weight compounds to macromolecular drugs. The swelling kinetics and permeability are two of the significant properties that govern the hydrogel performance. For oral delivery, the hydrogels loaded with drug are often used in the dehydrated state. As hydrogels undergo glass transition in the dehydrated state, the water-soluble drug is released by concomitant absorption of water and desorption of drug. This swelling and diffusion follows non-Fickian behaviour because of slow macromolecular relaxation process in the swollen region. Diverse release profiles are possible depending on the relative magnitude of the rate of polymer swelling and rate of drug diffusion. Precise control over water swelling and cross-linking density is a way to regulate the drug release from the hydrogels (Lee and Kim 1994).

The hydrogel nanocomposite beads are swelled in buffer solutions, and after a certain time, they were taken out, dried to remove excess water and weighed.

$$\text{Swelling rate} = \frac{W_S - W_D}{W_D}$$

where: W_D is the weight of the dry sample, W_S is the weight of swollen sample

In a study where GQDs were observed to direct the polymeric self-assembly (of copolymers) to exotic Kagome lattice and Janus nanostructures, the swelling ratio was measured using the above specified gravimetric method (Shakeel et al. 2019). On the contrary, the increase in GQD content was correlated to the decrease in swelling capacity of the nanocomposite where a single component (carboxymethyl cellulose, CMC) was used. Interestingly, the elevated swelling capacity brought about increased drug penetration and uptake (Rakhshaei et al. 2019). We believe that the swelling capacity, drug penetration and uptake depend on the morphology of the polymer nanocomposite which in turn depends on the interaction of the nanofiller with the monomeric/co-monomeric units governing the cross-linking density.

20.5.4 Optical Properties

It has been observed that the GQDs exhibit a strong optical absorption between 260 and 320 nm in the UV region. This characteristic absorbance is attributed to the π-π^* transition of the C=C bonds having sp^2 hybridization. Sometimes, a weak shoulder at 270–400 nm is observed which is due to the n-π^* transition of the C=O bonds, having a tail that extends into the visible wavelengths. The former peak implies that aromatic sp^2 domains undergo electronic transition, leading to the formation of delocalized π states on the basal planes which is similar to graphene (Sangam et al. 2018). In a recent work by Bhattacharya et al., fluorescent hydrogels containing carbon dots in branched polyethylenimine (PEI) were fabricated. The UV-vis absorbance spectra of the nanocomposite displayed broadened π-π^* transition peak along with the appearance of a new peak at 380 nm which is attributed to the imine absorption. This significant broadening and red shift of π-π^* demonstrate that C-dots are immobilized in the gel (Bhattacharya et al. 2019).

The absorption peak position and emission of GQDs is influenced synergistically by their size, edge defects and surface functional groups. When irradiated, GQDs show tunable PL properties which are found to range from blue, green, yellow, to red, endorsing them to be used in bioimaging arena. The PL emission of GQDs is also of keen interest to the researchers. PL intensity is influenced by excitation wavelength, with red shifts covering the visible range. The most acceptable model to explain this phenomenon is quantum confinement effect. The excitation-dependent PL emission was also explained with the help of surface traps, giant red-edge effect, edge states and electronegativity of heteroatom models. It is noteworthy that the uniformity in size and the emission sites of the sp^2 clustered in the GQDs also influence the excitation-independent PL.

20.5.5 Fourier Transform Infrared (FTIR) Spectroscopy

Fourier transform infrared (FTIR) spectroscopy is an easy and rapid analytical technique to unravel the fingerprint of organic, inorganic or polymeric material with absorption peaks that comply to the vibrational frequencies between the bonds of the atoms. This is a convenient technique for identifying the chemical structure of hydrogels or its composites based on the oscillations of atoms which identify the structure of a molecule which can be organic, inorganic or polymeric. The radiation absorbed by the sample molecules is transformed into rotational and vibrational energy. The detector produces a signal which will form a spectrum, from 4000 cm^{-1} to 400 cm^{-1}. Every molecule will have its own unique spectral fingerprint, and thus molecule can be easily identified by referring to the standard data. A change in the composition of the material is indicated by the change in the pattern of absorption bands. Furthermore, the size of the peaks corresponds to the amount of material present. The surface functionalization of the mesoporous structures of the fluorescence resonance energy transfer (FRET) nanocarrier was confirmed by FTIR spectroscopy (Zheng et al. 2015a, b).

The FTIR of hollow mesoporous carbon nanoparticle (HMCN)-COOH of HA-HMCN(DOX)/GQDs spectrum displayed the peaks of carbonyl groups at 1705 cm^{-1} and 3436 cm^{-1} which confirmed that GQD was successfully functionalized with carboxyl group (Fang et al. 2018). The FTIR spectrum of DOX-GQDs-RGD (arginine-glycine-aspartic acid) revealed that they have –OH, C=O and epoxy groups which improved the fluorescence and solubility of the GQDs (Dong et al. 2018). After comparison of the FTIR spectra of GQD, GQD-BTN (biotin) and GQD-BTN-DOX, GQD spectrum showed peaks around 3450 cm^{-1} (O–H), 1620 cm^{-1} (C=O) and 1072 cm^{-1} (C–O) indicating their hydrophilic nature. For GQD-BTN, a peak at 1660 cm^{-1} showed the presence of amide group. Peaks were found at 1450 cm^{-1} (O–H) and 3450–3550 cm^{-1} (N–H) in GQD-BTN-DOX (Iannazzo et al. 2017). The spectra of graphene oxide confirmed the presence of functional groups having oxygen due to the peaks at 1720, 3430 and 1097 cm^{-1}. In GQDs-PEG spectra, peaks at 1735 and 1580 cm^{-1} confirmed the presence of amide bond, whereas the peaks at ~2885 cm^{-1} and ~1106 cm^{-1} suggested that the PEG was conjugated with the GQDs. The reduction of GO is confirmed by the peak at 1720 cm^{-1} (Wang et al. 2013b). FTIR spectrum of CMC showed a stretching vibration at 3440 cm^{-1} (O–H) representing presence of hydrogen bonds in the CMC along with the appearance of peaks at of 1615 cm^{-1} (C–H), 1517 cm^{-1}, (C–O) 2920 cm^{-1} and 1250 cm^{-1} (COO^-). Spectrum of CMC/GQD had a peak at 1712 cm^{-1} due to carbonyl groups of the ester functionality which proved that the cellulosic materials are cross-linked by carboxylic acids via esterification (Rakhshaei et al. 2019). In the FTIR spectra, the peaks at 1735 cm^{-1} (C=O) and 1441 cm^{-1} (C–H) disappeared in GQD-PDEA (poly(N,N-diethylacrylamide)) revealing that GQDs have conjugated with PDEA. The appearance of broad –OH peak at 3466 cm^{-1} and small C–C peak at 1452 cm^{-1} in GQD-PDEA confirmed the presence of GQDs in PDEA. The disappearance of C=O and C–H bending peaks in the FTIR spectra confirmed the polymerization of N,N-diethylacrylamide (DEA), and the appearance of –OH and C–C peaks explained the grafting of GQDs with PDEA nanohydrogels (Havanur et al. 2019). FTIR analysis confirmed the existence of nitrogen in GQD. In nitrogen-doped GQDs (N-GQDs), the peaks of 3000 cm^{-1} (O–H), 3500 cm^{-1} (NH_2), 1640 cm^{-1} (COO^-) and 1360 cm^{-1} (CN) were present. The existence of NH and CN groups explained that N-GQDs are nitrogen functionalized. In the spectrum of methotrexate (MTX)-(N-GQDs) presence of bands of N-GQDs, MTX (at 1620 cm^{-1}) and N–H (at 3475 cm^{-1}) confirmed the loading of MTX to N-GQDs (Khodadadei et al. 2017). FTIR analysis showed that CMC/GQD hydrogel showed an enhanced drug delivery, water vapour permeability, in vitro swelling, degradation and toxicity for blood cancer cells (Javanbakht and Namazi 2018).

20.5.6 X-Ray Diffraction

X-ray diffraction is an analytical approach which is used for identification of amorphous or crystalline structure of a material and which displays the chemical composition, percentage of crystallinity and crystallite dimensions of the compound to unravel the crystal structure. Comprehensive peak-width investigation can tell about the information on the crystallite size and the microstrain established within a composite hydrogel. During the polymerization process in hydrogels, this quantitative technique may monitor the modification in the crystalline structure. The emergence of new diffraction peaks may designate the role of nanofillers, drug, etc. and their interaction with hydrogel matrix.

XRD analysis supported the fact that GQD gating was present in the compound. No diffraction peaks were observed for HMCN(DOX), while a broad diffraction peak corresponding to GQDs at $2\theta = 9°$ was observed for HMCN(DOX)@GQDs, which also indicated GQDs capping on the HMCN(DOX) nanoparticles of HMCN-COOH, thus confirming the success of carboxylation functionalization (Nasrollahi et al. 2019). The structure of GQD capped mesoporous silica nanoparticles (MSNs) was confirmed by XRD. The intensities of the XRD peaks became lower after capping with GQDs and decreased even more after DOX loading and the capping process (Chen et al. 2014). A wide (002) peak of around $2\theta = 26.98°$ in the XRD patterns of GQDs was detected which confirmed the carbonizing of citric acid to yield graphene structures. The XRD data revealed the crystalline nature of the drug and compatibility between drug and GQD in the DOX loaded CMC/GQD hydrogel (Javanbakht and Namazi 2018). XRD analysis confirmed that by doping of GQDs, specific surface area is increased (Khodadadei et al. 2017).

20.5.7 Thermal Analysis

Thermoporometry via differential scanning calorimetry (DSC) facilitates the evaluation of both size distribution and pore size of hybrid hydrogel/hydrogels in the liquid state which is based on the principle of crystallization depression or melting of a fluid limited in a pore. DSC thermoporometry discloses only pores below 200 nm.

The volume phase transition of water in hydrogel systems is used to scrutinize thermal cycling and the influence of cross-linker due to electrostatic hydration effects of amphoteric polymers by DSC and TGA analyses. The temperature was raised from 10–90 °C at a rate of 10 °C per minute. DSC is also used to control lower critical solution temperature (LCST) of hybrid GQD hydrogel and to quantify the bound water and free water amount. DSC analysis measures the endotherm of sublimated, vaporized and molten stage. The thermal stability of the hydrogel is conducted by thermogravimetry. TGA also helps in understanding of thermal stability of hydrogel and GQD-fluorescent smart hydrogel (Cheng et al. 2019).

20.5.8 Rheological Properties

The mechanical characteristics of hydrogel are estimated by deforming the hydrogel by applying a static compressive loading. The hydrogel counters the continuous load up to the breaking point where it flops. The resistance of the hybrid hydrogel to the applied force is calculated by the slope of the load deformation curve. The steeper the slope, the greater the resistance and hence the higher the modulus. The mechanical properties of hydrogels are evaluated to determine the aptness of the hydrogel for the required application. In non-biodegradable system, carrier hydrogel matrix should maintain mechanical and physical integrity to keep safe the biomolecules and drugs from adverse environment before reaching to the target site. The mechanical strength of the hydrogel can be optimized by incorporating the optimum level of cross-linking agents and co-monomers because a greater degree of cross-linking develops into more brittle and less elastic nature. Viscoelastic attributes of swollen hydrogels are analysed by a rheometer. Incorporation of GQD nanocrystal into hydrogel may affect the rheological properties influencing the biological behaviour. This modification can favor the use of formulations, namely, for drug delivery, as a little change in the viscosity can impact the occlusion, permeability and retention of these GQDs into biological.

20.5.9 In Vitro Drug Loading and Release Studies

The in vitro release profile analysis is a significant tool used to investigate drug release mechanism and kinetics performance during different phases of hydrogel fabrication. Furthermore, this also helps to design suitable system via a scientific and rational approach for an improvement to drug product. With the aim to unravel a potential carrier for controlled drug release, nanocomposite hydrogels are assessed for their encapsulation and loading capacity of the drug. At a specified time, the initial and final concentration of the drug in the solution is measured through UV-vis/NIR (near infrared) spectrophotometer which calculates the drug concentration in the GQD hybrid hydrogels.

The drug release percentage increased with increase in pH of the nanocomposite. When nanocomposite was encapsulated with CMC, the lowest drug release was found to occur in the gastric fluid pH range, wherein, chitosan favoured the drug release at pH 6.8 and 7.4. At pH 6.8, the deprotonation of carboxylate groups of CMC and repulsion between CMC and negatively charged sodium salicylate drug engendered the drug release (Javanbakht and Shaabani 2019). Rakhshaei et al. used GQDs as a cross-linker for CMC to fabricate hydrogels and studied the effects of

GQD concentration on the physicochemical properties of the synthesized CMC/GQDs films. The loading capacity of the DOX in the nanocomposite gels containing 30% of GQDs was higher than the ones containing 45%. With increase in the pH, the drug release concentration increased as a result of pH-sensitive swelling of CMC/GQD (Rakhshaei et al. 2019). Yue et al. studied the in vitro release profile of thermoresponsive GQDs-Dex/PNIPAM (poly(N-isopropylacrylamide)) nanocomposite loaded with buprenorphine. GQDs as nanofillers were found to modulate factors encompassing hydrophilic to hydrophobic component ratio of the polymer matrix, interrelationship of GQD and drug, GQD stability and many others (Yue et al. 2019).

20.5.10 Biocompatibility Evaluation

The advent of GQD/hydrogels/hybrid hydrogels challenges their utility in the biological field which needs to be addressed more comprehensively according to specific applications. In recent years, substantial efforts have been made to procure the toxic profile of GQDs, doped GQDs and its hybrid hydrogels to uncloak their tremendous biocompatible essence. In vitro and in vivo studies concoct two channels for GQDs' toxic behaviour evaluation. Cell viability testing, a perfect epitome of in vitro toxicity analysis, is employed via MTT, LDH and ATP assays (Wang et al. 2016). In a comparative study, Wu et al. reported the higher cytotoxicity of graphene oxide than GQD with MGC-803 cells (human gastric cancer) and MCF-7 cells (human breast cancer) (Wang et al. 2013a). Furthermore, Nurunnabi et al. in his comparative study manifested LDH release assay of KB (human epithermal cancer cells), MDA-MB231 (breast cancer cells) and A549 (human epithelial cancer cells) cells. Interestingly, human epidermal cancer cells showed lowest cell viability and highest LDH release due to membrane damage, under the same incubation time and GQD concentration (Nurunnabi et al. 2013). Additionally, cytotoxicity of GQDs to other mammalian cells was also addressed by several groups. For instance, Zhang et al. reported that even at high GQD concentration, the cell viability was greater than 80% after 3 days of culturing. Therefore, this long-term and cytocompatible labelling of stem cells is crucial for assessing their role in the field of regenerative medicine (Zhang et al. 2012). Nurunnabi et al. in his additional test on MDCK (kidney epithelial cells) presented 95% of cell viability over 48 h of incubation with GQDs at 500 μg/mL concentration (Nurunnabi et al. 2013). Chemically doped GQDs' toxic profile analysis is also significant and appertains in the category of the testing candidate. Wang et al. evaluated the toxicity of graphene oxide (GO) and nitrogen-doped GQDs (N-GQDs) for RBCs via haemolytic assay. GO leads to apparent haemolysis with the release of ATP, paradoxical to N-GQDs' toxic profile (Wang et al. 2015a). Hai et al. observed that boron-doped GQDs (at 4 mg/mL) are more biocompatible with HeLa cells (about 87% cell viability after 12 h of incubation) than undoped GQDs (Hai et al. 2015).

In vivo toxicity is pre-eminently concerned with three important factors: biodistribution of testing material among various organs, possible elimination route and any possible damage to vital organs. Animal model administered with a certain dosage of GQDs is monitored over a period of time for biodistribution, organic accumulation and excretion of GQDs. The toxic profile of GQDs was analysed by Nurunnabi et al. via in vivo imaging, biodistribution and ex vivo organic imaging in cancerous mice model. GQDs synthesized from the bottom-up method depicted low in vivo toxicity. Twelve hours after injection, GQD could be detected in tumour site located on the skin with no fluorescence signal detected from deep organs. However, after 24 h of injection, GQDs could not be detected, neither on the skin nor in the deep organs. Biodistribution of GQDs throughout the body is depicted via ex vivo imaging. Within the first 12 h, they accumulate in the liver and heart. However, in the next 12 h, it was observed that accumulation of GQDs increases gradually within the kidneys and decreases in the liver. Effective removal of GQDs via the excretory system after 24 h is confirmed through weakened fluorescence signals (Nurunnabi et al. 2013). Wang et al. studied the impact of GQD concentration on zebrafish embryos during development. The concentration greater than 50 μg/mL leads to a rise in the mortality rate and embryo disfigurement along with the decrease in heart rate and hatching process (Wang et al. 2015b). Another study showed GQD uptake in the zebrafish and its removal through the excretory system completed in 7 days, left with no obvious distinct physiology between the test group and the control group at a concentration lower than 2 mg/mL, indicating excellent biocompatibility of this nanodot. The contradictory results of the above two studies on the in vivo toxicity of GQD were upshot of the different synthetic routes adopted. The former adopted GQDs synthesized from cleaving graphene oxide, whereas the latter synthesized GQD via the hydrothermal treatment of leaf, according to its toxic features. From the above-mentioned literature, GQDs' high biocompatibility and low toxic profile are pellucid (Roy et al. 2014).

Similarly, the biocompatibility of GQD hydrogel has been assessed before complementing to its biomedical applications. Rakhshaei et al. in his comprehensive study showed CMC/GQD hydrogel nanocomposites' non-toxic nature at 45% GQD concentration and observed 80% cell viability. These nanocomposites in conjunction with DOX showed improved cell viability which could be attribute of π–π interactions between the GQDs and the drug (Rakhshaei et al. 2019). Another in vitro cell toxicity study was performed with DOX-loaded CMC/GQD nanocomposites on K562 cell lines. It was inferred that although the presence of GQD enhances the biocompatibility of the nanocomposites, higher toxicity was observed due to increased DOX dosage (Javanbakht and Namazi 2018).

GQD hydrogel nanocomposite is suitable platform for tissue compatibility testing due to its low toxicity and great biocompatibility. Yue et al. observed the non-toxic response of GQD fabricated dextran/glycol polymer-rich hydrogel (GQDsDex/PNIPAM) in some of the isolated animals' sciatic nerves and its adjacent tissues; it potentially did not harm the nerves and tissues as confirmed by haematoxylin and eosin (H&E) analysis further aided to pain management. Substantial infiltration of stromal cells (like macrophages, lymphocytes, neutrophils) was observed on the site of implantation of the GQDs coated nano matrix (Yue et al. 2019).

20.6 Application of GQDs and GQD Hydrogel Nanocomposites

The applications of GQD and their hybrid hydrogels are delineated below (Fig. 20.2).

20.6.1 Wound Healing

Hydrogel is non-adhesive to the wound, providing an ideal environment for wound cleaning as well as facilitating body defence in wound care management. Hydrogel nanocomposites with its enhanced physicochemical properties can shed new light for tailoring wound dressings in a better way compared to the native hydrogels. These GQD hybrid hydrogels enhance early wound bed restoration *via* re-epithelialization and fibroplasia, reinforcing angiogenesis, keeping the wound moist, protecting the wound against microorganism, desiccation, proficient of absorbing some exudate and promotes healing by the regeneration of complex layers of skin, including hair follicles and the glands. The accelerated wound healing ability of hybrid hydrogel with the optimal GQD mass ratio exhibits stronger antibacterial abilities than other hydrogels for effective wound dressing. Furthermore, GQD

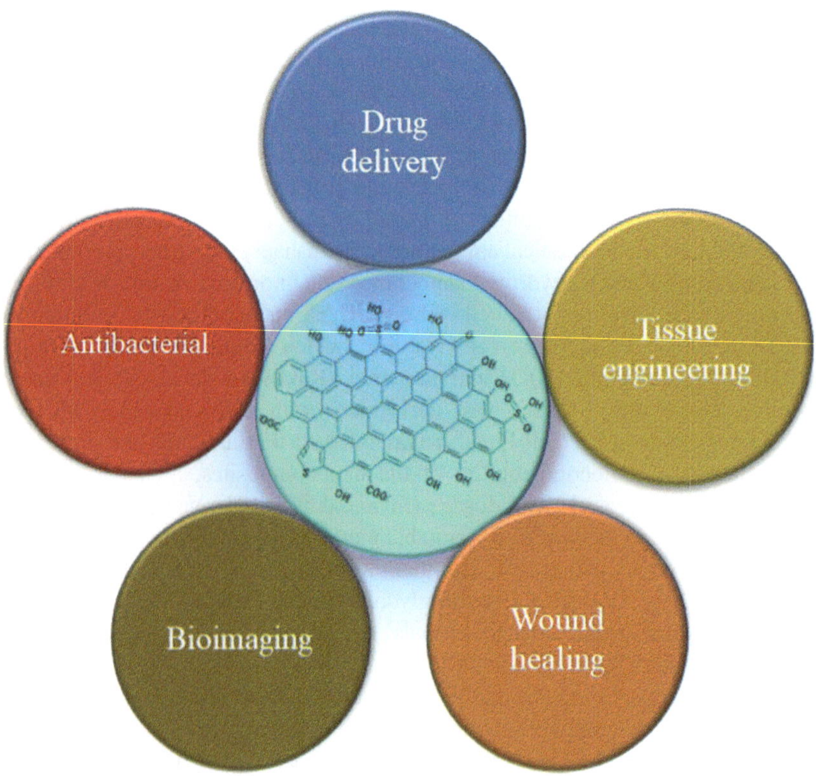

Fig. 20.2 Application of GQDs in theranostics

hybrid hydrogel demonstrates superb biocompatibility, great swelling ratio and decent extensibility (Fan et al. 2014).

GQD hybrid hydrogels formed via ex situ polymerization can be the promising and remarkable vehicle/medicament for wound healing compared to the conventional hydrogels since they act as a barrier to microorganisms, allow gaseous exchange, have great biocompatibility and remove surplus exudates without creating trauma. Most significantly, in in vivo experiments, graphene-derived hybrid hydrogel specifies acceleration in the wound healing process. Furthermore, histological evaluation delineates reconstruction of thickened and intact epidermis within few days of the wound creation. Thus, GQD-based composite hydrogel with antibacterial property encourages their applications in wound dressing.

20.6.2 Drug Delivery/Antitumour Drug Delivery Systems

The process of drug delivery includes the administration of the drug, the release of the active ingredients by the drug and the transport of the active ingredients to the site of action (Jain 2008). Over the years, various carriers were explored for an effective drug delivery system for therapeutic release to target sites, which includes synthetic and natural polymers, liposomes, dendrimers, metal oxide nanoparticles, carbon nanotubes, fullerenes and graphene. The polymers and copolymers that show reversible solgel transitions have applications like injectable drug transport systems. Hydrogels based on anionic or cationic polymers exhibit pH-sensitive swelling and the ability to govern diffusion in an on/off pattern. Similarly, lower critical solution temperature (LCST) polymer gels exhibit temperature sensitivity (Huang and Brazel 2001). Due to tunable physical properties, hydrogels can swell and hold large amount of water while retaining its original shape making them ideal compounds for the protection of drug in the adverse conditions. Moreover, due to the property of modifying their structure with the changes in environmental conditions, hydrogels have a command over discharge of drugs eliminating the fear and discomfort linked with intravenous chemotherapeutics.

Cancer has emanated as a frightening menace to humans despite the fact that medicine, science and technology have reached its zenith. Conventional chemotherapy often lacks specificity and selectivity for the cancerous cells, which leads to systemic toxicity. Furthermore, a person cannot get rid of whole tumour by a single chemotherapy. Although chemotherapy is one of the most accepted treatments against cancer, it leads to systemic toxicity due to lack of specificity and selectivity for the cancerous cells. To overcome these issues, researchers started exploring nanotechnology for the treatment of cancer due to the properties like drug solubility, reduced toxicity and site-specific targeting. Advantages of nanogels and other nanosized drug delivery systems over macro-sized delivery systems are as follows:

- Great stability, response to stimuli, permit for convenient functionalization and being easy to synthesize
- Control over drug release rates due to increased surface area

- The capability to target diseases due to easy flow into the body (Chacko et al. 2012)

Although bright results have been reported about nanoparticles, their intrinsic toxicity and low biological distribution is still a serious limitation. Thus, researchers started using new formulations like carbon nanotubes, graphene oxide and fullerenes to enhance the nano-drug efficiency while diminishing damage to the surrounding normal tissues.

Graphene nanomaterials with increased surface areas show higher drug loading ability. Reportedly, targeted delivery and controlled release of drugs become better with graphene-based hydrogels, although graphene oxide demonstrates very low fluorescence efficiency. GQDs, the next generation of graphene family, have shown even better results when compared with other carbon-based nanomaterials and conventional quantum dots (QDs), due to their exceptional hydrophilicity (results in high stability in physiological environments), low chemical inertness, effective delivery of partially water-soluble drugs like CDDP, water solubility, biocompatibility, high loading capacity and low toxicity to their ultrasmall size. By giving interaction anisotropy and shape, GQDs help in aiming the self-assembled complex. Furthermore, these compounds have improved the chemotherapeutic potential of anticancer agents by acting at the nuclear level, accelerating their nuclear accumulation and enhancing their DNA cleavage activity, which makes them a promising candidate for applications in drug delivery.

DOX (doxorubicin) is one of the chief anticancer drugs, which kills cells by incorporating with DNA and stops the cell division and the DNA replication process. GQD nanoparticles, conjunct with DOX, deliver it to the cancer tissues and dramatically increase cytotoxicity of DOX. Therefore, GQD-based carriers such as GQD hybrid hydrogels are promising and highly efficient carrier for delivery of DOX. Fang et al. synthesized a biocompatible HA-modified and GQD-gated HMCN pH-/NIR-responsive nanoparticle for CD44 cancer cells (Fang et al. 2018). Dong et al. covalently conjugated RGD with GQDs and loaded DOX on its surface via π-π stacking and hydrophobic interactions (Dong et al. 2018), whereas Iannazzo et al. functionalized GQDs with biotin (Iannazzo et al. 2017). Ding et al. developed a DOX@GQD-P-Cy, by loading anticancer drug DOX to GQD and attaching fluorescent cyanine (Cy) dye to GQD via cathepsin D-responsive P peptide (Ding et al. 2017). Su et al. reported a size-changeable GQD nanoaircraft (SCNA) that synergistically combines functions of hierarchically targeted nanoparticles and GQDs to overcome the low EPR effect and tumour heterogeneity (Su et al. 2017). Zheng et al. developed an anti-HER2 labelled GQD conjugated LPT nanoflower as a multifunctional nanocarrier for target drug delivery, in which the targeting ligand, imaging agent, and therapeutic compounds are all included into a single platform (Zheng et al. 2013). Chen et al. designed a biocompatible pH-responsive drug delivery system by grafting GQDs on the MSNs through acetal bonds. The combined system could effectively stop the leakage of drug molecules at neutral pH and release them at acidic pH (Wang et al. 2020). In another study, prepared GQDs-PEG (polyethylene glycol) is bestowed with high surface area, thus enabling them as a drug carrier

(Wang et al. 2013a). CMC/GQD hydrogel nanocomposite displayed enhanced controlled release of drug and pH sensitivity on addition of GQDs (Rakhshaei et al. 2019). Havanur et al. observed that addition of GQDs to poly(N, N-diethylacrylamide) (PDEA) enhanced its stimuli-responsiveness and other inherent characteristics (Havanur et al. 2019).

Methotrexate (MTX) is one of the most potent anticancer drugs for a wide range of hematologic malignancies, psoriasis, rheumatoid arthritis and tumours, but its non-selective nature and side effects like nephrotoxicity, leucopenia, bone marrow suppression to healthy cells and hepatotoxicity severely limit its efficacy. Thus, controlled and stimuli-responsive release of MTX is of extreme importance. Nasrollahi et al. conjugated GQDs to single-chain variable fragment of antibody, which was engineered to have high affinity B10 to epidermal growth factor receptor, through amide covalent linkages (Nasrollahi et al. 2019).

Zheng et al. fabricated an adenosine triphosphate (ATP)-responsive FRET nanocarrier for intracellular drug delivery and monitoring of drug release, where aptamer/GQD nanocomplexes acted as capping and quenching motifs (Zheng et al. 2015a, b). GQD-based nanovector was synthesized by Iannazzo et al. from multi-walled carbon nanotubes (MWCNT), for delivery of methyl 3,3-dimethyl-2-(3-methyl-2,3-dihydrobenzofuran-2-yl)butanoate (BFG) bearing riboflavin (RF; vitamin B2). After conjugation with the GQD, the water dispersibility of nanosystem increased which enhanced the pharmacological profile of the anticancer drug (Iannazzo et al. 2019). Gao et al. developed a photothermal, biocompatible and redox-responsive drug carrier where GQDs acted as gatekeepers to stop the release of rhodamine B (RhB) which was loaded into the mesopores of mesoporous silica nanoparticles. The nanocomposites have exquisite photothermal property originated from GQDs (Gao et al. 2019). Vatanparast et al. used the general dispersion correction (DFT-D3) method for the adsorption of 5-fluorouracil (FU) drug on pristine and doped GQDs. The results declared that doping of GQDs with aluminium and nitrogen (AlN) and aluminium and phosphorus (AlP) atoms remarkably increased the tendency of GQDs for adsorption in the order AlN-FU > AlP-FU > BP-FU (boron- and phosphorus-doped fluorouracil) > BNFU (boron- and nitrogen-doped fluorouracil) (Vatanparast and Shariatinia 2018). Hyaluronic acid (HA) and GQD-functionalized human serum albumin (HSA) nanoparticles were applied for bioimaging and targeted delivery of gemcitabine (Gem) to pancreatic cancer cells (Nigam et al. 2014). The extremely small GQD surface functionalized grafted dextran/PNIPAM (poly(*N*-isopropylacrylamide)) copolymeric hydrogel was synthesized by Yue et al. for the thermosensitive drug delivery of buprenorphine for pain management in tissue implantation (Yue et al. 2019).

20.6.3 Antibacterial Activity

Bacterial infection emerged as a hazardous obstacle in the medicinal sector which harvested a pivotal query related to its curability. Retaliation via the remarkable historic breakthrough of antibiotic invention in 1929, a promising curative candidate was utilized as a weapon against these infectious microbes. The ramifications of overusing antibiotics and the evolution of superbugs are one of the key factors in motivating researchers to come up with new treatment methods. Left with fewer effective armoury tools, high demand for new weapons against these drug-resistant pathogens is an ardent requirement (Harroun et al. 2017). Progression in nanotechnology instigates the metal and metal oxide nanoparticles with their biocidal activity (Hegab et al. 2016). Enhanced efficacy of nanoparticles by dint of multiple antimicrobial effects confers its antibacterial action, dwindling the antibacterial resistance. However, while designing a suitable treatment using nanoparticles, their toxicity to the mammalian cells should also be factored in. With higher biocompatibility than metal nanoparticles, graphene-based nanoparticles hold the potential to revolutionize the clinical sector (Harroun et al. 2017).

Owing to the inimitable properties of graphene-based nanomaterials, graphene oxide, appraised as a potent antibacterial, is an attribute of cutting-edge effect, cell entrapment and oxidative species effect (Xia et al. 2019). Higher toxicity of graphene oxide limits its biological applications entailing GQD exploration in the biomedical operations (Wang et al. 2016). Synthetic routes, heteroatoms and surface functionalization concomitant with GQDs alter their physicochemical properties and strongly influence its biocidal activity (Zheng et al. 2015a, b). Hui et al. illustrated the same in his comparative study of fullerene and GO-derived GQD antibacterial property encompassing Gram-positive *S. aureus* and *B. subtilis* and Gram-negative bacteria *E. coli* and *P. aeruginosa.* Species-selective killing of *S. aureus* persisters via C_{60}-GQD over GO-GQD captures the limelight. Engrossingly, GQD synthesized via chemical rupture of C_{60} cages shows significant bacterial loss of *S. aureus.* GO-GQD, prepared through a bottom-up approach, failed to elucidate its activity (Hui et al. 2016).

GQDs are responsible for redox species generation when illuminated with light, acting as photosensitizing agents in antimicrobial photodynamic therapy, exceptionally smart treatment of persisters (Sun et al. 2014). Under blue light irradiation, these photoactive nanodots deteriorate the bacterial cell count of multidrug-resistant clinical isolates (*E. coli, S. aureus*). Neither GQDs nor irradiation alone exhibits oxidative stress on bacterial cells and declines their viability (Marković et al. 2018). Ristic et al. delineated the same in their experimental analysis. This group irradiated the bacterial culture (*E. coli*, methicillin-resistant *Staphylococcus aureus* (MRSA)) with 470 nm and reported a noteworthy loss in the viability of bacterial persisters upon GQD photoexcitation. GQD toxicity to a mammalian cell, a noteworthy loss in bacterial persisters' viability, was reported. The cytotoxic effects of GQD were found to be concentration-dependent as well as on the duration of photo-exposure, with comparable efficiency in both *E. coli* and *S. aureus* bacterial species incubated with the fixed dose of GQD (Ristic et al. 2014). Barneck et al. found that 405 nm visible light sterilization (VLS) eradicated the bacterial isolates and was less toxic to human tissues. This study demonstrates the clinical potential of 405 nm VLS (Barneck et al. 2016).

Mitigation of reactive oxygen species (ROS) including superoxide radical (O_2^-), peroxide (O_2^{2-}) and hydroxyl radicals (•OH) with the antioxidant defence mechanism of bacteria contributing to antibiotic lethality incites pandrug-resistant bacteria encroachment. An elevated level of redox species considerably kills the bacteria (Courtney et al. 2016). Chong et al. confirmed the potentiality of photoactive GQD in increasing intracellular ROS generation via electron-hole pair upon light excitation. More importantly, they revealed about single oxygen species generation via photoexcited GQD, attributable to both energy transfer and electron transport (Chong et al. 2016). In a comprehensive study, MB-GQD (methylene blue-GQD) effectively enhanced the singlet oxygen species generation in the bacterial cell, and an increase in photoresponsivity induced the viability loss. No evident side effect of MB-GQD was observed on human cells. Moreover, MB-GQD deactivated the Gram-positive and Gram-negative bacteria within 5 min of irradiation. Furthermore, sulphur-doped GQD, when combined with MB, increased the antibacterial activity of GQD as a lower concentration of MB induced higher antibacterial activity (Kholikov et al. 2018).

Doping is one of the key factors directing semiconductors' physicochemical features to eventuate extraordinary optoelectronic property that is completely distinct from its intrinsic counterpart GQD. The line of the study carried out by Pourhajibagher et al. and Kuo et al. outlined the GQD enhanced antibacterial under illuminance when doped with curcumin and nitrogen atom, respectively. N-GQD employed as photosensitizer illuminated at 670 nm laser exposure eradicated *E. coli* cells efficiently. Kuo and his co-workers also showed that higher N content in N-GQD dominate bactericidal activity than the lower N content of N-GQD (Kuo et al. 2017). Pourhajibagher et al. in their comparative study showed 93%, 82.2% and 73.1% significant reduction in CFU/ml in perio-pathogen mixed biofilm of *A. actinomycetemcomitans, P. gingivalis and P. intermedia* via curcumin (cur)-GQD+blue LED, cur+blue LED and GQD+blue LED, respectively, in comparison to the control group (Pourhajibagher et al. 2019). In addition to distinguished photostability and excellent optical and electronic property, GQD possesses high peroxidase-like activity due to its excellent electron transport system, which arose from its hydrogen peroxide (H_2O_2) decomposition reaction. Sun et al. for the first time reported GQD peroxidase activity application in antibacterial field. A low dose of H_2O_2, otherwise toxic at high concentration, combined with GQD generates •OH radicals and contributes to potent antibacterial activity with a negligible toxic deed to a human cell. Dose-dependent antibacterial activity of H_2O_2 against *E. coli* and *S. aureus* was observed. 1 and 10 mM of H_2O_2 for *E. coli* and *S. aureus*, respectively, induce same effect on two bacterial species due to different membrane architecture and chemical composition of cell wall and "bacteria observer" effect (Sun et al. 2014).

In conclusion, combinatorial treatment, a bold move of modern science in the antibacterial field, enacts as a powerful weapon against multidrug-resistant (MDR) strain and assists in its eradication process. Also, tunable properties of nanomaterials such as size, surface moieties, and reduction, and oxidation potentials can help to gain control over redox perturbation for alleviating multidrug resistance and improving antibiotic efficacy (Courtney et al. 2017).

20.6.4 Tissue Engineering

Hydrogel and their nanocomposites have played a pivotal role in developing scaffolds for tissue engineering. Scaffolds are 3D porous solid bioactive systems which provide a physical surface for adsorption of biomolecules and immobilization of proteins, growth factors and other biologically active biomaterials. They show great potential in tissue engineering and regenerative medicine, catering bioactivity and specificity to the structure. These polymer scaffolds aid in organizing cells, render the stimuli to direct the formation of the desired tissues and function as delivery vehicles for carrying bioactive peptides, growth factors and drugs and space-filling agents. It is very essential to find a suitable material to deal with the vital biological blueprint inherent to each application. Scaffold matrices play a very fundamental role in the progress of new tissue morphogenesis by guiding the growth of cells seeded within them or to help cells migrate from surrounding tissues by communicating with them. Most of the human cell types necessitate sufficient anchorage to support tissue regeneration, and thus their deficiency can lead to cell necrosis and defective tissues. Hence, scaffolds act as a substrate and possess essential physical and chemical properties compulsory to promote cell attachment, proliferation, differentiation and migration (Mantha et al. 2019).

Recent advances in the stem cell arena and nanoscale engineering have been utilized to tailor innovative platforms, thus driving stem cell researchers to cater questions that have not been answered before and to overcome hurdles in regenerative medicine. Stem cells can be procured from a myriad of sources and subsequently diverge in their proliferation rates, ease of in vitro culture and adeptness to form specialized cell types. They possess the potential to self-renew and evolve into diverse specialized functional cell. Irrespective of the type of stem cell, the current research emphasizes on the expansion of stem cell, maintenance of stem cell state, differentiation and their clinical usage. Stem cells are categorized, on the basis of their developmental stage, into two broad classes, namely, embryonic stem cells (ESCs) and somatic stem cells (SSCs). Induced pluripotent stem cells (iPSCs) are derived from somatic cells by genetic reprogramming to an ESCs-like state by introducing the expression of certain genes and factors and have comparable attributes as ESCs. Though SSCs that are procured from adult individuals have an ease of access, they are less powerful than ESCs and iPSCs (Kingham and Oreffo 2013).

Now, the question arises such as “what are the major factors that guide the application of stem cells in the field of cell therapy and tissue engineering?”. The answer is the ability to regulate and sway the differentiation of cells into defined cell types. Researchers have been making an effort to understand the mechanism of stem cell differentiation and to manipulate it into abundant cell types encompassing cardiomyocytes, adipocytes, osteoblasts and neurocytes. Still, it could not be exploited to the fullest and hence curtails its usage in stem cell therapy. The ability

Fig. 20.3 Factors affecting cellular proliferation and viability

of the undifferentiated cells to trigger the threat of teratoma post-implantation is a major concern and hence only those stem cells that are committed to differentiate into particular lineages are often employed in the stem cell therapy (Fig. 20.3).

The extracellular milieu encodes molecular and physical information that can help us guide the development of new-generation scaffolds for tissue engineering. Nanomaterials present a great way to modulate the behaviour of cell due to their nanoscale dimensions, versatility of functionalization and ease of synthesis. Hitherto, graphene and its derivatives have been employed for various tissue engineering applications (Akhavan 2016; Han et al. 2019). The fact that the toxicity of graphene-based materials can be correlated with their particle size along with various other factors like oxygen content, surface charge, etc. supports the concept of higher biocompatibility of GQDs (Fasbender et al. 2019). Zero dimensional GQDs with quantum confinement and controllable particle size can be useful in exploring the interaction of GQDs and MSCs.

In a study by Qui et al., proliferation of mesenchymal stem cells (MSCs) upon exposure to GQDs was studied, and proliferation rate was found to improve in a time-dependent manner. The confirmation of the osteogenic differentiation potential of the GQDs was confirmed by the accretion of alkaline phosphatase (ALP) activity. The expression of phenotypically related genes in osteogenic differentiation was measured via quantitative polymerase chain reaction (q-PCR), and it was found that the expression of Runt-related transcription factor 2 (Runx-2), an important transcription factor that regulates a myriad of genes related to osteogenic differentiation, was increased along with the expression of bone extracellular proteins like osteopontin (OPN) and osteocalcin (OCN). The osteogenic differentiation of MSCs into osteoblasts was also assessed with the help of alizarin red-based assay of mineralization depicting that GQDs favoured the formation of calcium-rich

Fig. 20.4 Factors affecting the cellular attachment

deposits. Further, the microarray analysis of 21,000 global gene expression concluded that there were upregulation of 213 genes, with pivotal genes being bone morphogenetic protein-6 (BMP-6), transforming growth factor β-2 (TGFβ-2) and collagen V-α3 (COLV-α3), and downregulation of 263 genes. The influence of MSC differentiation into adipocytes was studied by staining the intracytoplasmic lipids with oil red O. The presence of intracytoplasmic lipids depicts that GQDs have no ill-effect on the differentiation of MSCs to adipogenic cells and the pluripotency ability of the stem cells was not affected even in the presence of GQDs (Qiu et al. 2016) (Fig. 20.4).

Yang et al. studied osteogenic differentiation of stem cells from human exfoliated deciduous teeth (SHEDs) in presence of graphene oxide (GO) and graphene oxide quantum dots (GOQDs). The proliferation was facilitated in presence of GO and GOQDs at a concentration of 1 μg/mL up to 7 days in vitro. Their excellent fluorescent properties helped to confirm their distribution in the cell cytoplasm. Significant increase in ALP activity, formation of calcium nodules and expression of RNA and protein in SHEDs treated GOQD containing osteogenic medium was observed. However, the trends were observed to be contrary when the osteogenic medium was supplemented with GO (Yang et al. 2019). The study conducted by Liu et al. showed that the proliferation of the rat bone marrow-derived mesenchymal stem cells (rBMSCs) in presence of 100 μg/mL N-GQDs was alleviated. N-GQDs have a good permeability engendered by their nanoscale size and superior dispersibility. The confirmation of their osteogenic potential was attributed to the upregulation in the expression levels of ALP, ECM, OPN and OCN in rBMSCs (Geng et al. 2018).

Thus, hydrogels have served to be the most common scaffold for tissue engineering due to their innate abilities including (i) potential to retain a distinct 3D morphology, (ii) mimicking and stimulating the native extracellular environment and (iii) acting as a mechanical support for cells in the engineered tissues (Qiu et al. 2016). Interaction of nanoparticles with MSCs affects their behaviour in terms of function, differentiation and self-renewal, and thus, we can harness this potential to gain new heights in the field of tissue engineering (Mantha et al. 2019). Therefore, incorporation of GQDs within the hydrogels needs to be explored and such scaffolds may hold promise by exhibiting complementary properties.

20.6.5 Bioimaging

One of the important approaches that aids in both research and clinical needs is bioimaging which helps in observing certain biological processes like cellular uptake, targeted delivery, biodistribution of therapeutics and many more (Chung et al. 2019). GQDs have made their place in the spectrum of bioimaging as they possess superior qualities like high aqueous solubility, good biocompatibility, stable photoluminescence (PL), superior resistance to photobleaching and low cytotoxicity (Rakhshaei et al. 2019). GQDs can be successfully used in the fields of NIR fluorescence imaging and MR imaging due to the stable photoluminescence property, making them favourable to be used as optical probes (Chung et al. 2019) and giving an edge over the use of other fluorescent dyces/nanoparticles.

Thus, the property of GQD, emitting fluorescence of different wavelengths, makes them one of the best suited imaging probes and hence is counted as a better option to be used in the field of bioimaging. There are many other carbon-based fluorescent nanomaterials like carbon dots (C-dots) that have emerged as a competitor of GQD, but the quasi zero-dimensional property of the GQDs entailing few atomic layers with strong PL property and without any passivation makes it an extraordinary class of bioimaging probe. Compared to the C-dots which are either crystalline in structure or amorphous in nature, GQDs have graphene lattices which are present inside the dots. These are produced from the starting materials containing graphene. Moreover, the special quality of GQDs having tunable PL by doping heteroatoms to the π-conjugated system proves beneficial. The wide range of emission and excitation properties of GQDs makes it appropriate to be used as imaging agents in tissue engineering, cancer therapy and cell imaging (Nurunnabi et al. 2015). Compared to the photostability of ODs, the stability of GQD is very high and is resistant against photobleaching and blinking (Shen et al. 2012).

To date, various probes made of GQD are actively used for the detection of biomolecules, ions and small organic molecules which contain aromatic structures. Various biomarkers have also been designed based on blue and green fluorescent GQDs. It has been observed that GQDs show very strong optical absorption in the UV region within a wavelength of around 230 nm which is due to the π-π^* transition of the C–C bonds. This irradiation boosts up the role of GQD in bioimaging. The biocompatibility of the GQDs is required to be assessed before using it for bioimaging. As a result of the recent advancements, GQDs have made great progress in the fields of both in vitro and in vivo bioimaging.

20.6.5.1 In Vitro Bioimaging

The application of GQDs in tumour cell imaging is possible by the researchers due to its properties like good biocompatibility, high photostability and tunable PL. Many cell lines have been implemented for the application of bioimaging. Huiting et al. showed in his studies that the HeLa cell line was mostly used for bioimaging with GQD alongside with T47D human breast cancer cells, MG-63 cells and MCF-7 cells. Due to pyrrole ring surface functionalization, GQDs were inserted inside the excited HeLa cells as they emit a strong bright yellow fluorescence that can be easily visualized under the inverted microscope at a wavelength of 495 nm (Lu et al. 2019). A type of PL-GQD testified by Zhou et al. was efficiently inserted inside the MCF-7 cells, and it showed better optical stability than the normal fluorescent dyes like Alexa Fluor 488 and fluorescein (Zhou et al. 2013). The blue luminescent GQDs that were prepared by combination of isopropanol along with photoreduced GQDs showed lower cytotoxicity but stronger PL properties in A549 cells. This is due to the presence of less negative surface charge on the photoreduced GQD. Supplementary to the tumour cells, the use of GQD for bioimaging is applied on other types of cells like stem cells, human hepatic cancer cells, MC3T3 cells, etc. Human cell line has continually been considered crucial for maintaining high research standards. It was observed that GQDs enter inside the human cells mainly by the process of endocytosis. Entry of GQDs inside the human cells, particularly the neural stem cells, does not affect the cell viability, proliferation, metabolism and differentiation. GQDs synthesized from *Mangifera indica* are useful NIR-responsive probes that can be used in bioimaging, and it can self-localize inside the cell cytoplasm. The use of GQDs has also been fortified in the multiphoton bioimaging field. Further, the adenine modified GQDs demonstrating high fluorescence activity with a quantum yield of about 21.63% rendering them suitable for two-photon fluorescent cell imaging (Kumawat et al. 2017). The absorption cross section of the two-photon path of the prepared GQDs reaches up to 31000 GM as observed by Wang et al., much higher than that of the traditional fluorescent materials. In the case of targeted imaging, GQDs play a very major role as it is difficult to get all the necessary information on a specific cell (Wang et al. 2014). As a solution to it, GQDs are surface modified with certain bioactive moieties like hyaluronic acid (HA), folic acid (FA) and arginine-glycine-aspartic acid (RGD) as the cell-targeting agents.

It is reported that the uptake of GQDs in cancer cell line was facilitated by receptor arbitrated endocytosis when affixed with HA, as in case of the A549 cancer cells. For the highly selective and specific tumour cell imaging, FA-conjugated GQDs were modelled (Wang et al. 2014). It was detected that the outcome of fluorescence caused by the GQDs was much intense in the HeLa cell lines as compared to the adenocarcinomic human alveolar basal epithelial cell line A549 as well as the human embryonic kidney cell line HEK293A, thus proving the fact that GQD-FA was successfully introduced by endocytosis. Targeted cancer fluorescence imaging technique leads to the detection of specific antitumour drugs.

The GQDs in the field of theranostic helps in integrated fluorescence imaging and even in some therapeutic strategies. The recent employment of amino-N-GQDs in the two-photon contrast probe which helps in tracking and in localization of analytes through in-depth two-photon imaging portrays higher PDT effect and also better two-photon properties other than unmodified GQDs. N-GQDs can also show high photothermal conversion efficiency for tumour dual-mode imaging along with NIR fluorescence and can be used for the diagnosis of tumour with advanced PTT effect (Xuan et al. 2018). Efficacious application of nitrogen-doped GQDs helps in the near-infrared second window imaging as well as photothermal therapy. The GQD-PEG-peptide exhibits good feasibility in the arena of intracellular cancer-related microRNA detection and also in the fluorescent imaging of the cancer cells (Lu et al. 2019).

20.6.5.2 In Vivo Bioimaging

Biological in vivo imaging plays a vital role in the advancements of the subfields of biomedicine and life sciences sector. Reports show that the role of GQD in in vivo bioimaging has been studied in mouse, zebrafish, etc. Biocompatible N-GQDs can be used as an efficient two-photon fluorescent probe for cellular and deep tissue imaging (Liu et al. 2013). The cross section of N-GQD in the two-photon absorption reached 48000 GM that was very high as compared to the organic dyes. The fluorescence imaging capability of the GQDs in animal studies can be demonstrated by the help of the GQD-labelled human adipose-derived stem cells study. It shows promising fluorescence signals in vivo compared to the untreated cells that were confined inside the hydrogels. Long wavelength emission is always much more preferred for bioimaging applications other than short wavelength emission. The excitation wavelength from 465 to 620 nm shows best results as it also contains little autofluorescence in the background.

Although we have discussed that GQDs have proved themselves to be highly promising materials for both in vitro and in vivo bioimaging, nevertheless GQDs have a long way to go for further development and for the improvement in certain features like short wavelength emission, low yield and quantum yield (QY). Till date, the highest QY reported for GQD with blue fluorescence imaging is only 54.5%, and that of other colour fluorescence is less than 20% (which is far lower than that of organic dyes and traditional semiconductor QDs). The requirement for longer wavelength is due to the fact that autofluorescence and light scattering background of biological specimens are in the short wavelength region and only long wavelength can infiltrate inside them with a higher imaging contrast. Fluorescent agents with long NIR region (1000–1700 nm) have gained importance because of their deep penetration power. So, for its advancement, GQDs with brighter fluorescent emission are required which fall mainly in the NIR-II spectral regions. Even the upconversion of the PL properties of GQDs proves advantageous, as it leads to the excitation in the NIR region, diminishes the use of harmful UV or blue excitations and prevents photodamage in the tissue. Even though these improvements are very challenging, engineered GQDs can help to fill the lacuna. Special biomolecule modified GQD can also be designed for targeted cancer imaging, diagnosis and therapy (Fan et al. 2015).

20.6.5.3 GQD Hydrogel in Bioimaging

The fluorescence property exhibited by GQD hydrogel nanocomposite evoked many possible applications of the GQD in the arena of bioimaging. GQDs that were cross-linked with carboxymethyl cellulose show photoluminescence with excitation-dependent emission and are also biocompatible and biodegradable. These GQDs when introduced into nano fibrillar gel act as a secondary cross-linker due to the π-π stacking with amino acids. The observed interactions lead to the electron transfer induced quenching of fluorescence of the anthracene moieties in the gel. Similar reports were observed with S- and N-doped GQDs in CNC hydrogel. It was found that upon swelling, the hydrogels still retain their thread-like pattern and shape and it also shows fluorescence upon excitation at 365 nm, proving that with increase in the GQD concentration, the blue light emission was decreased (Rakhshaei et al. 2019). Another study suggested that photoactivity of hydrogel with the presence of CMC was weak, whereas the presence of the cellulose nanocrystals (CNCs)-GQD hydrogels was high. The hydrogen bonding between carboxyl groups of GQDs and that of CNCs makes GQD hydrogel a promising tool in bioimaging owing to its fluorescence properties (Khabibullin et al. 2017).

20.7 Conclusion and Future Prospects

The development of versatile state-of-the-art GQD hybrid hydrogel bank on the complementation of the properties is arising from both, the polymer hydrogel network and GQDs that harness the potential of providing superior functionality with stimuli response and modulating the mechanical, electrical and biological properties of hydrogels, thus enabling the researchers to answer questions in diverse fields such as electronics, soft material catalysis, biosensing, nanomedicine, drug delivery and many more. This amalgamation stems in the synergistic augmentation of each component. Precise control over covalent and supramolecular interactions by design and prediction of ensuing properties of the hybrid hydrogels are prerequisites to scale new vistas in this arena. Such estimates, upon validation via experimental data, will pave the way to sketch hybrid hydrogel composites for anticipated outcomes. In the upcoming years, such designs will aid in predicting properties of novel nanocomposites from components. These systems can probably alter the efficacy, scale and cost of therapeutics. The enactment of the GQD hybrid hydrogel depends primarily on specific chemical and/or physical cross-linking interactions between the hydrogel matrix and the nanoparticles. Despite the great advances made in bioengineering and nanotechnology, there is still a long way to go to advance the utility of GQD hybrid hydrogel as a theranostic platform. This can be possible if we gain more insight on the interactions between the GQD hydrogel at nanoscale along with their potential to influence the tissue microenvironment. Their utility in drug delivery systems can help in limiting the number of doses necessary for long-term effects. They enhance the selectivity and potency of cancer drugs like doxorubicin and cisplatin. Hitherto studies have shown that GQDs along with metals have antibacterial activity; however, the reports with only GQDs are limited and impel further study in this domain. Newly exploited properties of GQDs and progression in the techniques of polymerization for synthetic polymers have unlocked new avenues in the arena of theranostic. However, there still exist some areas where the GQD hydrogel nanocomposites are yet to make an impact.

Acknowledgement Authors are thankful to Amity Institute of Click Chemistry Research and Studies (AICCRS), Amity University Uttar Pradesh (AUUP), Noida, Uttar Pradesh 201313, India, for providing research infrastructure.

References

Ahmed EM (2015) Hydrogel: preparation, characterization, and applications: a review. J Adv Res Cairo Univ 6(2):105–121. https://doi.org/10.1016/j.jare.2013.07.006

Akhavan O (2016) Graphene scaffolds in progressive nanotechnology/stem cell-based tissue engineering of the nervous system. J Mater Chem B 4(19):3169–3190. https://doi.org/10.1039/c6tb00152a

Annabi N et al (2014) 25th anniversary article: Rational design and applications of hydrogels in regenerative medicine. Adv Mater 26(1):85–124. https://doi.org/10.1002/adma.201303233

Argenta DF et al (2019) Hydrogel nanocomposite systems: physicochemical characterization and application for drug-delivery systems. Nanocarriers Drug Deliv. https://doi.org/10.1016/b978-0-12-814033-8.00003-5

Barneck MD et al (2016) Violet 405-nm light: a novel therapeutic agent against common pathogenic bacteria. J Surg Res Elsevier Inc 206(2):316–324. https://doi.org/10.1016/j.jss.2016.08.006

Bhattacharya S et al (2019) Fluorescent self-healing carbon dot/polymer gels. ACS Nano 13(2):1433–1442. https://doi.org/10.1021/acsnano.8b07087

Chacko RT et al (2012) Polymer nanogels: a versatile nanoscopic drug delivery platform. Adv Drug Deliv Rev Elsevier B.V. 64(9):836–851. https://doi.org/10.1016/j.addr.2012.02.002

Chai Q, Jiao Y, Yu X (2017) Hydrogels for biomedical applications: their characteristics and the mechanisms behind them. Gels MDPI 3(1). https://doi.org/10.3390/gels3010006

Chen T et al (2014) Graphene quantum dot-capped mesoporous silica nanoparticles through an acid-cleavable acetal bond for intracellular drug delivery and imaging. J Mater Chem B 2(31):4979–4982. https://doi.org/10.1039/c4tb00849a

Cheng C, Xing M, Wu Q (2019) Green synthesis of fluorescent carbon dots/hydrogel nanocomposite with stable Fe^{3+} sensing capability. J Alloys Compounds Elsevier B.V. 790:221–227. https://doi.org/10.1016/j.jallcom.2019.03.053

Chong Y et al (2016) Crossover between anti- and pro-oxidant activities of graphene quantum dots in the absence or presence of light. ACS Nano 10(9):8690–8699. https://doi.org/10.1021/acsnano.6b04061

Chung S, Revia RA, Zhang M (2019) Graphene quantum dots and their applications in bioimaging, biosensing, and therapy. Adv Mater 1904362:1–26. https://doi.org/10.1002/adma.201904362

Courtney CM et al (2016) Photoexcited quantum dots for killing multidrug-resistant bacteria. Nat Mater 15(5):529–534. https://doi.org/10.1038/nmat4542

Courtney CM et al (2017) Potentiating antibiotics in drug-resistant clinical isolates via stimuli-activated superoxide generation. Sci Adv 3(10):1–11. https://doi.org/10.1126/sciadv.1701776

Ding H et al (2017) Beyond a carrier: graphene quantum dots as a probe for programmatically monitoring anti-cancer drug delivery, release, and response. ACS Applied Materials and Interfaces 9(33):27,396–27,401. https://doi.org/10.1021/acsami.7b08824

Dong J et al (2018) Application of graphene quantum dots for simultaneous fluorescence imaging and tumor-targeted drug delivery. Sens Actuat B Chem Elsevier B.V. 256:616–623. https://doi.org/10.1016/j.snb.2017.09.200

Fan Z et al (2014) A novel wound dressing based on Ag/graphene polymer hydrogel: effectively kill bacteria and accelerate wound healing. Adv Funct Mater 24(25):3933–3943. https://doi.org/10.1002/adfm.201304202

Fan Z et al (2015) Fluorescent graphene quantum dots for biosensing and bioimaging. RSC Adv 3:10,715–10,722. https://doi.org/10.1039/C4RA17131D

Fang J et al (2018) Graphene quantum dots-gated hollow mesoporous carbon nanoplatform for targeting drug delivery and synergistic chemo-photothermal therapy. Int J Nanomed 13:5991–6007. https://doi.org/10.2147/IJN.S175934

Fasbender S et al (2019) The low toxicity of graphene quantum dots is reflected by marginal gene expression changes of primary human hematopoietic stem cells. Sci Rep 9(1):1–13. https://doi.org/10.1038/s41598-019-48567-6

Gao Y et al (2019) Mesoporous silica nanoparticles capped with graphene quantum dots as multifunctional drug carriers for photo-thermal and redox-responsive release. Microporous Mesoporous Mater Elsevier 278:130–137. https://doi.org/10.1016/j.micromeso.2018.11.030

Geng H (2018) Preparation and characterization of cellulose/N,N'-methylene bisacrylamide/graphene oxide hybrid hydrogels and aerogels. Carbohydrate Polym 196:289–298. https://doi.org/10.1016/j.carbpol.2018.05.058

Geng H et al (2018) Achieving stem cell imaging and osteogenic differentiation by using nitrogen doped graphene quantum dots. J Mater Sci Mater Med 29(6). https://doi.org/10.1007/s10856-018-6095-8

Guo Q et al (2014) Comparison of in situ and ex situ methods for synthesis of two-photon polymerization polymer nanocomposites. Polymers 6(7):2037–2050. https://doi.org/10.3390/polym6072037

Hai X et al (2015) An acid-free microwave approach to prepare highly luminescent boron-doped graphene quantum dots for cell imaging. J Mater Chem B 3(47):9109–9114. https://doi.org/10.1039/c5tb01954k

Han S et al (2019) The application of graphene-based biomaterials in biomedicine. Am J Transl Res 11(6):3246–3260

Harroun SG et al (2017) Reborn from the ashes: turning organic molecules to antimicrobial carbon quantum dots. ACS Infect Dis 3(11):777–779. https://doi.org/10.1021/acsinfecdis.7b00150

Havanur S, JagadeeshBabu PE (2018) Role of graphene quantum dots synthesized through pyrolysis in the release behavior of temperature responsive poly (N,N-diethyl acrylamide) hydrogel loaded with doxorubicin. Int J Polym Anal Characterization 23(7):606–620. https://doi.org/10.1080/1023666X.2018.1484207

Havanur S et al (2019) Poly(N,N-diethyl acrylamide)/functionalized graphene quantum dots hydrogels loaded with doxorubicin as a nano-drug carrier for metastatic lung cancer in mice. Mater Sci Eng C 105. https://doi.org/10.1016/j.msec.2019.110094

Hegab HM et al (2016) 'The controversial antibacterial activity of graphene-based materials. Carbon 105:362–376. https://doi.org/10.1016/j.carbon.2016.04.046

Huang X, Brazel CS (2001) On the importance and mechanisms of burst release in matrix-controlled drug delivery systems. J Control Release 73(2–3):121–136. https://doi.org/10.1016/S0168-3659(01)00248-6

Hui L et al (2016) Antibacterial property of graphene quantum dots (both source material and bacterial shape matter). ACS Appl Mater Interf 8(1):20–25. https://doi.org/10.1021/acsami.5b10132

Iannazzo D et al (2017) Graphene quantum dots for cancer targeted drug delivery. Int J Pharm 518(1–2):185–192. https://doi.org/10.1016/j.ijpharm.2016.12.060

Iannazzo D et al (2019) A smart nanovector for cancer targeted drug delivery based on graphene quantum dots. Nanomaterials 9(2):1–17. https://doi.org/10.3390/nano9020282

Jain K (2008) Drug delivery systems—an overview. Methods Mol Biol 437:1–50. https://doi.org/10.1007/978-1-59745-210-6_1

Javanbakht S, Namazi H (2018) Doxorubicin loaded carboxymethyl cellulose/graphene quantum dot nanocomposite hydrogel films as a potential anticancer drug delivery system. Mater Sci Eng C. https://doi.org/10.1016/j.msec.2018.02.010

Javanbakht S, Shaabani A (2019) Encapsulation of graphene quantum dot-crosslinked chitosan by carboxymethylcellulose hydrogel beads as a pH-responsive bio-nanocomposite for the oral delivery agent. Int J Biol Macromol 123:389–397. https://doi.org/10.1016/j.ijbiomac.2018.11.118

Joshi PN et al (2016) Graphene quantum dots—from emergence to nanotheranostic applications. Mater Sci. https://doi.org/10.5772/57353
Khabibullin A et al (2017) Injectable shear-thinning fluorescent hydrogel formed by cellulose nanocrystals and graphene quantum dots. Langmuir 33(43):12344–12350. https://doi.org/10.1021/acs.langmuir.7b02906
Khodadadei F, Safarian S, Ghanbari N (2017) Methotrexate-loaded nitrogen-doped graphene quantum dots nanocarriers as an efficient anticancer drug delivery system. Mater Sci Eng C 79:280–285. https://doi.org/10.1016/j.msec.2017.05.049
Kholikov K et al (2018) Improved singlet oxygen generation and antimicrobial activity of sulphur-doped graphene quantum dots coupled with methylene blue for photodynamic therapy applications. Photodiag Photodyn Ther 24:7–14. https://doi.org/10.1016/j.pdpdt.2018.08.011
Kingham E, Oreffo ROC (2013) Embryonic and induced pluripotent stem cells: understanding, creating, and exploiting the nano-niche for regenerative medicine. ACS Nano 7(3):1867–1881. https://doi.org/10.1021/nn3037094
Kumawat MK et al (2017) Graphene quantum dots from mangifera indica: application in near-infrared bioimaging and intracellular nanothermometry. ACS Sustain Chem Eng 5(2): 1382–1391. https://doi.org/10.1021/acssuschemeng.6b01893
Kuo WS et al (2017) Graphene quantum dots with nitrogen-doped content dependence for highly efficient dual-modality photodynamic antimicrobial therapy and bioimaging. Biomaterials 120: 185–194. https://doi.org/10.1016/j.biomaterials.2016.12.022
Lee PI, Kim C-J (1994) Probing the mechanisms of drug release from HPMC matrices. Pharm Res 11:1379–1384. https://doi.org/10.1023/A:1018975318805
Li Y et al (2018) Self-adapting hydrogel to improve the therapeutic effect in wound-healing. ACS Appl Mater Interf 10(31):26,046–26,055. https://doi.org/10.1021/acsami.8b08874
Lin S et al (2016) Stretchable hydrogel electronics and devices. Adv Mater 28(22):4497–4505. https://doi.org/10.1002/adma.201504152
Liu Q et al (2013) Strong two-photon-induced fluorescence from photostable, biocompatible nitrogen-doped graphene quantum dots for cellular and deep-tissue imaging. Nanoletters. https://doi.org/10.1021/nl400368v
Liu X et al (2017) Stretchable living materials and devices with hydrogel-elastomer hybrids hosting programmed cells. Proc Natl Acad Sci U S A 114(9):2200–2205. https://doi.org/10.1073/pnas.1618307114
Liu W et al (2019) Reduced graphene oxide (rGO) hybridized hydrogel as a near-infrared (NIR)/pH dual-responsive platform for combined chemo-photothermal therapy. J Colloid Interf Sci 536: 160–170. https://doi.org/10.1016/j.jcis.2018.10.050
Lu H et al (2019) Graphene quantum dots for optical bioimaging. Small 15(36):1–19. https://doi.org/10.1002/smll.201902136
Mantha S et al (2019) Smart hydrogels in tissue engineering and regenerative medicine. Materials 12(20). https://doi.org/10.3390/ma12203323
Marković ZM et al (2018) Photo-induced antibacterial activity of four graphene based nanomaterials on a wide range of bacteria. RSC Adv 8(55):31,337–31,347. https://doi.org/10.1039/c8ra04664f
Massoumi B, Mozaffari Z, Jaymand M (2018) A starch-based stimuli-responsive magnetite nanohydrogel as de novo drug delivery system. Int J Biol Macromol 117(2017):418–426. https://doi.org/10.1016/j.ijbiomac.2018.05.211
Nasrollahi F et al (2019) Targeting graphene quantum dots to epidermal growth factor receptor for delivery of cisplatin and cellular imaging. Mater Sci Eng C 94:247–257. https://doi.org/10.1016/j.msec.2018.09.020
Nigam P et al (2014) Graphene quantum dots conjugated albumin nanoparticles for targeted drug delivery and imaging of pancreatic cancer. J Mater Chem B 2(21):3190–3195. https://doi.org/10.1039/c4tb00015c
Nurunnabi M et al (2013) In-vivo biodistribution and toxicology of carboxylated graphene quantum dots. ACS Nano 7(8):6858–6867. https://doi.org/10.1021/nn402043c

Nurunnabi M et al (2015) Bioapplication of graphene oxide derivatives: Drug/gene delivery, imaging, polymeric modification, toxicology, therapeutics and challenges. RSC Adv 5(52): 42,141–42,161. https://doi.org/10.1039/c5ra04756k

Panchal N, Patel D, Shah N (2017) Synthesis of hydrogels. In: 4th international conference on multidisciplinary research & practice (4ICMRP-2017)

Paul A et al (2014) Injectable graphene oxide/hydrogel-based angiogenic gene delivery system for vasculogenesis and cardiac repair. ACS Nano 8(8):8050–8062. https://doi.org/10.1021/nn5020787

Pourhajibagher M et al (2019) Photoexcitation triggering via semiconductor Graphene Quantum Dots by photochemical doping with Curcumin versus perio-pathogens mixed biofilms. Photodiag Photodyn Ther 28:125–131. https://doi.org/10.1016/j.pdpdt.2019.08.025

Qiu J et al (2016) Effects of graphene quantum dots on the self-renewal and differentiation of mesenchymal stem cells. Adv Healthcare Mater 5(6):702–710. https://doi.org/10.1002/adhm.201500770

Rakhshaei R et al (2019) Graphene quantum dot cross-linked carboxymethyl cellulose nanocomposite hydrogel for pH-sensitive oral anticancer drug delivery with potential bioimaging properties. Int J Biol Macromol. https://doi.org/10.1016/j.ijbiomac.2019.10.118

Ranganathan N et al (2019) Synthesis and properties of hydrogels prepared by various polymerization reaction systems. In: Cellulose-based superabsorbent hydrogels, polymers and polymeric composites: a reference series. Springer International Publishing AG, Part of Springer Nature, Cham, pp 487–511. https://doi.org/10.1007/978-3-319-76573-0_18-1

Ristic BZ et al (2014) Photodynamic antibacterial effect of graphene quantum dots. Biomaterials 35(15):4428–4435. https://doi.org/10.1016/j.biomaterials.2014.02.014

Roy P et al (2014) Photoluminescent graphene quantum dots for in-vivo imaging of apoptotic cells. Nanoscale 3:10,715–10,722. https://doi.org/10.1039/b000000x

Sangam S et al (2018) Sustainable synthesis of single crystalline sulphur-doped graphene quantum dots for bioimaging and beyond. Green Chem Roy Soc Chem 20(18):4245–4259. https://doi.org/10.1039/c8gc01638k

Shakeel A et al (2019) Graphene quantum dots in the game of directing polymer self-assembly to exotic kagome lattice and janus nanostructures. ACS Nano 13(8):9397–9407. https://doi.org/10.1021/acsnano.9b04188

Shen J et al (2012) Graphene quantum dots: emergent nanolights for bioimaging, sensors, catalysis and photovoltaic devices. Chem Commun 48(31):3686–3699. https://doi.org/10.1039/c2cc00110a

Singh SK, Dhyani A, Juyal D (2017) Hydrogel: preparation, characterization and applications. Pharma Innov J 6(6):25–32

Su YL et al (2017) Hierarchically targeted and penetrated delivery of drugs to tumors by size-changeable graphene quantum dot nanoaircrafts for photolytic therapy. Adv Funct Mater 27(23):1–12. https://doi.org/10.1002/adfm.201700056

Subramanian KG, Vijayakumar V (2015) Hydrogels: classification, synthesis, characterization, and applications. Encyclopedia Biomed Polym Polym Biomater:3879–3892. https://doi.org/10.1081/e-ebpp-120049894

Sun H et al (2014) Graphene quantum dots-band-aids used for wound disinfection. ACS Nano 8(6): 6202–6210. https://doi.org/10.1021/nn501640q

Suzuki N et al (2016) Chiral graphene quantum dots. ACS Nano 10(2):1744–1755. https://doi.org/10.1021/acsnano.5b06369

Teo AJT et al (2016) Polymeric biomaterials for medical implants and devices. ACS Biomater Sci Eng 2(4):454–472. https://doi.org/10.1021/acsbiomaterials.5b00429

Unterman S et al (2017) Hydrogel nanocomposites with independently tunable rheology and mechanics. ACS Nano 11(3):2598–2610. https://doi.org/10.1021/acsnano.6b06730

Vatanparast M, Shariatinia Z (2018) AlN and AlP doped graphene quantum dots as novel drug delivery systems for 5-fluorouracil drug: theoretical studies. J Fluorine Chem 211:81–93. https://doi.org/10.1016/j.jfluchem.2018.04.003

Wang C et al (2013a) Enhancing cell nucleus accumulation and DNA cleavage activity of anti-cancer drug via graphene quantum dots. Sci Rep 3:1–8. https://doi.org/10.1038/srep02852
Wang Z et al (2013b) Synthesis of strongly green-photoluminescent graphene quantum dots for drug carrier. Colloids Surf B Biointerf 112:192–196. https://doi.org/10.1016/j.colsurfb.2013.07.025
Wang X et al (2014) Multifunctional graphene quantum dots for simultaneous targeted cellular imaging and drug delivery. Colloids Surf B Biointerf 122:638–644. https://doi.org/10.1016/j.colsurfb.2014.07.043
Wang T, Zhu S, Jiang X (2015a) Toxicity mechanism of graphene oxide and nitrogen-doped graphene quantum dots in RBCs revealed by surface-enhanced infrared absorption spectroscopy. Toxicol Res Royal Soc Chem 4(4):885–894. https://doi.org/10.1039/c4tx00138a
Wang ZG et al (2015b) Toxicity of graphene quantum dots in zebrafish embryo. Biomed Environ Sci 28(5):341–351. https://doi.org/10.3967/bes2015.048
Wang S, Cole IS, Li Q (2016) The toxicity of graphene quantum dots. RSC Adv 6(92): 89,867–89,878. https://doi.org/10.1039/c6ra16516h
Wang C et al (2020) Fabrication and characterization of novel cRGD modified graphene quantum dots for chemo-photothermal combination therapy. Sens Actuators B Chem 309:127,732. https://doi.org/10.1016/j.snb.2020.127732
Xia LW et al (2013) Nano-structured smart hydrogels with rapid response and high elasticity. Nat Commun 4:1–11. https://doi.org/10.1038/ncomms3226
Xia MY et al (2019) Graphene-based nanomaterials: the promising active agents for antibiotics-independent antibacterial applications. J Control Release 307:16–31. https://doi.org/10.1016/j.jconrel.2019.06.011
Xuan Y et al (2018) Targeting N-doped graphene quantum dot with high photothermal conversion efficiency for dual-mode imaging and therapy in-vitro. Nanotechnology:0–11. https://doi.org/10.1088/1361-6528/aacad0
Yang X et al (2019) Effects of graphene oxide and graphene oxide quantum dots on the osteogenic differentiation of stem cells from human exfoliated deciduous teeth. Artificial Cells Nanomed Biotechnol 47(1):822–832. https://doi.org/10.1080/21691401.2019.1576706
Yue J et al (2019) Facile design and development of photoluminescent graphene quantum dots grafted dextran/glycol-polymeric hydrogel for thermoresponsive triggered delivery of buprenorphine on pain management in tissue implantation. J Photochem Photobiol B Biol:197. https://doi.org/10.1016/j.jphotobiol.2019.111530
Zhang Y et al (2011) Synthesis of multiresponsive and dynamic chitosan-based hydrogels for controlled release of bioactive molecules. Biomacromolecules 12(8):2894–2901. https://doi.org/10.1021/bm200423f
Zhang M et al (2012) Facile synthesis of water-soluble, highly fluorescent graphene quantum dots as a robust biological label for stem cells. J Mater Chem 22(15):7461–7467. https://doi.org/10.1039/c2jm16835a
Zheng XT, He HL, Li CM (2013) Multifunctional graphene quantum dots-conjugated titanate nanoflowers for fluorescence-trackable targeted drug delivery. RSC Adv 3(47):24,853–24,857. https://doi.org/10.1039/c3ra44125c
Zheng FF et al (2015a) Aptamer/graphene quantum dots nanocomposite capped fluorescent mesoporous silica nanoparticles for intracellular drug delivery and real-time monitoring of drug release. Anal Chem 87(23):11739–11745. https://doi.org/10.1021/acs.analchem.5b03131
Zheng XT et al (2015b) Glowing graphene quantum dots and carbon dots: properties, syntheses, and biological applications. Small 11(14):1620–1636. https://doi.org/10.1002/smll.201402648
Zhou L, Geng J, Liu B (2013) Graphene quantum dots from polycyclic aromatic hydrocarbon for bioimaging and sensing of Fe^{3+} and hydrogen peroxide. Particle Particle Syst Characterization 30(12):1086–1092. https://doi.org/10.1002/ppsc.201300170

Functionalized Nanoparticles in Drug Delivery: Strategies to Enhance Direct Nose-to-Brain Drug Delivery via Integrated Nerve Pathways

21

Fakhara Sabir, Qurrat Ul Ain, Abbas Rahdar, Zhugen Yang, Mahmood Barani, Mauhammad Bilal, and Nikhil Bhalla

Abstract

Nose-to-brain drug delivery system is becoming a desirable alternative approach to conventional drug delivery systems used for the treatment of various neurological disorders. Trigeminal and olfactory routes are implicated to deliver drugs

The original version of this chapter was revised: a spelling error in the 2nd author's name and the affiliation of the 5th author have been corrected. The correction to this chapter can be found at https://doi.org/10.1007/978-981-16-6819-7_26

F. Sabir
University of Szeged, Faculty of Pharmacy, Institute of Pharmaceutical Technology and Regulatory Affairs, Szeged, Eötvös u. 6, Hungary

Q. U. Ain
Institute Teknologi Bandung, Bandung, Indonesia

A. Rahdar (✉)
Department of Physics, Faculty of Science, University of Zabol, Zabol, Iran
e-mail: a.rahdar@uoz.ac.ir

Z. Yang
Cranfield Water Science Institute, Cranfield University, Cranfield, Bedfordshire, UK

M. Barani
Medical Mycology and Bacteriology Research Center, Kerman University of Medical Sciences, Kerman, Iran

M. Bilal
School of Life Science and Food Engineering, Huaiyin Institute of Technology, Huaian, China

N. Bhalla (✉)
Nanotechnology and Integrated Bioengineering Centre (NIBEC), School of Engineering, Ulster University, Newtownabbey, Northern Ireland, UK

Healthcare Technology Hub, Ulster University, Newtownabbey, Northern Ireland, UK
e-mail: n.bhalla@ulster.ac.uk

A. Thakur et al. (eds.), *Synthesis and Applications of Nanoparticles*,
https://doi.org/10.1007/978-981-16-6819-7_21

from the nose-to-brain, which bypasses the blood-brain barrier and the first-pass metabolism. In this review, nanocarrier systems are evaluated, screened, and tested in order to evaluate its physiochemical features and configuration to enhance the bioavailability of drugs in the brain after intranasal intervention. The application of specific ligand, surface modifications, and use of permeation enhancers to increase brain targeting are discussed. Furthermore, we discuss the in vivo animal and in vitro cell line-based models, which are actively being employed to explore the nanomaterial-driven drug transport mechanisms via the intranasal route. These models can be used to study absorption, diffusion, permeation, and toxicological and pharmacokinetic profile of the active pharmaceutical ingredient. Our review provides evidence to conclude that the potential of nose-to-brain delivery and role of functionalization of nanomaterials enhance the drug efficacy in brain diseases. We also conclude that the biorecognitive surface modifiers have the ability to enhance and optimize the drug delivery to the brain, and we provided our insights and outlooks to address challenges and opportunities for nanosystems to speed up clinical translation.

Keywords

Nose-to-brain · Nanomaterials · Nerve pathways · Bioavailability · Neurological disorders

21.1 Introduction

About 1.5 billion people are suffering from different types of neurological ailments, and this situation will worsen by reaching 1.9 billion people by the end of 2021 (Pardeshi and Belgamwar 2018). Traditionally, a wide variety of therapeutic agents can be administered intranasally for topical, systemic, and central nervous system (CNS) action for the treatment of local ailments such as nasal allergy, sinusitis, nasal infections, and nasal congestion. This is because it is a reliable, safe, noninvasive, and convenient route for high-level and fast rate of drug absorption (Pardeshi and Belgamwar 2013). The target tissue for nasal drug delivery is the nasal mucosa as it offers many advantages over the oral route because of its large surface area, high blood flow, porous endothelial membrane, avoidance of hepatic first-pass metabolism, and its accessibility (Mittal et al. 2014). However, one of the challenges faced by CNS drugs is to enter into the brain due to the different physiological barriers such as the presence of blood-brain barrier (BBB), first-pass metabolism, enzymatic degradation, inadequate blood perfusion, systemic clearance, peripheral side effects, and reduced bioavailability (Agrawal et al. 2018). Previously, some drugs with high molecular weight and almost all the drugs with low molecular weight are unable to cross the BBB (Kulkarni et al. 2015). Moreover, the drugs that require large doses to enter into the CNS for their therapeutic effect may be effective at relatively lower doses with fewer adverse effects, when administered through the nose (Warnken et al. 2016). Drugs administered through the intranasal route avoid first-pass metabolism and degradation in the stomach. However, nasal mucosa provides benefits by increasing the safety and the rate of drug absorption and decreasing the onset of the therapeutic action (Piazzini et al. 2019). Diseases related to CNS include neurodegenerative diseases such as Parkinson's disease, Alzheimer's, multiple sclerosis, and cerebral ischemia representing a broad spectrum of pathological conditions that leads to a change in neural functions (Bonferoni et al. 2019). Various dosage forms, i.e., oral, topical, and IV injections, are available for the treatments of CNS disorders that reach CNS, but even after crossing the different physiological barriers, these methods have limited therapeutic efficacy. On the other hand, surgical intervention is a highly invasive short-term method of treatment (Agrawal et al. 2020). However, the ability of drugs to efficiently pass through the blood-brain barrier (BBB) is an important parameter in the development of drug products for the treatment of brain neurological disorders (Martins et al. 2019). Few novel drugs have been brought to market in the past four decades, and pharmaceutical companies are spending less on the development of psychiatric treatments. Indeed, only 7% of developed psychiatric treatments reach the market (Quintana et al. 2016).

Nasally administered drugs can adequately cross the BBB and reach the CNS through different pathways. The main and direct pathway for the nose-to-brain drug delivery is through olfactory nerves. There are also the trigeminal nerves, which have nerve endings in the respiratory epithelia and respiratory epithelium through which the drugs reach the systemic circulation by crossing through the BBB (Piazzini et al. 2019). A thick layer of mucous surrounds the human nasal cavity which is adequately supplied with the blood vessels responsible for removing and transporting foreign particles from the nasopharynx into the esophagus through mucociliary motion. The nasal cavity has a total volume and surface area of 15–20 mL and 150–200 cm^2, respectively (Feng et al. 2018). The nasal cavity is divided by the nasal septum into two halves which are open to the facial side and the rhino pharynx, *via* anterior and posterior nasal apertures, respectively. Each nasal cavity is further divided into three regions, i.e., the nasal vestibule, the olfactory region, and the respiratory region (Mistry et al. 2009).

Therefore, there are two pathways for nose-to-brain drug delivery, indirect pathway and direct pathway. The indirect pathway lies in the respiratory region mucosa, which is highly vascularized by blood capillaries that allow the drugs to be absorbed into the systemic circulation and then enter into CNS after crossing BBB (Martins et al. 2019). There are two primary pathways through which a drug can be transported, i.e., extracellular olfactory and trigeminal nerve fibers and intracellular pathway. Intracellular transport is also known as endocytosis in which the drug is absorbed by olfactory sensory neurons after depositing on the olfactory epithelium. Hydrophobic molecules with a lower molecular weight are more likely to use this mode of transport (Quintana et al. 2016). Figure 21.1 illustrates the nose-to-brain delivery pathway and intranasal administration. The most dominant path is an olfactory neuronal path which is linked to the nasal cavity from the olfactory bulb of the brain directly by crossing the cribriform plate. The drug transport takes place by passive diffusion in the case of small lipophilic molecules, whereas for hydrophilic drugs and large moieties, transport occurs by endocytosis (Agrawal et al. 2020).

Figure 21.1 also depicts the nasal administration of nanoparticles and pathways or fate of nanoparticles after administration. In the context of the aforementioned details and in continuation of efforts from our groups related to the synthesis of

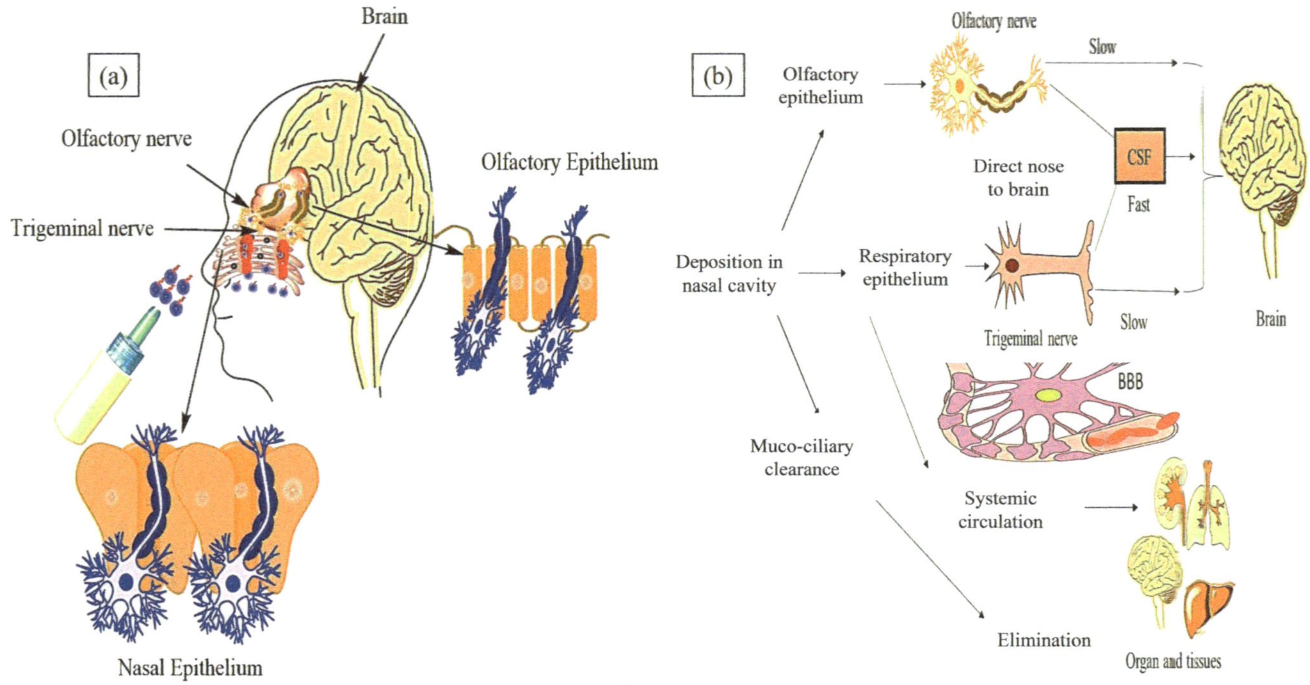

Fig. 21.1 (**a**) Nasal administration and (**b**) nose-to-brain delivery pathways of drug delivery or fate of nanoparticles after intranasal administration

nanomaterials and investigation about their potential bio applications (Barani et al. 2019a, b; 2020a, b, c, d, 2018; ; Bilal et al. 2020; Das et al. 2020; Davarpanah et al. 2018; Ebrahimi et al. 2018; Hajizadeh et al. 2019a; Hajizadeh et al. 2019b; Rahdar et al. 2020a, b, c, d, e, f; 2019a, b; Torkzadeh-Mahani et al. 2020; Taimoory et al. 2018; Nikazar et al. 2020a; Pillai et al. 2020; Saravani et al. 2020; Sivasankarapillai et al. 2020a, b; Sayadi et al. 2020; Davarpanah et al. 2019; Baranei et al. 2020; Nikazar et al. 2020b; Ghazy et al. 2020a, b; Sivasankarapillai et al. n.d.; Mukhtar et al. 2020), we here review different strategies to enhance the direct nose-to-brain delivery *via* integrated nerve pathways by nanomaterials.

21.2 Intranasal Barriers for Nose-to-Brain Drug Delivery System

By following simple formulation strategies, scientists can easily formulate a nasal drug delivery system, but various intranasal barriers impact the absorption of therapeutic drugs (Upadhyay et al. 2011). A thin lining of lipophilic mucous surrounds the nasal cavity that only allows the permeation of lipophilic molecules (Agrawal et al. 2018). One of the factors that limit the nasal absorption of polar drugs is that most of these drugs have a large molecular weight of the drugs which lead to low membrane permeability. However, polar drugs with low molecular weights, less than 1000 Da, prefer paracellular and transcellular routes by which the drugs cross the epithelial membrane through the concentration gradient by receptor-mediated or vesicular transport mechanisms *via* tight junctions between the cells (Jadhav et al. 2007). Therefore, the nasal route is suitable for the smaller-size, lipophilic drug molecules while restricting the permeation of hydrophilic drugs (Agrawal et al. 2018). Large molecules like peptides and proteins which are not permeable across the nasal cavity are destroyed in the lumen of the nasal cavity due to enzymatic degradation of the drug molecule in the lumen of the nasal cavity or while passing through the epithelial membrane (Upadhyay et al. 2011).

Enzymes such as epoxide hydroxylase, carboxylesterase, aldehyde dehydrogenases, glutathione S-transferases, cytochrome P-450, UDP-glucuronyltransferase, and glutathione transferase are responsible for the degradation of drugs in the human nasal mucosa, while cytochrome P450 metabolizes some drugs such as cocaine, nicotine, alcohols, decongestants, and progesterone in the nasal mucosa (Bahadur and Pathak 2012). The pH of the nasal secretions varies in adults and infants from 5.5 to 6.5 and 5.0 to 7.0, respectively. A drug is better absorbed at lower pH of the nasal mucus than the drug's pKa because the drug is predominantly in a unionized state. Therefore, changing the pH of nasal secretions can change the amount of drug absorbed by altering the drug ionization. As the pH of the nasal mucus changes the pH of the drug formulations and vice versa, therefore, the pH of a drug should ideally be within the range of 4.5–6.5 with a significant buffering capacity (Misra and Kher 2012). The nasal mucociliary clearance plays a vital role in removing foreign particles such as dust, allergens, and bacteria that are usually trapped during inhalation on the mucus blanket. This also affects the contact time between the drug and the nasal mucosa which leads to varied amounts of drug absorptions. On the other hand, the mucociliary clearance decreases the contact time, and thus, it is inversely related to the drug absorption (Tafaghodi et al. 2004). There are some limitations regarding nose-to-brain targeting. For instance, it is still an immense challenge to calculate the exact dose of the drugs to be administered intranasally, which is one of the main reasons that subcutaneous insulin injections are not yet replaced by insulin administered intranasally (Nasulin®) (Stützle et al. 2015).

As a result of which, administration methods depict a wide variety of variations, and they require adjustments according to anatomical characteristics of the individuals. Furthermore, due to a relatively small and confined nasal cavity, the dose quantity is limited. Moreover, while developing formulation, it must be in consideration to use such type of materials that do not cause mucosal toxicity or any such material that might cause an allergic reaction/irritation. Another key consideration should be the patient's health status. Essentially, if patients have flu symptoms or if they have any kind of allergies, they may face some issues while using intranasal drug delivery devices. Moreover, one must also consider that the nasal mucosal surface shouldn't be damaged by the use of such devices frequently. Drug reproducibility in the olfactory region also depicts that nose-to-brain targeting shows efficient and safe results (Pardeshi and Belgamwar 2013). pKa of drug and pH also affect the extent of nasal drug absorption at the specific absorption site. Hence, the pH of the formulation should be tuned accordingly to ensure drug stability for allowing a large amount of non-ionized drug absorption. Usually, the pH of the formulation should be similar to the human nasal mucosa, i.e., 5.0–6.5; otherwise, it can cause nasal mucosa irritation (Dae-Duk 2007). Moreover, the pH often prevents the growth of bacteria and also responsible for the integrity of nasal mucosa pH (Rahdar et al. 2019a). The viscosity of the formulation is directly proportional to the contact time between nasal mucosa, drug and any enhancers within it, which ultimately results in a potential increase of drug absorption. Moreover, increased viscosity of drug increases its permeability by interfering with the normal ciliary mechanisms (Dhakar et al. 2011). According to recent research study, it has also been reported that the viscosity of the solution may provide a larger therapeutic period of nasal formulations (Zaki et al. 2006).

21.3 Strategies to Improve Nasal Absorption

While considering systematic drug delivery in CNS, a vast range of drugs show promising results through nasal drug delivery systems. However, low bioavailability of drug remains a challenge for many drugs which are administered intranasally. There may be several reasons due to which dugs show low bioavailability which might be due to low solubility, rapid degradation of enzymes in the nasal cavity, their poor membrane permeability, or that the drugs might clear rapidly from a mucous membrane. To overcome these hindrances, several mechanisms have been adopted in the literature (Bahadur and Pathak 2012). Drugs that are administered nasally usually surpass the GIT tract and hepatic first-pass effect, but some drugs are metabolized by the enzymes existing in the lumen of the nasal cavity. To prevent the degradation of enzymes, several techniques have been used, and among them, the most recommended is by using proteases and peptidases inhibitors. For example, for aminopeptidase inhibitors, comostate amylase and bestatine have been used, and for calcitonin degradation, trypsin inhibitor aprotinin is used. Furthermore, to prevent enzymatic degradation of drugs like leucine, human growth hormone, enkephalin, amastatin, boroleucin, bacitracin, and puromycin are widely used. Moreover, few enzymatic inhibitors also show the ability as permeation enhancers. For example, for treating Alzheimer's disease, disodium ethylenediaminetetraacetic acid has been used which reduces the beta-sheet peptide enzyme degradation by enhancing drug absorption (Ying 2008). Several other excipients have been used to increase absorption and permeation of drugs which include fusidate derivatives, surfactants, bile salts, laureth-9 sulfate, fatty acids, hydrophilic polymers, cyclodextrins, etc.

Methylated β-cyclodextrins form the inclusion complexes with drugs, thus resulting in enhanced absorption of lipophilic drugs that have poor water solubility. Mucus interaction has been shown by hydrophobic and positively charged excipients; however, excipients that are hydrophilic and are negatively charged do not show mucus interaction. Poly(lactic-co-glycolic) acid (PLGA) is used as a nanoparticle/emulsions for developing a formulation that can penetrate the mucus membrane easily. Moreover, to eliminate the particle's hydrophobic interactions with mucus, PEGylation is used, and hydrophilic properties of the particles are ensured by coating processes (Gänger and Schindowski 2018). Usually, when the drugs are administered in solution form, they undergo dissolution before absorption. Despite poor water solubility, lipophilic drugs are absorbed easily through the nasal membrane. Therefore, to achieve an aqueous formulation of lipophilic drugs by using compounds of higher hydrophilic character, the prodrug technique can be used. One should also ensure that the prodrug must be converted into a patent drug when the finished formulation reaches the systematic circulation. To increase the retention time of drugs in the nasal cavity, several techniques have been adopted like the use of polymers chitosan or polycarbophil that act as bioadhesive or increase the formulation viscosity. Nasal drug absorption is also affected by the behavior of drugs, excipients, preservatives, or absorption enhancers that might inhibit or stimulate the nasal mucociliary clearance (Stevens et al. 2009).

21.4 Application of Nanomaterials for Nose-to-Brain Delivery

To overcome several difficulties to deliver the drug from the nose to the brain, various techniques are used to design formulations like solutions, mucoadhesives, microemulsions, and nanoparticles that may be polymeric or lipid base, including novel combinatorial therapies. These formulations are usually designed by considering the physicochemical properties of the drug (Warnken et al. 2016). The lipid-based nanocarriers are site-specific controlled drug delivery systems. Along with the targeting ability, there are several inherent traits of lipid nanocarriers which include excellent biodegradability, controlled drug release, improved stability, biocompatibility, and feasibility of carrying both lipophilic and hydrophilic drugs. Ideally, the lipid-based nanoemulsion-based carriers are suitable for therapeutic drug delivery. They are usually made up to omega-3 fatty acids which belong to the class of polyunsaturated fatty acid, and it easily permits the nanoemulsions to cross the membranes of the cell.

Neuro nanoemulsions are those types of nanoemulsions that are loaded with neurotherapeutic agents and are usually used for the treatment of neurological disorders. When neuro nanoemulsions are surface-modified with mucoadhesive agents, they are usually considered mucoadhesive neuro nanoemulsions. For the rapid clearance of drugs through the nose by mucociliary action, polymer-based mucoadhesive dosage forms are usually prepared (Pardeshi and Belgamwar 2018). It constitutes "soft" nanomaterials, which are prepared by several standard solution-based organic chemistry methods. A wide range of drugs which include proteins, chemotherapeutic drugs, nucleic acids, and contrast agents can be encapsulated in polymeric nanoparticles due to their low density and less rigid properties. For drug delivery, polymeric nanoparticles are considered as the most suitable carriers because of the presence of a wide range of different kinds of polymers along with their access to a greater surface area containing functional groups with which conjugation of biomolecules occurs (Illum 2007). Figure 21.2 demonstrates the application of different nanomaterials for enhanced brain uptake *via* intranasal administration.

21.5 Lipid Nanoparticles

Lipid nanoparticles are suitable drug carrier systems as they provide protection of embedded active pharmaceutical ingredient (API) from glycoproteins (P-gp) efflux/enzymatic transporters and from chemical destabilization. One of the key characteristics of developing the intranasal dosage form is their toxicological assessment. The prolonged contact of the formulations with the nasal epithelial may lead to the ciliotoxicity, tissue damage, and localized irritation that further leads to a microbial infection. Above all, the protection of the olfactory nerves and sense of smell is very critical while screening the carrier system, for intranasal delivery (Battaglia et al. 2018). The lipid nanoparticle system including liposomes, niosomes, and nanoemulsion are the most significant and efficient drug delivery approaches because of their biocompatible nature. The lipid nanoparticle average particle size is placed within the range of 50–1000 nm. Among lipid nanocarriers, solid-lipid nanoparticles (SLNs) are considered to be one of the efficacious vehicle systems. It comprises of lipid core surrounded by a layer of surfactants in an aqueous phase. Nanolipid carrier systems show low burst release and also have reduced drug expulsion as compared to the other nanoparticle system. However, almost all lipid nanoparticles have limited drug loading capacity; therefore, lipid-drug-conjugate nanoparticles have been fabricated (Yasir and Sara 2014). All categories of lipid nanoparticles can be effectively and remarkably used for scale-up production on the industrial level.

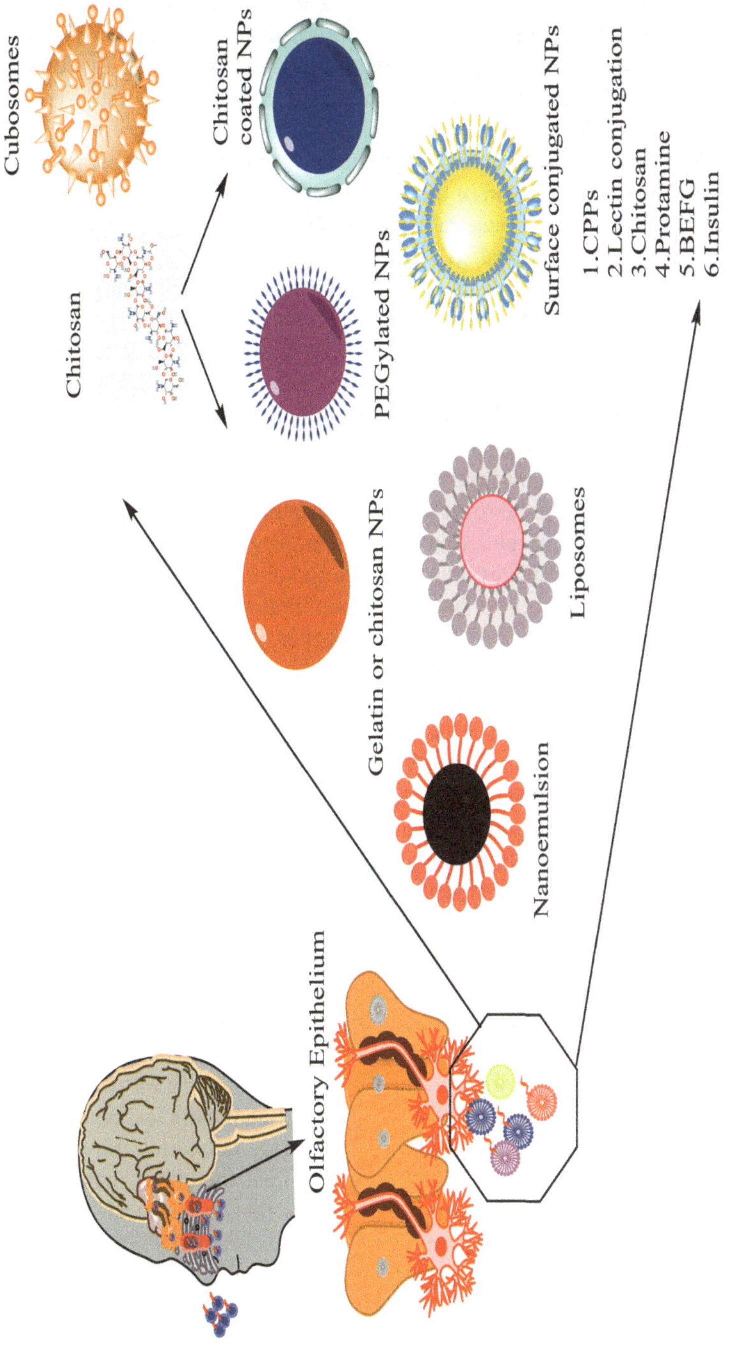

Fig. 21.2 Applications of different nanomaterials for intranasal delivery that can induce stability against enzymatic degradation, access to the CNS either through passive or active targeting, increased half-life, and enhanced concentration in brain

Moreover, several advantages including high scale-up, high biocompatibility, and ease of functionalization make lipid nanoparticles more efficient carrier systems than others. For instance, Giovanna Rassu et al. designed an efficient delivery system comprising of chitosan-coated and chitosan-uncoated solid-lipid nanoparticles which were subjected to intranasal delivery. Both of these formulations were tested to check the significance of olfactory and trigeminal nerve pathway for use in drug delivery to the brain. The system directed toward an optimal nose-to-brain delivery of BACE1siRNA in the treatment of the Alzheimer's disease. The short peptide also called as RVG-9R is a cell-penetrating peptide which is used to enhance the transcellular pathway (Rassu et al. 2017).

The coating process elucidates the significant effect of particle charge on siRNA protection. The cationic formulation ensured enhanced mucoadhesiveness to the particles and also enhanced the stay time in the nasal mucosa. The cellular transportation of siRNA released from the SLNs using Caco-2 as a model has also been studied. It is concluded that siRNA released from both uncoated and coated formulation permeates the monolayer to the large extent. Sonal Patel et al. developed risperidone (RSP)-loaded solid-lipid nanoparticles and measured the possibility of targeting the brain *via* intranasal route. The RSLNs paw test using perspex platforms has shown greater hind-limb retraction time as compared to the RSP solution. This suggests preference of RSLNs over the pure drug solution for brain delivery. The pharmacokinetics studies revealed that the value of brain-to-blood ratio of RSLNs in intranasal environment was five- to tenfold higher with respect to the RS and RSLNs. The results of gamma scintigraphy imaging of mice brain following intranasal and intravenous administration revealed the localization of risperidone in the brain. The results concluded that the existence of direct intranasal delivery for nanoparticle intervention to the brain (Patel et al. 2011). Laurent Salade et al. developed ghrelin encapsulated liposomes for treating cachexia. They compared anionic liposomes decorated with chitosan in either a dry powder formulation or in liquid. The dry chitosan powder formulation showed good adhesion to mucins and greater encapsulation efficiency. This also provides protection against enzymes with decreased ghrelin storage degradation at 25 °C. The chitosan-coated powder formulation has stronger adhesive property to mucins. They used device for deposition of a greater quantity of powder in the olfactory region. The combined evaluation of the device and powder could provide significant treatment against cachexia (Salade et al. 2017; de Barros et al. 2020; Aderibigbe and Naki 2019). S. Cunha et al. combine the literature evidences supporting the lipid nanoparticles suitability for targeting delivery. These lipid nanoparticles (liposomes, solid lipid nanoparticles, nanostructured lipid carriers) loaded with drug showed greater efficacy in treating brain disorders. The intranasal route also has greater potential to treat infections, cardiovascular diseases, and menopausal syndrome. It is obvious that in the near future, lipid nanoparticles play a vital role in targeting the intranasal route by circumvention of the blood-brain barrier (BBB) (Battaglia et al. 2018).

21.6 Mucoadhesive Nanoparticles

Mucociliary clearance is the most significant factor that can have drastic impact on intranasal transport. This defensive role of the respiratory tract efficiently eliminates the noxious substances and microbes in the mucus layer. These are limitations that hinder the delivery of traditional nasal formulations that could help by increasing the residence time, reduce variations in nasal absorption, increase the viscosity, and improve the bioavailability of nanoformulation. Different polymers have the ability to provide bioadhesion such as hydrophilic polymers which have intimate contact with mucus, electrostatic attractive forces, hydrogen, and van der Waals bonds (Sonvico et al. 2018). These polymers including natural, synthetic, and semi-synthetic, alginates, starch and gelatin, cellulose, and different acrylates, respectively, have been applied for nasal delivery of drugs. The nose-to-brain delivery via the olfactory region in humans has several applications of polymers like pectins and chitosan to prolong the residence time (Ugwoke et al. 2005). The sodium hyaluronate has been used for improved delivery of higher-molecular-weight model compounds like 4 kDa dextran after nasal administration to rats. These mucoadhesive polymers' adhesion mechanism when used in formulations of nanocarriers or for functionalization will remain the same. However, the higher surface area translates into a broader interface for more prolonged residence time. The particle size lower than 500 nm allows nanoparticles to squeeze in the nonviscous aqueous pores within the entangled mucin network. The hydrophobic and positive interactions with the mucin network further enhance the interaction with the mucus at a molecular level (Sosnik et al. 2014). However, surface functionalization can decrease the mucoadhesivity, and these characteristics are exploited by mucus-penetrating nanocarriers. There are several reasons for which mucoadhesivity features of nanocarriers have been broadly studied for intranasal delivery. Betbeder and his coworkers developed morphine-loaded nanoparticles for targeting the direct delivery pathway between the CNS and olfactory mucosa. The nanoparticles were composed of cationic polymer (maltodextrin)-coated particles surrounded by a lipid bilayer. These particles were studied for antinociceptive activity when co-encapsulated with the opioid in mice in comparison to morphine solution. The results concluded the enhanced transport of morphine-encapsulated nanoparticles via nose-to-brain delivery. However, the level of morphine or concentration of morphine in brain tissues was not that much high. The research find out that the cationic interaction of nanoparticles with nasal mucus layer may enhance the stay time and absorption from olfactory region in laboratory mice (Betbeder et al. 2000). Silvia Guterres et al. developed nanocapsule system encapsulated with olanzapine. This nanocapsule system comprises of methacrylic poly(ε-caprolactone) copolymer as a mucoadhesive carrier system for nose-to-brain delivery. Nanocapsule interaction with mucin was evaluated via NTA (nanoparticle tracking analysis test). The mucodhesivity was tested in terms of prolongation of residence time and mucoadhesion of olanzapine-encapsulated nanocapsules on porcine. In vivo studies ensured twice-fold enhanced accumulation of olanzapine in CNS tissues after intranasal administration. The authors reported that the brain accumulation of olanzapine was higher in the present study as compared to olanzapine-encapsulated PLGA nanoparticles and concluded the enhanced mucoadhesivity of coated nanocapsule (Fonseca et al. 2015). Among all synthetic polymers, polysaccharides are the most useful materials to design mucoadhesive polymers. Polysaccharides have prominent features that make them useful materials for nasal delivery including biodegradability, biocompatibility, and biomimetic recognition and ease of chemical functionalization. Applications of polysaccharides can be incorporated in three possible ways, by copolymerization, covalent attachment, or directly developing the polysaccharide comprising nanoparticles. Ondansetron-encapsulated nanolipid carriers were formulated for nose-to-brain delivery from *Delonix regia* gum. In vitro mucoadhesion testing was performed by evaluating the binding efficiency of DRG-NLC toward mucin. The study concluded the enhanced brain targeting of or direct brain targeting efficiency of about (DTE) 506% and direct transport percentage (DTP) of about 97% as compared to IV intervention of marketed ondansetron injection as a control. Alginate nanoparticles encapsulated with venlafaxine were developed for nose-to-brain delivery against treatment and management of depression. Ex vivo study results concluded venlafaxine-loaded nanoparticles have twice drug permeation via nasal mucosa compared to drug solution. Alginate nanoparticles performed good results as compared to controls in depressed animals. However, limitations including swallowing and inhalation of nanoparticle suspension are found when applied in relatively higher amount. An increase in absorption and reduction in nasal mucociliary clearance can increase mucosal permeation, and alteration of P-gp efflux transporters is a mechanism elaborated by pharmacokinetic results (Devkar et al. 2014). The results concluded the reduction of systemic side effects after direct nose-to-brain intervention. Chitosan polysaccharides have been concluded as a very advantageous material for nose-to-brain delivery. Wang et al. developed chitosan nanoparticle-encapsulated estradiol for nose-to-brain delivery in targeting Alzheimer's disease. In vivo study results showed DTP around about 68% and DTE about 320% in CSF (that was measured in cerebrospinal fluid) after intranasal administration. The results concluded enhanced chitosan potential to mucin binding along with paracellular transport. In several other studies, chitosan nanoparticles developed by ionotropic gelation have been suggested for nose-to-brain delivery. Chitosan nanoparticles were encapsulated with different drugs such as rasagiline or ropinirole for Alzheimer's disease and Parkinson's disease and tapentadol for pain management. There is one more study in which selegiline HCl was encapsulated in thiolated chitosan nanoparticles. The results concluded the enhanced anti-inflammatory and neuroprotective efficacy of thiolated polysaccharides in in vivo experiment (Wang et al. 2008). Di Gioia et al. evaluated and developed the dopamine-encapsulated chitosan nanoparticles for delivery to the striatum. The results concluded that the nanocarriers enhance the bioavailability of drugs that have good penetration in traditional formulation such as solution (Di Gioia et al. 2015). Another study developed chitosan functionalized PLGA nanoparticles encapsulated with chlorpromazine HCl for nose-to-brain delivery. The results concluded good mucoadhesion, enhanced permeation on sheep nasal mucuosa, and controlled release of the PLGA nanoparticles (Chalikwar et al. 2013).

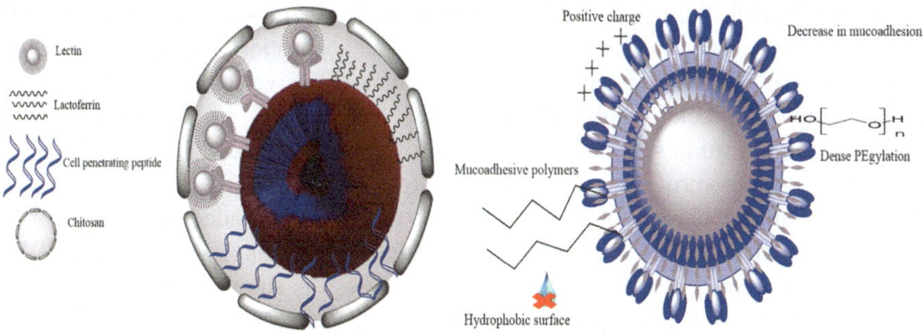

Fig. 21.3 Surface chemistry of nanomaterials affecting mucoadhesive behavior and conjugation of different peptidal moeities

Liposomes grafted with chitosan derivative and encapsulated with ghrelin were developed for intranasal administration. The results showed chitosan-coated liposomes attached to mucin have more mucoadhesivity in comparison to uncoated liposomes. These thiolated chitosan-decorated liposomes show more permeation through Calu-3 cell monolayer that was used as model for upper airways epithelial barrier (Mathieu 2019). For the nose-to-brain delivery of statins, Clementino et al. developed hybrid chitosan nanocapsule. The biodegradation of these nanoparticles provides a more efficient release of drug in the nasal mucosa. Gamma scintigraphy studies were significantly higher for 99mTc labelled simvastatin through intranasal administration. Maria Jose Alonso and her colleagues developed polyethylene glycol-coated nanocarriers that showed increased mucus permeation via nose-to-brain delivery. The results showed increased mucus diffusivity while minimizing interaction with mucins and providing close contact with epithelium (Clementino et al. 2016). Sekerdag et al. developed lipid/PEG-PLGA nanoparticles as mucus-penetrating carriers of farnesylthiosalicylic acid for nose-to-brain delivery. Intranasal administration of nanoparticles significantly decreased the tumor volume. Figure 21.3 shows the surface chemistry of mucoadhesive nanomaterials and surface grafting of peptides. The results concluded the dose reaching to the brain was 0.04% of the injected dose as part of drug accumulated into the liver was very low in comparison to IV administration (Sekerdag et al. 2017).

21.7 Penetration-Enhancing Nanocarriers

Besides developing the pegylated nanoparticles or stealth nanoparticles, many researchers developed particles whose composition showed enhanced penetration across the nasal mucosa. Chitosan acts as both mucoadhesive agent and penetration enhancer. However, penetration-enhancing particles are composed of components that have the ability to change the barrier property of the nasal mucosa such as surfactants. Examples of these types of carrier system encapsulating serotonin agonists (zolmitriptan, sumatriptan) comprise of PEG 400, vitamin E,

Transcutol P, and Pluronic® F127. Surfactants such as Transcutol P and vitamin E were previously used to enhance and solubilize the absorption of drugs through the nasal mucosa. Olanzapine encapsulated into nanocubic vesicles by utilizing the surface-active triblock copolymer poloxamer 188 or 407, in the lipid bilayer. Pharmacokinetic studies in rats showed improved and enhanced bioavailability as compared to the control nanoformulation (Guo et al. 2013; Anand et al. 2012). In relation to that, Albderahman et al. studied the nanovesicle-loaded risperidone. The following formulation used span 60, risperidone, and ethanol solution into PVA aqueous solution (Abdelrahman et al. 2017). The results showed enhanced permeation, greater viscosity, and improved CNS accumulation. Another study developed gelatin nanostructured lipid carriers for nose-to-brain delivery of fibroblast growth factor in Parkinson's disease. The intranasal administration enhanced the absorption of growth factor in the olfactory bulb without disturbing the integrity of the nasal mucosa. The reason for this improved attribute is because of poloxamer 188 and its ability to decrease the barrier property of the mucus layer. Other penetration enhancers or surfactants have similar effects, for example, SLNs encapsulated with rosmarinic acid applied for treating Huntington's disease (Kulkarni et al. 2015; Md et al. 2018). The improved level of oxidative stress and low behavioral abnormalities was observed in lab animals via application through intranasal route. Zolmitriptan-encapsulated novasomes were formulated (Abd-Elal et al. 2016). The following formulation showed enhanced brain concentration when compared with intravenous drug solution. Coumarin-encapsulated micelles were formulated that showed enhanced brain distribution after intranasal administration under in vivo experiment. The results stated enhanced penetration ability of these micelles intracellularly (Kanazawa et al. 2011). Another study showed micelles functionalized with CPP and encapsulated with camptothecin showing enhanced cytotoxic effect as compared to control group. The siRNA was used and condensed with Tat-mPEG-PCL to get the polyplexes. The brain distribution of the polyplexes after intranasal delivery results in increased accumulation of siRNA in the brain in comparison to controls (Sawant and Torchilin 2010). The results showed enhanced micelle permeation of the nasal mucosa within the trigeminal and olfactory nerves. However, these CPP conjugated formulations have few limitations including lack of efficient transport to the cytoplasm. Considering these limitations, Paolo Giunchedi developed a delivery system for nose-to-brain targeting. The RVG-siRNA complex was developed and encapsulated into SLNs. The results concluded the enhanced stability due to mucoadhesive nanocarrier that helps in crossing the mucosal barrier (Sonvico et al. 2018).

21.8 Lectin-Grafted Nanocarriers

Lectins are basically proteins of non-immunological origin found in all living organisms. Lectin interaction with transformed glycans can be used in the diagnosis and treatment of diseases. Selective targeting toward glycans has attracted the nanobiotechnology in designing the lectin functionalized nanoparticles. Conjugated lectins on the surface of nanocarriers can stimulate the development of the system with fewer side effects. The glycoprotein obtained from plants has high specificity to get attached at the surface receptors of the different cell types (Chen et al. 2012). These proteins have been proposed as targeting ligands for the drug delivery system. The high quantitative and qualitative work has been reported relevant to the significance of use about lectins by various researchers. The following data suggested that horseradish peroxidase conjugate can get attached to the receptor of the olfactory sensory cells. The wheat germ agglutinin (WGA) has tendency to aggregate at the olfactory nerve site via intranasal route at higher concentrations, that is, 100 times greater than IV administration (Muralidharan et al. 2014). This plant obtained glycoprotein from *Triticum vulgare* which binds specifically to sialic acid residue and N-acetyl-D-glucosamine in the nasal epithelium. WGA is the most studied ligand among nanocarriers for intranasal delivery. The PEG-PLA nanoparticles were developed with functionalization of WGA along with encapsulation of fluorescent dye 6-coumarin. When these conjugated nanoparticles were administered intranasally, they enhance the concentration of PEG-PLA at both sites in the brain and in blood. In several studies, WGA decorated PEG-PLA nanoparticles were encapsulated with different CNS agents via nasal administration to the laboratory animals. These studies concluded that nanoparticle transport to the brain was very high via olfactory, extracellular, and trigeminal nerves; however, the cerebrospinal fluid contributes very little part in this transport (Gao et al. 2008, 2006). A team of researchers also studied the PEG-PLA conjugated WGA encapsulated with neuroprotective peptide (VIP) via intranasal administration for targeting brain diseases. In pharmacokinetic study of brain drug distribution, it is found that WGA grafting improved brain concentration of VIP about sevenfold compared to non-conjugated nanoparticles. The WGA-ligated PEG-PLA nanoparticles were encapsulated with QDs (quantum dots) aiming to develop specific brain imaging agents for CNS targeting (Ong et al. 2014). The brain targeting ability of PEG-PLGA nanoparticles was grafted to other lectins such as Ulex europeus agglutinin I (UEA I), Solanum tuberosum lectin (STL), and odorranalectin (Gupta 2020). The STL-conjugated nanoparticles encapsulated with haloperidol were given through intranasal route resulting in enhanced concentration in the brain by threefold compared to non-functionalized nanoparticles. The immunogenic reaction is one of the main causes of lectins. However, odorranalectin is a peptide with lectin-like activity; this compound was evaluated as a potential ligand with low immunogenic response. The small peptide-loaded PEG-PLG nanoparticles were reported to have enhanced efficacy in Parkinson's disease (PD). The lectins somehow have potential toxicity issues; that's why it's compulsory to do toxicological assay of the lectins grafted carrier system at both local and systemic sites (Piazza et al. 2014). Xiaoling Gao et al. developed lectin functionalized biodegradable nanoparticles, a novel way to enhance the brain uptake of agents encapsulated into PEG-PLA nanoparticles by intranasal administration. Ulex europaeus agglutinin I(UEAI) belongs to the I-fucose that is at broad level located at the olfactory site (Guo et al. 2011; Xia et al. 2011). The in vitro study of this formulation suggested that UEA I conjugation at nanoparticles facilitated the permeation and absorption of coumarin fluorescent marker via intranasal administration. UEA I grafting also enhances the brain delivery of nanoparticles. The biodistribution of UEA I conjugated nanoparticles showed higher affinity of the olfactory mucosa than to the respiratory mucosa. Hence, results concluded that the UEA I grafted nanoparticles proved as potential carriers for neurodegenerative disorders (Gao et al. 2006, 2007a).

21.9 Cell-Penetrating Peptides as Surface Ligands

The positively charged cell penetrating peptides are more feasible for extracellular internalization and transport. This internalization and transport mechanism of these cell-specific proteins can be studied through model called as HIV transactivator of transcription (Tat) protein (Kanazawa et al. 2013). These CPPs are nonselective to the cell type and has the ability to transport the proteins, nucleic acids, small molecules, and nanocarriers across different barriers, for example, BBB, skin, etc. Several features like this make CPPs more efficient as functionalization component for intranasal delivery. In one study, researchers concluded that the grafting of CPPs to lipids and polymeric nanocargo clearly improved the ability to cross the in vitro olfactory model (Lin et al. 2016). Several other studies including PEG-PLA nanoparticles functionalized with low-molecular-weight protamine were found to aggregate in 16HBE14O cell line in higher concentration than the unmodified nanoparticles. Among all studies relevant to the CPPs, the HIV Tat peptide is the most commonly applied until now (Xia et al. 2011). The cell-penetrating features have been directly linked with the guanidinium groups of arginine that induce electrostatic bonding on the cell surface. T. Kanazawa et al. studied the significance of RNA-based active components for brain disorders in in vitro study. To significantly improve the siRNA transport to the brain, the intranasal delivery system comprises of CPP modified nanomicelles PEG-PCL copolymers. In this research, MPEG-PCL-Tat developed as model siRNA. The nose-to-brain delivery of dextran with MPEG-PCL-Tat enhanced brain targeting compared to IV delivery of dextran with or without conjugation of Tat. The results demonstrated the MPEG-PCL-Tat enhanced delivery along the trigeminal and olfactory pathway due to increased permeation across the nasal mucosa (Kanazawa et al. 2013, 2019; Samaridou et al. 2020).

Takanori Kanazawa et al. Tat conjugated MPEG-PCL amphiphilic copolymers via ester bond. The researchers evaluated the brain distribution of model compound coumarin after systemic and intranasal delivery of MPEG-PCL. The concentration of coumarin in the brain was very high after intranasal delivery. It is concluded that the application of nanosized micelles with Tat has proven efficacy for the direct targeting to the brain. In one study the authors developed nanolipid carrier system grafted with Tat-conjugated chitosan for nose-to-brain targeting. The results concluded the surface functionalization of Tat peptide chitosan encapsulated with GDNF (glial cell derived neurotropic factor) led to a potential increment of the therapeutic potential in PD. The PLGA nanoparticles surface functionalized with the cationic Tat peptide were able to accumulate approximately 7 times in the olfactory bulb in comparison to non-functionalized NPs. In one more study, surface conjugated MPEG-PCL nanomicelles with cationic surface charge were developed and studied for antitumor activity. The nanomicelles were encapsulated with camptothecin for targeting brain tumor via intranasal route. The MPEG-PCL micelles grafted with arginine based CPPs were developed for the study of nose-to-brain delivery. The in vivo study including imaging concluded that the formulation was reached to the olfactory bulb in no time or with zero delay (Kanazawa et al. 2011, 2012; Okada 2014). The short peptide fragments produced by enzymatic degradation known as protamine. In comparison to other peptides, these CPPs have low immunogenic response and very low toxicity. The grafting of short-chain peptide on PEG-PLGA nanoparticles led to a potential increase in nose-to-brain delivery of the model compound coumarin. All these following results of different studies including those mentioned in Table 21.1 describe the significant potential of nanoparticle mediated transport of peptides or larger molecules via intranasal route (Xia et al. 2011; Kanazawa 2015; Shamarekh et al. 2020; Kamei et al. 2016) (Table 21.2).

21.10 Mechanism to Enhance Direct Targeting of NPs via Neurological Pathway

There are many studies related to the enhanced brain delivery of nanoparticles in comparison to the free drug formulation. There are only few data regarding the mechanism of nanoparticles that can increase drug delivery in the brain. There are different mechanisms involved, but the simplest one in which the nanocarriers bind at the mucus layer and (Xia et al. 2011) release the drug in the mucus. The most significant mechanism is when drug-loaded nanoparticles cross the mucosal barrier. Moreover, these nanoparticles translocate along the nerve axons to reach the brain. Amid that, two scenarios involve nanoparticle uptake into the neuroepithelium. The free drug diffuse out into the perineural spaces of the CNS. Hence, the fate of nanoparticles rely on the physiochemical properties of nanoparticles (Nehoff et al. 2014). The size, morphology, and surface hydrophilicity can affect nanocarrier interaction with the biological environment (Yokoyama 2005). The properties of nanocarriers can influence the mucus interaction, the uptake by the neuroepithelial cells, and the transport to the brain by diffusion. Some authors study in vitro transport of nanoparticles along the olfactory monolayers, ex vivo study across the nasal mucosa, or in vivo in mammalian models like rodent model in order to elaborate the effect of charge, particle size, hydrophobicity, etc. Gartziandia et al. worked on and checked the permeation of nanoparticles across the olfactory mucosa having different physiochemical properties (Alexander et al. 2019). A fluorescent probe (DiR; 1-10dioctadecyl-3,3,30,30-tetranethylindotricarbocyanine) was encapsulated into nanoparticles to analyze their pathway. The results showed the material used for formulation has potential effect on permeation of nanoparticles. For instance, the lipid nanocarrier penetrated to the higher extent compared to the PLGA nanoparticles with same particle charge (−23 mV). The alteration of zeta

Table 21.1 Nanomaterial-based lipid carrier system for nose-to-brain delivery

Drug	Nanocargo system	Ligand molecule	Animals used	Therapeutic effects	Ref
Liposomes	H102 peptide	–	Rats	Threefold enhanced area under curve (AUC) in the brain than in solution	Zheng et al. (2015)
Oil in water nanoemulsion	CsA	–	Rats	Six-time increase in concentration of the brain in comparison to CsA solution	Yadav et al. (2015)
PEGylated cubosomes	Derivative of humanin	OL	Rats	1.7–3.5 times increase of coumarin distribution	Wu et al. (2012)
Cationic liposomes	OVA	–	Rats	Ten times enhanced efficacy. Four times greater AUC of brain-to-blood ratio	Migliore et al. (2010)
Gelatin NCL	SP		Rat	Significant increase of therapeutic effect	Lu et al. (2015)
NCL coated with chitosan	hIGF-I	–	Mice	Enhanced stay time in nasal epithelium and enhanced concentration in the brain	Gartziandia et al. (2015)
NCL coated with chitosan	GDNF	–	Rat	Significant increase of neuroprotective effect	Gartziandia et al. (2016)
NCL gelatin	bFGF	–	Rat	~Twofold increase of brain concentration of bFGF	Zhao et al. (2014)
NCL gelatin	bFGF	–	Rat	Increased concentration of peptide in the brain in comparison to simple solution	Zhao et al. (2016)

potential from negative to positive by coating with chitosan enhanced the intranasal transport 3 times compared to uncoated nanolipid carriers (Salem et al. 2020). Surface modifications using cell-penetrating peptides further increase the nanoparticle transport. The composition of nanocarrier system has different penetration effects, for example, formulation composed of poloxamer and polysorbate 80 PLGA nanoparticles in comparison to formulation without surfactant has low permeation, as well as high mucus-penetrating effect of PEG and the mucoadhesive features of PVA-coated particles. Musumeci and coworkers developed PLA and PLGA nanoparticles using Tween 80 surfactant and rhodamine as fluorescent dye. PLGA nanoparticles having surfactant were found with greater uptake in olfactory ensheathing cells as compared to other nanoparticles. Mistry et al. developed nanoparticles of different sizes and with different zeta potentials (Sonvico et al.

Table 21.2 Examples of targeting ligands and surface modification of nanocarriers for nose-to-brain delivery along with various in vivo models

Nanocarrier	Surface modification	Targeting ligand	Animal model used	Ref
PEG-PCL-NPs	Lactoferrin		Rats/mice	Samaridou and Alonso (2018), Liu et al. (2013), Hernando et al. (2018)
Chitosan coated NCL		Tat	Mice	Hernando et al. (2018)
PLGA NPs	Lactoferrin	Tat	Mice	Meng et al. (2018)
PEG-PLGA NPs	STL		Mice	Chen et al. (2012) Zhang et al. (2014)
PEG-PLA NPs	WGA		Rats/mice	Gao et al. (2006, 2007b)
PEG-PLGA NPs	OL		Rats	Gao et al. (2011)
Nanomicelles, MPEG-PCL		Tat	Rats	Kanazawa et al. (2011), Taki et al. (2012)
PEG-PLA NPs	UEA1		Rats	Xia et al. (2011), Kanazawa et al. (2017)
Nanomicelles		CH2R4H2C	Rats	Kanazawa et al. (2017)
PEG-PLA NPs		LMWP	Rats	Wu et al. (2012), Xia et al. (2011)
Pegylated cubosomes	OL		Rats	Wu et al. (2012)

2018; Musumeci et al. 2018). Among all these studied particles, the polysorbate 80-coated nanoparticles or PEGylated nanoparticles penetrated into the deeper layer of the nasal epithelium as compared to the uncoated one. In vivo study was performed on mouse model to check the nanoparticle uptake mechanism. A 15 μL of nanoparticles having particle size of 276, 163, and 107 nm chitosan-coated (+30 & +23 mV, respectively) or 180 and 107 nm Tween 80-coated NPs (−24 and −21 mV) were implicated for 3 days (Huckaby and Lai 2018). Among these studies the particles decorated with tween 80 have smaller size (between 107-180 nm) and showed more penetration as compared to larger size particles. The chitosan coated nanoparticles showed very little penetration in comparison to other pairs of coated particles. Ahmad et al. studied and compared nanoemulsions with droplet sizes of 900, 500,200, and 80 nm. It was evaluated that the nanoparticles with smaller droplet size have quite higher stay time in the nasal mucosa. The cationic-charged chitosan-grafted nanoparticles have higher stay in the nasal mucosa and in anionic-charged epithelial membranes (Ahmad et al. 2018). A recent study resulted in greater brain accumulation of PLGA nanoparticles after intranasal intervention in rats. The researchers developed rhodamine-loaded PLGA NPs coated with Tween 80 having zeta potential of −26 mV and particle size of 118 nm and chitosan PLGA-NPs 213 nm and +69 mV. The results showed that both types of nanoparticles either anion and cationic charged reached the neural cells. The exact site of particles

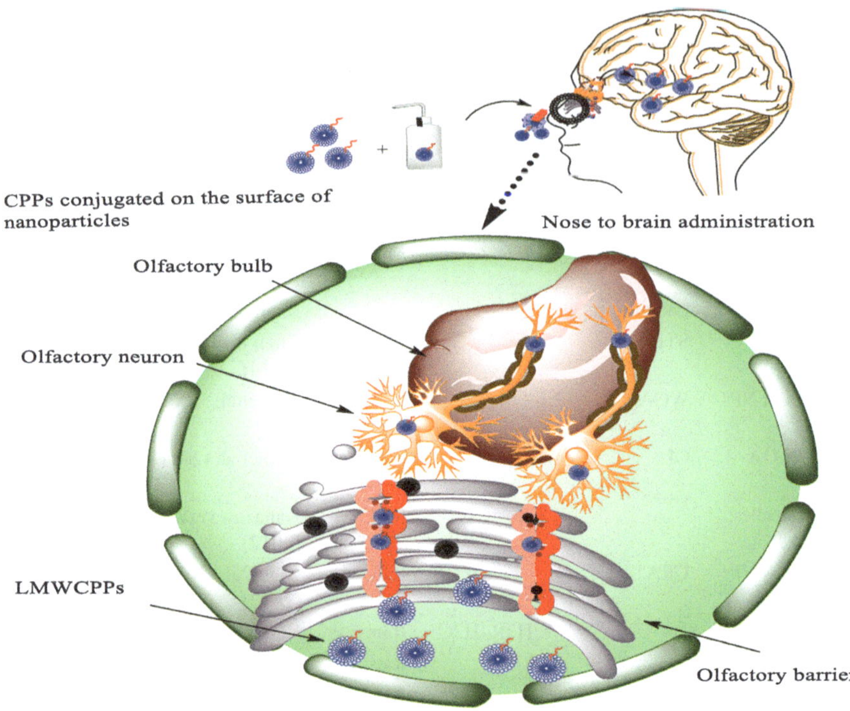

Fig. 21.4 Mechanism behind nose-to-brain delivery of CPP functionalized nanoparticles

deposition was different in brain cells depending on the surface charge on particles. The quite slower brain uptake of positively charged nanoparticles was seen in the nasal cavity after mucus nanoparticle interactions (Shamarekh et al. 2020). Kanazawa et al. also described the relationship between the brain distribution and nanocarrier characteristics by using peptide-based carriers. In this study, two stable micellar formulations were developed along with conjugation of arginine-rich oligopeptide. The particle size of both stable formulations including PEG-PCL peptide micelles and stearate peptide was 50 and 100 nm, and surface charge was +15 and +20 mV. The intranasal administration was performed along with controlled group of formulation. The results concluded the increase uptake of both nanoformulations into the brain. The hydrophobic stearate peptide showed higher penetration as compared to the PEG-PCL peptide (Kanazawa et al. 2019, 2017; Ghosh et al. 2019; Grossen et al. 2017). The results concluded that PEG-PCL peptide nanocarrier penetration was higher across the nasal mucosa through the trigeminal and olfactory nerves. Gabal et al. studied the nose-to-brain delivery of nanocarrier system depending on surface charge by developing cationic and anionic nanolipid carriers having particle sizes of 160 and 175 nm and zeta potential of +34 and −34 mV. The pharmacokinetic study of both nanoparticle formulations was performed in albino rats. Moreover, results show higher bioavailability for both charge particles due to enhanced stay time in the nasal mucosa (Gabal et al. 2014).

Figure 21.4 illustrated the mechanism behind the nose-to-brain delivery of nanomaterials functionalized with CPPs.

21.11 In Vivo/In Vitro Models for Testing Nose-to-Brain Delivery

The development of futuristic nanomedicines to treat malignant gliomas and central nervous system-related malignancies demands prospective approaches to improve drug availability. In vivo animal models and in vitro cell lines-based models are being employed to explore nanomaterial-driven drug transport mechanisms via the intranasal route. These models are used in different studies; for example, absorption and toxicological and pharmacokinetic profile of the active pharmaceutical ingredient are determined by in vivo animal-based models, whereas in vitro models can study diffusion, permeation, and mechanisms (Hornof et al. 2005; Urtti 2006; Sabir et al. 2020). The selection of in vivo models is highly important for the anatomical studies of the nasal cavity. The rat is among the foremost animal models applied to intranasal examination, and subsequently, many other animal models, such as monkeys, mice, rabbits, and sheep, have also been used. For preliminary absorption evaluation, rat and mouse models are generally recommended, whereas sheep, dog, rabbit, and monkey models are employed for adequate pharmacokinetic investigations (Sousa and Castro 2016). For example, (Azambuja et al. 2020) developed a cationic nanoemulsion to treat glioblastoma by delivering CD73siRNA through the nasal route. After nasal delivery, the NE-siRNA CD73R could be noticed in rat brain and serum resulting in a 60% reduction in tumor growth in glioma-carrying Wistar rats. This cationic nanoemulsion showed no toxicity in astrocytes or rat models. Male Sprague-Dawley rats were used to evaluate microemulsion efficacy for intranasal delivery of mebendazole for glioblastoma treatment. Results revealed a significant improvement in survival time in the C6 rat model than control groups without any damage in the epithelium (Mena-Hernández et al. 2020).

In vitro models were designed as a substitute for the in vivo models. The extrapolation of absorption and kinetics data recorded from the animal models to humans is quite challenging. Moreover, animal experiments in biomedical research are often criticized for high-cost, labor-intensive, time, and ethical concerns (Hornof et al. 2005). Many countries have imposed restrictive legislation to limit animal-based experimentations. These issues have rekindled researchers' attention to seeking in vitro methods as a potential alternative to swapping animal experiments and inspired investigators to recruit cell culture models that offer noteworthy platforms to investigate different cancers. Cell models are regarded as a suitable system for investigating barrier functions, cellular uptake, and transport mechanisms (Barar et al. 2009). They could be easily established and applied to various molecular and cellular studies, like biomarker detections, cellular metabolism, and the development of new therapeutic tools, including nanobodies, monoclonal antibodies, and genome-based nanotherapeutics. Being an extremely defined system, the utilization of cell culture models from human resources can provide reliable data and reproducible results (Barar et al. 2009). Nevertheless, it is important to choose adequate in vitro cell lines reproducing results at substantially low costs. A large number of in vitro cell culture models, such as Calu-3 (human lung adenocarcinoma), CaCo-2 cell lines (human colon carcinoma), BT (bovine turbinates), NAS2BL (rat nasal squamous carcinoma), 16HBE14o- (human normal bronchial epithelium of male heart-lung transplant patient), and RPMI 2650 (human nasal epithelial tissues) are available and tested (Sousa and Castro 2016). Among the cell lines mentioned above, RPMI 2650 and CaCo-2 are applied to study permeability and absorption via the nasal route. Nevertheless, some demerits are also associated with these cell lines; for instance, RPMI 2650 (undifferentiated cells) experience the restricted expression of goblet and ciliated cells. The lack of a well-developed monolayer renders this model cells unrealistic for transport studies. On the other hand, the Calu-3 cell line is suitable for transport evaluation because it grows in monolayers, but this does not originate from the normal epithelium of the nasal cavity. Due to high transepithelial electrical resistance (TEER), the 16HBE14o-cell line is appropriate for transportation study but originates from a male heart and lung transplant patient and normal bronchial epithelial cells. (Ullah et al. 2020) used human (U87MG) and rat (C6) glioblastoma cells for investigating the therapeutic effects of paclitaxel-incorporated nanoparticles against glioblastoma multiforme following nose-to-brain delivery. The intranasal administration of paclitaxel-incorporated nanoparticles significantly reduced tumor burden by impeding tumor proliferation and inducing apoptosis without impacting the healthy brain cells. Various formulations of low-molecular-weight heparins were administered via the nasal route to examine their absorption behavior in anesthetized rats. In vitro cell culture model was used for absorption-improving mechanism studies of cyclodextrins (Yang et al. 2004).

21.12 Comparative Analysis of In Vivo Studies of Nanomaterials vs Permeation Enhancer

Merkus and van den berg defined the selection criteria for animal study that aimed to increase evidences or justification to prove nose-to-brain delivery. These rules include (1) screening of particular administration quantity and volumes, (2) the pharmacokinetic evaluation in CNS and in plasma, (3) the comparison of the results after analysis with those obtained after intravenous intervention of the same product, and (4) the brain distribution evaluation of given product. Many other researchers elaborated on potential of various intranasal delivery strategies. These approaches included direct transport percentages (DTP%) and drug targeting efficiency % (DTE %). DTE% showed the potential of the drug to reach the brain in comparison to the blood following nose-to-brain delivery (Mistry et al. 2009; Merkus and van den Berg 2007). DTP% is defined as the percentage of the that delivered into the brain due to nose-to-brain route with respect to the total concentration of drug present in the brain. The following two equations elaborated on the mathematical expression of these two strategies (van Den Berg et al. 2002, 2004):

Drug targeting efficiency %(DTE)

$$\mathrm{DTE\%} = \frac{(\mathrm{AUCbrain/AUC\ blood})\mathrm{intranasal}}{(\mathrm{AUC\ brain/AUCblood})\mathrm{parentral}} * 100$$

Here, AUC brain and AUC blood are the area under the curve of the drug in the brain and blood, evaluated every time for respective route.

Direct transport percentage %(DTP)

$$\mathrm{DTP\%} = \frac{(\mathrm{Bin} - \mathrm{Bx})}{\mathrm{Bin}} * 100$$

Here, Bin is the AUC of the brain after nose-to-brain delivery and Bx is the AUC of brain fraction from the blood after intranasal application. The most data present regarding this do not clearly mentioned the pharmacokinetic and brain distribution but only represent quantitative bio-distribution results (Van Den Berg et al. 2003). Some studies did not elaborate on the comparison data systemic circulation that makes it difficult to conclude the direct nose-to-brain delivery pathway or mechanism. This is significant to mention that a number of studies showed ability of nanocarriers to pass through the olfactory pathway and deliver and release the encapsulated moiety at various brain targeting sites. However, the quantitative data related to nose-to-brain delivery of nanocarriers is not present in literature. The most intranasally applied nanocarrier used polymer and lipid-based systems. In related studies, the reported greater efficacy of peptide conjugated cationic liposomes of mean diameter up to 300nm showed 10 times enhanced bioavailability in the brain. PLA nanoparticles coated with chitosan were observed to have 9.5-fold increased amount of neurotoxin-i in the neuronal cells (Van Den Berg et al. 2003). Zhang et al. studied the PEG-PLGA particles grafted with STL, having negative zeta potential of −32 mV for intranasal delivery of growth hormone (bFGF). The tracing and screening method enabled tracing of protein molecule after its IV and IN intervention. The results showed greater DTE% and DTP% (~1050% and ~90.44% in olfactory bulb). In short, all study data until now showed that cationic or modified nanoparticles showed greater efficacy in comparison to anionic charge particles. However, there is no exact data regarding comparative studies for different nanomaterial intranasal deliveries (Li et al. 2001, 2011). Therefore, more extensive study in this arena will be more appreciated and needed in the near future. There are a huge list of studies regarding enhanced delivery of nanoparticles co-encapsulated with peptides via olfactory passage. The most common natural permeation enhancer used is chitosan, but Morishita's group of researchers used L-penetratin in comparison to free protein solution. Vaka et al. described ~14 times enhanced brain concentration of BDNF and NGF when applied intranasally. The peppermint oil is also used to have tight junction opening property. Zheng et al. studied the comparison between chitosan and PEGylated liposome formulation for intranasal delivery (Zhang et al. 2014). The results concluded 3 times higher concentration of the drug for liposomal formulation. It is concluded from the study that comparative studies are required to evaluate the conclusion between nanoparticles and permeation enhancers (Chen et al. 2012).

21.13 Conclusion and Future Perspectives

This review discussed on the physicochemical and delivery-related intranasal barriers that influence the absorption of therapeutic drugs, with attention to details relevant to novel, noninvasive, safe, and effective strategies that help overcome intranasal barriers, published in various research articles. The new nanomaterial and nanotechnology is more phenomenal in this emerging field, which may provide a significant potential in incapacitating the limitation and barriers in targeting this route. The present literature study also pointed out the enormous role of physiochemical features like nanoparticle size and surface potential in brain delivery. Materials including lipids, PEGylation, use of surfactant as permeation enhancer, and surface grafting of peptides (CPPs) were evaluated showing more compatible results. Both in vivo animal and in vitro cell line-based models are being employed to explore nanomaterial-mediated drug transport mechanisms via the intranasal route. In contrast to in vivo models, in vitro cell models are regarded as a suitable system for investigating barrier functions, cellular uptake, and transport mechanisms. They could be easily applied to various molecular and cellular studies and to develop new therapeutic tools, including nanobodies, monoclonal antibodies, and genome-based nanotherapeutics. Nevertheless, it is important to choose adequate in vitro cell lines reproducing results at substantially low costs. The chapter concludes and highlights the potential and significance of nanomaterial application in brain targeting. However, there is more pharmacokinetic and pharmacodynamic studies needed, and also mechanism behind uptake of nanoparticles needs to be more explicatory.

References

Abd-Elal RM et al (2016) Trans-nasal zolmitriptan novasomes: in-vitro preparation, optimization and in-vivo evaluation of brain targeting efficiency. Drug Deliv 23(9):3374–3386

Abdelrahman FE et al (2017) Response surface optimization, Ex vivo and In vivo investigation of nasal spanlastics for bioavailability enhancement and brain targeting of risperidone. Int J Pharm 530(1-2):1–11

Aderibigbe BA, Naki T (2019) Chitosan-based nanocarriers for nose to brain delivery. Appl Sci 9(11):2219

Agrawal M et al (2018) Nose-to-brain drug delivery: an update on clinical challenges and progress towards approval of anti-Alzheimer drugs. J Control Release 281:139–177

Agrawal M et al (2020) Stimuli-responsive In situ gelling system for nose-to-brain drug delivery. J Control Release

Ahmad N et al (2018) Impact of ultrasonication techniques on the preparation of novel Amiloride-nanoemulsion used for intranasal delivery in the treatment of epilepsy, Artificial Cells, Nanomedicine, and Biotechnology. 46(supp 3):S192–S207

Alexander A et al (2019) Recent expansions of novel strategies towards the drug targeting into the brain. Int J Nanomedicine 14:5895

Anand U, Feridooni T, Agu RU (2012) Novel mucoadhesive polymers for nasal drug delivery. Recent Adv Novel Drug Carrier Syst:315–330

Azambuja J et al (2020) Nasal administration of cationic nanoemulsions as CD73-siRNA delivery system for glioblastoma treatment: a new therapeutical approach. Mol Neurobiol 57(2):635–649

Bahadur S, Pathak K (2012) Physicochemical and physiological considerations for efficient nose-to-brain targeting. Expert Opin Drug Deliv 9(1):19–31

Baranei M et al (2020) Anticancer effect of green tea extract (GTE)-loaded pH-responsive niosome coated with PEG against different cell lines. Mater Today Commun:101751

Barani M et al (2018) Lawsone-loaded Niosome and its antitumor activity in MCF-7 breast Cancer cell line: a Nano-herbal treatment for Cancer. DARU J Pharm Sci 26(1):11–17

Barani M et al (2019a) Evaluation of carum-loaded niosomes on breast cancer cells: physicochemical properties, in vitro cytotoxicity, flow cytometric, DNA fragmentation and cell migration assay. Sci Rep 9(1):1–10

Barani M et al (2019b) In silico and in vitro study of magnetic niosomes for gene delivery: the effect of ergosterol and cholesterol. Mater Sci Eng C 94:234–246

Barani M et al (2020a) A new formulation of hydrophobin-coated niosome as a drug carrier to cancer cells. Mater Sci Eng C:110975

Barani M et al (2020b) Nanotreatment and nanodiagnosis of prostate cancer: recent updates. Nanomaterials 10(9):1696

Barani M et al (2020c) Comprehensive evaluation of gene expression in negative and positive trigger-based targeting niosomes in HEK-293 cell line. Iran J Pharm Res 19(1):166–180

Barani M et al (2020d) Nanotechnology in ovarian cancer: diagnosis and treatment. Life Sci:118,914

Barar J et al (2009) Ocular drug delivery; impact of in vitro cell culture models. J Ophthalmic Vis Res 4(4):238

de Barros T et al (2020) Cachexia: pathophysiology and ghrelin liposomes for nose-to-brain delivery. Int J Mol Sci 21(17):5974

Battaglia L et al (2018) Lipid nanoparticles for intranasal administration: application to nose-to-brain delivery. Expert Opin Drug Deliv 15(4):369–378

Betbeder D et al (2000) Biovector™ nanoparticles improve antinociceptive efficacy of nasal morphine. Pharm Res 17(6):743–748

Bilal M et al (2020) Nanomaterials for the treatment and diagnosis of Alzheimer's disease: an overview. NanoImpact:100,251

Bonferoni MC et al (2019) Nanoemulsions for "nose-to-brain" drug delivery. Pharmaceutics 11(2): 84

Chalikwar SS et al (2013) Self-assembled, chitosan grafted PLGA nanoparticles for intranasal delivery: design, development and ex vivo characterization. Polym-Plast Technol Eng 52(4): 368–380

Chen J et al (2012) Solanum tuberosum lectin-conjugated PLGA nanoparticles for nose-to-brain delivery: in vivo and in vitro evaluations. J Drug Target 20(2):174–184

Clementino A et al (2016) The nasal delivery of nanoencapsulated statins–an approach for brain delivery. Int J Nanomedicine 11:6575

Dae-Duk K (2007) Drug absorption studies: in situ, in vitro and in silico models. Springer, New York

Das SS et al (2020) Stimuli-responsive polymeric nanocarriers for drug delivery, imaging, and theragnosis. Polymers 12(6):1397

Davarpanah F et al (2018) Magnetic delivery of antitumor carboplatin by using PEGylated-Niosomes. DARU J Pharm Sci 26(1):57–64

Davarpanah AM et al (2019) (1-x) BaFe12O19/xCoFe2O4 hard/soft magnetic nanocomposites: Synthesis, physical characterization, and antibacterial activities study. J Mol Struct 1175:445–449 van Den Berg MP et al (2002) Serial cerebrospinal fluid sampling in a rat model to study drug uptake from the nasal cavity. J Neurosci Methods 116(1):99–107

Devkar TB, Tekade AR, Khandelwal KR (2014) Surface engineered nanostructured lipid carriers for efficient nose to brain delivery of ondansetron HCl using Delonix regia gum as a natural mucoadhesive polymer. Colloids Surf B Biointerfaces 122:143–150

Dhakar RC et al (2011) A review on factors affecting the design of nasal drug delivery system. Int J Drug Deliv 3(2):194

Di Gioia S et al (2015) Intranasal delivery of dopamine to the striatum using glycol chitosan/sulfobutylether-β-cyclodextrin based nanoparticles. Eur J Pharm Biopharm 94:180–193

Ebrahimi AK, Barani M, Sheikhshoaie I (2018) Fabrication of a new superparamagnetic metal-organic framework with core-shell nanocomposite structures: characterization, biocompatibility, and drug release study. Mater Sci Eng C 92:349–355

Feng Y et al (2018) An update on the role of nanovehicles in nose-to-brain drug delivery. Drug Discov Today 23(5):1079–1088

Fonseca FN et al (2015) Mucoadhesive amphiphilic methacrylic copolymer-functionalized poly (ε-caprolactone) nanocapsules for nose-to-brain delivery of olanzapine. J Biomed Nanotechnol 11(8):1472–1481

Gabal YM et al (2014) Effect of surface charge on the brain delivery of nanostructured lipid carriers in situ gels via the nasal route. Int J Pharm 473(1-2):442–457

Gänger S, Schindowski K (2018) Tailoring formulations for intranasal nose-to-brain delivery: a review on architecture, physico-chemical characteristics and mucociliary clearance of the nasal olfactory mucosa. Pharmaceutics 10(3):116

Gao X et al (2006) Lectin-conjugated PEG–PLA nanoparticles: preparation and brain delivery after intranasal administration. Biomaterials 27(18):3482–3490

Gao X et al (2007a) UEA I-bearing nanoparticles for brain delivery following intranasal administration. Int J Pharm 340(1-2):207–215

Gao X et al (2007b) Brain delivery of vasoactive intestinal peptide enhanced with the nanoparticles conjugated with wheat germ agglutinin following intranasal administration. J Control Release 121(3):156–167

Gao X et al (2008) Quantum dots for tracking cellular transport of lectin-functionalized nanoparticles. Biochem Biophys Res Commun 377(1):35–40

Gao M et al (2011) Synthesis and characterization of superparamagnetic Fe3O4@ SiO2 core-shell composite nanoparticles. World Jf Condensed Matter Phys 1(2):49–54

Gartziandia O et al (2015) Chitosan coated nanostructured lipid carriers for brain delivery of proteins by intranasal administration. Colloids Surf B Biointerfaces 134:304–313

Gartziandia O et al (2016) Intranasal administration of chitosan-coated nanostructured lipid carriers loaded with GDNF improves behavioral and histological recovery in a partial lesion model of Parkinson's disease. J Biomed Nanotechnol 12(12):2220–2280

Ghazy E et al (2020a) Scrutinizing the therapeutic and diagnostic potential of nanotechnology in thyroid cancer: edifying drug targeting by nano-oncotherapeutics. J Drug Delivery Sci Technol:102,221

Ghazy E et al (2020b) Nanomaterials for Parkinson disease: recent progress. J Mol Struct:129,698

Ghosh S et al (2019) Surface engineered liposomal delivery of therapeutics across the blood brain barrier: recent advances, challenges and opportunities. Expert Opin Drug Deliv 16(12): 1287–1311

Grossen P et al (2017) PEG-PCL-based nanomedicines: a biodegradable drug delivery system and its application. J Control Release 260:46–60

Guo J et al (2011) Aptamer-functionalized PEG–PLGA nanoparticles for enhanced anti-glioma drug delivery. Biomaterials 32(31):8010–8020

Guo Y et al (2013) The applications of Vitamin E TPGS in drug delivery. Eur J Pharm Sci 49(2): 175–186

Gupta A (2020) Emerging applications of lectins in cancer detection and biomedicine. Mater Today: Proc 31:651–661

Hajizadeh MR et al (2019a) In vitro cytotoxicity assay of D-limonene niosomes: an efficient nano-carrier for enhancing solubility of plant-extracted agents. Res Pharm Sci 14(5):448

Hajizadeh MR et al (2019b) Diosgenin-loaded niosome as an effective phytochemical nanocarrier: physicochemical characterization, loading efficiency, and cytotoxicity assay. DARU J Pharm Sci 27(1):329–339

Hernando S et al (2018) Intranasal administration of TAT-conjugated lipid nanocarriers loading GDNF for Parkinson's disease. Mol Neurobiol 55(1):145–155

Hornof M, Toropainen E, Urtti A (2005) Cell culture models of the ocular barriers. Eur J Pharm Biopharm 60(2):207–225

Huckaby JT, Lai SK (2018) PEGylation for enhancing nanoparticle diffusion in mucus. Adv Drug Deliv Rev 124:125–139

Illum L (2007) Nanoparticulate systems for nasal delivery of drugs: a real improvement over simple systems? J Pharm Sci 96(3):473–483

Jadhav KR et al (2007) Nasal drug delivery system-factors affecting and applications. Curr Drug Ther 2(1):27–38

Kamei N et al (2016) Visualization and quantitative assessment of the brain distribution of insulin through nose-to-brain delivery based on the cell-penetrating peptide noncovalent strategy. Mol Pharm 13(3):1004–1011

Kanazawa T (2015) Brain delivery of small interfering ribonucleic acid and drugs through intranasal administration with nano-sized polymer micelles. Med Devices (Auckland, NZ) 8:57

Kanazawa T et al (2011) Cell-penetrating peptide-modified block copolymer micelles promote direct brain delivery via intranasal administration. Pharm Res 28(9):2130–2139

Kanazawa T et al (2012) Suppression of tumor growth by systemic delivery of anti-VEGF siRNA with cell-penetrating peptide-modified MPEG–PCL nanomicelles. Eur J Pharm Biopharm 81(3):470–477

Kanazawa T et al (2013) Delivery of siRNA to the brain using a combination of nose-to-brain delivery and cell-penetrating peptide-modified nano-micelles. Biomaterials 34(36):9220–9226

Kanazawa T et al (2017) Enhancement of nose-to-brain delivery of hydrophilic macromolecules with stearate-or polyethylene glycol-modified arginine-rich peptide. Int J Pharm 530(1-2): 195–200

Kanazawa T et al (2019) Therapeutic effects in a transient middle cerebral artery occlusion rat model by nose-to-brain delivery of anti-TNF-alpha siRNA with cell-penetrating peptide-modified polymer micelles. Pharmaceutics 11(9):478

Kulkarni AD et al (2015) Nanotechnology-mediated nose to brain drug delivery for Parkinson's disease: a mini review. J Drug Target 23(9):775–788

Li Y-P et al (2001) PEGylated PLGA nanoparticles as protein carriers: synthesis, preparation and biodistribution in rats. J Control Release 71(2):203–211

Li J et al (2011) Targeting the brain with PEG–PLGA nanoparticles modified with phage-displayed peptides. Biomaterials 32(21):4943–4950

Lin T et al (2016) Nose-to-brain delivery of macromolecules mediated by cell-penetrating peptides. Acta Pharm Sin B 6(4):352–358

Liu Z et al (2013) Lactoferrin-modified PEG-co-PCL nanoparticles for enhanced brain delivery of NAP peptide following intranasal administration. Biomaterials 34(15):3870–3881

Lu C-T et al (2015) Gelatin nanoparticle-mediated intranasal delivery of substance P protects against 6-hydroxydopamine-induced apoptosis: an in vitro and in vivo study. Drug Des Devel Ther 9:1955

Martins PP, Smyth HD, Cui Z (2019) Strategies to facilitate or block nose-to-brain drug delivery. Int J Pharm 570:118635

Mathieu V (2019) Development and characterization of formulations for the nose-to-brain delivery of ghrelin and the management of cachexia. Université de Mons.

Md S et al (2018) Nano-carrier enabled drug delivery systems for nose to brain targeting for the treatment of neurodegenerative disorders. J Drug Delivery Sci Technol 43:295–310

Mena-Hernández J et al (2020) Preparation and evaluation of mebendazole microemulsion for intranasal delivery: an alternative approach for glioblastoma treatment. AAPS PharmSciTech 21(7):1–12
Meng Q et al (2018) Intranasal delivery of Huperzine A to the brain using lactoferrin-conjugated N-trimethylated chitosan surface-modified PLGA nanoparticles for treatment of Alzheimer's disease. Int J Nanomedicine 13:705
Merkus FW, van den Berg MP (2007) Can nasal drug delivery bypass the blood-brain barrier? Drugs in R & D 8(3):133–144
Migliore MM et al (2010) Brain delivery of proteins by the intranasal route of administration: a comparison of cationic liposomes versus aqueous solution formulations. J Pharm Sci 99(4): 1745–1761
Misra A, Kher G (2012) Drug delivery systems from nose to brain. Curr Pharm Biotechnol 13(12): 2355–2379
Mistry A, Stolnik S, Illum L (2009) Nanoparticles for direct nose-to-brain delivery of drugs. Int J Pharm 379(1):146–157
Mittal D et al (2014) Insights into direct nose to brain delivery: current status and future perspective. Drug Deliv 21(2):75–86
Mukhtar M et al (2020) Nanomaterials for diagnosis and treatment of brain cancer: recent updates. Chemsensors 8(4):117
Muralidharan P et al (2014) Inhalable PEGylated phospholipid nanocarriers and PEGylated therapeutics for respiratory delivery as aerosolized colloidal dispersions and dry powder inhalers. Pharmaceutics 6(2):333–353
Musumeci T et al (2018) Tangential flow filtration technique: an overview on nanomedicine applications. Pharm Nanotechnol 6(1):48–60
Nehoff H et al (2014) Nanomedicine for drug targeting: strategies beyond the enhanced permeability and retention effect. Int J Nanomedicine 9:2539
Nikazar S et al (2020a) Revisiting the cytotoxicity of quantum dots: an in-depth overview. Biophys Rev:1–16
Nikazar S et al (2020b) Photo-and magnetothermally responsive nanomaterials for therapy, controlled drug delivery and imaging applications. ChemistrySelect 5(40):12,590–12,609
Okada H (2014) Targeted siRNA therapy using cytoplasm-responsive nanocarriers and cell-penetrating peptides. J Pharm Investig 44(7):505–516
Ong W-Y, Shalini S-M, Costantino L (2014) Nose-to-brain drug delivery by nanoparticles in the treatment of neurological disorders. Curr Med Chem 21(37):4247–4256
Pardeshi CV, Belgamwar VS (2013) Direct nose to brain drug delivery via integrated nerve pathways bypassing the blood–brain barrier: an excellent platform for brain targeting. Expert Opin Drug Deliv 10(7):957–972
Pardeshi CV, Belgamwar VS (2018) N, N, N-trimethyl chitosan modified flaxseed oil based mucoadhesive neuronanoemulsions for direct nose to brain drug delivery. Int J Biol Macromol 120:2560–2571
Patel S et al (2011) Brain targeting of risperidone-loaded solid lipid nanoparticles by intranasal route. J Drug Target 19(6):468–474
Piazza J et al (2014) Haloperidol-loaded intranasally administered lectin functionalized poly (ethylene glycol)–block-poly (D, L)-lactic-co-glycolic acid (PEG–PLGA) nanoparticles for the treatment of schizophrenia. Eur J Pharm Biopharm 87(1):30–39
Piazzini V et al (2019) Chitosan coated human serum albumin nanoparticles: a promising strategy for nose-to-brain drug delivery. Int J Biol Macromol 129:267–280
Pillai AM et al (2020) Green synthesis and characterization of zinc oxide nanoparticles with antibacterial and antifungal activity. J Mol Struct:128,107
Quintana DS et al (2016) The promise and pitfalls of intranasally administering psychopharmacological agents for the treatment of psychiatric disorders. Mol Psychiatry 21(1):29–38

Rahdar A et al (2019a) Effect of tocopherol on the properties of Pluronic F127 microemulsions: physico-chemical characterization and in vivo toxicity. J Mol Liq 277:624–630

Rahdar A et al (2019b) Synthesis and characterization of highly efficacious Fe-doped ceria nanoparticles for cytotoxic and antifungal activity. Ceram Int 45(6):7950–7955

Rahdar A et al (2020a) The synthesis of methotrexate-loaded F127 microemulsions and their in vivo toxicity in a rat model. J Mol Liq:113449

Rahdar A et al (2020b) Synthesis, characterization, and intraperitoneal biochemical studies of zinc oxide nanoparticles in Rattus norvegicus. Appl Phys A 126:1–9

Rahdar A et al (2020c) Behavioral effects of zinc oxide nanoparticles on the brain of rats. Inorg Chem Commun 119:108,131

Rahdar A et al (2020d) Copolymer/graphene oxide nanocomposites as potential anticancer agents. Polymer Bull:1–22

Rahdar A et al (2020e) Gum-based cerium oxide nanoparticles for antimicrobial assay. Appl Phys A 126:1–9

Rahdar A et al (2020f) Deferasirox-loaded pluronic nanomicelles: synthesis, characterization, in vitro and in vivo studies. J Mol Liq:114,605

Rassu G et al (2017) Nose-to-brain delivery of BACE1 siRNA loaded in solid lipid nanoparticles for Alzheimer's therapy. Colloids Surf B Biointerfaces 152:296–301

Sabir F, Ismail R, Csoka I (2020) Nose-to-brain delivery of antiglioblastoma drugs embedded into lipid nanocarrier systems: status quo and outlook. Drug Discov Today 25(1):185–194

Salade L et al (2017) Development of coated liposomes loaded with ghrelin for nose-to-brain delivery for the treatment of cachexia. Int J Nanomedicine 12:8531

Salem LH et al (2020) Coated lipidic nanoparticles as a new strategy for enhancing nose-to-brain delivery of a hydrophilic drug molecule. J Pharm Sci

Samaridou E, Alonso MJ (2018) Nose-to-brain peptide delivery–the potential of nanotechnology. Bioorg Med Chem 26(10):2888–2905

Samaridou E et al (2020) Nose-to-brain delivery of enveloped RNA-cell permeating peptide nanocomplexes for the treatment of neurodegenerative diseases. Biomaterials 230:119657

Saravani R et al (2020) Newly crocin-coated magnetite nanoparticles induce apoptosis and decrease VEGF expression in breast carcinoma cells. J Drug Delivery Sci Technol 60:101,987

Sawant RR, Torchilin VP (2010) Multifunctionality of lipid-core micelles for drug delivery and tumour targeting. Mol Membr Biol 27(7):232–246

Sayadi K et al (2020) Atorvastatin-loaded SBA-16 nanostructures: synthesis, physical characterization, and biochemical alterations in hyperlipidemic rats. J Mol Struct 1202:127,296

Sekerdag E et al (2017) A potential non-invasive glioblastoma treatment: Nose-to-brain delivery of farnesylthiosalicylic acid incorporated hybrid nanoparticles. J Control Release 261:187–198

Shamarekh KS et al (2020) Development and evaluation of protamine-coated PLGA nanoparticles for nose-to-brain delivery of tacrine: in-vitro and in-vivo assessment. J Drug Delivery Sci Technol:101,724

Sivasankarapillai VS et al (2020a) Cancer theranostic applications of MXene nanomaterials: recent updates. Nano-Struct Nano-Objects 22:100,457

Sivasankarapillai VS et al (2020b) On facing the SARS-CoV-2 (COVID-19) with combination of nanomaterials and medicine: possible strategies and first challenges. Nanomaterials 10(5):852

Sivasankarapillai V et al Progress in natural polymer engineered biomaterials for transdermal drug delivery systems. Mater Today Chem 19:100,382

Sonvico F et al (2018) Surface-modified nanocarriers for nose-to-brain delivery: from bioadhesion to targeting. Pharmaceutics 10(1):34

Sosnik A, das Neves J, Sarmento B (2014) Mucoadhesive polymers in the design of nano-drug delivery systems for administration by non-parenteral routes: a review. Prog Polym Sci 39(12): 2030–2075

Sousa F, Castro P (2016) Cell-based in vitro models for nasal permeability studies. In: Concepts and models for drug permeability studies. Elsevier, pp 83–100

Stevens J et al (2009) A new minimal-stress freely-moving rat model for preclinical studies on intranasal administration of CNS drugs. Pharm Res 26(8):1911–1917

Stützle M et al (2015) Nose-to-brain delivery of insulin for Alzheimer's disease. ADMET and DMPK 3(3):190–202

Tafaghodi M et al (2004) Evaluation of the clearance characteristics of various microspheres in the human nose by gamma-scintigraphy. Int J Pharm 280(1-2):125–135

Taimoory SM et al (2018) The synthesis and characterization of a magnetite nanoparticle with potent antibacterial activity and low mammalian toxicity. J Mol Liq 265:96–104

Taki H et al (2012) Intranasal delivery of camptothecin-loaded tat-modified nanomicells for treatment of intracranial brain tumors. Pharmaceuticals 5(10):1092–1102

Torkzadeh-Mahani M et al (2020) A combined theoretical and experimental study to improve the thermal stability of recombinant D-lactate dehydrogenase immobilized on a novel superparamagnetic Fe3O4NPs@ metal–organic framework. Appl Organomet Chem 34(5): e5581

Ugwoke MI et al (2005) Nasal mucoadhesive drug delivery: background, applications, trends and future perspectives. Adv Drug Deliv Rev 57(11):1640–1665

Ullah I et al (2020) Nose-to-brain delivery of cancer-targeting paclitaxel-loaded nanoparticles potentiates antitumor effects in malignant glioblastoma. Mol Pharm 17(4):1193–1204

Upadhyay S et al (2011) Intranasal drug delivery system-a glimpse to become maestro. J Appl Pharm Sci 1(03):34–44

Urtti A (2006) Challenges and obstacles of ocular pharmacokinetics and drug delivery. Adv Drug Deliv Rev 58(11):1131–1135

Van Den Berg MP et al (2003) Hydroxocobalamin uptake into the cerebrospinal fluid after nasal and intravenous delivery in rats and humans. J Drug Target 11(6):325–331

Van Den Berg MP et al (2004) Uptake of melatonin into the cerebrospinal fluid after nasal and intravenous delivery: studies in rats and comparison with a human study. Pharm Res 21(5): 799–802

Wang X, Chi N, Tang X (2008) Preparation of estradiol chitosan nanoparticles for improving nasal absorption and brain targeting. Eur J Pharm Biopharm 70(3):735–740

Warnken ZN et al (2016) Formulation and device design to increase nose to brain drug delivery. J Drug Delivery Sci Technol 35:213–222

Wu H et al (2012) A novel small Odorranalectin-bearing cubosomes: preparation, brain delivery and pharmacodynamic study on amyloid-β25–35-treated rats following intranasal administration. Eur J Pharm Biopharm 80(2):368–378

Xia H et al (2011) Low molecular weight protamine-functionalized nanoparticles for drug delivery to the brain after intranasal administration. Biomaterials 32(36):9888–9898

Yadav S et al (2015) Comparative biodistribution and pharmacokinetic analysis of cyclosporine-a in the brain upon intranasal or intravenous administration in an oil-in-water nanoemulsion formulation. Mol Pharm 12(5):1523–1533

Yang T et al (2004) Cyclodextrins in nasal delivery of low-molecular-weight heparins: in vivo and in vitro studies. Pharm Res 21(7):1127–1136

Yasir M, Sara UVS (2014) Solid nanoparticles for nose to brain delivery of haloperidol: in vitro drug release and pharmacokinetics evaluation. Acta Pharm Sin B 4(6):454–463

Ying, W., The nose may help the brain: intranasal drug delivery for treating neurological diseases. 2008.

Yokoyama M (2005) Drug targeting with nano-sized carrier systems. J Artif Organs 8(2):77–84

Zaki N et al (2006) Rapid-onset intranasal delivery of metoclopramide hydrochloride: Part I. Influence of formulation variables on drug absorption in anesthetized rats. Int J Pharm 327(1-2):89–96

Zhang C et al (2014) Intranasal nanoparticles of basic fibroblast growth factor for brain delivery to treat Alzheimer's disease. Int J Pharm 461(1-2):192–202

Zhao Y-Z et al (2014) Gelatin nanostructured lipid carriers-mediated intranasal delivery of basic fibroblast growth factor enhances functional recovery in hemiparkinsonian rats. Nanomed Nanotechnol Biol Med 10(4):755–764

Zhao Y-Z et al (2016) Intranasal delivery of bFGF with nanoliposomes enhances in vivo neuroprotection and neural injury recovery in a rodent stroke model. J Control Release 224: 165–175

Zheng X et al (2015) Intranasal H102 peptide-loaded liposomes for brain delivery to treat Alzheimer's disease. Pharm Res 32(12):3837–3849

Environmental Nanomedicine

22

Nishant Srivastava, Gourav Mishra, and Shailendra K. Saxena

Abstract

A healthy environment is an utmost imperative requirement for the survival of human beings. Exaggerated deforestation, industrialization, waste generation, and negligent waste management contribute to severe pollution and high discharge of toxic hazardous materials in the environment. These hazardous environmental pollutants are the major cause of several severe diseases. Diseases associated with air, water, and soil pollution are arising exponentially. Scientists and medical professionals are witnessing the number of cases on a regular basis due to environmental problems. Availability of fresh air to breathe and clean water to drink seems like a dream nowadays. There is an urgent need to find a sustainable solution for these problems so that uncompromised development can be achieved. Nanotechnology seems like a ray of hope to overcome these environmental problems with a sustainable solution. Wastewater treatment, nanofiltration, nanocatalysis, photocatalysis, and nano nasal filters are some of the widely explored research areas to counter environmental challenges. The antimicrobial properties of nanoparticles can be further utilized for the complete remediation of wastewater. Waste decomposition and degradation using nanoparticles is of high use and is an important technological advancement in waste remediation and management. Nanosensors and biosensors provide better monitoring and diagnostic solutions for the management of environmental problems. In the present chapter, a brief description of environmental problems,

N. Srivastava · G. Mishra
Department of Biotechnology, Meerut Institute of Engineering and Technology, Meerut, India

S. K. Saxena (✉)
Centre for Advanced Research (CFAR), Faculty of Medicine, King George's Medical University (KGMU), Lucknow, India
e-mail: shailen@kgmcindia.edu

A. Thakur et al. (eds.), *Synthesis and Applications of Nanoparticles*,
https://doi.org/10.1007/978-981-16-6819-7_22

their effects on human health, and solutions by implying nanotechnology is provided.

Keywords

Environment · Diseases · Pollution · Nanotechnology · Nanomedicine

22.1 Introduction

Environment-related issues and their impact on health have become a major concern for scientists. Exponentially deteriorating environmental and climatic conditions are becoming a major contributor to various types of diseases. The continuous discharge of waste, toxic, nonbiodegradable, and organic-inorganic pollutants in water, air, and soil, is a serious concern for life on Earth. Since fresh air and clean water together make the core of a healthy environment for the existence of living organisms on Earth, conservation, sustainable obtainability, and limpidness of air and water cannot be disregarded. According to the World Health Organization (WHO), 24% of global diseases and 33% of diseases in children under 5 are caused due to environmental exposures which can be prevented (WHO 2006). The major noncommunicable diseases in the world are predominantly associated with detrimental environmental situations. The chief vulnerable population groups from poor environment situations are of age below 5 years and above 50 years. Approximately 25% of diseases can be prevented by applying proper environment prevention measures (Prüss-Ustün et al. 2016).

The growing harmful environmental situation is required to be addressed categorically for sustainable and efficient solutions. To achieve the efficient solution, five categories were identified to address the challenges and enhance global health (Whitmee et al. 2015).

i. **Awareness Challenge:** Lack of awareness and knowledge about sanitization, cleanliness, and purity, pollution hazard, and toxicity affect major sectors of population especially those living in underdeveloped and developing nations. Furthermore, the negligent and lethargic attitude of peoples in developed and developing countries is also responsible for the deteriorating environmental condition and health. Additionally, more focus on economic growth by improving gross domestic products is a major part of human development, but human health and nature are left behind.
ii. **Research and Education Challenge:** Lack of scientific interest, unavailability of requisite funds, less interest and inefficiency to understand and solve problems, very little interdisciplinary research, lack of scientific collaborations globally and locally, the gap between industries and educational institutions, knowledge gap, and lack of communication between the scientific community with educational institutes and students are major hurdles in addressing environmental issues efficiently.

iii. **Population Challenge:** The increasing population is also one of the major challenges which directly affect the environment and human health. To meet the demand for food, shelter, and other essential things, humans accelerate deforestation, industrialization, and other related activities like increasing transportation modes and others, causing heavy domestic discharge, pollution in the ocean, microplastic pollution, etc. predominantly damaging the health of humans itself.
iv. **Society Challenge:** The negligent and irresponsible behavior of society toward their environment for economic growth and comfortable life with lack of scientific knowledge and concepts of sustainable development and environment conservation systems leads the human race toward their extinction in the near future.
v. **Implementation Challenges:** The challenge of implementing remedies and scientific techniques for a prosperous environment and good health is also a very crucial task. The delay in recognition and response for environment related issues by governments and responsible organizations to intimidations causing severe health-related issues. The time gap between action and effect plays a very important role in controlling the situation. The dispassion and apathy of governments and related organizations toward environment-associated health impacts impose severe concern for our only habitable planet.

The adoption of green methods, elimination of harmful elements, and implementation of sustainable development to reduce environmental risks are the key to good health. The present chapter is a little effort to associate and provide a solution for environment-related issues with the help of nanotechnology for a better environment and good health.

22.2 Environmental Challenges and Effect on Human Health

The twenty-first century is believed to be the era of science and technology with the help of innovation and research for making human life more comfortable and better. But in this race of development, we left our environment behind and suffered immense environmental losses. The global race of industrialization to become an economic superpower makes our society so materialistic, selfish, and negligent that starts harming its species and whole planet harshly. From radioactive pollution on Earth to garbage of satellites and space crafts in outer space, human species are damaging the environment and nature without considering their survival. The exponentially growing industrialization, motor vehicles, industrial and domestic discharge into major water bodies, high use of chemical fertilizers, etc. are major contributors to the environmental crisis. Figure 22.1 is a depiction of some of the important sources contributing to the environmental crisis due to human activity.

The poor environmental condition promotes various diseases associated with various pollution-causing agents. Gastrointestinal diseases, diarrhea, anemia, various types of cancer like pulmonary cancer, liver cancer and failure, skin cancer,

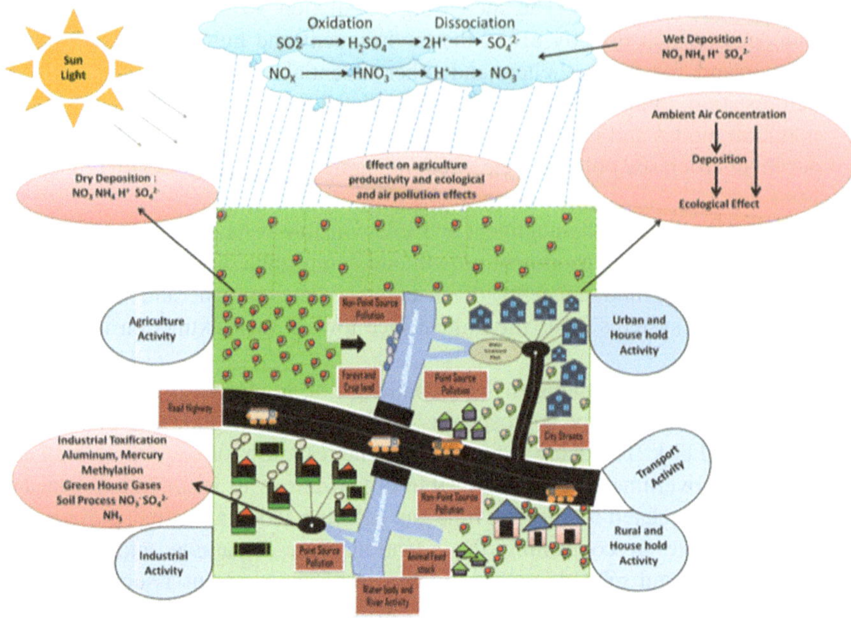

Fig. 22.1 Various sources of environmental pollution contributing to environment-associated illness

asthma, hepatitis, etc. are some of the health-related issues from a very large list that occurred due to pollution of water and air, radiation, hospital waste, inefficient waste management, etc. and responsible many deaths globally (Table 22.1).

22.3 Diseases Associated with Environmental Problems

As depicted in Table 22.1, various types of diseases are directly and indirectly associated with environmental pollution and depleting environmental conditions. Diseases associated with water, soil, and air pollution directly affect major populations globally and account for major deaths worldwide. Nature's life support system is under threat due to the unsustainable development activities like exploitation of natural resources, climate change, biodiversity forfeiture or damage, land degradation, deforestation, water insufficiency and demeaning quality of drinking water, pollution and acidification of ocean water, and radiation pollution impacting global health and posing serious health and life risks on the planet (Whitmee et al. 2015).

The environmental health hazards, namely, physical, chemical, and biological hazards, are respectively affecting human health majorly by using pollution as a vector. The physical hazards consist of hazardous impact on living organisms due to climate change, rise in temperature, forest fires, radioactive pollution, stress due to

Table 22.1 Diseases associated with environmental crises

Environmental crises	Associated diseases	Causative agent	References
Water	Gastrointestinal and noncommunicable diseases (NCDs)	Toxic insecticides and herbicides	Landrigan et al. (2019)
	Diarrhea, anemia, kidney damage, edema	Mercury (Hg), cadmium (Cd), arsenic (As), chromium (Cr), and lead (Pb)	Chouhan and Flora (2010), Alengebawy et al. (2021)
	Cancer	Nitrophenol compounds	Gopal et al. (2019)
	Cancer and Mutagens	Aromatic hydrocarbons	Andelman and Suess (1970), Moslen et al. (2019)
Air	Pulmonary cancer	Sulfur dioxide (SO_2), nitrogen oxides (NOx), carbon monoxide (CO), particulate matter (PM)	Shahadin et al. (2018)
	Asthma attacks	NO_2 and PM levels are highly associated with airflow obstruction	Burbank and Peden (2018)
	Chronic obstructive pulmonary disease—COPD	Second-hand smoke and dust, fumes	Jiang et al. (2016)
	Pneumonia	CO, black carbon (BC), and $PM_{2.5}$	Adaji et al. (2019)
Radioactive	Carcinogenic mutation at the genetic level	Radium and uranium: beta particles and gamma rays	Tsivoglou and Towne (1957)
Hospital and medical waste	Skin infections	Antibiotics, synthetic hormones, anti-inflammatories, statins, and cytotoxins	Quesada et al. (2019)
	Hepatitis B and hepatitis C	Contaminated infected blood, semen, or other bodily fluids, needles	Acharya (2013)
	Bacteremia	Sharps and other medical waste instruments	Moore et al. (2017)
	Meningitis and candida	Waste generated from hospitals	Alwabr et al. (2016)
Other factors	Myeloid leukemia	Gasoline, benzene	Mundt et al. (2021)
	Skin disease	Xenobiotic organic chemicals	Steffan et al. (2018), de Oliveira et al. (2020)

environmental change, and impact on behavioral and mental health due to change in the external atmosphere. The chemical hazard includes hazardous organic and inorganic chemical compounds and elements causing severe damage to health. Pollution of arsenic, fluoride, phenols, azo dyes, heavy metals, etc. in water from industrial wastewater discharge leads to several diseases like cancer, skin disease, kidney failure, liver diseases, reproductive issues, and gastrointestinal issues (Srivastava and Mukhopadhyay 2015a, 2015b, 2015c). Furthermore, the release of toxic fumes, toxic gases like carbon monoxide, carbon dioxide, sulfur dioxide, nitrogen oxides, particulate matters, etc., in the air causes various respiratory, cardiac, and urinary tract-related problems (Kampa and Castanas 2008). The biological hazards include microorganisms, parasites, fungi, viruses, and amoeba-related infections caused due to contaminated water, air, and surroundings. Open desiccation; open and frequent domestic discharge in rivers, ponds, lakes, and oceans; and contaminated groundwater cause various infections of biological origin. Viral and bacterial hepatitis; fibrosis; kidney diseases; typhoid; vector-borne diseases like malaria, dengue, chikungunya, filariasis, diarrhea, and cholera; and many more diseases occur due to consumption of contaminated water and food and lack of sanitization practices.

Diarrhea is the second important cause of death in children below 5 years majorly caused by two pathogens *Rotavirus* and *Escherichia coli*. Furthermore, cholera remains endemic in 69 countries infecting approximately 2.9 million people and claiming 95,000 lives per year worldwide. Additionally, 1 billion people worldwide suffer from neglected tropical diseases. As per WHO, approximately half of the population of the world is at risk due to vector (mosquito)-borne disease dengue which occurs due to stagnant water (WHO 2019).

Another cause of high mortality and hospitalization is air pollution. There are several natural and human-mediated activities causing air pollution. Volcanoes activity, forest fire, fire, toxic gases, etc. cause pollution to a small degree, but the major and devastating contributor of air pollution is human activity such as burning of fossil fuel. Particulate matters (PM_{10} and $PM_{2.5}$), volatile organic compounds, toxic gases, dioxins, ultrafine particles (>0.1 μM), and heavy metals are majorly released in the air from industrial activity, power plants, incinerators, burning of plastics, construction activity, and motor vehicles. These air pollutants cause illness via inhalation and ingestion in the human body. The disease occurs due to inhalation and ingestion of these pollutants and includes respiratory illness like asthma, fibrosis of lungs, silicosis, reduction in lung function, irritation in the throat, cough, cardiac problems, neurological disorders, liver cell damage, gastrointestinal and liver cancer, increase oxidative stress and inflammatory response at the cellular level, etc. that occur in humans.

Additionally, the pollution of air and water can cause severe problems in pregnant women and fetuses. Pollutants like heavy metals, dioxins, and arsenic can cause termination of pregnancy, slow or abridge growth of the fetus, damage newborns' motor and cognitive skills, and cause issues in the development of central nervous systems and disruption in endocrine activities (Kampa and Castanas 2008).

Environmental pollution of all kinds is very dangerous for all the life forms on Earth. This pollutant affects living organisms including humans from genetic to molecular to organ level and causes severe damages which may be permanent or irreversible. Few examples of man-led activity that turned into devastating accidents are the atomic bombing at Hiroshima and Nagasaki, the nuclear reactor explosion in

Chernobyl, Russia, and the gas tragedy of Bhopal due to leakage of methyl isocyanide from the plant of Union Carbide in Bhopal, India, showing their aftereffects even after several decades of their occurrence. The effect is so deep that newborns in these regions are stillborn with various mental and physical disabilities and people to date developing and suffering various diseases like cancer, neurological problems, cardiac issues, endocrine dysfunction, and lung-related diseases. The water, land, and air are badly polluted with toxic pollutants and radioactive substances.

On the other hand, various cases of radioactive poisoning are reported worldwide due to radioactive materials discharged in the ocean from nuclear power plants. These radioactive pollutants cause severe damage to aquatic life of the ocean and also affect humans directly and indirectly (through contaminated seafood).

22.4 Nanoparticles for Diagnostics and Treatment of Environmental and Pollution-Associated Diseases

Nanotechnology emerged as a hope of having a possible solution for the problems associated with health due to environmental crises. The advancements and development in the area of nanotechnology provide tremendous opportunities to the researchers for creating novel tools to analyze, address, and provide a sustainable solution for environmental remediation. The utilization of nanotechnology in the environment and medicine provides an opportunity to develop advanced tools that are smart, highly specific, accurate, easy to handle and carry, fast, and robust. Nanotechnology has a high potential to find the answer for countless environmental problems like wastewater remediation, pollutant degradation, desalination, air purification, and other pollutants and waste management. The small size and the high surface-to-volume ratio of nanomaterial lead to its proper applications in the environment and medicine (Srivastava and Saxena 2020). The small size of nanomaterial also eases its smooth and specific action toward the target like nanocatalysts, nanocomposites, and antimicrobials. Table 22.2 provides some of the nanotechnology-based solutions for the environmental crisis.

22.4.1 Advanced Oxidation Process (AOP) or Photocatalytic Oxidation

Nanomaterials are recognized as an efficient catalyst to their high surface area and high surface reactivity. These properties facilitate researchers to use nanomaterials for removal and/or degradation of pollutants (chlorophenols, dyes, nitrophenols, etc.) from water by applying techniques like advanced oxidation processes (Srivastava and Mukhopadhyay 2015a, 2015b, 2015c). Due to complex structure and high stability, pollutants like phenols and dyes are very difficult to remove or degrade completely from effluents (Srivastava and Mukhopadhyay 2014). As depicted in Fig. 22.2, the advanced oxidation process (AOP) or photocatalytic

Table 22.2 Nanotechnology-based solution for environmental crises

S. no.	Crises	Type of nanomaterial	Mechanism	Solution	References
1.	SARS-CoV-2	Titanium oxide (TiO_2) and silver (Ag)-zeolite	Metallic coatings (ROS mediated) and ROS generation on the surface of the NPs exposed to UV radiation	Disinfection-inhibition of interaction with viral glycoproteins	Talebian et al. (2020)
2.	Removal of bacteria	Halloysite nanotube (HNT) and TiO_2	Mixed matrix polymeric ultrafiltration membrane	*E. coli* bacteria removal efficiency approximately 99%	Mishra and Mukhopadhyay (2017), Mishra and Mukhopadhyay 2019)
3.	Detection of pesticides	Immobilization of Ag and TiO_2 on the surface of nanotubes	Fluorometric detection-based sensor	Dimethoate and parathion pesticide detection level approximately 0.01 mg/mL and 0.03 mg/L, respectively	Saraji et al. 2016, Hsu et al. (2017)
4.	Detection of dengue	Porous silicon microcavity (PSMC)	Bioconjugation of PSMC to label-free biosensor	Efficacy of dengue NS1 Ag detection ~90%	Mishra et al. (2016)
5.	Degradation of coloring dye	Tetragonal SnO_2 nanoparticles	ROS generation on the surface of the NPs exposed to UV radiation	Degradations of methylene blue, eriochrome black T, and methyl orange dye are ~ 93.3%, 97.8%, and 94.0%	Srivastava and Mukhopadhyay (2014)
6.	Detection of heavy metals	Arsenic-binding aptamer (Ars-3 aptamer) as a molecular recognition element	Label-free LC-based aptasensor, specific arsenic aptamer optical (polarized light microscope)	Aptasensor displayed a low detection limit of 50 nM (~3.7 ppb) toward arsenic	Nguyen and Jang (2020)
7.	Degradation of phenols	Gold nanoparticles as a catalyst	ROS generation on the surface of the NPs exposed to UV radiation	Degradation of 2,4-dichlorophenols	Srivastava and Mukhopadhyay (2015a, 2015b, 2015c)

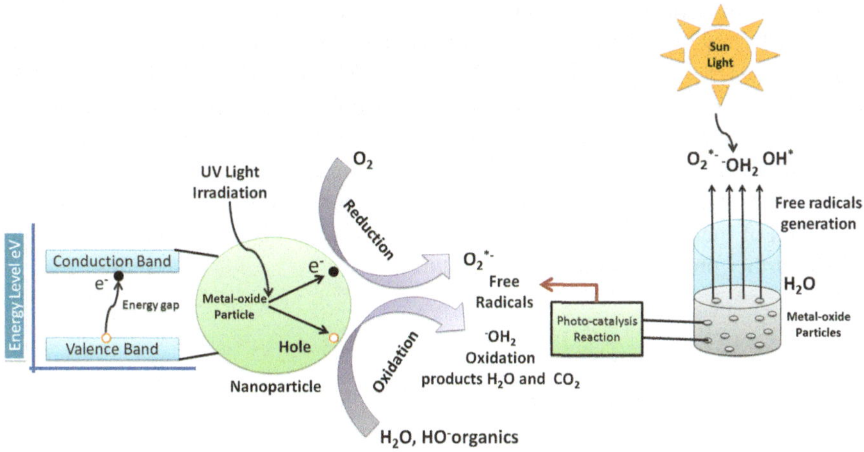

Fig. 22.2 Mechanism of photocatalytic oxidation of water pollutants

oxidation in presence of nanomaterials as a catalyst has the potential to remove these pollutants from the water completely (Smith et al. 2009). In the AOP, nanocatalyst is irradiated using UV radiation and activated for light-induced redox reaction by forming a redox environment in a water sample. The AOP for the degradation of pollutants is initiated by the photoexcitation process of nanoparticles. The surface of nanoparticles is irradiated with light energy (UV light) due to the irradiation conduction band electrons and valence band holes created. The valence band holes were trapped on the surface of the hydroxyl group at the nanocatalyst to provide highly oxidizing species OH^- radicals. Superoxide radical anions O_2^- are formed due to the reaction of dissolved oxygen molecules with conduction band electrons. On protonation, superoxide radical anions O_2^- produce hydroxyl radicals, HO_2. Finally, pollutants are found to be degraded completely because of the action of superoxide anions. The conduction band electrons and the valence band holes instigated the oxidation and reduction of pollutants in water (Smith et al. 2009; Tian et al. 2011; Lee et al. 2012; Srivastava and Mukhopadhyay 2014).

22.4.2 Antimicrobial Nanomaterials

Antimicrobial property of nanomaterials is well explored, and many researchers documented antimicrobial properties of nanoparticles such as silver, gold, selenium, copper, zinc, nickel, carbon and graphene, metal oxides, polymeric nanomaterials, and organic nanomaterials (Santoro et al. 2007; Singh et al. 2013; Srivastava and Mukhopadhyay 2015a, 2015b, 2015c; Raghunath and Perumal 2017; Shoeibi and Mashreghi 2017; Yousefi et al. 2017; Saha and Bandyopadhyay 2019; Rathinavel et al. 2020). The antimicrobial property of nanomaterials provides an opportunity to develop nonmaterial-based filtration membranes, composite materials, potential decontamination pills, water purifiers, and desalination plants. Researchers reported the *E. coli* bacteria removal efficiency of approximately 99% by using a mixed matrix polymeric ultrafiltration membrane containing halloysite nanotubes and TiO_2 nanoparticles (Mishra and Mukhopadhyay 2017; Mishra and Mukhopadhyay 2019). In another innovative research, scientists reported the disinfection and inhibition of viral glycoproteins by applying ROS-mediated metallic coatings and generating ROS on the surface of the nanoparticles by exposing them to UV radiation (Talebian et al. 2020). Nanomaterials were found to be very useful to prevent rusting and oxidation of metals by applying a coating of nanomaterials in the form of nanofilm which further prevents the formation of biofilms.

22.4.3 Nanomaterials for Air Pollution

Air pollution is one of the major environmental problems of the twenty-first century. The release of various pollutants in the air like particulate matter and toxic gases such as carbon monoxide, chloroform carbons, sulfur dioxide, nitrogen oxide, ozone, hydrocarbons, and volatile organic compounds causes a wide range of diseases (Yunus et al. 2012; Ibrahim et al. 2016). Another major problem that arises from air pollution is global warming due to the increase in greenhouse gases. Carbon dioxide is the major greenhouse gas (75% of all greenhouse gas), and various techniques have been under research for its emission control such as absorption on liquid, adsorption on solid, filtration, or a combination of these processes. The adsorption of carbon dioxide was found to be more efficient and economical on nanomaterial surfaces due to the high surface area and high surface-to-volume ratio. Nano adsorbents like TiO_2-coated nano calcium carbonate, Ca-based nano absorbents, lithium orthosilicate-based absorbents, and lithium orthosilicate-based (CRHA-Li_4SiO_4) adsorbents are used to adsorb carbon dioxide at a high temperature of 700 ° C. Moreover, the alkali nanotubes (e.g., potassium titanate (K-Ti-NT) and sodium titanate (Na-Ti-NT)) have been used to capture carbon dioxide at a temperature below 200 °C. Another greenhouse gas of concern is methane which can be thermally decomposed or degraded into hydrogen employing nickel or TiO_2 nanoparticles as catalyst (Ibrahim et al. 2016). Another major pollutant of air is NOx. The NOx can be treated by adsorption on carbon nanotubes, ruthenium nanoparticles, or zinc oxide nanoparticles or by the help of photocatalytic oxidation in presence of titanate nanotubes. Magnesium ferrite nanospheres, TiO_2 nanoparticles, magnetic nanoparticles, etc. were reported for remediation of sulfur dioxide from the air. Silica nanoparticles are reported to capture atmospheric lead, whereas TiO_2 nanoparticles can also be used for capturing volatile organic compounds (Baltrusaitis et al. 2011; Komvokis et al. 2011; Srisitthiratkul et al. 2011; Singh et al. 2012; Zhao et al. 2012; Yang et al. 2013; Hussein 2015; Celebioglu et al. 2016; Ibrahim et al. 2016). The innovation and efficient technological advancements in nanotechnology have the potential to control the pollution in the air and provide better air for the future generation.

22.4.4 Other Nano-Remediation Techniques

Nanofibers and nanocomposite membranes were found to be a better option for water filtration. Biofouling of membrane is a major problem in water purification membranes, but membrane composites of polyvinyl alcohol and polyacrylonitrile containing silver nanoparticles are found to have great antimicrobial activity of 91% and 99%, respectively (Yunus et al. 2012). These types of nanofiltration membranes are very beneficial for the removal of bacteria from contaminated water.

Another nanomaterial is the zerovalent iron nanoparticles which are utilized for in situ reductive immobilization of heavy metals in the soil as well as for treatment of groundwater in large quantities (Yunus et al. 2012; Ibrahim et al. 2016).

22.4.5 Nanosensors for Diagnosis and Monitoring

The continuous monitoring and analysis of the environment is another crucial yet important task. Nowadays nanosensors are taking much attention to monitoring the environment because of their portable size, low cost, accuracy, ultra-sensitivity, smart performance, and fast output. Nanomaterial-based electrochemical and gas sensors such as carbon-based nanomaterial, metal oxide-based nanosensors (ZnO, CdO, etc.), and optically active palladium nanoparticle-based volatile organic compound detectors are used for detection of pollutants in the air like NH_3, polyimide, volatile organic compound, humidity, etc. Similarly, optical-based nanosensors of gold nanoparticles, electrochemical sensors from fabricated carbon nanotubes, carbon nanotube-based sensors, etc. can be used for the detection of mercury and organophosphate pesticides in water, waterborne pathogens (pathogenic bacteria), and weather parameters respectively. Additionally, a microcantilever-based system can be used to determine the carbon in the soil. The nano-biosensors consist of two components:

i. **Receptor:** used to improve and increase detection sensitivity
ii. **Transducer:** sensing component (nanomaterial) either chemical or physical sense component that works with optical, thermal, biological, and electrochemical detection principles

Nanosensors can be connected with Bluetooth and GPS and facilitate the monitoring of requisite samples or sites remotely. These nanosensors can also be very useful for monitoring the parameters like pH, humidity, chemical pollutants, pathogens, toxic gases, and hazardous elements (Ibrahim et al. 2016; Debnath and Das 2020). The early and efficient analysis of the very minimal sample can revolutionize the way of environmental monitoring. The use of nanosensors for the environment will also help in timely detection of the problem, planning to deal with the problem, and timely execution of solution with further progress monitoring.

22.5 Conclusions

Human negligence and convulsive activities lead to a poor environment and severe health problems. The high amount of anthropogenic toxic pollutants in the environment causes severe water, air, and soil pollution. These pollutants further directly and indirectly affect human health. More than a quarter of deaths worldwide are occurring due to environmental crises. Many young children lost their lives due to poor air and water quality, lack of sanitation, and bad food. There is an urgent need to make people aware of the importance of a clean and healthy environment, implementation of laws and government policies needs to be executed timely, and extensive research must be carried out for solution and preservation of our environment. Nanotechnology emerges as one of the sustainable solutions to tackle present environmental problems. Nanotechnology has been providing answers for major environmental problems like waste management, wastewater remediation, availability of clean water, sanitation, air pollution remediation, improvement in soil quality for maximum fertility, etc. Nanotechnology shows promising features for fast, economical, efficient, and advanced processes for pollution handling and management. Nanotechnology can lead humanity toward a clean environment with zero waste generation and ensure the availability of clean and fresh water and air. The nanosensors in near future revolutionize the monitoring and diagnosis processes by making them more robust, accurate, noninvasive, economical, and rapid. Nanosensors provide analysis of toxic and harmful chemicals, gases, pathogenic microbes, and organic pollutants for proper management and remedy. Though nanotechnology has many positive aspects, its dark side also needs to be studied in detail. The impact of nanomaterials on the environment and human health needs to be explored deeply as long-term toxicity assessment has yet not been done.

Acknowledgments The authors are grateful to the Vice-Chancellor, King George's Medical University (KGMU), Lucknow, for the encouragement and support for this work. The authors have no other relevant affiliations or financial involvement with any organization or entity with a financial interest in or financial conflict with the subject matter or materials discussed in the manuscript apart from those disclosed.

References

Acharya SK (2013) This is hepatitis: know it, confront it. Indian J Med Res 138(1):8

Adaji EE, Ekezie W, Clifford M, Phalkey R (2019) Understanding the effect of indoor air pollution on pneumonia in children under 5 in low- and middle-income countries: a systematic review of evidence. Environ Sci Pollut Res Int 26(4):3208–3225

Alengebawy A, Abdelkhalek ST, Qureshi SR, Wang M-Q (2021) Heavy metals and pesticides toxicity in agricultural soil and plants: Ecological risks and human health implications. Toxics 9(3):42

Alwabr GM, Al-Mikhlafi AS, Al-Hakimi SA, Dughish MA (2016) Identification of bacteria and fungi in the solid waste generated in hospitals of Sana'a city, Yemen. Curr Life Sc 2(3):67–71

Andelman JB, Suess MJ (1970) Polynuclear aromatic hydrocarbons in the water environment. Bull World Health Organ 43(3):479

Baltrusaitis J, Jayaweera PM, Grassian VH (2011) Sulfur dioxide adsorption on TiO2 nanoparticles: influence of particle size, coadsorbates, sample pretreatment, and light on surface speciation and surface coverage. J Phys Chem C 115(2):492–500

Burbank AJ, Peden DB (2018) Assessing the impact of air pollution on childhood asthma morbidity: how, when and what to do. Curr Opin Allergy Clin Immunol 18(2):124

Celebioglu A, Sen HS, Durgun E, Uyar T (2016) Molecular entrapment of volatile organic compounds (VOCs) by electrospun cyclodextrin nanofibers. Chemosphere 144:736–744

Chouhan S, Flora S (2010) Arsenic and fluoride: two major groundwater pollutants. Indian J Exp Biol 48:666–678

Debnath N, Das S (2020) Nanobiosensor: current trends and applications. In: Saxena SK, Khurana SMP (eds) NanoBioMedicine. Springer Singapore, Singapore, pp 389–409

Gopal K, Mohd N, Raoov M, Suah F, Yahaya N, Zain N (2019) Development of a new efficient and economical magnetic sorbent silicone surfactant-based activated carbon for the removal of chloro- and nitro-group phenolic compounds from contaminated water samples. RSC Adv 9(63):36,915–36,930

Hsu C-W, Lin Z-Y, Chan T-Y, Chiu T-C, Hu C-C (2017) Oxidized multiwalled carbon nanotubes decorated with silver nanoparticles for fluorometric detection of dimethoate. Food Chem 224: 353–358

Hussein AK (2015) Applications of nanotechnology in renewable energies—a comprehensive overview and understanding. Renew Sustain Energy Rev 42:460–476

Ibrahim RK, Hayyan M, AlSaadi MA, Hayyan A, Ibrahim S (2016) Environmental application of nanotechnology: air, soil, and water. Environ Sci Pollut Res 23(14):13,754–13,788

Jiang X-Q, Mei X-D, Feng D (2016) Air pollution and chronic airway diseases: what should people know and do? J Thorac Dis 8(1):E31–E40

Kampa M, Castanas E (2008) Human health effects of air pollution. Environ Pollut 151(2):362–367

Komvokis VG, Marti M, Delimitis A, Vasalos IA, Triantafyllidis KS (2011) Catalytic decomposition of N2O over highly active supported Ru nanoparticles ($\leq$3nm) prepared by chemical reduction with ethylene glycol. Appl Catal Environ 103(1):62–71

Landrigan PJ, Fuller R, Fisher S, Suk WA, Sly P, Chiles TC, Bose-O'Reilly S (2019) Pollution and children's health. Sci Total Environ 650:2389–2394

Lee JS, Kwon OS, Jang J (2012) Facile synthesis of SnO2 nanofibers decorated with N-doped ZnO nanonodules for visible light photocatalysts using single-nozzle co-electrospinning. J Mater Chem 22(29):14,565–14,572

Mishra G, Mukhopadhyay M (2017) Flux improvement, rejection, surface energy and antibacterial properties of synthesized TiO2-Mo. HNTs/PVC nanocomposite ultrafiltration membranes. New J Chem 41(24):15,049–15,057

Mishra G, Mukhopadhyay M (2019) TiO 2 decorated functionalized halloysite nanotubes (TiO 2@ HNTs) and photocatalytic PVC membranes synthesis, characterization and its application in water treatment. Sci Rep 9(1):1–17

Mishra V, Patel P, Kumari S, Mishra G (2016) Dengue NS1 detection used chemically modified porous silicon microcavity (psmc). Silicon 8(3):401–407

Moore CA, Khalid MF, Patel PD, Goldstein JS (2017) Aeromonas salmonicida bacteremia associated with chronic well water consumption in a patient with diabetes. J Glob Infect 9(2):82

Moslen M, Miebaka C, Boisa N (2019) Bioaccumulation of Polycyclic Aromatic Hydrocarbon (PAH) in a bivalve (Arca senilis-blood cockles) and health risk assessment. Toxicol Rep 6:990–997

Mundt KA, Dell LD, Boffetta P, Beckett EM, Lynch HN, Desai VJ, Lin CK, Thompson WJ (2021) The importance of evaluating specific myeloid malignancies in epidemiological studies of environmental carcinogens. BMC Cancer 21(1):227–227

Nguyen DK, Jang C-H (2020) Label-free liquid crystal-based detection of As(III) ions using ssDNA as a recognition probe. Microchem J 156:104,834 de Oliveira M, Frihling BEF, Velasques J, Magalhães Filho FJC, Cavalheri PS, Migliolo L (2020) Pharmaceuticals residues and xenobiotics contaminants: occurrence, analytical techniques and sustainable alternatives for wastewater treatment. Sci Total Environ 705:135568

Prüss-Ustün A, Wolf J, Corvalán C, Neville T, Bos R, Neira M (2016) Diseases due to unhealthy environments: an updated estimate of the global burden of disease attributable to environmental determinants of health. J Public Health 39(3):464–475

Quesada HB, Baptista ATA, Cusioli LF, Seibert D, de Oliveira Bezerra C, Bergamasco R (2019) Surface water pollution by pharmaceuticals and an alternative of removal by low-cost adsorbents: a review. Chemosphere 222:766–780

Raghunath A, Perumal E (2017) Metal oxide nanoparticles as antimicrobial agents: a promise for the future. Int J Antimicrob Agents 49(2):137–152

Rathinavel L, Jothinathan D, Sivasankar V, Mylsamy P, Omine K, Selvarajan R (2020) Algal nanoparticles: boon for antimicrobial therapeutic applications. In: Prasad R, Siddhardha B, Dyavaiah M (eds) Nanostructures for antimicrobial and antibiofilm applications. Springer International Publishing, Cham, pp 127–143

Saha M, Bandyopadhyay PK (2019) Green biosynthesis of silver nanoparticle using garlic, allium sativum with reference to its antimicrobial activity against the pathogenic strain of Bacillus sp. and Pseudomonas sp. infecting Goldfish, Carassius auratus. Proc Zool Soc 72(2):180–186

Santoro CM, Duchsherer NL, Grainger DW (2007) Minimal in vitro antimicrobial efficacy and ocular cell toxicity from silver nanoparticles. NanoBiotechnology 3(2):55–65

Saraji M, Jafari MT, Mossaddegh M (2016) Halloysite nanotubes-titanium dioxide as a solid-phase microextraction coating combined with negative corona discharge-ion mobility spectrometry for the determination of parathion. Anal Chim Acta 926:55–62

Shahadin MS, Ab. Mutalib NS, Latif MT, Greene CM, Hassan T (2018) Challenges and future direction of molecular research in air pollution-related lung cancers. Lung Cancer 118:69–75

Shoeibi S, Mashreghi M (2017) Biosynthesis of selenium nanoparticles using Enterococcus faecalis and evaluation of their antibacterial activities. J Trace Elem Med Biol 39:135–139

Singh J, Mukherjee A, Sengupta SK, Im J, Peterson GW, Whitten JE (2012) Sulfur dioxide and nitrogen dioxide adsorption on zinc oxide and zirconium hydroxide nanoparticles and the effect on photoluminescence. Appl Surf Sci 258(15):5778–5785

Singh M, Kumar M, Kalaivani R, Manikandan S, Kumaraguru AK (2013) Metallic silver nanoparticle: a therapeutic agent in combination with the antifungal drug against human fungal pathogen. Bioprocess Biosyst Eng 36(4):407–415

Smith YR, Kar A, Subramanian V (2009) Investigation of physicochemical parameters that influence photocatalytic degradation of methyl orange over TiO2 nanotubes. Indus Eng Chem Res 48(23):10,268–10,276

Srisitthiratkul C, Pongsorrarith V, Intasanta N (2011) The potential use of nanosilver-decorated titanium dioxide nanofibers for toxin decomposition with antimicrobial and self-cleaning properties. Appl Surf Sci 257(21):8850–8856

Srivastava N, Mukhopadhyay M (2014) Biosynthesis of SnO2 nanoparticles using bacterium Erwinia herbicola and their photocatalytic activity for degradation of dyes. Indus Eng Chem Res 53(36):13,971–13,979

Srivastava N, Mukhopadhyay M (2015a) Biosynthesis and characterization of gold nanoparticles using Zooglea ramigera and assessment of its antibacterial property. J Cluster Sci 26(3): 675–692

Srivastava N, Mukhopadhyay M (2015b) Green synthesis and structural characterization of selenium nanoparticles and assessment of their antimicrobial property. Bioprocess Biosyst Eng 38(9):1723–1730

Srivastava N, Mukhopadhyay M (2015c) Ralstonia eutropha (Cupriavidus metallidurans) mediated biosynthesis of gold nanoparticles and catalytic treatment of 2, 4 dichlorophenol. Synthesis Reactivity Inorg Metal-Organic Nano-Metal Chem 45(2):238–247

Srivastava N, Saxena SK (2020) Opportunities in clinical translation and commercialization of nanomedicine. In: Saxena SK, Khurana SMP (eds) NanoBioMedicine. Springer Singapore, Singapore, pp 501–517

Steffan JJ, Brevik EC, Burgess LC, Cerdà A (2018) The effect of soil on human health: an overview. Eur J Soil Sci 69(1):159–171

Talebian S, Wallace GG, Schroeder A, Stellacci F, Conde J (2020) Nanotechnology-based disinfectants and sensors for SARS-CoV-2. Nat Nanotechnol 15(8):618–621

Tian Z, Liang C, Liu J, Zhang H, Zhang L (2011) Reactive and photocatalytic degradation of various water contaminants by laser ablation-derived SnOx nanoparticles in liquid. J Mater Chem 21(45):18,242–18,247

Tsivoglou EC, Towne WW (1957) Sources and control of radioactive water pollutants. Sewage Ind Waste 29(2):143–156

Whitmee S, Haines A, Beyrer C, Boltz F, Capon AG, de Souza Dias BF, Ezeh A, Frumkin H, Gong P, Head P, Horton R, Mace GM, Marten R, Myers SS, Nishtar S, Osofsky SA, Pattanayak SK, Pongsiri MJ, Romanelli C, Soucat A, Vega J, Yach D (2015) Safeguarding human health in the Anthropocene epoch: report of The Rockefeller Foundation; Lancet Commission on planetary health. Lancet 386(10007):1973–2028

WHO (2006) Almost a quarter of all disease caused by environmental exposure. https://www.who.int/mediacentre/news/releases/2006/pr32/en/

WHO (2019) Water, sanitation, hygiene, and health: a primer for health professionals. https://www.who.int/water_sanitation_health/en/

Yang X, Shen Z, Zhang B, Yang J, Hong W-X, Zhuang Z, Liu J (2013) Silica nanoparticles capture atmospheric lead: implications in the treatment of environmental heavy metal pollution. Chemosphere 90(2):653–656

Yousefi M, Dadashpour M, Hejazi M, Hasanzadeh M, Behnam B, de la Guardia M, Shadjou N, Mokhtarzadeh A (2017) Anti-bacterial activity of graphene oxide as a new weapon nanomaterial to combat multidrug-resistance bacteria. Mater Sci Eng C 74:568–581

Yunus IS, Harwin, Kurniawan A, Adityawarman D, Indarto A (2012) Nanotechnologies in water and air pollution treatment. Environ Technol Rev 1(1):136–148

Zhao W, Wang Z, Shen X, Li J, Xu C, Gan Z (2012) Hydrogen generation via photoelectrocatalytic water splitting using a tungsten trioxide catalyst under visible light irradiation. Int J Hydrogen Energy 37(1):908–915

Development of Nanostructured Material by Severe Plastic Deformation

23

Sanjeev Sharma and P. B. Sharma

Abstract

Recent statements of the Intergovernmental Panel on Climate Change (IPCC) have reported about 30 % rise in the CO_2 emissions from the transportation area since 1990. Weight reduction is thus indispensable; subsequently, growing demand for lightweight and strong materials has been generated. The core thrust of the transportation, particularly in the auto industries and aero industry, is to decrease the weight of vehicles because it will reduce manufacturing cost. However, users have various requirements to chase the completion with fulfilling the safety and luxury for human comfort. Some modules, for example, safety, refrigeration and air-conditioning, computer, and mechaelectronics, meet the user requirement, but then again, they increase weight.

Nanostructured metals with a standard grain size ranging from 100 to 250 nm have acknowledged substantial attention due to their improved and better mechanical properties, for example, elongation to failure and strength in assessment to the conventional grain size (CG) metals. In metal processing grain refinement is a significant area of research which provides the fundamental understanding of the grain growth or refinement. The famous Hall-Petch relation describes that "any material could be infinite strong if it's graine size infinite small." Severe plastic deformation is a top-down approach to produce the UFG/nanostructured metals; as the name indicates, severe plastic deformation has been carried out at room temperature or below the recrystallization temperature. It have some unique properties without change the original dimensions the sample or workpieces could be manufactured. Some of the most recognized severe plastic deformation techniques are ECAP (equal channel angular

S. Sharma (✉) · P. B. Sharma
Department of Mechanical Engineering, Amity University, Manesar, Haryana, India
e-mail: ssharma26@ggn.amity.edu

A. Thakur et al. (eds.), *Synthesis and Applications of Nanoparticles*,
https://doi.org/10.1007/978-981-16-6819-7_23

pressing), HPT (high-pressure torsion), and ARB (accumulative roll bonding). These are the most promising processes which are commonly used to fabricate the bulk UFG/nanostructured materials.

In this book chapter our work has been focused on various parameters which may affect to produce the bulk nanostructure metals by ARB, those are plastic deformation behavior, impact on its microstructures, and some other mechanical properties. The samples used in this study are commercial purity aluminum, OFHC copper, and DLP copper.

Keywords

Nanostructured material · Ultrafine grains · Severe plastic deformation · Accumulative roll bonding · Top-down approach

Abbreviations

ARB	Accumulative roll bonding
CHPT	Continuous high-pressure torsion
d	Grain size
dc	Diameter of ECAP sample
ECAP	Equal channel angular pressing
k	The channel displacement
MMC	Metal matrix composites
N	Number of passes
n	Number of cycles
n	Constant quantity which is greater than unity
P	Die pressure
Pav	Average die pressure during sintering forging
RPM	Rotation per minute
rt	Total reduction
SEM	Scanning electron microscope
SPD	Severe plastic deformation
S	Surface area
t	Thickness
t_0	Initial thickness on the strips
tr	The rolling time (steady)
UFG	Ultrafine grained
V	Volume
W_i	Internal energy dissipation
W_f	Frictional shear energy dissipation
W_v	Energy dissipation due to velocity discontinuation
W_a	Inertia energy dissipation
W_t	Energy dissipation due to traction
Wp	Plastic work done

XRD	X-ray diffraction
ε	Plastic strain
εeq	Magnitude of effective strain
γ	The total shear strain
Ø	Angle of die
ε	Strain

23.1 Introduction

Since the last three decades, nanostructured and UFG materials of grain size 10–250 nm have generated substantial attention because of their mechanical characteristic in terms of tensile strength, percentage failure to elongation, and ductility related to as-received CG materials. Those nanostructured and UFG materials can be used for future applications with high durability modules in the auto industries, aviation industries, and biomedical engineering. The UFG sheet metals can be successfully produced through recently developed technique accumulative roll bonding (ARB). This process comes in the category of top-down approach of the severe plastic deformation (SPD) process, presented by Japanese scholars Saito et al. (Saito et al. 1998). The ARB process has been conducted firstly on commercial purity aluminum. In addition to the ARB process, there are many other fabricating process of SPD techniques; by using these processes, nanostructured or UFG microstructure materials could be achieved. Later this achieved an enhancement in material strength and ductility. Equal channel angular pressing (ECAP) was originally known to Segal at the start of the 1980s (Valiev 2004, 1997), and previously recognized by Valiev and their researcher (Segal 1977) were high-pressure torsion (HPT) (Smirnova et al. 1986), folding and rolling (F&R) (Sagel et al. 1998), etc. However the major existing constraint of some SPD techniques is not to fabricate the bulk production for viable uses, for example, HPT and ECAP. However, the ARB technique is an acceptable process for tangible applications; later it can be scaled up and adapted to a conventional rolling procedure to produce bulk UFG sheets (Saito et al. 1999). In this technique, the surface of aluminum sheets is wire brushed first, and one must ensure that no oxidation layer should be there, arranged in such a way that faces of two sheets are toward each other and rolled together; the main feature of this technique is that the reduction of thickness should be 50%. Due to 50% thickness reduction of sheet, the length doubles; the next step is to cut the sheet in the middle and repeat the same procedure eight to ten times. With repeating this process again and again, the sheet undergoes plastic deformation; due to that the size of grains are becoming smaller and smaller depending on the number of passes. As a result this will make the mechanical properties extraordinary.

The properties of fabricated nanostructured aluminum sheet include extraordinary strength and have scientific importance particularly for strength weight ratio. Therefore it has established significant weightage as likely applicants for new materials in

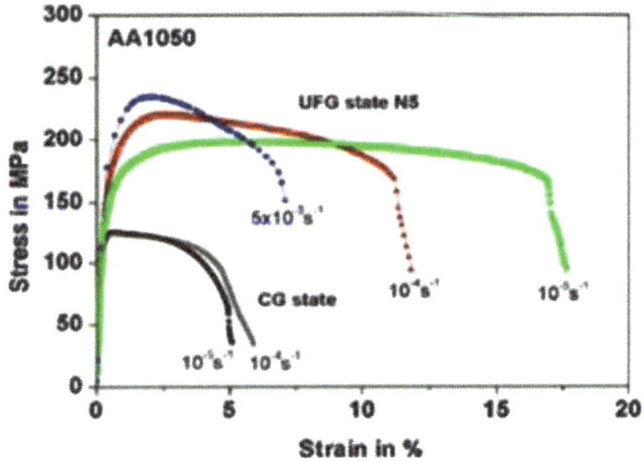

Fig. 23.1 Stress vs. strain for the as-received, cold rolled aluminum and the UFG aluminum (Höppel et al. 2004)

the auto and aerospace engineering (Gologranc 1975; Sharma et al. 2016a). The abovementioned roll bonding process is the important area, where there is more possibility of additional research on nanostructured aluminum sheets (Thomas et al. 1991). The friction stir welding (FSW) process has also been used to fabricate the UFG and known for joining technique. Due to several advancements, with more weight reduction perspective related to usual mechanical fasteners, easier joining of the hard-to-weld materials such as aluminum alloys, and a lesser heat input associated to fusion welding (Bodkhe et al. 2021a). UFG materials are also known for enhanced strain rate sensitivity in estimation to the as-received grained materials. Increase in strain rate develops a growth in yield strength and a decrease in elongation to failure; however reduction in strain rate causes lower strength and considerably advanced elongation to failure (Höppel et al. 2004) (see Fig. 23.1).

The outcome can be satisfactorily used for sheet metal work where the strain rate sensitivity affects the deformation ability and greater elongation to failure can be achieved at lesser strain rates. Though, with the object of increase the efficiency, it is important to reduction the time of the total metal work procedure. In this case, carrying out the sheet metal work of UFG materials at the higher temperature can be an extra suitable technique for accomplishing greater deformation and at the same time better procedure proficiency. In addition, compound strain rates are generally connected with crash occurrence and subsequently higher energy inclusion of a given constituent. From this position strain rate, responsive materials would also be valuable, as they demonstrate an increase in strength with growing strain rates. At this stage, the ductility of UFG aluminum sheets is still a constraint, which places additional significant challenge on ARB processed sheets.

Also, there have also been various details on the modest temperature *superplasticity* of the UFG metal sheets formed by the ARB process (Valiev et al. 2000; Tsuji et al. 1999a) and friction stir welding. The superplastic behavior of nanostructured sheets (Fig. 23.2) is likely for further focused application and hard to form workpiece with compound dimensions, due to the outstanding ductility acting a vital function.

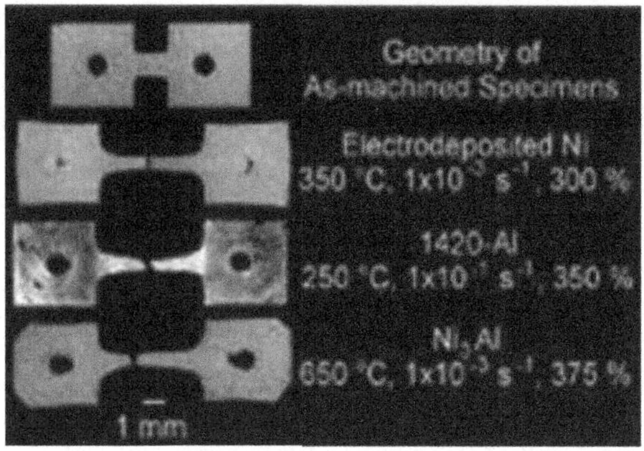

Fig 23.2 View of the Al 1420 alloy samples after tensile tests (Valiev et al. 2000)

23.2 General Aspects of Severe Plastic Deformation

SPD techniques demonstrate the capacity to fabricate the bulk UGF materials by applying huge plastic strain ($\varepsilon > 8$) and at the same time keeping the new geometry (Valiev et al. 2006; Tsuji et al. 2003a). The utmost important strengthening mechanisms, which demonstrate the way to the high strength of materials processed by any SPD technique, are grain improvement.

By employing the conventional manufacturing techniques such as cold rolling, total percentages of thickness reduction have occurred between 50 and 80%, and relatively big grain sizes of about 20 μm have been fabricated. The large plastic strain of above 6 is essential in order to obtain a nano-/microstructure from the CG metals. The most useful SPD comprise ECAP, ARB, and HPT used for sheet metal. ECAP is a discontinuous technique where a workpiece passed through a die having two channels of equal dimension with an intersecting angle of 90°. The metal undergoes to high strains and in the cross-sectional geometry. The process could be done a number of times in order to achieve the nanostructured grain (Horita et al. 1996). In HTP process, a coin-type metal workpiece is positioned in two dies. The lower dies rotate and produce very high torsional strain on the sample, while the upper dies apply a very high pressure of several GPa. Due to very high plastic strain attained while no change in dimension occurs throughout the process. The main obstacle of the HPT process is that the metal workpiece dimension is limited to the size of sample (Horita et al. 1996). All SPD techniques are costly and difficult and have low production rates and discontinuous processes. New methods of the bulk nanostructured material process have emerged. Some of these include the ARB (accumulative roll bonding) process, continuous confined strip shearing (CCSS) (Lee et al. 2002a), continuous cyclic bending (CCB) (Takayama et al. 1999), and repetitive corrugation and strengthening (RCS) process (Tsuji et al. 1999b). In spite of that, various SPD processes have been used, and several researchers established that there is a high probability to develop the nanostructure and an exceptional variety of mechanical properties by initiating very high strains. Nowadays these methods increased industrial production of nanostructured or UFG metals, and now it is commercialized and cost-effective. As a result, the modernization competence of the nanostructured and UFG metals for highly advanced purpose in engineering and the need for producing such materials are fascinating, more practical, and feasible.

23.3 Strengthening Mechanisms

The requirement of materials with valuable mechanical properties like strength, strain hardening, etc. could be achieved by refinement of grains. The dislocation and grain refinement mechanism is accountable for metal strengthening through SPD processes. Solid solution strengthening is dependent on the interface of solute atoms with dislocations and the subsequent endurance to dislocation movement. Determined by the size of the solute atoms, two forms of solid solutions are distinguished: one is substitutional and other is interstitial solid solution. Relationship among the solute molecules (c) and the strength ($\Delta\sigma_{\mathrm{ss}}$) is governed from the following equation:

$$\Delta\sigma_{\mathrm{ss}} \approx c^{n}, \tag{23.1}$$

where the value of n is about 0.5–0.6 (Sharma et al. 2016b). With the different directions and possible path, the solute atoms can work collectively with dislocations, together with paraelastic, dielastic, and chemical interface. In the case of the paraelastic interaction, solute atoms differ in size compared to the matrix molecules. Ding whether they are smaller or bigger than the matrix atoms, they basis moreover tensile or compressive stresses. The lattice distortion energy reason by the solute molecule can reduced if the solute atom segregates to a dislocation. However, if this dislocation moves, extra forces must be functional in order to conquer the backward stress from the solute molecule and separate the dislocation from it.

23.3.1 Strain Hardening

Work or strain hardening is a significant method to make stronger materials by using commercial, manufacturing process like deep drawing, cold rolling, extrusion, etc. In the process of work hardening, the dislocation starting place becomes triggered and the sum of dislocations rises. Subsequently the dislocations mix with strain fields due to lattice distortion. Thus, by raising the volume of plastic deformation, the quantity of dislocations rises, leading to a considerable increase of the yield strength. The rise in yield strength σ_{SH} is relative to the dislocation density (ρ):

$$\Delta\sigma_{SH} = (0.1\ldots0.2)\cdot M\cdot G\cdot b\cdot\sqrt{\rho}, \tag{23.2}$$

where G is the shear modulus, M stands for the Taylor factor, and b is the Burgers vector (Sharma et al. 2017).

23.3.2 Grain Boundary Strengthening

The mechanical characteristic of metals is greatly influenced by the grain size. Rise in yield strength with a reduction of size of grain was determined to apply to a wide range of metals which follows the Hall-Petch correlation:

$$\sigma_y = \sigma_0 + k \cdot d^{-\frac{1}{2}}, \tag{23.3}$$

where k is a constant for a given material, σ_y is the strength, σ_0 is the intrinsic yield stress, and d is the size of grain (Hall 1951). Dislocation cause inside one grain creates dislocations which accumulate at the succeeding grain boundary, the stress at the tip of the pile-up essential to go beyond normal stress to follow up the slip to continue into the adjoining grain. Thus, boundaries of the grain work as an efficient restriction to dislocation motion. As a result, it is repeatedly desirable to yield materials with UFG; therefore the more boundaries of grains performing as barriers to dislocation slide.

23.4 Accumulative Roll Bonding (ARB)

23.4.1 Conventional Sheet Metal Rolling

When a metal sheet is passed in between two rolls known as rolling (Dieter 1988), it is a process of plastic deformation. In this process, the metal sheet is enduring to excessive compressive stresses. Due to pressing process and execution of the rolls elevated stresses found from friction and the rolls, the primary purpose of roll mills is to decrease the thickness of the received sheet metals a long with to attain excellent mechanical properties (Fig. 23.3).

In the rolling process, the metal worksheet with an initial thickness t_0 enters in between rollers at starting velocity v_0. The worksheet is dragged out through the

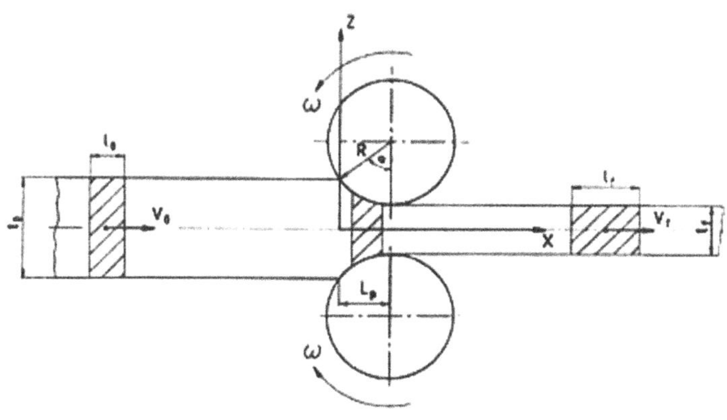

Fig 23.3 Sketch of a conventional rolling process after (Sharma et al. 2017)

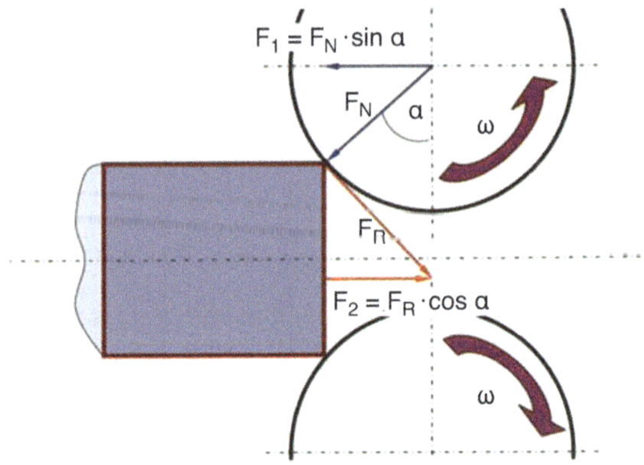

Fig. 23.4 Sketch of forces during typical rolling (Sharma et al. 2017)

rollers due to friction. At one point of contact, the worksheet achieves the roll velocity. That point is called the neutral point. The underlying conditions and boundaries, such as the percentage of thickness reduction, percentage amount of deformation thickness reduction, amount of deformation, velocity of the worksheet, and the length of the deformation zone, can all be taken into account by the following equations:

$$\text{Reduction of thickness}\Delta t = t_0 - t_f \tag{23.4}$$

$$\text{Deformation due to strain } \varepsilon_h = \frac{\Delta t}{t_0} \tag{23.5}$$

Worksheet are moderately big; it may be considered that by an increasing the length and not any raise in the width of a worksheet throughout rolling (Lange 1974). The increase in width can then be taken into account as insignificant as seen in Eq. (23.5).

$$b/_{L_p} \geq 20 \tag{23.6}$$

From the initial condition of the worksheet which can be rolled, there are two conditions, which have to be satisfied. The rolls must be capable of holding the incoming worksheet than out it during the rolling of roll mills (Geiger and Hennige 2001).

The two roll contacts can be derivates from Fig. 23.4.

$$\text{Worksheet is pulled in when } F_2 \geq F_1 \tag{23.7}$$

$$\text{where } F_1 = F_N \cdot sin\alpha \tag{23.8}$$

and

$$F_2 = F_R \cdot cos\alpha = \mu . F_N . cos\alpha \tag{23.9}$$

$$\text{Thus } \mu \geq \tan \alpha \tag{23.10}$$

$$\text{Holding state} : \mu \geq \tan \alpha \tag{23.11}$$

$$\text{Pull through condition} : 2\mu \geq \alpha \tag{23.12}$$

Hot rolling is generally performed above the recrystallization temperature of the metal; hence, it consists of small flow stresses. As a result, it improved the ductility of the sheet metal at the higher temperature and high deformation, while thickness reduction is accomplished in one pass. However, rolling is typically executed below the recrystallization temperature after hot rolling. The sheet material strain hardens considerably, and elevated stress is necessary for the deformation. The outcomes of cold rolling are enhanced thickness tolerance, smooth surface finish with no oxide formation on the layer of surface, and a possibility to obtain slimmer sheets (Kim et al. 2006; Lee et al. 2002b). During rolling process a number of defects may arise due to a compound contact between the rolls and the sheet metal such as sheet bending in the normal plane and the nonuniform thickness in direction of rolling. Normally, problems take place due to homogeneities in deformation in the rolling direction. If the gap in between the rolls is not entirely parallel, then one side of the worksheet is deformed and elongated more than the other side. As a result, sheet metal bending as well obtains no similar thickness; additionally, edge cracking of worksheets usually faces the problem. During the rolling usual trend of worksheets to increase in the transverse direction, but the flow of material is resisted by the frictional forces. This offers, inside towards the center of the sheet metal is under compression, and the boundaries are under tension leading to edge cracking.

23.4.2 Accumulative Roll Bonding Process

ARB process has firstly been presented and established by Saito et al. in 1998 (Saito et al. 1998); this is the novel severe plastic deformation technique which is illustrated in Fig. 23.5. In this process the first step is cleaning with the help of wire brush and removing the oxidized layer on the surface of sheet metal; after cleaning, the two equal and opposite faces each other; the metal sheets are joined and put in between the two rollers so that sheets are rolled up to 50% thickness reduction in every pass. This process is repeated several times; it might be ten times, depending on the materials' properties. After the first pass or cycle, the length of metal sheet doubles and it is cut half its length; once again it is cleaned and airbrushed and then again joined together for second pass of rolling; this process is repeated up to ten or more than ten times; after tenth cycle there is formation of 1024 layers which contains the UFG metal sheets. ARB process is shown in Fig. 23.5 (Sharma et al. 2017).

Plastic deformation has taken place mostly in the direction of rolling or in transverse direction of the metal worksheet. The change in length in worksheet in perpendicular direction to the rolling can be neglected. It is also described in

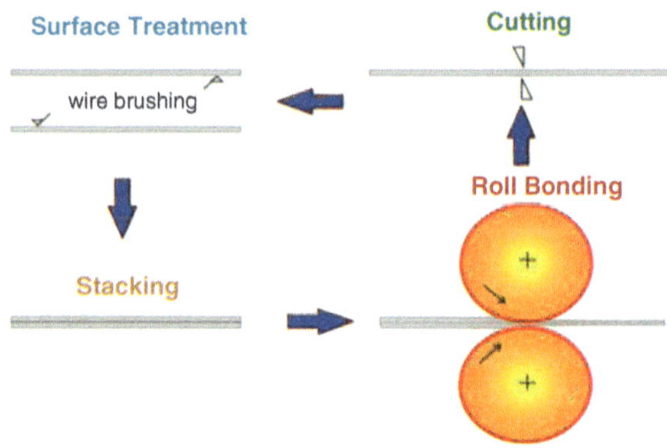

Fig. 23.5 Accumulative roll bonding (ARB) technique (Sharma et al. 2017)

Eq. (23.14). In a process of roll bonding, the number of separate layers within 1-mm-thick sheet increases according to Eq. (23.15), and the thickness of the separate layers can be found out from Eq. (23.16).

$$n = 2^N \tag{23.13}$$

$$t = \frac{t_0}{2^N} \tag{23.14}$$

where n = number of individual layers, N = number of passes/cycles, t_0 = original sheet thickness, t = ending/final thickness of layer

Therefore, the total reduction r_{tot} after N cycles can be calculated from Eq. (23.15).

$$r_{tot} = 1 - \frac{t}{t_0} = 1 - \frac{1}{2^N} \tag{23.15}$$

After the tenth pass of ARB cycles with 50% thickness reduction each cycle, the 1-mm-thick worksheet forms 1024 individual layers, each having a theoretical thickness of less than 1 μm, whereas the entire 100% reduction has been achieved. Take into consideration von Mises yield principle, where the entire corresponding strain up to next N passes is found out by using Eqs. (23.1)–(23.16) (Hansen and Jensen 1999) (Fig. 23.6).

$$\varepsilon_{tot} = \left\{\frac{2}{\sqrt{3}} \cdot In\left(\frac{1}{2}\right)\right\} \cdot N = 0.80 \cdot N \tag{23.16}$$

Through ARB technique of SPD, it has been possible to make a broad range of different ultrafine-grained materials including commercial purity aluminum, aluminum alloys, pure copper, titanium, and steels (Saito et al. 1998). Instead of it, many authors focused on the microstructural development and enhancement of the

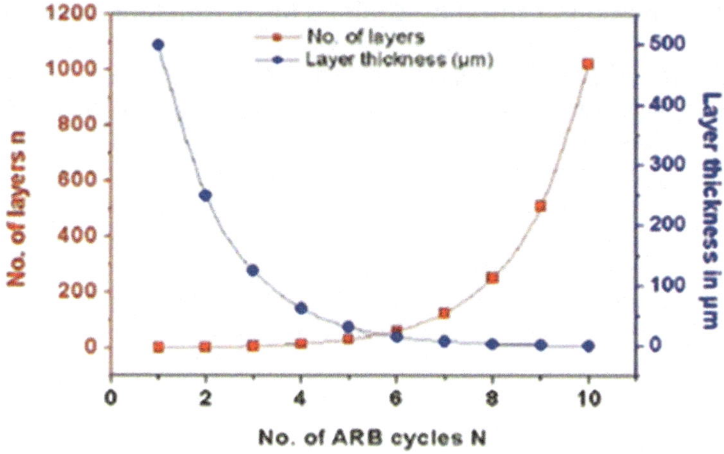

Fig. 23.6 Geometrical changes of the materials during ARB (Sharma et al. 2017)

Fig. 23.7 Geometrical changes of the materials during ARB where two metals sheets 1-mm-thick are roll bonded (Sharma et al. 2017)

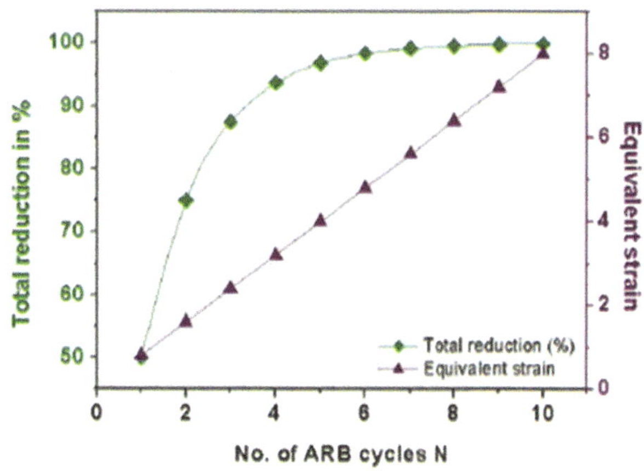

mechanical properties, whereas the actual ARB process advancement and upgrading are largely not considered (Fig. 23.7).

23.4.3 Microstructural Evolution of Accumulative Roll-Bonded Metal Sheets

The subsequent parameters for the microstructural evolution could be taken into account which are substantial for the evolution during rolling process of the ultrafine-grained metal sheets. The most important one will be reviewed in this section regarding various materials.

23.4.4 Formation of Ultrafine-Grained Microstructure During ARB

The nanostructure is produced after cyclic rolling which deforms a commercialized grain-sized sheet metal. Irrespective of the various kinds of material rolled by rolling mills, the different ranges of grain size have been achieved up to submicron and nanometer scale, ranging from 50 to 10,000 nm. The grains are developed and extended in the rolling direction through the ARB process. This was originally described by Tsuji et al. (Lee et al. 2002c) for aluminum alloys (Huang et al. 2003; Hansen and Jensen 1999). Established on the different investigations, cold working is divided by grain boundaries which are arranged in lamellar or subgrain arrangement; Hughes and Hansen (Hughes and Hansen 1997) recommended a promising technique for high-angle boundary development connecting to grain segment. Primarily, at small strains, the dislocations might be easily upended and might form grains or cells. The cells can isolate by dislocation boundaries for gap in dense dislocation edges. Due to rising of the strain level, the space between the HAGBs decreases in the normal direction, i.e., they are pushed together and they are separated by smaller cells or subgrains. There is a rising tendency for the dislocations to be stored within the subgrains, therefore subdividing the microstructure. The subgrain size reduces to sub-micrometer level, and the dislocation boundary misorientation has extensively distributed. Finally, the microstructure evolves into a lamellar structure, sandwiching in little dislocation boundaries, i.e., cells or subgrains, and is subdivided into high- and low-angle grain boundaries. The grain misorientation rises and as a result the number of HAGBs increases, representing that fall in grain size was induced by high plastic deformation (Hughes and Hansen 1997). An additional assumption about the formation of ultrafine-grained HAGB is based on recrystallization processes rather than the beforehand projected grain subdivision and rise in grain misorientation. Humphreys and Hatherly (Humphreys and Hatherly 1995) propose that the UFG HAGB microstructure develops by the mechanism of continuous recrystallization rather than the discontinuous convention recrystallization. The constant recrystallization classically arises afterwards annealing of severely deformed materials, wherever the grain edge relocation is greatly improved microstructure of small angle grain boundaries (LAGB) as well as high angle grain boundaries (HAGBs) (Gottstein 1998; Pérez-Prado et al. 2004). Bearing in mind both concepts into interpretation, it could be expressed that the evolution of HAGBs in severely deformed materials is still a strongly not clear-cut topic and does not ensue in a conventional nucleation and growth method. As a matter of fact, the significance of mechanism of HAGB growth in intensely deformed materials endures to be explained.

23.4.5 Influence of Second-Phase Particles

The effect of second-phase particles on the microstructural growth through ARB process has not been examined in detail (Humphreys and Hatherly 1995; Slamova et al. 2007). Both scientists have stated that the small subdivisions apply significant pinioning on grain borders, stabilize the deformed microstructure, and influence material hardening.

23.4.6 Conventional vs. ARB Rolling

The conventional rolling and the accumulative roll bonding technique are almost identical, but the piece of evidence is that UFG or nanograin can be developed by the ARB process; this the main differences are pointed out as follows.

- Fifty percentage reduction in each rolling cycles.
- Sizes of grain are very small (100–200 nm).
- High surface finishes due to additional shear stresses at the surface and complex shear strain distribution.
- The higher friction of high-angle grain boundaries (HAGBs).
- Grain misorientation.

In manufacturing industries, rolling mills carried out about 60%–80% of total thickness reduction (1.06–1.86 von Mises true strain). Once ensuring the recrystallization, the typical size of grains reaches nearly 10 μm (Hansen and Jensen 1999). But in the ARB technique, huge plastic strains in each pass have been accumulated, and the total percentage reduction can reach over 90%. One more important distinction is the consequence of friction. Normal rolling process is meant to cut the volume of friction among worksheet and the rolls because of the energy reduction, though the presence of friction in the ARB process has found to be beneficial and is required for bonding without lubrication, which is evident to take the lead to quicker grain refinement (Li et al. 2006). It has also some findings in a rapid expansion of an oxide layer on the roll mills (May 2004; Lee et al. 2002c). The grain misorientation processed by conventional rolling to 20% reduction has measured (Hughes et al. 2003), and explore those substantial angles of misorientation were previously present at comparatively small strains. The value of about 37° at a strain of 3.5 misorientation angle increases with strain in the low strain range but reaches a saturation level. During cold rolling, no penetration was observed; thus, the ARB technique appears to be a further efficient grain refining process as compared to normal rolling process for the equal percentage of reduction thickness. As expressed earlier, the ARB fabricated materials have strongly stretched out microstructure with elongate boundaries in the direction of rolling and are small, in transverse direction (Huang et al. 2003), whereas fraction of HAB is advanced as compared to as-received or usual deformed materials (Hansen and Jensen 1999). However, the sound structure and the subgrain differ and may be influenced by form of material being processed.

Solid-State Bonding Bonding through diffusion and bonding through mechanical process are the two types of main bonding procedures. Diffusion bonding is a joining process where interdiffusion of atoms through the boundary occurs. It needs adequate time and relevant temperature and pressure. Mechanical or conventional bonding takes place right away or in limited period of time and depends on additional factors like forces, temperature, and bonding force in between similar or dissimilar atoms between the atoms (Wu et al. 1998). Convectional roll bonding is a systematic solid-state bonding process, it is significant to emphasize some of its characteristics, and it is also important for the complete manufacturing process of accumulative roll bonding. ARB bonded worksheets are reliable in lamellar bonding and create more effective sheet metal forming process. In 1950 F.P. Bowden (Bowden and Tabor 1950) suggested that cold welding technique is a subject of adhesion force, analyzed by intermolecular forces between the interfaces of materials. The foremost condition for good adhesion is the contamination of free surfaces at an interatomic distance to each during the ARB process. In bonding during ARB process, one of the most important factors is microstructural textures which may affect the properties of sheet metal. During investigation, microstructural textures and some different aspects have been noticed which affect the whole process and the condition of the bond. Some examples include rolling speed, an oxide layer, surface roughness, normal pressure, annealing temperature, percentage reduction, and process temperature.

23.4.7 Accumulative Roll Bonding Parameters

The different range of parameters affects the success of the ARB process, for instance, the quality of the bond, microstructure, and the mechanical properties of the ARB prepared sheets. The important ARB process parameters are described below.

23.4.8 Process Temperature

Temperature is the fundamental factor in any manufacturing process, which influences the microstructure, thermal stability, and other mechanical properties such as the excellence of the worksheet bonding. It comprises good bonding among worksheets and an outstanding thermal stability (Homola et al. 2006). The maximum toughness was attained when sheet metal rolling has performed at the lower temperature; on the other hand, notable bonding has been achieved at about 250 °C. Work hardening occurred at elevated temperatures, at the first cycle material softens goes down up to after the next cycles. The reason for softening incorporates dynamic recovery and partial recrystallization. Increased temperatures also reduce the potential for a rapid grain refinement. The process temperature influences the bonding development as well as the division of the oxide layer which influences the roll bonding. At elevated temperatures the strength of the oxide layer has increased and bonding among the layers is reduced. Annealing is performed earlier to roll bonding and illustrates a substantial effect on the sheet metal bonding. Heat treatment at elevated temperate generally performs the strong bindings (Yan and Lenard 2004).

23.4.9 Percentage Reduction per Rolling Cycle

Usually, 50% reduction of thickness every cycle is essential for ARB process, which tends to an enhanced quality of bonding of aluminum alloy metal sheets and for different materials like titanium, IF steel, and Al-Mg alloys (Krallics and Lenard 2004). Roll bonding less than 50% reduction of thickness shows low bonding quality or unsuccessful bonding between the worksheets, and that bonding more than 50% has a tendency to edge cracking. Interlamellar bonding quality in terms of strength enhances with a higher percentage of thickness reduction for the aluminum alloy AA6111 (Fig. 23.8). Greater reduction of thickness at lower rolling speed is enough to produce a strong bond nearly 235 PMa. Therefore, the time of contact of rolling also plays a vital role.

23.4.10 Number of Rolling Cycles

The processing of a greater number of ARB passes/cycles results in the hardness, yield, tensile strength, and ductility increase. The initial ARB cycles might be one or two or in some cases three also, dominates to work hardening future, the involvement of grain boundary strengthening develops and dominating due to the development UFG microstructure (Krallics and Lenard 2004). It has reported that as that number of passes/cycles increased, the bonding strength of adjacent layers is enhanced.

23.4.11 Roll Diameter

Roll diameter in ARB process wholly controls the angular velocity of the rollers and the plastic deformation area of the worksheets (Krallics and Lenard 2004). Identical roll diameters guide and control good rolling speed, lesser area of contact for plastic deformation, and sufficient rolling pressures.

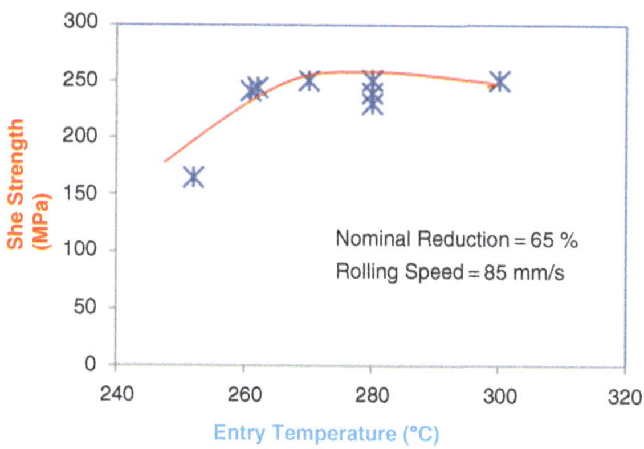

Fig. 23.8 Shear strength of the interlamellar bond of the AA6111 as a function of entry temperature and annealing temperature and the percentage reduction (Sharma et al. 2016b)

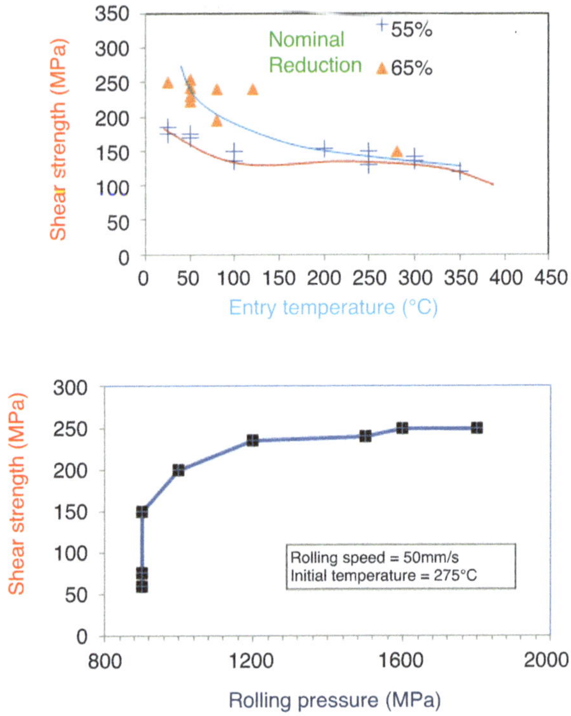

Fig. 23.9 Shear strength of interlamellar bonding of the AA6111 as a function of (**a**) rolling speed and percentage reduction and (**b**) roll pressure (Sharma et al. 2016b)

Rolling Speed In Fig. 23.9, it is noted that the quality of the bond declines by escalating the speed of rollers. This happens because of the fact that quicker strain hardening of the surface of worksheet and not sufficient contact time occurs among bonding sheets (Tsuji et al. 2003b). Primarily, it has projected that the elevated strain rate may give additional effective grain subdivisions, however the size of the grain increased with increase of the strain rate. This has an importance in the amount of heat generation during the process. On the other hand, it was also recommended that the high-speed plastic deformation has the possibility of efficient grain subdivision if the heat has to be controlled at minimum (Klösters 2006).

Roll Pressure Higher roll pressures generally show the better interlamellar bonding between the worksheets (Fig. 23.9). It has imagined that due to shear the bond move toward the base metal when the adequate quantity of energy the activation energy to start the process of roll bonding and has given for the two components to joined. Amount of energy may be produced by heating and by mechanical means (Le et al. 2004).

23.4.12 Surface Roughness

Eliminating the oxidized layer is very necessary for bonding of sheet; due to this reason, wire brushing has been done for cleaning the surfaces before stacking and rolling; plastic deformation, as well as surface roughness, is introduced. Adequate surface unevenness is one more constraint which impacts the strength of the sheet metal bonding, even though usually it might be a complex parameter to calibrate. It has been a known fact that the maximum level of surface finish attained by wire brushing provides the strongest bonding connected to the rolling sheets. Some peel tests were showed to compare interlamellar quality in terms of strength and the degree of surface roughness of peel. The ARB worksheets have produced in such manner that one part of the sheet was capable of being pulled apart from the other.

23.4.13 Oxide Layer

Investigators have been recognized that the elimination of the oxide coating before roll bonding and this can be meaningful for superior interlamellar bond strength between the worksheets contains oxide layers which might be act as the barrier to growth of grains (Saito et al. 1999). Earlier work of Milner and Rowe (Le et al. 2004) states that aluminum oxide is divided into small subdivisions as it is extended during normal cold rolling and leaves after a newly built metal surface. The fragmentation of the oxide coating has also been established by Barlow et al. (Barlow et al. 2004) in the verification of the roll bonding process of aluminum layers. They are associated in the direction of rolling and from stringers demarcate the new location of the layered surface (Barlow et al. 2004).

23.4.14 Friction

Rolling depends on the friction in between the rollers and the sheet metals. Sequentially, the metal sheet is to be grabbed and pulled by the rollers and pushed out through the rollers. The impact of friction on the ARB process was not measured in any publication till now.

23.4.15 Mechanical Properties of Accumulative Roll-Bonded Materials

Theoretically, a material could be made infinitely strong if the grains are made infinitely small. This theory was originally established by Hall (Huang et al. 2003) and Petch (Petch 1953) and well known as Hall-Petch relationship. Based on this inspiration, the aim is to produce nanostructured, and ultrafine-grained materials have noticeably improved over the last 30 years (Gleiter 1989). UFG/nanomaterials hold notable excellent strength in comparison to the CG-sized materials which have been recognized by various investigators (Lowe and Valiev 2000). The connection between ductility and strength has been firstly investigated by Valiev et al. on nanostructured titanium and copper processed by ECAP and told as the "paradox of strength and ductility" (Valiev et al. 2002). Therefore it is possible to increase the

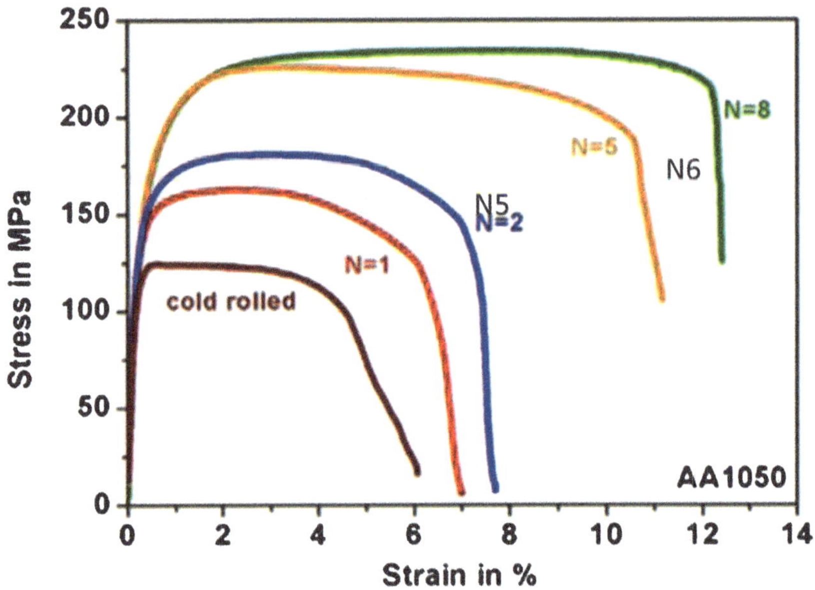

Fig. 23.10 Stress vs. strain for UFG aluminum at different numbers N of ARB cycles compared to the cold rolled state with approximately 3 μm grain size. Tensile tests were performed at 1.10-4 s-1 and room temperature (Valiev 2004)

elongation to failure and strength with processing more number of ARB cycle as related to the cold rolling of CG grained metal sheets.

Many types of research, those that examine and developed ultrafine grains of aluminum and aluminum alloys by using the ARB method, usually detected the increase in hardness and tensile strength with increasing number of ARB cycles. However, the earlier researcher suggested paradox of strength and ductility was not for all time evident (Sharma et al. 2016c). The characteristic curve of hardness and strain is seen in Fig. 23.10, which shows, after three to four ARB cycles, the hardness normally rises up to two to three times in evaluation to the as-received material; then, the hardness goes up to a saturation level and then does not increase additionally (Sharma et al. 2016c). On the other hand, depending on the many kinds of alloys and working condition, parameters, such as temperature, rollers' speed, or rollers' diameter, may come to softening after the maximum hardness has been reached.

The growth in the yield strength with an increase in the number of ARB cycles or strain has been examined for ARB-produced aluminum alloys shown in Fig. 23.11. Tensile strength increased by a factor of 3–3.5 in both cases. Approximately the same results have been reported by some other investigators (Xing et al. 2001). Nanostructured copper produced by the ARB process has been examined by Takata et al. (Sharma and Kumar 2016). Instead of that, the rate of elongation reduced notably after the first ARB cycle, though the overall elongation improved again with increasing number of the ARB cycles and touched 10% after eighth ARB cycle. It

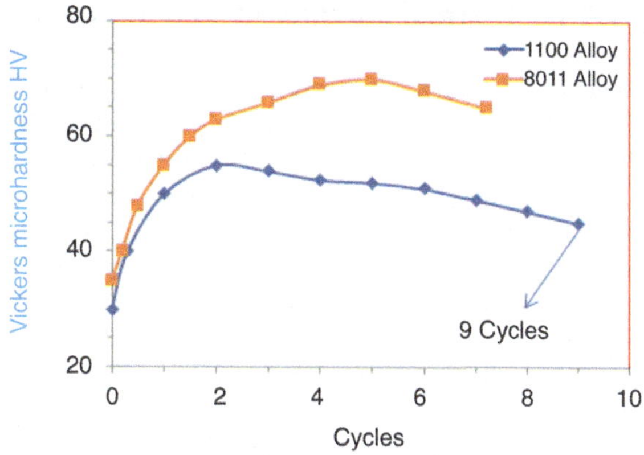

Fig. 23.11 Mechanical properties of AA1100 and AA60661 roll bonded at 200 °C and room temperature, respectively (Sharma et al. 2016b)

has been noticed that with cumulative numbers of passes, the value of strength constantly increased, while the total elongation to failure decreased (Takata et al. 2007). When the size of grains is decreased to the limit of nanoscale, the various dislocations influence work hardening which has lowered down; however the conventional understanding is that breaking of boundaries is limited, or it does not happen in UFG. The dislocations are formed through the process of severe plastic deformation, but they are also found at the grain boundaries Thus, these work as generators and sinks for dislocations (Terada et al. 2007). At normal temperature dynamic recovery turn into nanocrystalline or UFG materials (Meyers et al. 2006). This is generally indicated in the mechanical properties of nanostructured or UFG materials. The significance of ARB processed material is that during the work herding, the ductility goes down, but in this process some researcher has reported that the ductility has improved a lot. It is especially significant for those who have complicated geometries. Some other techniques are recently studied, which are worked to improve the ductility of UFG materials developed by severe plastic deformation process. Such process involves post annealing, which creates a bimodal microstructure, and the whole process undergoes cryogenic temperatures where it has reported that the strain rate sensitivity increased in most of ultrafine-grained materials. Annealing in two steps is more effective to fabricate the similar grain structures in all over the sheet metal as compared to earlier deformed grains (Kamikawa et al. 2006). Many researchers examined that a duplex microstructural grain improves the ductility quality. Further it has been found that grains smaller in size provide more yield strength; however the grains who are bigger in size lead to deformation. Optimal percentage of small-scale grain and large-scale grains starts to improve work hardening quality. The nanostructured duplex grain of copper and aluminum alloys (Wang et al. 2002; Witkins et al. 2003) has reduced yield strength, but then again a substantial increase in the elongation has been noticed. Rolling at cryogenic temperatures have two main advantages broadly those are put down the dynamic recovery and development of a UFG structure. The development of a duplex grain growth provides suitable blending of strength and ductility.

23.4.16 Thermal Stability of UFG Materials

Materials which are formed by severe plastic deformation process have been characterized by a nanoscale grain size and/or UFG scale size (Jin and Lloyd 2004). Earlier it has pointed out that ultrafine-grained materials have also generally relatively low elongation to failure due to prework hardening just before the severe plastic deformation process. It is meaningful to analyze the annealing operations of heavily plastic deformed materials which provide better ductility. Thermal stability is an additional most important factor which is also to be included for determining key material in SPD process. Ultrafine-grained materials are highly thermal sensitive; hence thermal stability has also been examined by some researchers (Bodkhe et al. 2021b). It has been proved that thermal stability of ultrafine-grained metals is subjected to the number of ARB or ECAP passes in addition to this material process temperature (Bodkhe et al. 2021b; Park et al. 2001).

References

Barlow CY, Nielsen P, Hansen N (2004) Multilayer roll bonded aluminium foil: processing, microstructure and flow stress. Acta Mater 52:3967–3972

Bodkhe M, Sharma S, Mourad A-HI, Sharma PB (2021a) An overview of the microstructure and mechanical properties of copper tube by SPD process, Materials Today: Proceedings

Bodkhe M, Sharma S, Mourad A-HI, Sharma PB (2021b) A review on SPD processes used to produce ultrafine-grained and multilayer nanostructured tubes, Materials Today Proc

Bowden FP, Tabor D (1950) The friction and lubrication of solids. Clarendon Press, Oxford

Dieter GE (1988) Mechanical metallurgy. McGraw-Hill Book Company, London

Geiger M, Hennige T (2001) SkriptzurVorlesungUmformtechnik 1. LehrstuhlfürFertigungstechnologie, Friedrich-Alexander Universität Erlangen-Nürnberg

Gleiter H (1989) Nanocrystalline materials. Progr Mater Sci 33:223–315

Gologranc F 1975 BeitragzurErmittlung von FließkurvenimkontinuierlichenhydraulischenTiefungsversuch (Evaluation of the flow stress curve with the continuous hydraulic bulge test), Dissertation, Institute for Metal Forming Technology, University of Stuttgart, Germany

Gottstein G (1998) PhysikalischeGrundlagen der Meterialkunde. Springer Verlag, Berlin Heidelberg

Hall EO (1951) The deformation and ageing of mild steel: III discussion of results. Proc Phys Soc Lond B 64:747–753

Hansen N, Jensen DJ (1999) Development of microstructure in FCC metals during cold work. Philos Trans R Soc Lond A 357:1447–1469

Homola P, Slámová M, Karlík M, Cizek J, Procházka I (2006) Effect of the temperature of accumulative roll bonding on the microstructure and properties of twin-roll cast AA8006 alloy. Mater Sci Forum 503–504:281–286

Höppel HW, May J, Göken M (2004) Enhanced strength and ductility in UFG aluminium produced by ARB. Adv Eng Mater 6:781–784

Horita Z, Smith DJ, Furukawa M, Nemoto M, Valiev RZ, Langdon TG (1996) Evolution of grain boundary structure in submicrometer-grained Al-Mg alloy. Mater Res Soc 11:1880–1890

Huang X, Tsuji N, Hansen N, Minamino Y (2003) Microstructural evolution during ARB of commercial purity Al. Mater Sci Eng A 340:265–271

Hughes DA, Hansen N (1997) High angle boundaries formed by grain subdivision mechanism. Acta Mater 45:3871–3886

Hughes DA, Hansen N, Bammann DJ (2003) Geometrically necessary boundaries, incidental dislocation boundaries and geometrically necessary dislocations. Acta Mater 48:147–153

Humphreys FJ, Hatherly M (1995) Recrystallization and related annealing aluminium phenomena. Pergamon Press, Oxford

Jin H, Lloyd DJ (2004) Effect of a duplex grain size on the tensile ductility of an ultra-fine grained Al–Mg alloy, AA5754, produced by asymmetric rolling and annealing. Scr Mater 50:1319–1323

Kamikawa N, Tsuji N, Huang X, Hansen N (2006) Quantification of annealed microstructures in ARB processed aluminum. Acta Mater 54:3055–3066

Kim YS, Kang SH, Shin DH (2006) Effect of rolling direction on the microstructure and mechanical properties of ARB processed commercially pure 1050 aluminium alloy. Mater Sci Forum 503–504:681–686

Klösters CC (2006) Einfluss der Oberflächenkonditionierung auf die mechanischenEigenschaften von ultrafeinkörnigenBlechenhergestelltmitkumulativemWalzen. Lehrstuhlfür Allgemeine Werkstoffeigenschaften, Friedrich-Alexander Universität Erlangen-Nürnberg, Diplomarbeit

Krallics G, Lenard JG (2004) An examination of the accumulative roll-bonding process. J Mater Process Technol 152:154–161

Lange K (1974) Lehrbuch der Umformtechnik, Band 2: Massivumformung. Springer Verlag, Berlin

Le HR, Sutcliffe MPF, Wang PZ, Burstein GT (2004) Surface oxide fracture in cold aluminium rolling. Acta Mater 52:911–920

Lee JC, Seok HK, Suh JY (2002a) Microstructural evolutions of the Al strip prepared by cold rolling and continuous equal channel angular pressing. Acta Mater 50:4005–4019

Lee SH, Saito Y, Sakai T, Utsunomiya H (2002b) Microstructures and mechanical properties of 6061 aluminium alloy processed by ARB. Mater Sci Eng A 325:228–235

Lee SH, Saito Y, Tsuji N, Utsunomiya H, Sakai T (2002c) Role of shear strain in ultragrain refinement by ARB process. Scr Mater 46:281–285

Li BL, Tsuji N, Kamikawa N (2006) Microstructure homogeneity in various metallic materials heavily deformed by accumulative roll-bonding. Mater Sci Eng A 423:331–342

Lowe TC, Valiev RZ (2000) Producing nanoscale microstructures through severe plastic deformation. J Miner Metals Mater Soc 52:27–28

May J (2004) Optimierung der Verfahrensparameterfür die ErzeugungultrafeinkörnigermetallischerWerkstoffemittels Severe Plastic Deformation (SPD). Diplomarbeit, Universität Erlangen-Nürnberg

Meyers MA, Mishra A, Benson DJ (2006) Mechanical properties of nanocrystalline materials. Progr Mater Sci 51:427–556

Park K-T, Kwon H-J, Kim W-J, Kim Y-S (2001) Microstructural characteristics and thermal stability of UFG 6061 Al alloy fabricated by ARB process. Mater Sci Eng A 316:124–152

Pérez-Prado MT, del Valle JA, Ruano OA (2004) Grain refinement of Mg–Al–Zn alloys via accumulative roll bonding. Scr Mater 51:1093–1097

Petch NJ (1953) The cleavage strength of polycrystals. J Iron Steel Inst 174:25–28

Sagel A, Sieber H, Fecht H-J, Perepezko JH (1998) Acta Mater 46:4233

Saito Y, Tsuji N, Utsunomiya H, Sakai T, Hong RG (1998) Ultra-fine grained bulk aluminium produced by accumulative roll-bonding (ARB) process. Scr Mater 39(9):1221–1227

Saito Y, Utsunomiya H, Tsuji N, Sakai T (1999) Novel ultra-high straining process for bulk materials development of the accumulative roll-bonding (ARB) process. Acta Mater 47:579–583

Segal VM (1977) Patent der UdSSR, Nr. 575892

Sharma S, Kumar S (2016) Accumulative roll bonding of Mg-Al Alloy. Int J Innov Eng Technol 6(3)

Sharma S, Singh RP, Kumar S (2016a) Development of nano-structured metals processed by severe plastic deformation. Indian J Sci Technol 9(27). https://doi.org/10.17485/ijst/2016/v9i27/94379

Sharma S, Singh RP, Kumar S (2016b) Bonding characteristics of Al and Al-alloy strips: processed by severe plastic deformation. Int J Eng Technol 8. https://doi.org/10.21817/ijet/2016/v8i5/160805403

Sharma S, Singh RP, Kumar S (2016c) Mechanical anisotropy of Aluminium AA1050 and Aluminium Alloy AA6014 produce by Accumulative roll bonding. Int J Innov Eng Technol 7(1)

Sharma S, Singh RP, Kumar S (2017) Development of nanostructure materials by severe plastic deformation for ductile metals. Doctoral Thesis, http://hdl.handle.net/10603/221470

Slamova M, Homola P, Karlik M (2007) Thermal stability of twin-roll cast Al–Fe–Mn–Si sheets accumulative roll bonded at different temperatures. Mater Sci Eng A 462:106–110

Smirnova NA, Levit VI, Pilyugin VI, Kuznetsov RI, Davidova LS, Sazonova VA (1986) Evolution of structure of fcc single crystals during strong plastic deformation. Phys Met Metallogr 61(6):127

Takata N, Lee S-H, Lim C-Y, Kim S-S, Tsuji N (2007) Nanostructured bulk copper fabricated by accumulative roll bonding. J Nanosci Nanotechnol 7:3985–3989

Takayama Y, Yamaguchi M, Tozaea T, Kato H, Watanabe H, Izawa T (1999) Proc. of the 4th Int. conf. on recrystallization and related phenomena (Rex'99), The Jap. Inst. Metals, Sendai, Japan 321

Terada D, Inoue S, Tsuji N (2007) Microstructure and mechanical properties of commercial purity titanium severely deformed by ARB process. J Mater Sci 42:1673–1681

Thomas WM, Nicholas ED, Needham JC, Murch MG, Templesmith P, Dawes CJ (1991) Friction stir butt welding, G. B. Patent: Application No. 9125978.8

Tsuji N, Shiotsuki K, Saito Y (1999a) Low temperature superplasticity of ultrafine-grained 5083 aluminium alloy produced by accumulative roll bonding. Mater Sci Forum 304–306:73–78

Tsuji N, Shiotsuki K, Saito Y (1999b) Superplasticity of ultrafine-grained Al-Mg alloy produced by accumulative roll bonding. Mater Trans JIM 40:765–771

Tsuji N, Saito Y, Lee SH, Minamino Y (2003a) ARB and other new techniques to produce bulk UFG materials. Adv Eng Mater 5:338–344

Tsuji N, Toyoda T, Minamino Y, Koizumi Y, Yamane T, Komatsu M, Kiritani M (2003b) Microstructural change of ultrafine-grained aluminum during high-speed plastic deformation. Mater Sci Eng A A350:108–116

Valiev RZ (1997) Structure and mechanical properties of ultrafine-grained metals. Mater Sci Eng A 234-236:59–66

Valiev RZ (2004) Nanostructuring of metals by severe plastic deformation for advanced properties. Nat Mater 3:511–515

Valiev RZ, Islamgaliev RK, Alexandrov IV (2000) Bulk nanostructured materials from severe plastic deformation. Progr Mater Sci 45:103–189

Valiev RZ, Alexandrov IV, Zhu YT, Lowe TC (2002) Paradox of strength and ductility in metals processed by severe plastic deformation. J Mater Res 17:5–8

Valiev RZ, Estrin Y, Horita Z, Langdon TG, Zehetbauer MJ, Zhu YT (2006) Producing bulk ultrafine-grained materials by severe plastic deformation. J Miner Metals Mater Soc 58(4): 33–39

Wang Y, Chen M, Zhou F, Ma E (2002) High tensile ductility in a nanostructured metal. Nature 419:912

Witkins D, Lee Z, Rodrigues R, Nutt S, Lavernia E (2003) Scr Mater 49:303

Wu H-Y, Lee S, Wang J-Y (1998) Solid state bonding of iron-based alloys, steel-brass and aluminium alloys. J Mater Process Technol 75:173–179

Xing ZP, Kang SB, Kim HW (2001) Softening behaviour of 8011 alloy produced by accumulative roll bonding process. Scr Mater 45:597–604

Yan H, Lenard JG (2004) A study of warm and cold roll-bonding of an aluminium alloy. Mater Sci Eng A 385:419–428

Nanotechnology for Functional/High-Performance/Smart Textiles

24

Sapna Gautam, Anupama Mishra, and Pooja Koundal

Abstract

Nanotechnology is an emerging area with the understanding and control of matter at the nanoscale, which is expected to have wide-ranging implications in all fields of science and technology such as materials science, materials processing technology, mechanics, electronics, optics, medicine, energy and aerospace, plastics, and textiles. Despite the fact that this technology is still in its early stages, it is already proven to be a useful tool for increasing textile performance and generating global attention. Nanotechnology's innovative application in textiles opens up a larger spectrum of qualities, as well as new product possibilities.

Keywords

Nanoparticles · Coating · Textiles · Antibacterial

24.1 Introduction

There are several newly developed coating techniques like soil-gel, layer-by-layer that can impart multi-functional, intelligent, excellent durability and waterproofness to fabrics (Musante and White 2012; Yetisen et al. 2016). The current

S. Gautam (✉)
Department of Textiles & Apparel Designing, College of Home Science, CSK HP Agricultural University, Palampur, India

A. Mishra
Central Agricultural University, Tura, Meghalaya, India

P. Koundal
Department of Fashion and Textile Design, NBSCFF, Subharti University, Meerut, Uttar Pradesh, India

A. Thakur et al. (eds.), *Synthesis and Applications of Nanoparticles*,
https://doi.org/10.1007/978-981-16-6819-7_24

review focuses on nanotechnology's prospective applications in generating multifunctional and smart nanocomposite-based fibres, nanofibres, and other innovative completed and coated textiles with nanotechnology-based ideas. Nanocoating is a relatively new technique that is still in the research and development stage. The manufacturing of nanocomposite fibres, in which the extraordinary capabilities of nanoparticles have been used to augment and impart many functionalities to traditional textile-based fibres, is the most explored area for producing multifunctional, intelligent fibres. Nanofibres with a diameter of less than a micron are also getting a lot of attraction in specialised technological applications including filter fabric, antibacterial patches, tissue engineering, and chemical protective suits. The textile sector has a lot of potential with nanotechnology owing to the fact that existing methods for imparting various qualities to fabrics frequently fail to provide long-lasting results and may lose their effectiveness after repeated laundering or use. Advances in nanotechnology have created considerable opportunities in textile finishing techniques, leading to the development of new finishes and application techniques. The most significant is to apply different types of nanoparticles or create structured surfaces based on nanotechnology to make chemical finishings more controllable and sustainable and dramatically increase functioning (Khan et al. 2012). Stain resistance, antimicrobial, controlled hydrophilicity/hydrophobicity, antistatic, UV-resistant, wrinkle-controlled, and shrink-proof abilities can be exploited for a wide range of technical textile applications including protective clothing, medical textiles, sportswear, and automotive textiles, among others.

24.2 Nanotechnology's Modern Application in the Textile Industry

Nanotechnology is gaining popularity around the world because it is commonly regarded as having enormous potential in a variety of applications. There is lot of potential for use of nanotechnology in textile industry due to the fact that traditional ways of imparting diverse qualities to fabrics frequently do not result in long-term impacts and lose their functionality after repeated use. Because nanoparticles have a higher surface area-to-volume ratio and a high surface energy, they have a better affinity to fabrics, resulting in increased function durability. Furthermore, a nanoparticle coating on the tissues has no effect on their breathability or hand sensation. Nanotechnology has already had an impact on the textile sector. Nanotechnology-based research to improve the performance of textile materials or to generate new functionalities is booming. Nanotechnology in composite fibre manufacture and in textile finishing are two examples of these applications (Ayatullah Hosne Asif and Hasan 2018).

1. Nanotechnology uses in the production of composite fibres are as follows:
   (a) Carbon nanofibres and carbon nanoparticles
   (b) Clay nanoparticles

    (c) Metal oxide nanoparticles
    (d) Carbon nanotubes
    (e) Nano cellular foam structures
2. Nanotechnology in textile finishing includes:
    (a) Upgrade of chemical finishes and resultant functions
    (b) Nanoparticles in finishing
    (c) Self-assembled nano layer

24.3 Nano Application in Apparel Industry

Sport, skin care, space, and apparel and materials technologies for increased protection in adverse situations are among the fields where nanotechnology-enhanced textiles are already being used. Using nanotechnological materials to process textiles improves the qualities of the textile, making it more durable, having more beautiful colours, and so on. Additional functions, such as energy storage and communications, can be added using nanotechnology. Here are few examples of nano-enhanced textiles now available in the market:

- Threads woven into textiles that are stain and wrinkle resistant.
- Phase change materials (PCMs) are used in body warmers to respond to changing body temperatures.
- Silver nanoparticles are used in the nanosocks which serve as an antiseptic and eliminate odour.

Swimwear: The shark skin suit worn at the world record Olympic swimming championship is the most well-known application. The suit, which incorporates a plasma layer strengthened by nanotechnology to reject water molecules, is designed to assist the swimmer in gliding through the water and has become a frequent part of major swimming competitions as all competitors want to improve their chances of winning.

Sports Items: Nanotechnology has also enhanced running shoes, tennis rackets, golf balls, skin creams, and a variety of other sporting equipment. In addition to designing fabrics that can resist harsh settings, scientists have sought for new building blocks for nanotechnology in naturally occurring viral microparticles that reside in some of the world's harshest environments. A garment that recognises its surroundings and interacts with its wearer is a major source of concern. These textile-based nanosensors might be used to create a tailored healthcare system, monitor your vital signs while you run up a hill or adapt to temporal changes.

Flexible Electronic Circuits: Nano ribbons, which are so flexible that they can wrap around the edge of a microscope cover slip and are elastic enough to be twisted into a corkscrew, serve as the foundation for the chips. These tiny, flexible electronic sheets could one day be used to line the brain to monitor activity in patients at risk of epilepsy or integrated into surgical gloes to monitor a patient's vital signs during operation, according to the researchers.

Lifestyle Applications: The earliest commercialised applications of nanotechnology are, maybe surprisingly, in lifestyle applications. Nanomaterials are being used in textiles and cosmetics for the first time. Bulletproof vests are an example of nanotechnology materials and technologies in a lifestyle application. Nanotubular fibres have been used to create a material that is 17 times stronger than Kevlar. Nanotechnology will be used in the future to make smart and interactive textiles (SMIT) that can detect electrical, thermal, chemical, magnetic, and other inputs (Kiron 2021).

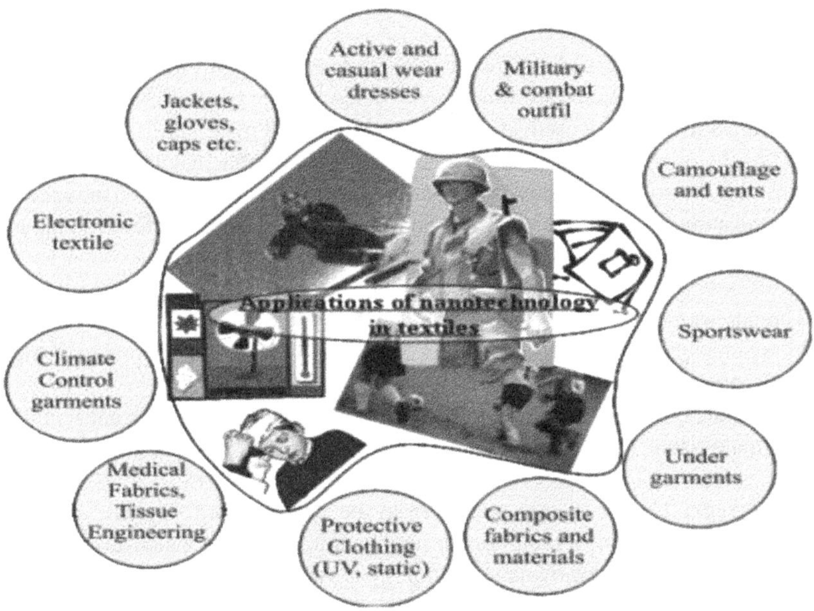

Applications of nanotechnology in textiles (Swahney et al. 2008)

24.4 Application Based on the Textile Material's Characteristics

Waterproofing; soil resistance; wrinkle resistance; antibacterial, antistatic, and UV protection; flame-retardant and increased dyeing capacity; and self-cleaning materials are just a few of the properties that nanotechnology can impart to textiles. The most important applications are briefly discussed below.

24.4.1 Water Repellence

Nanotechnology-based self-cleaning surfaces can be divided into two categories. First and foremost is incredibly water-repellent, microscopically rough surfaces: dirt particles cannot adhere to them and are thus washed away by rain or a simple water rinse. Photocatalytic layers are the second example: fouling organic material is destroyed by solar irradiation due to a layer of nanocrystalline titanium oxide.

(a) Nano-Tex enhances fabric's water-repellency by adding nano-whiskers, which are hydrocarbons 1/1000 the size of a conventional cotton fibre, to the fabric to give a peach fuzz effect without compromising the cotton's strength. Water remains on top of the whiskers and above the fabric surface because the spaces between the whiskers on the fabric are smaller than a typical drop of water but still larger than water molecules. If pressure is applied, the liquid can still permeate the tissue. While preserving breathability, the performance is permanent (Ayatullah Hosne Asif and Hasan 2018).

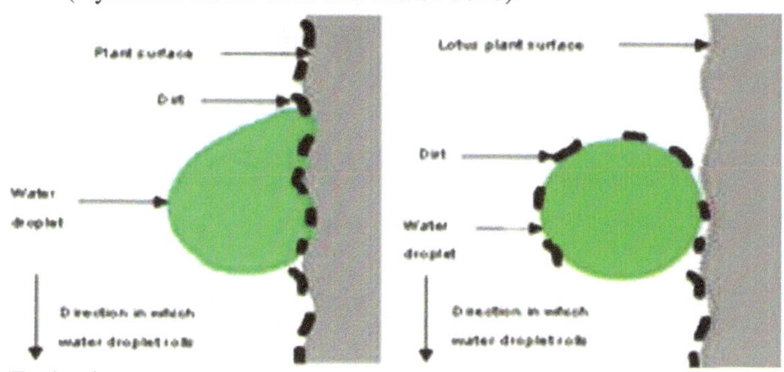

(b) To begin, we'll look at photocatalysts. Nanosized titanium dioxide and zinc oxide are employed in this process to confer self-cleaning and antibacterial properties. A thin layer of titanium dioxide particles with a diameter of 20 nanometres is applied to the fabric. Titanium dioxide is a photocatalyst, which means that when it is lit with light with a higher energy than its band gap, electrons in TiO_2 jump from the valence band to the conduction band, forming electron (e) and electric hole (h+) pairs on the photocatalysts' surface. Negative electrons mix with oxygen to form oxygen radical ions, whereas positive electric holes combine with water to form hydroxyl radicals OH. Because these products are chemically unstable, when an organic substance, such as dirt, pollutants, or microorganisms, falls on the photocatalysts' surface, it will mix with O_2- and OH– to form carbon dioxide (CO_2) and water (H_2O). Because titanium dioxide is utilised as a catalyst, it is never used. This is how the coating keeps decomposing stains over and over. Zinc oxide is likewise a photocatalyst, with a similar photocatalysis mechanism to titanium dioxide.

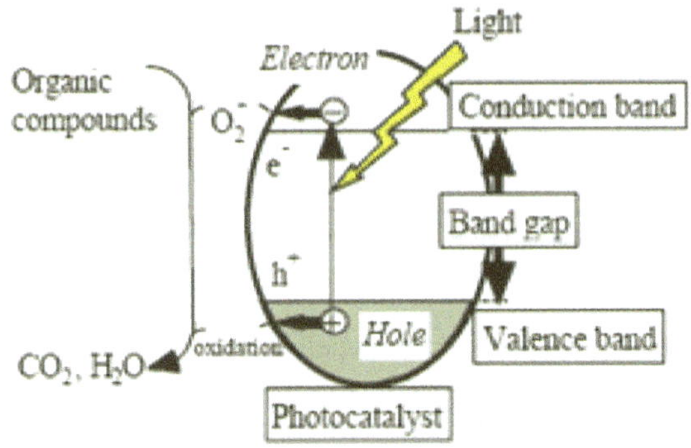

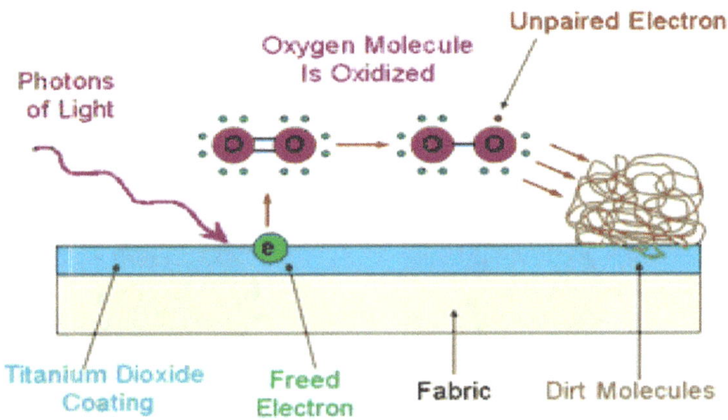

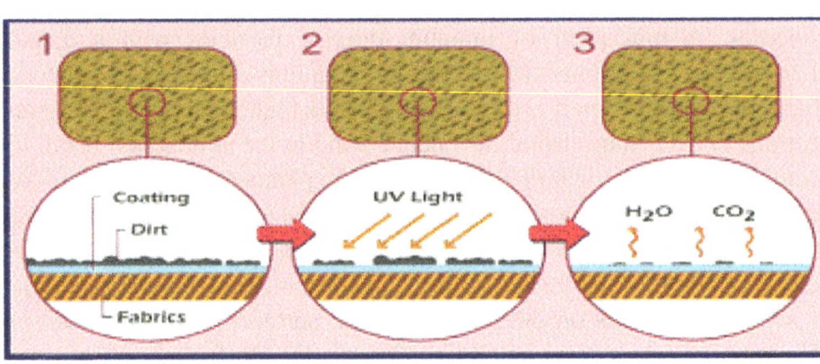

Working of self cleaning textiles

24.4.2 UV Protective Finish

The most crucial function of the garment is to keep the wearer protected from the elements. However, it also serves to shield the wearer from harmful UV rays. Ultraviolet radiation is defined as rays having wavelengths ranging from 150 to 400 nanometres. When a dye, pigment, delustrant, or ultraviolet absorber finish is present, ultraviolet radiation is absorbed and blocked from passing through the fabric to the skin; the fabric's UV-blocking quality is strengthened.

Metallic oxides, such as ZnO, are more stable as UV blockers than organic UV blockers. As a result of their enlarged surface area and high absorption into the UV zone, nanoZnO will significantly improve UV-blocking properties. In terms of profitability, brightness, and UV-blocking capabilities, ZnO nanoparticles outperform silver nanoparticles for antibacterial finishing. UV-absorbent fabric guarantees that clothing reflects harmful ultraviolet radiation from the sun, decreasing people's exposure to UV and protecting their skin from damage. The amount of UV protection required by different types of human skin varies based on geographic location, time of day, and season, as well as the intensity and distribution of UV radiation. The SPF (sun protection factor) value indicates how efficient the UV protection is; the higher the SPF value, the more effective the UV protection is (Kiron 2021).

24.4.3 Self-Cleaning Fabrics

Nanotex, an American firm, developed and commercialised a self-cleaning cotton fabric known as nano-care, and stain-resistant denim and khaki have been available since 1990. The cylindrical structure of the cotton fibres that make up the fabric is changed to generate nano-care materials. Cotton fibres mimic tree trunks at the nanoscale. These tree trunks are wrapped in microscopic whiskers that provide an air cushion around the fibre, thanks to nanotechnology. When water comes into contact with the whiskers on the cloth, the beads compress the air in the cavities between the whiskers, adding buoyancy. The cloth has been rendered super-non-wetting or super-hydrophobic on a technical level. Dirt has fewer contact places with whiskers. When water is applied to a dirty cloth, the dirt sticks to the water much better than it does to the textile's surface, and the water is wiped away as it beads up and rolls over the cloth's surface.

There are a number of plants in nature, including the lotus leaf, that exhibit remarkable superhydrophobicity wetting events. A surface that can totally bead off water droplets and has droplet advance angles of 150 degrees or greater at the surface area is referred to as a 'hyper hydrophobic surface' (Bozzi et al. 2005). As a result, a self-cleaning surface emerges, as water droplets rolling across the surface may easily capture dirt particles and leave a clean surface behind. Researchers have discovered numerous wonderful methods for creating ultra hydrophobic surfaces on textiles that imitate the nanostructured lotus leaf and so exhibit self-cleaning qualities in nature (Bhattacharyya and Joshi 2012).

24.4.4 Antistatic Finishes

Synthetic fibres, such as nylon and polyester, tend to generate static charge because they absorb little water. For handling static loads, cellulosic fibres have a higher moisture content, ensuring that no static loads build up. Because synthetic fibres have poor antistatic qualities, researchers have been looking at using nanotechnology to improve the antistatic properties of textiles. Antistatic characteristics have been demonstrated in synthetic fibres using nanometric titanium dioxide, zinc oxide whiskers, nano-antimony-doped tin oxide (ATO), and silanenano sol. Because TiO_2, ZnO, and ATO are electrically conductive materials, they have antistatic properties. The static load that accumulates on the fabric is effectively dispelled by such a material. Silanenano sol, on the other hand, improves fibre's antistatic characteristics by absorbing moisture and binding it to amino and hydroxyl groups (Kathiervelu 2003).

24.4.5 Wrinkle-Free Finishes

Nano-Tex claims that their new wrinkle-free treatment improves performance while keeping fabric strength and integrity, making it a viable alternative to harsh old procedures. Fabrics' tear and tensile strength are reduced by chemicals and treatment procedures. As a result, some materials and clothes, such as light fabrics or slim-fitting apparel, may not be suitable for wrinkle-free technology. Traditional materials are not ideal candidates for wrinkle-free technology. Wrinkle-free fabrics are popular and practical for time-pressed consumers, but traditional materials are not suitable candidates for wrinkle-free technology. To avoid fibre degradation caused by typical wrinkle-free methods, materials may need to be overcoated or strengthened.

In either case, today's technologies do not function on all fabrics, or the brand/retailer must invest additional costs solely to account for wrinkle-free chemistry's damaging qualities.

Nano-Tex's innovative Fortify DP technology penetrates deeper into the fibre for better wrinkle-free performance. Furthermore, it employs a longer and more flexible cross-linking chain that decreases stress on stressed fibres, resulting in a considerable reduction in strength compared to standard wrinkle-free chemistry.

24.4.6 Antibacterial Finishes

Antibacterial characteristics are conferred using nanosized silver, titanium dioxide, and zinc oxide. Metal ions and metal complexes have some sterilising properties. In theory, portion of the oxygen in the air or water is transformed to active oxygen by catalysis with the metallic ion, causing organic molecules to dissolve. The amount of particles per square inch can be raised when utilising nanoparticles, boosting the antibacterial characteristics. Nano-silver particles used in antimicrobial dressings have been shown to successfully inhibit the growth of microorganisms around wounds, resulting in faster wound healing. In addition, nano-silver medical fabrics for health and hygienic safety have been produced and marketed (Ali et al. 2011).

24.5 Economical and Environmental Aspects

Nanomaterials have piqued the interest of scientists and academics due to their unique features, but they have also piqued the interest of corporations due to their tremendous economic potential. Despite the COVID-19 dilemma, the global market for nanomaterials, which was anticipated to be worth US$7.1 billion in 2020, is expected to increase to US$12.1 billion by 2026, with a CAGR of 9.7% throughout the analysis period. Carbon nanotubes, one of the report's segments, are expected to grow at an 11 per cent compound annual growth rate (CAGR) and reach US $4.2 billion by the conclusion of the analysis period. After a detailed examination of the pandemic's commercial ramifications and the resulting economic crisis, the titanium dioxide nanoparticles' segment growth is updated to a revised 10.1 per cent CAGR over the next 7 years. The growing use of nanoparticles in the healthcare industry, as well as the growing use of nanomaterials in water treatment applications, is likely to increase global demand for nanomaterials. Nanotechnology is predicted to generate hundreds of billions of euros over the next 10 years (Anonymous n.d.). Nanotechnology may potentially have a positive environmental influence. Nanotechnology has the ability to save raw materials while improving quality of life by requiring fewer resources without sacrificing performance.

Nanotechnology has emerged as a fundamental technology that has resurrected materials science and has the potential to produce and advance a wide range of intelligent materials, including polymers and textiles. The following are the issues:

- The expense of producing nanoparticles in large quantities.
- Practical implications for determining mechanical fitness for a certain application.
- The impact of uncontrolled nanoparticle releases in the environment, as well as their effects on human health and the entire ecosystem, is comprehensively addressed by the field of 'nanotoxicology'.
- A realistic and ethical perspective on the widespread use of nanotechnology-based products (Ayatullah Hosne Asif and Hasan 2018; Kiron 2021; Ali et al. 2011).

References

Ali SW, Joshi M, Rajendran S (2011) Synthesis and characterization of chitosan nanoparticles with enhanced antimicrobial activity. Int J Nanosci 4&5:979–984

Anonymous (n.d.). https://www.prnewswire.com/news-releases/global-nanomaterials-market-to-reach-12-1-billion-by-2026%2D%2D301319895.html

Ayatullah Hosne Asif AKM, Hasan MZ (2018) Application of nanotechnology in modern textiles: a review. Int J Curr Eng Technol 8(2):227–231

Bhattacharyya, Joshi M (2012) Functional properties of microwave absorbent nanocomposite coatings based on thermoplastic polyurethane-based and hybrid carbon-based nanofillers. Polym Adv Technol 23(6):975–983

Bozzi A, Yuranova T, Kiwi J (2005) Self-cleaning of wool-polyamide and polyester textiles by TiO2-rutile modification under daylight irradiation at ambient temperature. J Photochem Photobiol A Chem 172:27–34

Kathiervelu SS (2003) Applications of nanotechnology in fibre finishing. Synthetic Fibres 32:20–22

Khan FR, Superb KM, Garcia AJ, Smith BD, Strekopytov S, Reinbow PS, Luoma SN, Valsami JE (2012) Bioaccumulation dynamics and modeling in an estuarine invertebrate following aqueous exposure to nanosized and dissolved silver. Environ Sci Technol 46:7621–7628

Kiron MI (2021) Sitemap of textile learner blog. https://textilelearner.net

Musante, White JC (2012) Toxicity of silver and copper to Cucurbita pepo: Differential effects of nano and bulk-size particles. Environ Toxicol 27(9):510–517

Swahney APS, Condon B, Singh KV, Pang SS, Li G, Hui D (2008) Modern applications of nanotechnology in textiles. Textile Res J 78(8):731–739

Yetisen AK, Qu H, Manbachi A, Butt H, Dokmeci MR, Hinestroza JP, Skorobogatiy M, Khademhosseini A, Yun SH (2016) Nanotechnology in textiles. ACS Nano 10(3):3042–3068

Toxicity of Nanomaterials: An Overview 25

Preeti Thakur, Yeshvandra Verma, and Atul Thakur

Abstract

Nanotechnology has attracted much attention due to its potential applications in diverse areas. Nanoparticles (NPs) have unique properties as compared to their bulk counterparts, and with their extremely small size as well as greater stability can persist in the human body for a long time period. Our understanding on the harmful effects of exposure of nanoparticles to the human environment is inadequate. Nanomaterials may enter the human body through inhalation, skin absorption or ingestion. So, in this context it is relevant to understand the toxic effects of nanoparticles. It has been generally agreed that NPs interfere with either assay materials or with detection systems. Also, the research so far done to investigate the toxic effects of nanoparticles is not coordinated, thereby generating data which is inconsistent. Therefore, it has become very difficult to categorize nanoparticles on the basis of their toxicity. It is the need of the hour to look for new toxicological models by concentrating more on in vivo studies.

Keywords

Nanoparticles · Toxicity · Exposure route

P. Thakur (✉)
Department of Physics, Amity School of Applied Sciences, Amity University Haryana, Gurugram, India

Y. Verma
Department of Toxicology, C.C.S. University Campus, Meerut, India

A. Thakur
Centre of Nanotechnology, Amity University Haryana, Gurugram, India

A. Thakur et al. (eds.), *Synthesis and Applications of Nanoparticles*, https://doi.org/10.1007/978-981-16-6819-7_25

25.1 Introduction

Nanotechnology is a developing new technology. It offers many benefits in several fields such as pharmaceuticals, medical technology, food technology, cosmetics, electronic devices and agricultural applications (Siegrist et al. 2008; Handford et al. 2014; Morganti 2010; Kahan et al. 2009; Mukhopadhyay 2014). Nanotechnology is basically defined as "the understanding and control of matter at dimensions between approximately 1 and 100 nm, where unique phenomena enable novel applications" (NNI 2017; 2014; NSTC 2011). The unique chemical, physical and biological properties of the nanoparticles are due to the variation in their size and shapes, which make them different from their bulk counterparts. Also, nanomaterials possess high surface area-to-volume ratio and, hence, are more reactive as compared to their coarse counterparts. Nanomaterials are known to possess physiochemical properties that can influence their effects in biological systems such as chemical reactivity, charge, shape, size, extent of agglomeration, solubility and surface area. The selectivity and reactivity achieved due to very small size have produced a wide variety of applications of nanomaterials including food packaging, pharmaceuticals, cosmetics, medicine, paints, coatings, electronics and fuel catalysts and water treatment. Also, the engineered nanomaterials have possible application in cancer treatment and its diagnosis wherein the nanoscale properties enable its intracellular transport to the specific target sites.

Whether these unique properties of engineered nanomaterials pose serious health concerns is still a matter which is not explored much. Exposure to these nanomaterials pose potential health risks which generally depend on the extent and duration of exposure; tenacity of material in the body; inherent toxicity of the material; and susceptibility or health status of the person (Nel et al. 2006). Inhalation, ingestion and penetration through the skin are some of the potential exposure routes for nanoparticles, whereas particle size, shape, surface area and surface chemistry collectively define the toxicity of nanoparticles. The intentional and unintentional increase in the use of nanomaterials consequently lead to their increased production and presence in the human environment, thereby potentially increasing human exposure. Therefore, it is vital to examine the transport of nanomaterials into and through the environment and also their impact on environmental health.

Essentially, nanomaterials found in human environment fall into four major types: carbon-based nanomaterials, metal-based nanomaterials, dendrimers and composites. The carbon-based nanomaterials such as fullerenes, carbon nanotubes and graphene quantum dots have unique chemical and physical properties including structural, mechanical, electrical, optical and thermal and hence are used in films, coatings, electronics, biosensing and biomedical applications. The metal-based nanomaterials have the ability to multi-bond and can be joined chemically with antibodies. Nanosilver, nanogold and metal oxides such as titanium dioxide (TiO_2) are hence used in food, cosmetics, healthcare and pharmaceutical industries. The dendrimers are nano-polymers. Their surface exhibits chain ends suitable for chemical manipulation as tools. They are used for drug delivery, anticancer drugs and pain management medications. Composites such as nano-clays combine one nanoparticle with other nanosized or larger particles. They are used in packaging materials. These four classes of nanomaterials differ in shape, size and chemical as well as biological properties. Once they enter the body, then by blood circulation, they may be translocated throughout the body. Their distribution in the body may depend on their size and surface characteristics such as hydrophilicity, polarity and catalytic activity (Wang et al. 2007; Yang et al. 2016). With the decrease in particle size, surface area per unit mass increases, and, hence, the nanoparticles exhibit increased chemical and biological activity in the body (Agnihotri et al. 2014; Ershov et al. 2016). So, it is generally hypothesized that smaller nanoparticles may be more toxic than their larger counterparts and therefore are taken up by the cells faster than the larger ones.

25.2 Potential Exposure Routes of Nanomaterials

Nanoparticles may pose a potential risk for the exposure of humans and the environment throughout their life cycle, starting from their manufacture to disposal. Permitted release of industrial effluents along with accidental spillage of nanoparticles in water bodies may result in its direct exposure to humans through skin contact, inhalation of water aerosols and direct ingestion of contaminated drinking water or particles adsorbed on vegetables or other foodstuffs. The smaller size eases the uptake and translocation of nanoparticles into the cells. These nanoparticles may transport to other places in the body, thereby interacting with tissues and prolonging their stay in the body. There is prime accumulation of nanoparticles in the organs such as the liver, kidney and spleen which have high phagocytic activity. Toxic effects have been observed at the pulmonary, cardiac, reproductive, renal, cutaneous and cellular levels.

Factors affecting exposure to nanoparticles include the quantity of nanomaterial being used, the susceptibility of nanomaterial to be easily dispersed and the degree of containment and the duration of use. Some of the prominent routes of exposure to a nanomaterial are through *inhalation, skin absorption and ingestion*. Inhalation is an important route of human exposure to the airborne nanomaterials (Borm et al. 2006; Oberdörster 2010). The surface of large airways is lined by ciliated bronchial epithelial cells and mucus-producing goblet cells. Particle's aerodynamic or thermodynamic diameter governs the deposition of distinct nanomaterials in the respiratory tract. Nanoparticles that are small in size can reach the lower airways and gain entry into the air-blood barrier, whereas the larger particles (>5 μm) remain trapped in the upper airways where the epithelial lining is thick and cells are incased by protective mucus. Another import route of human exposure to nanomaterials is skin absorption. The skin comprises the epidermis and dermis, with hair follicles and sweat glands providing pathways across these layers. Particles that are capable of being deposited in the gas exchange region of the lungs are considered respirable particles. Nanoparticles present in products such as paints, pastes and glues are the likely source of dermal exposure to nanoparticles. These nanomaterials are deposited in the lungs to a greater extent than the larger respirable particles. With an increase in the rate of breathing and also when there is a change from nasal to mouth breathing, the deposition of nanoparticles is found to increase. *Ingestion* is another route whereby nanomaterials may enter the body. Ingestion can occur from inadvertent hand-to-mouth transfer of materials. This can probably occur during handling of nanomaterials. Ingestion may also accompany inhalation exposure as it may happen that the nanoparticles that are cleared from the respiratory tract via the mucociliary escalator may be swallowed.

25.3 Need for Toxicity Evaluation of Nanomaterial

With the need and applications of nanomaterial continuously increasing, the probable exposure of nanoparticles into the human environment is also growing. Hence, to study and understand the toxicological effects of nanoparticles is of considerable interest. Appropriate characterization of nanomaterials is a critical factor for the toxicity evaluation of nanoparticles, so that potential risks to health and environment associated with their application can therefore be taken into account. Although prevailing toxicity testing procedures may be pertinent to recognize detrimental effects associated with nanomaterials, research to develop novel methods is essential to address the unusual properties of nanomaterials and their kinetics in living environment.

25.4 Nanoparticles of Iron Oxide

Iron oxide nanoparticles are widely used for hyperthermia, drug delivery and biomedical and diagnostic fields (Naqvi et al. 2010; Albukhaty et al. 2013). These nanoparticles are found to accumulate in the liver and other phagocytic cell organs such as the spleen, lungs and brain after inhalation (Liu et al. 2013). As observed during in vivo studies, once these iron oxide nanoparticles enter the cells, they remain in cell organelles for a long time and, then after decomposing, release into cytoplasm. The in vitro studies of iron oxide nanoparticles revealed their toxic effects which are observed in the form of inflammation, cell lysis, disturbing blood coagulation system and also reduced cell viability (Zhu et al. 2008). Naqvi et al. studied murine macrophage cells and reported the toxicity of Tween-coated supermagnetic iron oxide nanoparticles of size 30 nm (Naqvi et al. 2010). Low concentration of iron oxide nanoparticles is observed to exhibit more cell toxicity as compared to their high concentrations. However, in dextran-coated iron oxide nanoparticles (100–150 nm), after 7 days of incubation, reduced cell viability is observed in human macrophages by 20% (Pawelczyk et al. 2008). Another research to study the toxic effect of nanoparticles of iron oxide of size 25 nm was conducted on mouse neuroblastoma (Neuro-2A) cell line. It was found to exhibit less toxic effect in terms of change in mitochondrial function, cell morphology, permeability and apoptosis (Jeng and Swanson 2006). Using human hepatocellular carcinoma cells, chitosan-coated iron oxide NPs (13.8 nm) at concentration of 123.52 μg/mL have shown 10% cell viability after 12 h exposure (Jeng and Swanson 2006).

25.5 Nanoparticles of Zinc Oxide

Zinc oxide nanoparticles have applications in many personal care products, sunscreens, paints, UV detectors, gas sensors and wave filters (Huang et al. 2006). Because of its increased use in diverse areas, human exposure to nanoparticles of zinc oxide is imminent (Sharma et al. 2012). In vitro studies carried out to study the toxic effects of nanoparticles of zinc oxide on the bacteria and mammalian cells revealed cytotoxicity, damage of cell membrane and increased oxidative stress (Huang et al. 2010; Meyer et al. 2011). A study on these NPs revealed a change in cell viability and DNA damage (using MTT and comet assay, respectively) along with alteration in mitochondrial activity in human hepatocytes and embryonic kidney cells was observed (Guan et al. 2012). One more study using HEp-2 cell line confirmed genotoxic effect of zinc oxide nanoparticles via DNA damage employing comet assay and cytokinesis-blocked micronucleus assay techniques (Osman et al. 2010).

25.6 Nanoparticles of Aluminium Oxide

Nanoparticles of aluminium oxide have applications in paints, coatings, textiles, medical products and even as composite propellants in artillery. The phytotoxic activity of nanometric aluminium oxide is significantly lower as compared to nanoparticles of zinc oxide (Lin and Xing 2007; Hu et al. 2009). Aluminium oxide nanoparticles are found to stimulate reactive oxygen species (ROS), thereby producing deterioration of DNA, lipids, proteins and also quicker apoptosis (M'rad et al. 2018). A study on the toxic effects of these nanoparticles show that it causes a change in cell viability and mitochondrial function, increases oxidative stress and may modify tight junction protein expression of the blood-brain barrier (Chen et al. 2008). Another study indicated that aluminium oxide nanoparticles can cause cell depletion in the cortex and hippocampus, thereby affecting learning and memory ability (Wang et al. 2014). An experiment conducted with comet assay and micronucleus test using rat blood cells revealed that aluminium oxide nanoparticles (30–40 nm) own genotoxic properties that are dose-dependent (Balasubramanyam et al. 2009). Based on its tremendous applications in diverse fields and consequent human exposure, there is a dire need to study aluminium oxide nanoparticles for toxic health effects on the humans as per the standard protocols.

25.7 Nanoparticles of Titanium Oxide

Titanium oxide nanoparticles have low solubility and have applications in cosmetics, sunscreen, food additive, paints, waste-water treatment, sterilization and implant biomaterials and even as photocatalysts. Therefore, it is pertinent to study the toxicity effects of TiO_2 nanoparticles, and studies on animals have shown that these nanoparticles have toxic health effects resulting in DNA damage, genotoxicity and lung inflammation (Wu et al. 2010). A study on titanium oxide nanoparticles revealed a cytotoxic effect on glial cells when these nanoparticles are exposed to brain cells, thereby causing brain injury and contributing in the development of neurodegenerative diseases (Márquez-Ramírez et al. 2012).

25.8 Nanoparticles of Copper Oxide

Copper oxide is a semiconducting material with useful photovoltaic and photoconductive properties. Therefore, nanoparticles of copper oxide have tremendous applications in antimicrobial reagents, semiconductors, heat transfer fluids, solar cells, electronic chips and gas sensors (Singh et al. 2009; Shi et al. 2010). From the in vitro lung toxicity study of CuO nanoparticles, it is known that these nanoparticles transformed the regulation of several genes involved in cell cycle progression and genetic damage in human lung cells. Also, copper oxide nanoparticles have the likelihood to induce cytotoxicity and genotoxicity, facilitated through reactive oxygen species (ROS) generation and oxidative stress in human lung cells. Copper oxide nanomaterials have been found to have toxic effects resulting in severe impairment in the kidney, liver and spleen in experimental animals. Oral administration and interaction with gastric juice result in the development of highly reactive ionic copper which gets accumulated in the kidney of the exposed animals (Ahamed et al. 2015).

25.9 Nanoparticles of Gold

Nanoparticles of gold with unique physicochemical properties are relatively safe and non-toxic and have the ability of easy functionalization binding to amine and thiol groups and therefore lead the way for surface modification (Jia et al. 2017). These nanoparticles are used as drug delivery in cancer treatment and thermal therapy, and as contrast agents (Cheng et al. 2016). Gold nanoparticles show a size-dependent cellular toxicity wherein gold nanoparticles of size 1.4 nm were found to be the most cytotoxic, whereas gold nanoparticles of 15 nm were non-toxic even at a concentration which was 60-fold higher than that for the smaller particles (Pan et al. 2007). Also, cytotoxicity linked with gold nanoparticles depend on dose, physical/chemical properties, type of toxicity assay, cell line and the stabilizer used.

25.10 Nanoparticles of Silver

Silver nanoparticles are potential candidates known for their antibacterial activity and being used in a varied applications like wound dressings, coating of surgical instruments and prostheses. Experiments conducted after exposing the rats to Ag-based nanoparticles either through inhalation or by subcutaneous injection confirmed the presence of Ag nanoparticles in several organs, including the lungs, liver, spleen, kidney and brain. These nanoparticles are known to show more toxicity with regard to cell viability, production of reactive oxygen species (ROS) and lactate dehydrogenase (LDH) leakage (Beer et al. 2012). The production of ROS can be excited by the cell uptake of Ag nanoparticles, which results in oxidative stress and genotoxic effects. Also, the toxicity of silver nanoparticles depend both on the size and shape. In vitro studies have recognized the toxic effects of Ag nanoparticles on rat liver (BRL3A), neuronal cells, human lung cells and stem cells (Braydich-Stolle et al. 2005). The Ag^+ toxicity mechanism exhibits the formation of proteinaceous pores by Ag^+ ions in the mitochondrial membranes. Studies in rat liver mitochondria revealed an increase in permeability in mitochondrial membrane resulting in swelling, deviant metabolism and ultimately cellular apoptosis (Almofti et al. 2003). Silver nanoparticles exhibit stimulatory as well as suppressive effects on the creation of cytokines related with the inflammatory response and furthermore depends on case, dosage and type of the cell. Hence, it is important to study and address the toxicity effects of silver nanoparticles through suitable experimental models.

25.11 Nanoparticles of Cadmium

Cadmium nanoparticles (CdTe, CdSe) are primarily utilized for the construction of semiconductor quantum dots (Juzenas et al. 2008). They have notable electronic, optical properties specifically fluorescent properties having practical applications in optoelectronic devices, solar cells, optical interference coatings, biomedical imaging and cancer detection (Alivistas 2004; Hardman 2006). Enhancement in quantum yield of the quantum dots (QDs) has been observed after a silica coating is done (Zhang et al. 2009). However, cytotoxic effects of cadmium QDs have been reported by many researchers, causing mitochondrial injury, cellular impairment, apoptosis, oxidative stress and disorder in intracellular calcium signaling (Ambrosone et al. 2012).

25.12 Conclusions

In the last two decades, nanoparticles have demonstrated diverse applications in the fields as detection, sensing, catalysis, electronics and biomedical sciences. Humans are exposed to airborne nanosized particles from an early age, but exposure to such engineered nanoparticles has risen manyfold in the last century. Use of nanoparticles in human environment is fast growing. The wide application of nanomaterials in electronic industry, healthcare products and biomedical field has raised concerns regarding the potential toxicity of nanoparticles in humans. It is expected that human exposure to nanomaterials will continue to increase and, hence, extensive human exposure to nanomaterials is of main concern. Better tools to evaluate the safety of nanomaterials should be developed. A plan for storage and disposal of nanoparticles or nanoparticles contaminated waste should be developed, taking into account the hazardous nature of the particles and the quantities involved. Also, current understanding of the effects of human exposure to nanomaterials is very limited. The information collected from current literature indicates that nanomaterials found in human environment may have potential for toxicological effects. Our knowledge on the toxicological effects of nanomaterials is mainly dependent on the studies conducted through in vitro and in vivo test models using different sources and different experimental conditions and even different nanomaterial characterization techniques. Therefore, these data are hard to interpret. Extensive research on nanomaterial characterization, biological interaction, toxicity and health effects is required, and the test methods need to be validated along with harmonization of the data.

References

Agnihotri S, Mukherji S, Mukherji S (2014) Size-controlled silver nanoparticles synthesized over the range 5–100 nm using the same protocol and their antibacterial efficacy. RSCAdv 4:3974–3983

Ahamed M, Akhtar MJ, Alhadlaq HA, Alrokayan SA (2015) Assessment of the lung toxicity of copper oxide nanoparticles: current status. Nanomedicine 10(15):2365–2377

Albukhaty S, Naderi-Manesh H, Tiraihi T (2013) In vitro labeling of neural stem cells with poly-l-lysine coated super paramagnetic nanoparticles for green fluorescent protein transfection. Iran Biomed J 17(2):71–76

Alivistas P (2004) The use of nanocrystals in biological detection. Nat Biotechnol 22:47–52

Almofti MR, Ichikawa T, Yamashita K, Terada H, Shinohara Y (2003) Silver ion induces a cyclosporine a-insensitive permeability transition in rat liver mitochondria and release of apoptogenic cytochrome. J Biochem 134(1):43–49

Ambrosone A, Mattera L, Marchesano V, Quarta A, Susha AS, Tino A, Rogach AL, Tortiglione C (2012) Mechanisms underlying toxicity induced by CdTe quantum dots determined in an invertebrate model organism. Biomaterials 33:1991–2000

Balasubramanyam A, Sailaja N, Mahboob M, Rahman MF, Misra S, Hussain SM et al (2009) Evaluation of genotoxic effects of oral exposure to Aluminum oxide nanomaterials in rat bone marrow. Mutat Res 676:41–47

Beer C, Foldbjerg R, Hayashi Y, Sutherland DS, Autrup H (2012) Toxicity of silver nanoparticles—Nanoparticle or silver ion? Toxicol Lett 208(3):286–292

Borm PJ, Robbins D, Haubold S et al (2006) The potential risk of nanomaterials: a review. Part Fibre Toxicol 14:3–11

Braydich-Stolle L, Hussain S, Schlager JJ, Hofmann MC (2005) In vitro cytotoxicity of nanoparticles in mammalian germline stem cells. Toxicol Sci 88(2):412–419

Chen L, Yokel RA, Hennig B, Toborek M (2008) Manufactured aluminum oxide nanoparticles decrease expression of tight junction proteins in brain vasculature. J Neuroimmune Pharmacol 3:286–295

Cheng B, He H, Huang T, Berr SS, He J, Fan D, Zhang J, Xu P (2016) Gold nanosphere gated mesoporous silica nanoparticle responsive to near-infrared light and redox potential as a theranostic platform for cancer therapy. J Biomed Nanotechnol 12(3):435–449

Ershov BG, Abkhalimov EV, Solovov RD et al (2016) Gold nanoparticles in aqueous solutions: influence of size and pH on hydrogen dissociative adsorption and Au(III) ion reduction. Phys Chem Chem Phys 18:13,459–13,466

Guan R, Kang T, Lu F, Zhang Z, Shen H, Liu M (2012) Cytotoxicity, oxidative stress, and genotoxicity in human hepatocyte and embryonic kidney cells exposed to ZnO nanoparticles. Nanoscale Res Lett 7(1):602–607

Handford CE, Dean M, Henchion M et al (2014) Implications of nanotechnology for the agri-food industry: opportunities, benefits and risks. Trend Food Sci Technol 40(2):226–241

Hardman RA (2006) Toxicologic review of quantum dots: toxicity depends upon physic-chemical and environmental factors. Environ health Perspect 114:165–172

Hu X, Cook S, Wang P, Hwang H (2009) In vitro evaluation of cytotoxicity of engineered metal oxide nanoparticles. Sci Total Environ 407:3070–3072

Huang GG, Wang CT, Tang HT, Huang YS, Yang J (2006) ZnO nanoparticle-modified infrared internal reflection elements for selective detection of volatile organic compounds. Anal Chem 78(7):2397–2404

Huang CC, Aronstam RS, Chen DR, Huang YW (2010) Oxidative stress, calcium homeostasis, and altered gene expression in human lung epithelial cells exposed to ZnO nanoparticles. Toxicol In Vitro 24(1):45–55

Jeng HA, Swanson J (2006) Toxicity of metal oxide nanoparticles in mammalian cells. J Environ Sci Health Part A Tox Hazard Subst Environ Eng 41(12):2699–2711

Jia Y-P, Ma B-Y, Wei X-W, Qian Z-Y (2017) The in vitro and in vivo toxicity of gold nanoparticles. Chin Chem Lett 28(4):691–702

Juzenas P, Chen W, Sun YP, MAN C, Genralov R, Genraloves N, Christensen IL (2008) Quantum dots and nanoparticles for photodynamic and radiation therapies of cancer. Adv Drug Delivery Revs 60:1600–1614

Kahan DM, Braman D, Slovic P et al (2009) Cultural cognition of the risks and benefits of nanotechnology. Nat Nanotechnol 4:87–90

Lin D, Xing B (2007) Phytotoxicity of nanoparticles: inhibition of seed germination and root growth. Environ Pollut 150:243–250

Liu G, Gao J, Ai H, Chen X (2013) Applications and potential toxicity of magnetic iron oxide nanoparticles. Small 9(9–10):1533–1545

M'rad I, Jeljeli M, Rihane N, Hilber P, Sakly M, Amara S (2018) Aluminium oxide nanoparticles compromise spatial learning and memory performance in rats. EXCLI J 17:200

Márquez-Ramírez SG, Delgado-Buenrostro NL, Chirino YI, Iglesias GG, López-Marure R (2012) Titanium dioxide nanoparticles inhibit proliferation and induce morphological changes and apoptosis in glial cells. Toxicology 302(2-3):146–156

Meyer K, Rajanahalli P, Ahamed M, Rowe JJ, Hong Y (2011) ZnO nanoparticles induce apoptosis in human dermal fibroblasts via p53 and p38 pathways. Toxicol In Vitro 25(8):1721–1726

Morganti P (2010) Use and potential of nanotechnology in cosmetic dermatology. Clin Cosmet Investig Dermatol 3:5–13

Mukhopadhyay SS (2014) Nanotechnology in agriculture: prospects and constraints. Nanotechnol Sci Appl 7:63–71

Naqvi S, Samim M, Abdin M, Ahmed FJ, Maitra A, Prashant C, Dinda AK (2010) Concentration-dependent toxicity of iron oxide nanoparticles mediated by increased oxidative stress. Int J Nanomed 5:983–989

Nel A, Xia T, Madler L, Li N (2006) Toxic potential of materials at the nano-level. Science 311: 622–627

NNI (2014) National nanotechnology initiative strategic plan. http://nano.gov/sites/default/files/pub_resource/2014_nni_strategic_plan.pdf. Accessed 20 June 2017.

NNI (2017) NNI supplement to the president's 2017 budget—Obama White House. https://obamawhitehouse.archives.gov/sites/default/files/microsites/ostp/nni_fy17_budget_supplement.pdf Accessed 20 June 2017
NSTC (2011) National science and technology council, committee on technology, subcommittee on nanoscale science. National Technology Initiative Strategic Plan. https://www.nano.gov. Accessed 20 June 2017.
Oberdörster G (2010) Safety assessment for nanotechnology and nanomedicine: concepts of nanotoxicology. J Intern Med 267(1):89–105
Osman IF, Baumgartner A, Cemeli E, Fletcher JN, Anderson D (2010) Genotoxicity and cytotoxicity of zinc oxide and titanium dioxide in HEp-2 cells. Nanomedicine (Lond) 5(8):1193–1203
Pan Y, Neuss S, Leifert A, Fischler M, Wen F, Simon U, Schmid G, Brandau W (2007) Size-dependent cytotoxicity of gold nanoparticles. Nano Micro Small 3:1941–1949
Pawelczyk E, Arbab AS, Chaudhry A, Balakumaran A, Robey PG, Frank JA (2008) In vitro model of bromodeoxyuridine or iron oxide nanoparticle uptake by activated macrophages from labeled stem cells: implications for cellular therapy. Stem Cells 26(5):1366–1375
Sharma V, Anderson D, Dhawan A (2012) Zinc oxide nanoparticles induce oxidative DNA damage and ROS-triggered mitochondria mediated apoptosis in human liver cells (HepG2). Apoptosis 17(8):852–870
Shi J, Votruba AR, Farokhzad OC, Langer R (2010) Nanotechnology in drug delivery and tissue engineering: from discovery to applications. Nano Lett 10(9):3223–3230
Siegrist M, Stampfli N, Kastenholz H et al (2008) Perceived risks and perceived benefits of different nanotechnology foods and nanotechnology food packaging. Appetite 52(2):283–290
Singh N, Manshian B, Jenkins GJS et al (2009) NanoGenotoxicology: the DNA damaging potential of engineered nanomaterials. Biomaterials 30:3891–3914
Wang J, Zhoua G, Chena C et al (2007) Acute toxicity and biodistribution of different sized titanium dioxide particle sin mice after oral administration. Toxicol Lett 168:176–185
Wang L, Hu J, Zhao Y (2014) Effects of aluminium on βamyloid (1–42) and secretases (APP-cleaving enzymes) in rat brain. Neurochem Res 39:1338–1345
Wu J, Sun J, Xue Y (2010) Involvement of JNK and P53 activation in G2/M cell cycle arrest and apoptosis induced by titanium dioxide nanoparticles in neuron cells. Toxicol Lett 199:269–276
Yang C, Tang A, Li Z (2016) Reversible cardiac hypertrophy induced by PEG-coated gold nanoparticles in mice. Sci Rep 6:1–12. https://doi.org/10.1038/srep20203
Zhang Y, Gong SWY, Jin L, Li SM, Chen ZP, Zhong PC, Ma M, Gu N (2009) Magnetic nanocomposites of Fe3O4/SiO2-FITC with pH-dependent fluorescence emission. Chin Chem Lett 20:969–972
Zhu MT, Feng WY, Wang B, Wang TC, Gu YQ, Wang M, Wang Y, Ouyang H, Zhao YL, Chai ZF (2008) Comparative study of pulmonary responses to nano- and submicron-sized ferric oxide in rats. Toxicology 247(2-3):102–111

Correction to: Functionalized Nanoparticles in Drug Delivery: Strategies to Enhance Direct Nose-to-Brain Drug Delivery via Integrated Nerve Pathways

Fakhara Sabir, Qurrat Ul Ain, Abbas Rahdar, Zhugen Yang, Mahmood Barani, Mauhammad Bilal, and Nikhil Bhalla

Correction to:
chapter 21 in A. Thakur et al. (eds.),
***Synthesis and Applications of Nanoparticles*,**
https://doi.org/10.1007/978-981-16-6819-7_21

The original version of this chapter was inadvertently published with errors. The following errors have been corrected with this erratum

1. A spelling error in the 2nd author's name has been corrected to read as 'Qurrat Ul Ain'
2. The affiliation of the author Mahmood Barani has been updated as 'Medical Mycology and Bacteriology Research Center, Kerman University of Medical Sciences, Kerman, Iran'

The updated version of this chapter can be found at
https://doi.org/10.1007/978-981-16-6819-7_21

A. Thakur et al. (eds.), *Synthesis and Applications of Nanoparticles*,
https://doi.org/10.1007/978-981-16-6819-7_26

The manufacturer's authorised representative in the EU is Springer Nature Customer Service Centre GmbH, Europaplatz 3, 69115 Heidelberg, Germany. If you have any concerns regarding our products, please contact ProductSafety@springernature.com

Printed and bound by CPI Group (UK) Ltd, Croydon, CR0 4YY

30/09/2026

02225245-0018